Student's Solutions Manual

to accompany

Intermediate Algebra

Sixth Edition

Mark Dugopolski
Southeastern Louisiana University

Boston Burr Ridge, IL Dubuque, IA New York San Francisco St. Louis
Bangkok Bogotá Caracas Kuala Lumpur Lisbon London Madrid Mexico City
Milan Montreal New Delhi Santiago Seoul Singapore Sydney Taipei Toronto

The McGraw-Hill Companies

Student's Solutions Manual to accompany
INTERMEDIATE ALGEBRA, SIXTH EDITION
MARK DUGOPOLSKI

Published by McGraw-Hill Higher Education, an imprint of The McGraw-Hill Companies, Inc., 1221 Avenue of the Americas, New York, NY 10020.

This book is printed on recycled, acid-free paper containing 10% post consumer waste.

1 2 3 4 5 6 7 8 9 0 QPD/QPD 0 9 8

ISBN: 978-0-07-320614-1
MHID: 0-07-320614-8

www.mhhe.com

Table of Contents

1.1 WARM-UPS

1. False, since 5 is a counting number and $5 \notin A$.
2. False, since B has only 3 elements.
3. False, because the set has a specific number of members.
4. False, since $1 \notin B$.
5. True, since $3 \in A$.
6. True, because 3 and 4 are the only numbers that belong to both A and B.
7. True, because every member of C is also a member of B.
8. False, since $1 \in A$ but $1 \notin B$.
9. True, since $\emptyset$ is a subset of every set.
10. True, since $1 \in A$ and $1 \notin C$.

1.1 EXERCISES

1. A set is a collection of objects.
3. A Venn diagram is used to illustrate relationships between sets.
5. A is a subset of B if every member of set A is also a member of set B.
7. True, because 3 is one of the numbers listed in set A.
9. True, because 11 is not in the list of numbers for set A.
11. False, because C has only 5 numbers and N has infinitely many numbers.
13. True, because 2 is in B, but not in A.
15. True, because N is the set of all natural numbers.
17. Note that $A = \{1, 3, 5, 7, 9\}$ and $C = \{1, 2, 3, 4, 5\}$. Since $1, 2, 3, 4, 5, 7$, and 9 are in either A or C, $A \cup C = \{1, 2, 3, 4, 5, 7, 9\}$.
19. Note that $A = \{1, 3, 5, 7, 9\}$ and $C = \{1, 2, 3, 4, 5\}$. Since $1, 3$, and 5 are in both A and C, $A \cap C = \{1, 3, 5\}$.
21. The elements of B together with those of C give us $B \cup C = \{1, 2, 3, 4, 5, 6, 8\}$.
23. Since the empty set has no members, forming the union with another set does not add any new members. So $A \cup \emptyset = A$.
25. Since $\emptyset$ has no members in common with A, $A \cap \emptyset = \emptyset$.
27. Since every member of A is also a member of N, $A \cap N = A = \{1, 3, 5, 7, 9\}$.
29. Since the members of A are odd and the members of B are even, they have no members in common. So $A \cap B = \emptyset$.
31. $A \cup B = \{1, 2, 3, 4, 5, 6, 7, 8, 9\}$ from Exercise 20.
33. Take the elements that B has in common with C to get $B \cap C = \{2, 4\}$.
35. Since 3 is not in B, $3 \notin A \cap B$.
37. Since 4 is in both sets, $4 \in B \cap C$.
39. True, since each member of A is a counting number.
41. True, since both 2 and 3 are members of C.
43. True, since $6 \in B$ but $6 \notin C$.
45. True, since $\emptyset$ is a subset of every set.
47. False, since $1 \in A$ but $1 \notin \emptyset$.
49. True, since $A \cap B = \emptyset$ and $\emptyset$ is a subset of any set.
51. Using all numbers that belong to D or to E yields $D \cup E = \{2, 3, 4, 5, 6, 7, 8\}$.
53. Using only numbers that belong to both D and F gives $D \cap F = \{3, 5\}$.
55. Using all numbers that belong to E or to F gives $E \cup F = \{1, 2, 3, 4, 5, 6, 8\}$.
57. From Exercise 53, $D \cup E = \{2, 3, 4, 5, 6, 7, 8\}$ and $F = \{1, 2, 3, 4, 5\}$. Only 2, 3, 4, and 5 are in both sets. So $(D \cup E) \cap F = \{2, 3, 4, 5\}$.
59. Form the union of $E \cap F = \{2, 4\}$ with $D = \{3, 5, 7\}$ to get $D \cup (E \cap F) = \{2, 3, 4, 5, 7\}$.
61. Form the union of $D \cap F = \{3, 5\}$ with $E \cap F = \{2, 4\}$ to get $(D \cap F) \cup (E \cap F) = \{2, 3, 4, 5\}$.
63. Intersect $D \cup E = \{2, 3, 4, 5, 6, 7, 8\}$ with $D \cup F = \{1, 2, 3, 4, 5, 7\}$ to get $(D \cup E) \cap (D \cup F) = \{2, 3, 4, 5, 7\}$.
65. The set of even natural numbers less than 20 is $\{2, 4, 6, \ldots, 18\}$.
67. The set of odd natural numbers greater than 11 is $\{13, 15, 17, \ldots\}$.
69. The set of even natural numbers between 4 and 79 is $\{6, 8, 10, \ldots, 78\}$. Note that 4 is not in the set because 4 is not between 4 and 79.
71. Since $A = \{1, 2, 3, 4, 5\}$ and $B = \{4, 5, 6, 7, 8, 9\}$, use the numbers that are

in A or B or both to get
$A \cup B = \{1, 2, 3, 4, 5, 6, 7, 8, 9\}$.
73. Since $A = \{1, 2, 3, 4, 5\}$ and
$B = \{4, 5, 6, 7, 8, 9\}$, use the numbers that are in both A and B to get $A \cap B = \{4, 5\}$.
75. Since $A \cap B = \{4, 5\}$ and
$D = \{2, 4, 6, 8\}$, use the numbers that are in $A \cap B$ or in D to get
$(A \cap B) \cup D = \{2, 4, 5, 6, 8\}$.
77. Since $A \cup B = \{1, 2, 3, 4, 5, 6, 7, 8, 9\}$ and $D = \{2, 4, 6, 8\}$, use the numbers that are in $A \cup B$ or in D to get
$(A \cup B) \cup D = \{1, 2, 3, 4, 5, 6, 7, 8, 9\}$.
79. Since $B \cap C = \{5, 7\}$ and
$A = \{1, 2, 3, 4, 5\}$, use the numbers that are in $B \cap C$ and in A to get $(B \cap C) \cap A = \{5\}$.
81. Since $B \cap D = \{4, 6, 8\}$ and $C \cap D = \emptyset$, use the numbers that are in $B \cap D$ or in $C \cap D$ to get $(B \cap D) \cup (C \cap D) = \{4, 6, 8\}$.
83. Answers may vary. Two possible answers are $\{x \mid x$ is a natural number between 2 and 7$\}$ or $\{x \mid x$ is a natural number between 2.5 and 6.3$\}$.
85. Answers may vary. One possibility is $\{x \mid x$ is an odd natural number greater than 4$\}$.
87. Answers may vary. One possibility is $\{x \mid x$ is an even natural number between 5 and 83$\}$
89. Draw a Venn diagram for two sets, females and smokers. Since there are 5 female smokers, place 5 in the intersection of the two sets. Place 12 inside the smoker circle, but outside the female circle. Since there are 30 students in the union of the two sets, there must be 13 female nonsmokers.
91. If A has n elements and B has m elements, then the union of the two sets cannot have more members than $n + m$. So the union is also finite.
93. a) $3 \in \{1, 2, 3\}$ **b)** $\{3\} \subseteq \{1, 2, 3\}$
c) This is tricky. Since $\{\emptyset\}$ is a set containing the empty set, $\{\emptyset\}$ is not empty. So $\emptyset \neq \{\emptyset\}$

1.2 WARM-UPS

1. False, because π is irrational.
2. True, because the set of real numbers is the union of the rationals and the irrationals.
3. False, 0 is rational and not irrational.
4. False, because every rational number is a real number that is not in the set of irrational numbers.
5. True, since it is a repeating decimal.
6. False, it is not repeating.
7. True, because the points on the number line correspond to the real numbers.
8. False, because 6 is not in the interval $(2, 6)$. The interval $(2, 6)$ consists of real numbers between 2 and 6, not include the endpoints.
9. True, because 3 is not in the interval $(1, 3)$ but it is in the interval $[3, 4)$. So every number in $(1, 4)$ is in either $(1, 3)$ or $[3, 4)$.
10. False, because 2 is in the interval $(1, 5)$ and $[2, 8)$ but not in the interval $(2, 8)$.

1.2 EXERCISES

1. The integers consist of the positive and negative counting numbers and zero.
3. The repeating or terminating decimal numbers are rational numbers.
5. The set of real numbers is the union of the rational and irrational numbers.
7. False, because -2 is not a natural number.
9. True, since 0 is rational.
11. True, since a terminating decimal is rational.
13. False, because $1/2$ is a rational number that is not an integer.
15. False, since $1/2$ is not an integer.
17. The set of whole numbers smaller than 6 is the set {0, 1, 2, 3, 4, 5}. Note that 0 is a whole number.

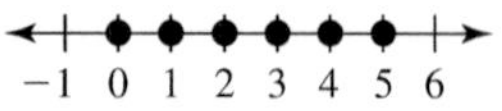

−1 0 1 2 3 4 5 6

19. The set of integers greater than -5 consists of every integer to the right of -5 on the number line:
{−4, −3, −2, −1, 0, 1, . . .}.

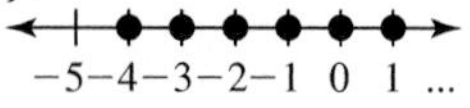

−5 −4 −3 −2 −1 0 1 ...

21. The set of natural numbers between 0 and 5 does not include either 0 or 5: {1, 2, 3, 4}.

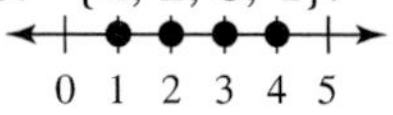

0 1 2 3 4 5

23. The set of integers between -3 and 5 does not include either -3 or 5: $\{-2, -1, 0, 1, 2, 3, 4\}$.

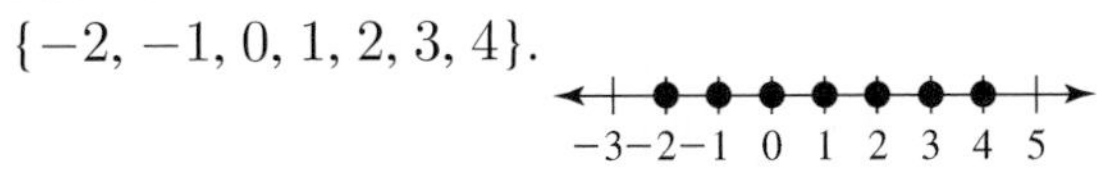

25. All of the numbers are real numbers.

27. The whole numbers include 0. $\{0, 8/2\}$

29. All of the numbers except $-\sqrt{10}$ and $\sqrt{2}$ are rational.
$\{-3, -5/2, -0.025, 0, 3\frac{1}{2}, 8/2\}$

31. True, since every rational number is also a real number.

33. False, since 0 is not irrational.

35. True, because all real numbers are either rational or irrational.

37. False, since nonrepeating decimals are irrational.

39. False, since repeating decimals are rational.

41. False, since repeating decimals are rational.

43. True, since π is irrational.

45. $N \subseteq W$, since every natural number is a whole number.

47. $Z \not\subseteq N$, since $-9 \in Z$ but $-9 \notin N$.

49. $Q \subseteq R$, since every rational number is a real number.

51. $\emptyset \subseteq I$, since $\emptyset$ is a subset of every set.

53. $N \subseteq R$, since every natural number is a real number.

55. $5 \in Z$, since 5 is an integer.

57. $7 \in Q$, since 7 is a rational number.

59. $\sqrt{2} \in R$, since $\sqrt{2}$ is a real number.

61. $0 \notin I$, since 0 is rational.

63. $\{2, 3\} \subseteq Q$, since both 2 and 3 are rational.

65. $\left\{3, \sqrt{2}\right\} \subseteq R$, since both numbers are real.

67. The set of real numbers greater than 1 is written in interval notation as $(1, \infty)$ and graphed as follows.

−1 0 1 2 3

69. The set of real numbers less than -1 is written in interval notation as $(-\infty, -1)$ and graphed as follows.

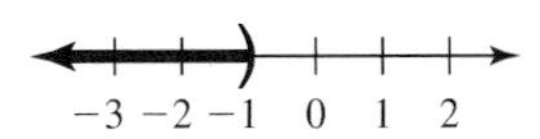

71. The set of real numbers between 3 and 4 is written in interval notation as $(3, 4)$ and graphed as follows.

2 3 4 5

73. The set of real numbers between 0 and 2 inclusive is written in interval notation as $[0, 2]$ and graphed as follows.

−1 0 1 2 3

75. The set of real numbers greater than or equal to 1 and less than 3 is written in interval notation as $[1, 3)$ and graphed as follows.

0 1 2 3 4

77. The graph shows the real numbers between 5 and 7 inclusive, which is written in interval notation as $[5, 7]$.

79. The graph shows the real numbers greater than -3 and less than or equal to 0, which is written in interval notation as $(-3, 0]$.

81. The graph shows the real numbers greater than or equal to 60, which is written in interval notation as $[60, \infty)$.

83. The graph shows the real numbers less than -5, which is written in interval notation as $(-\infty, -5)$.

85. Graph the intervals $(1, 5)$ and $(4, 9)$ on the same number line to see that the union of $(1, 5)$ and $(4, 9)$ is $(1, 9)$.

87. Graph $(0, 3)$ and $(2, 8)$ on the same number line to see that the numbers they have in common are in the interval $(2, 3)$.

89. Graph $(-2, 4)$ and $(0, \infty)$ on the same number line to see that the union of these two intervals is $(-2, \infty)$.

91. Graph $(-\infty, 2)$ and $(0, 6)$ on the same number line to see that the numbers they have in common are in the interval $(0, 2)$.

93. Graph $[2, 5)$ and $(4, 9]$ on the same number line to see that the union of these intervals is $[2, 9]$.

95. Graph $[2, 6)$ and $[2, 8)$ on the same number line to see that the numbers they have in common are in the interval $[2, 6)$.

97. Since 1.5 is in $(1, 3)$, but not in $[2, 4]$ the statement $(1, 3) \subseteq [2, 4]$ is false.

99. Since every number in the interval $(5,7)$ is a real number, the statement $(5,7) \subseteq R$ is true.
101. Since 0 is not in the interval $(0,1)$, the statement $0 \in (0,1)$ is false.
103. These intervals have no numbers in common. So the statement $(0,1) \cap [1,2] = \emptyset$ is true.
105. A rational number is a repeating or terminating decimal and an irrational number never repeats or terminates as a decimal number. Since $\sqrt{9} = 3$ it is rational. Since $\sqrt{3}$ is a nonterminating nonrepeating decimal number, it is irrational.
107. **a)** Find the decimal representations with a calculator. Note that each decimal is repeating.
b) The number of digits that repeats is the same as the number of nines in the denominator.

1.3 WARM-UPS

1. True, since $-6+6=0$.
2. True, -5 and 5 are opposites of each other.
3. False, $|\,6\,| = 6$.
4. True, since $b - a = b + (-a)$.
5. True, because the product of two numbers with opposite signs is negative.
6. False, since $6 + (-4) = 2$.
7. False, since $-3 - (-6) = -3 + 6 = 3$.
8. False, since $6 \div (-1/2) = 6(-2) = -12$.
9. False, since division by zero is undefined.
10. True, because 0 divided by any nonzero number is 0.

1.3 EXERCISES

1. The absolute value of a number is the number's distance from 0 on the number line.
3. Subtract their absolute values and use the sign of the number with the larger absolute value.
5. Multiply their absolute values, then affix a positive sign if the original numbers have the same sign or a negative sign if the original numbers have opposite signs.
7. Because -9 is a negative number, $|-9| = -(-9) = 9$.
9. Because 7 is a positive number, $|\,7\,| = 7$.
11. Since $|-4| = 4$, we have $-|-4| = -4$.
13. $-(-17) = 17$
15. $-(-(-5)) = -(5) = -5$
17. $(-5) + 9 = 9 - 5 = 4$
19. $(-4) + (-3) = -(4+3) = -7$
21. $-6 + 4 = -(6-4) = -2$
23. $7 + (-17) = -(17-7) = -10$
25. $(-11) + (-15) = -(11+15) = -26$
27. $18 + (-20) = -(20-18) = -2$
29. $-14 + 9 = -(14-9) = -5$
31. $-4 + 4 = 0$
33. $-\frac{1}{10} + \frac{1}{5} = \frac{1}{5} - \frac{1}{10} = \frac{2}{10} - \frac{1}{10} = \frac{1}{10}$
35. $\frac{1}{2} + \left(-\frac{2}{3}\right) = \frac{3}{6} + \left(-\frac{4}{6}\right) = -\frac{1}{6}$
37. $-15 + 0.02 = -(15.00 - 0.02) = -14.98$
39. $-2.7 + (-0.01) = -(2.70 + 0.01) = -2.71$
41. $47.39 + (-44.587) = 2.803$
43. $0.2351 + (-0.5) = -0.2649$
45. $7 - 10 = -(10-7) = -3$
47. $-4 - 7 = -4 + (-7) = -11$
49. $7 - (-6) = 7 + 6 = 13$
51. $-1 - 5 = -1 + (-5) = -6$
53. $-12 - (-3) = -12 + 3 = -9$
55. $20 - (-3) = 20 + 3 = 23$
57. $\frac{9}{10} - \left(-\frac{1}{10}\right) = \frac{9}{10} + \frac{1}{10} = 1$
59. $1 - \frac{3}{2} = \frac{2}{2} - \frac{3}{2} = -\frac{1}{2}$
61. $2.00 - 0.03 = 1.97$
63. $5.3 - (-2) = 5.3 + 2 = 7.3$
65. $-2.44 - 48.29 = -50.73$
67. $-3.89 - (-5.16) = 1.27$
69. $(25)(-3) = -(25 \cdot 3) = -75$
71. $\left(-\frac{1}{3}\right)\left(-\frac{1}{2}\right) = \frac{1}{6}$
73. $(0.3)(-0.3) = -0.09$
75. $(-0.02)(-10) = 0.02(10) = 0.2$
77. The reciprocal of 20 is $\frac{1}{20}$ or 0.05.
79. The reciprocal of $-\frac{6}{5}$ is $-\frac{5}{6}$.
81. The reciprocal of -0.3 is $-\frac{1}{0.3}$ or $-\frac{10}{3}$.
83. $-6 \div 3 = -2$

85. $30 \div (-0.8) = -37.5$
87. $(-0.8)(0.1) = -0.08$
89. $(-0.1) \div (-0.4) = 0.25$
91. $(0) \div (19) = 0$
93. Since division by zero is undefined, $-2 \div 0$ is undefined.
95. $9 \div \left(-\frac{3}{4}\right) = 9\left(-\frac{4}{3}\right) = -\frac{36}{3} = -12$
97. $-\frac{2}{3}\left(-\frac{9}{10}\right) = \frac{2}{3} \cdot \frac{9}{10} = \frac{18}{30} = \frac{3}{5}$
99. $(0.25)(-365) = -91.25$
101. $(-51) \div (-0.003) = 17{,}000$
103. $(0) \div (1.3422) = 0$
105. Since division by zero is undefined, $339.4 \div 0$ is undefined.
107. $-62 + 13 = -(62 - 13) = -49$
109. $-32 - (-25) = -32 + 25 = -7$
111. $|-15| = -(-15) = 15$
113. $\frac{1}{2}(-684) = -342$
115. $\frac{1}{2} - \left(-\frac{1}{4}\right) = \frac{2}{4} + \frac{1}{4} = \frac{3}{4}$
117. $-57 \div 19 = -3$
119. $|-17| + |-3| = 17 + 3 = 20$
121. $0 \div (-0.15) = 0$
123. $-63 + |8| = -63 + 8 = -55$
125. $-\frac{1}{2} + \left(-\frac{1}{2}\right) = -\frac{2}{2} = -1$
127. $-\frac{1}{2} - 19 = -\frac{1}{2} - \frac{38}{2} = -\frac{39}{2}$
129. $28 - 0.01 = 27.99$
131. $-29 - 0.3 = -29.3$
133. $(-2)(0.35) = -0.7$
135. $85{,}000 + (-45{,}000) + (-2300) + (-1500) + 1200 + 2(3500) = 44{,}400$
Net worth is \$44,400. The operation used here is addition.
137. $14° - (-6°) = 20°$F The operation used here is subtraction.
139. $-282 - (-1296) = 1014$ ft The operation used here is subtraction.
141. We learn addition first, because subtraction is defined in terms of addition.

1.4 WARM-UPS

1. False, $2^3 = 8$.
2. True, because $2^2 = 4$ and $-1 \cdot 2^2 = -4$.
3. True, since $-2^2 = -(2^2) = -4$.
4. False, since $6 + 3 \cdot 2 = 6 + 6 = 12$.
5. False, since $(6 + 3) \cdot 2 = 9 \cdot 2 = 18$.
6. False, since $(6 + 3)^2 = 9^2 = 81$.
7. True, since $6 + 3^2 = 6 + 9 = 15$.
8. True, since $(-3)^3 = -27$ and $-3^3 = -27$.
9. False, since $|-3 - (-2)| = |-1| = 1$.
10. False, since $|7 - 8| = 1$ and $|7| - |8| = 7 - 8 = -1$.

1.4 EXERCISES

1. An arithmetic expression is the result of writing numbers in a meaningful combination with the ordinary operations of arithmetic.
3. Grouping symbols are used to indicate the order in which operations are to be performed.
5. The order of operations tells us the order in which to perform operations when grouping symbols are omitted.
7. $(-3 \cdot 4) - (2 \cdot 5) = -12 - 10 = -22$
9. $4[5 - |3 - (2 \cdot 5)|] = 4[5 - 7] = -8$
11. $(6 - 8)(|2 - 3| + 6) = (-2)(7) = -14$
13. $2^5 = 2 \cdot 2 \cdot 2 \cdot 2 \cdot 2 = 32$
15. $(-1)^4 = (-1)(-1)(-1)(-1) = 1$
17. $\left(-\frac{1}{3}\right)^2 = \left(-\frac{1}{3}\right)\left(-\frac{1}{3}\right) = \frac{1}{9}$
19. $\sqrt{49} = 7$
21. $\sqrt{36 + 64} = \sqrt{100} = 10$
23. $\sqrt{4(7 + 9)} = \sqrt{64} = 8$
25. $\sqrt{3 + 13} + 9 = \sqrt{16} + 9 = 4 + 9 = 13$
27. $-2\sqrt{25 + 144} = -2\sqrt{169} = -2(13) = -26$
29. $4 - 6 \cdot 2 = 4 - 12 = -8$
31. $6/2 \cdot 3 = 3 \cdot 3 = 9$
33. $5 - 6(3 - 5) = 5 - 6(-2) = 5 + 12 = 17$
35. $\left(\frac{1}{3} - \frac{1}{2}\right)\left(\frac{1}{4} - \frac{1}{2}\right) = \left(-\frac{1}{6}\right)\left(-\frac{1}{4}\right) = \frac{1}{24}$
37. $-3^2 + (-8)^2 + 3 = -9 + 64 + 3 = 58$
39. $-(2 - 7)^2 = -(-5)^2 = -25$
41. $-5^2 \cdot 2^3 = -25 \cdot 8 = -200$
43. $(-5)(-2)^3 = -5(-8) = 40$
45. $-(3^2 - 4)^2 = -(9 - 4)^2 = -5^2 = -25$
47. $8 + 2\sqrt{5^2 - 3^2} = 8 + 2\sqrt{16} = 8 + 2 \cdot 4 = 8 + 8 = 16$

49. $-60 \div 10 \cdot 3 \div 2 \cdot 5 \div 6$
$= -6 \cdot 3 \div 2 \cdot 5 \div 6$
$= -18 \div 2 \cdot 5 \div 6$
$= -9 \cdot 5 \div 6$
$= -45 \div 6 = -7.5$

51. $5.5 - 2.3^4 = -22.4841$

53. $(1.3 - 0.31)(2.9 - 4.88) = -1.9602$

55. $-388.8 \div (13.5)(9.6) = -276.48$

57. $\frac{2-6}{9-7} = \frac{-4}{2} = -2$

59. $\frac{-3-5}{6-(-2)} = \frac{-8}{8} = -1$

61. $\frac{4+2\cdot 7}{3\cdot 2-9} = \frac{18}{-3} = -6$

63. $\frac{-3^2-(-9)}{2-3^2} = \frac{-9+9}{2-9} = \frac{0}{-7} = 0$

65. $\frac{4-7}{-3-(-3)} = \frac{-3}{-3+3} = \frac{-3}{0}$
The expression is undefined.

67. $3^2 - 4(-1)(-4) = 9 - 16 = -7$

69. $\frac{-1-3}{-1-(-4)} = \frac{-4}{-1+4} = -\frac{4}{3}$

71. $(-1-3)(-1+3) = (-4)(2) = -8$

73. $\sqrt{(-4)^2 - 2(-4) + 1} = \sqrt{16+8+1}$
$= 5$

75. $\frac{2}{-1} + \frac{3}{-4} - \frac{1}{-4} = -2 - \frac{3}{4} + \frac{1}{4}$
$= -\frac{8}{4} - \frac{3}{4} + \frac{1}{4} = -\frac{10}{4} = -\frac{5}{2}$

77. $|-1-3| = |-4| = 4$

79. $\frac{-6-4}{-7-2} = \frac{-10}{-9} = \frac{10}{9}$

81. $\frac{2-(-1)}{1-(-3)} = \frac{3}{4}$

83. $\frac{5.6-2.4}{4.7-5.9} = -\frac{8}{3} \approx -2.67$

85. $-2^2 + 5(3)^2 = -4 + 45 = 41$

87. $(-2+5)3^2 = 3 \cdot 9 = 27$

89. $\sqrt{5^2 - 4(1)(6)} = \sqrt{25-24} = 1$

91. $[13 + 2(-5)]^2 = 3^2 = 9$

93. $\frac{4-(-1)}{-3-2} = \frac{5}{-5} = -1$

95. $3(-2)^2 - 5(-2) + 4 = 12 + 10 + 4 = 26$

97. $-4\left(\frac{1}{2}\right)^2 + 3\left(\frac{1}{2}\right) - 2$
$= -4\left(\frac{1}{4}\right) + \frac{3}{2} - 2$
$= -\frac{2}{2} + \frac{3}{2} - \frac{4}{2} = -\frac{3}{2}$

99. $|6 - 3\cdot 7| + |7-5|$
$= |-15| + |2|$
$= 17$

101. $3 - 7[4 - (2-5)] = 3 - 7[4+3]$
$= 3 - 49 = -46$

103. $3 - 4(2 - |4-6|) = 3 - 4(2-2)$
$= 3$

105. $[3-(-1)]^2 + [-1-4]^2 = 4^2 + (-5)^2$
$= 16 + 25 = 41$

107. Find the target heart rates for both women and subtract them:
$0.65(220-25) - 0.65(220-65) = 26$
The target heart rate for a 25-yr old woman is 26 beats per minute larger than the target heart rate for a 65 yr-old woman. From the graph, a woman's target heart rate is 115 at about 43 years of age.

109. The perimeter is $2(34) + 2(18)$ or 104 feet.

111. a) Using the graph, the investment amounts to approximately \$60,000.
b) The investment amounts to $10{,}000(1+0.062)^{30}$ or \$60,776.47.

113. According to the graph she owes approximately \$5500.
The actual amount owed when the payments start is $4000(1+0.08)^4$ or \$5441.96.

115. a) Use $\frac{n(n+1)}{2}$ with $n = 50$ to get $\frac{50(51)}{2} = 1275$.
b) Use $\frac{n(n+1)(2n+1)}{6}$ with $n = 40$ to get $\frac{40(41)(81)}{6} = 22{,}140$.
c) Use $\frac{n^2(n+1)^2}{4}$ with $n = 30$ to get $\frac{30^2(31)^2}{4} = 216{,}225$.
d) To find the square of the sum, use $\left(\frac{n(n+1)}{2}\right)^2$ with $n = 20$ to get $\left(\frac{20(21)}{2}\right)^2 = 210^2 = 44{,}100$.
e) To find the cube of the sum, use $\left(\frac{n(n+1)}{2}\right)^3$ with $n = 10$ to get $\left(\frac{10(11)}{2}\right)^3 = 55^3 = 166{,}375$.

1.5 WARM-UPS

1. True, because of the commutative properties.
2. False, since $8 \div (4 \div 2) = 4$ and $(8 \div 4) \div 2 = 1$.
3. False, since $10 \div 2 = 5$ and $2 \div 10 = 0.2$.
4. False, since $5 - 3 = 2$ and $3 - 5 = -2$.
5. False, since $10 - (7 - 3) = 6$ and $(10 - 7) - 3 = 0$.
6. False, since $4(6 \div 2) = 12$ and $(4 \cdot 6) \div (4 \cdot 2) = 3$.
7. True, since $(0.02)(50) = 1$.
8. True, because of Warm-up number 2.
9. False, because if $x = 0$ we get $3 = 0$.
10. False, 1 is the multiplicative identity.

1.5 EXERCISES

1. The commutative property of addition says that $a + b = b + a$ and the commutative property of multiplication says that $a \cdot b = b \cdot a$.
3. The commutative property of addition says that you get the same result when you add two numbers in either order. The associative property of addition has to do with which two numbers are added first when adding three numbers.
5. Zero is the additive identity because adding zero to a number does not change the number.
7. $9 - 4 + 6 - 10 = 15 - 14 = 1$
9. $6 - 10 + 5 - 8 - 7 = 11 - 25 = -14$
11. $-4 - 11 + 6 - 8 + 13 - 20 = -43 + 19 = -24$
13. $-3.2 + 1.4 - 2.8 + 4.5 - 1.6 = -7.6 + 5.9 = -1.7$
15. $3.27 - 11.41 + 5.7 - 12.36 - 5 = 8.97 - 28.77 = -19.8$
17. $4(x - 6) = 4 \cdot x - 4 \cdot 6 = 4x - 24$
19. $a(3 + t) = 3a + at$
21. $-2(w - 5) = -2w - (-10) = -2w + 10$
23. $-1(-2x - y) = (-1)(-2x) - (-1)y = 2x + y$
25. $\frac{1}{2}(4x + 8) = \frac{4}{2}x + \frac{8}{2} = 2x + 4$
27. $2m + 10 = 2 \cdot m + 2 \cdot 5 = 2(m + 5)$
29. $5x - 5 = 5 \cdot x - 5 \cdot 1 = 5(x - 1)$
31. $3y - 15 = 3 \cdot y - 3 \cdot 5 = 3(y - 5)$
33. $3a + 9 = 3 \cdot a + 3 \cdot 3 = 3(a + 3)$
35. $bw + w = b \cdot w + w \cdot 1 = w(b + 1)$
37. The reciprocal of $\frac{1}{2}$ is 2 because $2 \cdot \frac{1}{2} = 1$.
39. The reciprocal of 1 is 1 because $1 \cdot 1 = 1$.
41. The reciprocal of 6 is $\frac{1}{6}$, because $6 \cdot \frac{1}{6} = 1$.
43. Since $0.25 = \frac{1}{4}$, its reciprocal is 4.
45. Since $-0.7 = -\frac{7}{10}$, its reciprocal is $-\frac{10}{7}$.
47. Since $-1.8 = -\frac{18}{10}$ or $-\frac{9}{5}$, the reciprocal of -1.8 is $-\frac{5}{9}$.
49. $\frac{1}{2.3} + \frac{1}{5.4} \approx 0.6200$
51. $\dfrac{\frac{1}{4.3}}{\frac{1}{5.6} + \frac{1}{7.2}} \approx 0.7326$
53. Because $3 + x = x + 3$ has the same form as $a + b = b + a$, the commutative property of addition is illustrated here.
55. Because $5(x - 7) = 5x - 35$ has the same form as $a(b + c) = ab + ac$, the distributive property is illustrated.
57. Because $3(xy) = (3x)y$ has the same form as $a(bc) = (ab)c$, the associative property of multiplication is illustrated.
59. Because 4 and 0.25 have a product of 1, the multiplicative inverse property is illustrated.
61. Because $y^3x = xy^3$ has the same form as $ab = ba$, the commutative property of multiplication is illustrated.
63. Because multiplying x by 1 does not change x, the multiplicative identity property is illustrated.
65. Because $2x + 3x = (2 + 3)x$ has the same form as $a(b + c) = ab + ac$, the distributive property is illustrated.
67. Because 7 and -7 have a sum of 0, the additive inverse property is illustrated.
69. Because the product of 0 and any number is 0, the multiplication property of zero is illustrated.

71. Because $xy + x = x(y+1)$ has the same form as $a(b+c) = ab + ac$, the distributive property is illustrated.
73. Because $a + b = b + a$ for any real numbers, $5 + w = w + 5$.
75. Because $a(bc) = (ab)c$ for any real numbers, $5(xy) = (5x)y$.
77. Because $ab + ac = a(b+c)$ for any real numbers, $\frac{1}{2}x - \frac{1}{2} = \frac{1}{2}x - \frac{1}{2} \cdot 1 = \frac{1}{2}(x-1)$.
79. Because $ab + ac = a(b+c)$ for any real numbers, $6x + 9 = 3 \cdot 2x + 3 \cdot 3 = 3(2x+3)$.
81. Since 8 and 0.125 are reciprocals, $8(0.125) = 1$.
83. $0 = 5(0)$
85. $0.25(4) = 1$
87. Fortunately, the order in which the groceries are placed does not affect the total bill, because of the commutative and associative properties of addition.

1.6 WARM-UPS

1. True by the distributive property.
2. False, because $-4x + 8 = -4(x-2)$.
3. True, because multiplying by -1 is equivalent to finding the opposite.
4. True, by the distributive property.
5. False, $(2x)(5x) = 10x^2$
6. True because of the distributive property .
7. False, $a + a = 2a$.
8. False, $b \cdot b = b^2$.
9. False, because 1 and $7x$ are not like terms.
10. True, because the like terms are combined correctly.

1.6 EXERCISES

1. A term is a single number or a product of a number and one or more variables.
3. The coefficient of a term is the number preceding the variables.
5. You can multiply and divide unlike terms.
7. $(45 \cdot 2) \cdot 100 = 90 \cdot 100 = 9000$
9. $\frac{4}{3}(0.75) = \frac{4}{3} \cdot \frac{3}{4} = 1$
11. $427 + (68 + 32) = 427 + 100 = 527$
13. $47 \cdot 4 + 47 \cdot 6 = 47(4+6) = 470$
15. $19 \cdot 2 \cdot 5 \cdot \frac{1}{5} = 19 \cdot 2 \cdot 1 = 19 \cdot 2 = 38$
17. $120 \cdot 4 \cdot 100 = 480 \cdot 100 = 48{,}000$
19. $13 \cdot 377 \cdot 0 = 0$
21. $348 + (5 + 45) = 348 + 50 = 398$
23. $\frac{2}{3} \cdot 1.5 = \frac{2}{3} \cdot \frac{3}{2} = 1$
25. $17 \cdot 101 - 17 \cdot 1 = 17(101 - 1) = 1700$
27. $354 + (7+3) + (8+2)$
$= 354 + 10 + 10 = 374$
29. $(567 + 874)(0) = 0$
31. $-4n + 6n = (-4+6)n = 2n$
33. $3w - (-4w) = 3w + 4w = 7w$
35. $4mw^2 - 15mw^2 = (4 - 15)mw^2$
$= -11mw^2$
37. $-5x - (-2x) = -5x + 2x = -3x$
39. $4ay + 5ya = 4ay + 5ay = 9ay$
41. $9mn - mn = 9mn - 1mn = 8mn$
43. $-kz^6 - kz^6 = -1kz^6 + (-1kz^6)$
$= -2kz^6$
45. $4(7t) = (4 \cdot 7)t = 28t$
47. $(-2x)(-5x) = (-2)(-5)x \cdot x = 10x^2$
49. $(-h)(-h) = (-1)(-1)h \cdot h = h^2$
51. $7w(-4) = -4 \cdot 7w = -28w$
53. $-x(1 - x) = -x \cdot 1 - (-x)(x)$
$= -x + x^2$
55. $5k \cdot 5k = 5 \cdot 5 \cdot k \cdot k = 25k^2$
57. $3 \cdot \frac{y}{3} = 3 \cdot \frac{1}{3} \cdot y = 1 \cdot y = y$
59. $9 \cdot \frac{2y}{9} = 9 \cdot \frac{1}{9} \cdot 2y = 1 \cdot 2y = 2y$
61. $\frac{6x^3}{2} = \frac{6}{2}x^3 = 3x^3$
63. $\frac{3x^2y + 15x}{3} = \frac{3x^2y}{3} + \frac{15x}{3} = x^2y + 5x$
65. $\frac{2x - 4}{-2} = \frac{2x}{-2} - \frac{4}{-2} = -x + 2$
67. $\frac{-xt + 10}{-2} = \frac{-xt}{-2} + \frac{10}{-2} = \frac{1}{2}xt - 5$
69. $a - (4a - 1) = a - 4a + 1 = -3a + 1$
71. $6 - (x - 4) = 6 - x + 4 = 10 - x$
73. $4m + 6 - m - 5 = 3m + 1$
75. $-5b + at - 7b = -12b + at$
77. $t^2 - 5w + 2w + t^2 = 2t^2 - 3w$
79. $x^2 - x^2 + y^2 + z = y^2 + z$
81. $(2x + 3) + (7x + 5) = 9x + 8$
83. $(-3x + 4) - (6x - 6) = -9x + 10$
85. $3(5x + 2) - 2(2x + 4) + x$
$= 15x + 6 - 4x - 8 + x = 12x - 2$

87. $3x^2 - 2(x^2 - 5) - 5(2x^2 + 1)$
$= 3x^2 - 2x^2 + 10 - 10x^2 - 5$
$= -9x^2 + 5$

89. $5(x^2 - 6x + 4) + 4(x^2 + 3x - 1)$
$= 5x^2 - 30x + 20 + 4x^2 + 12x - 4$
$= 9x^2 - 18x + 16$

91. $8 - 7k^3 - 21 - 4 = -7k^3 - 17$

93. $x - 0.04x - 0.04(50) = 0.96x - 2$

95. $0.10x + 0.5 - 0.04x - 2 = 0.06x - 1.5$

97. $3k + 5 - 6k + 8 - k + 3 = -4k + 16$

99. $3(1 - xy) - 2(xy - 5) - (35 - xy)$
$= 3 - 3xy - 2xy + 10 - 35 + xy$
$= -4xy - 22$

101. $w \cdot 3w + 5w(-6w) - w(2w)$
$= 3w^2 - 30w^2 - 2w^2 = -29w^2$

103. $3a^2w^2 - 5a^2w^2 - 4a^2w^2 = -6a^2w^2$

105. $\frac{1}{6} - \frac{1}{3}\left(-6x^2y - \frac{1}{2}\right) = \frac{1}{6} + 2x^2y + \frac{1}{6}$
$= 2x^2y + \frac{1}{3}$

107. $-\frac{1}{2}m\left(-\frac{1}{2}m\right) - \frac{1}{2}m - \frac{1}{2}m$
$= \frac{1}{4}m^2 - m$

109. $\frac{-8t^3}{-2} - \frac{6t^2}{-2} + \frac{2}{-2} = 4t^3 + 3t^2 - 1$

111. $\frac{-6xyz}{-3} - \frac{3xy}{-3} + \frac{9z}{-3}$
$= 2xyz + xy - 3z$

113. Perimeter is $s + s + 2 + s + 4$ or $3s + 6$ ft.

115. Perimeter is $2x + 2\left(x + \frac{1}{6}x\right)$ or $\frac{13}{3}x$ meters. Area is $x(x + \frac{1}{6}x) = \frac{7}{6}x^2$ m^2.

Chapter 1 Wrap-Up

Enriching Your Mathematical Word Power

1. a **2.** c **3.** a **4.** d **5.** a
6. c **7.** a **8.** d **9.** b **10.** a
11. b **12.** a **13.** b **14.** c

CHAPTER 1 REVIEW

1. True, because 3 is the only number common to A and B.

3. False, because $A \cup B = \{1, 2, 3, 4, 5\}$.

5. True, because $B \subseteq C$.

7. False, $A \cap \emptyset = \emptyset$.

9. True, because $A \cap B = \{3\}$.

11. True, because every member of B is also a member of C.

13. False, because $1 \in A$ and $1 \notin B$.

15. True, because 3 is a member of D.

17. False, because $0 \notin E$.

19. True, because $\emptyset$ is a subset of every set.

21. The set of whole numbers in the given set is {0, 1, 31}

23. The set of integers contained in the given set is {−1, 0, 1, 31}

25. The set of irrational numbers in the given set is $\left\{-\sqrt{2}, \sqrt{3}, \pi\right\}$.

27. The set of real numbers greater than 0 is written in interval notation as $(0, \infty)$ and graphed as follows:

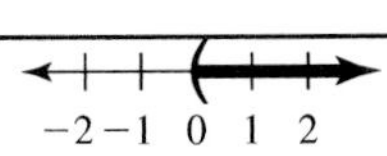

29. The set of real numbers between 5 and 6 is written in interval notation as $(5, 6)$ and graphed as follows:

3 4 5 6 7 8

31. The set of real numbers greater than or equal to -1 and less than 2 is written in interval notation as $[-1, 2)$ and graphed as follows:

−2 −1 0 1 2 3 4

33. Graph the intervals $(0, 2)$ and $(1, 5)$ on the same number line to see that their union is $(0, 5)$.

35. Graph the intervals $(2, 4)$ and $(3, \infty)$ on the same number line to see that the numbers they have in common are in the interval $(3, 4)$.

37. Graph the intervals $[2, 6)$ and $(4, 8)$ on the same number line to see that their union is the interval $[2, 8)$.

39. $-4 + 9 = 9 - 4 = 5$

41. $25 - 37 = -12$

43. $(-4)(6) = -24$

45. $(-8) \div (-4) = 2$

47. $-\frac{1}{4} + \frac{1}{12} = -\frac{3}{12} + \frac{1}{12} = -\frac{2}{12} = -\frac{1}{6}$

49. $\frac{-20}{-2} = 10$

51. $-0.04 + 10 = 10.00 - 0.04 = 9.96$

53. $-6-(-2)=-6+2=-4$

55. $-0.5+0.5=0$

57. $3.2\div(-0.8)=-4$

59. $0\div(-0.3545)=0$

61. $4+7\cdot 5=4+35=39$

63. $20/2\cdot 5=10\cdot 5=50$

65. $(4+7)^2=11^2=121$

67. $6-(7-8)=6-(-1)=6+1=7$

69. $5-6-8-10=5-24=-19$

71. $4^2-9+3^2=16-9+9=16$

73. $5+3\cdot|6-4\cdot 3|=5+3\cdot|6-12|$
$=5+3\cdot 6=5+18$
$=23$

75. $\sqrt{3^2+4^2}=\sqrt{9+16}=\sqrt{25}=5$

77. $\dfrac{-4-5}{7-(-2)}=\dfrac{-9}{7+2}=\dfrac{-9}{9}=-1$

79. $\dfrac{-12+2(6)}{4-(-3)}=\dfrac{0}{7}=0$

81. $\dfrac{-1-(-6)}{-4-(-4)}=\dfrac{5}{-4+4}=\dfrac{-5}{0}$
This expression is undefined.

83. $1-(0.8)(0.3)=1.00-0.24=0.76$

85. $(-3)^2-(4)(-1)(-2)=9-8=1$

87. $\sqrt{3^2-4(-2)(-1)}=\sqrt{9-8}$
$=\sqrt{1}=1$

89. $(-1-3)(-1+3)=-4\cdot 2=-8$

91. $(-2)^2+2(-2)(3)+3^2=4-12+9$
$=1$

93. $(-2)^3-3^3=-8-27=-35$

95. $\dfrac{3+(-1)}{-2+3}=\dfrac{2}{1}=2$

97. $|-2-3|=|-5|=5$

99. $(-2+3)(-1)=(1)(-1)=-1$

101. Commutative property of addition

103. Distributive property

105. Associative property of multiplication

107. Multiplicative identity property

109. Multiplicative inverse property

111. Multiplication property of zero

113. Additive identity property

115. Additive inverse property

117. $3(w+1)=3\cdot w+3\cdot 1=3w+3$

119. $-1(x+5)=-1\cdot x+(-1)\cdot 5$
$=-x-5$

121. $-3(-2x-5)=-3\cdot(-2x)-(-3)(5)$
$=6x+15$

123. $3x-6a=3x-3\cdot 2a=3(x-2a)$

125. $7x+7=7x+7\cdot 1=7(x+1)$

127. $p-pt=p\cdot 1-pt=p(1-t)$

129. $ab+b=ab+b\cdot 1=b(a+1)$

131. $3a+7+4a-5=7a+2$

133. $5(t-4)-3(2t-6)$
$=5t-20-6t+18=-t-2$

135. $-(a-2)+2-a=-a+2+2-a$
$=-2a+4$

137. $5-3(x-2)+7(x+4)$
$=5-3x+6+7x+28-6$
$=4x+33$

139. $0.2(x+0.1)-(x+0.5)$
$=0.2x+0.02-x-0.50$
$=-0.8x-0.48$

141. $0.05(x+3)-0.1(x+20)$
$=0.05x+0.15-0.10x-2$
$=-0.05x-1.85$

143. $\frac{1}{2}(x+4)-\frac{1}{4}(x-8)$
$=\frac{1}{2}x+2-\frac{1}{4}x+2=\frac{1}{4}x+4$

145. $\dfrac{-9x^2}{3}-\dfrac{6x}{3}+\dfrac{3}{3}=-3x^2-2x+1$

147. $32(4)(-6+6)=32(4)(0)=0$, additive inverse property, multiplication property of 0

149. $768(4)+768(6)=768(10)=7680$, distributive property

151. $(12\cdot 4+(-6))+6$
$=12\cdot 4+(-6+6)=48+0=48$, associative property of addition, additive inverse property, additive identity property

153. $752(-6)+752(6)=752(-6+6)$
$=752(0)=0$, distributive property, additive inverse property

155. $(47\cdot 6)\frac{4}{24}=47\left(6\cdot\frac{1}{6}\right)=47\cdot 1=47$, associative property of multiplication, multiplicative inverse property, multiplicative identity property

157. $(-6\cdot 24)\frac{1}{6}=(24\cdot -6)\frac{1}{6}$
$=24(-6\cdot\frac{1}{6})=24(-1)=-24$, commutative property of multiplication, associative property of multiplication, multiplicative inverse property

159. $5(-6+6)(4+24)=5\cdot 0\cdot 28=0$, additive inverse property, multiplication property of 0

161. The perimeter is $x + x + 3 + x + x + 3$ or $4x + 6$ feet. The area is $x(x + 3)$ or $x^2 + 3x$ square feet.

163. a) $53{,}625(1.05)^6 \approx 71{,}863$
The Hummer will cost \$71,863 in 2013.

b) From the graph, the first year in which the car is over \$80,000 will be 2016. Check using the formula: $53{,}625(1.05)^8 \approx 79{,}229$ and $53{,}625(1.05)^9 \approx 83{,}190$.

CHAPTER 1 TEST

1. List the elements in A or in B:
$A \cup B = \{2, 3, 4, 5, 6, 7, 8, 10\}$

2. Only 6 and 7 are in both B and C. So $B \cap C = \{6, 7\}$.

3. The intersection of $A = \{2, 4, 6, 8, 10\}$ with $B \cup C = \{3, 4, 5, 6, 7, 8, 9, 10\}$ is $\{4, 6, 8, 10\}$.

4. $\{0, 8\}$

5. $\{-4, 0, 8\}$

6. $\{-4, -\frac{1}{2}, 0, 1.65, 8\}$

7. $\left\{-\sqrt{3}, \sqrt{5}, \pi\right\}$

8. The integers between -3 and 5 are $-2, -1, 0, 1, 2, 3,$ and 4. The graph follows.

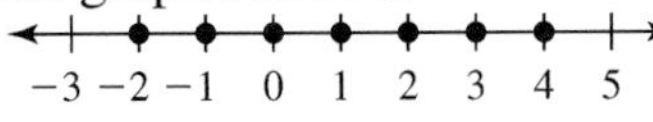

9. The interval of real numbers $(-3, 5]$ consists of all points on the number line greater than -3 and less than or equal to 5. The graph follows.

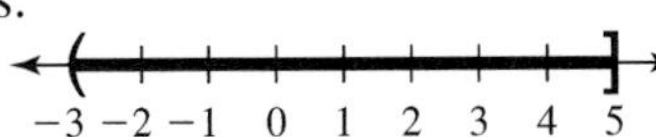

10. Graph $(-\infty, 2)$ and $(1, 4)$ on the same number line to see that their union is $(-\infty, 4)$.

11. Graph $(2, 8)$ and $[4, 9)$ on the same number line to see that the numbers they have in common are in the interval $[4, 8)$.

12. $6 + 3(-5) = 6 + (-15) = -9$

13. $\sqrt{(-2)^2 - 4(3)(-5)} = \sqrt{4 - (-60)}$
$= \sqrt{64} = 8$

14. $-5 + 6 - 12 = -17 + 6 = -11$

15. $0.02 - 2 = -1.98$

16. $\dfrac{-3 - (-7)}{3 - 5} = \dfrac{-3 + 7}{-2} = \dfrac{4}{-2} = -2$

17. $\dfrac{-6 - 2}{4 - 2} = \dfrac{-8}{2} = -4$

18. $\left(\frac{2}{3} - 1\right)\left(\frac{1}{3} - \frac{1}{2}\right) = \left(-\frac{1}{3}\right)\left(-\frac{1}{6}\right)$
$= \dfrac{1}{18}$

19. $-\frac{4}{7} - \frac{1}{2}\left(24 - \frac{8}{7}\right) = -\frac{4}{7} - 12 + \frac{4}{7}$
$= -12$

20. $|\,3 - 5 \cdot 2\,| = |\,3 - 10\,| = |\,-7\,| = 7$

21. $5 - 2\,|\,6 - 10\,| = 5 - 2\,|\,-4\,|$
$= 5 - 2 \cdot 4 = 5 - 8 = -3$

22. $(452 + 695)[-8 + 8] = (452 + 695)[0]$
$= 0$

23. $478(8 + 2) = 478(10) = 4780$

24. $-8 \cdot 3 - 4(6 - 9 \cdot 2^3)$
$= -24 - 4(6 - 9 \cdot 8)$
$= -24 - 4(-66) = -24 + 264$
$= 240$

25. $(-4)^2 - 4(-3)(2) = 16 + 24 = 40$

26. $\dfrac{(-3)^2 - (-4)^2}{-4 - (-3)} = \dfrac{9 - 16}{-1} = 7$

27. $\dfrac{(-3)(-4) - 6(2)}{(-4)^2 - 2^2} = \dfrac{12 - 12}{16 - 4} = \dfrac{0}{12} = 0$

28. Distributive property

29. Commutative property of multiplication

30. Associative property of addition

31. Additive inverse property

32. Commutative property of multiplication

33. $3(m - 5) - 4(-2m - 3)$
$= 3m - 15 + 8m + 12 = 11m - 3$

34. $x + 3 - 0.05x - 0.1 = 0.95x + 2.9$

35. $\frac{1}{2}x - \frac{4}{2} + \frac{1}{4}x + \frac{3}{4} = \frac{2}{4}x + \frac{1}{4}x - \frac{8}{4} + \frac{3}{4}$
$= \frac{3}{4}x - \frac{5}{4}$

36. $-3x^2 + 6y - 6y + 8x^2 = 5x^2$

37. $\dfrac{-6x^2}{-2} - \dfrac{4x}{-2} + \dfrac{2}{-2} = 3x^2 + 2x - 1$

38. $5x - 40 = 5x - 5 \cdot 8 = 5(x - 8)$

39. $7t - 7 = 7t - 7 \cdot 1 = 7(t - 1)$

40. The perimeter is $2x + 2(x - 4)$ or $4x - 8$ feet. If $x = 9$, the perimeter is $4 \cdot 9 - 8$ or 28 feet. The area is $x^2 - 4x$. If $x = 9$, then the area is $9^2 - 4 \cdot 9$, or 45 ft^2.

41. $6(1.03)^{25} \approx 12.6$ billion

2.1 WARM-UPS

1. False, it is equivalent to $-2x = 5$.
2. True, because
$x - (x - 3) = x - x + 3 = 3$.
3. False, multiply each side by $\frac{4}{3}$.
4. True, because we can multiply each side by -1.
5. True, because all of the denominators divide into the LCD.
6. True, subtract 5 from each side and then divide each side by 3.
7. True, because it simplifies to $8 = 12$.
8. True, it is equivalent to $x = -3$.
9. True, it is equivalent to $0.8x = 0.8x$.
10. True, since it is equivalent to $3x = 12$.

2.1 EXERCISES

1. An equation is a sentence that expresses the equality of two algebraic expressions.
3. Equivalent equation are equations that have the same solution set.
5. If the equation involves fractions then multiply each side by the LCD.
7. A conditional equation is an equation that has at least one solution but is not an identity.
9. Yes, because $3 \cdot (-4) + 7 = -5$ is correct.
11. Yes, because $\frac{1}{2} \cdot 12 - 4 = \frac{1}{3} \cdot 12 - 2$.
13. No, because $0.2(200 - 50) = 30$ and $20 - 0.05 \cdot 200 = 10$.

15.
$$\begin{aligned} x + 3 &= 24 \\ x + 3 - 3 &= 24 - 3 \\ x &= 21 \end{aligned}$$
The solution set is $\{21\}$.

17.
$$\begin{aligned} 5x &= 20 \\ \frac{1}{5} \cdot 5x &= \frac{1}{5} \cdot 20 \\ x &= 4 \end{aligned}$$
The solution set is $\{4\}$.

19.
$$\begin{aligned} 2x - 3 &= 25 \\ 2x - 3 + 3 &= 25 + 3 \\ 2x &= 28 \\ \frac{1}{2} \cdot 2x &= \frac{1}{2} \cdot 28 \\ x &= 14 \end{aligned}$$
The solution set is $\{14\}$.

21.
$$\begin{aligned} -72 - x &= 15 \\ -72 - x + 72 &= 15 + 72 \\ -x &= 87 \\ (-1)(-x) &= -1 \cdot 87 \\ x &= -87 \end{aligned}$$
The solution set is $\{-87\}$.

23.
$$\begin{aligned} -3x - 19 &= 5 - 2x \\ -3x &= 24 - 2x \\ -3x + 2x &= 24 \\ -x &= 24 \\ x &= -24 \end{aligned}$$
The solution set is $\{-24\}$.

25.
$$\begin{aligned} 2x - 3 &= 0 \\ 2x - 3 + 3 &= 0 + 3 \\ 2x &= 3 \\ \frac{2x}{2} &= \frac{3}{2} \\ x &= \frac{3}{2} \end{aligned}$$
The solution set is $\left\{\frac{3}{2}\right\}$.

27.
$$\begin{aligned} -2x + 5 &= 7 \\ -2x + 5 - 5 &= 7 - 5 \\ -2x &= 2 \\ \frac{-2x}{-2} &= \frac{2}{-2} \\ x &= -1 \end{aligned}$$
The solution set is $\{-1\}$.

29.
$$\begin{aligned} -12x - 15 &= 21 \\ -12x - 15 + 15 &= 21 + 15 \\ -12x &= 36 \\ \frac{-12x}{-12} &= \frac{36}{-12} \\ x &= -3 \end{aligned}$$
The solution set is $\{-3\}$.

31.
$$\begin{aligned} 26 &= 4x + 16 \\ 26 - 16 &= 4x + 16 - 16 \\ 10 &= 4x \\ \frac{10}{4} &= \frac{4x}{4} \\ \frac{5}{2} &= x \end{aligned}$$
The solution set is $\left\{\frac{5}{2}\right\}$.

33.
$$\begin{aligned} -3(x - 16) &= 12 - x \\ -3x + 48 &= 12 - x \\ -3x &= -36 - x \\ -2x &= -36 \\ x &= 18 \end{aligned}$$
The solution set is $\{18\}$.

35.
$$\begin{aligned} 2(x + 9) - x &= 36 \\ 2x + 18 - x &= 36 \\ x + 18 &= 36 \end{aligned}$$

$x+18-18=36-18$
$x=18$
The solution set is $\{18\}$.

37. $2+3(x-1)=x-1$
$2+3x-3=x-1$
$3x-1=x-1$
$3x-x=-1+1$
$2x=0$
$x=0$
The solution set is $\{0\}$.

39. $-\frac{3}{7}x=4$
$-\frac{7}{3}\left(-\frac{3}{7}x\right)=-\frac{7}{3}(4)$
$x=-\frac{28}{3}$
The solution set is $\left\{-\frac{28}{3}\right\}$.

41. $-\frac{5}{7}x-1=3$
$-\frac{5}{7}x=4$
$-\frac{7}{5}\left(-\frac{5}{7}x\right)=-\frac{7}{5}(4)$
$x=-\frac{28}{5}$
The solution set is $\left\{-\frac{28}{5}\right\}$.

43. $\frac{x}{3}+\frac{1}{2}=\frac{7}{6}$
$6\left(\frac{x}{3}+\frac{1}{2}\right)=6\cdot\frac{7}{6}$
$2x+3=7$
$2x=4$
$x=2$
The solution set is $\{2\}$.

45. $\frac{2}{3}x+5=-\frac{1}{3}x+17$
$3\left(\frac{2}{3}x+5\right)=3\left(-\frac{1}{3}x+17\right)$
$2x+15=-x+51$
$3x+15=51$
$3x=36$
$x=12$
The solution set is $\{12\}$.

47. $\frac{1}{2}x+\frac{1}{4}=\frac{1}{4}\left(x-6\right)$
$4\left(\frac{1}{2}x+\frac{1}{4}\right)=4\cdot\frac{1}{4}\left(x-6\right)$
$4\left(\frac{1}{2}x+\frac{1}{4}\right)=4\cdot\frac{1}{4}\left(x-6\right)$
$2x+1=x-6$
$x+1=-6$
$x=-7$
The solution set is $\{-7\}$.

49. $8-\frac{x-2}{2}=\frac{x}{4}$
$4\left(8-\frac{x-2}{2}\right)=4\left(\frac{x}{4}\right)$
$32-2(x-2)=x$
$32-2x+4=x$
$-3x=-36$
$x=12$
The solution set is $\{12\}$.

51. $\frac{y-3}{3}-\frac{y-2}{2}=-1$
$6\left(\frac{y-3}{3}\right)-6\left(\frac{y-2}{2}\right)=6(-1)$
$2y-6-3y+6=-6$
$-y=-6$
$y=6$
The solution set is $\{6\}$.

53. $x-0.2x=72$
$10(x-0.2x)=10(72)$
$10x-2x=720$
$8x=720$
$x=90$
The solution set is $\{90\}$.

55. $0.03(x+200)+0.05x=86$
$0.03x+0.03(200)+0.05x=86$
$0.08x+6=86$
$0.08x=80$
$x=1000$
The solution set is $\{1000\}$.

57. $0.1x+0.05(x-300)=105$
$0.1x+0.05x-0.05(300)=105$
$0.15x-15=105$
$0.15x=120$
$x=800$
The solution set is $\{800\}$.

59. $2(x+1)=2(x+3)$
$2x+2=2x+6$
$2=6$
The solution set is $\emptyset$. The equation is inconsistent.

61. $x+x=2x$
$2x=2x$
The solution set is R. The equation is an identity.

63. $x+x=2$
$2x=2$
$x=1$
The solution set is $\{1\}$ and the equation is conditional.

65. $\frac{4x}{4}=x$
$x=x$

The solution set is R and the equation is an identity.

67. $x \cdot x = x^2$

$x^2 = x^2$

The solution set is R and the equation is an identity.

69. $2(x+3) - 7 = 5(5-x) + 7(x+1)$

$2x + 6 - 7 = 25 - 5x + 7x + 7$

$2x - 1 = 32 + 2x$

$-1 = 32$

The solution set is $\emptyset$. The equation is inconsistent.

71. $2\left(\frac{1}{2}x + \frac{3}{2}\right) - \frac{7}{2} = \frac{3}{2}(x+1) - \left(\frac{1}{2}x + 2\right)$

$x + 3 - \frac{7}{2} = \frac{3}{2}x + \frac{3}{2} - \frac{1}{2}x - 2$

$2x + 6 - 7 = 3x + 3 - x - 4$

$2x - 1 = 2x - 1$

The solution set to this identity is R.

73. $2(0.5x + 1.5) - 3.5 = 3(0.5x + 0.5)$

$x + 3 - 3.5 = 1.5x + 1.5$

$x - 0.5 = 1.5x + 1.5$

$10x - 5 = 15x + 15$

$-20 = 5x$

$-4 = x$

The solution set is to this conditional equation is $\{-4\}$.

75. $4 - 6(2x-3) + 1 = 3 + 2(5-x)$

$4 - 12x + 18 + 1 = 3 + 10 - 2x$

$-12x + 23 = 13 - 2x$

$-10x = -10$

$x = 1$

The solution set is $\{1\}$.

77. $5x - 2(3x+6) = 4 - (2+x) + 7$

$5x - 6x - 12 = 4 - 2 - x + 7$

$-x - 12 = -x + 9$

$-12 = 9$

The solution set is the empty set, $\emptyset$.

79. $\frac{2x-5}{4} - \frac{3x-1}{6} = -\frac{13}{12}$

$12 \cdot \frac{2x-5}{4} - 12 \cdot \frac{3x-1}{6} = -\frac{13}{12} \cdot 12$

$6x - 15 - 6x + 2 = -13$

$-13 = -13$

All real numbers satisfy this equation. The solution set is R or $(-\infty, \infty)$.

81. $\frac{1}{2}\left(y - \frac{1}{6}\right) + \frac{2}{3} = \frac{5}{6} + \frac{1}{3}\left(\frac{1}{2} - 3y\right)$

$\frac{1}{2}y - \frac{1}{12} + \frac{2}{3} = \frac{5}{6} + \frac{1}{6} - y$

$6y - 1 + 8 = 10 + 2 - 12y$

$6y + 7 = 12 - 12y$

$18y = 5$

$y = \frac{5}{18}$

The solution set is $\left\{\frac{5}{18}\right\}$.

83. $\frac{40x-5}{2} + \frac{5}{2} = \frac{33-2x}{3} - 11$

$20x - \frac{5}{2} + \frac{5}{2} = 11 - \frac{2}{3}x - 11$

$20x = -\frac{2}{3}x$

$60x = -2x$

$62x = 0$

$x = 0$

The solution set is $\{0\}$.

85. $1.3 - 0.2(6 - 3x) = 0.1(0.2x + 3)$

$1.3 - 1.2 + 0.6x = 0.02x + 0.3$

$130 - 120 + 60x = 2x + 30$

$10 + 60x = 2x + 30$

$58x = 20$

$x = \frac{20}{58} = \frac{10}{29}$

The solution set is $\left\{\frac{10}{29}\right\}$.

87. $3x - 9 = 0$

$3x = 9$

$x = 3$

The solution set is $\{3\}$.

89. $7 - z = -9$

$16 = z$

The solution set is $\{16\}$.

91. $\frac{2}{3}x = \frac{1}{2}$

$x = \frac{3}{2} \cdot \frac{1}{2} = \frac{3}{4}$

The solution set is $\left\{\frac{3}{4}\right\}$.

93. $-\frac{3}{5}y = 9$

$y = -\frac{5}{3} \cdot 9 = -15$

The solution set is $\{-15\}$.

95. $3y + 5 = 4y - 1$

$6 = y$

The solution set is $\{6\}$.

97. $5x + 10(x+2) = 110$

$5x + 10x + 20 = 110$

$15x = 90$

$x = 6$

The solution set is $\{6\}$.

99. $\frac{P+7}{3} - \frac{P-2}{5} = \frac{7}{3} - \frac{P}{15}$

$15\left(\frac{P+7}{3}\right) - 15\left(\frac{P-2}{5}\right)$

$= 15\left(\frac{7}{3}\right) - 15\left(\frac{P}{15}\right)$

$$5P + 35 - 3P + 6 = 35 - P$$
$$2P + 41 = 35 - P$$
$$3P = -6$$
$$P = -2$$

The solution set is $\{-2\}$.

101. $x - 0.06x = 50{,}000$
$$0.94x = 50{,}000$$
$$x = \frac{50{,}000}{0.94} \approx 53{,}191.49$$

The solution set is $\{53{,}191.49\}$.

103. $2.365x + 3.694 = 14.8095$
$$2.365x = 14.8095 - 3.694$$
$$2.365x = 11.1155$$
$$x = \frac{11.1155}{2.365} = 4.7$$

The solution set is $\{4.7\}$.

105. a) For 1992, $x = 7$.
$$0.45(7) + 39.05 = 42.2$$

Public school enrollment in 1992 was 42.2 million.

b) $0.45x + 39.05 = 50$
$$0.45x = 10.95$$
$$x \approx 24.3$$

So in 24.3 years (or during the 25th year) public school enrollment will reach 50 million. $1985 + 25 = 2010$.

c) Judging from the graph, enrollment is increasing.

107. To eliminate decimals multiply each side by an appropriate power of 10.

2.2 WARM-UPS

1. False, P is on both sides of the equation.
2. False, because we use the distributive property to factor out P on the right side.
3. False, because we divide each side by Pr to get $t = \frac{I}{Pr}$.
4. True, because $\frac{5 \cdot 6}{2} = 15$.
5. True, because $P = 2L + 2W$.
6. True, because $V = LWH$.
7. False, because $A = \frac{1}{2}h(b_1 + b_2)$.
8. True, $x - y = 5$ is equivalent to $x = 5 + y$, or $y = x - 5$.
9. True, because $-2(-3) - 4 = 6 - 4 = 2$.
10. False, perimeter is the distance around the outside edge.

2.2 EXERCISES

1. A formula is an equation involving two or more variables.

3. Solving for a variable means to rewrite the formula with the indicated variable isolated.

5. To find the value a variable, solve for that variable, then replace all other variables with the given numbers.

7. $I = Prt$
$$\frac{I}{Pr} = \frac{Prt}{Pr}$$
$$\frac{I}{Pr} = t$$
$$t = \frac{I}{Pr}$$

9. $F = \frac{9}{5}C + 32$
$$F - 32 = \frac{9}{5}C$$
$$\frac{5}{9}(F - 32) = \frac{5}{9} \cdot \frac{9}{5}C$$
$$C = \frac{5}{9}(F - 32)$$

11. $A = LW$
$$\frac{A}{L} = \frac{LW}{L}$$
$$\frac{A}{L} = W$$
$$W = \frac{A}{L}$$

13. $A = \frac{1}{2}(b_1 + b_2)$
$$2A = 2 \cdot \frac{1}{2}(b_1 + b_2)$$
$$2A = b_1 + b_2$$
$$2A - b_2 = b_1$$
$$b_1 = 2A - b_2$$

15. $P = 2L + 2W$
$$P - 2W = 2L$$
$$\frac{P - 2W}{2} = L$$
$$L = \frac{P - 2W}{2} \text{ or } L = \frac{1}{2}P - W$$

17. $V = \pi r^2 h$
$$\frac{V}{\pi r^2} = \frac{\pi r^2 h}{\pi r^2}$$
$$h = \frac{V}{\pi r^2}$$

19. $2x + 3y = 9$
$$3y = -2x + 9$$
$$\frac{3y}{3} = \frac{-2x}{3} + \frac{9}{3}$$
$$y = -\frac{2}{3}x + 3$$

21.
$$\begin{aligned} x - y &= 4 \\ -y &= -x + 4 \\ \frac{-y}{-1} &= \frac{-x}{-1} + \frac{4}{-1} \\ y &= x - 4 \end{aligned}$$

23.
$$\begin{aligned} \tfrac{1}{2}x - \tfrac{1}{3}y &= 2 \\ -\tfrac{1}{3}y &= -\tfrac{1}{2}x + 2 \\ -3\left(-\tfrac{1}{3}y\right) &= -3\left(-\tfrac{1}{2}x + 2\right) \\ y &= \tfrac{3}{2}x - 6 \end{aligned}$$

25.
$$\begin{aligned} y - 2 &= \tfrac{1}{2}(x - 3) \\ y - 2 &= \tfrac{1}{2}x - \tfrac{3}{2} \\ y &= \tfrac{1}{2}x - \tfrac{3}{2} + 2 \\ y &= \tfrac{1}{2}x + \tfrac{1}{2} \end{aligned}$$

27.
$$\begin{aligned} A &= P + Prt \\ A - P &= Prt \\ \frac{A - P}{Pr} &= \frac{Prt}{Pr} \\ \frac{A - P}{Pr} &= t \\ t &= \frac{A - P}{Pr} \end{aligned}$$

29.
$$\begin{aligned} ab + a &= 1 \\ a(b + 1) &= 1 \\ \frac{a(b + 1)}{b + 1} &= \frac{1}{b + 1} \\ a &= \frac{1}{b + 1} \end{aligned}$$

31.
$$\begin{aligned} xy + 5 &= y - 7 \\ xy - y &= -5 - 7 \\ y(x - 1) &= -12 \\ y &= \frac{-12}{x - 1} \\ y &= \frac{-12(-1)}{(x - 1)(-1)} \\ y &= \frac{12}{1 - x} \end{aligned}$$

33.
$$\begin{aligned} xy^2 + xz^2 &= xw^2 - 6 \\ xy^2 + xz^2 - xw^2 &= -6 \\ x(y^2 + z^2 - w^2) &= -6 \\ x &= \frac{-6}{y^2 + z^2 - w^2} \\ x &= \frac{6}{w^2 - y^2 - z^2} \end{aligned}$$

35.
$$\begin{aligned} 3.35x - 54.6 &= 44.3 - 4.58x \\ 3.35x + 4.58x &= 44.3 + 54.6 \\ x(3.35 + 4.58) &= 44.3 + 54.6 \\ x &= \frac{44.3 + 54.6}{3.35 + 4.58} \\ x &\approx 12.472 \end{aligned}$$

37.
$$\begin{aligned} 4.59x - 66.7 &= 3.2(x - 5.67) \\ 4.59x - 66.7 &= 3.2x - 3.2(5.67) \\ x(4.59 - 3.2) &= 66.7 - 3.2(5.67) \\ x &= \frac{66.7 - 3.2(5.67)}{4.59 - 3.2} \\ x &\approx 34.932 \end{aligned}$$

39.
$$\begin{aligned} \frac{x}{19} - \frac{3}{23} &= \frac{4}{31} + \frac{3x}{7} \\ \frac{x}{19} + \frac{3x}{7} &= \frac{4}{31} + \frac{3}{23} \\ x\left(\frac{1}{19} + \frac{3}{7}\right) &= \frac{4}{31} + \frac{3}{23} \\ x &= \frac{\frac{4}{31} + \frac{3}{23}}{\frac{1}{19} + \frac{3}{7}} \approx 0.539 \end{aligned}$$

41.
$$\begin{aligned} 2x - 3y &= 5 \\ 2(3) - 3y &= 5 \\ -3y &= -1 \\ y &= \tfrac{1}{3} \end{aligned}$$

43.
$$\begin{aligned} -4x + 2y &= 1 \\ -4(3) + 2y &= 1 \\ 2y &= 13 \\ y &= \tfrac{13}{2} \end{aligned}$$

45.
$$\begin{aligned} y &= -2x + 5 \\ y &= -2(3) + 5 \\ y &= -6 + 5 \\ y &= -1 \end{aligned}$$

47.
$$\begin{aligned} -x + 2y &= 5 \\ -3 + 2y &= 5 \\ 2y &= 8 \\ y &= 4 \end{aligned}$$

49.
$$\begin{aligned} y - 1.046 &= 2.63(x - 5.09) \\ y - 1.046 &= 2.63(3 - 5.09) \\ y - 1.046 &= -5.4967 \\ y &= -4.4507 \end{aligned}$$

51.
$$\begin{aligned} wxy &= 5 \\ 4x(2) &= 5 \\ 8x &= 5 \\ x &= \tfrac{5}{8} \end{aligned}$$

53.
$$\begin{aligned} x + xz &= 7 \\ x + x(-3) &= 7 \\ -2x &= 7 \\ x &= -\tfrac{7}{2} \end{aligned}$$

55.
$$\begin{aligned} w(x - z) &= y(x - 4) \\ 4[x - (-3)] &= 2(x - 4) \\ 4x + 12 &= 2x - 8 \\ 2x &= -20 \\ x &= -10 \end{aligned}$$

57.
$$\begin{aligned} w &= \tfrac{1}{2}xz \\ 4 &= \tfrac{1}{2}x(-3) \end{aligned}$$

$$4 = -\frac{3}{2}x$$
$$-\frac{2}{3}(4) = -\frac{2}{3}\left(-\frac{3}{2}x\right)$$
$$-\frac{8}{3} = x$$

59.
$$\frac{1}{w} + \frac{1}{x} = \frac{1}{y}$$
$$\frac{1}{4} + \frac{1}{x} = \frac{1}{2}$$
$$4x \cdot \frac{1}{4} + 4x \cdot \frac{1}{x} = 4x \cdot \frac{1}{2}$$
$$x + 4 = 2x$$
$$4 = x$$

61.
$$I = Prt$$
$$300 = 1000 \cdot 2 \cdot \mathrm{r}$$
$$r = \frac{300}{2000} = 15\%$$

63.
$$I = Prt$$
$$20 = 500 \cdot \mathrm{r} \cdot \frac{2}{52}$$
$$r = \frac{20}{500 \cdot \frac{2}{52}} = 104\%$$

65. Use $P = 2000$, $r = 0.18$, and $I = 180$ in the formula for simple interest $I = Prt$.
$$180 = 2000(0.18)t$$
$$180 = 360t$$
$$0.5 = t$$
The time is one-half year.

67. The formula for the area of a circle is $A = \pi r^2$.

69. Since $C = 2\pi r$, we can divide each side by 2π to get $r = \frac{C}{2\pi}$.

71. Because $P = 2L + 2W$, we can subtract $2L$ from each side to get $2W = P - 2L$, and then divide each side by 2 to get $W = \frac{P - 2L}{2}$ or $W = \frac{1}{2}P - L$.

73.
$$A = LW$$
$$23 = L(4)$$
$$L = \frac{23}{4} = 5.75 \text{ yards}$$

75.
$$V = LWH$$
$$36 = 2(2.5)H$$
$$36 = 5H$$
$$H = 7.2 \text{ feet}$$

77. $V = 900 \text{ gal.}\left(\frac{1 \text{ ft}^3}{7.5 \text{ gal}}\right) = 120 \text{ ft}^3$
$$V = LWH$$
$$120 = 4 \cdot 6 \cdot H$$
$$H = \frac{120}{24} = 5 \text{ feet}$$

79.
$$A = \frac{1}{2}bh$$
$$30 = \frac{1}{2}4h$$
$$30 = 2h$$
$$h = 15 \text{ feet}$$

81.
$$A = \frac{1}{2}h(b_1 + b_2)$$
$$300 = \frac{1}{2}(20)(16 + b_2)$$
$$300 = 10(16 + b_2)$$
$$30 = 16 + b_2$$
$$b_2 = 14 \text{ inches}$$

83.
$$P = 2L + 2W$$
$$600 = 2L + 2(132)$$
$$336 = 2L$$
$$L = 168 \text{ feet}$$

85.
$$C = 2\pi r$$
$$3\pi = 2\pi r$$
$$r = \frac{3\pi}{2\pi}$$
$$r = 1.5 \text{ meters}$$

87.
$$C = 2\pi r$$
$$r = \frac{C}{2\pi}$$
$$r = \frac{25{,}000}{2\pi}$$
$$r \approx 3979 \text{ miles}$$

89. If the diameter is 3 in., then the radius is 1.5 in.
$$V = \pi r^2 h$$
$$30 = \pi(1.5)^2 h$$
$$30 = 2.25\pi h$$
$$h = \frac{30}{2.25\pi} \approx 4.24 \text{ inches}$$

91. Let $W = 62 \text{ lb/ft}^3$, $D = 32$ ft, and $A = 48 \text{ ft}^2$ in $F = WDA$:
$$F = 62 \text{ lb/ft}^3 \cdot 32 \text{ ft} \cdot 48 \text{ ft}^2 = 95{,}232 \text{ lb}$$

93. $\$1000 \div \$2 = 500$ feet
$$b_1 + b_2 = 500$$
$$A = \frac{1}{2}h(b_1 + b_2)$$
$$50{,}000 = \frac{1}{2}h \cdot 500$$
$$h = 200 \text{ feet}$$

95.
$$A = bh$$
$$60{,}000 = b \cdot 200$$
$$b = 300 \text{ feet}$$
There is 300 ft on each street. So 600 ft at \$2 each is a \$1200 assessment.

97. Since August has 31 days, the total is $1 + 2 + 3 + 4 + \ldots + 31$. Use $n = 31$ in the formula $S = \frac{n(n+1)}{2}$:
$$S = \frac{31(32)}{2} = 31 \cdot 16 = 496$$
So her total jogging time is 496 minutes.

99. a) $N = B + S - 1$
$N = 2003 + 455 - 1 = 2457$

b) $1452 = 1033 + S - 1$
$420 = S$

2.3 WARM-UPS

1. False, first identify what the variable stands for.
2. True, we must know what the letters represent.
3. False, you may have the wrong equation.
4. False, odd integers differ by 2.
5. True, because $x + 6 - x = 6$.
6. True, because $x + 7 - x = 7$.
7. False, $5x - 2 = 3(x + 20)$ since $5x$ is larger than $3(x + 20)$.
8. True, because 8% of x is $0.08x$.
9. False, because 10% of \$88,000 is \$8,800 and \$88,000 – \$8,800 is not \$80,000.
10. False, because the acid in the mixture must be between 10% and 14%.

2.3 EXERCISES

1. Three unknown consecutive integers are represented by x, $x + 1$, and $x + 2$.
3. The formula $P = 2L + 2W$ expresses the perimeter in terms of the length and width.
5. The commission is a percentage of the selling price.
7. Since two consecutive even integers differ by 2, we can use x and $x + 2$ to represent them.
9. The expressions x and $10 - x$ have a sum of 10, since $x + 10 - x = 10$.
11. Two numbers with a difference of 2 can be represented as x and $x + 2$ or x and $x - 2$.
13. If x is the selling price, then eighty-five percent of the selling price is $0.85x$.
15. Since $D = RT$, the distance is $3x$ miles.
17. Since the perimeter is twice the length $(x + 5)$ plus twice the width (x), we can represent the perimeter by $2(x + 5) + 2(x)$ or $4x + 10$.
19. Let $x =$ the first integer, $x + 1 =$ the second integer, and $x + 2 =$ the third integer. Their sum is 84:

$$x + x + 1 + x + 2 = 84$$
$$3x + 3 = 84$$
$$3x = 81$$
$$x = 27$$

If $x = 27$, then $x + 1 = 28$, and $x + 2 = 29$. The integers are 27, 28, and 29.

21. Let $x =$ the first even integer, $x + 2 =$ the second, and $x + 4 =$ the third. Their sum is 252:

$$x + x + 2 + x + 4 = 252$$
$$3x + 6 = 252$$
$$3x = 246$$
$$x = 82$$

If $x = 82$, then $x + 2 = 84$ and $x + 4 = 86$. The integers are 82, 84, and 86.

23. Let $x =$ the first odd integer and $x + 2 =$ the second. Their sum is 128:

$$x + x + 2 = 128$$
$$2x = 126$$
$$x = 63$$

If $x = 63$, then $x + 2 = 65$. The integers are 63 and 65.

25. Let $x =$ the number.

$$x + 5 = -8$$
$$x = -13$$

The number is -13.

27. Let $x =$ the number.

$$2x + 6 = 52$$
$$2x = 46$$
$$x = 23$$

The number is 23.

29. Let $x =$ the number.

$$\frac{1}{6}x - \frac{1}{7}x = 1$$
$$42\left(\frac{1}{6}x - \frac{1}{7}x\right) = 42$$
$$7x - 6x = 42$$
$$x = 42$$

The number is 42.

31. Let $x =$ the length and $x - 2 =$ the width. Since $P = 2L + 2W$ we can write the following equation.

$$2x + 2(x - 2) = 16$$
$$4x - 4 = 16$$
$$4x = 20$$
$$x = 5$$
$$x - 2 = 3$$

So the length is 5 ft and the width is 3 ft.

33. Let $x =$ the width and $2x + 4 =$ the length. Since the perimeter is 26 in., we can write the following equation.

$$2x + 2(2x + 4) = 26$$
$$2x + 4x + 8 = 26$$
$$6x = 18$$
$$x = 3$$
$$2x + 4 = 10$$

The width of the glass is 3 in. and the length of the glass is 10 in.

35. Let $x =$ the length and $\frac{1}{2}x - 2 =$ the width. Since the perimeter is 44 cm, we can write the following equation.

$$2x + 2(\tfrac{1}{2}x - 2) = 44$$
$$2x + x - 4 = 44$$
$$3x = 48$$
$$x = 16$$
$$\tfrac{1}{2}x - 2 = 6$$

The width of the sign is 6 cm and the length is 16 cm.

37. Let $x =$ the width and $x + 2 =$ the length. Since he is fencing two width and one length, we can write the following equation.

$$2x + 1(x + 2) = 14$$
$$2x + x + 2 = 14$$
$$3x = 12$$
$$x = 4$$
$$x + 2 = 6$$

The width of the region is 4 ft and the length is 6 ft.

39. Let $x =$ the width, and $2x + 5 =$ the length. To fence the 3 sides, we use 2 widths and 1 length:

$$2(x) + (2x + 5) = 50$$
$$4x = 45$$
$$x = 11.25$$
$$2x + 5 = 27.5$$

Width is 11.25 feet and the length is 27.5 feet.

41. Let $x =$ the length of the first side, $2x - 10 =$ the length of the second side, and $x + 50 =$ the length of the third side. Since the perimeter is 684, we have the following equation.

$$x + 2x - 10 + x + 50 = 684$$
$$4x + 40 = 684$$
$$4x = 644$$
$$x = 161$$

If $x = 161$ feet, then $2x - 10 = 312$ feet, and $x + 50 = 211$ feet.

43. Let $x =$ the amount invested at 5% and $2x =$ the amount invested at 9%. Since the interest on the investments is $0.05x$ and $0.09(2x)$ we can write the equation

$$0.05x + 0.09(2x) = 920$$
$$0.05x + 0.18x = 920$$
$$0.23x = 920$$
$$x = 4000$$
$$2x = 8000$$

He invested \$4000 at 5% and \$8000 at 9%.

45. Let $x =$ the amount invested at 6% and $x + 1000 =$ the amount invested at 10%. Since the interest on the investments is $0.06x$ and $0.10(x + 1000)$ we can write the equation

$$0.06x + 0.10(x + 1000) = 340$$
$$0.16x + 100 = 340$$
$$0.16x = 240$$
$$x = 1500$$
$$x + 1000 = 2500$$

He invested \$1500 at 6% and \$2500 at 10%.

47. Let $x =$ the amount of his inheritance. He invests $\frac{1}{2}x$ at 10% and $\frac{1}{4}x$ at 12%. His total income of \$6400 can be expressed as

$$0.10\left(\tfrac{1}{2}x\right) + 0.12\left(\tfrac{1}{4}x\right) = 6400$$
$$0.05x + 0.03x = 6400$$
$$0.08x = 6400$$
$$x = 80{,}000$$

His inheritance was \$80,000.

49. Let $x =$ the amount of Claudette's inheritance, $\frac{1}{2}x =$ the amount he invested at 5%, and $\frac{1}{3}x =$ the amount invested at 6%.

$$\tfrac{1}{2}x \cdot 0.05 + \tfrac{1}{3}x \cdot 0.06 = 9000$$

Multiply by 6:

$$3x \cdot 0.05 + 2x \cdot 0.06 = 54000$$
$$0.15x + 0.12x = 54000$$
$$0.27x = 54000$$
$$x = 200,000$$

The inheritance was \$200,000.

51. Let $x =$ the number of gallons of 5% solution. In the 5% solution there are $0.05x$ gallons of acid, and in the 20 gallons of 10% solution there are $0.10(20)$ gallons of acid. The final mixture consists of $x + 20$ gallons of which 8% is acid. The total acid in the mixture is the sum of the acid from each solution mixed together:

$$0.05x + 0.10(20) = 0.08(x + 20)$$
$$0.05x + 2 = 0.08x + 1.6$$
$$0.4 = 0.03x$$
$$x = \frac{0.4}{0.03} = \frac{40}{3}$$

Use $\frac{40}{3}$ gallons of 5% solution.

53. Let $x =$ the amount of \$8 per pound cherries and $x + 12 =$ the amount of the \$7 per pound mixture.

$$12(5) + x(8) = (x + 12)7$$
$$60 + 8x = 7x + 84$$
$$x = 24$$

He should use 24 pounds of cherries.

55. Let $x =$ the amount of 5% solution and $6 - x =$ the amount of 13% solution.

$$0.05x + 0.13(6 - x) = 0.08(6)$$
$$0.05x + 0.78 - 0.13x = 0.48$$
$$-0.08x = -0.30$$
$$x = 3.75$$
$$6 - x = 2.25$$

He should us 3.75 gallons of 5% solution and 2.25 gallons of 13% solution.

57. Let $x =$ the number of gallons of pure acid. After mixing we will have $1 + x$ gallons of 6% solution. The original gallon has 0.05(1) gallons of acid in it. The equation totals up the acid:

$$0.05(1) + x = 0.06(1 + x)$$
$$0.05 + x = 0.06 + 0.06x$$
$$0.94x = 0.01$$
$$x = 0.010638 \text{ gallons}$$

Use 1 gallon $=$ 128 ounces, to get approximately 1.36 ounces of pure acid.

59. Let $x =$ his speed in the fog and $x + 30 =$ his increased speed. Since $D = RT$, his distance in the fog was $3x$ and his distance later was $6(x + 30)$. The equation gives the total distance:

$$3x + 6(x + 30) = 540$$
$$9x + 180 = 540$$
$$9x = 360$$
$$x = 40$$

His speed in the fog was 40 mph.

61. Let $x =$ the speed of the commuter bus and $x + 25 =$ the speed of the express bus. Use $D = RT$ to get the distance traveled by each as $2x$ and $\frac{3}{4}(x + 25)$. The equation expresses the fact that they travel the same distance:

$$2x = \frac{3}{4}(x + 25)$$
$$4 \cdot 2x = 4 \cdot \frac{3}{4}(x + 25)$$
$$8x = 3x + 75$$
$$5x = 75$$
$$x = 15$$

The speed of the commuter bus was 15 mph.

63. Let $x =$ the speed before lunch and $x + 15 =$ the speed after lunch. The distance before is $3x$ and the distance after is $4(x + 15)$. Write the equation about the total distance:

$$3x + 4(x + 15) = 410$$
$$3x + 4x + 60 = 410$$
$$7x = 350$$
$$x = 50$$
$$x + 15 = 65$$

His average speed before lunch was 50 mph.

65. Let $x =$ Candy's speed and $x + 10 =$ Fran's speed. The distance for Candy is $\frac{1}{2}x$ and the distance for Fran is $\frac{1}{3}(x + 10)$. Write the equation about the fact that the distances are the same:

$$\frac{1}{2}x = \frac{1}{3}(x + 10)$$

Multiply each side by 6:

$$3x = 2(x + 10)$$
$$3x = 2x + 20$$
$$x = 20$$

Candy's speed is 20 mph.

67. If $x =$ the selling price, then the commission is $0.08x$. The owner gets the selling price minus the commission:

$$x - 0.08x = 80{,}000$$
$$0.92x = 80{,}000$$
$$x \approx \$86{,}957$$

69. If $x =$ the selling price, then $0.07x =$ the amount of sales tax. The selling price plus the sales tax is the total amount paid:

$$x + 0.07x = 9041.50$$
$$1.07x = 9041.50$$
$$x = \$8450$$

71. Let $x - 3 =$ the distance from the base line to the service line and $x =$ the distance from the service line to the net. The total is 39 feet:

$$x + x - 3 = 39$$
$$2x = 42$$
$$x = 21$$

The distance from service line to the net is 21 ft.

73. Let $x =$ the width, and $2x + 1 =$ the length. Since $P = 2L + 2W$ we can write the following equation.

$$2(x) + 2(2x + 1) = 38$$
$$6x + 2 = 38$$
$$6x = 36$$
$$x = 6$$

If $x = 6$, then $2x + 1 = 13$. The width is 6 m and the length is 13 m.

75. Let $x =$ the driving time for Suzie and $x + 3 =$ the driving time for Scott. The distance for Suzie is $54x$ and the distance for Scott is $58(x + 3)$. Write the equation about the total distance:

$$54x + 58(x + 3) = 734$$
$$54x + 58x + 174 = 734$$
$$112x = 560$$
$$x = 5$$
$$x + 3 = 8$$

So Scott drove for 8 hours.

77. Let $x =$ the price per pound of the mixture, which contains 5 pounds of nuts. Write an equation about the total cost:

$$5(x) = 3(8) + 2(6)$$
$$5x = 36$$
$$x = 7.20$$

The mixture should cost \$7.20 per pound.

79. Let $x =$ the length and $\frac{1}{2}x - 3 =$ the width. Use $P = 2L + 2W$:

$$2(x) + 2\left(\frac{1}{2}x - 3\right) = 48$$
$$2x + x - 6 = 48$$
$$3x = 54$$
$$x = 18$$
$$\frac{1}{2}x - 3 = 6$$

So the length is 18 cm and the width is 6 cm.

81. Let $x =$ the number of points scored by the Packers and $x - 25 =$ the number of points scored by the Chiefs.

$$x + x - 25 = 45$$
$$2x = 70$$
$$x = 35$$

The score was Packers 35, Chiefs 10.

83. Let $x =$ the price per pound for the blended coffee. The price of 0.75 lb of Brazilian coffee at \$10 per lb is \$7.50, and the price of 1.5 lb of Colombian coffee at \$8 per lb is \$12. The total price for 2.25 lb of blended coffee at x dollars per kg is $2.25x$ dollars. We write an equation expressing the total cost:

$$7.50 + 12 = 2.25x$$
$$19.50 = 2.25x$$
$$x \approx 8.67$$

The blended coffee should sell for \$8.67 per lb to the nearest cent.

85. Let $x =$ the number of pounds of apricots. The total cost of 10 pounds of bananas at \$3.20 per pound is \$32, the total cost of x pounds of apricots at \$4 per pound is $4x$, and the total cost $x + 10$ pounds of mix at \$3.80 per pound is $3.80(x + 10)$. Write an equation expressing the total cost:

$$32 + 4x = 3.80(x + 10)$$
$$32 + 4x = 3.80x + 38$$
$$0.20x = 6$$
$$x = 30$$

The mix should contain 30 pounds of apricots.

87. Let $x =$ the number of quarts to be drained out. In the original 20 qt radiator there are 0.30(20) qts of antifreeze. In the x qts that are drained there are $0.30x$ qts of antifreeze, but in the x qts put back in there are x qts of antifreeze. The equation accounts for all of the antifreeze:

$$0.30(20) - 0.30x + x = 0.50(20)$$
$$6 + 0.70x = 10$$
$$0.7x = 4$$
$$x = \frac{4}{0.7} = \frac{40}{7}$$

The amount drained should be $\frac{40}{7}$ qts.

89. If $x =$ the profit in the third quarter of the previous year.

$$x + 0.158x = 44$$
$$1.158x = 44$$
$$x \approx 38$$

The profit per share in the previous year was 38 cents.

91. Let $x =$ Brian's inheritance, $\frac{1}{2}x =$ Daniel's inheritance, and $\frac{1}{3}x - 1000 =$ Raymond's inheritance. The sum of the three amounts is \$25,400:

$$x + \frac{1}{2}x + \frac{1}{3}x - 1000 = 25400$$
$$\frac{11}{6}x = 26400$$
$$x = \frac{6}{11} \cdot 26400$$
$$= 14{,}400$$
$$\frac{1}{2}x = 7200$$
$$\frac{1}{3}x - 1000 = 3800$$

Brian gets \$14,400, Daniel \$7200, and Raymond \$3800.

93. Let x = the first integer and $x + 1$ = the second. Subtract the larger from twice the smaller to get 21:

$$2x - (x + 1) = 21$$
$$x - 1 = 21$$
$$x = 22$$
$$x + 1 = 23$$

The integers are 22 and 23.

95. Let x = Berenice's time and $x - 2$ = Jarrett's time. Berenice's distance is $50x$ and Jarrett's distance is $56(x - 2)$.

$$50x + 56(x - 2) = 683$$
$$106x - 112 = 683$$
$$106x = 795$$
$$x = 7.5$$

Berenice drove for 7.5 hours.

97. Let x = the length of a side of the square. She will use x meters of fencing for each of 3 sides, but only $\frac{1}{2}x$ meters for the side with the opening.

$$3x + \frac{1}{2}x = 70$$
$$\frac{7}{2}x = 70$$
$$x = \frac{2}{7} \cdot 70 = 20$$

The square will by 20 meters by 20 meters.

99. Let x = the amount invested at 8% and $3000 - x$ = the amount invested at 10%. Income on the first investment is $0.08x$ and income on the second is $0.10(3000 - x)$. The total income is \$290:

$$0.08x + 0.10(3000 - x) = 290$$
$$0.08x + 300 - 0.10x = 290$$
$$-0.02x = -10$$
$$x = \frac{-10}{-0.02} = 500$$
$$3000 - x = 2500$$

She invested \$500 at 8% and \$2500 at 10%.

101. Let x = the number of gallons of 5% alcohol and $5 - x$ = the number of gallons of 10% alcohol. The alcohol in the final 5 gallons is the sum of the alcohol in the two separate quantities that are mixed together:

$$0.05x + 0.10(5 - x) = 0.08(5)$$
$$0.05x + 0.5 - 0.10x = 0.4$$
$$-0.05x = -0.1$$
$$x = \frac{-0.1}{-0.05} = 2$$
$$5 - x = 3$$

Use 2 gallons of 5% solution and 3 gallons of 10% solution.

103. Let x = the number of gallons of ethanol.

$$x = 0.85(x + 90)$$
$$x = 0.85x + 76.5$$
$$0.15x = 76.5$$
$$x = 510$$

Use 510 gallons of ethanol.

105. Let x = Darla's age now and $78 - x$ = Todd's age now. In 6 years Todd will be $78 - x + 6$ or $84 - x$, and 6 years ago Darla was $x - 6$. Todd's age in 6 years is twice what Darla's age was 6 years ago:

$$84 - x = 2(x - 6)$$
$$84 - x = 2x - 12$$
$$-3x = -96$$
$$x = 32$$
$$78 - x = 46$$

Todd is 46 now and Darla is 32 now.

2.4 WARM-UPS

1. False, $0 = 0$.
2. False, $-300 < -2$.
3. True, because $-60 = -60$.
4. False, since $6 < x$ is equivalent to $x > 6$.
5. False, $-2x < 10$ is equivalent to $x > -5$.
6. False, $3x \geq -12$ is equivalent to $x \geq -4$.
7. True, multiply each side by -1.
8. True, because of the trichotomy property.
9. True, because of the trichotomy property.
10. True, because $3 - 4(-2) \leq 11$ is correct.

2.4 EXERCISES

1. An inequality is a statement that expresses inequality between two algebraic expressions.

3. If a is less than b, then a lies to the left of b on the number line.

5. When you multiply or divide by a negative number, the inequality symbol is reversed.

7. False, because $-3 > -9$.

9. True, because $0 < 8$.

11. True, because $-60 > -120$.

13. True, because $9 - (-3) = 12$.

15. Yes, because $2(-3) - 4 < 8$ simplifies to $-10 < 8$.

17. No, because $2(5) - 3 \leq 3(5) - 9$ simplifies to $7 \leq 6$.

19. No, because $5 - (-1) < 4 - 2(-1)$ simplifies to $6 < 6$.

21. The solution set is the interval $(-\infty, -1]$. To graph it, shade the numbers to the left of -1 and including -1.

23. The solution set is the interval $(20, \infty)$. To graph it, shade numbers to the right of 20.

25. The solution set is the interval $[3, \infty)$ because $3 \leq x$ is equivalent to $x \geq 3$. We shade the numbers to the right of 3, including 3.

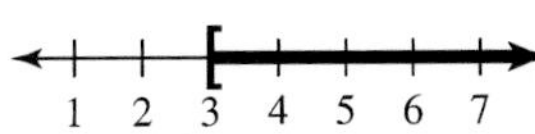

27. The solution set is the interval $(-\infty, 2.3)$. Shade to the left of 2.3.

29. $x + 5 > 12$ is equivalent to $x > 7$.

31. $-x < 6$ is equivalent to $x > -6$.

33. $-2x \geq 8$ is equivalent to $x \leq -4$.

35. $4 < x$ is equivalent to $x > 4$.

37. $-9 \leq -x$ is equivalent to $x \leq 9$.

39. $x + 3 < 5$

$x < 2$

Solution set is $(-\infty, 2)$.

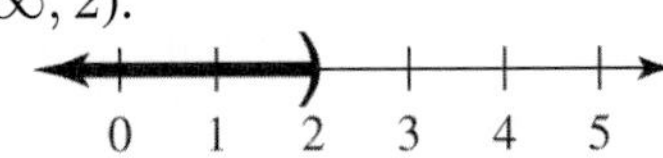

41. $7x > -14$

$x > -2$

Solution set is $(-2, \infty)$.

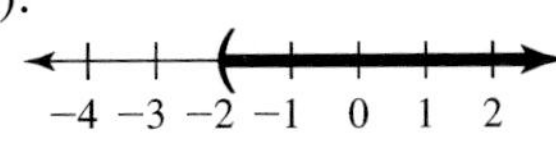

43. $-3x \leq 12$

$x \geq -4$

Solution set is $[-4, \infty)$.

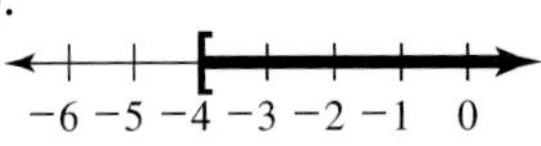

45. $-x > 2$

$x < -2$

Solution set is $(-\infty, -2)$.

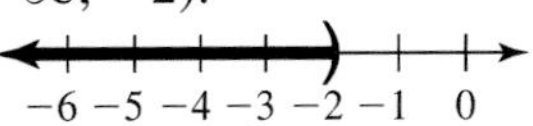

47.

$$\begin{aligned} 2x - 3 &> 7 \\ 2x &> 10 \\ x &> 5 \end{aligned}$$

Solution set is $(5, \infty)$.

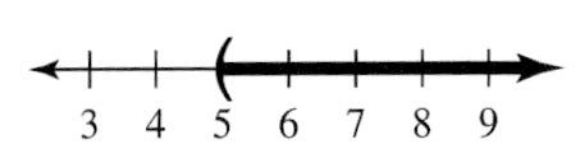

49.

$$\begin{aligned} 4 - x &\leq 3 \\ -x &\leq -1 \\ x &\geq 1 \end{aligned}$$

Solution set is $[1, \infty)$.

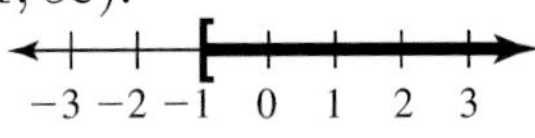

51.

$$\begin{aligned} 18 &\geq 3 - 5x \\ 15 &\geq -5x \\ -3 &\leq x \\ x &\geq -3 \end{aligned}$$

Solution set is $[-3, \infty)$.

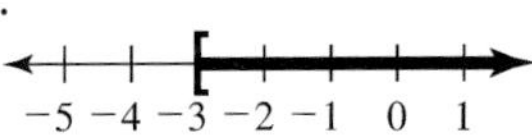

53.

$$\begin{aligned} \frac{x-3}{-5} &< -2 \\ x - 3 &> 10 \\ x &> 13 \end{aligned}$$

Solution set is $(13, \infty)$.

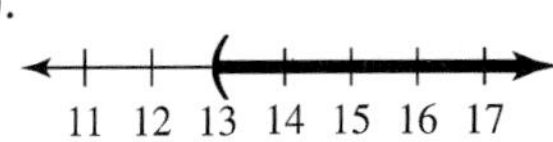

55.

$$\begin{aligned} 2 &\geq \frac{5-3x}{4} \\ 8 &\geq 5 - 3x \\ 3 &\geq -3x \\ -1 &\leq x \\ x &\geq -1 \end{aligned}$$

Solution set is $[-1, \infty)$.

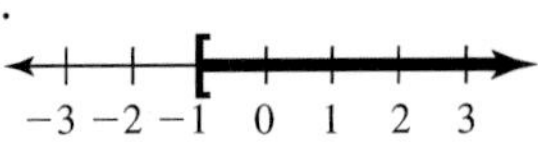

57.

$$\begin{aligned} 3 - \frac{1}{4}x &\geq 2 \\ 4\left(3 - \frac{1}{4}x\right) &\geq 4(2) \\ 12 - x &\geq 8 \\ -x &\geq -4 \\ x &\leq 4 \end{aligned}$$

Solution set is $(-\infty, 4]$.

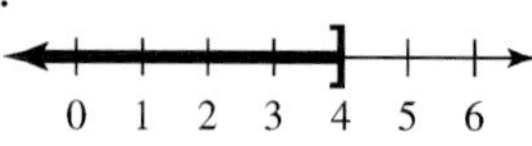

59. $\frac{1}{4}x - \frac{1}{2} < \frac{1}{2}x - \frac{2}{3}$

$12\left(\frac{1}{4}x - \frac{1}{2}\right) < 12\left(\frac{1}{2}x - \frac{2}{3}\right)$

$3x - 6 < 6x - 8$

$-3x < -2$

$x > \frac{2}{3}$

Solution set is $(2/3, \infty)$.

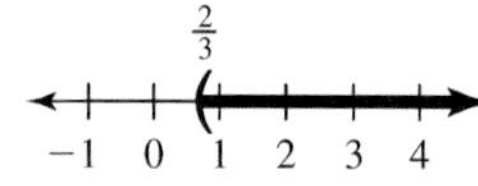

61. $\frac{y-3}{2} > \frac{1}{2} - \frac{y-5}{4}$

$4 \cdot \frac{y-3}{2} > 4 \cdot \frac{1}{2} - 4 \cdot \frac{y-5}{4}$

$2y - 6 > 2 - y + 5$

$3y > 13$

$y > \frac{13}{3}$

Solution set is $(13/3, \infty)$.

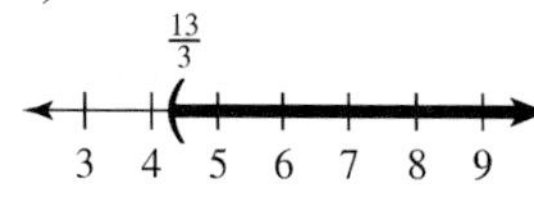

63. $x - 3 > x$

$-3 > 0$

Since $-3 > 0$ is false, the inequality has no solutions. The solution set is the empty set, $\emptyset$.

65. $x \geq x$

Subtract x from each side:

$0 \geq 0$

Since $0 \geq 0$ is true, all real numbers satisfy the original inequality. The solution set is R or $(-\infty, \infty)$.

67. $3(x + 2) \leq 9 + 3x$

$3x + 6 \leq 9 + 3x$

$6 \leq 9$

Solution set is $(-\infty, \infty)$.

69. $-2(5x - 1) \leq -5(5 + 2x)$

$-10x + 2 \leq -25 - 10x$

$2 \leq -25$

Solution set is $\emptyset$.

71. $3x - (4 - 2x) < 5 - (2 - 5x)$

$3x - 4 + 2x < 5 - 2 + 5x$

$5x - 4 < 3 + 5x$

$-4 < 3$

Solution set is $(-\infty, \infty)$.

73. $\frac{1}{2}x + \frac{1}{4}x < \frac{1}{8}(6x - 4)$

$\frac{3}{4}x < \frac{3}{4}x - \frac{1}{2}$

$0 < -\frac{1}{2}$

Solution set is $\emptyset$.

75. If $x =$ Tony's height in feet, then $x > 6$ feet.

77. If $s =$ Wilma's salary in dollars, then $s < 80{,}000$.

79. If $v =$ speed of the Concorde in mph, then $v \leq 1450$ mph.

81. If $a =$ amount in dollars that Julie can afford, then $a \leq \$400$.

83. If $b =$ Burt's height in feet, then $b \leq 5$ feet.

85. If $t =$ Tina's hourly wage in dollars, then $t \leq \$8.20$.

87. Let $x =$ the price of the car in dollars and $0.08x =$ the amount of tax. To spend less than \$10,000 we must satisfy the inequality

$x + 0.08x + 172 < 10{,}000$

$1.08x < 9828$

$x < 9100$

The price range for the car is $x < \$9100$.

89. Let $x =$ the price of the truck in dollars and $0.09x =$ the amount of sales tax. The total cost of at least \$10,000 is expressed as

$x + 0.09x + 80 \geq 10{,}000$

$1.09x \geq 9920$

$x \geq 9100.9174$

The price range for the truck is $x \geq \$9100.92$ to the nearest cent.

91. a) Decreasing

b) $-0.52n + 71.1 < 55$

$-0.52n < -16.1$

$n > 30.96$

Round to 31 years and add 31 to 1980 to get 2011 as the first year in which the number of births will be less than 55 per 1000 women.

93. Let $x =$ the final exam score. One-third of the midterm plus two-thirds of the final must be at least 70:

$\frac{1}{3}(56) + \frac{2}{3}x \geq 70$

$3\left(\frac{1}{3}(56) + \frac{2}{3}x\right) \geq 3(70)$

$56 + 2x \geq 210$

$$2x \geq 154$$
$$x \geq 77$$

The final exam score must satisfy $x \geq 77$.

95. Let $x =$ the price of a pair of A-Mart jeans and $x + 50 =$ the price of a pair of designer jeans. Four pairs of A-Mart jeans cost less than one pair of designer jeans is written as follows.

$$4x < x + 50$$
$$3x < 50$$
$$x < 16.6666$$

The price range for A-Mart jeans is $x < \$16.67$.

97. a) Write the inequality $x \geq 2$ and subtract -6 from each side to get $x - (-6) \geq 2 - (-6)$, or $x + 6 \geq 8$. So the results are all greater than or equal to 8, which is the interval $[8, \infty)$.
b) Write the inequality $x < -3$ and multiply each side by 2 to get $2x < -6$. So the results are all less than -6, which is the interval $(-\infty, -6)$.
c) If every number in $(8, \infty)$ is divided by 4 the result is the interval $(2, \infty)$.
d) If every number in $(6, \infty)$ is multiplied by -2, the result is the interval $(-\infty, -12)$
e) If every number in $(-\infty, -10)$ is divided by -5, the result is the interval $(2, \infty)$.

2.5 WARM-UPS

1. True, because both inequalities are true.
2. True, because both inequalities are correct.
3. False, because $3 > 5$ is incorrect.
4. True, because $3 \leq 10$ is correct.
5. True, because both inequalities are correct.
6. True, because both are correct.
7. False, because $0 < -2$ is incorrect.
8. True, because only numbers larger than 8 are larger than 3 and larger than 8.
9. False, because $(3, \infty) \cup [8, \infty) = (3, \infty)$.
10. True, because the numbers greater than -2 and less than 9 are between -2 and 9.

2.5 EXERCISES

1. A compound inequality consists of two inequalities joined with the words "and" or "or."
3. A compound inequality using or is true when either one or the other or both inequalities is true.
5. The inequality $a < b < c$ means that $a < b$ and $b < c$.
7. No, because $-6 > -3$ is incorrect.
9. Yes, because both inequalities are correct.
11. No, because both inequalities are incorrect.
13. No, because $-4 > -3$ is incorrect.
15. Yes, because even though $-4 > -3$ is incorrect, $-4 < 5$ is correct.
17. Yes, because $-4 - 3 \geq -7$ is correct.
19. Yes, because $2(-4) - 1 < -7$ is correct.
21. The solution set is the set of numbers between -1 and 4:

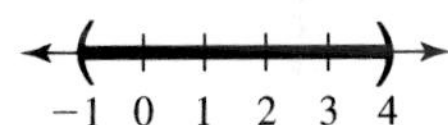

23. Numbers that satisfy both $x \leq 3$ and $x \leq 0$ must be less than or equal to 0. Graph the intersection of the two solution sets.

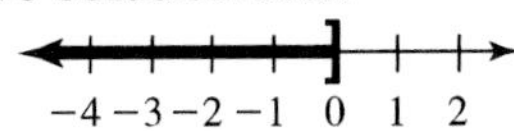

25. Numbers that satisfy $x \geq 2$ or $x \geq 5$ are greater than or equal to 2. Graph the union of the two solution sets:

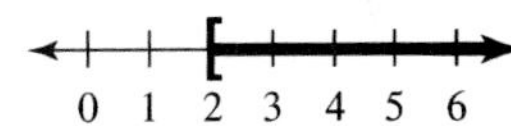

27. The union of $(-\infty, 6]$ with $(-2, \infty)$ is the interval $(-\infty, \infty)$. The union of the two solution sets consists of all real numbers:

−3 −2 −1 0 1 2 3

29. The solution set is $\emptyset$, because no number is greater than 9 and less than or equal to 6. There is no graph.
31. The union of the two solution sets is graphed as follows:

4 5 6 7 8 9 10 11

33. The solution set is $\emptyset$, because there is no intersection to the two solution sets. There is no graph.

35. $x - 3 > 7$ or $3 - x > 2$

$x > 10$ or $-x > -1$

$x > 10$ or $x < 1$

$(-\infty, 1) \cup (10, \infty)$

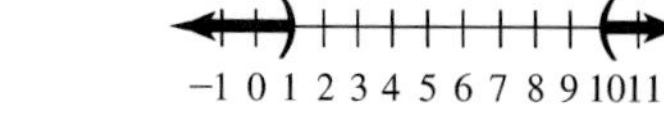

37. $3 < x$ and $1 + x > 10$

$x > 3$ and $x > 9$

$(9, \infty)$

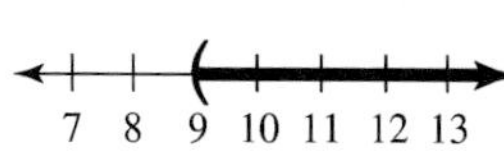

39. $\frac{1}{2}x > 5$ or $-\frac{1}{3}x < 2$

$x > 10$ or $x > -6$

$(-6, \infty)$

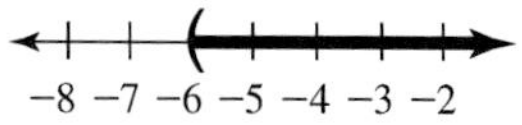

41. $2x - 3 \leq 5$ and $x - 1 > 0$

$2x \leq 8$ and $x > 1$

$x \leq 4$ and $x > 1$

$(1, 4]$

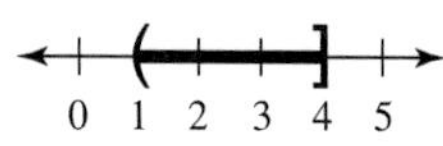

43. $\frac{1}{2}x - \frac{1}{3} \geq -\frac{1}{6}$ or $\frac{2}{7}x \leq \frac{1}{10}$

$3x - 2 \geq -1$ or $x \leq \frac{7}{2} \cdot \frac{1}{10}$

$x \geq \frac{1}{3}$ or $x \leq \frac{7}{20}$

$(-\infty, \infty)$

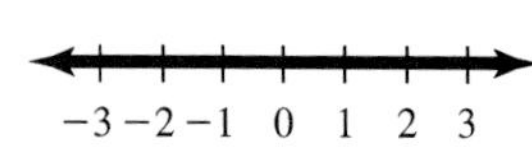

45. $0.5x < 2$ and $-0.6x < -3$

$x < 4$ and $x > 5$

The solution set is $\emptyset$, because there are no numbers that are less than 4 and greater than 5. There is no graph.

47. $-3 < x + 1 < 3$

$-3 - 1 < x + 1 - 1 < 3 - 1$

$-4 < x < 2$

$(-4, 2)$

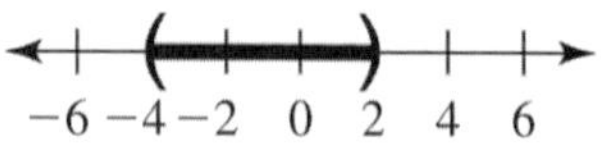

49. $5 < 2x - 3 < 11$

$5 + 3 < 2x - 3 + 3 < 11 + 3$

$8 < 2x < 14$

$4 < x < 7$

$(4, 7)$

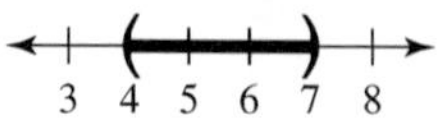

51. $-1 < 5 - 3x \leq 14$

$-6 < -3x \leq 9$

$\frac{-6}{-3} > \frac{-3x}{-3} \geq \frac{9}{-3}$

$2 > x \geq -3$

$-3 \leq x < 2$

$[-3, 2)$

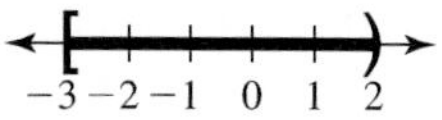

53. $-3 < \frac{3m + 1}{2} \leq 5$

$2(-3) < 2 \cdot \frac{3m + 1}{2} \leq 2 \cdot 5$

$-6 < 3m + 1 \leq 10$

$-7 < 3m \leq 9$

$-\frac{7}{3} < m \leq 3$

$(-7/3, 3]$

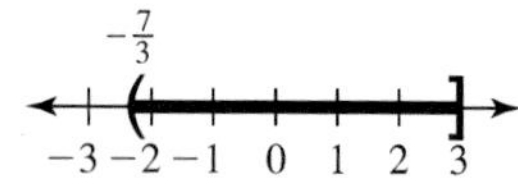

55. $-2 < \frac{1 - 3x}{-2} < 7$

$-2(-2) > -2 \cdot \frac{1 - 3x}{-2} > -2(7)$

$4 > 1 - 3x > -14$

$3 > -3x > -15$

$-1 < x < 5$

$(-1, 5)$

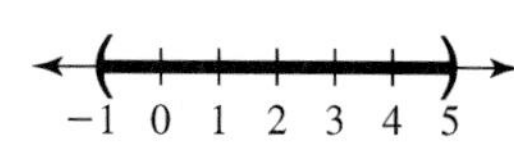

57. $3 \leq 3 - 5(x - 3) \leq 8$

$3 \leq 3 - 5x + 15 \leq 8$

$3 \leq 18 - 5x \leq 8$

$-15 \leq -5x \leq -10$

$3 \geq x \geq 2$

$2 \leq x \leq 3$

$[2, 3]$

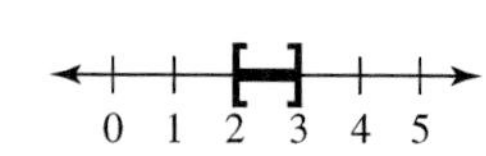

59. $(2, \infty) \cup (4, \infty) = (2, \infty)$

61. $(-\infty, 5) \cap (-\infty, 9) = (-\infty, 5)$

63. $(-\infty, 4] \cap [2, \infty) = [2, 4]$

65. $(-\infty, 5) \cup [-3, \infty) = (-\infty, \infty)$

67. $(3, \infty) \cap (-\infty, 3] = \emptyset$

69. $(3, 5) \cap [4, 8) = [4, 5)$

71. $[1, 4) \cup (2, 6] = [1, 6]$

73. The graph shows real numbers to the right of 2: $x > 2$

75. The graph shows the real numbers to the left of 3: $x < 3$

77. This graph is the union of the numbers greater than 2 with the numbers less than or equal to -1: $x > 2$ or $x \leq -1$

79. This graph shows real numbers between -2 and 3, including -2: $-2 \leq x < 3$

81. The graph shows real numbers greater than or equal to -3: $x \geq -3$

83. Let $x =$ the final exam score. We write an inequality expressing the fact that $1/3$ of the midterm plus $2/3$ of the final must be between 70 and 79 inclusive.

$$70 \leq \frac{1}{3} \cdot 64 + \frac{2}{3}x \leq 79$$
$$210 \leq 64 + 2x \leq 237$$
$$146 \leq 2x \leq 173$$
$$73 \leq x \leq 86.5$$

85. The car is replaced when $0.0004x + 20 > 40$ and $20{,}000 - 0.2x < 12{,}000$. Solve these inequalities to get $x > 50{,}000$ and $x > 40{,}000$, which is equivalent to $x > 50{,}000$. So the car is replaced when x is in the interval $(50{,}000, \infty)$.

87. The president worries if $20 + 0.1x < 22$ or $30 - 0.5x < 15$. Solve each inequality to get $x < 20$ or $x > 30$. So the president worries if x is in the union $(-\infty, 20) \cup (30, \infty)$.

89. Let $x =$ the price of the truck. The total spent will be $x + 0.08x + 84$.

$$12{,}000 \leq x + 0.08x + 84 \leq 15{,}000$$
$$12{,}000 \leq 1.08x + 84 \leq 15{,}000$$
$$11{,}916 \leq 1.08x \leq 14{,}916$$
$$\$11{,}033 \leq x \leq \$13{,}811$$

91. Let $x =$ the number of cigarettes smoked on the run, giving the equivalent of $\frac{1}{2}x$ cigarettes smoked. Thus, she smokes $3 + \frac{1}{2}x$ whole cigarettes per day and this number is between 5 and 12 inclusive:

$$5 \leq 3 + \frac{1}{2}x \leq 12$$
$$10 \leq 6 + x \leq 24$$
$$4 \leq x \leq 18$$

She smokes from 4 to 18 cigarettes on the run.

93. a) In 2000, we have $n = 10$.
$16.45(10) + 1062.45 = 1226.95$
In 2000, there were 1,226,950 bachelors degrees awarded.

b)
$$16.45n + 1062.45 = 1{,}400$$
$$16.45n = 337.55$$
$$n \approx 21$$

$1990 + 21 = 2011$

c)
$$16.45n + 1062.45 > 1{,}400$$
$$16.45n > 337.55$$
$$n > 20.52$$

$$7.79n + 326.82 > 550$$
$$7.79n > 223.18$$
$$n > 28.64$$

Both happen in the year $1990 + 29$, or 2019.

d) Either happens in the year $1990 + 21$, or 2011.

95. If $a < b$ and $a < -x < b$, we can multiply each part of this inequality by -1 to get $-a > x > -b$ or $-b < x < -a$. In words, x is between $-b$ and $-a$.

97. a) If $3 < x < 8$, then $12 < 4x < 32$.
$(12, 32)$

b) If $-2 \leq x < 4$, then

$$(-5)(-2) \geq -5x > (-5)(4)$$
$$10 \geq -5x > -20$$
$$-20 < -5x \leq 10.$$

$(-20, 10]$

c) If $-3 < x < 6$, then $0 < x + 3 < 9$.
$(0, 9)$

d) If $3 \leq x \leq 9$, then

$$\frac{3}{-3} \geq \frac{x}{-3} \geq \frac{9}{-3}$$
$$-1 \geq \frac{x}{-3} \geq -3$$
$$-3 \leq \frac{x}{-3} \leq -1.$$

$[-3, -1]$

2.6 WARM-UPS

1. True, because both 2 and -2 have absolute value 2.

2. False, because $|x| = 0$ has only one solution and $|x| = -1$ has no solutions.

3. False, because it is equivalent to $2x - 3 = 7$ or $2x - 3 = -7$.

4. True, because $|x| > 5$ means that x is more than 5 units away from 0 and that is true to the right of 5 or to the left of -5.

5. False, because this equation has no solution.

6. True, because only 3 satisfies the equation.

7. False, because only inequalities that express x between two numbers are written this way.

8. False, because $|x| < 7$ is equivalent to $-7 < x < 7$.
9. True, because 2 is subtracted from each side.
10. False, because if $x = 0$, then $|x| = 0$ and 0 is not positive.

2.6 EXERCISES

1. Absolute value of a number is the number's distance from 0 on the number line.
3. Since both 4 and -4 are four units from 0, $|x| = 4$ has two solutions.
5. Since the distance from 0 for every number on the number line is greater than or equal to 0, $|x| \geq 0$.
7. $|a| = 5$
$a = 5$ or $a = -5$
Solution set: $\{-5, 5\}$
9. $|x - 3| = 1$
$x - 3 = 1$ or $x - 3 = -1$
$x = 4$ or $x = 2$
Solution set: $\{2, 4\}$
11. $|3 - x| = 6$
$3 - x = 6$ or $3 - x = -6$
$-x = 3$ or $-x = -9$
$x = -3$ or $x = 9$
Solution set: $\{-3, 9\}$
13. $|3x - 4| = 12$
$3x - 4 = 12$ or $3x - 4 = -12$
$3x = 16$ or $3x = -8$
$x = \frac{16}{3}$ or $x = -\frac{8}{3}$
Solution set: $\left\{-\frac{8}{3}, \frac{16}{3}\right\}$
15. $\left|\frac{2}{3}x - 8\right| = 0$
$\frac{2}{3}x - 8 = 0$
$\frac{2}{3}x = 8$
$x = 12$
Solution set: $\{12\}$
17. $|5x + 2| = -3$
The solution set is $\emptyset$, because the absolute value of any quantity is nonnegative.
19. $6 - 0.2x = 10$ or $6 - 0.2x = -10$
$-0.2x = 4$ or $-0.2x = -16$
$x = -20$ or $x = 80$
Solution set: $\{-20, 80\}$
21. $|2(x - 4) + 3| = 5$
$2(x - 4) + 3 = 5$ or $2(x - 4) + 3 = -5$
$2x - 5 = 5$ or $2x - 5 = -5$
$2x = 10$ or $2x = 0$
$x = 5$ or $x = 0$
Solution set: $\{0, 5\}$
23. $|7.3x - 5.26| = 4.215$
$7.3x - 5.26 = 4.215$
$7.3x = 9.475$
$x \approx 1.298$
or $7.3x - 5.26 = -4.215$
$7.3x = 1.045$
$x \approx 0.143$
Solution set: $\{0.143, 1.298\}$
25. $3 + |x| = 5$
$|x| = 2$
$x = 2$ or $x = -2$
Solution set: $\{-2, 2\}$
27. $3|a| - 6 = 21$
$3|a| = 27$
$|a| = 9$
$a = 9$ or $a = -9$
The solution set is $\{-9, 9\}$.
29. $3|w + 1| - 2 = 7$
$3|w + 1| = 9$
$|w + 1| = 3$
$w + 1 = 3$ or $w + 1 = -3$
$w = 2$ or $w = -4$
The solution set is $\{-4, 2\}$.
31. $2 - |x + 3| = -6$
$-|x + 3| = -8$
$|x + 3| = 8$
$x + 3 = 8$ or $x + 3 = -8$
$x = 5$ or $x = -11$
Solution set: $\{-11, 5\}$
33. $5 - \frac{|3 - 2x|}{3} = 4$
$15 - |3 - 2x| = 12$
$-|3 - 2x| = -3$
$|3 - 2x| = 3$
$3 - 2x = 3$ or $3 - 2x = -3$
$-2x = 0$ or $-2x = -6$
$x = 0$ or $x = 3$
The solution set is $\{0, 3\}$.
35. $x - 5 = 2x + 1$ or $x - 5 = -(2x + 1)$
$-6 = x$ or $x - 5 = -2x - 1$
$3x = 4$
$x = 4/3$
Solution set: $\left\{-6, \frac{4}{3}\right\}$

37. $\left|\frac{5}{2} - x\right| = \left|2 - \frac{x}{2}\right|$

$\frac{5}{2} - x = 2 - \frac{x}{2}$ or $\frac{5}{2} - x = -\left(2 - \frac{x}{2}\right)$

$5 - 2x = 4 - x$ or $\frac{5}{2} - x = -2 + \frac{x}{2}$

$1 = x$ or $5 - 2x = -4 + x$

$-3x = -9$

$x = 3$

Solution set: $\{1, 3\}$

39. $|x - 3| = |3 - x|$

$x - 3 = 3 - x$ or $x - 3 = -(3 - x)$

$2x = 6$ or $x - 3 = -3 + x$

$x = 3$ or $x - 3 = x - 3$

The second equation is an identity. So the solution set is the set of real numbers, which is written as R or $(-\infty, \infty)$.

41. The graph shows the real numbers between -2 and 2: $|x| < 2$

43. The graph shows numbers greater than 3 or less than -3: $|x| > 3$

45. The graph shows numbers between -1 and 1 inclusive: $|x| \leq 1$

47. The graph shows numbers two or more units away from 0: $|x| \geq 2$

49. No, because $|x| < 3$ is equivalent to $-3 < x < 3$.

51. Yes.

53. No, because $|x - 3| \geq 1$ is equivalent to $x - 3 \geq 1$ or $x - 3 \leq -1$.

55. $|x| > 6$

$x > 6$ or $x < -6$

$(-\infty, -6) \cup (6, \infty)$

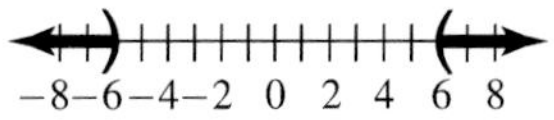

57. $|t| \leq 2$

$-2 \leq t \leq 2$

$[-2, 2]$

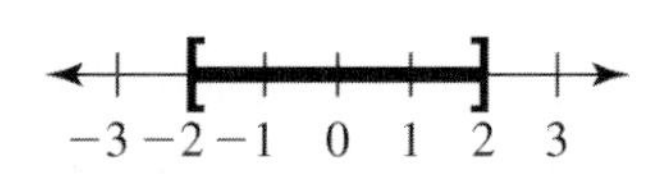

59. $|2a| < 6$

$-6 < 2a < 6$

$-3 < a < 3$

$(-3, 3)$

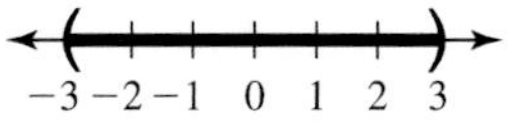

61. $x - 2 \geq 3$ or $x - 2 \leq -3$

$x \geq 5$ or $x \leq -1$

$(-\infty, -1] \cup [5, \infty)$

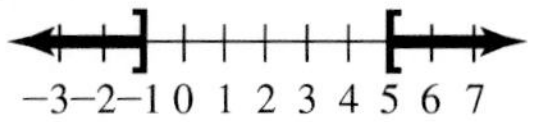

63. $3|a| - 3 > 21$

$3|a| > 24$

$|a| > 8$

$a > 8$ or $a < -8$

$(-\infty, -8) \cup (8, \infty)$

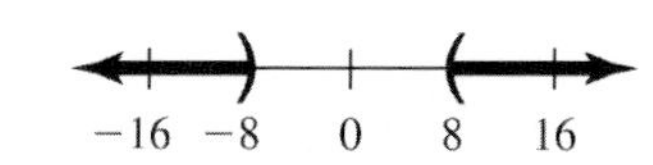

65. $3|w + 1| - 5 \leq 7$

$3|w + 1| \leq 12$

$|w + 1| \leq 4$

$-4 \leq w + 1 \leq 4$

$-5 \leq w \leq 3$

$[-5, 3]$

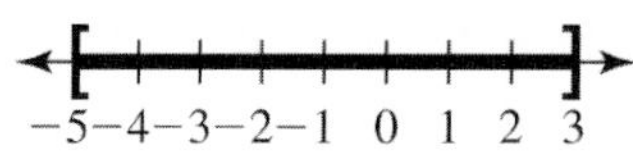

67. $\frac{1}{5}|2x - 4| < 1$

$|2x - 4| < 5$

$-5 < 2x - 4 < 5$

$-1 < 2x < 9$

$-\frac{1}{2} < x < \frac{9}{2}$ $\left(-\frac{1}{2}, \frac{9}{2}\right)$

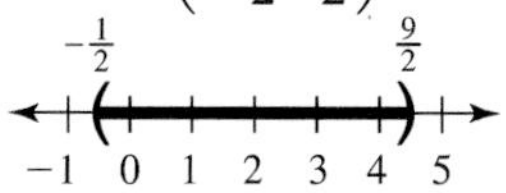

69. $-2|5 - x| \geq -14$

$|5 - x| \leq 7$

$-7 \leq 5 - x \leq 7$

$-12 \leq -x \leq 2$

$12 \geq x \geq -2$

$[-2, 12]$

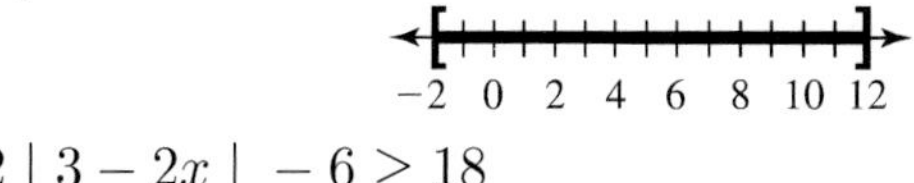

71. $2|3 - 2x| - 6 \geq 18$

$2|3 - 2x| \geq 24$

$|3 - 2x| \geq 12$

$3 - 2x \geq 12$ or $3 - 2x \leq -12$

$-2x \geq 9$ or $-2x \leq -15$

$x \leq -\frac{9}{2}$ or $x \geq \frac{15}{2}$

$(-\infty, -\frac{9}{2}] \cup [\frac{15}{2}, \infty)$

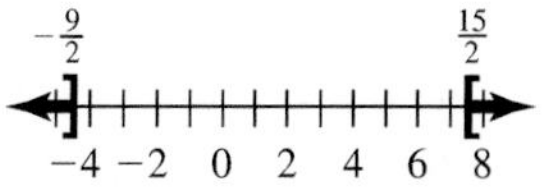

73. $|x| > 0$ is true except when $x = 0$ so the solution set is the set of all real numbers except 0, which is written in interval notation as $(-\infty, 0) \cup (0, \infty)$ and graphed as follows:

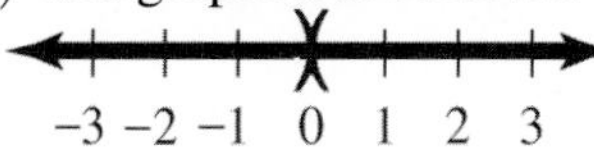

75. $|x| \leq 0$ is true only when $x = 0$ because the absolute value of any nonzero real number is positive. The solution set is $\{0\}$ and it is a single point on the number line:

−3 −2 −1 0 1 2 3

77. The inequality $|x - 5| \geq 0$ is satisfied by every real number because the absolute value of any real number is nonnegative. The solution set is R, or $(-\infty, \infty)$.

−3 −2 −1 0 1 2 3

79. $-2|3x - 7| > 6$ is equivalent to
$$|3x - 7| < -3.$$
Absolute value of an expression cannot be less than a negative number. Solution set is $\emptyset$.

81.
$$|2x + 3| + 6 > 0$$
$$|2x + 3| > -6$$
Since the absolute value of any expression is greater than or equal to zero, it is greater than any negative number. The solution set is R or $(-\infty, \infty)$.

−3 −2 −1 0 1 2 3

83.
$$1 < |x + 2|$$
$$|x + 2| > 1$$
$$x + 2 > 1 \quad \text{or} \quad x + 2 < -1$$
$$x > -1 \quad \text{or} \quad x < -3$$
$(-\infty, -3) \cup (-1, \infty)$

85.
$$5 > |x + 1|$$
$$|x| + 1 < 5$$
$$|x| < 4$$
$$-4 < x < 4$$
$(-4, 4)$

87.
$$3 - 5|x| > -2$$
$$-5|x| > -5$$
$$|x| < 1$$
$$-1 < x < 1$$
$(-1, 1)$

89.
$$|5.67x - 3.124| < 1.68$$
$$-1.68 < 5.67x - 3.124 < 1.68$$
$$1.444 < 5.67x < 4.804$$
$$0.255 < x < 0.847$$
$(0.255, 0.847)$

91. Let $x =$ the year of the battle of Orleans. Since the difference between x and 1415 is 14 years, we have $|x - 1415| = 14$.
$$x - 1415 = 14 \quad \text{or} \quad x - 1415 = -14$$
$$x = 1429 \quad \text{or} \quad x = 1401$$
The battle Agincourt was either in 1401 or 1429.

93. Let $x =$ the weight of Kathy. The difference between their weights is less than 6 pounds is expressed by the absolute value inequality
$$|x - 127| < 6$$
$$-6 < x - 127 < 6$$
$$121 < x < 133$$
Kathy weighs between 121 and 133 pounds.

95. a) If x is the percentage that approve of the president then x is between $39\% + 5$ and $39\% - 5\%$. That is, the percentage is between 34% and 39%.

b) The difference between x and 39% or 0.39 is less than 0.05.
$|x - 0.39| < 0.05$

97. a) From the graph it appears that the balls are at the same height when $t = 1$ second.

b) Height of first ball is $S = -16t^2 + 50t$ and height of second ball is
$S = -16t^2 + 40t + 10$.
When the balls are at the same height we have
$$-16t^2 + 50t = -16t^2 + 40t + 10$$
$$50t = 40t + 10$$
$$10t = 10$$
$$t = 1$$
The balls are at the same height when $t = 1$ sec.

c) The difference between the heights is less than 5 feet when
$$|-16t^2 + 50t - (-16t^2 + 40t + 10)| < 5$$
$$|10t - 10| < 5$$
$$-5 < 10t - 10 < 5$$
$$5 < 10t < 15$$
$$0.5 < t < 1.5$$

99. a) The equation $|m - n| = |n - m|$ is satisfied for all real numbers, because $m - n$ and $n - m$ are opposites of each other and opposites always have the same absolute value. So both m and n can be in the interval $(-\infty, \infty)$.

b) $|mn| = |m| \cdot |n|$ is satisfied for all real numbers, because of the rules for multiplying real numbers. So both m and n can be in the the interval $(-\infty, \infty)$.

c) Since you cannot have 0 in a denominator, the equation is satisfied by all real numbers except if $n = 0$.

Chapter 2 Wrap-Up

Enriching Your Mathematical Word Power

1. c **2.** b **3.** d **4.** c
5. c **6.** d **7.** d **8.** a
9. b **10.** d **11.** d **12.** d
13. c **14.** d

CHAPTER 2 REVIEW

1.
$$2x - 7 = 9$$
$$2x = 16$$
$$x = 8$$
Solution set: $\{8\}$

3.
$$5 - 4x = 11$$
$$-4x = 6$$
$$x = \frac{6}{-4} = -\frac{3}{2}$$
Solution set: $\left\{-\frac{3}{2}\right\}$

5. $x - 6 - (x - 6) = 0$
$$x - 6 - x + 6 = 0$$
$$0 = 0$$
Solution set: R or $(-\infty, \infty)$

7. $2(x - 3) - 5 = 5 - (3 - 2x)$
$$2x - 6 - 5 = 5 - 3 + 2x$$
$$2x - 11 = 2 + 2x$$
$$-11 = 2$$
Solution set: $\emptyset$

9. $\frac{3}{17}x = 0$
$$x = 0$$
Solution set: $\{0\}$

11. $\frac{1}{4}x - \frac{1}{5} = \frac{1}{5}x + \frac{4}{5}$
$$20\left(\frac{1}{4}x - \frac{1}{5}\right) = 20\left(\frac{1}{5}x + \frac{4}{5}\right)$$
$$5x - 4 = 4x + 16$$
$$x = 20$$
Solution set: $\{20\}$

13.
$$\frac{t}{2} - \frac{t-2}{3} = \frac{3}{2}$$
$$6\left(\frac{t}{2}\right) - 6\left(\frac{t-2}{3}\right) = 6\left(\frac{3}{2}\right)$$
$$3t - 2t + 4 = 9$$
$$t + 4 = 9$$
$$t = 5$$
Solution set: $\{5\}$

15. $1 - 0.4(x - 4) + 0.6(x - 7) = -0.6$
$$1 - 0.4x + 1.6 + 0.6x - 4.2 = -0.6$$
$$0.2x - 1.6 = -0.6$$
$$0.2x = 1$$
$$x = 5$$
The solution set is $\{5\}$.

17. $ax + b = 0$
$$ax = -b$$
$$x = \frac{-b}{a}$$
$$x = -\frac{b}{a}$$

19.
$$ax + 2 = cx$$
$$ax - cx = -2$$
$$x(a - c) = -2$$
$$\frac{x(a-c)}{a-c} = \frac{-2}{a-c}$$
$$x = \frac{2}{c-a}$$

21.
$$mwx = P$$
$$\frac{mwx}{mw} = \frac{P}{mw}$$
$$x = \frac{P}{mw}$$

23.
$$\frac{x}{2} + \frac{a}{6} = \frac{x}{3}$$
$$6\left(\frac{x}{2} + \frac{a}{6}\right) = 6\left(\frac{x}{3}\right)$$
$$3x + a = 2x$$
$$x = -a$$

25.
$$3x - 2y = -6$$
$$-2y = -3x - 6$$
$$y = \frac{-3x - 6}{-2}$$
$$y = \frac{3}{2}x + 3$$

27.
$$y - 2 = -\frac{1}{3}x + 2$$
$$y = -\frac{1}{3}x + 4$$

29. $\frac{1}{2}x - \frac{1}{4}y = 5$
$$4\left(\frac{1}{2}x - \frac{1}{4}y\right) = 4(5)$$
$$2x - y = 20$$
$$-y = -2x + 20$$
$$y = 2x - 20$$

31. Let $W =$ the width and $W + 5.5 =$ the length. Since $2L + 2W = P$, we have
$$2(W + 5.5) + 2W = 45$$
$$2W + 11 + 2W = 45$$
$$4W = 34$$
$$W = 8.5$$
Length is 14 inches and width is 8.5 inches.

33. Let $x =$ the wife's income and $x + 8000 =$ Roy's income. Roy saves $0.10(x + 8000)$ and his wife saves $0.08x$. Since the total saved is \$5660, we can write
$$0.10(x + 8000) + 0.08x = 5660$$
$$0.10x + 800 + 0.08x = 5660$$

$$0.18x = 4860$$
$$x = 27{,}000$$

Roy's wife earns \$27,000 and Roy earns \$35,000.

35. Let $x =$ the list price and $0.20x =$ the discount. The list price minus the discount is equal to \$7,600.

$$x - 0.20x = 7600$$
$$0.80x = 7600$$
$$x = 9500$$

The list price was \$9500.

37. Let $x =$ the number of nickels and $15 - x =$ the number of dimes. The value of x nickels is $5x$ cents and the value of $15 - x$ dimes is $10(15 - x)$ cents. Since she has a total of 95 cents, we can write the equation

$$5x + 10(15 - x) = 95$$
$$5x + 150 - 10x = 95$$
$$-5x = -55$$
$$x = 11$$
$$15 - x = 4$$

She has 11 nickels and 4 dimes.

39. Let $x =$ her walking speed and $x + 9 =$ her riding speed. Since she walked for 3 hours, $3x$ is the number of miles that she walked. Since she rode for 5 hours, $5(x + 9)$ is the number of miles that she rode. Her total distance was 85 miles.

$$3x + 5(x + 9) = 85$$
$$3x + 5x + 45 = 85$$
$$8x = 40$$
$$x = 5$$

She walked 5 miles per hour for 3 hours, so she walked 15 miles.

41. $3 - 4x < 15$
$$-4x < 12$$
$$x > -3$$

$(-3, \infty)$

43. $-3 - x > 2$
$$-x > 5$$
$$x < -5$$

$(-\infty, -5)$

45. $2(x - 3) > -6$
$$2x - 6 > -6$$
$$2x > 0$$
$$x > 0$$

$(0, \infty)$

47. $-\frac{3}{4}x \geq 6$
$$-3x \geq 24$$
$$x \leq -8$$

$(-\infty, -8]$

49. $3(x + 2) > 5(x - 1)$
$$3x + 6 > 5x - 5$$
$$-2x > -11$$
$$x < \frac{11}{2}$$

$(-\infty, 11/2)$

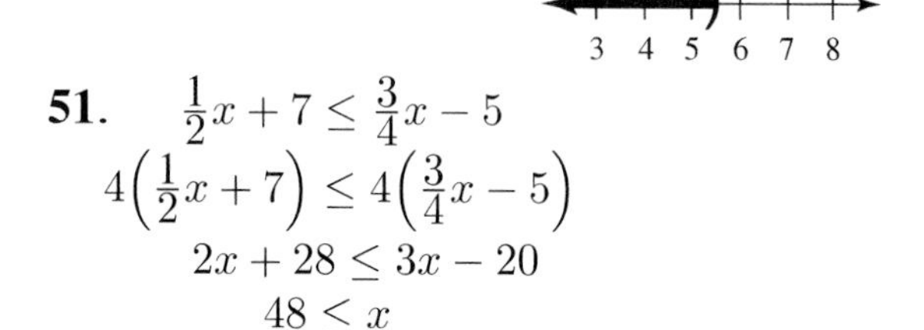

51. $\frac{1}{2}x + 7 \leq \frac{3}{4}x - 5$
$$4\left(\frac{1}{2}x + 7\right) \leq 4\left(\frac{3}{4}x - 5\right)$$
$$2x + 28 \leq 3x - 20$$
$$48 \leq x$$

$[48, \infty)$

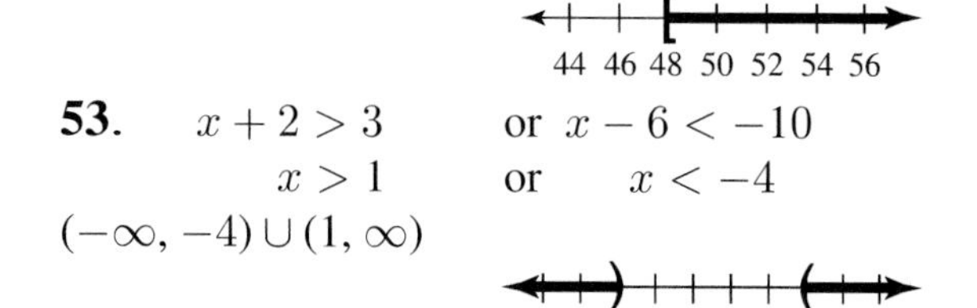

53. $x + 2 > 3$ or $x - 6 < -10$

$x > 1$ or $x < -4$

$(-\infty, -4) \cup (1, \infty)$

55. $x > 0$ and $x - 6 < 3$

$x > 0$ and $x < 9$

$(0, 9)$

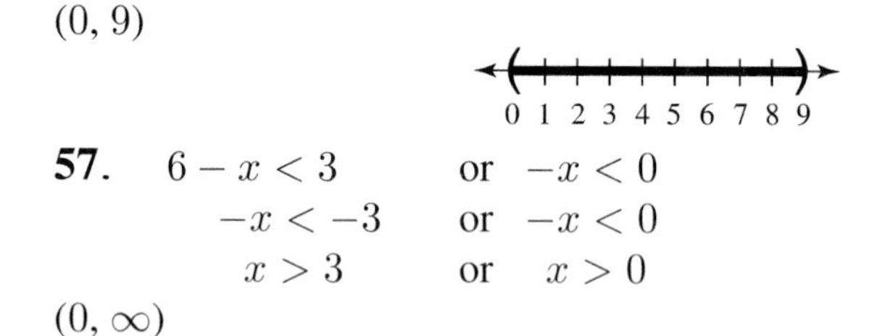

57. $6 - x < 3$ or $-x < 0$

$-x < -3$ or $-x < 0$

$x > 3$ or $x > 0$

$(0, \infty)$

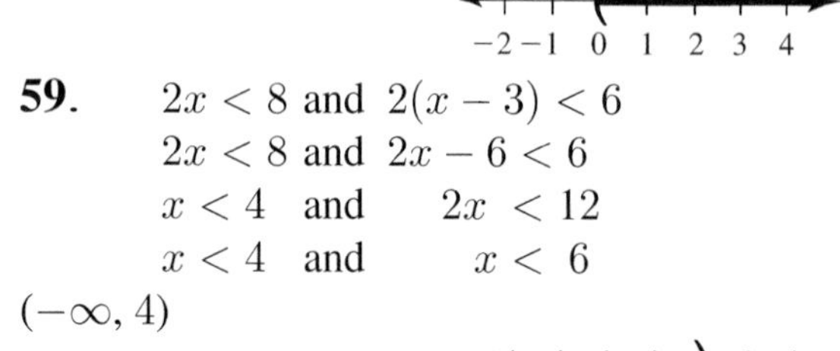

59. $2x < 8$ and $2(x - 3) < 6$

$2x < 8$ and $2x - 6 < 6$

$x < 4$ and $2x < 12$

$x < 4$ and $x < 6$

$(-\infty, 4)$

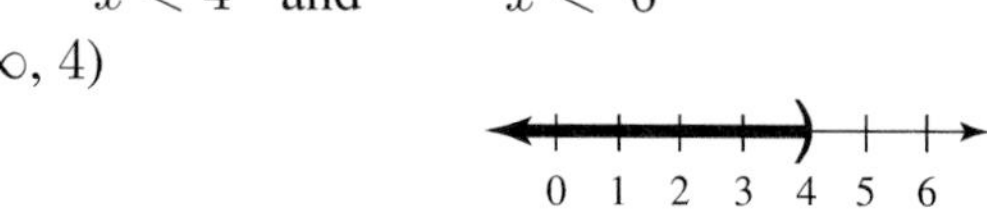

61. $x - 6 > 2$ and $6 - x > 0$

$x > 8$ and $-x > -6$

$x > 8$ and $x < 6$

No number is greater than 8 and less than 6.

$\emptyset$

63. $0.5x > 10$ or $0.1x < 3$

$x > 20$ or $x < 30$

Every number is either greater than 20 or less than 30. Solution set is R or $(-\infty, \infty)$

65. $-2 \le \frac{2x - 3}{10} \le 1$

$10(-2) \le 10 \cdot \frac{2x - 3}{10} \le 10(1)$

$-20 \le 2x - 3 \le 10$

$-17 \le 2x \le 13$

$-\frac{17}{2} \le x \le \frac{13}{2}$

$\left[-\frac{17}{2}, \frac{13}{2}\right]$

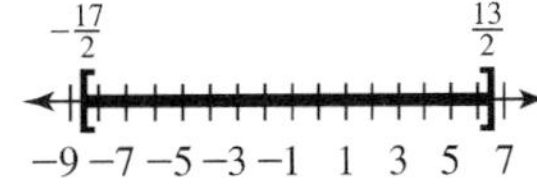

67. $[1, 4) \cup (2, \infty) = [1, \infty)$

69. $(3, 6) \cap [2, 8] = (3, 6)$

71. $(-\infty, 5) \cup [5, \infty) = (-\infty, \infty)$

73. $(-3, -1] \cap [-2, 5] = [-2, -1]$

75. $|x| + 2 = 16$

$|x| = 14$

$x = 14$ or $x = -14$

$\{-14, 14\}$

−21 −14 −7 0 7 14 21

77. $|4x - 12| = 0$

$4x - 12 = 0$

$4x = 12$

$x = 3$

$\{3\}$

−1 0 1 2 3 4

79. Since $|x| \ge 0$ for any real number, the solution set is $\emptyset$.

81. $|2x - 1| - 3 = 0$

$|2x - 1| = 3$

$2x - 1 = 3$ or $2x - 1 = -3$

$2x = 4$ or $2x = -2$

$x = 2$ or $x = -1$

$\{-1, 2\}$

−2 −1 0 1 2 3 4 5

83. $|2x| \ge 8$

$2x \ge 8$ or $2x \le -8$

$x \ge 4$ or $x \le -4$

$(-\infty, -4] \cup [4, \infty)$

−6 −4 −2 0 2 4 6

85. $\left|1 - \frac{x}{5}\right| > \frac{9}{5}$

$1 - \frac{x}{5} > \frac{9}{5}$ or $1 - \frac{x}{5} < -\frac{9}{5}$

$5 - x > 9$ or $5 - x < -9$

$-x > 4$ or $-x < -14$

$x < -4$ or $x > 14$

$(-\infty, -4) \cup (14, \infty)$

−8 −4 0 4 8 12 16

87. Since $|x - 3| \ge 0$ for any value of x, the solution set is $\emptyset$.

89. Since $|x + 4| \ge 0$ for any value of x, $|x + 4| \ge -1$ for any x. Solution set: R

−3 −2 −1 0 1 2 3

91. $1 - \frac{3}{2}|x - 2| < -\frac{1}{2}$

$2 - 3|x - 2| < -1$

$-3|x - 2| < -3$

$|x - 2| > 1$

$x - 2 > 1$ or $x - 2 < -1$

$x > 3$ or $x < 1$

$(-\infty, 1) \cup (3, \infty)$

−1 0 1 2 3 4 5

93. Let x = the rental price, $0.45x$ = the overhead per tape, and $x - 0.45x$ or $0.55x$ = the profit per video. The rental price must be less than or equal to \$5 and satisfy the inequality

$0.55x \ge 1.65$

$x \ge 3$

The range of the rental price is $\$3 \le x \le \$5$.

95. Since $150 < h < 180$, we have

$150 < 60.089 + 2.238F < 180$

$89.911 < 2.238F < 119.911$

$40.2 < F < 53.6$

The length of the femur is in $(40.2, 53.6)$.

97. Let x = the number on the mile marker where Dane was picked up. We can write the absolute value equation

$|x - 86| = 5$

$x - 86 = 5$ or $x - 86 = -5$

$x = 91$ or $x = 81$

He was either at 81 or 91.

99. $b = 0.20(300{,}000 - b)$

$b = 60{,}000 - 0.20b$

$1.2b = 60{,}000$

$$b = 50{,}000$$

Bonus is \$50,000 according to the accountant, and \$60,000 according the employees.

101. Let $x =$ the number of cows in Washington County and $3600 - x =$ the number of cows in Cade County. We have the equation

$$0.30x + 0.60(3600 - x) = 0.50(3600)$$
$$-0.30x + 2160 = 1800$$
$$-0.30x = -360$$
$$x = 1200$$
$$3600 - x = 2400$$

There are 1200 cows in Washington County and 2400 in Cade County.

103. The numbers to the right of 1 are described by the inequality $x > 1$.

105. The number 2 satisfies the equation $|x - 2| = 0$.

107. The numbers 3 and -3 both satisfy the equation $|x| = 3$.

109. The numbers to the left of and including -1 satisfy $x \leq -1$.

111. The numbers between -2 and 2 including the endpoints satisfy $|x| \leq 2$.

113. $x \leq 2$ or $x \geq 7$

115. The numbers greater than 3 or less than -3 satisfy $|x| > 3$.

117. $5 < x < 7$ or $|x - 6| < 1$

119. Every number except 0 has a positive absolute value and satisfies $|x| > 0$.

CHAPTER 2 TEST

1. $-10x - 5 + 4x = -4x + 3$

$$-6x - 5 = -4x + 3$$
$$-2x = 8$$
$$x = -4$$

Solution set: $\{-4\}$

2. $\frac{y}{2} - \frac{y-3}{3} = \frac{y+6}{6}$

$$6\left(\frac{y}{2}\right) - 6\left(\frac{y-3}{3}\right) = 6\left(\frac{y+6}{6}\right)$$
$$3y - 2y + 6 = y + 6$$
$$y + 6 = y + 6$$

The equation is an identity. Solution set: R

3. $|w| + 3 = 9$

$$|w| = 6$$
$$w = 6 \text{ or } w = -6$$

Solution set: $\{-6, 6\}$

4. $|3 - 2(5 - x)| = 3$

$$|-7 + 2x| = 3$$
$$-7 + 2x = 3 \text{ or } -7 + 2x = -3$$
$$2x = 10 \quad \text{or} \quad 2x = 4$$
$$x = 5 \quad \text{or} \quad x = 2$$

Solution set: $\{2, 5\}$

5. $2x - 5y = 20$

$$-5y = -2x + 20$$
$$y = \frac{2}{5}x - 4$$

6. $y = 3xy + 5$

$$y - 3xy = 5$$
$$y(1 - 3x) = 5$$
$$y = \frac{5}{1 - 3x}$$

7. $|m - 6| \leq 2$

$$-2 \leq m - 6 \leq 2$$
$$4 \leq m \leq 8$$

[4, 8]

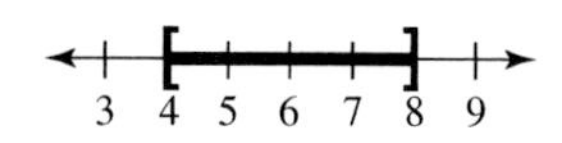

8. $2|x - 3| - 5 > 15$

$$2|x - 3| > 20$$
$$|x - 3| > 10$$
$$x - 3 > 10 \quad \text{or} \quad x - 3 < -10$$
$$x > 13 \quad \text{or} \quad x < -7$$

$(-\infty, -7) \cup (13, \infty)$

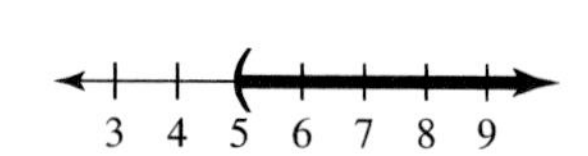

9. $2 - 3(w - 1) < -2w$

$$2 - 3w + 3 < -2w$$
$$-w < -5$$
$$w > 5$$

$(5, \infty)$

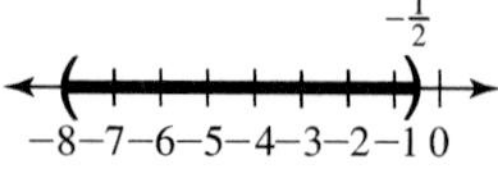

10. $2 < \frac{5 - 2x}{3} < 7$

$$3(2) < 3\left(\frac{5 - 2x}{3}\right) < 3(7)$$
$$6 < 5 - 2x < 21$$
$$1 < -2x < 16$$
$$-\frac{1}{2} > x > -8$$

$\left(-8, -\frac{1}{2}\right)$

11. $3x - 2 < 7$ and $-3x \leq 15$
$3x < 9$ and $x \geq -5$
$x < 3$ and $x \geq -5$

$[-5, 3)$

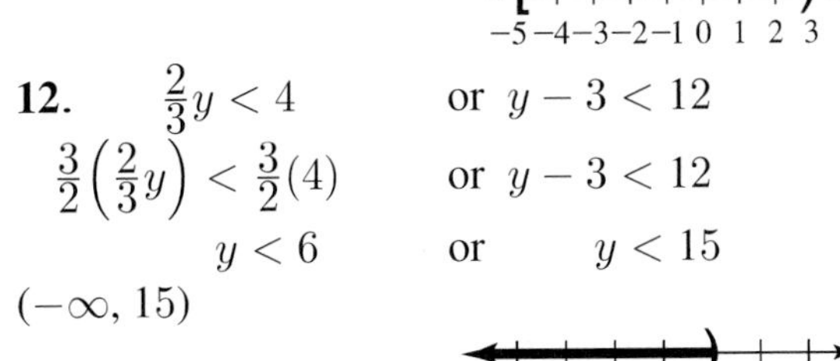

12. $\frac{2}{3}y < 4$ or $y - 3 < 12$
$\frac{3}{2}\left(\frac{2}{3}y\right) < \frac{3}{2}(4)$ or $y - 3 < 12$
$y < 6$ or $y < 15$

$(-\infty, 15)$

11 12 13 14 15 16 17

13. The equation $|2x - 7| = -3$ has no solution because the absolute value of any real number is greater than or equal to zero. The solution set is $\emptyset$.

14. $x - 4 > 1$ or $x < 12$
$x > 5$ or $x < 12$

$(-\infty, \infty)$

15. $3x < 0$ and $x - 5 > 2$
$x < 0$ and $x > 7$

No real number is both less than 0 and greater than 7. Solution set: $\emptyset$

16. Since no real number satisfies $|2x - 5| < 0$, we need only solve $|2x - 5| = 0$, which is equivalent to $2x - 5 = 0$, or $x = 2.5$.
Solution set: $\{2.5\}$

17. Since no real number satisfies $|x - 3| < 0$, the solution set is $\emptyset$.

18. Since $x + 3x = 4x$ is equivalent to $4x = 4x$, the solution set is R or $(-\infty, \infty)$.

19. $2(x + 7) = 2x + 9$
$2x + 14 = 2x + 9$
$14 = 9$

Solution set is $\emptyset$.

20. Since $|x - 6| \geq 0$ for any real number x, $|x - 6| \geq -6$ for any real number x. The solution set is R or $(-\infty, \infty)$.

21. $x - 0.04(x - 10) = 96.4$
$x - 0.04x + 0.4 = 96.4$
$0.96x = 96$
$x = 100$

$\{100\}$

22. Let W = the width and $W + 16$ = the length. Use $2W + 2L = P$ to write the equation

$2W + 2(W + 16) = 84$
$4W + 32 = 84$
$4W = 52$
$W = 13$

The width is 13 meters.

23. Let h = the height. Use the formula $A = \frac{1}{2}bh$ to write the following equation.

$\frac{1}{2} \cdot 3h = 21$
$3h = 42$
$h = 14$

The height is 14 inches.

24. Let x = the original price and $0.30x$ = the amount of discount. The original price minus the discount is equal to the price she paid.

$x - 0.30x = 210$
$0.70x = 210$
$x = 300$

The original price was \$300.

25. Let x = the number of liters of 11% alcohol. The mixture will be $x + 60$ liters of 7% alcohol. There are $0.11x$ liters of alcohol in the 11% solution, 0.05(60) liters of alcohol in the 5% solution, and $0.07(x + 60)$ liters of alcohol in the 7% solution.

$0.11x + 0.05(60) = 0.07(x + 60)$
$0.11x + 3 = 0.07x + 4.2$
$0.04x = 1.2$
$x = 30$

Use 30 liters of 11% alcohol solution.

26. Let b = Brenda's salary.

$|b - 28{,}000| > 3000$
$b - 28{,}000 > 3000$ or $b - 28{,}000 < -3000$
$b > 31{,}000$ or $b < 25{,}000$

Brenda's salary is either greater than \$31,000 or less than \$25,000.

Making Connections

Chapters 1-2

1. $5x + 6x = 11x$
2. $5x \cdot 6x = 30x^2$
3. $\frac{6x + 2}{2} = \frac{6x}{2} + \frac{2}{2} = 3x + 1$
4. $5 - 4(2 - x) = 5 - 8 + 4x = 4x - 3$
5. $(30 - 1)(30 + 1) = 29 \cdot 31 = 899$
6. $(30 + 1)^2 = 31^2 = 961$
7. $(30 - 1)^2 = 29^2 = 841$
8. $(2 + 3)^2 = 5^2 = 25$

9. $2^2 + 3^2 = 4 + 9 = 13$

10. $(8 - 3)(3 - 8) = 5(-5) = -25$

11. $(-1)(3 - 8) = -1(-5) = 5$

12. $-2^2 = -(2^2) = -4$

13. $3x + 8 - 5(x - 1) = 3x + 8 - 5x + 5$
$= -2x + 13$

14. $(-6)^2 - 4(-3)2 = 36 + 24 = 60$

15. $3^2 \cdot 2^3 = 9 \cdot 8 = 72$

16. $4(-6) - (-5)(3) = -24 + 15 = -9$

17. $-3x \cdot x \cdot x = -3x^3$

18. $(-1)^6 = 1$

19.
$$\begin{aligned} 5x + 6x &= 8x \\ 11x &= 8x \\ 3x &= 0 \\ x &= 0 \end{aligned}$$
Solution set: $\{0\}$

20.
$$\begin{aligned} 5x + 6x &= 11x \\ 11x &= 11x \end{aligned}$$
This equation is an identity. Solution set is R or $(-\infty, \infty)$.

21.
$$\begin{aligned} 5x + 6x &= 0 \\ 11x &= 0 \\ x &= 0 \end{aligned}$$
Solution set: $\{0\}$

22.
$$\begin{aligned} 5x + 6 &= 11x \\ -6x &= -6 \\ x &= 1 \end{aligned}$$
Solution set: $\{1\}$

23.
$$\begin{aligned} 3x + 1 &= 0 \\ 3x &= -1 \\ x &= -\frac{1}{3} \end{aligned}$$
Solution set: $\left\{-\frac{1}{3}\right\}$

24.
$$\begin{aligned} 5 - 4(2 - x) &= 1 \\ 5 - 8 + 4x &= 1 \\ 4x &= 4 \\ x &= 1 \end{aligned}$$
Solution set: $\{1\}$

25.
$$\begin{aligned} 3x + 6 &= 3(x + 2) \\ 3x + 6 &= 3x + 6 \end{aligned}$$
This equation is an identity. Solution set is R or $(-\infty, \infty)$.

26.
$$\begin{aligned} x - 0.01x &= 990 \\ 0.99x &= 990 \\ x &= 1000 \end{aligned}$$
Solution set: $\{1000\}$

27. $|\,5x + 6\,| = 11$
$$\begin{aligned} 5x + 6 &= 11 & \text{or}\quad 5x + 6 &= -11 \\ 5x &= 5 & \text{or}\quad 5x &= -17 \\ x &= 1 & \text{or}\quad x &= -17/5 \end{aligned}$$
Solution set: $\{-17/5, 1\}$

28. a) From the graph it appears that the cost of renting and buying are equal at around 87,500 copies.
b) If $x =$ the number of copies made in 5 years, then the cost for renting is
$R = 60(75) + 0.06x$ dollars
or $R = 4500 + 0.06x$ dollars.
The cost if the copier is purchased is
$P = 8000 + 0.02x$ dollars.
c)
$$\begin{aligned} 60(75) + 0.06x &= 8000 + 0.02x \\ 0.04x &= 3500 \\ x &= 87{,}500 \end{aligned}$$
Five-year cost is same for 87,500 copies.

d) If 120,000 copies are made, then renting cost is \$11,700 and buying cost is \$10,400. So buying is \$1300 cheaper.

e)
$$\begin{aligned} |60(75) + 0.06x - (8000 + 0.02x)| &< 500 \\ |-3500 + 0.04x| &< 500 \\ -500 < -3500 + 0.04x &< 500 \\ 3000 < 0.04x &< 4000 \\ 75{,}000 < x &< 100{,}000 \end{aligned}$$
If the number of copies is between 75,000 and 100,000, then the plans differ by less than \$500.

3.1 WARM-UPS

1. False, because $3(5) - 2(2) = -4$ is not correct.
2. False, the vertical axis is called the y-axis.
3. False, because the point $(0, 0)$ is not in any quadrant.
4. True, because its first coordinate is 0.
5. True, because the graph of $x = k$ for any real number k is a vertical line.
6. True, because $8 - y = 0$ is equivalent to $y = 8$ and the graph of $y = k$ for any real number k is a horizontal line.
7. True, because if $x = 0$, then $y = -3$.
8. True, because if $n = 2$, then $C = 3(2) + 4 = 10$.
9. False, because if $P = 12$, then $3x = 12$ and $x = 4$.
10. True, because we usually put the dependent variable on the vertical axis.

3.1 EXERCISES

1. The origin is the point where the x-axis and y-axis intersect.
3. Intercepts are points where a graph crosses the axes.
5. The graph of an equation of the type $x = k$ where k is a fixed number is a vertical line.
7. To plot $(2, 5)$, start at the origin and go 2 units to the right and up 5. The point is in quadrant I.
9. To plot $(-3, -1/2)$, start at the origin and go 3 units to the left and down $1/2$ unit. The point is in quadrant III.
11. To plot $(0, 4)$, start at the origin and go 4 units upward. The point is on the y-axis.
13. To plot $(\pi, -1)$, start at the origin and go approximately 3.14 units to the right and 1 unit downward. The point is in quadrant IV.
15. To plot $(-4, 3)$, start at the origin and go 4 units to the left and 3 units upward. The point is in quadrant II.
17. To plot $(3/2, 0)$, start at the origin and go approximately $3/2$ units to the right. The point is on the x-axis.
19. To plot $(0, -7/3)$, start at the origin and go $7/3$ units downward. The point is on the y-axis.

21. $y = x + 1$
If $x = 0$, $y = 0 + 1 = 1$.
If $x = 1$, $y = 1 + 1 = 2$.
If $x = 2$, $y = 2 + 1 = 3$.
If $x = -1$, $y = -1 + 1 = 0$.
Plot $(0, 1)$, $(1, 2)$, $(2, 3)$, $(-1, 0)$, and draw a line through the points.

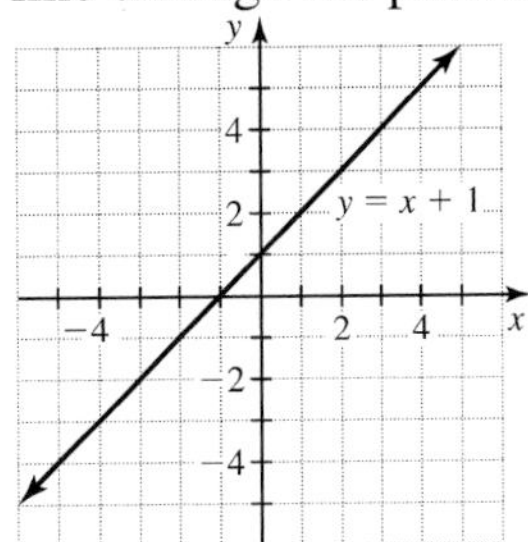

23. $y = -2x + 3$
If $x = 0$, $y = -2(0) + 3 = 3$.
If $x = 1$, $y = -2(1) + 3 = 1$.
If $x = 2$, $y = -2(2) + 3 = -1$.
If $x = -1$, $y = -2(-1) + 3 = 5$.
Plot $(0, 3)$, $(1, 1)$, $(2, -1)$, and $(-1, 5)$ and draw a line through the points.

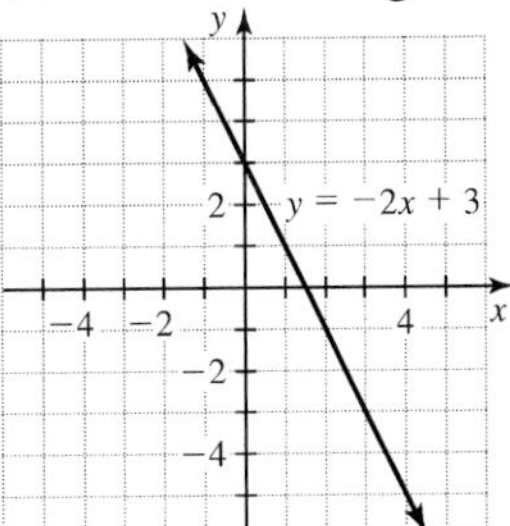

25. Since the x and y-coordinates are equal in $y = x$, plot $(0, 0)$, $(1, 1)$, $(2, 2)$, $(-1, -1)$ and draw a line through the points.

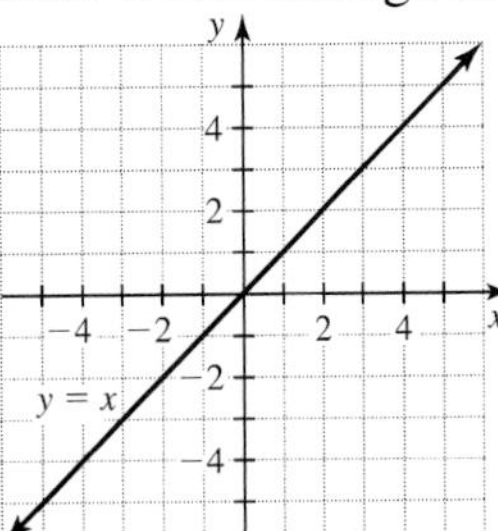

27. In $y = 3$, the x-coordinate can be any number as long as the y-coordinate is 3. Plot $(0, 3)$, $(1, 3)$, $(2, 3)$, $(-1, 3)$, and draw a horizontal line through the points.

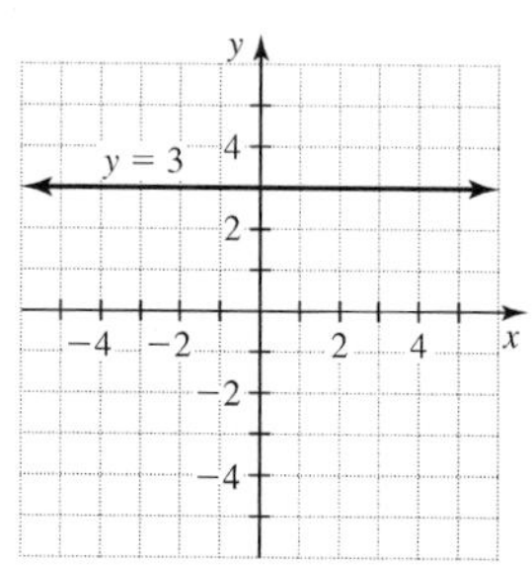

29. For the equation $y = 1 - x$, plot the points $(0, 1)$, $(1, 0)$, $(2, -1)$, and $(-2, 3)$. Draw a line through the points.

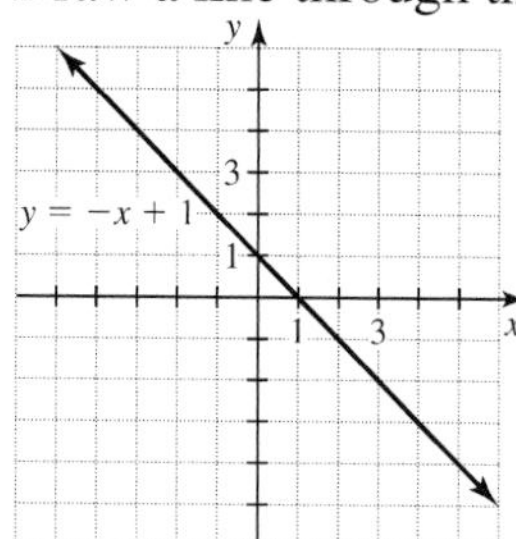

31. For $x = 2$, the y-coordinate can be any number, but the x-coordinate is 2. Plot $(2, 0)$, $(2, 1)$, $(2, 3)$, $(2, -1)$, and draw a vertical line through the points.

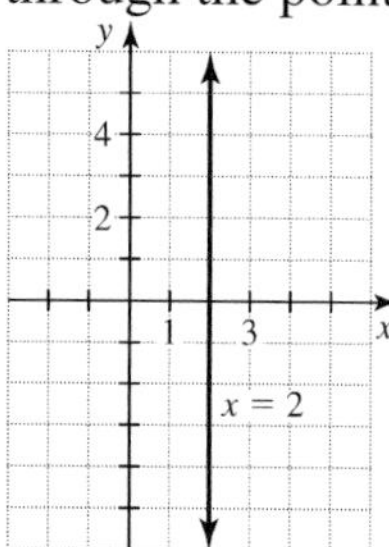

33. For the equation $y = \frac{1}{2}x - 1$, plot the points $(0, -1)$, $(2, 0)$, $(4, 1)$, and $(-4, -3)$. Draw a line through the points.

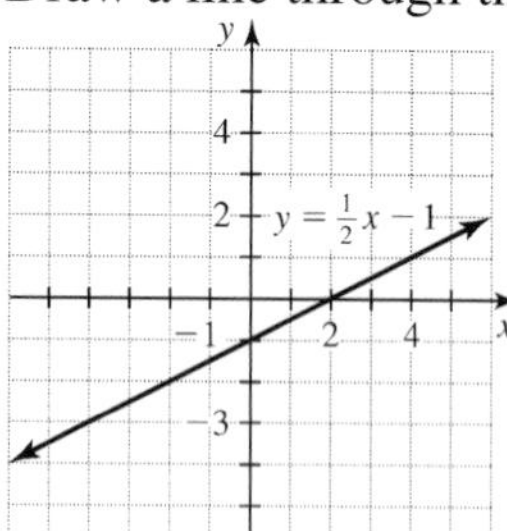

35.

$$y + 3 = 0$$
$$y = -3$$

Plot $(-2, -3)$, $(0, -3)$, $(1, -3)$, and $(3, -3)$ and draw a line through the points.

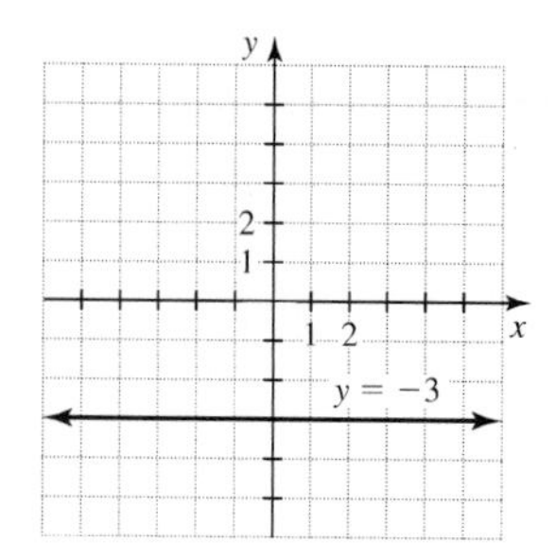

37.

$$x - 4 = 0$$
$$x = 4$$

Plot $(4, 5)$, $(4, 2)$, $(4, -1)$, $(4, 1)$, and draw a line through the points.

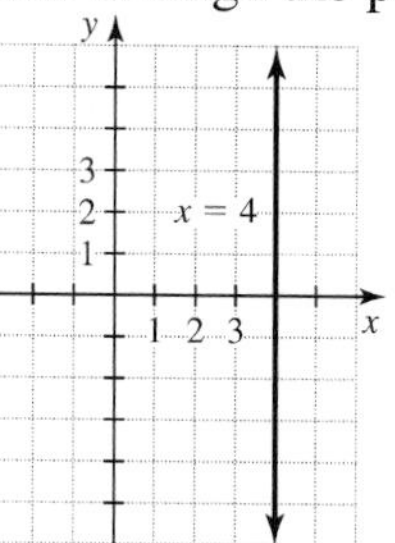

39. For $y = \frac{1}{2}x$ plot $(0, 0)$, $(2, 1)$, $(4, 2)$, $(6, 3)$ and draw a line through the points.

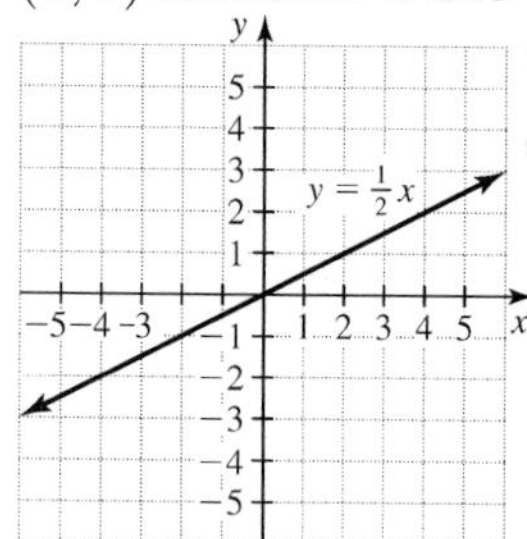

41.

$$3x + y = 5$$
$$y = -3x + 5$$

Plot $(0, 5)$, $(1, 2)$, $(2, -1)$, $(-1, 8)$, and draw a line through the points.

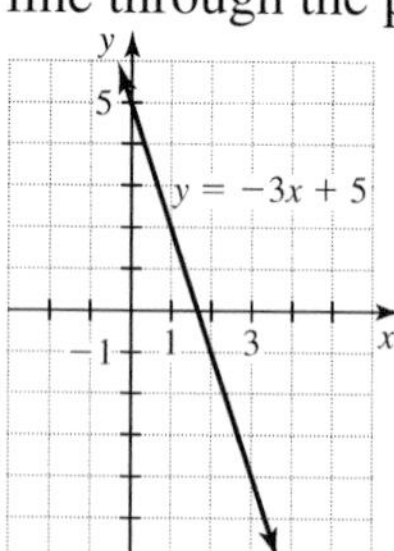

43.

$$6x + 3y = 0$$
$$3y = -6x$$
$$y = -2x$$

Plot the points $(0, 0)$, $(1, -2)$, $(2, -4)$, $(-2, 4)$ and draw a line through the points.

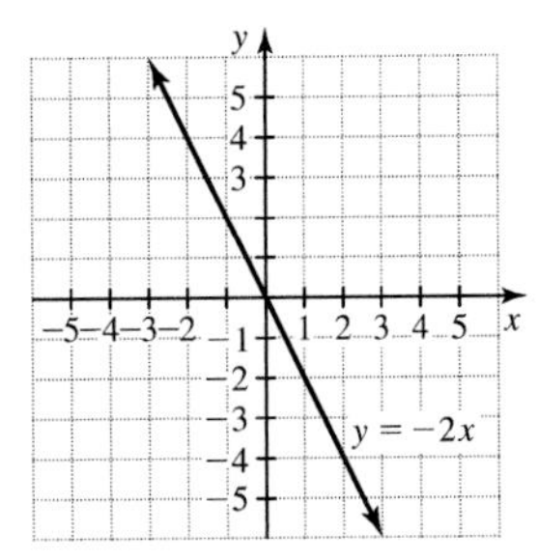

45. For $y = 2x - 20$ plot the points $(0, -20)$, $(5, -10)$, $(10, 0)$, $(-5, -30)$ and draw a line through them.

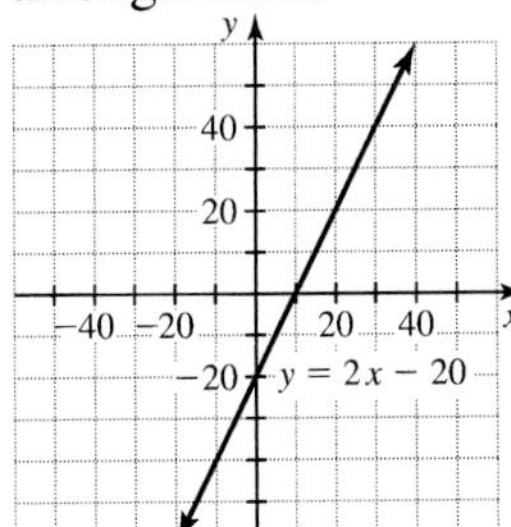

47. If $x = 0$ in $4x - 3y = 12$, then $4(0) - 3y = 12, -3y = 12$, or $y = -4$.
If $y = 0$, in $4x - 3y = 12$, then $4x - 3(0) = 12$, $4x = 12$, or $x = 3$.
The intercepts are $(0, -4)$ and $(3, 0)$. Draw a line through these points.

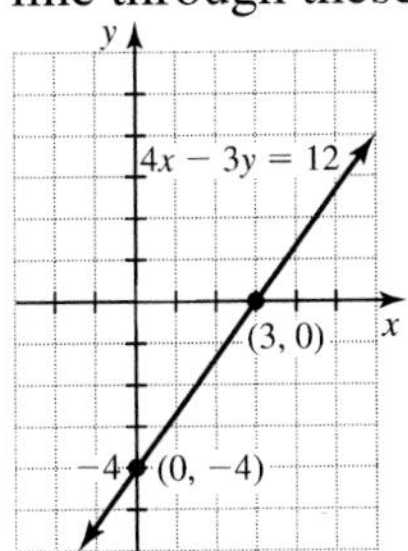

49. If $x = 0$, in $x - y + 5 = 0$, then $0 - y + 5 = 0$ or $5 = y$. If $y = 0$, then $x - 0 + 5 = 0$ or $x = -5$. The intercepts are $(0, 5)$ and $(-5, 0)$. Draw a line through these points.

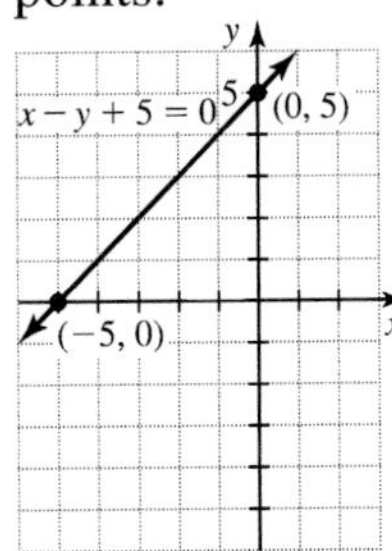

51. If $x = 0$, in $2x + 3y = 5$, then $2(0) + 3y = 5$ or $y = 5/3$. If $y = 0$, $2x + 3(0) = 5$ or $x = 5/2$. Plot the intercepts $(0, 5/3)$ and $(5/2, 0)$, and draw a line through the points.

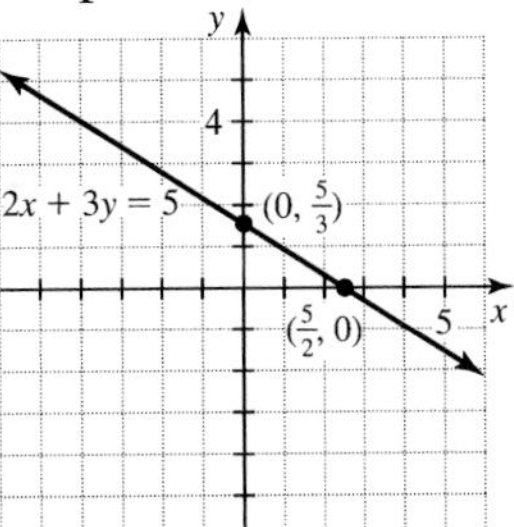

53. If $x = 0$, in $2x - 3y = 60$, then $2(0) - 3y = 60$, or $y = -20$. If $y = 0$, then $2x = 60$ or $x = 30$. Plot $(0, -20)$ and $(30, 0)$, and draw a line through the points.

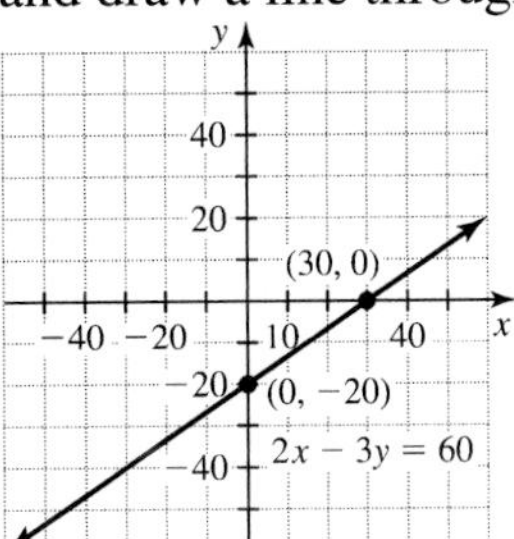

55. If $x = 0$ in $y = 2x - 4$, then $y = -4$. If $y = 0$, then $2x - 4 = 0$, or $x = 2$. Draw a line through the intercepts $(0, -4)$ and $(2, 0)$.

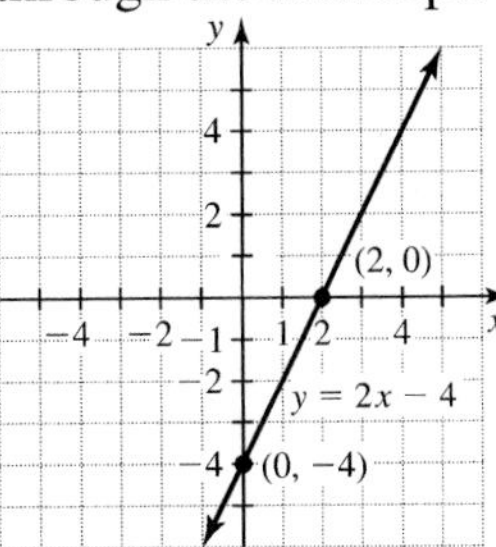

57. If $x = 0$ in $y = -\frac{1}{2}x - 20$, then $y = -20$. If $y = 0$, then $-\frac{1}{2}x - 20 = 0$, or $x = -40$. Draw a line through the intercepts $(0, -20)$ and $(-40, 0)$.

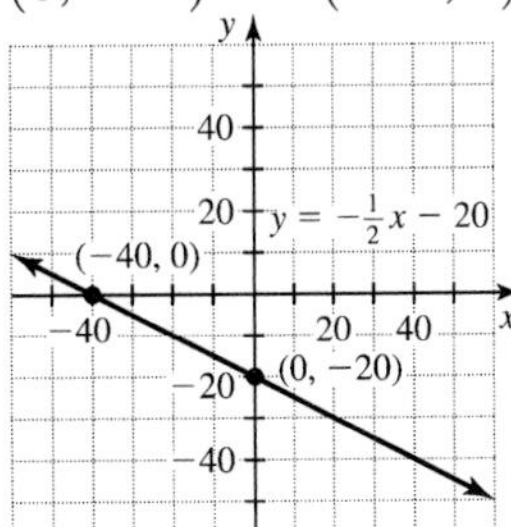

59. From the calculator graph the intercepts are $(0, 1)$ and $(1/3, 0)$.

61. From the calculator graph the intercepts are $(0, -2)$ and $(0.005, 0)$.
63. From the calculator graph the intercepts are $(0, 300)$ and $(600, 0)$.
65. From the calculator graph the intercepts are $(0, 23.54)$ and $(-5.53, 0)$.
67. If $x = 0$ in $x + y = 50$, then $0 + y = 50$ or $y = 50$. If $y = 0$, then $x + 0 = 50$ or $x = 50$. The intercepts are $(0, 50)$ and $(50, 0)$.
69. If $x = 0$ in $3x - 5y = 15$, then $-5y = 15$ or $y = -3$. If $y = 0$, then $3x = 15$ or $x = 5$. The intercepts are $(0, -3)$ and $(5, 0)$.
71. If $x = 0$ in $y = 5x$, then $y = 5(0) = 0$. If $y = 0$, then $5x = 0$ or $x = 0$. The intercept is $(0, 0)$.
73. The equation $6x + 3 = 0$ is equivalent to $x = -\frac{1}{2}$, which is a vertical line. It has no y-intercept. The x-intercept is $\left(-\frac{1}{2}, 0\right)$.
75. The equation $12 + 18y = 0$ is equivalent to $y = -\frac{2}{3}$, which is a horizontal line. It has no x-intercept. The y-intercept is $\left(0, -\frac{2}{3}\right)$.
77. If $x = 0$ in $2 - 4y = 8x$ then $2 - 4y = 0$ or $y = \frac{1}{2}$. If $y = 0$ then $2 = 8x$ or $x = \frac{1}{4}$. The intercepts are $\left(0, \frac{1}{2}\right)$ and $\left(\frac{1}{4}, 0\right)$.
79. If $x = 2$, then $y = -3(2) + 6 = 0$. If $y = -3$, then $-3 = -3x + 6$, or $x = 3$. The points are (2, 0) and (3, -3).
81. If $x = -4$, then $\frac{1}{2}(-4) - \frac{1}{3}y = 9$, or $y = -33$. If $y = 6$, then $\frac{1}{2}x - \frac{1}{3} \cdot 6 = 9$, or $x = 22$.
The points are (-4, -33) and (22, 6).
83. a) Use $n = 9$ in $P = 793n + 15{,}950$ to get $P = 793(9) + 15{,}950 = 23{,}087$. In 2009 the car will cost \$23,087.
b) Every year the price goes up by \$793.
c)

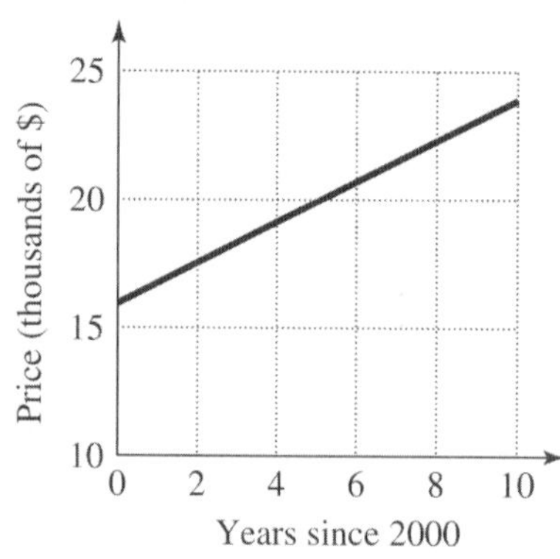

85. a) If $m = 400$ in $C = 0.26m + 42$, then $C = 0.26(400) + 42 = 146$. The charge for a car driven 400 miles is \$146.
b) If $m = 0$, $C = 0.26(0) + 42 = 42$. If $m = 1000$, $C = 0.26(1000) + 42 = 302$. Plot (0, 42) and (1000, 302), and draw a line segment with those endpoints.

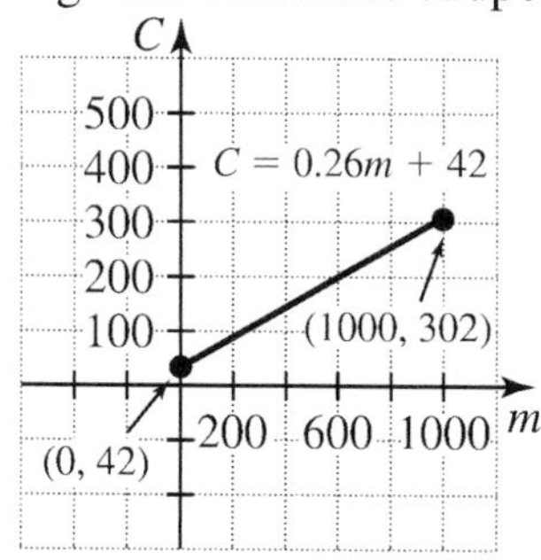

87. a) $C = 0.50(5) + 8.95 = \$11.45$.
b) $0.50t + 8.95 = 14.45$
$$0.50t = 5.50$$
$$t = 11$$
A pizza that sells for \$14.45 has 11 toppings.
89. Let $n =$ then number of note pads and $b =$ the number of binders. Since the total cost is \$100, we have $1n + 2b = 100$ or $n + 2b = 100$. If he orders 30 note pads, then $30 + 2b = 100$ or $b = 35$. The graph follows. Note that either n or b could be placed on the horizontal axis.

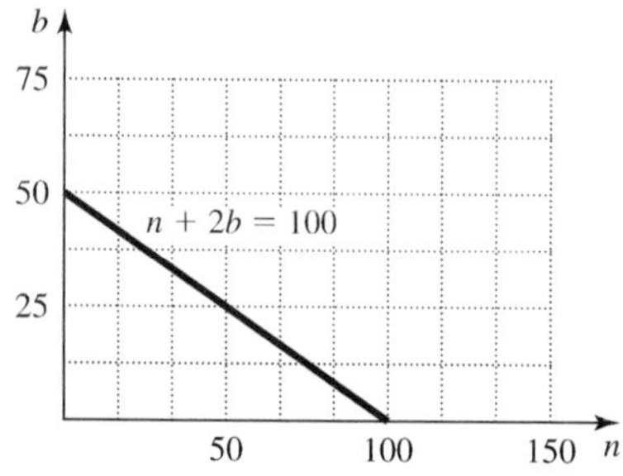

91. a) If $x = 850$, then $C = 0.55(850) + 50 = \$517.50$

$R = 1.50(850) = \$1275$
$P = 0.95(850) - 50 = \$757.50$. We have found her cost, revenue, and profit if she sells 850 roses.
b) If $P = 995$, then

$$0.95x - 50 = 995$$
$$0.95x = 1045$$
$$x = 1100$$

Here profit was \$995 when she sold 1100 roses.
c) If $x = 1100$, then $R = \$1650$ and $C = \$655$. So $R - C = \$995$. The difference between revenue and cost is \$995, which is her profit and this agrees with part (b).
velocity of the ball when it hits the ground.

3.2 WARM-UPS

1. True, because the closer the line is to vertical, the larger its slope.
2. False, slope is rise divided by run.
3. False, because it is horizontal line with 0 slope.
4. True, because it is a vertical line.
5. False, slope of a line can be any real number.
6. False, the slope is $2/5$.
7. False, it has slope 1.
8. True, because it is rising as you go from left to right.
9. False, lines with slope $2/3$ and $-3/2$ are perpendicular.
10. False, because two vertical parallel lines do not have equal slopes.

3.2 EXERCISES

1. Slope measures the steepness of a line.
3. A horizontal line has zero slope because the rise is zero.
5. If m_1 and m_2 are the slopes of perpendicular lines, then $m_1 = -1/m_2$.
7. In going from $(-3, 0)$ to $(0, 2)$ we rise 2 and run 3. The slope is $2/3$.
9. The slope for this line is undefined because it is a vertical line.
11. The slope of any horizontal line is 0.
13. In going from $(-2, 3)$ to $(3, -3)$, we rise -5 and run 5. The slope is $-5/5$ or -1.
15. In going from $(2, 0)$ to $(4, 3)$ we rise 3 and run 2. The slope is $3/2$.
17. In going from $(-2, 4)$ to $(4, -2)$ we rise -6 and run 6. The slope is $-6/6$ or -1.
19. $m = \frac{6-1}{2-5} = -\frac{5}{3}$
21. $m = \frac{-1-3}{-3-4} = \frac{-4}{-7} = \frac{4}{7}$
23. $m = \frac{2-7}{-2-(-1)} = \frac{-5}{-1} = 5$
25. $m = \frac{-5-0}{3-0} = \frac{-5}{3} = -\frac{5}{3}$
27. $m = \frac{3-0}{0-5} = -\frac{3}{5}$
29. $m = \frac{-1-(-\frac{1}{2})}{\frac{3}{4}-(-\frac{1}{2})} = \frac{-\frac{1}{2}}{\frac{5}{4}}$
$= -\frac{1}{2} \cdot \frac{4}{5} = -\frac{2}{5}$
31. $m = \frac{212-209}{6-7} = \frac{3}{-1} = -3$
33. $m = \frac{7-7}{4-(-12)} = \frac{0}{16} = 0$
35. $\frac{6-(-6)}{2-2} = \frac{12}{0}$ The slope is undefined because 0 is in the denominator.
37. $m = \frac{11.9-8.4}{24.3-3.57} = \frac{3.5}{20.73} \approx 0.169$

39. The line through $(2, 3)$ and $(4, 9)$ has slope

$$m = \frac{9-3}{4-2} = \frac{6}{2} = 3.$$

Any line parallel to this line has slope 3. Line l has slope 3.

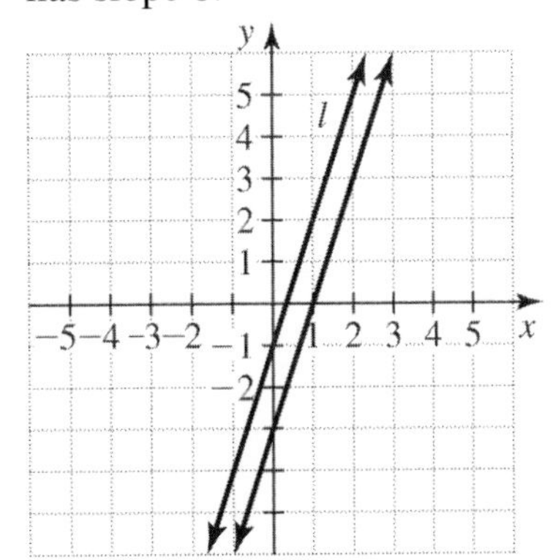

41. The line through $(-5, 1)$ and $(3, -2)$ has slope

$$m = \frac{1-(-2)}{-5-3} = \frac{3}{-8} = -\frac{3}{8}.$$

Any line perpendicular to this line has slope $8/3$ because $8/3$ is the opposite of the reciprocal of $-3/8$. Line l has slope $8/3$.

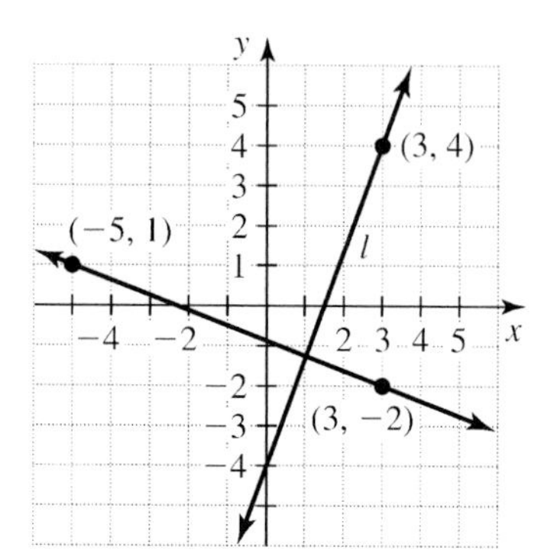

43. The slope of the line through $(-3, -2)$ and $(4, 1)$ is

$$m = \frac{-2-1}{-3-4} = \frac{3}{7}.$$

Any line parallel to this line also has slope $3/7$. So line l has slope $3/7$.

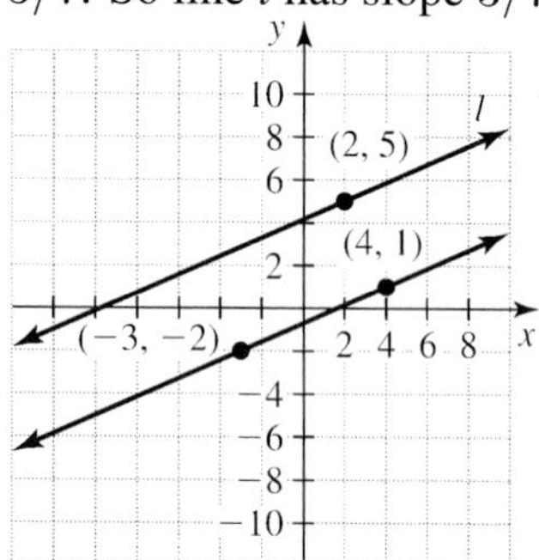

45. The slope of the line through $(0, 0)$ and $(2, 4)$ is

$$m = \frac{4-0}{2-0} = \frac{4}{2} = 2.$$

The slope of any line perpendicular to a line with slope 2 has slope $-1/2$. Line l has slope $-1/2$.

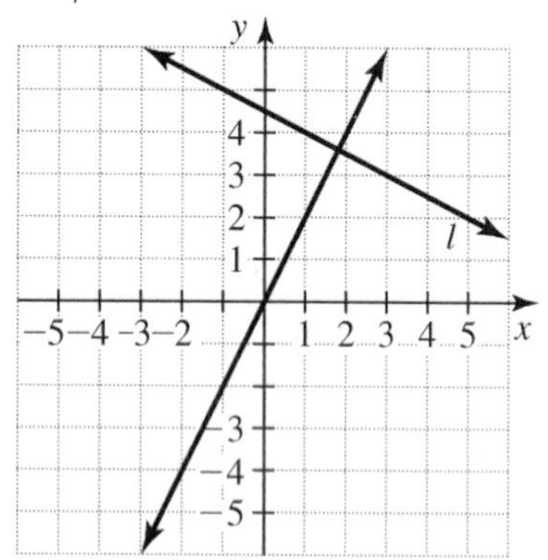

47. The opposite of the reciprocal of $4/5$ is $-5/4$. Line l has slope $-5/4$.

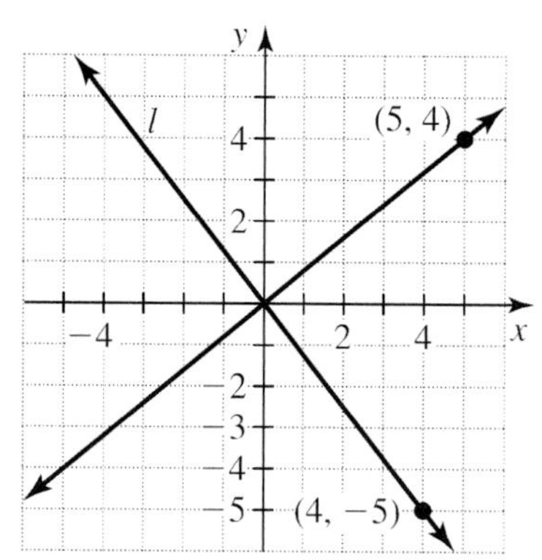

49. Any line parallel to a line with slope 2 also has slope 2. So line l has slope 2. Draw one line through the origin with slope 2 and line l through $(0, 4)$ with slope 2.

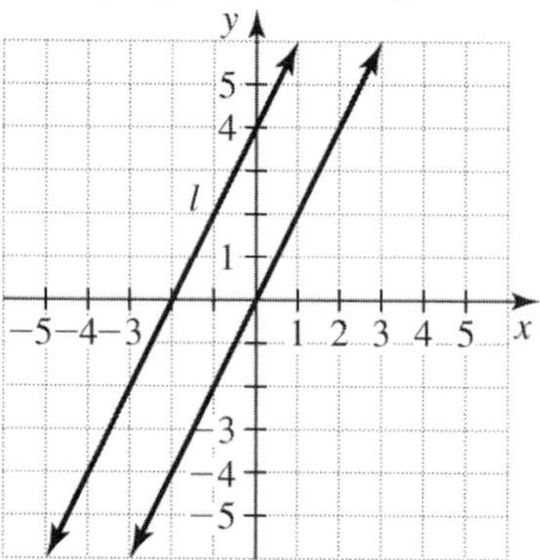

51. The slope for l_1 is $m_1 = \frac{8-2}{4-1} = 2.$
The slope for l_2 is $m_2 = \frac{2-3}{2-0} = -\frac{1}{2}.$
So the lines are perpendicular.

53. The slope for l_1 is $m_1 = \frac{6-4}{-1-0} = -2.$
The slope for l_2 is $m_2 = \frac{9-7}{8-7} = 2.$
So the lines are neither parallel nor perpendicular.

55. The slope for l_1 is $m_1 = \frac{9-2}{7-0} = 1.$
The slope for l_2 is $m_2 = \frac{-2-(-3)}{1-0} = 1.$
So the lines are parallel.

57. The slope for l_1 is $m_1 = \frac{5-0}{-2-0} = -\frac{5}{2}.$
The slope for l_2 is $m_2 = \frac{6-1}{2-0} = \frac{5}{2}.$
So the lines are neither parallel nor perpendicular.

59. The slope for l_1 is $m_1 = \frac{1-(-3)}{4-(-2)} = \frac{2}{3}.$
The slope for l_2 is $m_2 = \frac{4-7}{1-(-1)} = -\frac{3}{2}.$
So the lines are perpendicular.

61. Draw a quadrilateral with the four given points as vertices. Find the slope of each side.

$m_1 = \frac{3-1}{0-(-6)} = \frac{1}{3}$ $m_2 = \frac{3-1}{0-4} = -\frac{1}{2}$
$m_3 = \frac{1-(-1)}{4-(-2)} = \frac{1}{3}$
$m_4 = \frac{-1-1}{-2-(-6)} = -\frac{1}{2}$
Since opposite sides of the quadrilateral have equal slopes, the quadrilateral is a parallelogram.

63. Draw a quadrilateral with the four given points as vertices. Find the slope of each side.
$m_1 = \frac{6-2}{3-(-3)} = \frac{2}{3}$
$m_2 = \frac{6-4}{3-6} = -\frac{2}{3}$
$m_3 = \frac{4-(-1)}{6-(-1)} = \frac{5}{7}$
$m_4 = \frac{-1-2}{-1-(-3)} = -\frac{3}{2}$
Since there are no parallel sides, this quadrilateral is not a trapezoid.

65. Draw a triangle with the three given points as vertices. Find the slope of each side.
$m_1 = \frac{6-3}{-1-(-3)} = \frac{3}{2}$ $m_2 = \frac{6-0}{-1-0} = -6$
$m_3 = \frac{3-0}{-3-0} = -1$
From the slopes we see that none of the line segments are perpendicular. The triangle is not a right triangle.

67. a) $m = \frac{27.505 - 21.135}{9-0} \approx 0.708$
$708 dollars per year
b) Using the graph we might guess that the price in 2010 will be about $29,000.
c) The slope is the average yearly increase in price. So from 2007 to 2010 there will be 3 increases of $708 each year, for an increase of $2124. So the price in 2010 should be $27,505 + $2124, or $29,629.

69. Use graph paper to solve this problem. The slope 4 means that we can rise 4 and run 1 to get additional points on this line. From $(2, 1)$ rise 4 and run 1 to get to $(3, 5)$. From $(2, 1)$ we can rise -4 and run -1 to get to $(1, -3)$. From $(1, -3)$ we can rise -4 and run -1 to get to $(0, -7)$.

71. Find the slope:
$$\frac{k-(-5)}{2-(-3)} = \frac{1}{2}$$
$$\frac{k+5}{5} = \frac{1}{2}$$
$$2k + 10 = 5$$
$$k = -\frac{5}{2}$$

73. The slope of the line perpendicular to the line with slope 0.247 is the opposite of the reciprocal of 0.247:
$$-\frac{1}{0.247} \approx -4.049$$

75. A horizontal line has zero slope and a vertical line has undefined slope.

77. Draw a quadrilateral with the four given vertices. The slope of the diagonal joining $(-3, -1)$ and $(5, 3)$ is
$$m_1 = \frac{3-(-1)}{5-(-3)} = \frac{1}{2}.$$
The slope of the diagonal joining $(0, 3)$ and $(2, -1)$ is
$$m_2 = \frac{3-(-1)}{0-2} = -2.$$
Since the opposite of the reciprocal of $1/2$ is -2, the diagonals are perpendicular.

79. Increasing m makes the graph increase faster. The slopes of these lines are 1, 2, 3, and 4.

3.3 WARM-UPS

1. True, because there is a one-to-one correspondence between the slopes of lines through a fixed point and the real numbers.
2. False, the line through (a, b) with slope m has equation $y - b = m(x - a)$.
3. False, $y = mx + b$ goes through $(0, b)$ and has slope m.
4. True, because the y-intercept has x-coordinate 0.
5. True, because the x-intercept has y-coordinate 0.
6. False, because vertical lines do not have equations in slope intercept form.
7. True, because if we solve the equation for y, we get $y = (-3/2)x + (7/2)$.
8. False, because a line perpendicular to a line with slope 3 has slope $-1/3$.
9. False, because if $x = 0$ in this equation, we get $y = 5/2$.

10. True, because even the vertical lines can be expressed in standard form.

3.3 EXERCISES

1. Slope-intercept form is $y = mx + b$ where m is the slope and $(0, b)$ is the y-intercept.

3. Standard form is $Ax + By = C$ where A, B, and C are real numbers with A and B not both zero.

5. Point-slope form is $y - y_1 = m(x - x_1)$ where m is the slope and (x_1, y_1) is a point on the line.

7. The line goes through (0, 2) and $(-4, 0)$. The slope is $\frac{1}{2}$ and the y-intercept is (0, 2). Using slope-intercept form, we can write $y = \frac{1}{2}x + 2$.

9. This line has 0 slope and y-intercept $(0, -2)$. So its equation in slope-intercept form is $y = 0 \cdot x - 2$, or $y = -2$.

11. This is a vertical line with an x-intercept of $(1, 0)$ and so its equation is $x = 1$.

13. This line has a y-intercept of $(0, 0)$ and also goes through $(3, -3)$. Its slope is -1. Its equation in slope-intercept form is $y = -x$.

15. This line goes through (2, 0) and $(0, -3)$. Its slope is $3/2$ and its equation is $y = \frac{3}{2}x - 3$.

17. This line goes through (0, 2) and (2, 0). Since its slope is -1, its equation is $y = -x + 2$.

19.
$$\begin{aligned} 2x + 5y &= 1 \\ 5y &= -2x + 1 \\ y &= -\frac{2}{5}x + \frac{1}{5} \end{aligned}$$

The slope is $-\frac{2}{5}$ and the y-intercept is $(0, 1/5)$.

21.
$$\begin{aligned} 3x - y - 2 &= 0 \\ 3x - 2 &= y \\ y &= 3x - 2 \end{aligned}$$

The slope is 3 and the y-intercept is $(0, -2)$.

23.
$$\begin{aligned} y + 3 &= 5 \\ y &= 2 \end{aligned}$$

The slope is 0 and the y-intercept is (0, 2).

25.
$$\begin{aligned} y - 2 &= 3(x - 1) \\ y - 2 &= 3x - 3 \\ y &= 3x - 1 \end{aligned}$$

The slope is 3 and the y-intercept is $(0, -1)$.

27.
$$\begin{aligned} y - \frac{1}{2} &= \frac{1}{3}\left(x + \frac{1}{4}\right) \\ y - \frac{1}{2} &= \frac{1}{3}x + \frac{1}{12} \\ y &= \frac{1}{3}x + \frac{1}{12} + \frac{1}{2} \\ y &= \frac{1}{3}x + \frac{7}{12} \end{aligned}$$

The slope is $\frac{1}{3}$ and the y-intercept is $\left(0, \frac{7}{12}\right)$.

29.
$$\begin{aligned} y - 6000 &= 0.01(x + 5700) \\ y - 6000 &= 0.01x + 57 \\ y &= 0.01x + 6057 \end{aligned}$$

The slope is 0.01 and the y-intercept is $(0, 6057)$.

31. The slope of $y = (1/2)x$ is $1/2$ and the y-intercept is (0, 0). To graph the line start at $(0, 0)$ and go up 1 unit and 2 units to the right to find another point on the line, (2, 1). Draw the line through (0, 0) and (2, 1).

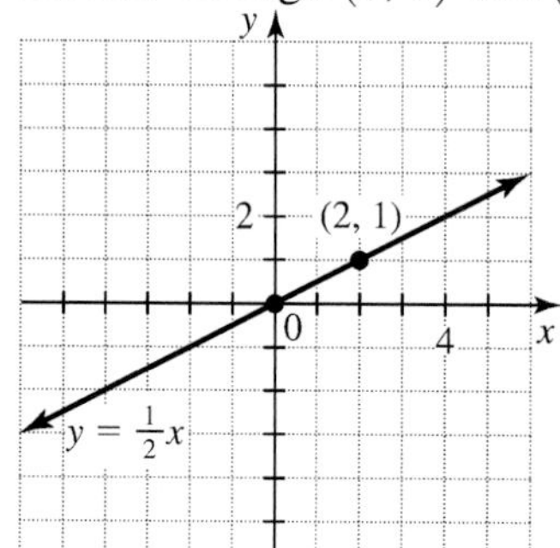

33. The slope of $y = 2x - 3$ is 2 and the y-intercept is $(0, -3)$. The slope of 2 is obtained from $2/1$. To graph the line start at $(0, -3)$ and rise 2 units and go 1 unit to the right to get to the point $(1, -1)$. Draw a line through the points $(0, -3)$ and $(1, -1)$.

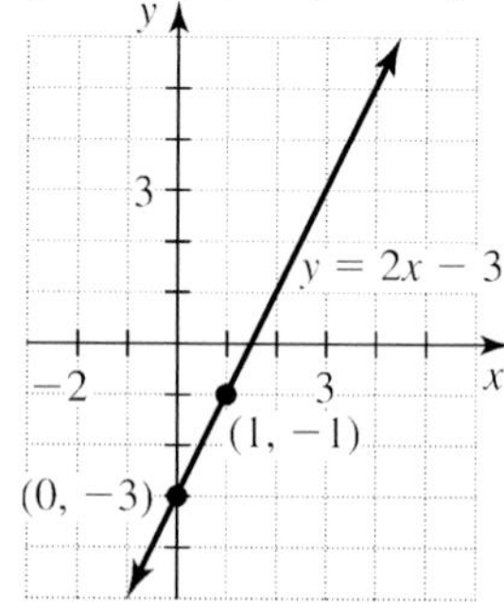

35. The slope of $y = (-2/3)x + 2$ is $-2/3$ and the y-intercept is (0, 2). To graph the line start at (0, 2) and go down 2 units and then 3 units to the right to locate the point (3, 0). Draw a line through (0, 2) and (3, 0).

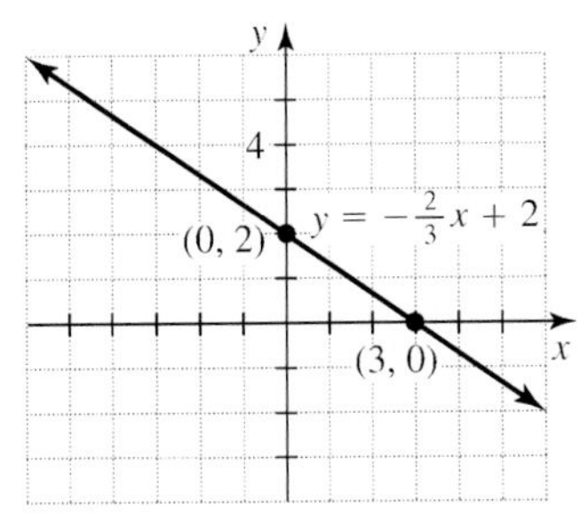

37. The equation $3y + x = 0$ can be written as $y = (-1/3)x$. Its slope is $-1/3$ and its y-intercept is $(0, 0)$. To graph the line start at $(0, 0)$ and go 1 unit down and 3 units to the right to locate the point $(3, -1)$. Draw a line through $(0, 0)$ and $(3, -1)$.

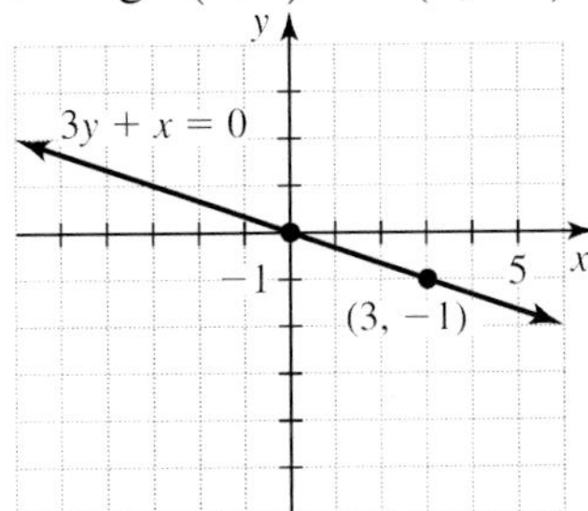

39. The graph of $y = x + 3$ has y-intercept $(0, 3)$ and slope 1. The graph of $y = x + 2$ has y-intercept $(0, 2)$ and slope 1. The lines are parallel.

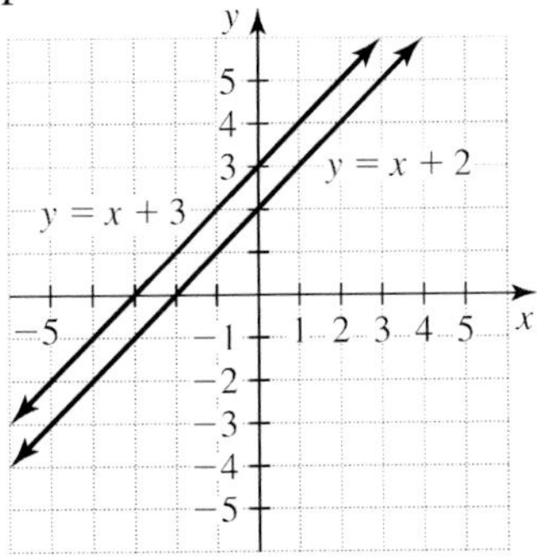

41. The graph of $y = 3x + 1$ has y-intercept $(0, 1)$ and slope 3. The graph of $y = -\frac{1}{3}x + 1$ has y-intercept $(0, 1)$ and slope $-\frac{1}{3}$. The lines are perpendicular.

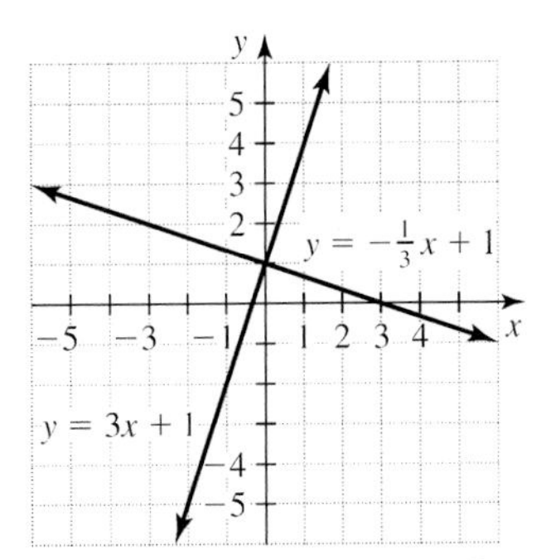

43. The graph of $y = \frac{2}{3}x - 1$ has y-intercept $(0, -1)$ and slope $\frac{2}{3}$. The graph of $y = \frac{2}{3}x + 1$ has y-intercept $(0, 1)$ and slope $\frac{2}{3}$. The lines are parallel.

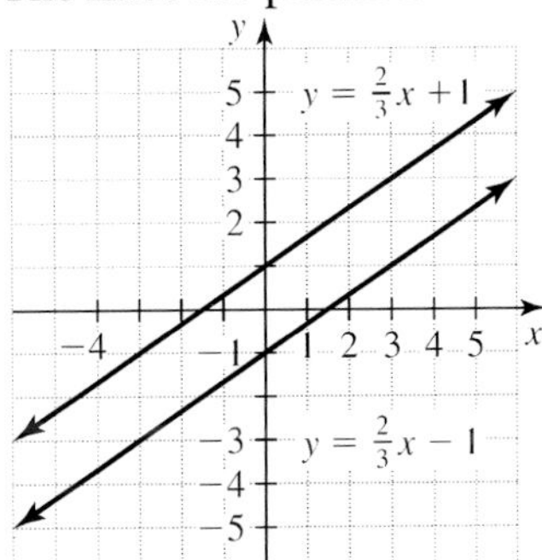

45. The graph of $y = \frac{3}{4}x + 3$ has y-intercept $(0, 3)$ and slope $\frac{3}{4}$. The graph of $y = -\frac{4}{3}x + 1$ has y-intercept $(0, 1)$ and slope $-\frac{4}{3}$. The lines are perpendicular.

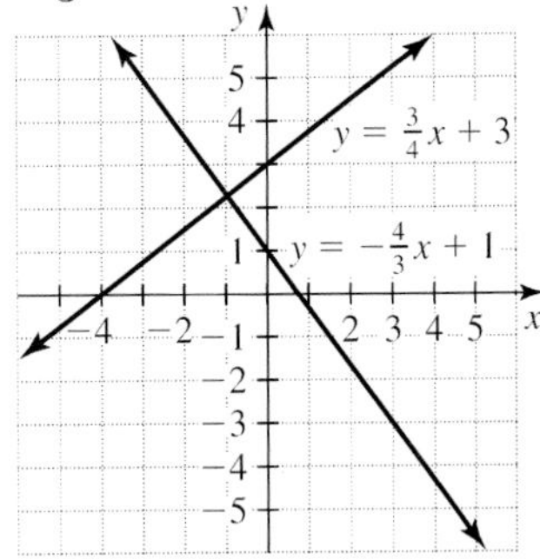

47.
$$\begin{aligned} y &= \tfrac{1}{3}x - 2 \\ 3(y) &= 3(\tfrac{1}{3}x - 2) \\ 3y &= x - 6 \\ -x + 3y &= -6 \\ x - 3y &= 6 \end{aligned}$$

49.
$$\begin{aligned} y - 5 &= \tfrac{1}{2}(x + 3) \\ 2(y - 5) &= 2 \cdot \tfrac{1}{2}(x + 3) \end{aligned}$$

$$2y - 10 = x + 3$$
$$-x + 2y = 13$$
$$x - 2y = -13$$

51. $y + \frac{1}{2} = \frac{1}{3}(x - 4)$

$$6\left(y + \frac{1}{2}\right) = 6 \cdot \frac{1}{3}(x - 4)$$
$$6y + 3 = 2(x - 4)$$
$$-2x + 6y = -11$$
$$2x - 6y = 11$$

53. $0.05x + 0.06y - 8.9 = 0$

$$100(0.05x + 0.06y - 8.9) = 100(0)$$
$$5x + 6y - 890 = 0$$
$$5x + 6y = 890$$

55. Use the given point and the given slope in the point-slope formula:

$$y - (-3) = 2(x - 2)$$
$$y + 3 = 2x - 4$$
$$y = 2x - 7$$

57. Use the point $(-2, 3)$ and the slope $-1/2$ in the point-slope formula:

$$y - 3 = -\frac{1}{2}(x - (-2))$$
$$y - 3 = -\frac{1}{2}x - 1$$
$$y = -\frac{1}{2}x + 2$$

59. Use the point (5, 12) and slope 0 in the point-slope formula.

$$y - 12 = 0(x - 5)$$
$$y = 12$$

61. Use the point (3, 60) and slope 20 in the point-slope formula.

$$y - 60 = 20(x - 3)$$
$$y - 60 = 20x - 60$$
$$y = 20x$$

63. The slope of the line through $(-1, -11)$ and $(4, 4)$ is

$$m = \frac{4 - (-11)}{4 - (-1)} = 3.$$

Use $(4, 4)$ and slope 3 in the point- slope formula:

$$y - 4 = 3(x - 4)$$
$$y - 4 = 3x - 12$$
$$y = 3x - 8$$

65. The slope of the line through $(2, 2)$ and $(-1, 1)$ is

$$m = \frac{1 - 2}{-1 - 2} = \frac{1}{3}.$$

Use $(2, 2)$ and slope $1/3$ in the point- slope formula:

$$y - 2 = \frac{1}{3}(x - 2)$$
$$y - 2 = \frac{1}{3}x - \frac{2}{3}w$$
$$y = \frac{1}{3}x - \frac{2}{3} + 2$$
$$y = \frac{1}{3}x + \frac{4}{3}$$

67. The slope of the line through $(8, 0)$ and $(-6, 7)$ is

$$m = \frac{7 - 0}{-6 - 8} = -\frac{1}{2}.$$

Use $(8, 0)$ and slope $-1/2$ in the point- slope formula:

$$y - 0 = -\frac{1}{2}(x - 8)$$
$$y = -\frac{1}{2}x + 4$$

69. The slope of the line through $(2, 13)$ and $(4, 26)$ is

$$m = \frac{26 - 13}{4 - 2} = \frac{13}{2}.$$

Use $(2, 13)$ and slope $13/2$ in the point- slope formula:

$$y - 13 = \frac{13}{2}(x - 2)$$
$$y - 13 = \frac{13}{2}x - 13$$
$$y = \frac{13}{2}x$$

71. The slope of the line through $(-5, 6)$ and $(14, 6)$ is

$$m = \frac{6 - 6}{14 - (-5)} = 0.$$

Use $(-5, 6)$ and slope 0 in the point- slope formula:

$$y - 6 = 0(x - (-5))$$
$$y = 6$$

73. First find slope of the line through $(-3, 1)$ and $(5, -1)$.

$$m = \frac{-1 - 1}{5 - (-3)} = -\frac{1}{4}$$

Any line perpendicular to a line with slope $-1/4$ has slope 4. Use the point $(-1.-12)$ and slope 4 in the point-slope formula.

$$y - (-12) = 4(x - (-1))$$
$$y + 12 = 4x + 4$$
$$y = 4x - 8$$

75. The line through $(9, -3)$ and $(-3, 6)$ has slope

$$m = \frac{6 - (-3)}{-3 - 9} = -\frac{3}{4}$$

Use the same slope $-3/4$ and $(0, 0)$ in the point-slope formula:

$$y - 0 = -\frac{3}{4}(x - 0)$$
$$y = -\frac{3}{4}x$$

77. First find the slope of the line $3x - 12y = 1$:

$$-12y = -3x + 1$$
$$y = \frac{1}{4}x - \frac{1}{12}$$

The slope is $1/4$. Any line perpendicular to a line with slope $1/4$ has slope -4. Use the point $(3, 2)$ and slope -4 in the point-slope formula.

$$y - 2 = -4(x - 3)$$
$$y - 2 = -4x + 12$$
$$4x + y = 14$$

79. First find the slope of the line $4x + 2y = 5$:

$$2y = -4x + 5$$
$$y = -2x + \frac{5}{2}$$

The slope is -2. Any line parallel to a line with slope -2 also has slope -2. Use the point $(4, -2)$ and slope -2 in the point-slope formula.

$$y - (-2) = -2(x - 4)$$
$$y + 2 = -2x + 8$$
$$2x + y = 6$$

81. Since the y-intercept is $(0, 5)$ and the slope is $1/2$, we can use slope-intercept form to write the equation.

$$y = \frac{1}{2}x + 5$$

To get integral coefficients multiply by 2.

$$2y = x + 10$$
$$-x + 2y = 10$$
$$x - 2y = -10 \quad \text{(Standard form)}$$

83. The slope of the line through $(2, 0)$ and $(0, 4)$ is

$$m = \frac{4 - 0}{0 - 2} = -2.$$

Use slope-intercept form to write the equation of the line with slope -2 and y-intercept $(0, 4)$:

$$y = -2x + 4$$
$$2x + y = 4 \quad \text{(Standard form)}$$

85. Any line parallel to $y = 2x + 6$ has slope 2. Use point-slope formula to find the equation of a line with slope 2 and going through $(-3, -1)$:

$$y - (-1) = 2(x - (-3))$$
$$y + 1 = 2x + 6$$
$$-2x + y = 5$$
$$2x - y = -5 \quad \text{(Standard form)}$$

87. Write $2x + 4y = 1$ in slope-intercept form to identify the slope:

$$4y = -2x + 1$$
$$y = -\frac{1}{2}x + \frac{1}{4}$$

Any line parallel to $2x + 4y = 1$ has slope $-1/2$. Use the point-slope formula to find the equation of the line with slope $-1/2$ through $(-3, 5)$:

$$y - 5 = -\frac{1}{2}(x - (-3))$$
$$y - 5 = -\frac{1}{2}x - \frac{3}{2}$$
$$2y - 10 = -x - 3$$
$$x + 2y = 7 \quad \text{(Standard form)}$$

89. A line perpendicular to $y = (1/2)x - 3$ has slope -2. Use the point-slope formula to find the equation of a line with slope -2 going through $(1, 1)$:

$$y - 1 = -2(x - 1)$$
$$y - 1 = -2x + 2$$
$$2x + y = 3 \quad \text{(Standard form)}$$

91. Write $x + 3y = 4$ in slope-intercept form to determine its slope:

$$3y = -x + 4$$
$$y = -\frac{1}{3}x + \frac{4}{3}$$

The slope of $x + 3y = 4$ is $-1/3$. Any line perpendicular to $x + 3y = 4$ has slope 3. Use the point-slope formula to find the equation of the line with slope 3 through $(-4, -3)$.

$$y - (-3) = 3(x - (-4))$$
$$y + 3 = 3x + 12$$
$$-3x + y = 9$$
$$3x - y = -9 \quad \text{(Standard form)}$$

93. Any line parallel to the x-axis has slope 0. If it goes through $(2, 5)$, it has y-intercept $(0, 5)$. Using slope-intercept form we get

$$y = 0 \cdot x + 5$$
$$y = 5$$

95. Solve $x + 3y = 7$ for y to get $y = -\frac{1}{3}x + \frac{7}{3}$. The lines are perpendicular because there slopes are 3 and $-1/3$.

97. Solve $2x - 4y = 9$ for y to get $y = \frac{1}{2}x - \frac{9}{4}$. Solve $\frac{1}{3}x = \frac{2}{3}y - 8$ for y to get $y = \frac{1}{2}x + 12$. The lines are parallel because there slopes are equal.

99. Solve $2y = x + 6$ for y to get $y = \frac{1}{2}x + 3$. Solve $y - 2x = 4$ for y to get

$y = 2x + 4$. The lines are neither parallel nor perpendicular.

101. Solve $x - 6 = 9$ for x to get $x = 15$. Solve $y - 4 = 12$ for y to get $y = 16$. The lines are perpendicular, because one is horizontal and the other is vertical.

103. a) The slope of the line through $(0, 60)$ and $(120, 200)$ is

$$m = \frac{200 - 60}{120 - 0} = \frac{140}{120} = \frac{7}{6}.$$

Using slope-intercept form with a t-intercept of $(0, 60)$ we get $t = \frac{7}{6}s + 60$.

b) After 30 seconds the temperature will be

$$t = \frac{7}{6}(30) + 60 = 35 + 60 = 95°\text{F}.$$

c)

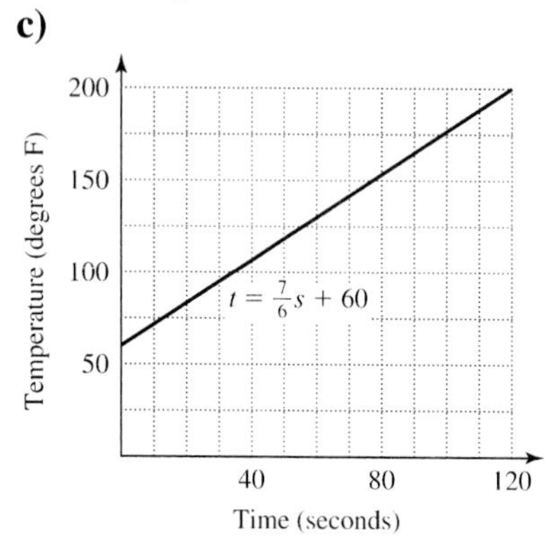

105. a) $m = \dfrac{26 - 14}{2000 - 1970} = \dfrac{12}{30} = \dfrac{2}{5} = 0.4$

$$y - 14 = 0.4(x - 1970)$$
$$y - 14 = 0.4x - 788$$
$$y = 0.4x - 774$$

b) If $x = 2010$, then

$$y = 0.4(2010) - 774 = 30.$$

So in 2010 worldwide emission of CO_2 will be 30 billion tons

107. a) The slope of the line through $(9.14, 1230)$ and $(7.84, 826)$ is

$$m = \frac{1230 - 826}{9.14 - 7.84} \approx 310.77$$

Use the point-slope formula for the equation of a line with slope 310.77 and the point $(7.84, 826)$.

$$w - 826 = 310.77(d - 7.84)$$
$$w - 826 = 310.77d - 2436.4368$$
$$w = 310.77d - 1610.44$$

b) Use $d = 8.25$ in the formula to find w.

$$w = 310.77(8.25) - 1610.44 \approx 953$$

When depth is 8.25 ft, the flow is 953 ft^3/sec.

c) As the depth increases, we can see from the graph that the flow increases.

109. a) If $y = 0$, then $x/4 = 1$ or $x = 4$. If $x = 0$, then $y/6 = 1$ or $y = 6$. So the intercepts are $(4, 0)$ and $(0, 6)$.

b) If $y = 0$, then $x/a = 1$ or $x = a$. If $x = 0$, then $y/b = 1$ or $y = b$. So the intercepts are $(a, 0)$ and $(0, b)$.

c) Since the intercepts are $(0, 3)$ and $(-5, 0)$ the equation is $\frac{x}{-5} + \frac{y}{3} = 1$.

d) To be written in intercept form a line must have two intercepts. So horizontal, vertical, and lines through $(0, 0)$ cannot be written in intercept form.

111. The x-intercept is $(3000, 0)$ and the y-intercept is $(0, -3000)$. So a viewing window such as x in $[-4000, 4000]$ and y in $[-4000, 4000]$ would show both intercepts. The are many other possibilities.

113. The lines intersect at $(50, 97)$. This intersection can be seen if x is in $[0, 100]$ and y is in $[0, 200]$.

3.4 WARM-UPS

1. True, because $-3 > -3(2) + 2$.

2. False, because $3x - y > 2$ is equivalent to $y < 3x - 2$, which is below the line $3x - y = 2$.

3. True, because $3x + y < 5$ is equivalent to $y < -3x + 5$.

4. False, the region $x < -3$ is to the left of the vertical line $x = -3$.

5. True, because the word "and" is used.

6. True, because the word "or" is used.

7. False, because $(2, -5)$ does not satisfy the inequality $y > -3x + 5$. Note that $-5 > -3(2) + 5$ is incorrect.

8. True, because $(-3, 2)$ satisfies $y \leq x + 5$.

9. False, it is equivalent to the compound inequality $2x - y \leq 4$ and $2x - y \geq -4$.

10. True, because in general $|x| > k$ (for a positive k) is equivalent to $x > k$ or $x < -k$.

3.4 EXERCISES

1. A linear inequality is an inequality of the form $Ax + By \leq C$ (or using $<$, $>$, or $\geq$) where A, B, and C are real numbers and A and B are not both zero.

3. If the inequality includes equality then the line should be solid.

5. The test point method is used to determine which side of the boundary line to shade.

7. Graph the equation $y = x + 2$. Use its slope of 1 and y-intercept of (0, 2).

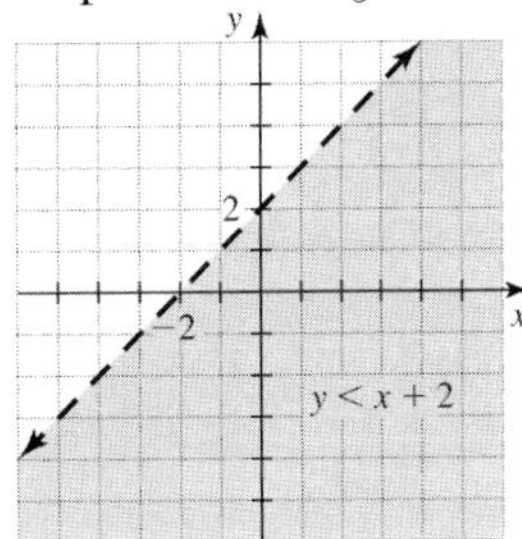

The graph of $y < x + 2$ is the region below the line $y = x + 2$. The line is dashed because of the inequality symbol. Points on the line satisfy the equation but not the inequality.

9. First graph the line $y = -2x + 1$. Start at its y-intercept (0, 1) and use a slope of -2. Go down 2 units and 1 to the right to locate a second point on the line.

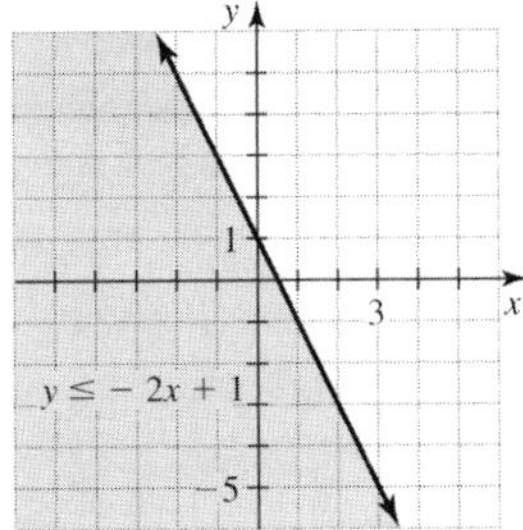

The graph of the inequality $y \leq -2x + 1$ is the region below the line and including the line. For this reason the line is drawn solid.

11. The inequality $x + y > 3$ is equivalent to $y > -x + 3$. First use slope of -1 and a y-intercept of (0, 3) to graph $y = -x + 3$.

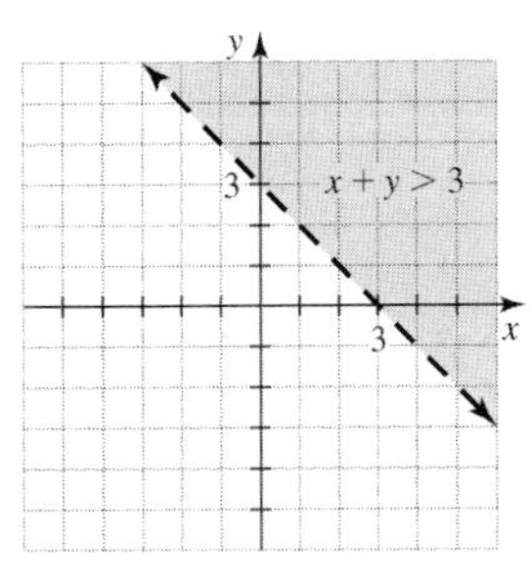

The graph of $y > -x + 3$ is the region above the line. The line is drawn dashed because of the inequality symbol.

13. The inequality $2x + 3y < 9$ is equivalent to $y < -\frac{2}{3}x + 3$. Use a slope of $-2/3$ and a y-intercept of 3 to graph $y = -\frac{2}{3}x + 3$.
The graph of the inequality is the region below the line. The line is not included in the graph and so it is drawn dashed.

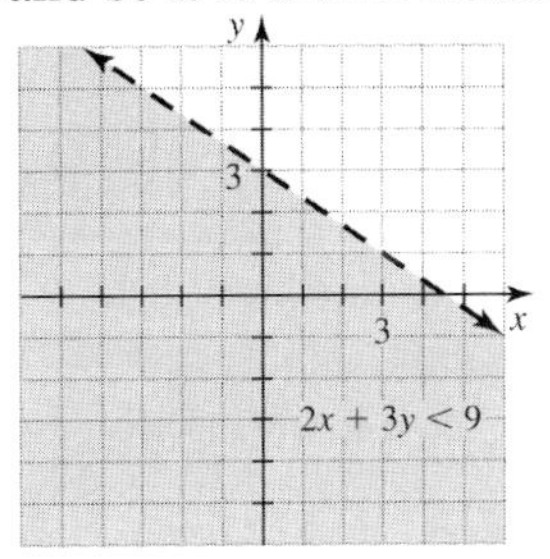

15. Solve the inequality for y:

$$3x - 4y \leq 8$$
$$-4y \leq -3x + 8$$
$$y \geq \frac{3}{4}x - 2$$

To graph the equation $y = \frac{3}{4}x - 2$, draw a solid line with y-intercept $(0, -2)$ and slope $3/4$. Since the inequality symbol is $\geq$, shade the region above the line.

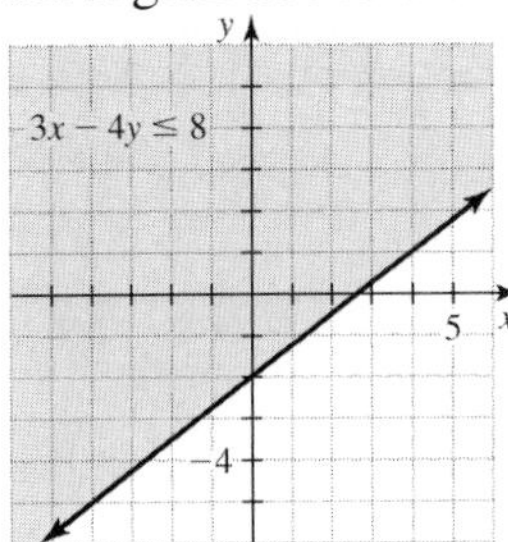

17. If we solve $x - y > 0$ for y we get $y < x$. To graph $y = x$, use a y-intercept of (0, 0) and a slope of 1. Draw a dashed line and shade below the line for the graph of $y < x$.

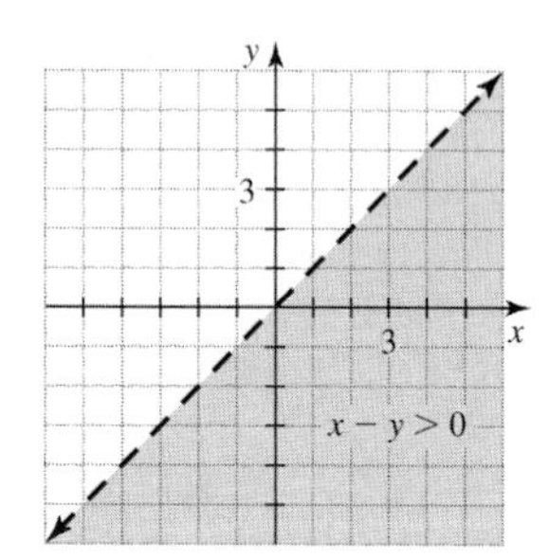

19. The graph of $x \geq 1$ consists of the vertical line $x = 1$ together with the region to the right of the line. Draw a solid vertical line through $(1, 0)$ and shade the region to the right.

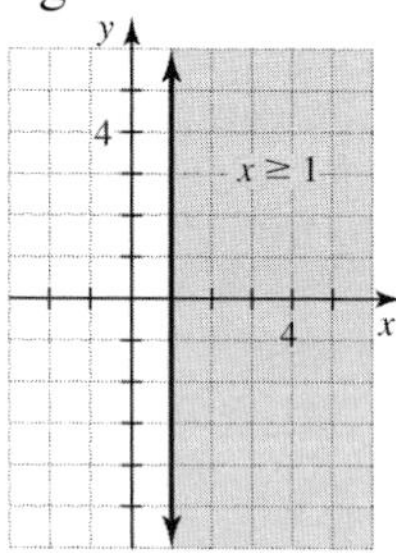

21. The graph of $y < 3$ is the region below the horizontal line $y = 3$. Draw a dashed horizontal line through $(0, 3)$ and shade the region below.

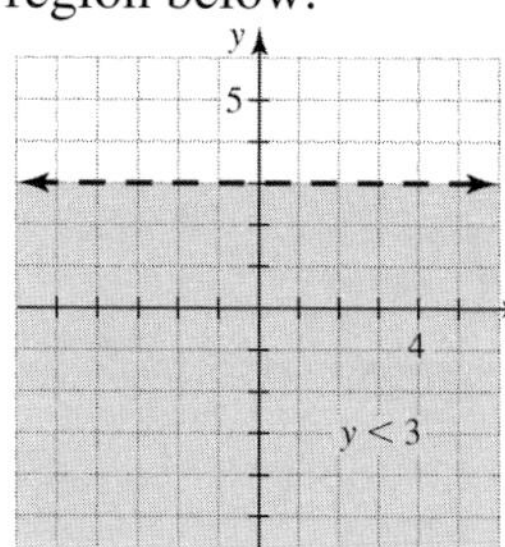

23. To graph $2x - 3y < 5$ using a test point, we can use the x and y-intercepts to determine the graph of the line $2x - 3y = 5$.
If $x = 0$, $2(0) - 3y = 5$ or $y = -5/3$.
If $y = 0$, $2x - 3(0) = 5$ or $x = 5/2$.
Locate the points $(0, -5/3)$ and $(5/2, 0)$ and draw a dashed line through them. Test the point $(0, 0)$ in the inequality:

$$2(0) - 3(0) < 5$$

Since $(0, 0)$ satisfies the inequality, we shade the side of the line that includes $(0, 0)$.

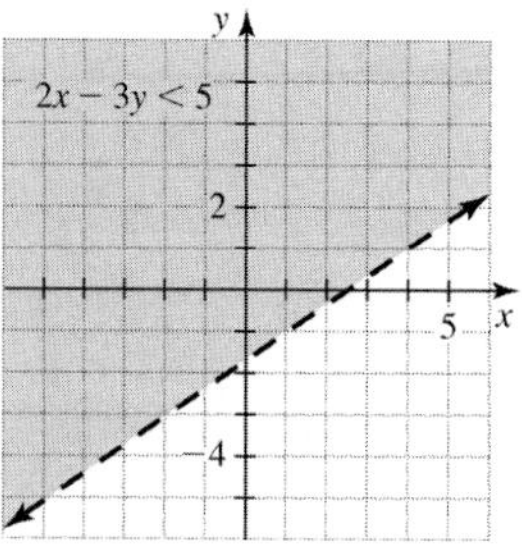

25. Graph $x + y + 3 = 0$ by using the x and y-intercepts, $(0, -3)$ and $(-3, 0)$. Draw a solid line through these two points. Test $(0, 0)$ in the inequality: $\quad 0 + 0 + 3 \geq 0$
Since $(0, 0)$ satisfies the inequality, we shade the side of the line that includes $(0, 0)$.

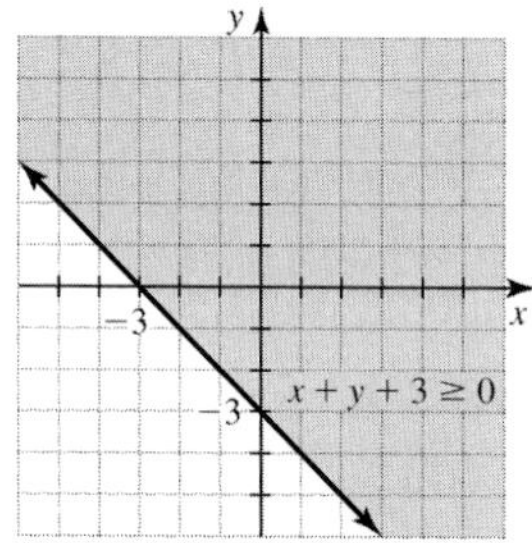

27. First graph the line $y - 2x = 0$ or $y = 2x$, which has slope 2 and y-intercept $(0, 0)$. Now test a point, say $(3, 0)$ in $y - 2x \leq 0$. Since $0 - 2(3) \leq 0$ is correct, we shade the region containing $(3, 0)$.

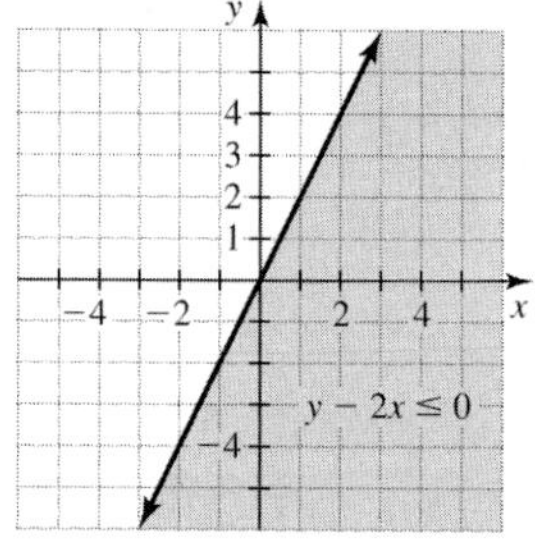

29. First graph the dashed line $3x - 2y = 0$ or $y = \frac{3}{2}x$, which has slope $3/2$ and y-intercept $(0, 0)$. Now test a point, say $(4, 0)$ in $3x - 2y > 0$. Since $3(4) - 2(0) > 0$ is correct, we shade the region containing $(4, 0)$.

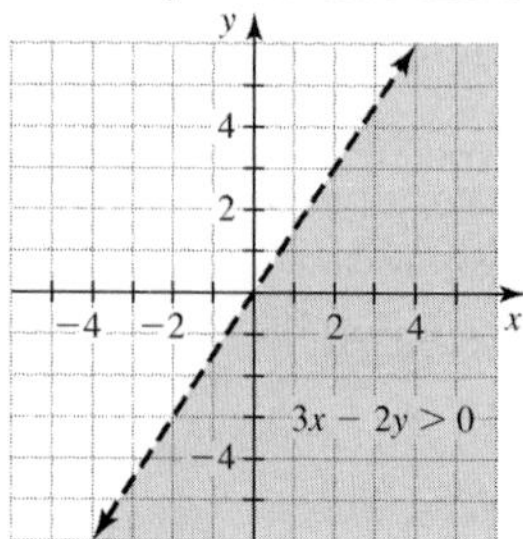

31. Find x and y-intercepts for $\frac{1}{2}x + \frac{1}{3}y = 1$:
If $x = 0$, we get $\frac{1}{3}y = 1$ or $y = 3$.
If $y = 0$, we get $\frac{1}{2}x = 1$ or $x = 2$.
Draw a dashed line through the intercepts $(0, 3)$ and $(2, 0)$. Test the point $(0, 0)$ in the inequality:

$$\frac{1}{2}(0) + \frac{1}{3}(0) < 1$$

Since $(0, 0)$ satisfies the inequality, we shade the region that includes $(0, 0)$.

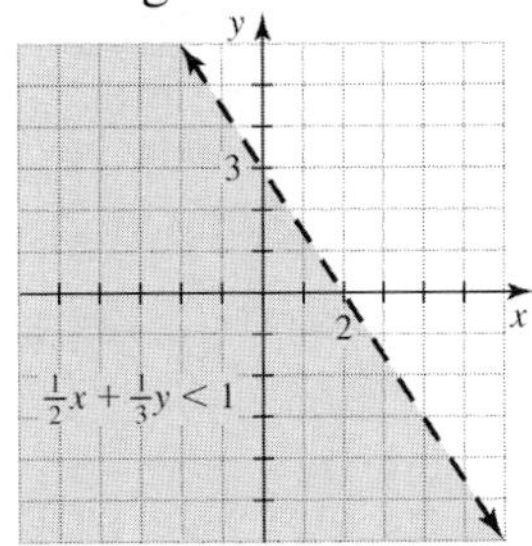

33. Check each ordered pair in
$y > 4$ or $x < 1$:

$(1, 3)$	$3 > 4$ or $1 < 1$	False
$(-2, 5)$	$5 > 4$ or $-2 < 1$	True
$(-6, -4)$	$-4 > 4$ or $-6 < 1$	True
$(7, -8)$	$-8 > 4$ or $7 < 1$	False

Only $(-2, 5)$ and $(-6, -4)$ satisfy the compound inequality.

35. Check each ordered pair in
$y > 4$ and $x < 1$:

$(1, 3)$	$3 > 4$ and $1 < 1$	False
$(-2, 5)$	$5 > 4$ and $-2 < 1$	True
$(-6, -4)$	$-4 > 4$ and $-6 < 1$	False
$(7, -8)$	$-8 > 4$ and $7 < 1$	False

Only $(-2, 5)$ satisfies the compound inequality.

37. Check each ordered pair in
$y > 5x$ and $y < -x$:

$(1, 3)$	$3 > 5(1)$ and $3 < -1$	
	$3 > 5$ and $3 < -1$	False
$(-2, 5)$	$5 > 5(-2)$ and $5 < -(-2)$	
	$5 > -10$ and $5 < 2$	False
$(-6, -4)$	$-4 > 5(-6)$ and $-4 < -(-6)$	
	$-4 > -30$ and $-4 < 6$	True
$(7, -8)$	$-8 > 5(7)$ and $-8 < -(7)$	
	$-8 > 35$ and $-8 < -7$	False

Only $(-6, -4)$ satisfies the compound inequality.

39. Check each ordered pair in
$y > -x + 1$ or $y > 4x$:

$(1, 3)$	$3 > -1 + 1$ or $3 > 4(1)$	
	$3 > 0$ or $3 > 4$	True
$(-2, 5)$	$5 > -(-2) + 1$ or $5 > 4(-2)$	
	$5 > 3$ or $5 > -8$	True
$(-6, -4)$	$-4 > -(-6) + 1$ or $-4 > 4(-6)$	
	$-4 > 7$ or $-4 > -24$	True
$(7, -8)$	$-8 > -(7) + 1$ or $-8 > 4(7)$	
	$-8 > -6$ or $-8 > 28$	False

So $(1, 3)$, $(-2, 5)$, and $(-6, -4)$ satisfy the compound inequality.

41. Check each ordered pair in
$|x + y| < 3$:

$(1, 3)$	$\|1 + 3\| < 3$	
	$4 < 3$	False
$(-2, 5)$	$\|-2 + 5\| < 3$	
	$3 < 3$	False
$(-6, -4)$	$\|-6 + (-4)\| < 3$	
	$10 < 3$	False
$(7, -8)$	$\|7 + (-8)\| < 3$	
	$1 < 3$	True

So $(7, -8)$ satisfies the absolute value inequality.

43. To graph $y > x$ and $y > -2x + 3$, we first draw dashed lines for $y = x$ and for $y = -2x + 3$. Test one point in each of the four regions to see if it satisfies the compound inequality. Test $(5, 0)$, $(0, 5)$, $(-5, 0)$ and $(0, -5)$. Only $(0, 5)$ satisfies both inequalities. So shade the region containing $(0, 5)$.

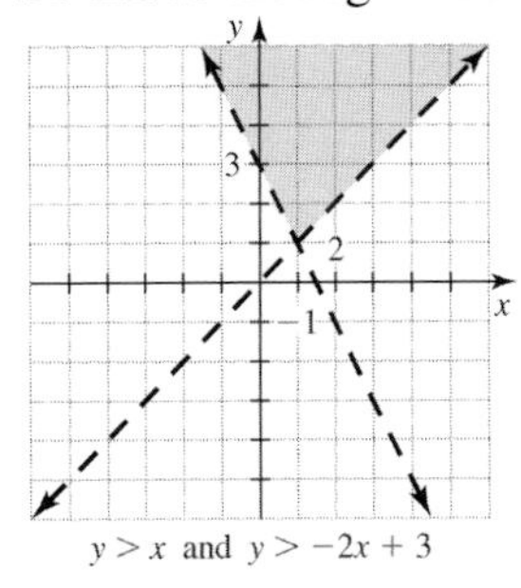

45. First graph the equations $y = x + 3$ and $y = -x + 2$. Test the points $(0, 5)$, $(5, 0)$, $(0, -5)$, and $(-5, 0)$ in the compound inequality. Only $(-5, 0)$ fails to satisfy the compound inequality. So shade all regions except the one containing $(-5, 0)$.

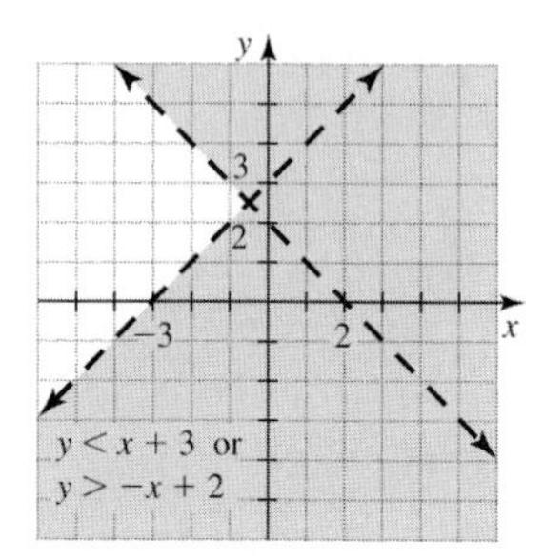

47. First graph the lines $x - 4y = 0$ and $3x + 2y = 6$. Use a dashed line for the first and a solid line for the second. Select a point in each region as a test point. Use $(0, 5)$, $(0, 1)$, $(0, -5)$, and $(6, 0)$. Only $(0, 5)$ satisfies the compound sentence $x - 4y < 0$ and $3x + 2y \geq 6$. So shade the region that contains $(0, 5)$.

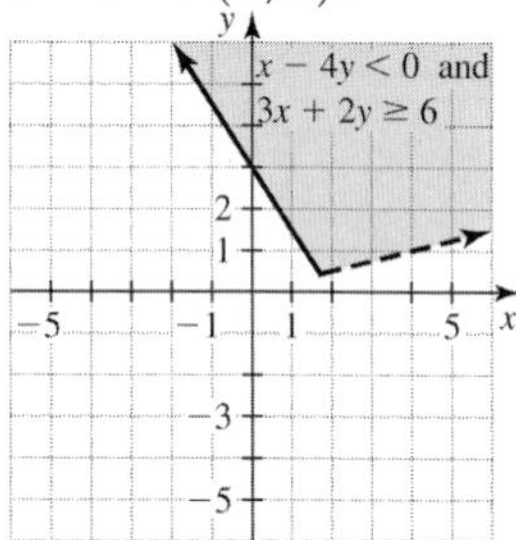

49. First graph the equations $x + y = 5$ and $x - y = 3$. Test one point in each of the 4 regions. Only points in the region containing $(0, 0)$ satisfy both inequalities. Shade that region including the boundary lines.

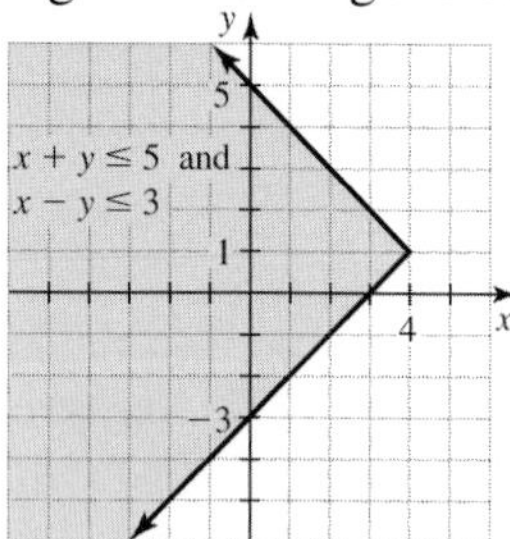

51. Graph the equations $x - 2y = 4$ and $2x - 3y = 6$. Only the region containing the point $(0, -5)$ fails to satisfy the compound inequality. Shade the other three regions including the boundary lines.

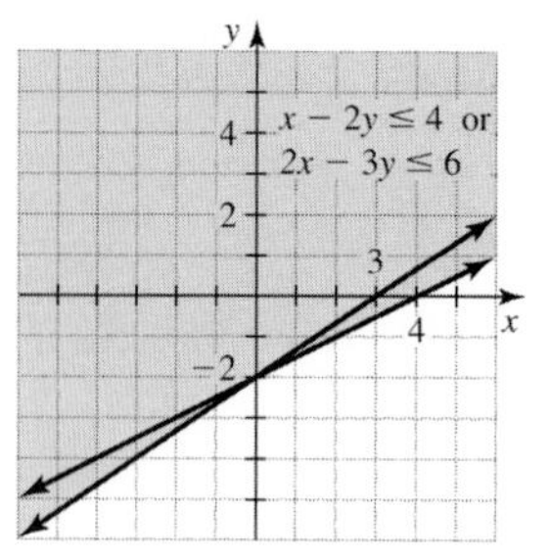

53. Graph the horizontal line $y = 2$ and the vertical line $x = 3$. Only points in the region containing $(0, 5)$ satisfy both inequalities. Shade that region with dashed boundary lines.

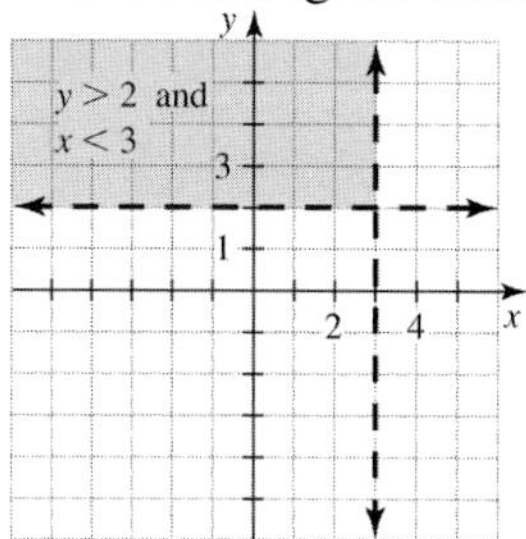

55. Graph $y = x$ and $x = 2$. Only points in the region containing $(0, 5)$ satisfy both inequalities. Shade that region and include the boundary lines.

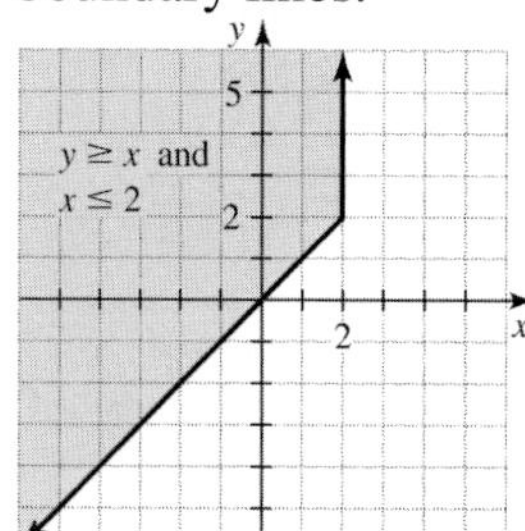

57. Graph $2x = y + 3$ and $y = 2 - x$. Only points in the region containing $(0, -5)$ fail to satisfy the compound inequality. Shade all regions except the one containing $(0, -5)$. Use dashed lines for the boundaries because of the inequality symbols.

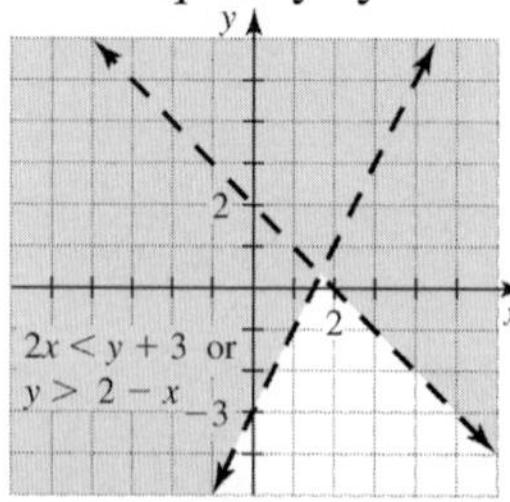

59. Graph the lines $y = x - 1$ and $y = x + 3$. Only points in the region containing $(0, 0)$ satisfy the compound inequality. Shade that region and use dashed boundary lines.

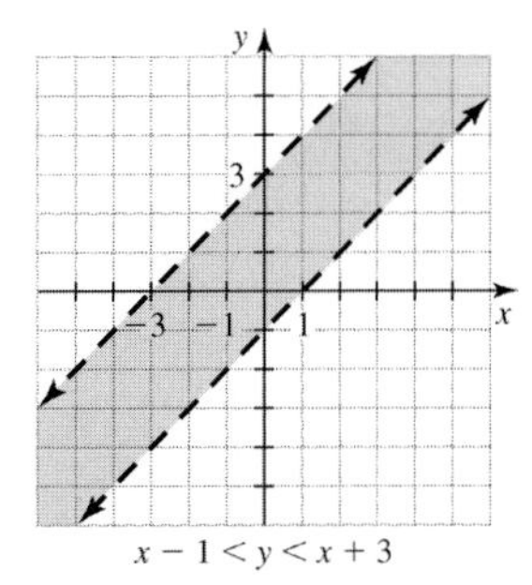

$x - 1 < y < x + 3$

61. Graph the lines $y = 0$, $y = x$, and $x = 1$. Only points inside the triangular region bounded by the lines satisfy the compound inequality. Shade that region and use solid boundary lines.

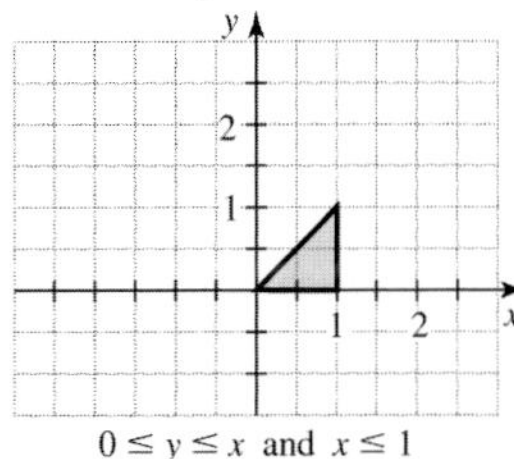

$0 \le y \le x$ and $x \le 1$

63. Graph $x = 1$, $x = 3$, $y = 2$, and $y = 5$.

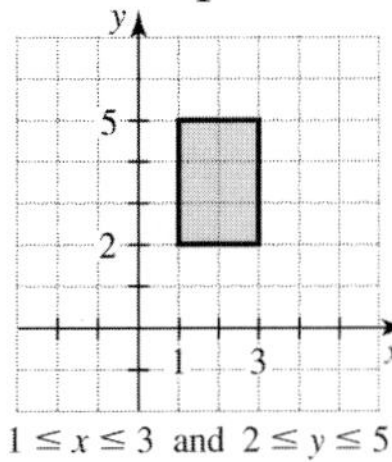

$1 \le x \le 3$ and $2 \le y \le 5$

Only points inside the rectangular region satisfy the compound inequality. Shade that region and include the boundary lines.

65. Graph the equations $x + y = 2$ and $x + y = -2$. Only points between these two parallel lines satisfy the absolute value inequality.

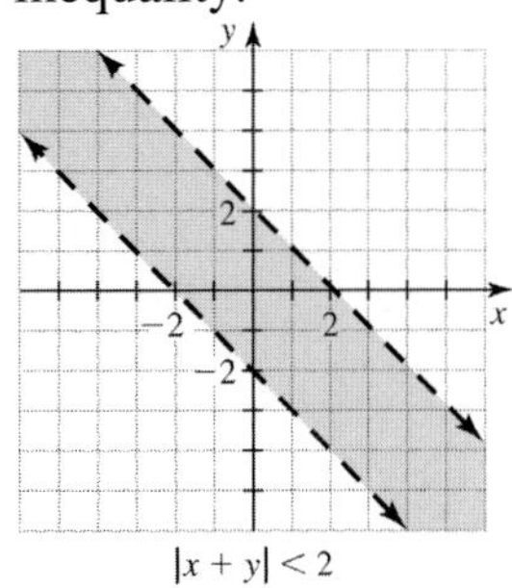

$|x + y| < 2$

67. Graph the parallel lines $2x + y = 1$ and $2x + y = -1$. Points between the lines do not satisfy the absolute value inequality. So shade the other two regions.

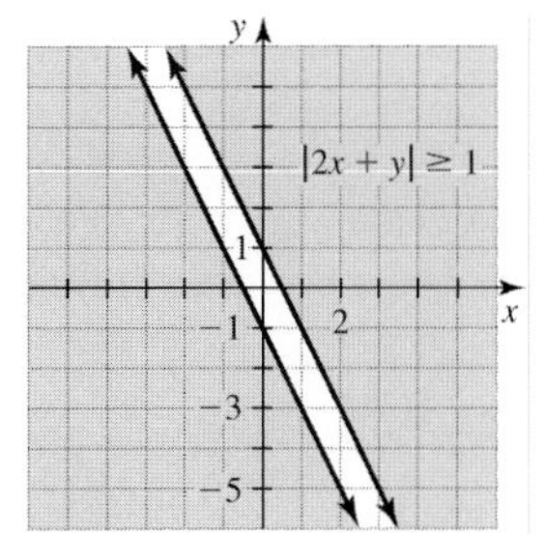

69. The inequality $|y - x| > 2$ is equivalent to

$$y - x > 2 \quad \text{or} \quad y - x < -2$$
$$y > x + 2 \quad \text{or} \quad y < x - 2$$

Graph the lines $y = x + 2$ and $y = x - 2$. Points above $y = x + 2$ together with points below $y = x - 2$ satisfy the absolute value inequality.

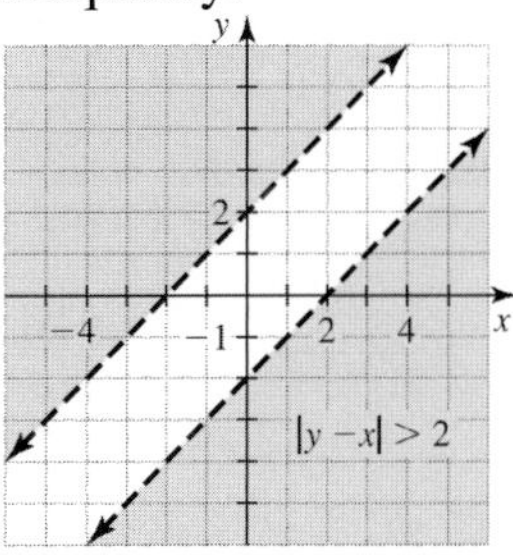

71. Graph the parallel lines $x - 2y = 4$ and $x - 2y = -4$. Points in the region between the lines satisfy the inequality.

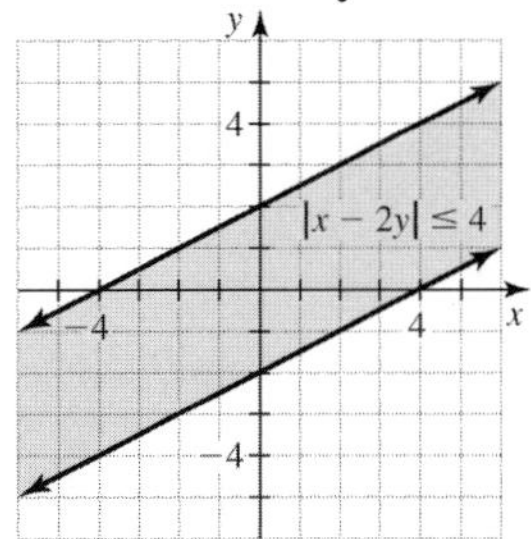

73. Graph the vertical lines $x = 2$ and $x = -2$. Points in the region between the lines do not satisfy the inequality but points in the other two regions do.

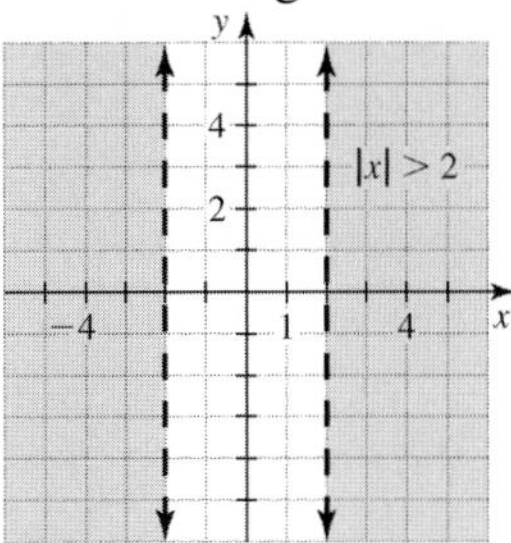

75. Graph the horizontal lines $y = 1$ and $y = -1$. Only points in the region between the lines satisfy $|\,y\,| < 1$.

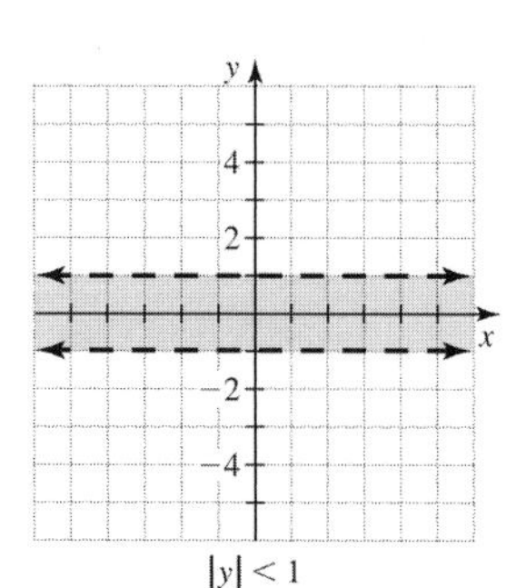

$|y| < 1$

77. Graph the lines $x = 2$, $x = -2$, $y = 3$, and $y = -3$. Only points inside the rectangular region bounded by the lines satisfy the compound inequality $|x| < 2$ and $|y| < 3$.

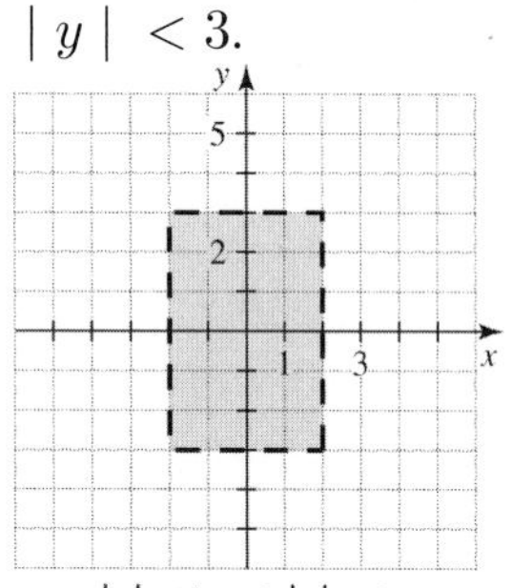

$|x| < 2$ and $|y| < 3$

79. The inequality $|x - 3| < 1$ is equivalent to $-1 < x - 3 < 1$, or $2 < x < 4$. The inequality $|y - 2| < 1$ is equivalent to $-1 < y - 2 < 1$, or $1 < y < 3$. Graph the lines $x = 2$, $x = 4$, $y = 1$, and $y = 3$. Only points inside the square region bounded by these lines satisfy the inequalities $|x - 3| < 1$ and $|y - 2| < 1$.

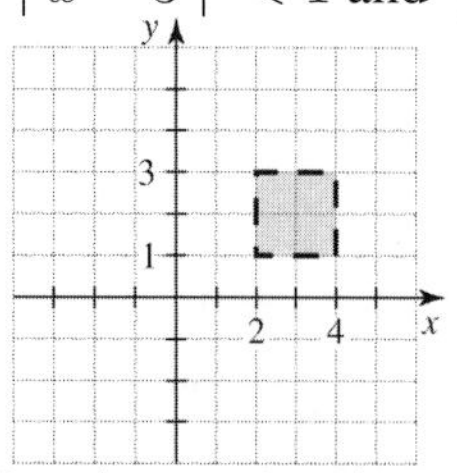

$|x - 3| < 1$ and $|y - 2| < 1$

81. Since $(0, 5)$ satisfies $y > x$ and $x < 1$ the solution set is not the empty set.

83. Points that satisfy $y < 2x - 5$ and $y > 2x + 5$ would be above the line $y = 2x + 5$ and below the line $y = 2x - 5$. Since these are parallel lines, there are no such points. The solution set is the empty set.

85. Since $(10, 0)$ satisfies $y < 2x - 5$ and the connecting word is "or", the solution set is not the empty set.

87. Points that satisfy $y < 2x$ lie below the line $y = 2x$ and points that satisfy $y > 3x$ lie above the line $y = 3x$. Since these lines are not parallel there are points that satisfy both inequalities. For example $(-5, -12)$ satisfies both. So the solution set is not the empty set.

89. This compound inequality indicates that $y < x$ and $y > x$. Points that satisfy both would be above the line $y = x$ and below the line $y = x$. There are no such points. The solution set is the empty set, $\emptyset$.

91. Since no real number has an absolute value that is less than 0, there are no points that satisfy $|y + 2x| < 0$. The solution set is the empty set, $\emptyset$.

93. Since no real number has an absolute value that is less than 0, there are no points that satisfy $|3x + 2y| \leq -4$. The solution set is the empty set, $\emptyset$.

95. Since the absolute value of any real number is nonnegative, all ordered pairs of real numbers satisfy $|x + y| > -4$. The solution set is not the empty set.

97. Let $x =$ the number of compact cars and $y =$ the number of full-size cars. We have

$$15{,}000x + 20{,}000y \leq 120{,}000$$

$$3x + 4y \leq 24$$

We also have $x \geq 0$ and $y \geq 0$ because they cannot purchase a negative number of cars. Graph $3x + 4y \leq 24$, $x \geq 0$, and $y \geq 0$.

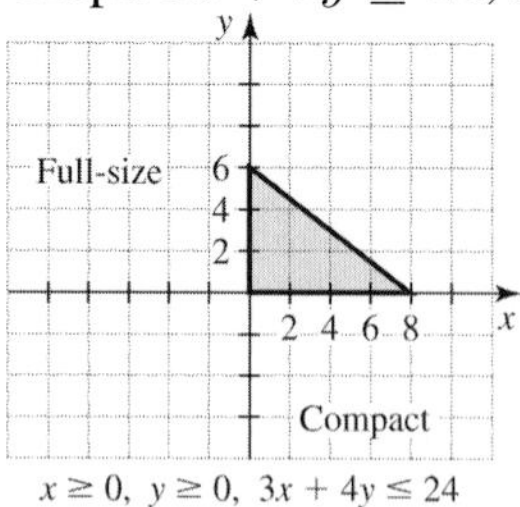

$x \geq 0,\ y \geq 0,\ 3x + 4y \leq 24$

99. Graph $3x + 4y \leq 24$, $x \geq 0$, $y \geq 0$, and $y > x$:

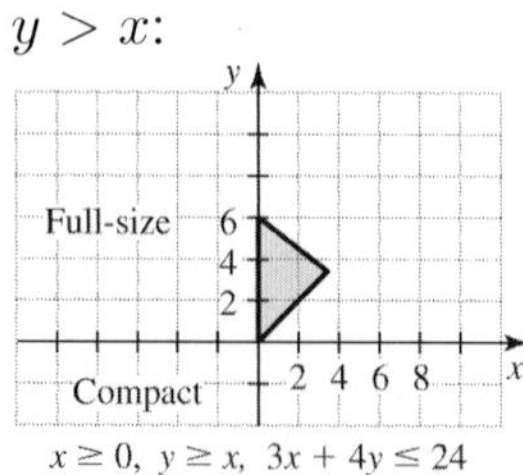

$x \geq 0,\ y \geq x,\ 3x + 4y \leq 24$

101. Graph $h \leq 187 - 0.85a$ and $h \geq 154 - 0.70a$ for $20 \leq a \leq 75$.

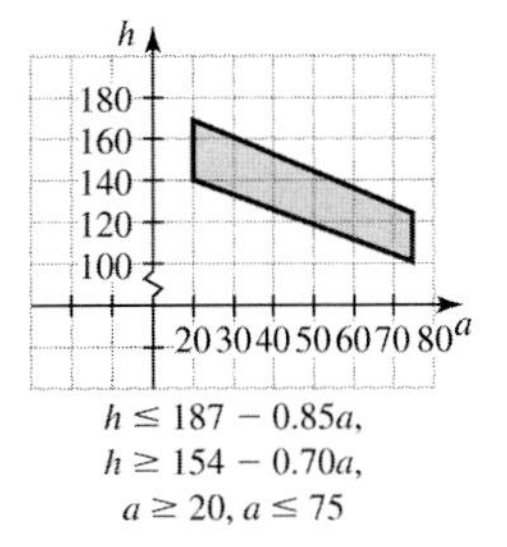

103. Let $d =$ the number of days of the *Daily Chronicle* advertising and $t =$ the number times an ad is aired on TV.

$$300d + 1000t \leq 9000$$
$$3d + 10t \leq 90$$

Graph $3d + 10t \leq 90$, $d \geq 0$, and $t \geq 0$:

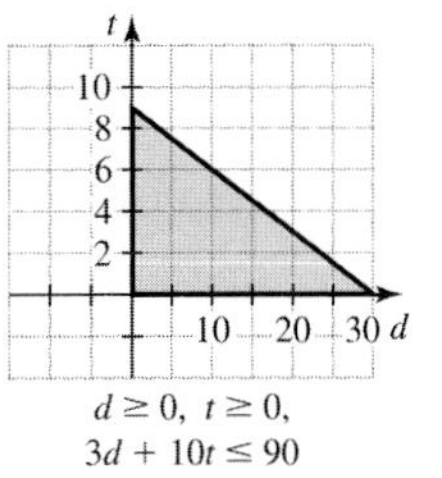

105. The solution set to a compound inequality using "and" is the intersection of the two solution sets. The solution set to a compound inequality using "or" is the union of the two solution sets.

3.5 WARM-UPS

1. False, $\{(1, 2), (1, 3)\}$ is not a function.
2. True, because $C = \pi D$.
3. True, because no two ordered pairs have the same first coordinate and different second coordinates.
4. False, because (1, 5) and (1, 7) have the same first coordinate and different second coordinates.
5. True, because every x-coordinate in $y = x^2$ corresponds to a unique y-coordinate.
6. False, because $\{(1,5),(3,6),(1,7)\}$ is a relation that is not a function.
7. True, because of the definition of domain.
8. False, because the domain of a relation or function is the set of first coordinates.
9. True, because $\sqrt{x}$ is a real number only if $x \geq 0$.
10. True, because if $h(x) = x^2 - 3$, then $h(-2) = (-2)^2 - 3 = 4 - 3 = 1$.

3.5 EXERCISES

1. To say that b is a function of a means that b is uniquely determined by a.
3. A relation is any set of ordered pairs.
5. The range of relation is the set of all possible second coordinates.
7. The number of gallons y that you get for \$10 is uniquely determined by the price per gallon x. So y is a function of x.
9. A student's test score y is not determined by the number of hours spent studying, because two students can study for the same amount of time and score differently on the test. So y is not a function of x.
11. The Fahrenheit temperature y is uniquely determined by the Celsius temperature x by a formula. So y is a function of x.
13. Two items with the same price x can have different universal product codes y. So the product code cannot be determined from the cost and y is not a function of x.
15. The total cost, C, is found by multiplying the number of toppings, t, by the cost for each topping, \$0.50, and adding on the \$5 base charge: $C = 0.50t + 5$
17. The total cost T is the price of the groceries S plus $0.09S$ or $T = S + 0.09S$. Combining like terms yields $T = 1.09S$.
19. The circumference of a circle C is found using the well-known formula $C = 2\pi r$, where r is the radius.
21. Since a square has 4 equal sides, the perimeter is 4 times the length of a side, $P = 4s$.
23. Since $A = \frac{1}{2}bh$ and $b = 10$, $A = 5h$ expresses the area as a function of the height.
25. This table has no ordered pairs with the same first coordinate and different second coordinates. So the table does give y as a function of x.
27. This table has no ordered pairs with the same first coordinate t and different second coordinates v. So the table does give v as a function of t.

29. Since the a-coordinate of 2 corresponds to both 2 and -2, P is not a function of a.

31. This table has no ordered pairs with the same first coordinate b and different second coordinates q. So the table does give q as a function of b

33. A set of ordered pairs is a function unless there are two pairs with the same first coordinate and different second coordinates. So this set is a function.

35. This set of ordered pairs is not a function because $(-2, 4)$ and $(-2, 6)$ have the same x-coordinate and different y-coordinates.

37. This set of ordered pairs is not a function because $(\pi, -1)$ and $(\pi, 1)$ have the same x-coordinate and different y-coordinates.

39. This set is a function because no two ordered pairs have the same first coordinate and different second coordinates.

41. Note that the value of y^2 is the same for a number and its opposite. So if $y = \pm 1$, $x = 2(\pm 1)^2 = 2$. So $(2, 1)$ and $(2, -1)$ both satisfy the equation.

43. Note that the value of $|2y|$ is the same for a number and its opposite. So if $y = \pm 4$, $x^2 = |2(\pm 4)| = 8$. So $(8, 4)$ and $(8, -4)$ both satisfy the equation.

45. Note that the value of x^2 or y^2 is the same for a number and its opposite. So if $y = \pm 1$, $x^2 + (\pm 1)^2 = 1$, or $x^2 = 0$. Now 0 satisfies $x^2 = 0$. So $(0, 1)$ and $(0, -1)$ both satisfy the equation.

47. Note that the value of y^4 is the same for a number and its opposite. So if $y = \pm 2$, $x = (\pm 2)^4 = 16$. So $(16, 2)$ and $(16, -2)$ both satisfy the equation.

49. Note that the value of $|y|$ is the same for a number and its opposite. So if $y = \pm 1$, $x - 2 = |\pm 1| = 1$, or $x = 3$. So $(3, 1)$ and $(3, -1)$ both satisfy the equation.

51. This relation is a function because for each value of x, $y = x^2$ determines only one value of y.

53. This relation is not a function because $(2, 1)$ and $(2, -1)$ both satisfy $x = |y| + 1$.

55. This relation is a function because for each value of x, $y = x$ determines only one value of y.

57. This relation is not a function because $(2, 1)$ and $(2, -1)$ both satisfy $x = y^4 + 1$.

59. This relation is a function because for each value of x, $y = \sqrt{x}$ determines only one value of y.

61. This relation is not a function because $(2, 1)$ and $(2, -1)$ both satisfy $|x| = |2y|$.

63. This relation is not a function because $(0, 3)$ and $(0, -3)$ both satisfy $x^2 + y^2 = 9$.

65. This relation is a function because for each value of x, $x = 2\sqrt{y}$ determines only one value of y.

67. This relation is a function because $(0, 5)$ and $(0, -5)$ both satisfy $x + 5 = |y|$.

69. Since a vertical line can be drawn to cross this graph more than once, this graph is not the graph of a function.

71. Since no vertical line can be drawn to cross this graph more than once, this graph is the graph of a function.

73. Since a vertical line can be drawn to cross this graph more than once, this graph is not the graph of a function.

75. The domain is the set of first coordinates, $\{4, 7\}$, and the range is the set of second coordinates, $\{1\}$.

77. The domain is the set of first coordinates, $\{2\}$, and the range is the set of second coordinates, $\{3, 5, 7\}$.

79. Since any number can be use in place of x in $y = x + 1$, the domain is the set of all real numbers, $(-\infty, \infty)$. Since any number can be used in place of y the range is also $(-\infty, \infty)$.

81. Since any number can be use in place of x in $y = 5 - x$, the domain is the set of all real numbers, $(-\infty, \infty)$. Since any number can be used in place of y the range is also $(-\infty, \infty)$.

83. Since $y = \sqrt{x - 2}$, we must have $x - 2 \geq 0$ or $x \geq 2$. So the domain is $[2, \infty)$. The values of y must be nonnegative. The range is $[0, \infty)$.

85. Since $y = \sqrt{2x}$, we must have $2x \geq 0$ or $x \geq 0$. So the domain is $[0, \infty)$. The values of y must be nonnegative. The range is $[0, \infty)$.

87. $f(0) = 3(0) - 2 = -2$

89. $f(4) = 3(4) - 2 = 12 - 2 = 10$

91. $g(-2) = -(-2)^2 + 3(-2) - 2$
$= -4 - 6 - 2 = -12$

93. $h(-3) = |-3 + 2| = |-1| = 1$

95. $h(-4.236) = |-4.236 + 2|$
$= |-2.236| = 2.236$

97. $f(2) = 3(2) - 2 = 4$
$g(3) = -3^2 + 3(3) - 2 = -2$
$f(2) + g(3) = 4 + (-2) = 2$

99. $g(2) = -2^2 + 3(2) - 2 = 0$
$h(-3) = |-3 + 2| = 1$
$\frac{g(2)}{h(-3)} = \frac{0}{1} = 0$

101. $f(-1) = 3(-1) - 2 = -5$
$h(-4) = |-4 + 2| = 2$
$f(-1) \cdot h(-4) = -5 \cdot 2 = -10$

103. a) $h(2) = 256 - 16(2)^2 = 192$ ft
b) $h(4) = 256 - 16(4)^2 = 0$ ft

105. The area of a square is found by squaring the length of a side: $A = s^2$ or $A(s) = s^2$.

107. The cost of the purchase, C, is the price per yard times the number of yards purchased: $C(x) = 3.98x$, $C(3) = 3.98(3) = \$11.94$.

109. The total cost, C, is found by multiplying the number of toppings, n, by the cost for each topping, \$0.50, and adding on the \$14.95 base charge: $C(n) = 0.50n + 14.95$
$C(6) = 0.50(6) + 14.95 = \$17.95$

111. The relation $y = x + 2$ is a function because there is only one y for any value of x. The relation $y > x + 2$ is not a function because $(2, 5)$ and $(2, 6)$ both satisfy $y > x + 2$.

Chapter 3 Wrap-Up

Enriching Your Mathematical Word Power

1. d **2.** a **3.** a **4.** b **5.** c
6. a **7.** b **8.** b **9.** a **10.** c
11. b **12.** c **13.** b **14.** a **15.** d
16. b

CHAPTER 3 REVIEW

1. Since both coordinates are negative, the point $(-3, -2)$ lies in quadrant III.

3. The point $(\pi, 0)$ lies on the x-axis. The quadrants do not include points on any axis.

5. The point $(0, -1)$ is on the y-axis.

7. Since the first coordinate is positive and the second negative, the point $\left(\sqrt{2}, -3\right)$ is in quadrant IV.

9. If $x = 0$, $y = -3(0) + 2 = 2$.
The ordered pair is $(0, 2)$.
If $y = 0$, then $0 = -3x + 2$.
$3x = 2$
$x = \frac{2}{3}$
The ordered pair is $\left(\frac{2}{3}, 0\right)$.
If $x = 4$, $y = -3(4) + 2 = -10$.
The ordered pair is $(4, -10)$.
If $y = -3$, then $-3 = -3x + 2$.
$3x = 5$
$x = \frac{5}{3}$
The ordered pair is $\left(\frac{5}{3}, -3\right)$.

11. $m = \frac{9 - 6}{-2 - (-5)} = \frac{3}{3} = 1$

13. $m = \frac{-2 - 1}{-3 - 4} = \frac{-3}{-7} = \frac{3}{7}$

15. $m = \frac{-1 - (-4)}{5 - (-3)} = \frac{3}{8}$
The slope of any line parallel to one with slope 3/8 is also 3/8.

17. $m = \frac{-6 - 5}{4 - (-3)} = \frac{-11}{7} = -\frac{11}{7}$
The slope of any line perpendicular to one with slope $-11/7$ is $7/11$.

19. The slope of the line through $(2, 2)$ and $(3, 3)$ is

$$m = \frac{3 - 2}{3 - 2} = 1.$$

Any line perpendicular to a line with slope 1 has slope -1. So l goes through the origin and has slope 1. Both lines are graphed as follows.

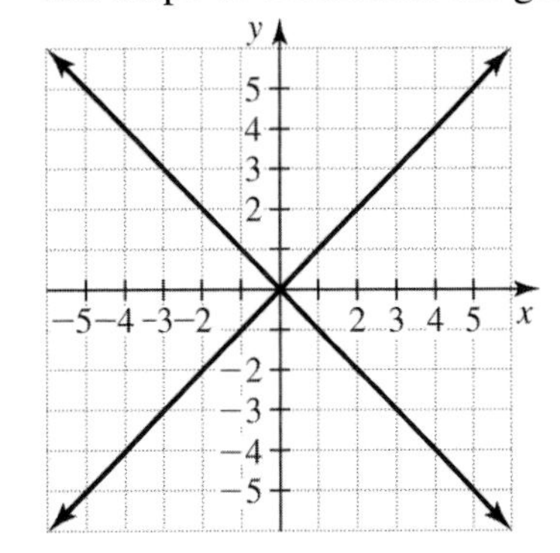

21. Any line parallel to a line with slope 2 also has slope 2. So l has slope 2. Both lines are graphed as follows.

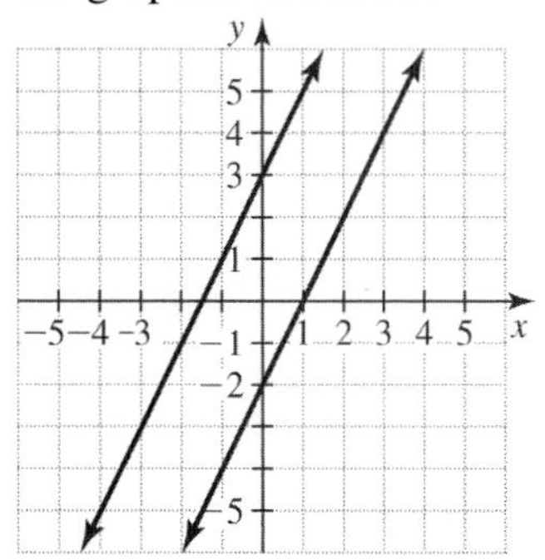

23. Any line perpendicular to a line with slope 2/3 has slope $-3/2$. So l has slope $-3/2$. Both lines are graphed through the origin as follows.

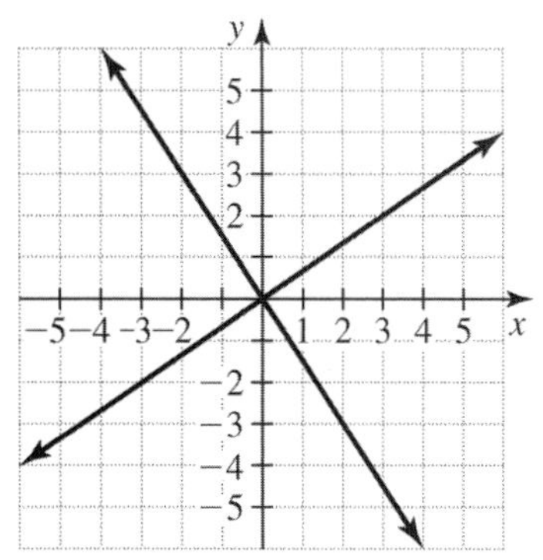

25. The slope for $y = -3x + 4$ is -3, and the y-intercept is $(0, 4)$.

27. Write the equation in slope-intercept form.

$$y - 3 = \frac{2}{3}(x - 1)$$
$$y - 3 = \frac{2}{3}x - \frac{2}{3}$$
$$y = \frac{2}{3}x + \frac{7}{3}$$

The slope is $\frac{2}{3}$ and the y-intercept is $\left(0, \frac{7}{3}\right)$.

29.
$$y = \frac{2}{3}x - 4$$
$$3(y) = 3(\tfrac{2}{3}x - 4)$$
$$3y = 2x - 12$$
$$-2x + 3y = -12$$
$$2x - 3y = 12 \quad \text{(Standard form)}$$

31.
$$y - 1 = \frac{1}{2}(x + 3)$$
$$2(y - 1) = 2 \cdot \frac{1}{2}(x + 3)$$
$$2y - 2 = x + 3$$
$$-x + 2y = 5$$
$$x - 2y = -5 \quad \text{(Standard form)}$$

33. Start with point-slope form:

$$y - (-3) = \frac{1}{2}(x - 1)$$
$$y + 3 = \frac{1}{2}x - \frac{1}{2}$$
$$2y + 6 = x - 1$$
$$-x + 2y = -7$$
$$x - 2y = 7$$

35. Start with point-slope form:

$$y - 6 = -\frac{3}{4}(x - (-2))$$
$$y - 6 = -\frac{3}{4}x - \frac{3}{2}$$
$$4(y - 6) = 4(-\tfrac{3}{4}x - \tfrac{3}{2})$$
$$4y - 24 = -3x - 6$$
$$3x + 4y = 18$$

37. Any line with slope 0 is a horizontal line. Since it contains $(3, 5)$ its equation is $y = 5$.

39. To graph $y = 2x - 3$ note that the y-intercept is $(0, -3)$ the slope is $2 = 2/1$. Start at $(0, -3)$ and go up 2 units and 1 unit to the right to locate the second point $(1, -1)$. Draw a line through the two points.

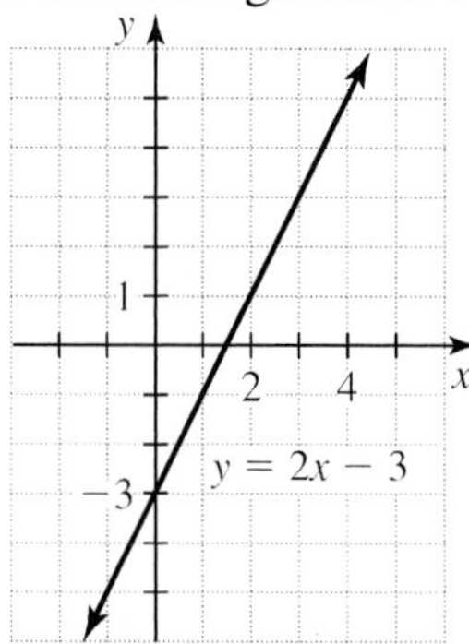

41. If $x = 0$, $3(0) - 2y = -6$, or $y = 3$
If $y = 0$, then $3x - 2(0) = -6$, $x = -2$.
Draw a line through the intercepts $(0, 3)$ and $(-2, 0)$.

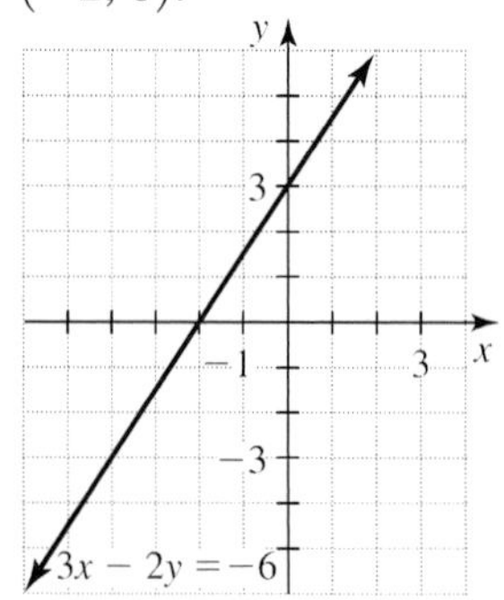

43. The equation $y - 3 = 10$ is equivalent to $y = 13$. Its graph is a horizontal line with a y-intercept of $(0, 13)$.

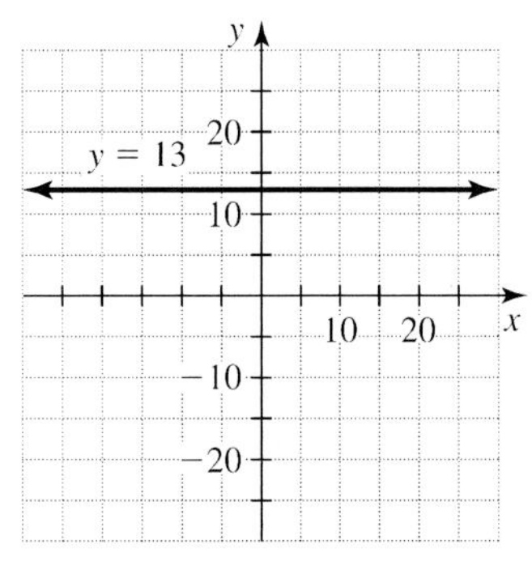

45. If $x = 0$, then $5(0) - 3y = 7$, or $y = -7/3$. If $y = 0$, then $5x - 3(0) = 7$, or $x = 7/5$. Draw a line through the intercepts $(0, -7/3)$ and $(7/5, 0)$.

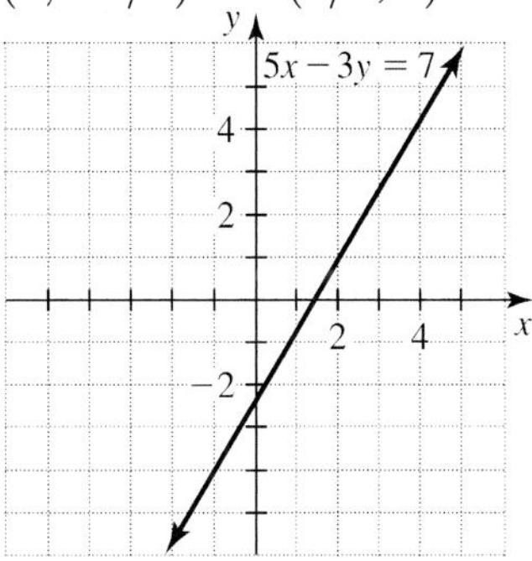

47. For $5x + 4y = 100$, the intercepts are $(0, 25)$ and $(20, 0)$.

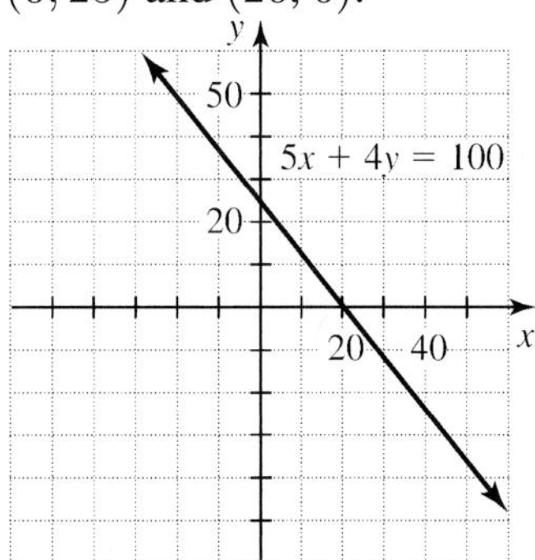

49. For $x - 80y = 400$ the intercepts are $(0, -5)$ and $(400, 0)$.

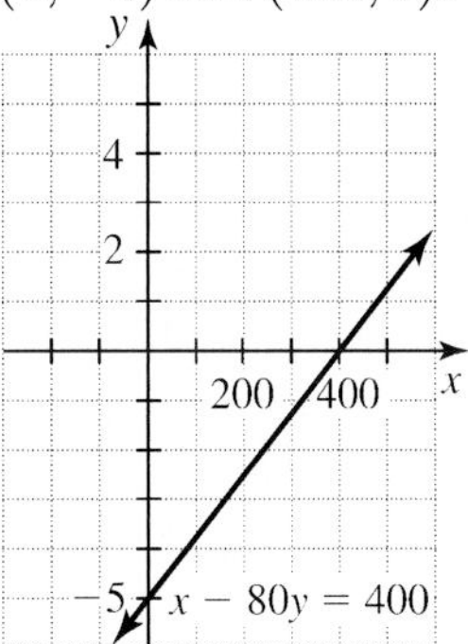

51. Use the slope $3 = 3/1$ and y-intercept $(0, -2)$ to graph the equation $y = 3x - 2$. Start at $(0, -2)$ and go up 3 units and 1 to the right. Draw a dashed line through the two points $(0, -2)$ and $(1, 1)$. Shade the region above the line to indicate the graph of $y > 3x - 2$.

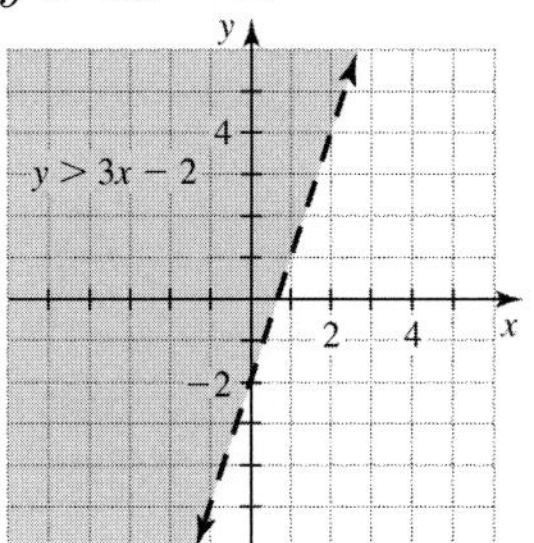

53. First graph $x - y = 5$ using the intercepts $(0, -5)$ and $(5, 0)$. The line should be solid because of the inequality symbol $\leq$. Test the point $(0, 0)$ in the inequality $x - y \leq 5$:

$$0 - 0 \leq 5$$

Since $(0, 0)$ satisfies the inequality, shade the region containing $(0, 0)$.

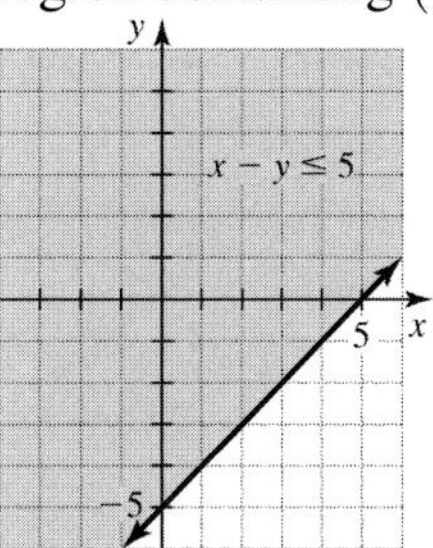

55. The inequality $3x > 2$ is equivalent to $x > \frac{2}{3}$. First graph the vertical line $x = \frac{2}{3}$ as a dashed line through $\left(\frac{2}{3}, 0\right)$. The graph of $x > \frac{2}{3}$ is the region to the right of this line.

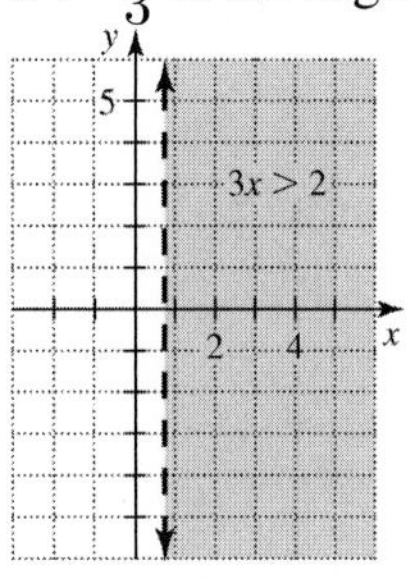

57. The inequality $4y \leq 0$ is equivalent to $y \leq 0$. Graph the line $y = 0$. The line $y = 0$ coincides with the x-axis. Draw the line solid. Points on the line and in the region below the line satisfy the inequality $y \leq 0$.

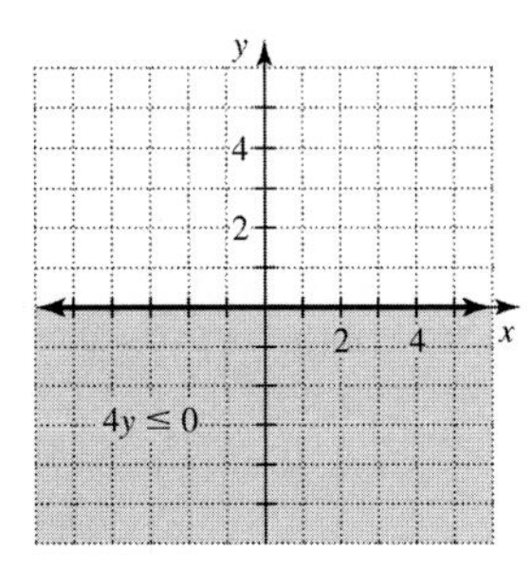

59. Solve $4x - 2y \geq 6$ for y:

$$-2y \geq -4x + 6$$
$$y \leq 2x - 3$$

All points on the line $y = 2x - 3$ together with the points below the line satisfy the inequality. Use y-intercept $(0, -3)$ and slope 2 to graph the solid line.

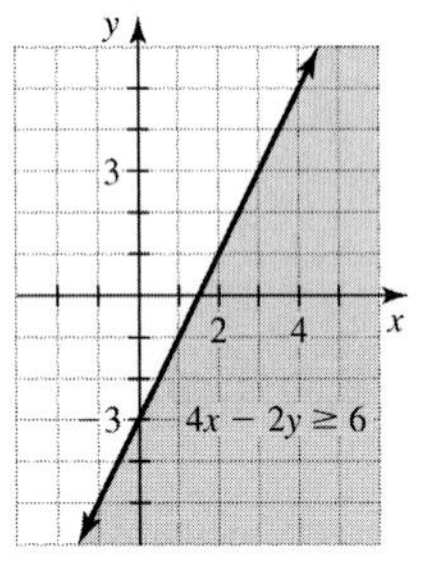

61. The x and y-intercepts for the line $5x - 2y = 9$ are $\left(0, -\frac{9}{2}\right)$ and $\left(\frac{9}{5}, 0\right)$. Draw a dashed line through the intercepts. Check $(0, 0)$ in the inequality $5x - 2y < 9$:

$5(0) - 2(0) < 9$ is correct.

Shade the region containing $(0, 0)$.

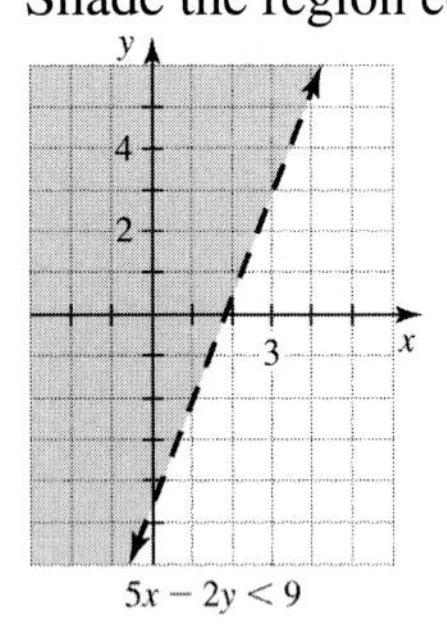

63. First graph the lines $y = 3$ and $y - x = 5$.

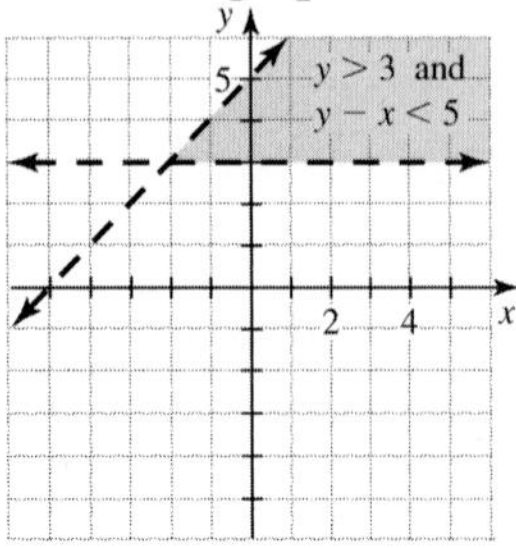

Test the points $(0, 0)$, $(0, 4)$, $(0, 6)$, and $(-6, 0)$. Only $(0, 4)$ satisfies $y > 3$ and $y - x < 5$. Shade the region containing $(0, 4)$.

65. First graph the lines $3x + 2y = 8$ and $3x - 2y = 6$. Test the points $(0, 0)$, $(0, 5)$, $(0, -6)$, and $(5, 0)$. The points $(0, 0)$, $(0, 5)$, and $(5, 0)$ satisfy the compound inequality $3x + 2y \geq 8$ or $3x - 2y \leq 6$. Shade the regions containing those points.

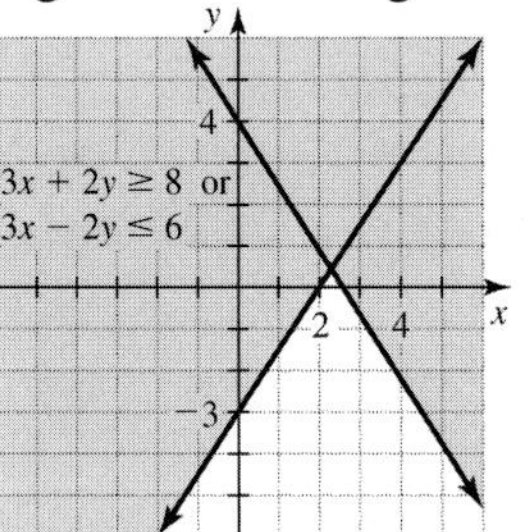

67. First graph the equations $x + 2y = 10$ and $x + 2y = -10$. Test the points $(0, 0)$, $(0, 8)$, and $(0, -8)$ in the inequality $|x + 2y| < 10$. Only $(0, 0)$ satisfies the inequality. So shade the region containing $(0, 0)$.

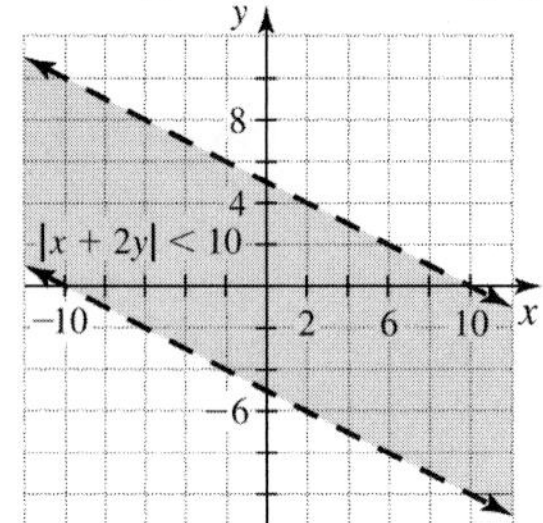

69. First graph the two vertical lines $x = 5$ and $x = -5$.

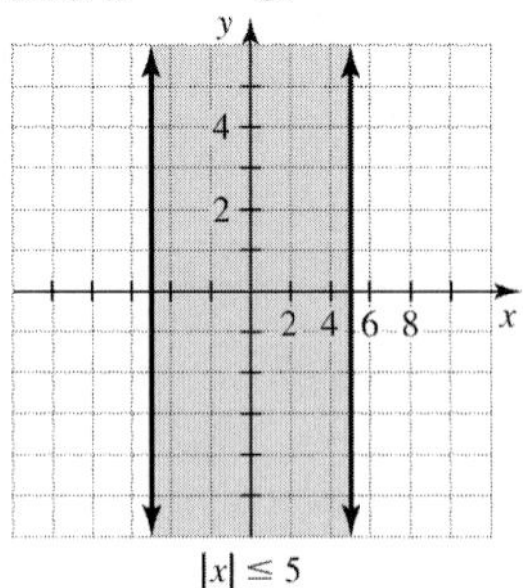

Test the points $(0, 0)$, $(8, 0)$, $(-8, 0)$ in the inequality $|x| \leq 5$. Only $(0, 0)$ satisfies the inequality. So shade the region containing $(0, 0)$, the region between the parallel lines.

71. First graph the lines $y - x = 2$ and $y - x = -2$. Test the points $(0, 0)$, $(0, 4)$, and $(0, -4)$ in the inequality $|y - x| > 2$. The points $(0, 4)$ and $(0, -4)$ satisfy the inequality. Shade the regions containing those points.

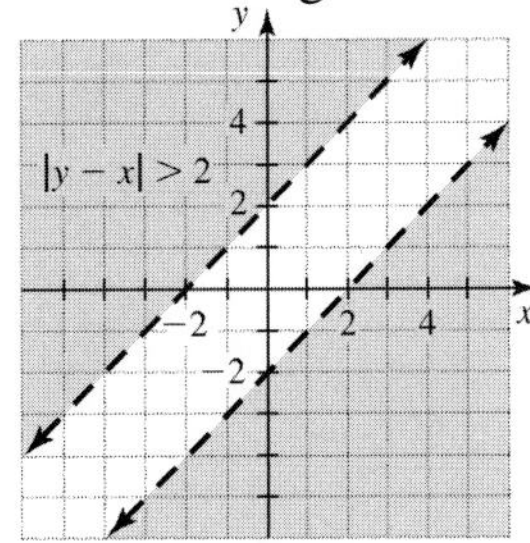

73. The relation is not a function because two ordered pairs $(5, 7)$ and $(5, 10)$ have the same first coordinate and different second coordinates.

75. The relation is a function because no two ordered pairs have the same x-coordinate and different y-coordinates.

77. The relation is a function because the condition that $y = x^2$ guarantees that no two ordered pairs will have the same x-coordinate and different y-coordinates.

79. This relation is not a function because the ordered pairs (16, 2) and (16, −2) are both in the relation.

81. The domain is the set of x-coordinates $\{3, 4, 5\}$, and the range is the set of y-coordinates $\{1, 5, 9\}$.

83. Since we can use any number for x in the equation $y = x + 1$, the domain is $(-\infty, \infty)$. Since y can be any real number in $y = x + 1$, the range is $(-\infty, \infty)$.

85. In the equation $y = \sqrt{x + 5}$, the value of $x + 5$ must be nonnegative. So $x + 5 \geq 0$ or $x \geq -5$. The domain is $[-5, \infty)$. Since y is equal to a square root, y must be nonnegative. The range is $[0, \infty)$.

87. $f(0) = 2(0) - 5 = -5$

89. $g(0) = 0^2 + 0 - 6 = -6$

91. $g\left(\frac{1}{2}\right) = \left(\frac{1}{2}\right)^2 + \frac{1}{2} - 6$
$= \frac{1}{4} + \frac{2}{4} - \frac{24}{4} = -\frac{21}{4}$

93. This line contains the points $(2, 0)$ and $(0, -6)$. The line has slope

$$m = \frac{-6 - 0}{0 - 2} = 3.$$

In slope-intercept form its equation is

$$y = 3x - 6,$$

and in standard form $3x - y = 6$.

95. Use the point-slope formula.

$$y - 4 = -\frac{1}{2}(x - (-1))$$
$$y - 4 = -\frac{1}{2}x - \frac{1}{2}$$
$$2y - 8 = -x - 1$$
$$x + 2y = 7$$

97. The line through $(2, -6)$ and $(2, 5)$ is a vertical line. All vertical lines are of the form $x = k$. So the equation is $x = 2$.

99. Since $x = 5$ is a vertical line, a line perpendicular to it is horizontal. The equation of a horizontal line through $(0, 0)$ is $y = 0$.

101. Any line parallel to $y = 2x + 1$ has slope 2. Use the point-slope formula with the point $(-1, 4)$ and slope 2:

$$y - 4 = 2(x - (-1))$$
$$y - 4 = 2x + 2$$
$$-2x + y = 6$$
$$2x - y = -6$$

103. The line with y-intercept $(0, 6)$ and slope 3 has equation $y = 3x + 6$. Written in standard form it is $3x - y = -6$.

105. A horizontal line has slope 0. If it goes through $(2, 5)$, its y-intercept is $(0, 5)$. Its equation is $y = 0 \cdot x + 5$ or simply $y = 5$.

107. Draw a quadrilateral with the 4 points as vertices. Find the slope of each side:

$$m_1 = \frac{-5 - (-1)}{-5 - (-3)} = \frac{-4}{-2} = 2$$

$$m_2 = \frac{-1 - 2}{-3 - 6} = \frac{-3}{-9} = \frac{1}{3}$$

$$m_3 = \frac{2 - (-2)}{6 - 4} = \frac{4}{2} = 2$$

$$m_4 = \frac{-2 - (-5)}{4 - (-5)} = \frac{3}{9} = \frac{1}{3}$$

Since the slopes of the opposite sides are equal, it is a parallelogram.

109. Plot the four points and find the slope of each side of the quadrilateral.

$$m_1 = \frac{2 - 6}{-2 - 2} = 1 \quad m_2 = \frac{6 - 4}{2 - 4} = -1$$

$$m_3 = \frac{4 - 0}{4 - 0} = 1 \quad m_4 = \frac{2 - 0}{-2 - 0} = -1$$

Since opposite sides have the same slope, the quadrilateral is a parallelogram. Since the slopes of the adjacent sides are 1 and −1, the adjacent sides are perpendicular. Therefore, the quadrilateral is a rectangle.

111. a) Find the slope of the line through $(20, 200)$ and $(70, 150)$.

$$m = \frac{200 - 150}{20 - 70} = \frac{50}{-50} = -1$$

Use the point-slope formula with $m = -1$ and the point $(20, 200)$:

$$h - h_1 = m(a - a_1)$$
$$h - 200 = -1(a - 20)$$
$$h - 200 = -a + 20$$
$$h = 220 - a$$

b) If $a = 40$, then $h = 220 - 40 = 180$ beats per minute.

c) The maximum heart rate decreases as you get older.

113. To determine the number of days it would take for the rental charge to equal \$1080, we solve the equation

$$1080 = 26 + 17d.$$
$$1054 = 17d$$
$$62 = d$$

It will take 62 days for the rental charge to equal the cost of the air hammer.
If $d = 1$, then $C = 26 + 17(1) = 43$.
If $d = 30$, then $C = 26 + 17(30) = 536$.
The graph of this function for d ranging from 1 to 30 is a straight line segment joining the two points $(1, 43)$ and $(30, 536)$.

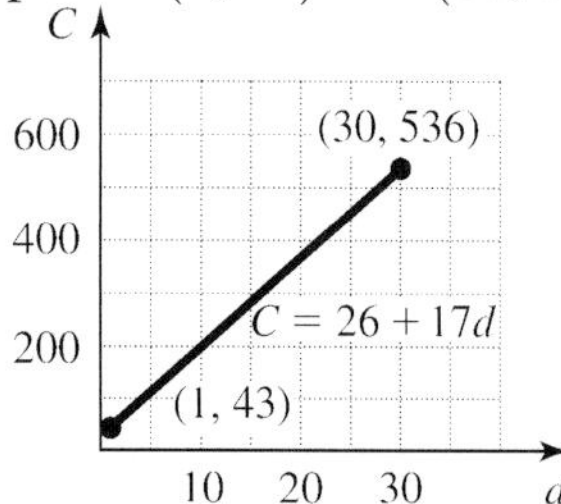

CHAPTER 3 TEST

1. If $x = 0$, then $2(0) + y = 5$.
$y = 5$

If $y = 0$, then $2x + 0 = 5$.
$x = 5/2$

If $x = 4$, then $2(4) + y = 5$.
$y = -3$

If $y = -8$, then $2x + (-8) = 5$.
$2x = 13$
$x = 13/2$

The pairs are $(0, 5)$, $\left(\frac{5}{2}, 0\right)$, and $\left(\frac{13}{2}, -8\right)$.

2. $m = \frac{7 - 1}{-3 - 2} = \frac{6}{-5} = -\frac{6}{5}$

3. Solve the equation $8x - 5y = -10$ for y:

$$-5y = -8x - 10$$
$$y = \frac{8}{5}x + 2$$

The slope is $\frac{8}{5}$ and the y-intercept is $(0, 2)$.

4. Plot the four points and find the slopes of the sides of the quadrilateral.

$m_1 = \frac{2 - 0}{6 - 0} = \frac{1}{3}$ $\quad m_2 = \frac{2 - 0}{6 - 5} = 2$

$m_3 = \frac{-2 - 0}{-1 - 5} = \frac{1}{3}$ $\quad m_4 = \frac{-2 - 0}{-1 - 0} = 2$

Since the opposite sides have equal slopes, the quadrilateral is a parallelogram.

5. The first coordinate is a and the second coordinate is V. We want the equation of the line containing the two points $(0, 22{,}000)$ and $(3, 16{,}000)$. The slope of the line is

$$m = \frac{22{,}000 - 16{,}000}{0 - 3} = -2000.$$

Using the slope and y-intercept, we can write the equation $V = -2000a + 22{,}000$.

6. The equation of the line with y-intercept $(0, 3)$ and slope $-1/2$ is

$$y = -\frac{1}{2}x + 3$$
$$2y = -x + 6$$
$$x + 2y = 6$$

7. Use the point-slope form to write the equation of the line through $(-3, 5)$ with slope -4:

$$y - 5 = -4(x - (-3))$$
$$y - 5 = -4(x + 3)$$
$$y - 5 = -4x - 12$$
$$4x + y = -7$$

8. Solve $3x - 5y = 7$ for y:

$$-5y = -3x + 7$$
$$y = \frac{3}{5}x - \frac{7}{5}$$

Since this line has slope $3/5$, any line perpendicular to it has slope $-5/3$. The line through $(2, 3)$ with slope $-5/3$ has equation

$$y - 3 = -\frac{5}{3}(x - 2)$$
$$y - 3 = -\frac{5}{3}x + \frac{10}{3}$$
$$3y - 9 = -5x + 10$$
$$5x + 3y = 19$$

9. The line shown goes through the points $(0, 2)$ and $(-4, 0)$. We can see from the graph that its slope is $2/4$ or $1/2$ and its y-intercept is $(0, 2)$. In slope-intercept form its equation is

$$\begin{aligned} y &= \frac{1}{2}x + 2. \\ 2y &= x + 4 \\ -x + 2y &= 4 \\ x - 2y &= -4 \end{aligned}$$

10. The graph of $y = 4$ is a horizontal line through $(0, 4)$.

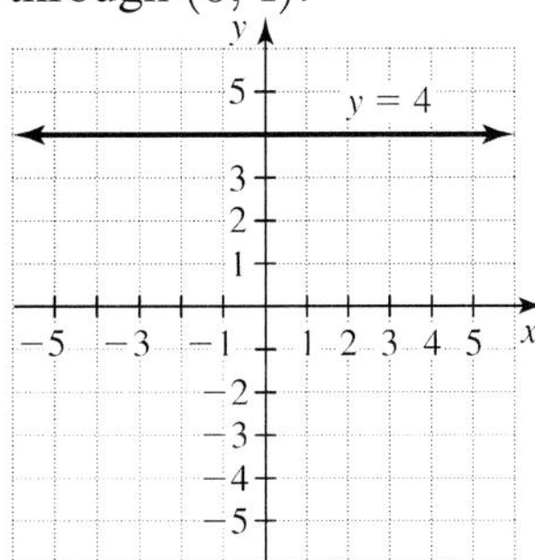

11. The graph of $x = 3$ is a vertical line through $(3, 0)$.

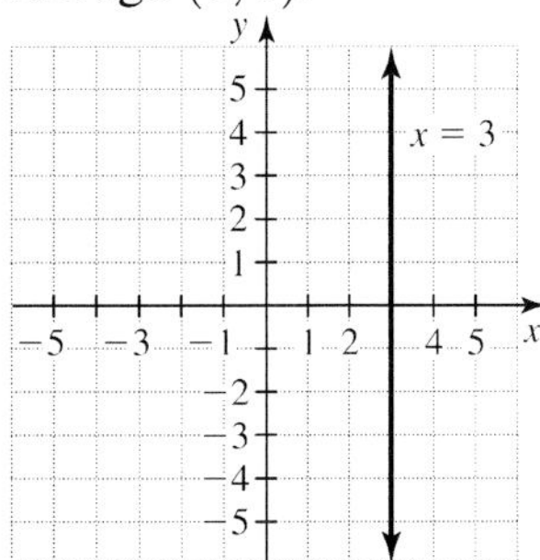

12. The graph of $3x + 4y = 12$ has x-intercept $(4, 0)$ and y-intercept $(0, 3)$.

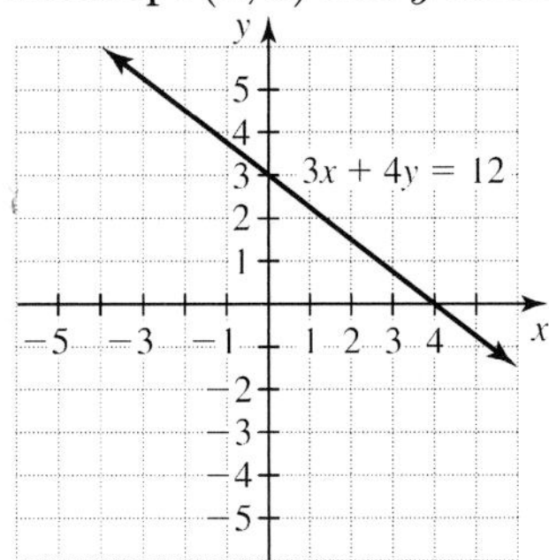

13. The graph of $y = \frac{2}{3}x - 2$ has y-intercept $(0, -2)$ and slope $2/3$.

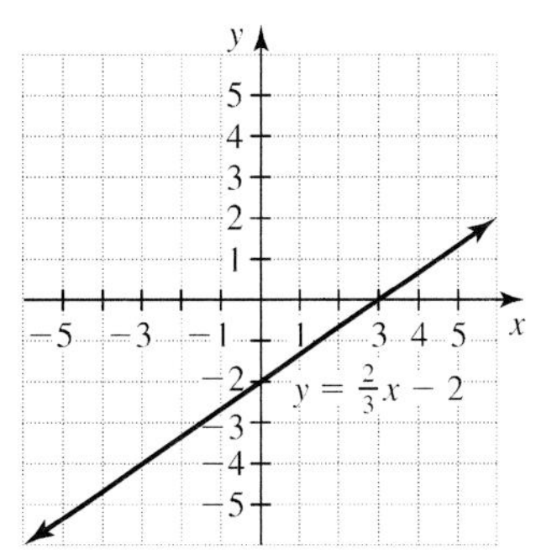

14. First graph the equation $y = (-1/2)x + 3$, using y-intercept of $(0, 3)$, a slope of $-1/2$, and a dashed line. The graph of $y > (-1/2)x + 3$ is the region above this line.

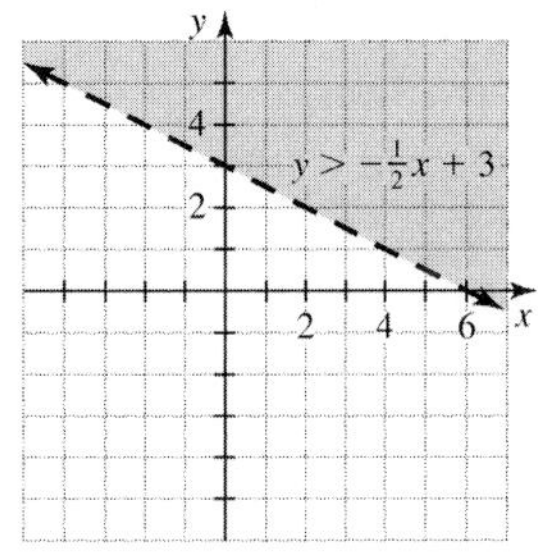

15. Graph the vertical line $x = 2$ using a dashed line. Graph $x + y = 0$ using a dashed line. Test the points $(-2, 0)$, $(1, 0)$, $(5, 0)$, and $(3, -5)$. Only $(5, 0)$ satisfies $x > 2$ and $x + y > 0$. So shade the region containing $(5, 0)$.

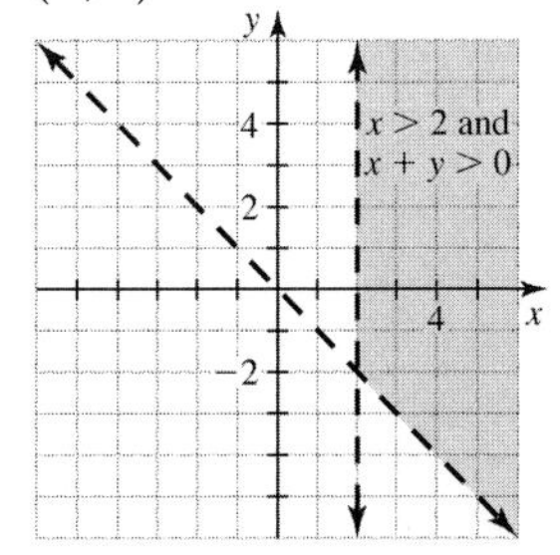

16. First graph the parallel lines $2x + y = 3$ and $2x + y = -3$, using solid lines. Test the points $(-5, 0)$, $(0, 0)$ and $(5, 0)$ in $|2x + y| \geq 3$. Both $(-5, 0)$ and $(5, 0)$ satisfy the inequality, so shade the regions containing those points.

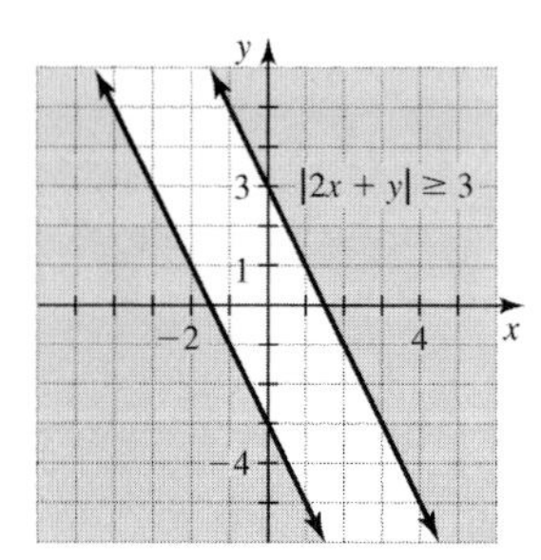

17. This set of ordered pairs is a function because no two of the ordered pairs have the same first coordinate and different second coordinates.

18. If $f(x) = -2x + 5$, then $f(-3) = -2(-3) + 5 = 11$.

19. Since $x - 7$ must be nonnegative, we have $x - 7 \geq 0$, or $x \geq 7$. So the domain is $[7, \infty)$. Since y is equal to a square root, y must be nonnegative also. So the range is $[0, \infty)$.

20. The shipping and handling fee S is a linear function of the weight of the order n, $S = 0.50n + 3$.

21. $A(2) = -16(2)^2 + 32(2) + 6 = 6$ ft

Making Connections

Chapters 1 - 3

1. $2^3 \cdot 4^2 = 8 \cdot 16 = 128$

2. $2^7 - 2^6 = 128 - 64 = 64$

3. $3^2 - 4(5)(-2) = 9 + 40 = 49$

4. $3 - 2|5 - 7 \cdot 3| = 3 - 2 \cdot 16 = -29$

5. $\dfrac{2-(-3)}{5-6} = \dfrac{5}{-1} = -5$

6. $\dfrac{-3-7}{-1-(-3)} = \dfrac{-10}{2} = -5$

7. $3t \cdot 4t = 12t^2$

8. $3t + 4t = 7t$

9. $\dfrac{4x+8}{4} = x + 2$

10. $\dfrac{-8y}{-4} - \dfrac{10y}{-2} = 2y - (-5y) = 7y$

11. $3(x-4) - 4(5-x)$
$= 3x - 12 - 20 + 4x = 7x - 32$

12. $-2(3x^2 - x) + 3(2x - 5x^2)$
$= -6x^2 + 2x + 6x - 15x^2$
$= -21x^2 + 8x$

13.
$$15(b - 27) = 0$$
$$15b - 405 = 0$$
$$15b = 405$$
$$b = 27$$
The solution set is $\{27\}$.

14.
$$0.05a - 0.04(a - 50) = 4$$
$$0.01a + 2 = 4$$
$$0.01a = 2$$
$$a = 200$$
The solution set is $\{200\}$.

15.
$$|\,3v - 7\,| = 0$$
$$3v - 7 = 0$$
$$3v = 7$$
$$v = \frac{7}{3}$$
The solution set is $\left\{\frac{7}{3}\right\}$.

16.
$$|\,3u - 7\,| = 3$$
$$3u - 7 = 3 \text{ or } 3u - 7 = -3$$
$$3u = 10 \text{ or } 3u = 4$$
$$u = \frac{10}{3} \text{ or } u = \frac{4}{3}$$
The solution set is $\{4/3,\ 10/3\}$.

17.
$$|\,3x - 7\,| = -77$$
The absolute value of any quantity is nonnegative. So the equation has no solution. The solution set is $\emptyset$.

18.
$$|\,3x - 7\,| + 1 = 8$$
$$|\,3x - 7\,| = 7$$
$$3x - 7 = 7 \text{ or } 3x - 7 = -7$$
$$3x = 14 \text{ or } 3x = 0$$
$$x = \frac{14}{3} \text{ or } x = 0$$
The solution set is $\{0,\ 14/3\}$.

19.
$$2x - 1 > 7$$
$$2x > 8$$
$$x > 4$$

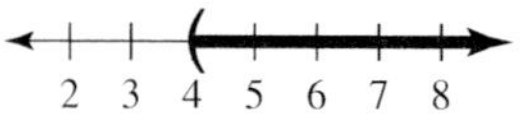

20.
$$5 - 3x \leq -1$$
$$-3x \leq -6$$
$$x \geq 2$$

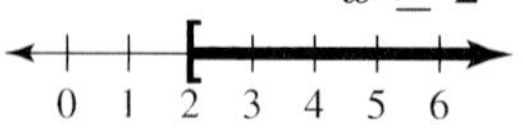

21.
$$x - 5 \leq 4 \text{ and } 3x - 1 < 8$$
$$x \leq 9 \text{ and } 3x < 9$$
$$x \leq 9 \text{ and } x < 3$$
The last compound inequality is equivalent to the simple inequality $x < 3$.

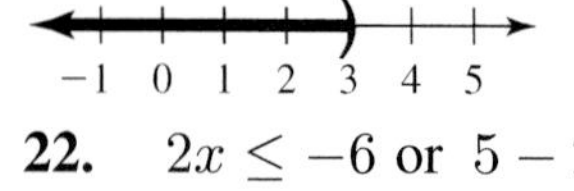

22.
$$2x \leq -6 \text{ or } 5 - 2x < -7$$
$$x \leq -3 \text{ or } -2x < -12$$
$$x \leq -3 \text{ or } x > 6$$

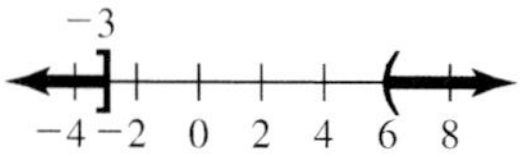

23. $|\,x-3\,| < 2$

$-2 < x-3 < 2$

$1 < x < 5$

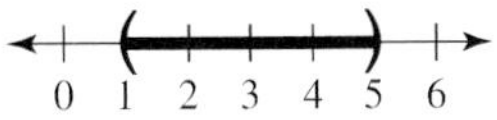

24. $|\,1-2x\,| \geq 7$

$1-2x \geq 7$ or $1-2x \leq -7$

$-2x \geq 6$ or $-2x \leq -8$

$x \leq -3$ or $x \geq 4$

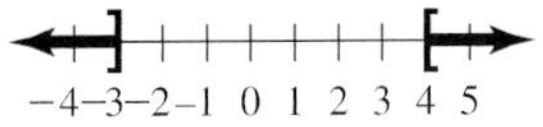

25. First graph the dashed line $y = 2x - 1$ through $(0, -1)$ with slope 2. Now shade below the line for $y < 2x - 1$.

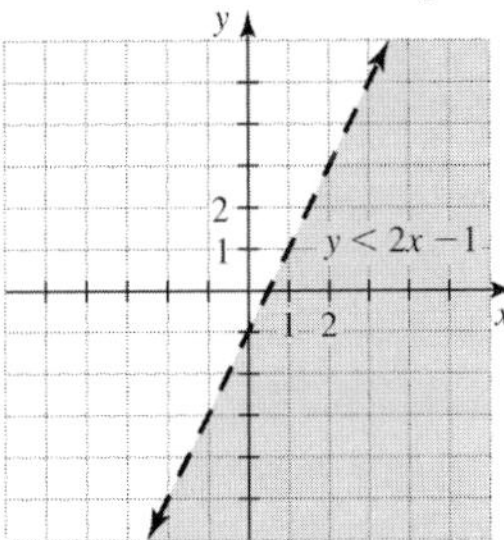

26. First graph the solid line $3x - y = 2$ or $y = 3x - 2$ through $(0, -2)$ with slope 3. Now test $(0, 0)$ in $3x - y \leq 2$. Since $(0, 0)$ satisfies $3x - y \leq 2$, we shade the side of the line containing $(0, 0)$.

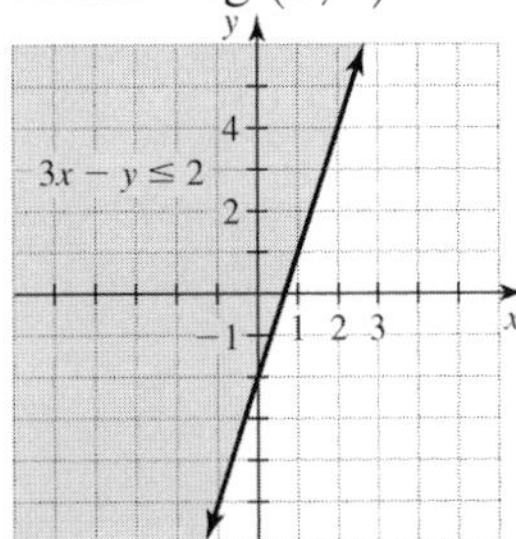

27. Graph the dashed line $y = x$ through $(0, 0)$ and $(1, 1)$. Next graph the dashed lin $y = -3x + 5$ through $(0, 5)$ with slope -3. Now test a point in each of the four regions. Only points that lie above $y = x$ and below $y = -3x + 5$ satisfy both inequalities. So shade that region.

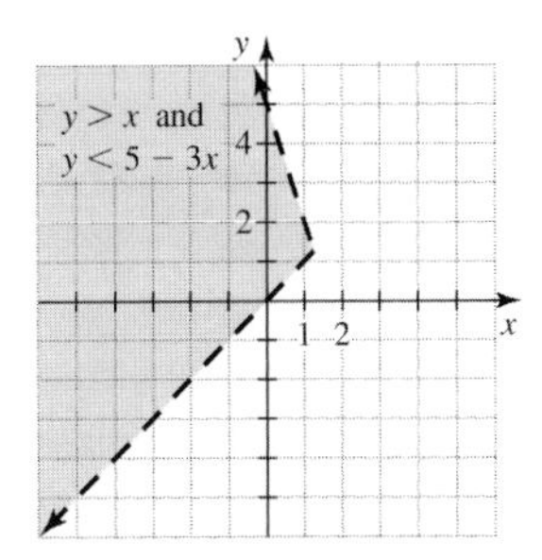

28. The graph of $y < 2$ consists of the points below the horizontal line $y = 2$. The graph of $x > -3$ consists of the points to the right of the vertical line $x = -3$. The region that lies below $y = 2$ and to the right of $x = -3$ is shaded on the graph. The boundary lines are drawn solid because of the $\leq$ and $\geq$ symbols.

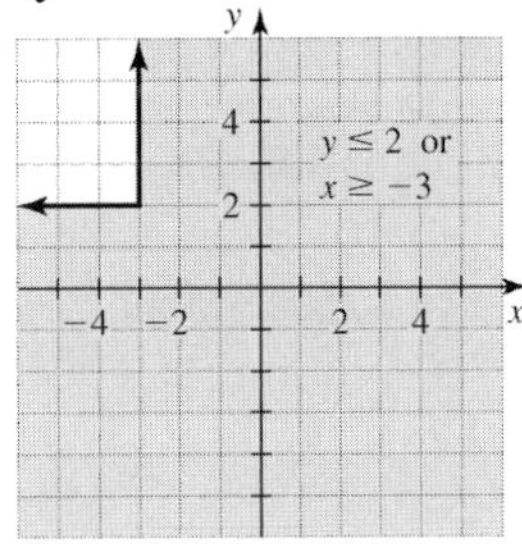

29. a) For ages 62 through 64
$b = 500a - 24{,}000$.
For ages 65 through 67
$b = 667a - 34{,}689$.
For ages 68 through 70
$b = 800a - 43{,}600$.

b) At age 64 $b = 500(64) - 24{,}000 = \$8000$

c) If a person get \$11,600 then

$11{,}600 = 800a - 43{,}600$

$55{,}200 = 800a$

$69 = a$

d) The slopes 500, 667, and 800 indicate the additional amount per year received beyond the basic amount in each category.

4.1 WARM-UPS

1. True, because if $x = 1$ and $y = 2$, then $2(1) + 2 = 4$ is correct.
2. False, because if $x = 1$ and $y = 2$, then $3(1) - 2 = 6$ is incorrect. Both equations must be satisfied for the compound statement using "and" to be true.
3. False, because (2, 3) does not satisfy $4x - y = -5$.
4. True.
5. True, because when we substitute, one of the variables is eliminated.
6. True, because each of these lines has slope 3 and they are parallel.
7. True, because the lines are not parallel and they have different y-intercepts.
8. True, because the lines are parallel and have different y-intercepts.
9. True, because dependent equations have the same solution sets.
10. True, because a system of independent equations has one point in its solution set.

4.1 EXERCISES

1. The intersection point of the graphs is the solution to an independent system.
3. The graphing method can be very inaccurate.
5. If the equation you get after substituting turns out to be incorrect, such as $0 = 9$, then the system has no solution.
7. The graph of $y = 2x$ is a straight line with y-intercept $(0, 0)$ and slope 2. The graph of $y = -x + 3$ is a straight line with y-intercept $(0, 3)$ and slope -1.

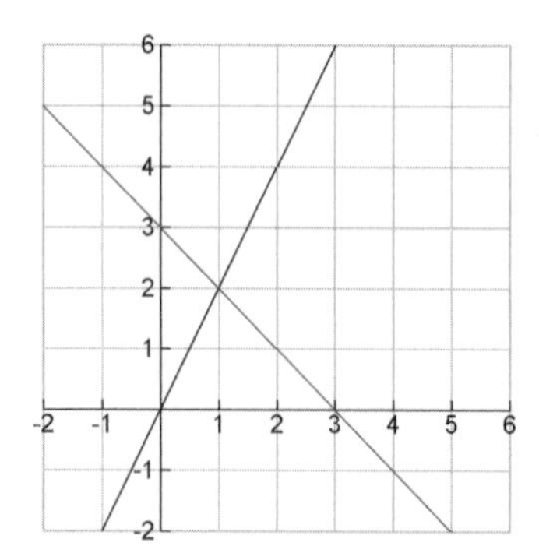

The graphs appear to intersect at (1, 2). Check that (1, 2) satisfies both equations. The solution set is $\{(1, 2)\}$.
9. The graph of $y = 2x - 1$ is a straight line with y-intercept $(0, -1)$ and slope 2. The graph of $2y = x - 2$ or $y = \frac{1}{2}x - 1$ is a straight line with y-intercept $(0, -1)$ and slope $\frac{1}{2}$.

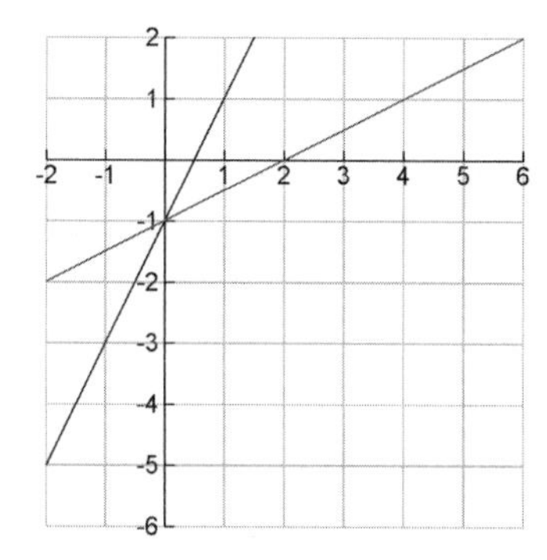

The graphs intersect at $(0, -1)$. The solution set to the system is $\{(0, -1)\}$.
11. The graph of $y = x - 3$ is a line with y-intercept $(0, -3)$ and slope 1. The graph of $x - 2y = 4$ or $y = \frac{1}{2}x - 2$ is a line with y-intercept $(0, -2)$ and slope $\frac{1}{2}$.

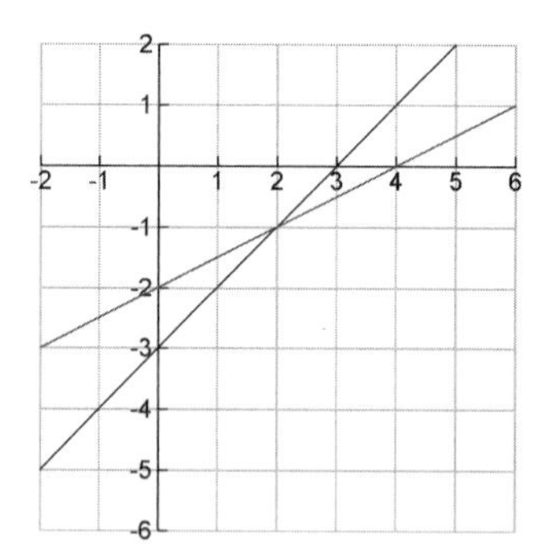

The lines appear to intersect at $(2, -1)$. To be sure, check that $(2, -1)$ satisfies both of the original equations. The solution set is $\{(2, -1)\}$.
13. The graph of $y = 2x + 4$ is a line with slope 2 and y-intercept $(0, 4)$. The graph of $3x + y = -1$ or $y = -3x - 1$ is a line with slope -3 and y-intercept $(0, -1)$.

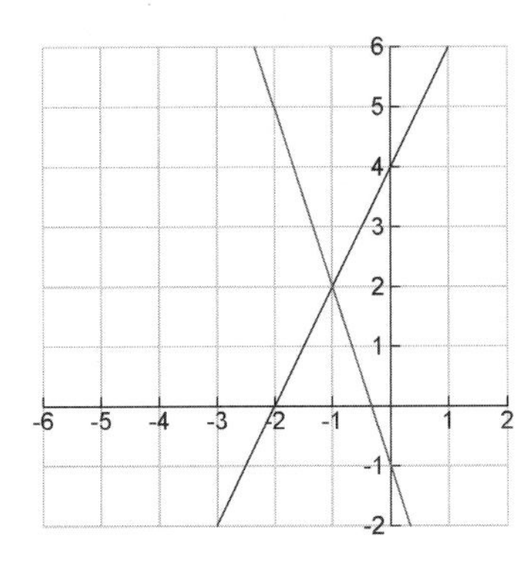

The lines appear to intersect at $(-1, 2)$. Check that $(-1, 2)$ satisfies both equations. The solution set is $\{(-1, 2)\}$

15. Solving $x + 2y = 8$ for y yields $y = -\frac{1}{2}x + 4$, which is the same as the first equation. So the equations have the same graph. All points on the line satisfy both equations. The solution set is $\{(x, y) \mid x + 2y = 8\}$.

17. If we rewrite these equations in slope-intercept form, we get $y = x + 1$ and $y = x + 3$. The graphs are parallel lines with slopes of 1 and y-intercepts of (0, 1) and (0, 3).

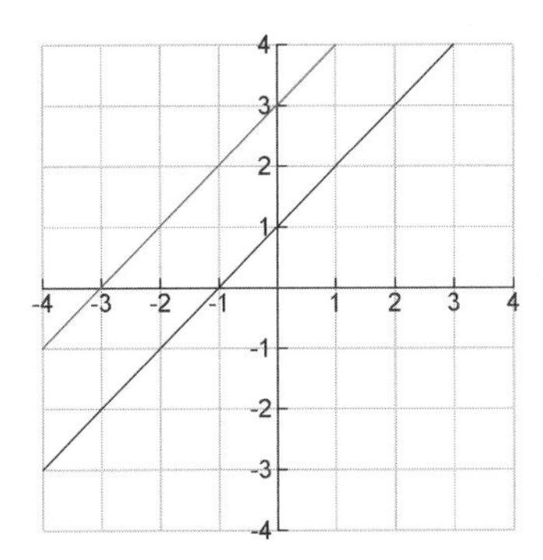

Since the lines do not intersect, the solution set is the empty set, $\emptyset$.

19. The graph of $y = -\frac{1}{4}x$ is a line with slope $-1/4$ and y-intercept (0, 0). The graph of $x + 4y = 8$ or $y = -\frac{1}{4}x + 2$ is a line with slope $-1/4$ and y-intercept (0, 2).

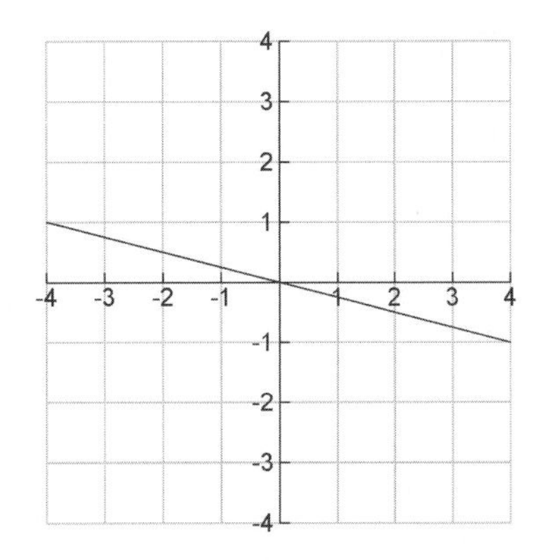

The lines have the same slope and do not intersect. The solution set is the empty set, $\emptyset$.

21. The lines in graph (c) intersect at (3, -2) and (3, -2) satisfies both equations of the given system.

23. The lines in graph (b) intersect at $(-3, -2)$ and $(-3, -2)$ satisfies both equations.

25. Enter the equations in your calculator and use the intersect feature to find the point of intersection.

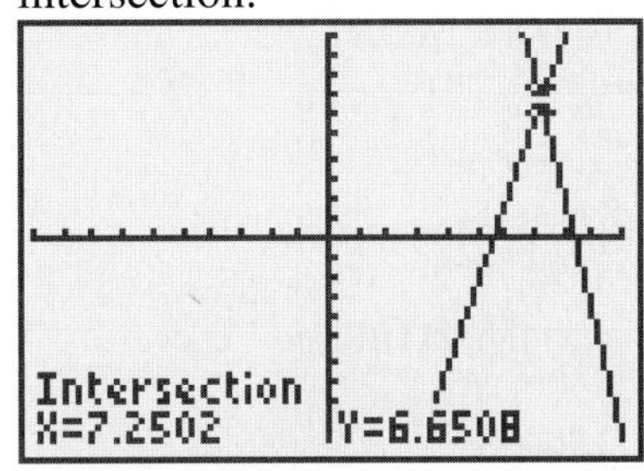

The solution set is $\{(7.25, 6.65)\}$.

27.

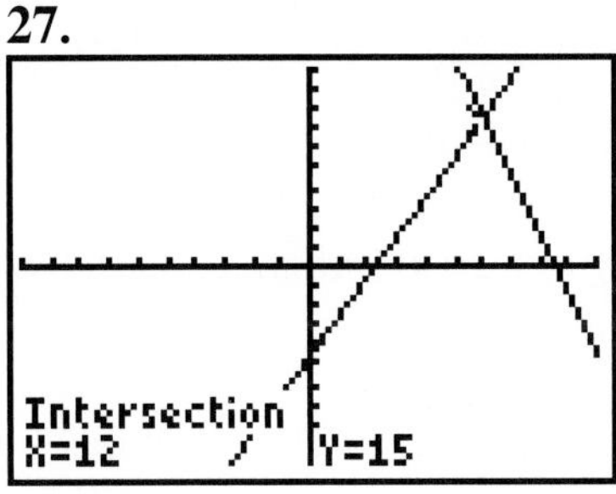

The solution set is $\{(12, 15)\}$.

29.

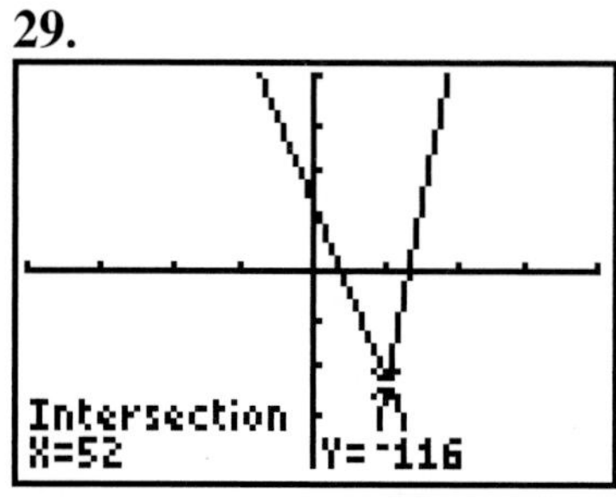

The solution set is $\{(52, -116)\}$.

31.

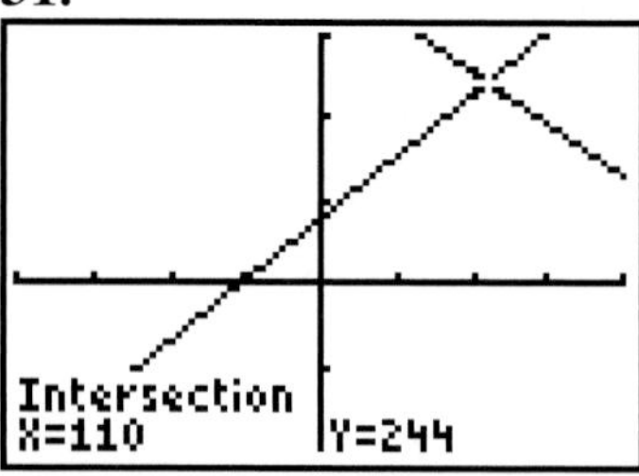

The solution set is $\{(110, 244)\}$.

33.

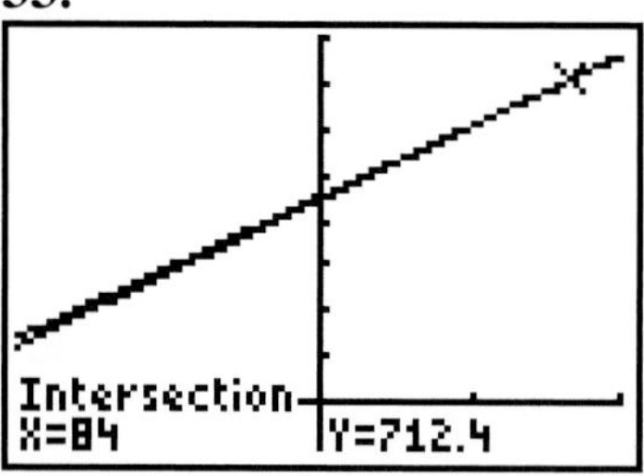

The solution set is $\{(84, 712.4)\}$.

35. Substitute $y = 4x - 1$ into $y = x + 8$.

$$4x - 1 = x + 8$$
$$3x = 9$$
$$x = 3$$

Use $x = 3$ in $y = x + 8$ to find y.

$$y = 3 + 8$$
$$y = 11$$

The solution set is $\{(3, 11)\}$. The equations are independent.

37. Substitute $y = -2x$ into $y = 4x + 12$.

$$-2x = 4x + 12$$
$$-6x = 12$$
$$x = -2$$

Use $x = -2$ in $y = -2x$ to find y.

$$y = -2(-2)$$
$$y = 4$$

The solution set is $\{(-2, 4)\}$. The equations are independent.

39. Substitute $y = \frac{1}{3}x + 2$ into $y = -\frac{1}{2}x + 7$.

$$\frac{1}{3}x + 2 = -\frac{1}{2}x + 7$$
$$2x + 12 = -3x + 42$$
$$5x = 30$$
$$x = 6$$

Use $x = 6$ in $y = \frac{1}{3}x + 2$ to find y.

$$y = \frac{1}{3}(6) + 2$$
$$y = 4$$

The solution set is $\{(6, 4)\}$. The equations are independent.

41. Substitute $y = x - 5$ into $2x - 5y = 1$.

$$2x - 5(x - 5) = 1$$
$$2x - 5x + 25 = 1$$
$$-3x = -24$$
$$x = 8$$

Use $x = 8$ in $y = x - 5$ to find y.

$$y = 8 - 5$$
$$y = 3$$

The solution set is $\{(8, 3)\}$. The equations are independent.

43. Substitute $x = 2y - 7$ into $3x + 2y = -5$.

$$3(2y - 7) + 2y = -5$$
$$6y - 21 + 2y = -5$$
$$8y = 16$$
$$y = 2$$

Use $y = 2$ in $x = 2y - 7$ to find x.

$$x = 2(2) - 7 = -3$$

The solution set is $\{(-3, 2)\}$. The equations are independent.

45. Substitute $y = 2x - 30$ into the other equation.

$$\frac{1}{5}x - \frac{1}{2}(2x - 30) = -1$$
$$\frac{1}{5}x - x + 15 = -1$$
$$5\left(\frac{1}{5}x - x + 15\right) = 5(-1)$$
$$x - 5x + 75 = -5$$
$$-4x = -80$$
$$x = 20$$

If $x = 20$, then

$$y = 2x - 30 = 2(20) - 30 = 10.$$

The solution set is $\{(20, 10)\}$ and the equations are independent.

47. Write $2x + y = 9$ as $y = -2x + 9$ and substitute into $2x - 5y = 15$.

$$2x - 5(-2x + 9) = 15$$
$$2x + 10x - 45 = 15$$
$$12x - 45 = 15$$
$$12x = 60$$
$$x = 5$$

Use $x = 5$ in $y = -2x + 9$ to find y.

$$y = -2(5) + 9 = -1$$

The solution set is $\{(5, -1)\}$. The equations are independent.

49. Write $x - y = 0$ as $y = x$ and substitute into $2x + 3y = 35$.

$$2x + 3x = 35$$
$$5x = 35$$
$$x = 7$$

Since $y = x$, $y = 7$ also. The solution set is $\{(7, 7)\}$ and the equations are independent.

51. Write $x + y = 40$ as $x = 40 - y$ and substitute into $0.2x + 0.8y = 23$.

$$0.2(40 - y) + 0.8y = 23$$
$$8 - 0.2y + 0.8y = 23$$
$$0.6y = 15$$
$$y = 25$$

Since $x = 40 - y$, we get $x = 40 - 25 = 15$. The solution set is $\{(15, 25)\}$ and the equations are independent.

53. Substitute $y = 2x - 5$ into $y + 1 = 2(x - 2)$.

$$2x - 5 + 1 = 2(x - 2)$$
$$2x - 4 = 2x - 4$$

Since the last equation is an identity, the equations are dependent. Any ordered pair that satisfies one equation will satisfy the other. The solution set is $\{(x, y) \mid y = 2x - 5\}$.

55. Write $x - y = 5$ as $x = y + 5$ and substitute into $2x = 2y + 14$.

$$2(y + 5) = 2y + 14$$
$$2y + 10 = 2y + 14$$
$$10 = 14$$

Since this last equation is incorrect no matter what values are used for x and y, the equations are inconsistent and the solution set is $\emptyset$.

57. Substitute $y = \frac{5}{7}x$ into $x = -\frac{2}{3}y$:

$$x = -\frac{2}{3}\left(\frac{5}{7}x\right)$$
$$x = -\frac{10}{21}x$$
$$21x = -10x$$
$$31x = 0$$
$$x = 0$$

$$y = \frac{5}{7}x$$
$$y = \frac{5}{7} \cdot 0 = 0$$

The solution set is $\{(0, 0)\}$. The system is independent.

59. First simplify.

$$3(y - 1) = 2(x - 3)$$
$$3y - 3 = 2x - 6$$
$$3y - 2x = -3$$

By simplifying the first equation we see that it is the same as the second equation. There is no need to substitute now. The equations are dependent and the solution set is $\{(x, y) \mid 3y - 2x = -3\}$.

61. Substitute $y = 3x$ into $y = 3x + 1$.

$$3x = 3x + 1$$
$$0 = 1$$

Since the last equation is false, the equations are inconsistent. The solution set to the system is the empty set, $\emptyset$.

63. Substitute $y = \frac{5}{2}x$ into $x + 3y = 3$:

$$x + 3\left(\frac{5}{2}x\right) = 3$$
$$x + \frac{15}{2}x = 3$$
$$2x + 15x = 6$$
$$17x = 6$$
$$x = \frac{6}{17}$$

$$y = \frac{5}{2}x$$
$$y = \frac{5}{2} \cdot \frac{6}{17} = \frac{15}{17}$$

The solution set is $\left\{\left(\frac{6}{17}, \frac{15}{17}\right)\right\}$. The system is independent.

65. Write $x - y = 5$ as $x = y + 5$ and substitute into $x + y = 4$.

$$y + 5 + y = 4$$
$$2y = -1$$
$$y = -1/2$$

If $y = -1/2$, then

$$x = y + 5 = -1/2 + 5 = 9/2.$$

The solution set is $\left\{\left(\frac{9}{2}, -\frac{1}{2}\right)\right\}$ and the equations are independent.

67. Write $2x - 4y = 0$ as $y = \frac{1}{2}x$. Substitute $y = \frac{1}{2}x$ into $6x + 8y = 5$.

$$6x + 8\left(\frac{1}{2}x\right) = 5$$
$$6x + 4x = 5$$
$$10x = 5$$
$$x = \frac{1}{2}$$

$$y = \frac{1}{2}\left(\frac{1}{2}\right) = \frac{1}{4}$$

The solution set is $\left\{\left(\frac{1}{2}, \frac{1}{4}\right)\right\}$.

69. Write $3x + y = 2$ as $y = -3x + 2$.
Substitute $y = -3x + 2$ into $-x - 3y = 6$.

$$-x - 3(-3x + 2) = 6$$
$$-x + 9x - 6 = 6$$
$$8x = 12$$
$$x = \frac{3}{2}$$
$$y = -3\left(\frac{3}{2}\right) + 2 = -\frac{9}{2} + \frac{4}{2} = -\frac{5}{2}$$

The solution set is $\left\{\left(\frac{3}{2}, -\frac{5}{2}\right)\right\}$.

71. Write $-9x + 6y = 3$ as $y = \frac{3}{2}x + \frac{1}{2}$.
Substitute $y = \frac{3}{2}x + \frac{1}{2}$ into $18x + 30y = 1$.

$$18x + 30\left(\frac{3}{2}x + \frac{1}{2}\right) = 1$$
$$18x + 45x + 15 = 1$$
$$63x = -14$$
$$x = -\frac{2}{9}$$
$$y = \frac{3}{2}\left(-\frac{2}{9}\right) + \frac{1}{2} = \frac{1}{6}$$

The solution set is $\left\{\left(-\frac{2}{9}, \frac{1}{6}\right)\right\}$.

73. Substitute $y = -2x$ into $3y - x = 1$.

$$3(-2x) - x = 1$$
$$-7x = 1$$
$$x = -\frac{1}{7}$$
$$y = -2\left(-\frac{1}{7}\right) = \frac{2}{7}$$

The solution set is $\left\{\left(-\frac{1}{7}, \frac{2}{7}\right)\right\}$.

75. Substitute $x = -6y + 1$ into $2y = -5x$.

$$2y = -5(-6y + 1)$$
$$2y = 30y - 5$$
$$-28y = -5$$
$$y = \frac{5}{28}$$
$$x = -6\left(\frac{5}{28}\right) + 1 = -\frac{1}{14}$$

The solution set is $\left\{\left(-\frac{1}{14}, \frac{5}{28}\right)\right\}$.

77. Write $x - y = 0.1$ as $x = y + 0.1$ and substitute into $2x - 3y = -0.5$.

$$2(y + 0.1) - 3y = -0.5$$
$$2y + 0.2 - 3y = -0.5$$
$$-y = -0.7$$
$$y = 0.7$$

If $y = 0.7$, then $x = 0.7 + 0.1 = 0.8$. The solution set is $\{(0.8, 0.7)\}$.

79. Let x represent the width and y represent the length, then $2x + 2y = 84$ and $y = x + 12$.

$$2x + 2(x + 12) = 84$$
$$4x + 24 = 84$$
$$4x = 60$$
$$x = 15$$
$$y = 15 + 12 = 27$$

The length is 27 feet and the width is 15 feet.

81. Let x represent the width and y represent the length, then $2x + 2y = 28$ and $x = \frac{1}{2}y - 1$.

$$2\left(\frac{1}{2}y - 1\right) + 2y = 28$$
$$y - 2 + 2y = 28$$
$$3y = 30$$
$$y = 10$$
$$x = \frac{1}{2}(10) - 1 = 4$$

The length is 10 ft and the width is 4 ft.

83. Let x and y represent the numbers.

$$x + y = 10$$
$$x - y = 3$$

Solve $x - y = 3$ for x to get $x = y + 3$.
Substitute into $x + y = 10$.

$$y + 3 + y = 10$$
$$2y = 7$$
$$y = \frac{7}{2}$$
$$x = \frac{7}{2} + 3 = \frac{13}{2}$$

The numbers are $7/2$ and $13/2$ or as decimals 3.5 and 6.5.

85. Let x and y represent the numbers.

$$x + y = 1$$
$$x - y = 20$$

Solve $x - y = 20$ for x to get $x = y + 20$.
Substitute into $x + y = 1$.

$$y + 20 + y = 1$$
$$2y = -19$$
$$y = -9.5$$
$$x = -9.5 + 20 = 10.5$$

The numbers are -9.5 and 10.5.

87. Let x represent the number of \$200 tickets and y represent the number of \$250 tickets.

$$x + y = 200$$
$$200x + 250y = 44{,}000$$

Solve $x + y = 200$ for x to get $x = 200 - y$.
Substitute into $200x + 250y = 44,000$.

$$200(200 - y) + 250y = 44,000$$
$$40000 - 200y + 250y = 44,000$$
$$50y = 4000$$
$$y = 80$$
$$x = 200 - 80 = 120$$

There were 120 tickets for \$200 each and 80 tickets for \$250 each.

89. Let x represent the number of student tickets (\$6) and y represent the number of nonstudent tickets (\$11).

$$y = 2x$$
$$6x + 11y = 1540$$

Substitute $y = 2x$ into $6x + 11y = 1540$.

$$6x + 11(2x) = 1540$$
$$28x = 1540$$
$$x = 55$$
$$y = 2(55) = 110$$

There were 55 tickets for \$6 each and 110 tickets for \$11 each.

91. Let x represent the amount invested at 5% and y represent the amount invested at 8%.

$$x + y = 40,000$$
$$0.05x + 0.08y = 2300$$

Substitute $y = 40,000 - x$ into $0.05x + 0.08y = 2300$.

$$0.05x + 0.08(40,000 - x) = 2300$$
$$0.05x + 3200 - 0.08x = 2300$$
$$-0.03x = -900$$
$$x = 30,000$$
$$y = 40,000 - 30,000 = 10,000$$

She invested \$30,000 at 5% and \$10,000 at 8%.

93. Let x represent the amount of 5% acid solution and y represent the amount of 25% acid solution.

$$x + y = 50$$
$$0.05x + 0.25y = 0.20(50)$$

Substitute $y = 50 - x$ into $0.05x + 0.25y = 10$.

$$0.05x + 0.25(50 - x) = 10$$
$$0.05x + 12.5 - 0.25x = 10$$
$$-0.20x = -2.5$$
$$x = 12.5$$
$$y = 50 - 12.5 = 37.5$$

The chemist should use 12.5 L of 5% solution and 37.5 L of 25% solution.

95. Let $x =$ the amount invested at 5% and $y =$ the amount invested at 10%. The interest earned on the investments is $0.05x$ and $0.10y$ respectively. We can write two equations, one expressing the total amount invested, and the other expressing the total of the interest earned.

$$x + y = 30{,}000$$
$$0.05x + 0.10y = 2{,}300$$

Write the first equation as $y = 30{,}000 - x$ and substitute into the second.

$$0.05x + 0.10(30{,}000 - x) = 2{,}300$$
$$0.05x + 3{,}000 - 0.10x = 2{,}300$$
$$-0.05x = -700$$
$$x = 14{,}000$$

If $x = 14{,}000$, then $y = 30{,}000 - 14{,}000 = 16{,}000$. She invested \$14,000 at 5% and \$16,000 at 10%.

97. Let x and y be the numbers. The fact that their sum is 2 and their difference is 26 gives us a system of equations:

$$x + y = 2$$
$$x - y = 26$$

Write $x + y = 2$ as $x = 2 - y$ and substitute.

$$2 - y - y = 26$$
$$-2y = 24$$
$$y = -12$$

If $y = -12$, then $x = 2 - y = 2 - (-12) = 14$. The numbers are -12 and 14.

99. Let $x =$ the number of toasters and $y =$ the number of vacation coupons. We write one equation expressing the fact that the total number of prizes is 100 and the other expressing the total bill of \$708.

$$x + y = 100$$
$$6x + 24y = 708$$

Write $x + y = 100$ as $x = 100 - y$ and substitute.

$$6(100 - y) + 24y = 708$$
$$600 - 6y + 24y = 708$$
$$18y = 108$$
$$y = 6$$

If $y = 6$, then $x = 100 - y = 100 - 6 = 94$. He gave away 94 toasters and 6 vacation coupons.

101. $s = 0.05(100{,}000 - f)$

$f = 0.30(100{,}000 - s)$

Simplify:

$$s = 5000 - 0.05f$$
$$f = 30{,}000 - 0.30s$$

Substitute:

$$f = 30{,}000 - 0.30(5000 - 0.05f)$$
$$f = 30{,}000 - 1500 + 0.015f$$

$$0.985f = 28{,}500$$
$$f \approx 28{,}934$$
$$s = 5{,}000 - 0.05(28{,}934)$$
$$\approx 3553$$

Rounding to the nearest dollar, the state tax is \$3,553 and federal tax is \$28,934.

103. $B = 0.20N$
$$N = 120{,}000 - B$$

Substitute:
$$N = 120{,}000 - 0.20N$$
$$1.2N = 120{,}000$$
$$N = 100{,}000$$
$$B = 0.20 \cdot 100{,}000 = 20{,}000$$

The bonus is \$20,000.

105. a) Use $x = 10{,}000$ in $C = 10x + 400{,}000$ to get the cost of 10,000 textbooks. The cost is \$500,000.

b) Use $x = 10.000$ in $F = 30x$ to get the revenue for 10,000 textbooks. The revenue for 10,000 textbooks is \$300,000.

c) The cost and revenue are equal when
$$30x = 10x + 400{,}000$$
$$20x = 400{,}000$$
$$x = 20{,}000.$$

The cost is equal to the revenue for 20,000 textbooks.

d) Let $x = 0$ in $C = 10x + 400{,}000$ to get $C = 400{,}000$. The fixed cost is \$400,000.

107. Solve each equation for y to get $y = \frac{2}{3}x - 2$ in every case except (a), which is $y = \frac{2}{3}x + 2$.

4.2 WARM-UPS

1. True, because when we add the equations, the y-terms add up to 0 and one variable is eliminated.

2. False, multiply the first by -2 and the second by 3 to eliminate x.

3. True, multiplying the second by 2 and adding will eliminate both x and y.

4. True, because both ordered pairs satisfy both equations.

5. False. The solution set is $\{(x, y) \mid 4x - 2y = 20\}$, which is a set of ordered pairs and it is not the same as the set of all real numbers.

6. True, because the equations are inconsistent.

7. True.

8. False, as long as we add the left sides and the right sides, the form does not matter.

9. True.

10. True.

4.2 EXERCISES

1. In this section we learned the addition method.

3. In some cases we multiply one or both of the equations on each side to change the coefficients of the variable that we are trying to eliminate.

5. If an identity (such as $0 = 0$) results form addition of the equations, then the equations are dependent.

7.
$$\begin{array}{r} x - y = 1 \\ x + y = 7 \\ \hline 2x \quad = 8 \\ x = 4 \end{array}$$

Use $x = 4$ in $x + y = 7$ to find y.
$$4 + y = 7$$
$$y = 3$$

The solution set is $\{(4, 3)\}$.

9.
$$\begin{array}{r} 3x - 4y = 11 \\ -3x + 2y = -7 \\ \hline -2y = 4 \\ y = -2 \end{array}$$

Use $y = -2$ in $3x - 4y = 11$.
$$3x - 4(-2) = 11$$
$$3x + 8 = 11$$
$$3x = 3$$
$$x = 1$$

The solution set is $\{(1, -2)\}$.

11.
$$\begin{array}{r} x - y = 12 \\ 2x + y = 3 \\ \hline 3x \quad = 15 \\ x = 5 \end{array}$$

Use $x = 5$ in $x - y = 12$.
$$5 - y = 12$$
$$-7 = y$$

The solution set is $\{(5, -7)\}$.

13.
$$\begin{aligned} 3x - y &= 5 \\ 5x + y &= -2 \\ \hline 8x \quad &= 3 \\ x &= \frac{3}{8} \end{aligned}$$

Use $x = \frac{3}{8}$ in $5x + y = -2$.

$$5 \cdot \frac{3}{8} + y = -2$$
$$y = -2 - \frac{15}{8} = -\frac{31}{8}$$

The solution set is $\left\{\left(\frac{3}{8}, -\frac{31}{8}\right)\right\}$.

15. If we multiply $2x - y = -5$ by 2, we get $4x - 2y = -10$. Add this equation to the second equation.

$$\begin{aligned} 4x - 2y &= -10 \\ 3x + 2y &= 3 \\ \hline 7x \quad &= -7 \\ x &= -1 \end{aligned}$$

Use $x = -1$ in $2x - y = -5$.

$$\begin{aligned} 2(-1) - y &= -5 \\ -2 - y &= -5 \\ 3 &= y \end{aligned}$$

The solution set is $\{(-1, 3)\}$.

17. If we multiply the first equation $-3x + 5y = 1$ by 3, we get $-9x + 15y = 3$. Add this equation to the second equation.

$$\begin{aligned} -9x + 15y &= 3 \\ 9x - 3y &= 5 \\ \hline 12y &= 8 \\ y &= \frac{2}{3} \end{aligned}$$

Use $y = 2/3$ in $9x - 3y = 5$:

$$\begin{aligned} 9x - 3 \cdot \frac{2}{3} &= 5 \\ 9x - 2 &= 5 \\ 9x &= 7 \\ x &= \frac{7}{9} \end{aligned}$$

The solution set is $\left\{\left(\frac{7}{9}, \frac{2}{3}\right)\right\}$.

19. Multiply the first equation by 4 and the second by 5.

$$\begin{aligned} 4(2x - 5y) &= 4(13) \\ 5(3x + 4y) &= 5(-15) \end{aligned}$$

$$\begin{aligned} 8x - 20y &= 52 \\ 15x + 20y &= -75 \\ \hline 23x \quad &= -23 \\ x &= -1 \end{aligned}$$

Use $x = -1$ in $2x - 5y = 13$.

$$\begin{aligned} 2(-1) - 5y &= 13 \\ -2 - 5y &= 13 \\ -5y &= 15 \\ y &= -3 \end{aligned}$$

The solution set is $\{(-1, -3)\}$.

21. Rewrite the first equation to match the form of the second.

$$\begin{aligned} 2x - 3y &= 11 \\ 7x - 4y &= 6 \end{aligned}$$

Multiply the first equation by -7 and the second by 2.

$$\begin{aligned} -7(2x - 3y) &= -7(11) \\ 2(7x - 4y) &= 2(6) \end{aligned}$$

$$\begin{aligned} -14x + 21y &= -77 \\ 14x - 8y &= 12 \\ \hline 13y &= -65 \\ y &= -5 \end{aligned}$$

Use $y = -5$ in $2x = 3y + 11$.

$$\begin{aligned} 2x &= 3(-5) + 11 \\ 2x &= -4 \\ x &= -2 \end{aligned}$$

The solution set is $\{(-2, -5)\}$.

23.
$$\begin{aligned} -12(x + y) &= -12(48) \\ 12x + 14y &= 628 \end{aligned}$$

$$\begin{aligned} -12x - 12y &= -576 \\ 12x + 14y &= 628 \\ \hline 2y &= 52 \\ y &= 26 \\ x + 26 &= 48 \\ x &= 22 \end{aligned}$$

The solution set is $\{(22, 26)\}$.

25.
$$\begin{aligned} 3x - 4y &= 9 \\ -3x + 4y &= 12 \\ \hline 0 &= 21 \end{aligned}$$

Since we obtained an incorrect equation by addition, the equations are inconsistent and the solution set is $\emptyset$.

27. Multiply $5x - y = 1$ by -2, to get $-10x + 2y = -2$. Add this equation to the second equation.

$$\begin{aligned} -10x + 2y &= -2 \\ 10x - 2y &= 2 \\ \hline 0 &= 0 \end{aligned}$$

Since we obtained an identity by adding the equations, the equations are dependent and the solution set is $\{(x, y) \mid 5x - y = 1\}$.

29.
$$\begin{array}{r} 2x - y = 5 \\ \underline{2x + y = 5} \\ 4x \quad = 10 \\ x = 10/4 = 5/2 \end{array}$$

Use $x = 5/2$ in $2x + y = 5$ to find y.
$$2(5/2) + y = 5$$
$$5 + y = 5$$
$$y = 0$$

The solution set is $\{(5/2, 0)\}$ and the equations are independent.

31. Multiplying the first equation by 12 to eliminate the fractions gives us the following system.
$$3x + 4y = 60$$
$$x - y = 6$$
Multiply the second equation by 4 and then add.

$$\begin{array}{r} 3x + 4y = 60 \\ \underline{4x - 4y = 24} \\ 7x \quad = 84 \\ x = 12 \end{array}$$

Use $x = 12$ in $x - y = 6$ to find y.
$$12 - y = 6$$
$$6 = y$$
The solution set is $\{(12, 6)\}$.

33. If we multiply the first equation by 12 and the second by 24 to eliminate fractions, we get the following system.
$$\begin{array}{r} 3x - 4y = -48 \\ \underline{3x + 4y = 0} \\ 6x \quad = -48 \\ x = -8 \end{array}$$
Use $x = -8$ in $3x + 4y = 0$ to find y.
$$3(-8) + 4y = 0$$
$$4y = 24$$
$$y = 6$$
The solution set is $\{(-8, 6)\}$.

35. Multiply the first equation by 8 and the second by -16 to get the following system.
$$\begin{array}{r} x + 2y = 40 \\ \underline{-x - 8y = -112} \\ -6y = -72 \\ y = 12 \\ x + 2(12) = 40 \\ x = 16 \end{array}$$
The solution set is $\{(16, 12)\}$.

37. Multiply the first equation by 6 and the second by 12 to get the following system.
$$2x + 3y = 2$$
$$10x - 9y = 2$$

Now multiply the first equation by -5 to get the following system:

$$\begin{array}{r} -10x - 15y = -10 \\ \underline{10x - 9y = 2} \\ -24y = -8 \\ y = \frac{1}{3} \end{array}$$
Use $y = 1/3$ in $\frac{1}{3}x + \frac{1}{2}y = \frac{1}{3}$
$$\frac{1}{3}x + \frac{1}{2} \cdot \frac{1}{3} = \frac{1}{3}$$
$$2x + 1 = 2$$
$$2x = 1$$
$$x = \frac{1}{2}$$
The solution set is $\left\{\left(\frac{1}{2}, \frac{1}{3}\right)\right\}$.

39. Multiply the first equation by 100 and the second by -10 to get the following system.
$$\begin{array}{r} 5x + 10y = 130 \\ \underline{-10x - 10y = -190} \\ -5x \quad = -60 \\ x = 12 \end{array}$$
Use $x = 12$ in $x + y = 19$ to get $y = 7$. The solution set is $\{(12, 7)\}$.

41. Multiply the first equation by -9 and the second by 100 to get the following system.
$$\begin{array}{r} -9x - 9y = -10{,}800 \\ \underline{12x + 9y = 12{,}000} \\ 3x \quad = 1200 \\ x = 400 \end{array}$$
Use $x = 400$ in $x + y = 1200$ to get $y = 800$. The solution set is $\{(400, 800)\}$.

43. Multiply the first equation by -2 to get the following system.

$$\begin{array}{r} -3x + 4y = 0.5 \\ \underline{3x + 1.5y = 6.375} \\ 5.5y = 6.875 \\ y = 1.25 \end{array}$$

Use $y = 1.25$ in $3x + 1.5y = 6.375$ to find x.
$$3x + 1.5(1.25) = 6.375$$
$$3x + 1.875 = 6.375$$

$$3x = 4.5$$
$$x = 1.5$$

The solution set is $\{(1.5, 1.25)\}$.

45. Multiply both equations by 100 to get the following system.

$$24x + 60y = 58$$
$$80x - 12y = 52$$

Multiply the second equation by 5 and add:

$$\begin{aligned} 24x + 60y &= 58 \\ 400x - 60y &= 260 \\ \hline 424x &= 318 \end{aligned}$$

$$x = \frac{318}{424} = \frac{3}{4}$$

Use $x = 3/4$ in $24x + 60y = 58$ to find y.

$$24 \cdot \frac{3}{4} + 60y = 58$$
$$60y = 40$$
$$y = \frac{40}{60} = \frac{2}{3}$$

The solution set is $\left\{\left(\frac{3}{4}, \frac{2}{3}\right)\right\}$.

47. Substitute $y = x + 1$ into $2x - 5y = -20$:

$$2x - 5(x + 1) = -20$$
$$-3x - 5 = -20$$
$$-3x = -15$$
$$x = 5$$

If $x = 5$, then $y = 5 + 1 = 6$. The solution set is $\{(5, 6)\}$.

49.
$$\begin{aligned} x - y &= 19 \\ 2x + y &= -13 \\ \hline 3x \quad &= 6 \end{aligned}$$
$$x = 2$$

Use $x = 2$ in $x - y = 19$.

$$2 - y = 19$$
$$-17 = y$$

The solution set is $\{(2, -17)\}$.

51. Substitute $x = y - 1$ into $2y = x + 2$:

$$2y = y - 1 + 2$$
$$y = 1$$

If $y = 1$, then $x = 1 - 1 = 0$. The solution set is $\{(0, 1)\}$.

53.
$$\begin{aligned} 2y - 3x &= -1 \\ 5y + 3x &= 29 \\ \hline 7y \quad &= 28 \end{aligned}$$
$$y = 4$$

Use $y = 4$ in $5y + 3x = 29$.

$$5(4) + 3x = 29$$
$$3x = 9$$
$$x = 3$$

The solution set is $\{(3, 4)\}$.

55. Substitute $y = \frac{2}{3}x$ into $6x + 3y = 4$:

$$6x + 3\left(\frac{2}{3}x\right) = 4$$
$$6x + 2x = 4$$
$$8x = 4$$
$$x = \frac{1}{2}$$

$$y = \frac{2}{3}x$$
$$y = \frac{2}{3} \cdot \frac{1}{2} = \frac{1}{3}$$

The solution set is $\left\{\left(\frac{1}{2}, \frac{1}{3}\right)\right\}$.

57. Substitute $y = 3x + 1$ into $x = \frac{1}{3}y + 5$:

$$x = \frac{1}{3}(3x + 1) + 5$$
$$x = x + \frac{1}{3} + 5$$
$$0 = \frac{16}{3}$$

The solution set is the empty set, $\emptyset$.

59.
$$\begin{aligned} x - y &= 0 \\ x + y &= 2x \\ \hline 2x \quad &= 2x \end{aligned}$$

The system is dependent and the solution set is $\{(x, y) \mid y = x\}$.

61. If $(2, 3)$ satisfies the system then it satisfies $x + y = 5$ and $x - y = a$. So $2 + 3 = 5$ and $2 - 3 = a$. For these equations to be correct, a must be -1.

63. If $(5, 12)$ satisfies the system then it satisfies $y = ax + 2$ and $y = bx + 17$. So $12 = a(5) + 2$ and $12 = b(5) + 17$. For the first equation to be correct, we must have $5a + 2 = 12$, $5a = 10$, and $a = 2$. For the second equation to be correct, we must have $5b + 17 = 12$, $5b = -5$, and $b = -1$.

65. Let $x =$ the price of one doughnut and $y =$ the price of one cup of coffee. His bills for Monday and Tuesday give us the following system of equations.

$$3x + 2y = 3.40$$
$$2x + 3y = 3.60$$

Multiply the first equation by -2 and the second by 3.

$$\begin{aligned} -6x - 4y &= -6.80 \\ 6x + 9y &= 10.80 \\ \hline 5y &= 4.00 \\ y &= 0.80 \end{aligned}$$

Use $y = 0.80$ in $3x + 2y = 3.40$ to find x.

$$\begin{aligned} 3x + 2(0.80) &= 3.40 \\ 3x + 1.60 &= 3.40 \\ 3x &= 1.80 \\ x &= 0.60 \end{aligned}$$

Doughnuts are \$0.60 each and a cup of coffee is \$0.80. So his total bill on Wednesday was \$1.40.

67. Let $x =$ the number of boys and $y =$ the number of girls at Freemont High. We get a system of equations from the number of boys and girls who attended the dance and game.

$$\begin{aligned} \frac{1}{2}x + \frac{1}{3}y &= 570 \\ \frac{1}{3}x + \frac{1}{2}y &= 580 \end{aligned}$$

Multiply the first equation by 12 and the second equation by -18.

$$\begin{aligned} 6x + 4y &= 6840 \\ -6x - 9y &= -10440 \\ \hline -5y &= -3600 \\ y &= 720 \end{aligned}$$

Use $y = 720$ in $6x + 4y = 6840$ to find x.

$$\begin{aligned} 6x + 4(720) &= 6840 \\ 6x + 2880 &= 6840 \\ 6x &= 3960 \\ x &= 660 \end{aligned}$$

There are 660 boys and 720 girls at Freemont High, or 1380 students.

69. Let $x =$ the number of dimes and $y =$ the number of nickels. We write one equation about the total number of the coins and the other about the value of the coins (in cents).

$$\begin{aligned} x + y &= 35 \\ 10x + 5y &= 330 \end{aligned}$$

Multiply the first equation by -5.

$$\begin{aligned} -5x - 5y &= -175 \\ 10x + 5y &= 330 \\ \hline 5x \qquad &= 155 \\ x &= 31 \end{aligned}$$

Use $x = 31$ in $x + y = 35$ to find that $y = 4$. He has 31 dimes and 4 nickels.

71. a) From the graph it appears that there should be about 20 pounds of Chocolate fudge and 30 pounds of Peanut Butter fudge.

b) Let $x =$ the number of pounds of Chocolate fudge and $y =$ the number of pounds of Peanut Butter fudge.

$$\begin{aligned} x + y &= 50 \\ 0.35x + 0.25y &= 0.29(50) \end{aligned}$$

$$\begin{aligned} y &= 50 - x \\ 0.35x + 0.25(50 - x) &= 14.5 \\ 0.10x &= 2 \\ x &= 20 \\ 20 + y &= 50 \\ y &= 30 \end{aligned}$$

Use 20 pounds of Chocolate fudge and 30 pounds of Peanut Butter fudge.

73. Let $a =$ the time from Allentown to Harrisburg and $h =$ the time from Harrisburg to Pittsburgh.

$$\begin{aligned} a + h &= 6 \\ 42a + 51h &= 288 \end{aligned}$$

$$\begin{aligned} -51a - 51h &= -306 \\ 42a + 51h &= 288 \\ \hline -9a \qquad &= -18 \\ a &= 2 \\ h &= 6 - 2 = 4 \end{aligned}$$

It took 4 hours to go from Harrisburg to Pittsburgh.

75. Let $p =$ the probability of rain and $q =$ the probability that it doesn't rain:

$$\begin{aligned} p + q &= 1 \\ p &= 4q \end{aligned}$$

$$\begin{aligned} 4q + q &= 1 \\ 5q &= 1 \\ q &= 0.20 = 20\% \\ p &= 80\% \end{aligned}$$

The probability of rain is 80%.

77. Let $W =$ the width and $L =$ the length.

$$\begin{aligned} W &= 0.75L \\ 2W + 2L &= 700 \end{aligned}$$

$$\begin{aligned} 2(0.75L) + 2L &= 700 \\ 1.5L + 2L &= 700 \\ 3.5L &= 700 \\ L &= 200 \\ W &= 0.75(200) = 150 \end{aligned}$$

The width is 150 m and the length is 200 m.

79. If one equation is already solved for x or y then substitution is usually easier. If both equations are in standard form then addition is usually easier.

81. a) Find the equation of the line through the two points. A system satisfied by both

$(-1, 2)$ and $(4, 5)$ is $y = \frac{3}{5}x + \frac{13}{5}$ and $5y = 3x + 13$.
b) The equations are dependent.
c) There is no independent system satisfied by both lines, because two points determines a unique line.

4.3 WARM-UPS

1. False, because $1 + (-2) - 3 = 4$ is incorrect.
2. False, because there are infinitely many ordered triples that satisfy $x + y - z = 4$.
3. True, because if $x = 1$, $y = -1$, and $z = 2$ then each of the equations is satisfied.
4. False, we can use substitution and addition to eliminate variables.
5. False, because two planes never intersect in a single point.
6. True, because if we multiply the second one by -1 and add, then we get $0 = 2$.
7. True, because -2 times the first equation is the same as the second equation.
8. False, because the graph is a plane.
9. False, because the value of x nickels, y dimes, and z quarters is $5x + 10y + 25z$ cents.
10. False, because $x = -2$, $z = 3$, and $-2 + y + 3 = 6$ implies that $y = 5$.

4.3 EXERCISES

1. A linear equation in three variables is an equation of the form $Ax + By + Cz = D$ where $A, B,$ and C cannot all be zero.
3. A solution to a system of linear equations in three variable is an ordered triple that satisfies all of the equations in the system.
5. The graph of a linear equation in three variables is a plane in a three-dimensional coordinate system.
7. Substituting $z = 4$ into the first two equations yields the following system.
$$x + y + 4 = 9$$
$$y + 4 = 7$$
Simplify.
$$x + y = 5$$
$$y = 3$$
Substitute.
$$x + 3 = 5$$
$$x = 2$$
The solution set is $\{(2, 3, 4)\}$.
9. Add the second and third equations to eliminate y.
$$\begin{array}{r} x - y = -1 \\ \underline{x + y = 5} \\ 2x \quad = 4 \\ x = 2 \end{array}$$
Substitute $x = 2$ into $x + y = 5$.
$$2 + y = 5$$
$$y = 3$$
Substitute $x = 2$ and $y = 3$ into $x + y + z = 10$.
$$2 + 3 + z = 10$$
$$z = 5$$
The solution set is $\{(2, 3, 5)\}$.
11. Add the first and last equations.
$$\begin{array}{r} x + y + z = 6 \\ \underline{x - y - z = -4} \\ 2x \qquad = 2 \\ x = 1 \end{array}$$
Replace x with 1 in the first two equations.
$$1 + y + z = 6$$
$$1 - y + z = 2$$
Simplify and add.
$$\begin{array}{r} y + z = 5 \\ \underline{-y + z = 1} \\ 2z = 6 \\ z = 3 \end{array}$$
If $z = 3$ and $y + z = 5$, then $y = 2$. The solution set is $\{(1, 2, 3)\}$.
13. Add the first two equations.
$$\begin{array}{rl} x + y + z = 2 & \\ \underline{x + 2y - z = 6} & \\ 2x + 3y \quad = 8 & \text{A} \end{array}$$
Add the first and third equations.
$$\begin{array}{rl} x + y + z = 2 & \\ \underline{2x + y - z = 5} & \\ 3x + 2y \quad = 7 & \text{B} \end{array}$$
Multiply equation A by 3 and equation B by -2.
$$\begin{array}{rl} 6x + 9y = 24 & 3 \times \text{A} \\ \underline{-6x - 4y = -14} & -2 \times \text{B} \\ 5y = 10 & \\ y = 2 & \end{array}$$

Use $y = 2$ in the equation $3x + 2y = 7$.

$$3x + 2(2) = 7$$
$$3x = 3$$
$$x = 1$$

Use $x = 1$ and $y = 2$ in $x + y + z = 2$.

$$1 + 2 + z = 2$$
$$z = -1$$

The solution set is $\{(1, 2, -1)\}$.

15. Multiply the first equation by -1 to get $-x + 2y - 4z = -3$. Now add this equation and the second, and this equation and the third.

$$\begin{array}{r} -x + 2y - 4z = -3 \\ x + 3y - 2z = 6 \\ \hline 5y - 6z = 3 \end{array} \qquad \begin{array}{r} -x + 2y - 4z = -3 \\ x - 4y + 3z = -5 \\ \hline -2y - z = -8 \end{array}$$

Multiply $-2y - z = -8$ by -6 and add the result to $5y - 6z = 3$.

$$\begin{array}{r} 12y + 6z = 48 \\ 5y - 6z = 3 \\ \hline 17y \quad = 51 \end{array}$$
$$y = 3$$

Use $y = 3$ in $5y - 6z = 3$.

$$5(3) - 6z = 3$$
$$-6z = -12$$
$$z = 2$$

Use $y = 3$ and $z = 2$ in $x + 3y - 2z = 6$.

$$x + 3(3) - 2(2) = 6$$
$$x = 1$$

The solution set is $\{(1, 3, 2)\}$.

17. Multiply $2x - y + z = 10$ by 2 to get $4x - 2y + 2z = 20$. Add this equation and the second and this equation and the third.

$$\begin{array}{r} 4x - 2y + 2z = 20 \\ 3x - 2y - 2z = 7 \\ \hline 7x - 4y \quad = 27 \end{array} \qquad \begin{array}{r} 4x - 2y + 2z = 20 \\ x - 3y - 2z = 10 \\ \hline 5x - 5y \quad = 30 \\ x - y \quad = 6 \end{array}$$

Multiply $x - y = 6$ by -4 to get $-4x + 4y = -24$. Add this to $7x - 4y = 27$.

$$\begin{array}{r} -4x + 4y = -24 \\ 7x - 4y = 27 \\ \hline 3x \quad = 3 \end{array}$$
$$x = 1$$

Use $x = 1$ in $7x - 4y = 27$ to find y.

$$7(1) - 4y = 27$$
$$-4y = 20$$
$$y = -5$$

Use $x = 1$ and $y = -5$ in $2x - y + z = 10$.

$$2(1) - (-5) + z = 10$$
$$z = 3$$

The solution set is $\{(1, -5, 3)\}$.

19. Multiply $x - y + 2z = -5$ by 2 to get $2x - 2y + 4z = -10$. Add this equation and the second equation, and add the first and second equations.

$$\begin{array}{r} 2x - 2y + 4z = -10 \\ -2x + y - 3z = 7 \\ \hline -y + z = -3 \end{array} \qquad \begin{array}{r} 2x - 3y + z = -9 \\ -2x + y - 3z = 7 \\ \hline -2y - 2z = -2 \\ y + z = 1 \end{array}$$

Add the last two equations to eliminate y.

$$\begin{array}{r} -y + z = -3 \\ y + z = 1 \\ \hline 2z = -2 \end{array}$$
$$z = -1$$

Use $z = -1$ in $y + z = 1$ to get $y = 2$. Now use $z = -1$ and $y = 2$ in $x - y + 2z = -5$.

$$x - 2 + 2(-1) = -5$$
$$x = -1$$

The solution set is $\{(-1, 2, -1)\}$.

21. Multiply the first equation by -2 and add the result to the last equation.

$$\begin{array}{r} -4x + 10y - 4z = -32 \\ 4x - 3y + 4z = 18 \\ \hline 7y \quad = -14 \\ y \quad = -2 \end{array}$$

Multiply the first equation by 3 and the second by 2 and add the results.

$$\begin{array}{r} 6x - 15y + 6z = 48 \\ 6x + 4y - 6z = -38 \\ \hline 12x - 11y \quad = 10 \end{array}$$

Use $y = -2$ in the last equation to find x.

$$12x - 11(-2) = 10$$
$$12x = -12$$
$$x = -1$$

Use $x = -1$ and $y = -2$ in $2x - 5y + 2z = 16$.

$$2(-1) - 5(-2) + 2z = 16$$
$$2z = 8$$
$$z = 4$$

The solution set is $\{(-1, -2, 4)\}$.

23. If we add the last two equations we get $x + 2y = 7$. Multiply the first equation by -1 to get $-x - y = -4$. Add these two equations.

$$\begin{array}{r} x + 2y = 7 \\ -x - y = -4 \\ \hline y = 3 \end{array}$$

Use $y = 3$ in $x + y = 4$ to get $x = 1$. Use $y = 3$ in $y - z = -2$.

$$3 - z = -2$$
$$5 = z$$

The solution set is $\{(1, 3, 5)\}$.

25. Multiply the first equation by -1 and add the result to the second equation.

$$\begin{array}{r} -x - y \quad = -7 \\ y - z = -1 \\ \hline -x \quad - z = -8 \end{array}$$

Add this result to the last equation.

$$\begin{array}{r} -x - z = -8 \\ x + 3z = 18 \\ \hline 2z = 10 \\ z = 5 \end{array}$$

Use $z = 5$ in $x + 3z = 18$ to get $x + 15 = 18$ or $x = 3$. Use $x = 3$ in $x + y = 7$ to get $3 + y = 7$ or $y = 4$. The solution set is $\{(3, 4, 5)\}$.

27. Note that the second equation is obtained from multiplying the first by -1 and the third is obtained from multiplying the first by 2. So all three equations are equivalent. So any point that satisfies one of them satisfies all of them. The solution set is $\{(x, y, z) | x + y - z = 2\}$.

29. Add the first two equations to get $2x + 2y = 10$ or $x + y = 5$. Add the second and third equations to get $2x + 2y = 14$ or $x + y = 7$. By substitution we get $5 = 7$, which is false. So there is no solution to the system. The solution set is the empty set $\emptyset$.

31. Use $z = 1$ two write the first equation as $x + y = 8$. Now the second equation indicates that $x + y = 5$. By substitution we have $8 = 5$, which is false. So there is no solution to the system. The solution set is the empty set $\emptyset$.

33. Add the first two equations.

$$\begin{array}{r} x - y + 2z = 3 \\ 2x + y - z = 5 \\ \hline 3x \quad + z = 8 \end{array} \quad \text{A}$$

Multiply the second equation by 3 and add the result to the last equation.

$$\begin{array}{r} 6x + 3y - 3z = 15 \\ 3x - 3y + 6z = 4 \\ \hline 9x \quad + 3z = 19 \end{array} \quad \text{B}$$

Multiply A by -3 and add the result to B.

$$\begin{array}{r} -9x - 3z = -24 \\ 9x + 3z = 19 \\ \hline 0 = -5 \end{array}$$

Since the last equation is false no matter what values the variables have, the solution set is $\emptyset$.

35. Multiply the last equation $-x + 2y - 3z = -6$ by -2 to get $2x - 4y + 6z = 12$, the first equation. Multiply the last equation $-x + 2y - 3z = -6$ by -6 to get $6x - 12y + 18z = 36$, the second equation. Since the first and second equations are multiples of the last equation, the system is dependent. The solution set is $\{(x, y, z) \mid -x + 2y - 3z = -6\}$.

37. Add the first and second equation to get $x + z = 11$. Multiply this equation by -2 and then add the result to the last equation.

$$\begin{array}{r} -2x - 2z = -22 \\ 2x + 2z = 7 \\ \hline 0 = -15 \end{array}$$

Since the last result is false, the solution set is $\emptyset$.

39. Multiply the first equation by 300 to get $30x + 24y - 12z = 900$. Multiply the second equation by 6 to get $30x + 24y - 12z = 900$. Multiply the third equation by 100 to get $30x + 24y - 12z = 900$. Since all three of these equations are different forms of the same equation, the system is dependent. The solution set is $\{(x, y, z) \mid 5x + 4y - 2z = 150\}$.

41. Multiply the second equation by 10 and add the result to the first equation.

$$\begin{array}{r} 37x - 2y + 0.5z = 4.1 \\ 3x + 2y - 0.4z = 0.1 \\ \hline 40x \quad + 0.1z = 4.2 \end{array} \quad \text{A}$$

Multiply the second equation by 19 and add the result to the last equation.

$$\begin{array}{r} 70.3x - 3.8y + 0.95z = 7.79 \\ -2x + 3.8y - 2.1z = -3.26 \\ \hline 68.3x \quad - 1.15z = 4.53 \end{array} \quad \text{B}$$

Multiply equation A by 11.5 and add the result to equation B.

$$\begin{array}{r} 460x + 1.15z = 48.3 \\ 68.3x - 1.15z = 4.53 \\ \hline 528.3x \quad = 52.83 \\ x = 0.1 \end{array}$$

Use $x = 0.1$ in $40x + 0.1z = 4.2$.

$$40(0.1) + 0.1z = 4.2$$
$$0.1z = 0.2$$
$$z = 2$$

Use $x = 0.1$ and $z = 2$ in
$3x + 2y - 0.4z = 0.1$.

$$3(0.1) + 2y - 0.4(2) = 0.1$$
$$0.3 + 2y - 0.8 = 0.1$$
$$2y = 0.6$$
$$y = 0.3$$

The solution set is $\{(0.1, 0.3, 2)\}$.

43. Let x = the price of the Chevy, y = the price of the Ford, and z = the price of the Toyota. Write the following system.

$$x + y + z = 66{,}000$$
$$y = x + 2000$$
$$z = y + 2000$$

By substitution $z = x + 4000$ and
$x + (x + 2000) + (x + 4000) = 66{,}000$

$$3x = 60{,}000$$
$$x = 20{,}000$$
$$y = x + 2{,}000 = 22{,}000$$
$$z = y + 2{,}000 = 24{,}000$$

The Chevrolet was \$20,000, the Ford was \$22,000, and the Toyota was \$24,000.

45. Let x = the number of hours driven on the first day, y = the number of hours for the second day, and z = the number of hours for the third day. His distance each day was $64x$, $62y$, and $58z$.

$$x + y + z = 36$$
$$64x + 62y + 58z = 2196$$
$$z = x + 4$$

Substitute:

$$x + y + (x + 4) = 36$$
$$64x + 62y + 58(x + 4) = 2196$$

Simplify:

$$2x + y = 32$$
$$122x + 62y = 1964$$

Multiply the first equation by -62:

$$-124x - 62y = -1984$$
$$122x + 62y = 1964$$
$$\overline{-2x \qquad = -20}$$
$$x = 10$$
$$z = x + 4 = 10 + 4 = 14$$
$$x + y + z = 36$$
$$10 + y + 14 = 36$$
$$y = 12$$

So he drove 10 hours on the first day, 12 hours on the second day, and 14 hours on the third day.

47. Let x = her investment in stocks, y = her investment in bonds, and z = her investment in a mutual funds. We can write 3 equations concerning x, y, and z.

$$x + y + z = 12{,}000$$
$$0.10x + 0.08y + 0.12z = 1230$$
$$x + y = z$$

Substitute $z = x + y$ into the first equation.

$$x + y + x + y = 12000$$
$$2x + 2y = 12000$$
$$x + y = 6000 \quad \text{A}$$

Substitute $z = x + y$ into the second equation.

$$0.10x + 0.08y + 0.12(x + y) = 1230$$
$$0.22x + 0.20y = 1230$$
$$-5(0.22x + 0.20y) = -5(1230)$$
$$-1.1x - y = -6150 \quad \text{B}$$

Add equations A and B.

$$x + y = 6000$$
$$-1.1x - y = -6150$$
$$\overline{-0.1x \qquad = -150}$$
$$x = 1500$$

Use $x = 1500$ in $x + y = 6000$ to get $y = 4500$. Since $z = x + y$, we must have $z = 6000$. So Ann invested \$1500 in stocks, \$4500 in bonds, and \$6000 in a mutual fund.

49. Let x = Anna's weight, y = Bob's weight, and z = Chris's weight. We are given three equations.

$$x + y \qquad = 226$$
$$y + z = 210$$
$$x \qquad + z = 200$$

From the first equation we get $y = 226 - x$. Substitute this equation into the second equation to eliminate y.

$$226 - x + z = 210$$
$$-x + z = -16$$

Add this last equation to $x + z = 200$ to eliminate x.

$$-x + z = -16$$
$$x + z = 200$$
$$\overline{2z = 184}$$
$$z = 92$$

If $z = 92$ and $x + z = 200$, we get $x = 108$.
If $z = 92$ and $y + z = 210$, we get $y = 118$.
So Anna weighs 108 pounds, Bob weighs 118 pounds, and Chris weighs 92 pounds.

51. Let $x =$ the number of nickels, $y =$ the number of dimes, and $z =$ the number of quarters. Since he has 13 coins altogether, $x + y + z = 13$. Since the value of the coins in cents is 175, we have $5x + 10y + 25z = 175$ or $x + 2y + 5z = 35$. Since the number of dimes is twice the number of nickels, $y = 2x$.

$$\begin{aligned} x + y + z &= 13 \\ x + 2y + 5z &= 35 \\ y &= 2x \end{aligned}$$

Substitute $y = 2x$ into the other two equations to get the following system.

$$\begin{aligned} 3x + z &= 13 \\ 5x + 5z &= 35 \end{aligned}$$

Divide the second equation by -5 and add:

$$\begin{aligned} 3x + z &= 13 \\ -x - z &= -7 \\ \hline 2x \quad &= 6 \\ x &= 3 \end{aligned}$$

If $x = 3$, then $y = 6$. If $x = 3$ and $y = 6$, then $3 + 6 + z = 13$, or $z = 4$. So he used 3 nickels, 6 dimes, and 4 quarters.

53. Let $x =$ her income from teaching, $y =$ her income from house painting, and $z =$ her royalties.

$$\begin{aligned} x + y + z &= 48{,}000 \\ x - y &= 6000 \\ z &= \tfrac{1}{7}(x + y) \end{aligned}$$

The last equation can be written as $x + y = 7z$. Replacing $x + y$ in the first equation by $7z$ gives us $7z + z = 48{,}000$, or $z = 6000$. If $z = 6000$, then $x + y = 42{,}000$. Add $x + y = 42000$ and the second equation.

$$\begin{aligned} x + y &= 42{,}000 \\ x - y &= 6000 \\ \hline 2x \quad &= 48{,}000 \\ x &= 24{,}000 \end{aligned}$$

Use $x = 24{,}000$ in $x - y = 6000$ to get $y = 18{,}000$. So she made \$24,000 teaching, \$18,000 house painting, and \$6,000 from royalties.

55. Let $x =$ Edwin's age, $y =$ his father's age and $z =$ his grandfather's age. Since their average age is 53, we can write

$$\frac{x + y + z}{3} = 53$$

or $$x + y + z = 159.$$

We can also write

$$\frac{1}{2}z + \frac{1}{3}y + \frac{1}{4}x = 65$$

or $$12\left(\frac{1}{2}z + \frac{1}{3}y + \frac{1}{4}x\right) = 12(65)$$

or $$6z + 4y + 3x = 780.$$

Four years ago Edwin was $x - 4$ years old and his grandfather was $z - 4$ years old. Since 4 years ago the grandfather was 4 times as old as Edwin, we can write

$$\begin{aligned} z - 4 &= 4(x - 4) \\ z - 4 &= 4x - 16 \\ z &= 4x - 12. \end{aligned}$$

Substitute $z = 4x - 12$ into the two equations above.

$$\begin{aligned} x + y + 4x - 12 &= 159 \\ 6(4x - 12) + 4y + 3x &= 780 \end{aligned}$$

$$\begin{aligned} 5x + y &= 171 \\ 27x + 4y &= 852 \end{aligned}$$

Substitute $y = 171 - 5x$ into $27x + 4y = 852$.

$$\begin{aligned} 27x + 4(171 - 5x) &= 852 \\ 7x + 684 &= 852 \\ 7x &= 168 \\ x &= 24 \end{aligned}$$

Since $z = 4x - 12$ we have $z = 84$. Since $x + y + z = 159$, we have $24 + y + 84 = 159$, or $y = 51$. So Edwin is 24, his father is 51, and his grandfather is 84 years old.

4.4 WARM-UPS

1. True
2. True.
3. True. Replace R_2 of (a) by $R_1 + R_2$ to get matrix (b).
4. False, because matrix (c) corresponds to an inconsistent system and (d) corresponds to a dependent system.
5. True, because the last row represents the equation $0 = 7$.
6. False. Replace R_2 by $2R_1 + R_2$ to get $0 = -3$ which is inconsistent.
7. False, because the system corresponding to (d) consists of two equations equivalent to $x + 3y = 5$.
8. False, because the augmented matrix is a 2×3 matrix.

9. True.
10. False, because it means to interchange R_1 and R_2.

4.4 EXERCISES

1. A matrix is a rectangular array of numbers.
3. The size of a matrix is the number of rows and columns.
5. An augmented matrix is a matrix where the entries in the first column are the coefficients of x, the entries in the second column are the coefficients of y, and the entries in the third column are the constants from a system of two linear equations in two unknowns.
7. The matrix is a 2×2 matrix because it has 2 rows and 2 columns.
9. The matrix is a 3×2 matrix because it has 3 rows and 2 columns.
11. The matrix is a 1×3 matrix because it has 1 row and 3 columns.
13. The matrix is a 2×1 matrix because it has 2 rows and 1 column.
15. Use the coefficients 2 and -3, and the constant 9 as the first row. Use the coefficients -3 and 1, and the constant -1 as the second row.

$$\begin{bmatrix} 2 & -3 & 9 \\ -3 & 1 & -1 \end{bmatrix}$$

17. Use the coefficients 1, -1, and 1, and the constant 1 as the first row. Use the coefficients 1, 1, and -2, and the constant 3 as the second row. Use the coefficients 0, 1, and -3, and the constant 4 as the third row.

$$\begin{bmatrix} 1 & -1 & 1 & 1 \\ 1 & 1 & -2 & 3 \\ 0 & 1 & -3 & 4 \end{bmatrix}$$

19. The entries in the first row $(5, 1, -1)$ represent the equation $5x + y = -1$. The entries in the second row $(2, -3, 0)$ represent the equation $2x - 3y = 0$. So the matrix represents the following system.

$$5x + y = -1$$
$$2x - 3y = 0$$

21. The entries in the first row $(1, 0, 0, 6)$ represent the equation $x = 6$. The entries in the second row $(-1, 0, 1, -3)$ represent the equation $-x + z = -3$. The entries in the third row $(1, 1, 0, 1)$ represent the equation $x + y = 1$. So the matrix represents the following system.

$$x = 6$$
$$-x + z = -3$$
$$x + y = 1$$

23. $R_1 \leftrightarrow R_2$ means that R_1 and R_2 are interchanged. So the new matrix is

$$\begin{bmatrix} 1 & 0 & 6 \\ 0 & 2 & 4 \end{bmatrix}.$$

25. $\frac{1}{4}R_1 \to R_1$ means that $\frac{1}{4}R_1$ replaces R_1. So multiply R_1 by $\frac{1}{4}$ and write the result in place of R_1:

$$\begin{bmatrix} 1 & 3 & 4 \\ 2 & -4 & 3 \end{bmatrix}.$$

27. $R_1 + R_2 \to R_2$ means that the sum of rows 1 and 2 is used in place of R_2. So the new matrix is

$$\begin{bmatrix} 1 & 0 & -3 \\ 0 & 2 & 1 \end{bmatrix}.$$

29. $2R_1 + R_2 \to R_2$ means that the sum of twice row 1 and row 2 is used in place of R_2. So the new matrix is

$$\begin{bmatrix} 1 & 2 & 3 \\ 0 & 7 & 11 \end{bmatrix}.$$

31. Row 1 and row 2 are interchanged. This operation is written in symbols as $R_1 \leftrightarrow R_2$.
33. The second row is multiplied by $\frac{1}{5}$. In symbols $\frac{1}{5}R_2 \to R_2$

35. $\begin{bmatrix} 1 & -1 & -3 \\ 0 & 1 & 4 \end{bmatrix}$

$$\begin{bmatrix} 1 & 0 & 1 \\ 0 & 1 & 4 \end{bmatrix} \quad R_2 + R_1 \to R_1$$

The solution set is $\{(1, 4)\}$.

37. $\begin{bmatrix} 1 & 1 & -6 \\ 0 & 3 & 6 \end{bmatrix}$

$$\begin{bmatrix} 1 & 1 & -6 \\ 0 & 1 & 2 \end{bmatrix} \quad \tfrac{1}{3}R_2 \to R_2$$

$$\begin{bmatrix} 1 & 0 & -8 \\ 0 & 1 & 2 \end{bmatrix} \quad -R_2 + R_1 \to R_1$$

The solution set is $\{(-8, 2)\}$.

39. $\begin{bmatrix} 1 & -1 & 7 \\ -1 & -1 & -3 \end{bmatrix}$

$\begin{bmatrix} 1 & -1 & 7 \\ 0 & -2 & 4 \end{bmatrix}$ $R_1 + R_2 \rightarrow R_2$

$\begin{bmatrix} 1 & -1 & 7 \\ 0 & 1 & -2 \end{bmatrix}$ $-\frac{1}{2}R_2 \rightarrow R_2$

$\begin{bmatrix} 1 & 0 & 5 \\ 0 & 1 & -2 \end{bmatrix}$ $R_2 + R_1 \rightarrow R_1$

The solution set is $\{(5, -2)\}$.

41. $\begin{bmatrix} 1 & 1 & 3 \\ -3 & 1 & -1 \end{bmatrix}$

$\begin{bmatrix} 1 & 1 & 3 \\ 0 & 4 & 8 \end{bmatrix}$ $3R_1 + R_2 \rightarrow R_2$

$\begin{bmatrix} 1 & 1 & 3 \\ 0 & 1 & 2 \end{bmatrix}$ $\frac{1}{4}R_2 \rightarrow R_2$

$\begin{bmatrix} 1 & 0 & 1 \\ 0 & 1 & 2 \end{bmatrix}$ $-R_2 + R_1 \rightarrow R_1$

The solution set is $\{(1, 2)\}$.

43. $\begin{bmatrix} 2 & -1 & 3 \\ 1 & 1 & 9 \end{bmatrix}$

$\begin{bmatrix} 1 & 1 & 9 \\ 2 & -1 & 3 \end{bmatrix}$ $R_1 \leftrightarrow R_2$

$\begin{bmatrix} 1 & 1 & 9 \\ 0 & -3 & -15 \end{bmatrix}$ $-2R_1 + R_2 \rightarrow R_2$

$\begin{bmatrix} 1 & 1 & 9 \\ 0 & 1 & 5 \end{bmatrix}$ $R_2 \div (-3) \rightarrow R_2$

$\begin{bmatrix} 1 & 0 & 4 \\ 0 & 1 & 5 \end{bmatrix}$ $-R_2 + R_1 \rightarrow R_1$

The solution set is $\{(4, 5)\}$.

45. $\begin{bmatrix} 3 & -1 & 4 \\ 2 & 1 & 1 \end{bmatrix}$

$\begin{bmatrix} 1 & -2 & 3 \\ 2 & 1 & 1 \end{bmatrix}$ $-R_2 + R_1 \rightarrow R_1$

$\begin{bmatrix} 1 & -2 & 3 \\ 0 & 5 & -5 \end{bmatrix}$ $-2R_1 + R_2 \rightarrow R_2$

$\begin{bmatrix} 1 & -2 & 3 \\ 0 & 1 & -1 \end{bmatrix}$ $R_2 \div 5 \rightarrow R_2$

$\begin{bmatrix} 1 & 0 & 1 \\ 0 & 1 & -1 \end{bmatrix}$ $2R_2 + R_1 \rightarrow R_1$

The solution set is $\{(1, -1)\}$.

47. $\begin{bmatrix} 6 & -7 & 0 \\ 2 & 1 & 20 \end{bmatrix}$

$\begin{bmatrix} 0 & -10 & -60 \\ 2 & 1 & 20 \end{bmatrix}$ $-3R_2 + R_1 \rightarrow R_1$

$\begin{bmatrix} 0 & 1 & 6 \\ 2 & 1 & 20 \end{bmatrix}$ $R_1 \div (-10) \rightarrow R_1$

$\begin{bmatrix} 2 & 1 & 20 \\ 0 & 1 & 6 \end{bmatrix}$ $R_1 \leftrightarrow R_2$

$\begin{bmatrix} 2 & 0 & 14 \\ 0 & 1 & 6 \end{bmatrix}$ $-R_2 + R_1 \rightarrow R_1$

$\begin{bmatrix} 1 & 0 & 7 \\ 0 & 1 & 6 \end{bmatrix}$ $R_1 \div 2 \rightarrow R_1$

The solution set is $\{(7, 6)\}$.

49. $\begin{bmatrix} 1 & 1 & -1 & 4 \\ 0 & 1 & 1 & 6 \\ 0 & 0 & 1 & 2 \end{bmatrix}$

$\begin{bmatrix} 1 & 0 & -2 & -2 \\ 0 & 1 & 1 & 6 \\ 0 & 0 & 1 & 2 \end{bmatrix}$ $-R_2 + R_1 \rightarrow R_1$

$\begin{bmatrix} 1 & 0 & 0 & 2 \\ 0 & 1 & 1 & 6 \\ 0 & 0 & 1 & 2 \end{bmatrix}$ $2R_3 + R_1 \rightarrow R_1$

$\begin{bmatrix} 1 & 0 & 0 & 2 \\ 0 & 1 & 0 & 4 \\ 0 & 0 & 1 & 2 \end{bmatrix}$ $-R_3 + R_2 \rightarrow R_2$

The solution set is $\{(2, 4, 2)\}$.

51. $\begin{bmatrix} 1 & 1 & 1 & 6 \\ 1 & -1 & 1 & 2 \\ 0 & 2 & -1 & 1 \end{bmatrix}$

$\begin{bmatrix} 1 & 1 & 1 & 6 \\ 0 & -2 & 0 & -4 \\ 0 & 2 & -1 & 1 \end{bmatrix}$ $-R_1 + R_2 \rightarrow R_2$

$\begin{bmatrix} 1 & 1 & 1 & 6 \\ 0 & 1 & 0 & 2 \\ 0 & 2 & -1 & 1 \end{bmatrix}$ $R_2 \div (-2) \rightarrow R_2$

$-R_2 + R_1 \rightarrow R_1$

$\begin{bmatrix} 1 & 0 & 1 & 4 \\ 0 & 1 & 0 & 2 \\ 0 & 0 & -1 & -3 \end{bmatrix}$

$-2R_2 + R_3 \rightarrow R_3$

$\begin{bmatrix} 1 & 0 & 1 & 4 \\ 0 & 1 & 0 & 2 \\ 0 & 0 & 1 & 3 \end{bmatrix}$

$R_3 \div (-1) \rightarrow R_3$

$\begin{bmatrix} 1 & 0 & 0 & 1 \\ 0 & 1 & 0 & 2 \\ 0 & 0 & 1 & 3 \end{bmatrix}$ $-R_3 + R_1 \rightarrow R_1$

The solution set is $\{(1, 2, 3)\}$.

53. $\begin{bmatrix} 2 & 1 & 1 & 4 \\ 1 & 1 & -1 & 1 \\ 1 & -1 & 2 & 2 \end{bmatrix}$

$\begin{bmatrix} 1 & 1 & -1 & 1 \\ 2 & 1 & 1 & 4 \\ 1 & -1 & 2 & 2 \end{bmatrix}$ $R_1 \leftrightarrow R_2$

$\begin{bmatrix} 1 & 1 & -1 & 1 \\ 0 & -1 & 3 & 2 \\ 0 & -2 & 3 & 1 \end{bmatrix}$ $-2R_1 + R_2 \rightarrow R_2$

$-R_1 + R_3 \rightarrow R_3$

$\begin{bmatrix} 1 & 1 & -1 & 1 \\ 0 & 1 & -3 & -2 \\ 0 & -2 & 3 & 1 \end{bmatrix}$ $-1 \cdot R_2 \rightarrow R_2$

$\begin{bmatrix} 1 & 0 & 2 & 3 \\ 0 & 1 & -3 & -2 \\ 0 & 0 & -3 & -3 \end{bmatrix}$ $-1 \cdot R_2 + R_1 \rightarrow R_1$

$2R_2 + R_3 \rightarrow R_3$

$\begin{bmatrix} 1 & 0 & 2 & 3 \\ 0 & 1 & -3 & -2 \\ 0 & 0 & 1 & 1 \end{bmatrix}$ $-\frac{1}{3} \cdot R_3 \rightarrow R_3$

$-2R_3 + R_1 \rightarrow R_1$

$\begin{bmatrix} 1 & 0 & 0 & 1 \\ 0 & 1 & 0 & 1 \\ 0 & 0 & 1 & 1 \end{bmatrix}$ $3R_3 + R_2 \rightarrow R_2$

The solution set is $\{(1, 1, 1)\}$.

55. $\begin{bmatrix} 1 & 1 & -3 & 3 \\ 2 & -1 & 1 & 0 \\ 1 & -1 & 1 & -1 \end{bmatrix}$

$\begin{bmatrix} 1 & 1 & -3 & 3 \\ 0 & -3 & 7 & -6 \\ 0 & -2 & 4 & -4 \end{bmatrix}$ $-2R_1 + R_2 \rightarrow R_2$

$-R_1 + R_3 \rightarrow R_3$

$\begin{bmatrix} 1 & 1 & -3 & 3 \\ 0 & 1 & -1 & 2 \\ 0 & -2 & 4 & -4 \end{bmatrix}$ $-2R_3 + R_2 \rightarrow R_2$

$\begin{bmatrix} 1 & 0 & -2 & 1 \\ 0 & 1 & -1 & 2 \\ 0 & 0 & 2 & 0 \end{bmatrix}$ $-R_2 + R_1 \rightarrow R_1$

$2R_2 + R_3 \rightarrow R_3$

$\begin{bmatrix} 1 & 0 & -2 & 1 \\ 0 & 1 & -1 & 2 \\ 0 & 0 & 1 & 0 \end{bmatrix}$ $R_3 \div 2 \rightarrow R_3$

$2R_3 + R_1 \rightarrow R_1$

$\begin{bmatrix} 1 & 0 & 0 & 1 \\ 0 & 1 & 0 & 2 \\ 0 & 0 & 1 & 0 \end{bmatrix}$ $R_3 + R_2 \rightarrow R_2$

The solution set is $\{(1, 2, 0)\}$.

57. $\begin{bmatrix} 1 & -1 & -4 & -3 \\ -1 & 3 & 1 & 0 \\ 1 & 1 & 2 & 3 \end{bmatrix}$

$\begin{bmatrix} 1 & -1 & -4 & -3 \\ 0 & 2 & -3 & -3 \\ 0 & 2 & 6 & 6 \end{bmatrix}$ $R_1 + R_2 \rightarrow R_2$

$-R_1 + R_3 \rightarrow R_3$

$$\begin{bmatrix} 1 & -1 & -4 & -3 \\ 0 & 2 & -3 & -3 \\ 0 & 0 & 9 & 9 \end{bmatrix} \quad -R_2 + R_3 \rightarrow R_3$$

$$\begin{bmatrix} 1 & -1 & -4 & -3 \\ 0 & 2 & -3 & -3 \\ 0 & 0 & 1 & 1 \end{bmatrix} \quad R_3 \div 9 \rightarrow R_3$$

$$\begin{bmatrix} 1 & -1 & 0 & 1 \\ 0 & 2 & 0 & 0 \\ 0 & 0 & 1 & 1 \end{bmatrix} \quad \begin{matrix} 4R_3 + R_1 \rightarrow R_1 \\ 3R_3 + R_2 \rightarrow R_2 \end{matrix}$$

$$\begin{bmatrix} 1 & -1 & 0 & 1 \\ 0 & 1 & 0 & 0 \\ 0 & 0 & 1 & 1 \end{bmatrix} \quad R_2 \div 2 \rightarrow R_2$$

$$\begin{bmatrix} 1 & 0 & 0 & 1 \\ 0 & 1 & 0 & 0 \\ 0 & 0 & 1 & 1 \end{bmatrix} \quad R_2 + R_1 \rightarrow R_1$$

The solution set is $\{(1, 0, 1)\}$.

59. $\begin{bmatrix} 1 & -5 & 11 \\ -2 & 10 & -22 \end{bmatrix}$

$$\begin{bmatrix} 1 & -5 & 11 \\ 0 & 0 & 0 \end{bmatrix} \quad 2R_1 + R_2 \rightarrow R_2$$

Since the second equation is a multiple of the first equation, the equations are equivalent. The solution set is $\{(x, y) | x - 5y = 11\}$.

61. $\begin{bmatrix} 2 & -3 & 4 \\ -2 & 3 & 5 \end{bmatrix}$

$$\begin{bmatrix} 2 & -3 & 4 \\ 0 & 0 & 9 \end{bmatrix} \quad R_1 + R_2 \rightarrow R_2$$

Since the second row represents the equation $0 = 9$, the solution set is the empty set, $\emptyset$.

63. $\begin{bmatrix} 1 & 2 & 1 \\ 3 & 6 & 3 \end{bmatrix}$

$$\begin{bmatrix} 1 & 2 & 1 \\ 0 & 0 & 0 \end{bmatrix} \quad -3R_1 + R_2 \rightarrow R_2$$

Since the system is equivalent to the single equation $x + 2y = 1$, the equations are dependent and the solution set is $\{(x, y) \mid x + 2y = 1\}$.

65. $\begin{bmatrix} 1 & -1 & 1 & 1 \\ 2 & -2 & 2 & 2 \\ -3 & 3 & -3 & -3 \end{bmatrix}$

$$\begin{bmatrix} 1 & -1 & 1 & 1 \\ 0 & 0 & 0 & 0 \\ 0 & 0 & 0 & 0 \end{bmatrix} \quad \begin{matrix} -2R_1 + R_2 \rightarrow R_2 \\ 3R_1 + R_3 \rightarrow R_3 \end{matrix}$$

Since the system of equations is equivalent to the first equation, the solution set is $\{(x, y, z) \mid x - y + z = 1\}$.

67. $\begin{bmatrix} 1 & 1 & -1 & 2 \\ 2 & -1 & 1 & 1 \\ 3 & 3 & -3 & 8 \end{bmatrix}$

$$\begin{bmatrix} 1 & 1 & -1 & 2 \\ 0 & -3 & 3 & -3 \\ 0 & 0 & 0 & 2 \end{bmatrix} \quad \begin{matrix} -2R_1 + R_2 \rightarrow R_2 \\ -3R_1 + R_3 \rightarrow R_3 \end{matrix}$$

Since the third equation is $0 = 2$, there is no solution to the system. The solution set is $\emptyset$.

69. Let x and y be the numbers.

$$x + y = 12$$
$$x - y = 2$$

$$\begin{bmatrix} 1 & 1 & 12 \\ 1 & -1 & 2 \end{bmatrix}$$

$$\begin{bmatrix} 1 & 1 & 12 \\ 0 & -2 & -10 \end{bmatrix} \quad -R_1 + R_2 \rightarrow R_2$$

$$\begin{bmatrix} 1 & 1 & 12 \\ 0 & 1 & 5 \end{bmatrix} \quad -\frac{1}{2}R_2 \rightarrow R_2$$

$$\begin{bmatrix} 1 & 0 & 7 \\ 0 & 1 & 5 \end{bmatrix} \quad -R_2 + R_1 \rightarrow R_1$$

So the numbers are 5 and 7.

71. Let $x =$ the length and $y =$ the width.

$$x - y = 2.5$$
$$2x + 2y = 39$$

$$\begin{bmatrix} 1 & -1 & 2.5 \\ 2 & 2 & 39 \end{bmatrix}$$

$$\begin{bmatrix} 1 & -1 & 2.5 \\ 0 & 4 & 34 \end{bmatrix} \quad -2R_1 + R_2 \rightarrow R_2$$

$$\begin{bmatrix} 1 & -1 & 2.5 \\ 0 & 1 & 8.5 \end{bmatrix} \quad \tfrac{1}{4}R_2 \rightarrow R_2$$

$$\begin{bmatrix} 1 & 0 & 11 \\ 0 & 1 & 8.5 \end{bmatrix} \quad R_2 + R_1 \rightarrow R_1$$

So the length is 11 in. and the width is 8.5 in.

73. Let $x =$ the selling price and $y =$ the buying price..

$$x - y = 2$$
$$49x - 56y = 0$$

$$\begin{bmatrix} 1 & -1 & 2 \\ 49 & -56 & 0 \end{bmatrix}$$

$$\begin{bmatrix} 1 & -1 & 2 \\ 0 & -7 & -98 \end{bmatrix} \quad -49R_1 + R_2 \rightarrow R_2$$

$$\begin{bmatrix} 1 & -1 & 2 \\ 0 & 1 & 14 \end{bmatrix} \quad -\tfrac{1}{7}R_2 \rightarrow R_2$$

$$\begin{bmatrix} 1 & 0 & 16 \\ 0 & 1 & 14 \end{bmatrix} \quad R_2 + R_1 \rightarrow R_1$$

So he buys for \$14 and sells for \$16.

75. Let $x =$ the number of four-wheeled cars, $y =$ the number of two-wheeled cars, and $z =$ the number of two-wheeled motorcycles.

$$x + y + z = 50$$
$$4x + 3y + 2z = 192$$
$$x = 9(y + z)$$

Write $x = 9(y + z)$ as $x - 9y - 9z = 0$:

$$\begin{bmatrix} 1 & 1 & 1 & 50 \\ 4 & 3 & 2 & 192 \\ 1 & -9 & -9 & 0 \end{bmatrix}$$

$$\begin{bmatrix} 1 & 1 & 1 & 50 \\ 0 & -1 & -2 & -8 \\ 0 & -10 & -10 & -50 \end{bmatrix}$$

$-4R_1 + R_2 \rightarrow R_2$
$-4_1 + R_3 \rightarrow R_3$

$$\begin{bmatrix} 1 & 1 & 1 & 50 \\ 0 & 1 & 2 & 8 \\ 0 & -10 & -10 & -50 \end{bmatrix}$$

$-R_2 \rightarrow R_2$

$$\begin{bmatrix} 1 & 0 & -1 & 42 \\ 0 & 1 & 2 & 8 \\ 0 & 0 & 10 & 30 \end{bmatrix}$$

$-R_2 + R_1 \rightarrow R_1$
$10R_2 + R_3 \rightarrow R_3$

$$\begin{bmatrix} 1 & 0 & -1 & 42 \\ 0 & 1 & 2 & 8 \\ 0 & 0 & 1 & 3 \end{bmatrix}$$

$\tfrac{1}{10}R_3 \rightarrow R_3$

$$\begin{bmatrix} 1 & 0 & 0 & 45 \\ 0 & 1 & 0 & 2 \\ 0 & 0 & 1 & 3 \end{bmatrix}$$

$R_3 + R_1 \rightarrow R_1$
$-2R_3 + R_2 \rightarrow R_2$

There were 45 four-wheeled cars, 2 three-wheeled cars, and 3 two-wheeled motorcycles.

4.5 WARM-UPS

1. True. because the determinant is $(-1)(-5) - (2)(3) = -1$.

2. False, because the determinant is $2 \cdot 8 - (-4)(4) = 32$.

3. False, because if $D = 0$, then Cramer's rule fails to give us the solution.

4. True, because the determinant is the value of $ad - bc$.

5. True, because this is the case where Cramer's rule fails to give the precise solution.

6. True.

7. True, because this is precisely when Cramer's rule works.

8. True, because of the definition of determinant of a 3×3 matrix.

9. True.

10. True.

4.5 EXERCISES

1. A determinant is a real number associated with a square matrix.

3. Cramer's rule works on systems that have exactly one solution.

5. A minor for an element is obtained by deleting the row and column of the element and then finding the determinant of the 2×2 matrix that remains.

7. $\begin{vmatrix} 2 & 5 \\ 3 & 7 \end{vmatrix} = 2 \cdot 7 - 3 \cdot 5 = -1$

9. $\begin{vmatrix} 0 & 3 \\ 1 & 5 \end{vmatrix} = 0 \cdot 5 - 1 \cdot 3 = -3$

11. $\begin{vmatrix} -3 & -2 \\ -4 & 2 \end{vmatrix} = -3 \cdot 2 - (-4)(-2) = -14$

13. $\begin{vmatrix} 0.05 & 0.06 \\ 10 & 20 \end{vmatrix} = 0.05(20) - 0.06(10)$

$= 0.4$

15. $D = \begin{vmatrix} 1 & -1 \\ 0 & 2 \end{vmatrix} = 2$

$D_x = \begin{vmatrix} -4 & -1 \\ 12 & 2 \end{vmatrix} = 4$

$D_y = \begin{vmatrix} 1 & -4 \\ 0 & 12 \end{vmatrix} = 12$

$x = \frac{D_x}{D} = \frac{4}{2} = 2 \quad y = \frac{D_y}{D} = \frac{12}{2} = 6$

The solution set is $\{(2, 6)\}$.

17. $D = \begin{vmatrix} 1 & 1 \\ 2 & 0 \end{vmatrix} = -2$

$D_x = \begin{vmatrix} 0 & 1 \\ -16 & 0 \end{vmatrix} = 16$

$D_y = \begin{vmatrix} 1 & 0 \\ 2 & -16 \end{vmatrix} = -16$

$x = \frac{D_x}{D} = \frac{16}{-2} = -8 \quad y = \frac{D_y}{D} = \frac{-16}{-2} = 8$

The solution set is $\{(-8, 8)\}$.

19. $D = \begin{vmatrix} 2 & -1 \\ 3 & 2 \end{vmatrix} = 7$

$D_x = \begin{vmatrix} 5 & -1 \\ -3 & 2 \end{vmatrix} = 7$

$D_y = \begin{vmatrix} 2 & 5 \\ 3 & -3 \end{vmatrix} = -21$

$x = \frac{D_x}{D} = \frac{7}{7} = 1 \quad y = \frac{D_y}{D} = \frac{-21}{7} = -3$

The solution set is $\{(1, -3)\}$.

21. $D = \begin{vmatrix} 3 & -5 \\ 2 & 3 \end{vmatrix} = 19$

$D_x = \begin{vmatrix} -2 & -5 \\ 5 & 3 \end{vmatrix} = 19$

$D_y = \begin{vmatrix} 3 & -2 \\ 2 & 5 \end{vmatrix} = 19$

$x = \frac{D_x}{D} = \frac{19}{19} = 1 \quad y = \frac{D_y}{D} = \frac{19}{19} = 1$

The solution set is $\{(1, 1)\}$.

23. $D = \begin{vmatrix} 4 & -3 \\ 2 & 5 \end{vmatrix} = 26$

$D_x = \begin{vmatrix} 5 & -3 \\ 7 & 5 \end{vmatrix} = 46 \quad D_y = \begin{vmatrix} 4 & 5 \\ 2 & 7 \end{vmatrix} = 18$

$x = \frac{D_x}{D} = \frac{46}{26} = \frac{23}{13} \quad y = \frac{D_y}{D} = \frac{18}{26} = \frac{9}{13}$

The solution set is $\left\{\left(\frac{23}{13}, \frac{9}{13}\right)\right\}$.

25. $D = \begin{vmatrix} 0.5 & 0.2 \\ 0.4 & -0.6 \end{vmatrix} = -0.38$

$D_x = \begin{vmatrix} 8 & 0.2 \\ -5 & -0.6 \end{vmatrix} = -3.8$

$D_y = \begin{vmatrix} 0.5 & 8 \\ 0.4 & -5 \end{vmatrix} = -5.7$

$x = \frac{D_x}{D} = \frac{-3.8}{-0.38} = 10$

$y = \frac{D_y}{D} = \frac{-5.7}{-0.38} = 15$

The solution set is $\{(10, 15)\}$.

27. Multiply the first equation by 4 and the second by 6 to eliminate the fractions.

$$2x + y = 20$$
$$2x - 3y = -6$$

$$D = \begin{vmatrix} 2 & 1 \\ 2 & -3 \end{vmatrix} = -8$$

$$D_x = \begin{vmatrix} 20 & 1 \\ -6 & -3 \end{vmatrix} = -54$$

$$D_y = \begin{vmatrix} 2 & 20 \\ 2 & -6 \end{vmatrix} = -52$$

$$x = \frac{D_x}{D} = \frac{-54}{-8} = \frac{27}{4} \; y = \frac{D_y}{D} = \frac{-52}{-8} = \frac{13}{2}$$

The solution set is $\left\{\left(\frac{27}{4}, \frac{13}{2}\right)\right\}$.

29. Enter the matrix into the calculator and use the determinant feature to get

$$\begin{vmatrix} 2.3 & -1.6 \\ 4.8 & 5.1 \end{vmatrix} = 19.41.$$

31. Enter the matrix into the calculator and use the determinant feature to get

$$\begin{vmatrix} 1/3 & 1/4 \\ 1/8 & 1/6 \end{vmatrix} 7/288.$$

33. Use the determinant feature of your calculator to find the determinants:

$$D = \begin{vmatrix} 3.2 & -5.7 \\ 4.6 & 7.1 \end{vmatrix} = 48.94$$

$$D_x = \begin{vmatrix} 6.24 & -5.7 \\ 33.44 & 7.1 \end{vmatrix} = 234.912$$

$$D_y = \begin{vmatrix} 3.2 & 6.24 \\ 4.6 & 33.44 \end{vmatrix} = 78.304$$

$$x = \frac{D_x}{D} = \frac{234.912}{48.94} = 4.8$$

$$y = \frac{D_y}{D} = \frac{78.304}{48.94} = 1.6$$

The solution set is $\{(4.8, 1.6)\}$.

35. Use the determinant feature of your calculator to find the determinants:

$$D = \begin{vmatrix} \sqrt{2} & -\sqrt{3} \\ \sqrt{8} & \sqrt{12} \end{vmatrix} = 2\sqrt{24} \approx 9.79796$$

$$D_x = \begin{vmatrix} -11 & -\sqrt{3} \\ 38 & \sqrt{12} \end{vmatrix} \approx 27.712813$$

$$D_y = \begin{vmatrix} \sqrt{2} & -11 \\ \sqrt{8} & 38 \end{vmatrix} \approx 84.85281$$

$$x = \frac{D_x}{D} = \frac{27.712813}{9.79796} \approx 2.83$$

$$y = \frac{D_y}{D} = \frac{84.85281}{9.79796} \approx 8.66$$

The solution set is $\{(2.83, 8.66)\}$.

37. The minor for 3 is the determinant of the matrix obtained by deleting the row and column containing 3, the first row and first column.

$$\begin{vmatrix} -3 & 7 \\ 1 & -6 \end{vmatrix} = 11$$

39. The minor for 5 is the determinant of the matrix obtained by deleting the first row and third column.

$$\begin{vmatrix} 4 & -3 \\ 0 & 1 \end{vmatrix} = 4$$

41. The minor for 7 is the determinant of the matrix obtained by deleting the second row and third column.

$$\begin{vmatrix} 3 & -2 \\ 0 & 1 \end{vmatrix} = 3$$

43. The minor for 1 is the determinant of the matrix obtained by deleting the third row and second column.

$$\begin{vmatrix} 3 & 5 \\ 4 & 7 \end{vmatrix} = 1$$

45. $1\begin{vmatrix} 3 & 1 \\ 1 & 5 \end{vmatrix} - 2\begin{vmatrix} 1 & 2 \\ 1 & 5 \end{vmatrix} + 3\begin{vmatrix} 1 & 2 \\ 3 & 1 \end{vmatrix}$

$$= 1(14) - 2(3) + 3(-5) = -7$$

47. $2\begin{vmatrix} 0 & 1 \\ 1 & 2 \end{vmatrix} - 1\begin{vmatrix} 1 & 0 \\ 1 & 2 \end{vmatrix} + 3\begin{vmatrix} 1 & 0 \\ 0 & 1 \end{vmatrix}$

$$= 2(-1) - 1(2) + 3(1) = -1$$

49. $-2\begin{vmatrix} 3 & 1 \\ 4 & 0 \end{vmatrix} + 3\begin{vmatrix} 1 & 2 \\ 4 & 0 \end{vmatrix} - 5\begin{vmatrix} 1 & 2 \\ 3 & 1 \end{vmatrix}$

$$= -2(-4) + 3(-8) - 5(-5) = 9$$

51. $1\begin{vmatrix} 3 & 2 \\ 2 & 3 \end{vmatrix} - 0\begin{vmatrix} 1 & 5 \\ 2 & 3 \end{vmatrix} + 0\begin{vmatrix} 1 & 5 \\ 3 & 2 \end{vmatrix}$

$$= 1(5) - 0(-7) + 0(-13) = 5$$

53. Expand by minors about the second column because it has two zeros in it.

$$-1\begin{vmatrix}2 & 6\\4 & 1\end{vmatrix}+0\begin{vmatrix}3 & 5\\4 & 1\end{vmatrix}-0\begin{vmatrix}3 & 5\\2 & 6\end{vmatrix}$$
$$=-1(-22)=22$$

55. Expand by minors about the first column because it has one zero in it.

$$-2\begin{vmatrix}1 & -1\\-4 & -3\end{vmatrix}-0\begin{vmatrix}1 & 3\\-4 & -3\end{vmatrix}+2\begin{vmatrix}1 & 3\\1 & -1\end{vmatrix}$$
$$=-2(-7)+2(-4)=6$$

57. Expand by minors about the third column because it has two zeros in it.

$$0\begin{vmatrix}4 & -1\\0 & 3\end{vmatrix}-0\begin{vmatrix}-2 & -3\\0 & 3\end{vmatrix}+5\begin{vmatrix}-2 & -3\\4 & -1\end{vmatrix}$$
$$=5(14)=70$$

59. Expand by minors about the second column.

$$-1\begin{vmatrix}0 & 5\\5 & 4\end{vmatrix}+0\begin{vmatrix}2 & 1\\5 & 4\end{vmatrix}-0\begin{vmatrix}2 & 1\\0 & 5\end{vmatrix}$$
$$=-1(-25)=25$$

61. $D=\begin{vmatrix}1 & 1 & 1\\1 & -1 & 1\\2 & 1 & 1\end{vmatrix}=2$

$$D_x=\begin{vmatrix}6 & 1 & 1\\2 & -1 & 1\\7 & 1 & 1\end{vmatrix}=2$$

$$D_y=\begin{vmatrix}1 & 6 & 1\\1 & 2 & 1\\2 & 7 & 1\end{vmatrix}=4$$

$$D_z=\begin{vmatrix}1 & 1 & 6\\1 & -1 & 2\\2 & 1 & 7\end{vmatrix}=6$$

$x=\frac{D_x}{D}=\frac{2}{2}=1,\ y=\frac{D_y}{D}=\frac{4}{2}=2,$

$z=\frac{D_z}{D}=\frac{6}{2}=3$

The solution set is $\{(1, 2, 3)\}$.

63. $D=\begin{vmatrix}1 & -3 & 2\\1 & 1 & 1\\1 & -1 & 1\end{vmatrix}=-2$

$$D_x=\begin{vmatrix}0 & -3 & 2\\2 & 1 & 1\\0 & -1 & 1\end{vmatrix}=2$$

$$D_y=\begin{vmatrix}1 & 0 & 2\\1 & 2 & 1\\1 & 0 & 1\end{vmatrix}=-2$$

$$D_z=\begin{vmatrix}1 & -3 & 0\\1 & 1 & 2\\1 & -1 & 0\end{vmatrix}=-4$$

$x=\frac{D_x}{D}=\frac{2}{-2}=-1,\ y=\frac{D_y}{D}=\frac{-2}{-2}=1,$

$z=\frac{D_z}{D}=\frac{-4}{-2}=2$

The solution set is $\{(-1, 1, 2)\}$.

65. $D=\begin{vmatrix}1 & 1 & 0\\0 & 2 & -1\\1 & 1 & 1\end{vmatrix}=2$

$$D_x=\begin{vmatrix}-1 & 1 & 0\\3 & 2 & -1\\0 & 1 & 1\end{vmatrix}=-6$$

$$D_y=\begin{vmatrix}1 & -1 & 0\\0 & 3 & -1\\1 & 0 & 1\end{vmatrix}=4$$

$$D_z=\begin{vmatrix}1 & 1 & -1\\0 & 2 & 3\\1 & 1 & 0\end{vmatrix}=2$$

$x=\frac{D_x}{D}=\frac{-6}{2}=-3,\ y=\frac{D_y}{D}=\frac{4}{2}=2,$

$z=\frac{D_z}{D}=\frac{2}{2}=1$

The solution set is $\{(-3, 2, 1)\}$.

67. $D=\begin{vmatrix}1 & 1 & -1\\2 & 2 & 1\\1 & -3 & 0\end{vmatrix}=12$

$$D_x=\begin{vmatrix}0 & 1 & -1\\6 & 2 & 1\\0 & -3 & 0\end{vmatrix}=18$$

$$D_y=\begin{vmatrix}1 & 0 & -1\\2 & 6 & 1\\1 & 0 & 0\end{vmatrix}=6$$

$$D_z=\begin{vmatrix}1 & 1 & 0\\2 & 2 & 6\\1 & -3 & 0\end{vmatrix}=24$$

$x = \frac{D_x}{D} = \frac{18}{12} = \frac{3}{2}, \quad y = \frac{D_y}{D} = \frac{6}{12} = \frac{1}{2},$

$z = \frac{D_z}{D} = \frac{24}{12} = 2$

The solution set is $\left\{\left(\frac{3}{2}, \frac{1}{2}, 2\right)\right\}$.

69. $D = \begin{vmatrix} 1 & 1 & 1 \\ 0 & 2 & 2 \\ 3 & -1 & 0 \end{vmatrix} = 2$

$D_x = \begin{vmatrix} 0 & 1 & 1 \\ 0 & 2 & 2 \\ -1 & -1 & 0 \end{vmatrix} = 0$

$D_y = \begin{vmatrix} 1 & 0 & 1 \\ 0 & 0 & 2 \\ 3 & -1 & 0 \end{vmatrix} = 2$

$D_z = \begin{vmatrix} 1 & 1 & 0 \\ 0 & 2 & 0 \\ 3 & -1 & -1 \end{vmatrix} = -2$

$x = \frac{D_x}{D} = \frac{0}{2} = 0, \quad y = \frac{D_y}{D} = \frac{2}{2} = 1,$

$z = \frac{D_z}{D} = \frac{-2}{2} = -1$

The solution set is $\{(0, 1, -1)\}$.

71. $D = \begin{vmatrix} 1.3 & -1.4 & 1.5 \\ 2.4 & 3.1 & -5.6 \\ 3.7 & -1.5 & 4.8 \end{vmatrix} = 30.955$

$D_x = \begin{vmatrix} 1.7 & -1.4 & 1.5 \\ -0.92 & 3.1 & -5.6 \\ 8.51 & -1.5 & 4.8 \end{vmatrix} = 34.0505$

$D_y = \begin{vmatrix} 1.3 & 1.7 & 1.5 \\ 2.4 & -0.92 & -5.6 \\ 3.7 & 8.51 & 4.8 \end{vmatrix} = 37.146$

$D_z = \begin{vmatrix} 1.3 & -1.4 & 1.7 \\ 2.4 & 3.1 & -0.92 \\ 3.7 & -1.5 & 8.51 \end{vmatrix} = 40.2415$

$x = \frac{D_x}{D} = \frac{34.0505}{30.955} = 1.1,$

$y = \frac{D_y}{D} = \frac{37.146}{30.955} = 1.2,$

$z = \frac{D_z}{D} = \frac{40.2415}{30.955} = 1.3$

The solution set is $\{(1.1, 1.2, 1.3)\}$.

73. a) From the graph it appears that there should be approximately 9 servings of peas and 11 servings of beets.

b) Let $x =$ the number of servings of canned peas and $y =$ the number of servings of canned beets. In the first equation we find the total grams of protein and in the second the total grams of carbohydrates.

$$3x + y = 38$$
$$11x + 8y = 187$$

$\text{D} = \begin{vmatrix} 3 & 1 \\ 11 & 8 \end{vmatrix} = 13 \quad D_x = \begin{vmatrix} 38 & 1 \\ 187 & 8 \end{vmatrix} = 117$

$D_y = \begin{vmatrix} 3 & 38 \\ 11 & 187 \end{vmatrix} = 143$

$x = \frac{D_x}{D} = \frac{117}{13} = 9 \quad y = \frac{D_y}{D} = \frac{143}{13} = 11$

To get the required grams of protein and carbohydrates we need 9 servings of peas and 11 servings of beets.

75. Let $x =$ the price of a gallon of milk and $y =$ the price of the magazine. Note that she paid \$0.30 tax. The first equation expresses the total price of the goods and the second expresses the total tax.

$$x + y = 4.65$$
$$0.05x + 0.08y = 0.30$$

$D = \begin{vmatrix} 1 & 1 \\ 0.05 & 0.08 \end{vmatrix} = 0.03$

$D_x = \begin{vmatrix} 4.65 & 1 \\ 0.3 & 0.08 \end{vmatrix} = 0.072$

$D_y = \begin{vmatrix} 1 & 4.65 \\ 0.05 & 0.3 \end{vmatrix} = 0.0675$

$x = \frac{D_x}{D} = \frac{0.072}{0.03} = 2.4$

$y = \frac{D_y}{D} = \frac{0.0675}{0.03} = 2.25$

The price of the milk was \$2.40 and the price of the magazine was \$2.25.

77. Let $x =$ the number of singles and $y =$ the number of doubles. Write one equation for the patties and one equation for the tomato slices.

$$x + 2y = 32$$
$$2x + y = 34$$

$$D = \begin{vmatrix} 1 & 2 \\ 2 & 1 \end{vmatrix} = -3 \quad D_x = \begin{vmatrix} 32 & 2 \\ 34 & 1 \end{vmatrix} = -36$$

$$D_y = \begin{vmatrix} 1 & 32 \\ 2 & 34 \end{vmatrix} = -30$$

$$x = \frac{D_x}{D} = \frac{-36}{-3} = 12 \quad y = \frac{D_y}{D} = \frac{-30}{-3} = 10$$

He must sell 12 singles and 10 doubles.

79. Let $x =$ Gary's age and $y =$ Harry's age. Since Gary is 5 years older than Harry, $x = y + 5$. Twenty-nine years ago Gary was $x - 29$ and Harry was $y - 29$. Gary was twice as old as Harry (29 years ago) is expressed as $x - 29 = 2(y - 29)$. These equations can be rewritten as follows.

$$x - y = 5$$
$$x - 2y = -29$$

$$D = \begin{vmatrix} 1 & -1 \\ 1 & -2 \end{vmatrix} = -1$$

$$D_x = \begin{vmatrix} 5 & -1 \\ -29 & -2 \end{vmatrix} = -39$$

$$D_y = \begin{vmatrix} 1 & 5 \\ 1 & -29 \end{vmatrix} = -34$$

$$x = \frac{D_x}{D} = \frac{-39}{-1} = 39 \quad y = \frac{D_y}{D} = \frac{-34}{-1} = 34$$

So Gary is 39 and Harry is 34.

81. Let $x =$ the length of a side of the square and $y =$ the length of a side of the equilateral triangle. Since the perimeters are to be equal, $4x = 3y$. Since the total of the two perimeters is 80, $4x + 3y = 80$. Rewrite the equations as follows.

$$4x - 3y = 0$$
$$4x + 3y = 80$$

$$D = \begin{vmatrix} 4 & -3 \\ 4 & 3 \end{vmatrix} = 24$$

$$D_x = \begin{vmatrix} 0 & -3 \\ 80 & 3 \end{vmatrix} = 240$$

$$D_y = \begin{vmatrix} 4 & 0 \\ 4 & 80 \end{vmatrix} = 320$$

$$x = \frac{D_x}{D} = \frac{240}{24} = 10 \quad y = \frac{D_y}{D} = \frac{320}{24} = \frac{40}{3}$$

The length of the side of the square should be 10 feet and the length of the side of the triangle should be $40/3$ feet.

83. Let $x =$ the number of gallons of 10% solution and $y =$ the number of gallons of 25% solution. Since the total mixture is to be 30 gallons, $x + y = 30$. The next equation comes from the fact that the chlorine in the two parts is equal to the chlorine in the 30 gallons, $0.10x + 0.25y = 0.20(30)$. Rewrite the equations as follows.

$$x + \quad y = 30$$
$$0.10x + 0.25y = 6$$

$$D = \begin{vmatrix} 1 & 1 \\ 0.10 & 0.25 \end{vmatrix} = 0.15$$

$$D_x = \begin{vmatrix} 30 & 1 \\ 6 & 0.25 \end{vmatrix} = 1.5$$

$$D_y = \begin{vmatrix} 1 & 30 \\ 0.10 & 6 \end{vmatrix} = 3$$

$$x = \frac{D_x}{D} = \frac{1.5}{0.15} = 10$$

$$y = \frac{D_y}{D} = \frac{3}{0.15} = 20$$

Use 10 gallons of 10% solution and 20 gallons of 25% solution.

85. Let $x =$ Mimi's weight, $y =$ Mitzi's weight, and $z =$ Cassandra's weight. We can write 3 equations.

$$\begin{aligned} x + y + z &= 175 \\ x \quad + z &= 143 \\ y + z &= 139 \end{aligned}$$

$$D = \begin{vmatrix} 1 & 1 & 1 \\ 1 & 0 & 1 \\ 0 & 1 & 1 \end{vmatrix} = -1$$

$$D_x = \begin{vmatrix} 175 & 1 & 1 \\ 143 & 0 & 1 \\ 139 & 1 & 1 \end{vmatrix} = -36$$

$$D_y = \begin{vmatrix} 1 & 175 & 1 \\ 1 & 143 & 1 \\ 0 & 139 & 1 \end{vmatrix} = -32$$

$$D_z = \begin{vmatrix} 1 & 1 & 175 \\ 1 & 0 & 143 \\ 0 & 1 & 139 \end{vmatrix} = -107$$

$$x = \frac{D_x}{D} = \frac{-36}{-1} = 36,$$

$$y = \frac{D_y}{D} = \frac{-32}{-1} = 32,$$

$$z = \frac{D_z}{D} = \frac{-107}{-1} = 107$$

So Mimi weights 36 pounds, Mitzi weighs 32 pounds, and Cassandra weighs 107 pounds.

87. Let $x =$ the number of degrees in the larger of the two acute angles, $y =$ the number of degrees in the smaller acute angle, and $z =$ the number of degrees in the right angle. We can write 3 equations about the sizes of the angles.

$$\begin{aligned} x + y + z &= 180 \\ x - y \quad &= 12 \\ z &= 90 \end{aligned}$$

$$D = \begin{vmatrix} 1 & 1 & 1 \\ 1 & -1 & 0 \\ 0 & 0 & 1 \end{vmatrix} = -2$$

$$D_x = \begin{vmatrix} 180 & 1 & 1 \\ 12 & -1 & 0 \\ 90 & 0 & 1 \end{vmatrix} = -102$$

$$D_y = \begin{vmatrix} 1 & 180 & 1 \\ 1 & 12 & 0 \\ 0 & 90 & 1 \end{vmatrix} = -78$$

$$D_z = \begin{vmatrix} 1 & 1 & 180 \\ 1 & -1 & 12 \\ 0 & 0 & 90 \end{vmatrix} = -180$$

$$x = \frac{D_x}{D} = \frac{-102}{-2} = 51,$$

$$y = \frac{D_y}{D} = \frac{-78}{-2} = 39,$$

$$z = \frac{D_z}{D} = \frac{-180}{-2} = 90$$

The measures of the three angles of the triangle are 39°, 51°, and 90°.

89. If $D = 0$, then use one of the other methods.

91. No, because Cramer's rule only works on linear systems and the given system is not linear.

4.6 WARM-UPS

1. False, because $x \geq 0$ consists of the points on or to the right of the y-axis.

2. False, because $y \geq 0$ consists of the points on or above the x-axis.

3. False, because $x + y \leq 6$ consists of the points on or below the line $x + y = 6$.

4. False, because the x-intercept is (15, 0) and the y-intercept is (0, 10).

5. False, because the solution set to a system is the intersection of the solution sets.

6. True, because that is the definition of constraint.

7. False, because a linear function does not have an x^2 in it.

8. True, because $R(2, 4) = 3(2) + 5(4) = 26$.

9. False, because $C(0, 5) = 12(0) + 10(5) = 50$.

10. True, because the maximum or minimum of a linear function occurs at a vertex.

4.6 EXERCISES

1. A constraint is an inequality that restricts the values of the variables.

3. Constraints may be limitations on the amount of available supplies, money, or other resources.

5. The maximum or minimum of a linear function subject to linear constraints occurs at a vertex of the region determined by the constraints.

7. The graph of $x \geq 0$ is the region on or to the right of the y-axis. The graph of $y \geq 0$ is the region on or above the x-axis. The graph of $x + y \leq 5$ is the region on or below the line $x + y = 5$. The intersection of these 3 regions is shaded here.

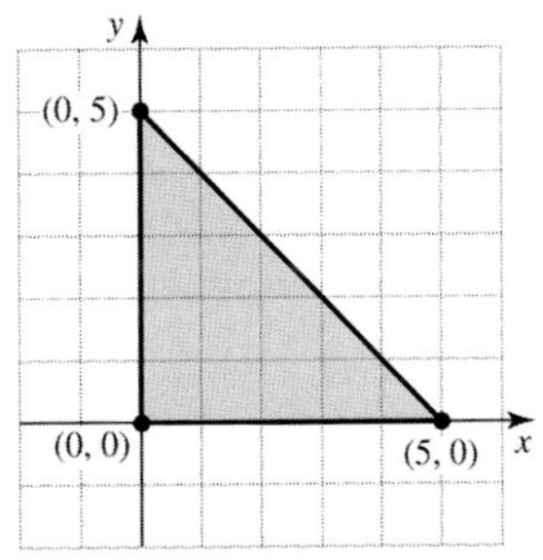

9. The graph of $x \geq 0$ is the region on or to the right of the y-axis. The graph of $y \geq 0$ is the region on or above the x-axis. The graph of $2x + y \leq 4$ is the region on or below the line $2x + y = 4$. The graph of $x + y \leq 3$ is the region on or below the line $x + y = 3$. The intersection of these regions is shaded below.

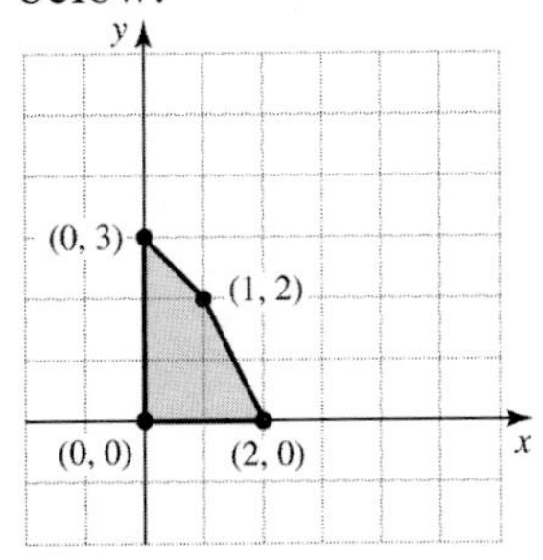

11. The graph of $x \geq 0$ is the region on or to the right of the y-axis. The graph of $y \geq 0$ is the region on or above the x-axis. The graph of $2x + y \geq 3$ is the region on or above the line $2x + y = 3$. The graph of $x + y \geq 2$ is the region on or above the line $x + y = 2$. The intersection of these regions is shaded below.

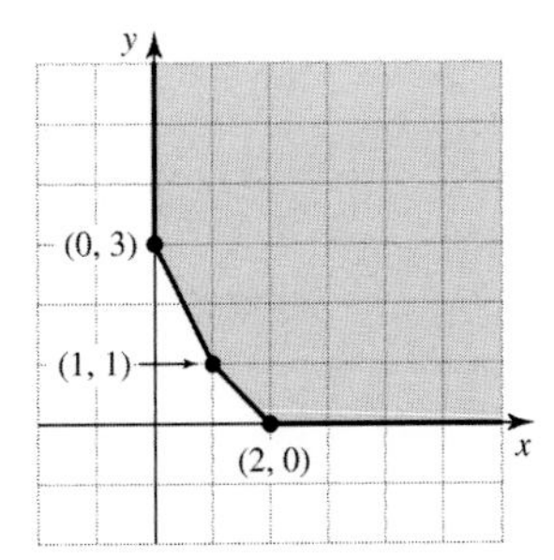

13. The graph of $x \geq 0$ is the region on or to the right of the y-axis. The graph of $y \geq 0$ is the region on or above the x-axis. The graph of $x + 3y \leq 15$ is the region on or below the line $x + 3y = 15$. The graph of $2x + y \leq 10$ is the region on or below the line $2x + y = 10$. The intersection of these regions is shaded below.

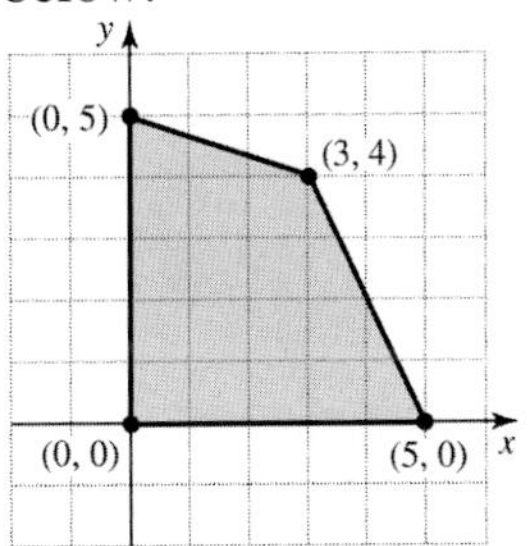

15. The graph of $x \geq 0$ is the region on or to the right of the y-axis. The graph of $y \geq 0$ is the region on or above the x-axis. The graph of $x + y \geq 4$ is the region on or above the line $x + y = 4$. The graph of $3x + y \geq 6$ is the region on or above the line $3x + y = 6$. The intersection of these regions is shaded below.

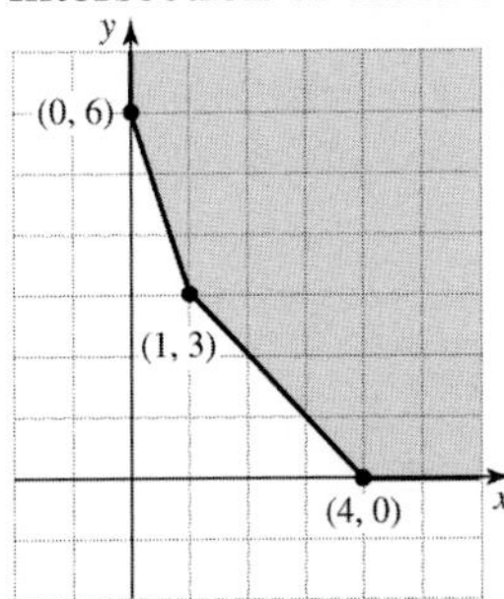

17. If x represents the number of acoustic and y represents the number of electric guitars, then we can write an inequality about the available money for material and available hours of labor.

$$100x + 200y \leq 3000$$
$$20x + 15y \leq 300$$

Simplify:

$x + 2y \le 30$
$4x + 3y \le 60$
Graphing these along with $x \ge 0$ and $y \ge 0$ gives the region shown here.

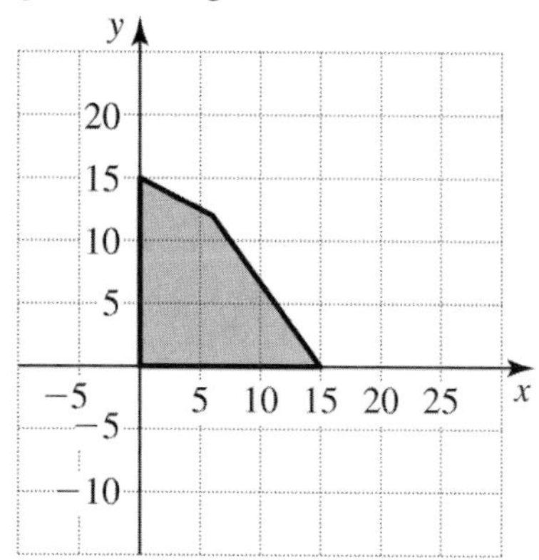

19. Since $P(x, y) = 6x + 8y$,
$P(1, 5) = 6(1) + 8(5) = 46$.
21. Since $R(x, y) = 11x + 20y$,
$R(8, 0) = 11(8) + 20(0) = 88$.
23. Since $C(x, y) = 5x + 12y$,
$C(4, 9) = 5(4) + 12(9) = 128$.
25. Evaluate $P(x, y) = 2x + 3y$ at each vertex:
$P(0, 3) = 9 \quad P(1, 2) = 6$
$P(2, 0) = 4 \quad P(0, 0) = 0$
So the maximum value is 9.
27. Evaluate $R(x, y) = 9x + 8y$ at each vertex:
$R(0, 5) = 40 \quad R(3, 4) = 59$
$R(5, 0) = 45 \quad R(0, 0) = 0$
So the maximum value is 59.
29. Evaluate $C(x, y) = 11x + 10y$ at each vertex:
$C(0, 3) = 30 \quad C(1, 1) = 21$
$C(2, 0) = 22$
So the minimum value is 21.
31. Evaluate $A(x, y) = 9x + 3y$ at each vertex:
$A(0, 6) = 18 \quad A(1, 3) = 18$
$A(4, 0) = 36$
So the minimum value is 18.
33. a) $A(0, 0) = 0$
$A(0, 80) = 9000(0) + 4000(80) = 320{,}000$
$A(50, 0) = 9000(50) + 4000(0) = 450{,}000$
$A(30, 60) = 9000(30) + 4000(60) = 510{,}000$
b) To maximize the audience they should use 30 TV ads and 60 radio ads.

35. Let $x =$ the number of doubles and $y =$ the number of triples. We can write four inequalities.
$x \ge 0, \quad y \ge 0$
$2x + 3y \le 24$
$6x + 3y \le 48$
Graph the system of inequalities. The vertices of the region are (0, 8), (6, 4), (8, 0), and (0, 0). Since Doubles are \$1.20 each and Triples are \$1.50 each, the revenue in dollars from x Doubles and y Triples is given by
$R(x, y) = 1.2x + 1.5y$.
Find the revenue at each vertex.
$R(0, 0) = 0$
$R(0, 8) = 1.2(0) + 1.5(8) = \$12.00$
$R(6, 4) = 1.2(6) + 1.5(4) = \$13.20$
$R(8, 0) = 1.2(8) + 1.5(0) = \$9.60$
The maximum revenue occurs at (6, 4). So they should make 6 Doubles and 4 Triples to maximize their revenue.
37. Let $x =$ the number of Doubles and $y =$ the number of Triples. The constraints and the graph are the same as in Exercise 19. Since the Doubles now sell for \$1.00 each and the Triples now sell for \$2.00 each, the revenue in dollars is given by the function is $R(x, y) = x + 2y$. Find the revenue at each vertex.
$R(0, 0) = 0$
$R(0, 8) = 0 + 2(8) = \$16.00$
$R(6, 4) = 6 + 2(4) = \$14.00$
$R(8, 0) = 8 + 2(0) = \$8.00$
The maximum revenue occurs at (0, 8). So to maximize the revenue they should make no Doubles and 8 Triples.
39. Let $x =$ the number of cups of Doggie Dinner, and $y =$ the number of cups of Puppie Power. We can write four inequalities.
$x \ge 0, \quad y \ge 0$
$20x + 30y \ge 200$
$40x + 20y \ge 180$
Graph the region that satisfies the constraints. Since one cup of Doggie Dinner costs 16 cents and one cup of Puppie Power costs 20 cents, the total cost in dollars for x cups of DD and y cups of PP is $C(x, y) = 0.16x + 0.20y$.

Find the cost at each vertex.

$C(0, 9) = 0.16(0) + 0.20(9) = \$1.80$
$C(1.75, 5.5) = 0.16(1.75) + 0.20(5.5) = \$1.38$
$C(10, 0) = 0.16(10) + 0.20(0) = \$1.60$

The minimum cost occurs at (1.75, 5.5). To minimize the cost and satisfy the constraints she should use 1.75 cups of DD and 5.5 cups of PP.

41. If the cost of one cup of DD is 4 cents and one cup of PP is 10 cents then the total cost in dollars of x cups of DD and y cups of PP is $C(x, y) = 0.04x + 0.10y$. The inequalities and the region are the same as in Exercise 39. Find the cost at each vertex of the region.

$C(0, 9) = 0.04(0) + 0.10(9) = \$0.90$
$C(1.75, 5.5) = 0.04(1.75) + 0.10(5.5) = \$0.62$
$C(10, 0) = 0.04(10) + 0.10(0) = \$0.40$

The minimum cost occurs at (10, 0). She should use 10 cups of DD and 0 cups of PP to satisfy the constraints and minimize the cost.

43. Let $x =$ the amount invested in the laundromat and $y =$ the amount invested in the car wash. We can write four inequalities.
$x \geq 0, \quad y \geq 0$
$x + y \leq 24{,}000$
$2x \leq y \leq 3x$

Graph the region that satisfies the system of inequalities. Since she makes 18% on the money invested in the laundry and 12% on the money invested in the car wash, the income in dollars is given by the function
$I(x, y) = 0.18x + 0.12y$.
Find the income at each vertex.

$I(0, 0) = 0$
$I(8000, 16000) = 0.18(8000) + 0.12(16000)$
$= \$3{,}360$
$I(6000, 18000) = 0.18(6000) + 0.12(18000)$
$= \$3{,}240$

The maximum income occurs at (8000, 16000). So she should invest \$8000 in the laundromat and \$16,000 in the car wash.

Chapter 4 Wrap-Up

Enriching Your Mathematical Word Power

1. c **2.** a **3.** a **4.** d **5.** b **6.** c
7. a **8.** c **9.** d **10.** b **11.** a **12.** d
13. a **14.** b

CHAPTER 4 REVIEW

1. The graph of $y = 2x - 1$ has y-intercept $(0, -1)$ and slope 2. the graph of $y = -x + 2$ has y-intercept $(0, 2)$ and slope -1. The graphs appear to intersect at (1, 1). Check that (1, 1) satisfies both equations. The solution set is $\{(1, 1)\}$ and the system is independent.

3. Solve $x + 2y = 4$ for y to get $y = -\frac{1}{2}x + 2$, which is the same as the second equation. So the two equations have the same graph and all points on that graph satisfy the system. The solution set to this dependent system is $\{(x, y) | x + 2y = 4\}$.

5. The graph of $y = -x$ is a line with slope -1 and y-intercept $(0, 0)$. The graph of $y = -x + 3$ is a line with slope -1 and y-intercept $(0, 3)$. Since these lines are parallel, there is no solution to the system. The system is inconsistent and the solution set is $\emptyset$.

7. Substitute $y = 3x + 11$ into $2x + 3y = 0$.
$2x + 3(3x + 11) = 0$
$11x + 33 = 0$
$11x = -33$
$x = -3$
$y = 3(-3) + 11 = 2$
The solution set is $\{(-3, 2)\}$ and the system is independent.

9. Substitute $x = y + 5$ into $2x - 2y = 12$.
$2(y + 5) - 2y = 12$
$2y + 10 - 2y = 12$
$10 = 12$
The solution set is $\emptyset$ and the system is inconsistent.

11. Write $2x - y = 3$ as $y = 2x - 3$.
Substitute into $6x - 9 = 3y$.
$6x - 9 = 3(2x - 3)$
$6x - 9 = 6x - 9$

Since the last equation is an identity, the solution set is $\{(x, y) \mid 2x - y = 3\}$. The system is dependent.

13. Substitute $y = \frac{1}{2}x - 3$ into $y = \frac{1}{3}x + 2$.

$$\frac{1}{3}x + 2 = \frac{1}{2}x - 3$$
$$2x + 12 = 3x - 18$$
$$30 = x$$

$$y = \frac{1}{3}x + 2$$
$$y = \frac{1}{3} \cdot 30 + 2 = 12$$

The solution set to this independent system is $\{(30, 12)\}$.

15. Substitute $x = 1 - 2y$ into $8x + 6y = 4$:

$$8(1 - 2y) + 6y = 4$$
$$8 - 16y + 6y = 4$$
$$-10y = -4$$
$$y = \frac{2}{5}$$

$$x = 1 - 2y$$
$$x = 1 - 2 \cdot \frac{2}{5} = \frac{1}{5}$$

The solution set to this independent system is $\left\{\left(\frac{1}{5}, \frac{2}{5}\right)\right\}$.

17. Multiply the first equation by 2 and the second by 3, and then add the resulting equations.

$$\begin{array}{r} 10x - 6y = -40 \\ \underline{9x + 6y = 21} \\ 19x \qquad = -19 \end{array}$$
$$x = -1$$

Use $x = -1$ in $3x + 2y = 7$.

$$3(-1) + 2y = 7$$
$$2y = 10$$
$$y = 5$$

The solution set is $\{(-1, 5)\}$ and the system is independent.

19. Rewrite the first equation.

$$2(y - 5) + 4 = 3(x - 6)$$
$$2y - 10 + 4 = 3x - 18$$
$$-3x + 2y = -12$$

Add this last equation to the original second equation.

$$\begin{array}{r} -3x + 2y = -12 \\ \underline{3x - 2y = 12} \\ 0 = 0 \end{array}$$

The two equations are just different forms of an equation for the same straight line. The solution set is $\{(x, y) \mid 3x - 2y = 12\}$ and the system is dependent.

21. Rewrite the first equation in the same form as the last so that they can be added.

$$3x - 4(y - 5) = x + 2$$
$$3x - 4y + 20 = x + 2$$
$$2x - 4y = -18$$
$$x - 2y = -9$$
$$-2y + x = -9$$

Add this last equation with the second equation.

$$\begin{array}{r} -2y + x = -9 \\ \underline{2y - x = 7} \\ 0 = -2 \end{array}$$

Since the result of the addition is a false statement, the system is inconsistent. The solution set is $\emptyset$.

23. Multiply the first equation by 8 and the second equation by 2 to clear the fractions:

$$2x + 3y = 3$$
$$5x - 12y = 14$$

Multiply the first equation by 4 and add:

$$\begin{array}{r} 8x + 12y = 12 \\ \underline{5x - 12y = 14} \\ 13x \qquad = 26 \end{array}$$
$$x = 2$$

Use $x = 2$ in $2x + 3y = 3$ to find y.

$$2(2) + 3y = 3$$
$$4 + 3y = 3$$
$$3y = -1$$
$$y = -\frac{1}{3}$$

The solution set is $\left\{\left(2, -\frac{1}{3}\right)\right\}$. The system is independent.

25. Multiply the first equation by 100 and the second equation by 100 to clear the decimals:

$$40x + 6y = 1160$$
$$80x - 5y = 1300$$

Multiply the first equation by -2 and add:

$$\begin{array}{r} -80x - 12y = -2320 \\ \underline{80x - 5y = 1300} \\ -17y = -1020 \end{array}$$
$$y = 60$$

Use $y = 60$ in $40x + 6y = 1160$ to find x.

$$40x + 6(60) = 1160$$
$$40x = 800$$
$$x = 20$$

The solution set is $\{(20, 60)\}$. The system is independent.

27. Add the first and second equations.

$$\begin{array}{r} x - y + z = 4 \\ -x + 2y - z = 0 \\ \hline y \qquad = 4 \end{array}$$

Add the first and third equations:

$$\begin{array}{r} x - y + z = 4 \\ -x + y - 3z = -16 \\ \hline -2z = -12 \\ z = 6 \end{array}$$

Use $y = 4$ and $z = 6$ in $x - y + z = 4$.

$$x - 4 + 6 = 4$$
$$x = 2$$

The solution set is $\{(2, 4, 6)\}$.

29. Add the first and second equations.

$$\begin{array}{l} 2x - y - z = 3 \\ 3x + y + 2z = 4 \\ \hline 5x \qquad + z = 7 \quad \text{A} \end{array}$$

Multiply the first equation by 2 and add the result to the last equation.

$$\begin{array}{l} 4x - 2y - 2z = 6 \\ 4x + 2y - \ z = -4 \\ \hline 8x \qquad - 3z = 2 \quad \text{B} \end{array}$$

Multiply equation A by 3 and add the result to equation B.

$$\begin{array}{l} 15x + 3z = 21 \\ 8x - 3z = 2 \\ \hline 23x \qquad = 23 \\ \qquad x = 1 \end{array}$$

Use $x = 1$ in $8x - 3z = 2$.

$$8(1) - 3z = 2$$
$$-3z = -6$$
$$z = 2$$

Use $x = 1$ and $z = 2$ in $3x + y + 2z = 4$.

$$3(1) + y + 2(2) = 4$$
$$y = -3$$

The solution set is $\{(1, -3, 2)\}$.

31. Add the first two equations.

$$\begin{array}{r} x + y - z = 4 \\ y + z = 6 \\ \hline x + 2y \quad = 10 \end{array}$$

Substituting $x + 2y = 10$ into the third equation $x + 2y = 8$ yields $10 = 8$, which is false. The solution set is $\emptyset$.

33. If the first equation is multiplied by -1 the result is the second equation and if the first is multiplied by 2 the result is the third equation. So all three equations are equivalent. Any point that satisfies one of them satisfies all of them. The system is dependent. The solution set is $\{(x, y, z) \mid x - 2y + z = 8\}$.

35. $\begin{bmatrix} 1 & 1 & 7 \\ -1 & 2 & 5 \end{bmatrix}$

$\begin{bmatrix} 1 & 1 & 7 \\ 0 & 3 & 12 \end{bmatrix}$ $R_1 + R_2 \rightarrow R_2$

$\begin{bmatrix} 1 & 1 & 7 \\ 0 & 1 & 4 \end{bmatrix}$ $R_2 \div 3 \rightarrow R_2$

$\begin{bmatrix} 1 & 0 & 3 \\ 0 & 1 & 4 \end{bmatrix}$ $-1R_2 + R_1 \rightarrow R_1$

The solution set is $\{(3, 4)\}$.

37. $\begin{bmatrix} 1 & -3 & 14 \\ 2 & 1 & 0 \end{bmatrix}$

$\begin{bmatrix} 1 & -3 & 14 \\ 0 & 7 & -28 \end{bmatrix}$ $-2R_1 + R_2 \rightarrow R_2$

$\begin{bmatrix} 1 & -3 & 14 \\ 0 & 1 & -4 \end{bmatrix}$ $R_2 \div 7 \rightarrow R_2$

$\begin{bmatrix} 1 & 0 & 2 \\ 0 & 1 & -4 \end{bmatrix}$ $3R_2 + R_1 \rightarrow R_1$

The solution set is $\{(2, -4)\}$.

39. $\begin{bmatrix} 1 & 1 & -1 & 0 \\ 1 & -1 & 2 & 4 \\ 2 & 1 & -1 & 1 \end{bmatrix}$

$\begin{bmatrix} 1 & 1 & -1 & 0 \\ 0 & -2 & 3 & 4 \\ 0 & -1 & 1 & 1 \end{bmatrix}$ $-R_1 + R_2 \rightarrow R_2$

$-2R_1 + R_3 \rightarrow R_3$

$$\begin{bmatrix} 1 & 1 & -1 & 0 \\ 0 & -1 & 1 & 1 \\ 0 & -2 & 3 & 4 \end{bmatrix} \quad R_2 \leftrightarrow R_3$$

$$\begin{bmatrix} 1 & 1 & -1 & 0 \\ 0 & 1 & -1 & -1 \\ 0 & -2 & 3 & 4 \end{bmatrix} \quad -R_2 \to R_2$$

$$\begin{bmatrix} 1 & 0 & 0 & 1 \\ 0 & 1 & -1 & -1 \\ 0 & 0 & 1 & 2 \end{bmatrix} \quad \begin{matrix} -R_2 + R_1 \to R_1 \\ 2R_2 + R_3 \to R_3 \end{matrix}$$

$$\begin{bmatrix} 1 & 0 & 0 & 1 \\ 0 & 1 & 0 & 1 \\ 0 & 0 & 1 & 2 \end{bmatrix} \quad R_3 + R_2 \to R_2$$

The solution set is $\{(1, 1, 2)\}$.

41. $\begin{vmatrix} 1 & 3 \\ 0 & 2 \end{vmatrix} = 1 \cdot 2 - 0 \cdot 3 = 2$

43. $\begin{vmatrix} 0.01 & 0.02 \\ 50 & 80 \end{vmatrix} = 0.01(80) - 0.02(50)$
$= 0.8 - 1 = -0.2$

45. $D = \begin{vmatrix} 2 & -1 \\ 3 & 1 \end{vmatrix} = 5$

$$D_x = \begin{vmatrix} 0 & -1 \\ -5 & 1 \end{vmatrix} = -5$$

$$D_y = \begin{vmatrix} 2 & 0 \\ 3 & -5 \end{vmatrix} = -10$$

$$x = \frac{D_x}{D} = \frac{-5}{5} = -1$$

$$y = \frac{D_y}{D} = \frac{-10}{5} = -2$$

The solution set is $\{(-1, -2)\}$.

47. Write the system in standard form.
$-2x + y = -3$
$3x - 2y = 4$

$$D = \begin{vmatrix} -2 & 1 \\ 3 & -2 \end{vmatrix} = 1$$

$$D_x = \begin{vmatrix} -3 & 1 \\ 4 & -2 \end{vmatrix} = 2$$

$$D_y = \begin{vmatrix} -2 & -3 \\ 3 & 4 \end{vmatrix} = 1$$

$$x = \frac{D_x}{D} = \frac{2}{1} = 2 \qquad y = \frac{D_y}{D} = \frac{1}{1} = 1$$

The solution set is $\{(2, 1)\}$.

49. Expand by minors using the first column.

$$2\begin{vmatrix} 2 & 4 \\ 1 & 1 \end{vmatrix} - (-1)\begin{vmatrix} 3 & 1 \\ 1 & 1 \end{vmatrix} + 6\begin{vmatrix} 3 & 1 \\ 2 & 4 \end{vmatrix}$$
$$= 2(-2) + 2 + 6(10) = 58$$

51. Expand by minors using the second column.

$$-3\begin{vmatrix} 2 & 4 \\ -1 & 3 \end{vmatrix} + 0\begin{vmatrix} 2 & -2 \\ -1 & 3 \end{vmatrix} - 0\begin{vmatrix} 2 & -2 \\ 2 & 4 \end{vmatrix}$$
$$= -3(10) = -30$$

53. $D = \begin{vmatrix} 1 & 1 & 0 \\ 1 & 1 & 1 \\ 1 & -1 & -1 \end{vmatrix} = 2$

$$D_x = \begin{vmatrix} 3 & 1 & 0 \\ 0 & 1 & 1 \\ 2 & -1 & -1 \end{vmatrix} = 2$$

$$D_y = \begin{vmatrix} 1 & 3 & 0 \\ 1 & 0 & 1 \\ 1 & 2 & -1 \end{vmatrix} = 4$$

$$D_z = \begin{vmatrix} 1 & 1 & 3 \\ 1 & 1 & 0 \\ 1 & -1 & 2 \end{vmatrix} = -6$$

$$x = \frac{D_x}{D} = \frac{2}{2} = 1, \quad y = \frac{D_y}{D} = \frac{4}{2} = 2,$$

$$z = \frac{D_z}{D} = \frac{-6}{2} = -3$$

The solution set is $\{(1, 2, -3)\}$.

55. The graph of $x \geq 0$ consists of points on and to the right of the y-axis. The graph of $y \geq 0$ consists of points on and above the x-axis. The graph of $x + 2y \leq 6$ consists of points on and below the line $x + 2y = 6$. The graph of $x + y \leq 5$ consists of points on and below the line $x + y = 5$. Points that satisfy all 4 inequalities are shown in the following graph.

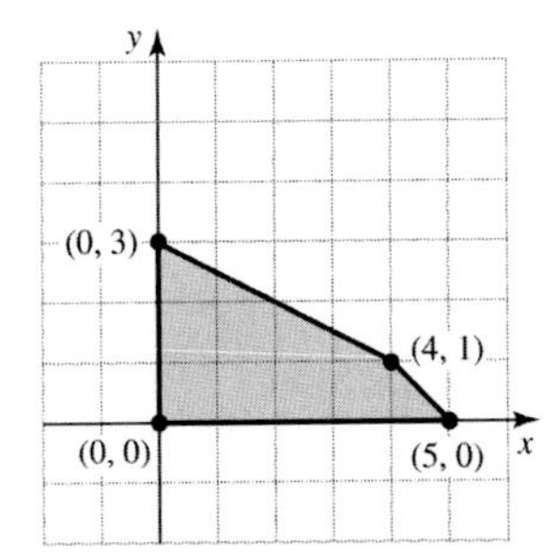

The vertices of the region are (0, 3), (4, 1), (5, 0), and (0, 0).

57. First graph the region that satisfies the four inequalities. The vertices of the region are $(0, 0)$, $(0, 3)$, $(2, 2)$, and $(3, 0)$.
Evaluate $R(x, y) = 6x + 9y$ at each vertex.
$R(0, 0) = 0$
$R(0, 3) = 6(0) + 9(3) = 27$
$R(2, 2) = 6(2) + 9(2) = 30$
$R(3, 0) = 6(3) + 9(0) = 18$
The maximum value of $R(x, y)$ is 30.

59. Let x represent the width and y represent the length, then $2x + 2y = 82$ and $y = x + 15$.

$$2x + 2(x + 15) = 82$$
$$4x + 30 = 82$$
$$4x = 52$$
$$x = 13$$
$$y = 13 + 15 = 28$$

The length is 28 feet and the width is 13 feet.

61. Let $x =$ the tens digit and $y =$ the ones digit. The value of the number is $10x + y$. If the digits are reversed the value will be $10y + x$.
We can write the following two equations.

$$x + y = 15$$
$$10x + y = 10y + x - 9$$

$$x + y = 15$$
$$9x - 9y = -9$$

$$x + y = 15$$
$$x - y = -1$$
$$2x = 14$$
$$x = 7$$
$$y = 8 \quad \text{Since } x + y = 15$$

The number is 78.

63. Let $b =$ his boat's speed in still water and $c =$ the speed of the current. Let $t =$ his time to go the same distance in the lake.

	Distance	Rate	Time
Down	$\frac{1}{2}(b + c)$	$b + c$	$\frac{1}{2}$
Back	$\frac{3}{4}(b - c)$	$b - c$	$\frac{3}{4}$
Lake	bt	b	t

Since the distance down the stream is the same as the distance going back, we can write the following equation.

$$\frac{1}{2}(b + c) = \frac{3}{4}(b - c)$$
$$\frac{1}{2}b + \frac{1}{2}c = \frac{3}{4}b - \frac{3}{4}c$$
$$2b + 2c = 3b - 3c$$
$$5c = b$$

Since $bt = \frac{1}{2}(b + c)$, we can substitute $b = 5c$ into this equation.

$$(5c)t = \frac{1}{2}(5c + c)$$
$$5ct = 3c$$
$$t = \frac{3c}{5c} = \frac{3}{5}$$

So the time in the lake is $3/5$ of an hour or 36 minutes.

65. Let $x =$ the number of liters of solution A, $y =$ the number of liters of solution B, and $z =$ the number of liters of solution C. We can write 3 equations.

$$x + y + z = 20$$
$$0.30x + 0.20y + 0.60z = 0.38(20)$$
$$z = 2x$$

Substitute $z = 2x$ into the other two equations.

$$x + y + 2x = 20$$
$$0.3x + 0.2y + 0.6(2x) = 7.6$$

$$3x + y = 20$$
$$1.5x + 0.2y = 7.6$$

Substitute $y = 20 - 3x$ into the second equation.

$$1.5x + 0.2(20 - 3x) = 7.6$$
$$0.9x + 4 = 7.6$$
$$0.9x = 3.6$$
$$x = 4$$

If $x = 4$, then $y = 20 - 3(4) = 8$ and $z = 2(4) = 8$. She should use 4 liters of 30% solution (A), 8 liters of 20% solution (B), and 8 liters of 60% solution (C).

67. Let $x =$ the number of servings of beets and $y =$ the number of servings of beans. We can write two equations.

$$x + 6y = 21$$
$$6x + 20y = 78$$

Substitute $x = 21 - 6y$ into $6x + 20y = 78$.

$$6(21 - 6y) + 20y = 78$$
$$126 - 36y + 20y = 78$$
$$-16y = -48$$
$$y = 3$$

If $y = 3$, then $x = 21 - 6(3) = 3$. It would take 3 servings of each.

Chapter 4 TEST

1. The graph of $y = -x + 4$ is a straight line with y-intercept (0, 4) and slope -1. The graph of $y = 2x + 1$ is a straight line with y-intercept (0, 1) and slope 2. The graphs appear to intersect at (1, 3). After checking that (1, 3) satisfies both equations, we can be sure that the solution set is $\{(1, 3)\}$.

2. Substitute $y = 2x - 8$ into $4x + 3y = 1$.

$$4x + 3(2x - 8) = 1$$
$$10x = 25$$
$$x = \frac{5}{2}$$

If $x = 5/2$, then $y = 2(5/2) - 8 = -3$. The solution set is $\left\{\left(\frac{5}{2}, -3\right)\right\}$.

3. Substitute $y = x - 5$ into the second equation.

$$3x - 4(y - 2) = 28 - x$$
$$3x - 4(x - 5 - 2) = 28 - x$$
$$3x - 4(x - 7) = 28 - x$$
$$28 - x = 28 - x$$

Since the last equation is an identity, the solution set is $\{(x, y) \mid y = x - 5\}$.

4. Multiply the first equation by 3 and the second by 2 and add the results.

$$9x + 6y = 9$$
$$8x - 6y = -26$$
$$17x \quad = -17$$
$$x = -1$$

Use $x = -1$ in $3x + 2y = 3$ to find y.

$$3(-1) + 2y = 3$$
$$2y = 6$$
$$y = 3$$

The solution set is $\{(-1, 3)\}$.

5. Multiply the first equation by 2 and then add the result to the second equation.

$$6x - 2y = 10$$
$$-6x + 2y = 1$$
$$0 = 11$$

The solution set to the system is $\emptyset$.

6. The lines both have slope 3 and different y-intercepts, so they are parallel. There is no solution to the system. The system is inconsistent.

7. If we multiply the second equation by 2, the result is the same as the first equation. These are two equations for the same straight line. The system is dependent.

8. The lines have different slopes and different y-intercepts. They will intersect in exactly one point. The system is independent.

9. Add the first two equations.

$$x + y - z = 2$$
$$2x - y + 3z = -5$$
$$3x \quad + 2z = -3 \quad \text{A}$$

Multiply the first equation by 3 and add the result to the last equation.

$$3x + 3y - 3z = 6$$
$$x - 3y + z = 4$$
$$4x \quad - 2z = 10 \quad \text{B}$$

Add equations A and B.

$$3x + 2z = -3$$
$$4x - 2z = 10$$
$$7x \quad = 7$$
$$x = 1$$

Use $x = 1$ in $3x + 2z = -3$.

$$3(1) + 2z = -3$$
$$2z = -6$$
$$z = -3$$

Use $x = 1$ and $z = -3$ in $x + y - z = 2$.

$1 + y - (-3) = 2$

$y = -2$

The solution set is $\{(1, -2, -3)\}$.

10. $\begin{bmatrix} 1 & 2 & 12 \\ 3 & -1 & 1 \end{bmatrix}$

$\begin{bmatrix} 1 & 2 & 12 \\ 0 & -7 & -35 \end{bmatrix}$ $-3R_1 + R_2 \rightarrow R_2$

$\begin{bmatrix} 1 & 2 & 12 \\ 0 & 1 & 5 \end{bmatrix}$ $R_2 \div (-7) \rightarrow R_2$

$\begin{bmatrix} 1 & 0 & 2 \\ 0 & 1 & 5 \end{bmatrix}$ $-2R_2 + R_1 \rightarrow R_1$

The solution set is $\{(2, 5)\}$.

11. $\begin{bmatrix} 1 & -1 & -1 & 1 \\ -1 & -1 & 2 & -2 \\ -1 & -3 & 1 & -5 \end{bmatrix}$

$\begin{bmatrix} 1 & -1 & -1 & 1 \\ 0 & -2 & 1 & -1 \\ 0 & -4 & 0 & -4 \end{bmatrix}$ $R_1 + R_2 \rightarrow R_2$

$R_1 + R_3 \rightarrow R_3$

$\begin{bmatrix} 1 & -1 & -1 & 1 \\ 0 & -4 & 0 & -4 \\ 0 & -2 & 1 & -1 \end{bmatrix}$ $R_2 \leftrightarrow R_3$

$\begin{bmatrix} 1 & -1 & -1 & 1 \\ 0 & 1 & 0 & 1 \\ 0 & -2 & 1 & -1 \end{bmatrix}$ $R_2 \div (-4) \rightarrow R_2$

$\begin{bmatrix} 1 & 0 & -1 & 2 \\ 0 & 1 & 0 & 1 \\ 0 & 0 & 1 & 1 \end{bmatrix}$ $R_2 + R_1 \rightarrow R_1$

$2R_2 + R_3 \rightarrow R_3$

$\begin{bmatrix} 1 & 0 & 0 & 3 \\ 0 & 1 & 0 & 1 \\ 0 & 0 & 1 & 1 \end{bmatrix}$ $R_3 + R_1 \rightarrow R_1$

The solution set is $\{(3, 1, 1)\}$.

12. $\begin{vmatrix} 2 & 3 \\ 4 & -3 \end{vmatrix} = 2(-3) - 4(3) = -18$

13. Expand by minors using the third column.

$-1\begin{vmatrix} 2 & 3 \\ 1 & 1 \end{vmatrix} - 1\begin{vmatrix} 1 & -2 \\ 1 & 1 \end{vmatrix} + 0\begin{vmatrix} 1 & -2 \\ 2 & 3 \end{vmatrix}$

$= -1(-1) - 1(3) + 0(7) = -2$

14. $D = \begin{vmatrix} 2 & -1 \\ 3 & 1 \end{vmatrix} = 5$

$D_x = \begin{vmatrix} -4 & -1 \\ -1 & 1 \end{vmatrix} = -5$

$D_y = \begin{vmatrix} 2 & -4 \\ 3 & -1 \end{vmatrix} = 10$

$x = \frac{D_x}{D} = \frac{-5}{5} = -1 \quad y = \frac{D_y}{D} = \frac{10}{5} = 2$

The solution set is $\{(-1, 2)\}$.

15. $D = \begin{vmatrix} 1 & 1 & 0 \\ 1 & -1 & 2 \\ 2 & 1 & -1 \end{vmatrix} = 4$

$D_x = \begin{vmatrix} 0 & 1 & 0 \\ 6 & -1 & 2 \\ 1 & 1 & -1 \end{vmatrix} = 8$

$D_y = \begin{vmatrix} 1 & 0 & 0 \\ 1 & 6 & 2 \\ 2 & 1 & -1 \end{vmatrix} = -8$

$D_z = \begin{vmatrix} 1 & 1 & 0 \\ 1 & -1 & 6 \\ 2 & 1 & 1 \end{vmatrix} = 4$

$x = \frac{D_x}{D} = \frac{8}{4} = 2, \quad y = \frac{D_y}{D} = \frac{-8}{4} = -2,$

$z = \frac{D_z}{D} = \frac{4}{4} = 1$ The solution set is $\{(2, -2, 1)\}$.

16. Let x = the price of a single and y = the price of a double. We can write an equation for each night.

$$5x + 12y = 390$$
$$9x + 10y = 412$$

Multiply the first equation by -9 and the second by 5, then add the results.

$$\begin{aligned} -45x - 108y &= -3510 \\ 45x + 50y &= 2060 \\ \hline -58y &= -1450 \\ y &= 25 \end{aligned}$$

Use $y = 25$ in $5x + 12y = 390$ to find x.

$$\begin{aligned} 5x + 12(25) &= 390 \\ 5x + 300 &= 390 \\ 5x &= 90 \\ x &= 18 \end{aligned}$$

So singles rent for \$18 per night and doubles rent for \$25 per night.

17. Let $x =$ Jill's study time, $y =$ Karen's study time, and $z =$ Betsy's study time. We can write three equations.

$$\begin{aligned} x + y + z &= 93 \\ x + y &= \tfrac{1}{2}z \\ x &= y + 3 \end{aligned}$$

Rewrite the equations as follows.

$$\begin{aligned} x + y + z &= 93 \\ 2x + 2y - z &= 0 \\ x - y \quad &= 3 \end{aligned}$$

Adding the first two equations to eliminate z gives us $3x + 3y = 93$, or $x + y = 31$. Add this result to the third equation.

$$\begin{aligned} x + y &= 31 \\ x - y &= 3 \\ \hline 2x \quad &= 34 \\ x &= 17 \end{aligned}$$

Since $x + y = 31$, we get $y = 14$. Use $x = 17$ and $y = 14$ in the first equation.

$$\begin{aligned} 17 + 14 + z &= 93 \\ z &= 62 \end{aligned}$$

So Jill studied 17 hours, Karen studied 14 hours, and Betsy studied 62 hours.

18. First graph the system of inequalities. The vertices of the region are $(0, 0)$, $(0, 4)$, $(3, 2)$, and $(5, 0)$. Evaluate the function $P(x, y) = 8x + 10y$ at each vertex.

$$\begin{aligned} P(0, 0) &= 0 \\ P(0, 4) &= 8(0) + 10(4) = 40 \\ P(3, 2) &= 8(3) + 10(2) = 44 \\ P(5, 0) &= 8(5) + 10(0) = 40 \end{aligned}$$

The maximum value of the function $P(x, y)$ is 44.

Making Connections

Chapters 1 - 4

1. $-3^4 = -(3^4) = -81$

2. $\frac{1}{3}(3) + 6 = 1 + 6 = 7$

3. $(-5)^2 - 4(-2)(6) = 25 - (-48) = 73$

4. $6 - (0.2)(0.3) = 6 - 0.06 = 5.94$

5. $5(t - 3) - 6(t - 2) = 5t - 15 - 6t + 12$
$= -t - 3$

6.
$0.1(x - 1) - (x - 1) = 0.1x - 0.1 - x + 1$
$= -0.9x + 0.9$

7. $\dfrac{-9x^2 - 6x + 3}{-3} = 3x^2 + 2x - 1$

8. $\dfrac{4y - 6}{2} - \dfrac{3y - 9}{3} = 2y - 3 - (y - 3)$
$= y$

9.
$$\begin{aligned} 3x - 5y &= 7 \\ -5y &= -3x + 7 \\ y &= \tfrac{3}{5}x - \tfrac{7}{5} \end{aligned}$$

10.
$$\begin{aligned} Cx - Dy &= W \\ -Dy &= -Cx + W \\ y &= \tfrac{C}{D}x - \tfrac{W}{D} \end{aligned}$$

11.
$$\begin{aligned} Cy &= Wy - K \\ K &= Wy - Cy \\ K &= y(W - C) \\ y &= \frac{K}{W - C} \end{aligned}$$

12.
$$\begin{aligned} A &= \tfrac{1}{2}b(w - y) \\ 2A &= bw - by \\ by &= bw - 2A \\ y &= \frac{bw - 2A}{b} \end{aligned}$$

13. $2x + 3(x - 5) = 5$
$$5x - 15 = 5$$
$$5x = 20$$
$$x = 4$$
$$y = x - 5 = 4 - 5 = -1$$
The solution set is $\{(4, -1)\}$.

14. $y = 1200 - x$
$$0.05x + 0.06(1200 - x) = 67$$
$$-0.01x + 72 = 67$$
$$-0.01x = -5$$
$$x = 500$$
$$y = 1200 - x = 1200 - 500 = 700$$
The solution set is $\{(500, 700)\}$.

15. $x = 5y - 17$
$$3(5y - 17) - 15y = -51$$
$$15y - 51 - 15y = -51$$
$$-51 = -51$$
The equations are equivalent and the solution set is $\{(x, y) \mid x + 17 = 5y\}$.

16. Multiply the first equation by 100 to get $7a + 30b = 670$. This equation is inconsistent with the equation $7a + 30b = 67$. So the solution set is the empty set $\emptyset$.

17. The slope of the line is $55/99$ or $5/9$. Since the y-intercept is $(0, 55)$, the equation is $y = \frac{5}{9}x + 55$.

18. The slope of the line through $(2, -3)$ and $(-4, 8)$ is $11/-6$ or $-11/6$. Use the point-slope formula:
$$y - 8 = -\frac{11}{6}(x - (-4))$$
$$y - 8 = -\frac{11}{6}x - \frac{44}{6}$$
$$y = -\frac{11}{6}x - \frac{22}{3} + \frac{24}{3}$$
$$y = -\frac{11}{6}x + \frac{2}{3}$$

19. Any line parallel to $y = 5x$ has slope 5.
$$y - 6 = 5(x - (-4))$$
$$y - 6 = 5x + 20$$
$$y = 5x + 26$$

20. Any line perpendicular to $y = -2x + 1$ has slope $1/2$.
$$y - 7 = \frac{1}{2}(x - 4)$$
$$y - 7 = \frac{1}{2}x - 2$$
$$y = \frac{1}{2}x + 5$$

21. Any line parallel to the x-axis has slope 0 and is of the form $y = k$. Since it must go through $(3, 5)$ it is $y = 5$.

22. Any line perpendicular to the x-axis is of the form $x = k$. So the equation is $x = -7$.

23. a) From the y-intercept we see that the purchase price for A is \$4000 and for B is \$2000. So A has the larger purchase price.

b) Machine B makes 300,000 copies for \$12,000, or \$0.04 per copy. Machine A makes 300,000 copies for \$9,000 or \$0.03 per copy.

c) The slope of the line for machine B is 0.04 and the slope of the line for machine A is 0.03, which is the per copy cost for each machine.

d) Machine B: $y = 0.04x + 2000$
Machine A: $y = 0.03x + 4000$

e) $0.04x + 2000 = 0.03x + 4000$
$$0.01x = 2000$$
$$x = 200{,}000$$
The total costs are equal at 200,000 copies.

5.1 WARM-UPS

1. True, because $3^5 \cdot 3^4 = 3^{5+4} = 3^9$.
2. False, because $2x^{-4} = 2 \cdot \frac{1}{x^4} = \frac{2}{x^4}$.
3. False, because $10^{-3} = \frac{1}{1000} = 0.001$.
4. False, because $\frac{x^5}{x^{-2}} = x^{5-(-2)} = x^7$.
5. True, because $\frac{2^5}{2^{-2}} = 2^{5-(-2)} = 2^7$.
6. False, because $2^3 \cdot 5^2 = 8 \cdot 25 = 200$.
7. True, because $-2^{-2} = -\frac{1}{2^2} = -\frac{1}{4}$.
8. True, $46.7 \times 10^5 = 4.67 \times 10^1 \times 10^5$.
9. True, $0.512 \times 10^{-3} = 5.12 \times 10^{-1} \times 10^{-3}$.
10. False, because $\frac{8 \times 10^{30}}{2 \times 10^{-5}} = 4 \times 10^{35}$.

5.1 EXERCISES

1. An exponential expression is an expression of the form a^n.
3. The product rule says that $a^m a^n = a^{m+n}$.
5. To convert a number in scientific notation to standard notation move the decimal point n places to the left if the exponent on 10 is $-n$ or move the decimal point n places to the right if the exponent on 10 is n, assuming that n is a positive integer.
7. $2^2 = 4, -2^2 = -(2^2) = -4, (-2)^2 = 4,$
$2^{-2} = \frac{1}{2^2} = \frac{1}{4}, -2^{-2} = -\frac{1}{2^2} = -\frac{1}{4},$
$(-2)^{-2} = \frac{1}{(-2)^2} = \frac{1}{4}$
9. $2^3 = 8, -2^3 = -(2^3) = -8, (-2)^3 = -8,$
$2^{-3} = \frac{1}{2^3} = \frac{1}{8}, -2^{-3} = -\frac{1}{2^3} = -\frac{1}{8},$
$(-2)^{-3} = \frac{1}{(-2)^3} = -\frac{1}{8}$
11. $\frac{1}{5^2} = \frac{1}{25}, \frac{1}{-5^2} = -\frac{1}{25},$
$\frac{1}{(-5)^2} = \frac{1}{25}, \frac{1}{5^{-2}} = 5^2 = 25,$
$\frac{1}{-5^{-2}} = -5^2 = -25, \frac{1}{(-5)^{-2}} = (-5)^2 = 25$
13. $7^1 = 7, -7^1 = -7, (-7)^1 = -7,$
$7^{-1} = \frac{1}{7}, -7^{-1} = -\frac{1}{7^1} = -\frac{1}{7},$
$(-7)^{-1} = \frac{1}{(-7)^1} = -\frac{1}{7}$
15. $\left(\frac{1}{2}\right)^2 = \frac{1}{4}, \left(-\frac{1}{2}\right)^2 = \frac{1}{4},$
$-\left(\frac{1}{2}\right)^2 = -\frac{1}{4}, \left(\frac{1}{2}\right)^{-2} = \left(\frac{2}{1}\right)^2 = 4,$
$\left(-\frac{1}{2}\right)^{-2} = \left(-\frac{2}{1}\right)^2 = 4,$
$-\left(\frac{1}{2}\right)^{-2} = -\left(\frac{2}{1}\right)^2 = -4$
17. $\left(\frac{2}{3}\right)^3 = \frac{8}{27}, \left(-\frac{2}{3}\right)^3 = -\frac{8}{27},$
$-\left(\frac{2}{3}\right)^3 = -\frac{8}{27}, \left(\frac{2}{3}\right)^{-3} = \left(\frac{3}{2}\right)^3 = \frac{27}{8},$
$\left(-\frac{2}{3}\right)^{-3} = \left(-\frac{3}{2}\right)^3 = -\frac{27}{8},$
$-\left(\frac{2}{3}\right)^{-3} = -\left(\frac{3}{2}\right)^3 = -\frac{27}{8}$
19. $2^5 \cdot 2^{12} = 2^{5+12} = 2^{17}$
21. $-2x^{-7} \cdot 3x = -6x^{-7+1} = -6x^{-6} = -\frac{6}{x^6}$
23. $-7b^{-7}(-3b^{-3}) = 21b^{-10} = \frac{21}{b^{10}}$
25. $3^0 = 1, -3^0 = -(3^0) = -1, (-3)^0 = 1,$
$-(-3)^0 = -1$
27. $(2+3)^0 = 5^0 = 1,$
$2^0 + 3^0 = 1 + 1 = 2,$
$(2^0 + 3)^0 = (1+3)^0 = 1$
29. $3st^0 = 3s \cdot 1 = 3s,$
$3(st)^0 = 3 \cdot 1 = 3, (3st)^0 = 1$
31. $2w^{-3}(w^7 \cdot w^{-4}) = 2w^0 = 2$
33. $\frac{2}{4^{-2}} = 2 \cdot 4^2 = 2 \cdot 16 = 32$
35. $\frac{3^{-1}}{10^{-2}} = \frac{10^2}{3^1} = \frac{100}{3}$
37. $\frac{2x^{-3}(4x)}{5y^{-2}} = \frac{8x^{-2}}{5y^{-2}} = \frac{8y^2}{5x^2}$
39. $\frac{4^{-2}x^3x^{-6}}{3x^{-3}x^2} = \frac{x^{-3}}{16 \cdot 3x^{-1}} = \frac{x^{-2}}{48} = \frac{1}{48x^2}$
41. $x^{5-3} = x^2$
43. $3^{6-(-2)} = 3^8$
45. $\frac{4a^{-5-(-2)}}{12} = \frac{a^{-3}}{3} = \frac{1}{3a^3}$
47. $\frac{-6w^{-5}}{2w^3} = -3w^{-5-3} = -3w^{-8} = \frac{-3}{w^8}$
49. $\frac{3^3w^{-2}w^5}{3^{-5}w^{-3}} = \frac{3^3w^3}{3^{-5}w^{-3}} = 3^8w^6$
51. $\frac{3x^{-6} \cdot x^2y^{-1}}{6x^{-5}y^{-2}} = \frac{3x^{-4}y}{6x^{-5}}$
$= \frac{x^{-4-(-5)}y}{2} = \frac{xy}{2}$
53. $3^{-1} \cdot \left(\frac{1}{3}\right)^{-3} = \frac{1}{3} \cdot 3^3 = 9$
55. $-2^4 + \left(\frac{1}{2}\right)^{-1} = -16 + 2 = -14$
57. $-(-2)^{-3} \cdot 2^{-1} = -\left(-\frac{1}{8}\right)\frac{1}{2} = \frac{1}{16}$
59. $7 \cdot 2^{-3} - 2 \cdot 4^{-1} = \frac{7}{8} - \frac{1}{2} = \frac{3}{8}$

61. $(1+2^{-1})^{-2} = \left(1+\frac{1}{2}\right)^{-2}$
$= \left(\frac{3}{2}\right)^{-2} = \left(\frac{2}{3}\right)^{2} = \frac{4}{9}$

63. $2x^2 \cdot 5x^{-5} = 10x^{-3} = \frac{10}{x^3}$

65. $\frac{-3a^5(-2a^{-1})}{6a^3} = \frac{6a^4}{6a^3} = a$

67. $\frac{(-3x^3y^2)(-2xy^{-3})}{-9x^2y^{-5}} = \frac{6x^4y^{-1}}{-9x^2y^{-5}}$
$= \frac{-2x^2y^4}{3}$

69. $8 = 2^3$

71. $\frac{1}{4} = 2^{-2}$

73. $16 = \left(\frac{1}{2}\right)^{-4}$

75. $10^{-3} = 0.001$

77. Since the exponent in 4.86×10^8 is positive, we move the decimal point 8 places to the right: 486,000,000

79. Since the exponent in 2.37×10^{-6} is negative, we move the decimal point 6 places to the left: 0.00000237

81. Since the exponent in 4×10^6 is positive, we move the decimal point 6 places to the right: 4,000,000

83. Since the exponent in 5×10^{-6} is negative, we move the decimal point 6 places to the left: 0.000005

85. The decimal point in 320,000 must be moved 5 places to the left for scientific notation. Since the original number is larger than 10 the exponent is positive: 3.2×10^5

87. The decimal point in 0.00000071 must be moved 7 places to the right to get scientific notation. Since the original number is smaller than 1, the exponent is negative: 7.1×10^{-7}

89. Move the decimal point 5 places to the right: 7.03×10^{-5}

91. $205 \times 10^5 = 2.05 \times 10^2 \times 10^5$
$= 2.05 \times 10^7$

93. $(4000)(5000)(0.0003)$
$= 4 \times 10^3 \cdot 5 \times 10^3 \cdot 3 \times 10^{-4} = 60 \times 10^2$
$= 6 \times 10^1 \times 10^2 = 6 \times 10^3$

95. $\frac{(5{,}000{,}000)(0.0003)}{2000}$
$= \frac{5 \times 10^6 \cdot 3 \times 10^{-4}}{2 \times 10^3} = \frac{7.5 \times 10^2}{10^3}$
$= 7.5 \times 10^{-1}$

97. By the quotient rule, we subtract exponents when we divide exponential expressions: $3 \times 10^{40-18} = 3 \times 10^{22}$

99. $\frac{(-4 \times 10^5)(6 \times 10^{-9})}{2 \times 10^{-16}}$
$= -12 \times 10^{5+(-9)-(-16)}$
$= -12 \times 10^{12}$
$= -1.2 \times 10^1 \times 10^{12} = -1.2 \times 10^{13}$

101. Enter the numbers into a calculator using scientific notation. The calculator will give answers to the computations in scientific notation. 1.578×10^5

103. Use a calculator to get 9.187×10^{-5}.

105. Use a calculator to get 3.828×10^{30}.

107. $9.3 \times 10^7 \text{miles} \times \frac{5280 \text{ feet}}{1 \text{ mile}}$
$= 4.910 \times 10^{11}$ ft

109. Since $T = \frac{D}{R}$, $T = \frac{4.6 \times 10^{12} \text{ km}}{1.2 \times 10^5 \text{ kps}}$
$\approx 3.833 \times 10^7$ seconds

111. $\frac{8.71 \times 10^7 \text{ ton} \times \frac{2000 \text{ lb}}{1 \text{ ton}}}{1.80863 \times 10^8 \text{ people}} \div 365 \text{ day} \approx$
2.639 lb/person/day

113. a) The only integers for which $2^m \cdot 3^n = 6^{m+n}$ is correct are $m = 0$ and $n = 0$. There are other values that would work, but they are not integers and it would require logarithms to find them.
b) If m and n are nonzero integers, then $2^m \cdot 3^n \neq 6^{m+n}$.

115. Suppose that a is *positive.* The value of $-a^n$ is always negative because $-a^n = -(a^n)$. The value of $(-a)^n$ is negative and equal to $-a^n$ when n is odd. The value of $(-a)^n$ is positive and not equal to $-a^n$ when n is even.
Similar statement can be made if a is negative.

5.2 WARM-UPS

1. False, because $(2^2)^3 = 2^6$.
2. True, because $(2^{-3})^{-1} = 2^3 = 8$.
3. True, because $(x^{-3})^3 = x^{-3 \cdot 3} = x^{-9}$.
4. False, because $2^3 \cdot 2^3 = 2^6$ and $(2^3)^3 = 2^9$.
5. False, because $(2x)^3 = 2^3x^3 = 8x^3$.
6. False, because $(-3y^3)^2 = 9y^6$.
7. True, the reciprocal of $\frac{2}{3}$ is $\frac{3}{2}$.

8. True, because $\left(\frac{2}{3}\right)^3 = \frac{2^3}{3^3} = \frac{8}{27}$.

9. True, because $\left(\frac{x^2}{2}\right)^3 = \frac{x^6}{2^3} = \frac{x^6}{8}$.

10. True, because $\left(\frac{2}{x}\right)^{-2} = \left(\frac{x}{2}\right)^2 = \frac{x^2}{4}$.

5.2 EXERCISES

1. The power of a power rule says that $(a^m)^n = a^{mn}$.

3. The power of a quotient rule says that $(a/b)^m = a^m/b^m$.

5. To compute the amount A when interest is compounded annually use $A = P(1+i)^n$ where P is the principal, i is the annual interest rate, and n is the number of years.

7. $(2^2)^3 = 2^6 = 64$

9. $(y^2)^5 = y^{2 \cdot 5} = y^{10}$

11. $(x^2)^{-4} = x^{(2)(-4)} = x^{-8} = \frac{1}{x^8}$

13. $(m^{-3})^{-6} = m^{(-3)(-6)} = m^{18}$

15. $(x^{-2})^3(x^{-3})^{-2} = x^{-6}x^6 = x^0 = 1$

17. $\frac{(x^3)^{-4}}{(x^2)^{-5}} = \frac{x^{-12}}{x^{-10}} = x^{-2} = \frac{1}{x^2}$

19. $(-9y)^2 = (-9)^2y^2 = 81y^2$

21. $(-5w^3)^2 = (-5)^2w^6 = 25w^6$

23. $(x^3y^{-2})^3 = x^9y^{-6} = \frac{x^9}{y^6}$

25. $(3ab^{-1})^{-2} = 3^{-2}a^{-2}b^2 = \frac{b^2}{9a^2}$

27. $\frac{2xy^{-2}}{(3x^2y)^{-1}} = \frac{2xy^{-2}}{3^{-1}x^{-2}y^{-1}}$
$= 2 \cdot 3x^3y^{-1} = \frac{6x^3}{y}$

29. $\frac{(2ab)^{-2}}{2ab^2} = \frac{2^{-2}a^{-2}b^{-2}}{2ab^2}$
$= 2^{-3}a^{-3}b^{-4} = \frac{1}{8a^3b^4}$

31. $\left(\frac{w}{2}\right)^3 = \frac{w^3}{2^3} = \frac{w^3}{8}$

33. $\left(-\frac{3a}{4}\right)^3 = \frac{(-3)^3a^3}{4^3} = -\frac{27a^3}{64}$

35. $\left(\frac{2x^{-1}}{y}\right)^{-2} = \frac{2^{-2}x^2}{y^{-2}} = \frac{x^2y^2}{4}$

37. $\left(\frac{-3x^3}{y}\right)^{-2} = \frac{(-3)^{-2}(x^3)^{-2}}{y^{-2}} = \frac{x^{-6}y^2}{(-3)^2}$
$= \frac{y^2}{9x^6}$

39. $\left(\frac{2}{5}\right)^{-2} = \left(\frac{5}{2}\right)^2 = \frac{25}{4}$

41. $\left(-\frac{1}{2}\right)^{-2} = (-2)^2 = 4$

43. $\left(-\frac{2x}{3}\right)^{-3} = \left(-\frac{3}{2x}\right)^3$
$= -\frac{3^3}{2^3x^3} = -\frac{27}{8x^3}$

45. $\left(\frac{2x^2}{3y}\right)^{-3} = \left(\frac{3y}{2x^2}\right)^3 = \frac{3^3y^3}{2^3(x^2)^3} = \frac{27y^3}{8x^6}$

47. $5^{2t} \cdot 5^{4t} = 5^{6t}$

49. $(2^{-3w})^{-2w} = 2^{6w^2}$

51. $\frac{7^{2m+6}}{7^{m+3}} = 7^{2m+6-m-3} = 7^{m+3}$

53. $8^{2a-1}(8^{a+4})^3 = 8^{2a-1} \cdot 8^{3a+12}$
$= 8^{5a+11}$

55. Add the exponents: $6x^9$

57. Multiply the exponents: $-8x^6$

59. Eliminate negative exponents: $\frac{3z}{x^2y}$

61. The reciprocal of $-\frac{2}{3}$ is $-\frac{3}{2}$.

63. Multiply the exponents: $\frac{4x^6}{9}$

65. $(-2x^{-2})^{-1} = (-2)^{-1}x^2 = -\frac{x^2}{2}$

67. $\left(\frac{2x^2y}{xy^2}\right)^{-3} = \left(\frac{xy^2}{2x^2y}\right)^3 = \frac{x^3y^6}{8x^6y^3} = \frac{y^3}{8x^3}$

69. $\frac{(5a^{-1}b^2)^3}{(5ab^{-2})^4} = \frac{5^3a^{-3}b^6}{5^4a^4b^{-8}}$
$= 5^{-1}a^{-7}b^{14} = \frac{b^{14}}{5a^7}$

71. $\frac{(2x^{-2}y)^{-3}}{(2xy^{-1})^2} \cdot 2x^2y^{-7}$
$= \frac{2^{-3}x^6y^{-3}}{2^2x^2y^{-2}} \cdot 2x^2y^{-7}$
$= 2^{-4}x^6y^{-8} = \frac{x^6}{16y^8}$

73. $\left(\frac{6a^{-2}b^3}{2c^4}\right)^{-2}(3a^{-1}b^2)^3$
$= \frac{6^{-2}a^4b^{-6}}{2^{-2}c^{-8}} \cdot 3^3a^{-3}b^6 = 3a^1b^0c^8 = 3ac^8$

75. $32 \cdot 64 = 2^5 \cdot 2^6 = 2^{11}$

77. $81 \cdot 6^{-4} = 3^4(2 \cdot 3)^{-4} = 3^42^{-4}3^{-4} = 2^{-4}$

79. $4^{3n} = (2^2)^{3n} = 2^{6n}$

81. $\frac{1}{5^{-2}} = 25$

83. $2^{-1} + 2^{-2} = 3/4$

85. $(0.036)^{-2} + (4.29)^3 \approx 850.559$

87. $\frac{(5.73)^{-1} + (4.29)^{-1}}{(3.762)^{-1}} \approx 1.533$

89. $40{,}000(1 + 0.12)^3 = 40{,}000(1.12)^3$
$= \$56{,}197.12$

91. $10{,}000(1 + 0.07)^{-18} \approx \$2{,}958.64$

93. a) $L = 72.2(1.002)^{20} \approx 75.1$ yr
b) $L = 72.2(1.002)^{60} \approx 81.4$ yr
95. If both a and b are nonzero, then $(ab)^{-1} = a^{-1}b^{-1}$. If $a = b = 2$, then $2 + 2)^{-1} = \frac{1}{4}$ and $2^{-1} + 2^{-1} = 1$.
97. All expressions have value -1 except (d) which has value 1.
99. a)

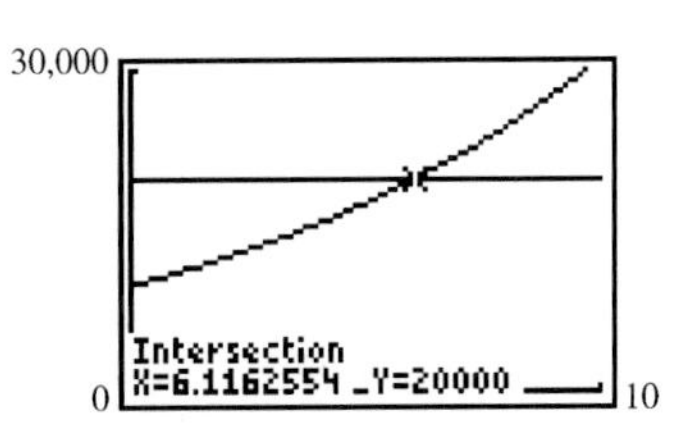

b) (6.116, 20,000)
c) 6.116 years

5.3 WARM-UPS

1. False, because it has a negative exponent.
2. False, because the coefficient of x is -5.
3. False, because the highest power of x is 3.
4. True, because $5^2 - 3 = 22$.
5. False, because $P(0) = 30(0) + 10 = 10$.
6. True, because the two trinomials are added correctly.
7. False, because
$$(x^2 - 5x) - (x^2 - 3x) = -2x$$
for any value of x.
8. True, because the monomial and binomial are correctly multiplied.
9. True, because $-(a - b) = b - a$ for any values of a and b.
10. False, because $-(y + 5) = -y - 5$ for any value of y.

5.3 EXERCISES

1. A term of a polynomial is a single number or the product of a number and one or more variables raised to whole number powers.
3. A constant is simply a number.
5. The degree of a polynomial in one variable is the highest power of the variable in the polynomial.
7. Yes, it is a monomial.
9. No, because polynomials do not have negative exponents.
11. Yes, it is a trinomial.
13. No, because $\frac{1}{x}$ is not allowed in a polynomial.
15. The degree is 4 because 4 is the highest power. The coefficient of x^3 is -8. It is a binomial because it has two terms.
17. The degree of -8 is 0 and the coefficient of x^3 is 0. It is a monomial.
19. The degree is 7 because 7 is the highest power. The coefficient of x^3 is 0, because x^3 is missing. It is a monomial because it has one term.
21. The degree is 6 because 6 is the highest power. The coefficient of x^3 is 1. It is a trinomial because it has three terms.
23. Replace x by 3 in $P(x) = x^4 - 1$.
$P(3) = 3^4 - 1 = 81 - 1 = 80$
25. Replace x by -2:
$M(-2) = -3(-2)^2 + 4(-2) - 9 = -29$
27. Replace x by 1:
$R(1) = 1^5 - 1^4 + 1^3 - 1^2 + 1 - 1 = 0$
29. $(2a - 3) + (a + 5) = 2a + a - 3 + 5$
$= 3a + 2$
31. $(7xy + 30) - (2xy + 5)$
$= 7xy + 30 - 2xy - 5 = 5xy + 25$
33. $(x^2 - 3x) + (-x^2 + 5x - 9)$
$= x^2 - x^2 - 3x + 5x - 9 = 2x - 9$
35. $(2x^3 - 4x - 3) - (x^2 - 2x + 5)$
$= 2x^3 - x^2 - 4x + 2x - 3 - 5$
$= 2x^3 - x^2 - 2x - 8$
37. Add the like terms to get $11x^2 - 2x - 9$.
39. $5x - 4x + 2 - (-3) = x + 5$.
41. Change the sign of every term on the bottom and add it to the appropriate like term on the top to get $-6x^2 + 5x + 2$.
43. Add the like terms to get $2x$.
45. Add exponents to get $-3x^2 \cdot 5x^4 = -15x^6$.
47. $x^2(x - 2) = x^3 - 2x$
49. $-1(3x - 2) = -3x + 2$

51. $5x^2y^3(3x^2y - 4x) = 15x^4y^4 - 20x^3y^3$

53. $(x - 2)(x + 2) = (x - 2)x + (x - 2)2$
$= x^2 - 2x + 2x - 4 = x^2 - 4$

55. $(x^2+x+2)(2x-3)$
$= (x^2+x+2)2x + (x^2+x+2)(-3)$
$= 2x^3+2x^2+4x-3x^2-3x-6$
$= 2x^3-x^2+x-6$

57. Multiply $-5x$ by $2x-3$ to get $-10x^2+15x$.

59.
$$\begin{array}{r} x+5 \\ \underline{x+5} \\ 5x+25 \\ \underline{x^2+5x \quad\;\;} \\ x^2+10x+25 \end{array}$$

61.
$$\begin{array}{r} x+6 \\ \underline{2x-3} \\ -3x-18 \\ \underline{2x^2+12x \quad\;\;} \\ 2x^2+9x-18 \end{array}$$

63.
$$\begin{array}{r} x^2+xy+y^2 \\ \underline{x-y} \\ -x^2y-xy^2-y^3 \\ \underline{x^3+x^2y+xy^2 \quad\quad\;\;} \\ x^3 \quad\quad\quad\quad -y^3 \end{array}$$

65. Add like terms to get $2x-5$.

67. Add like terms: $4a^2-11a-4$

69. $(w^2-7w-2)-(w-3w^2+5)$
$= w^2-7w-2-w+3w^2-5$
$= 4w^2-8w-7$

71. $(x-2)(x^2+2x+4)$
$= (x-2)x^2+(x-2)2x+(x-2)4$
$= x^3-2x^2+2x^2-4x+4x-8$
$= x^3-8$

73. $(x-w)(z+2w)$
$= (x-w)z+(x-w)2w$
$= xz-wz+2xw-2w^2$

75. $(x^2-x+2)(x^2+x-2)$
$= (x^2-x+2)(x^2)+(x^2-x+2)(x)$
$\quad + (x^2-x+2)(-2)$
$= x^4-x^3+2x^2+x^3-x^2+2x$
$\quad -2x^2+2x-4$
$= x^4-x^2+4x-4$

77. $(2.31x-5.4)6.25x+(2.31x-5.4)1.8$
$= 14.4375x^2-29.592x-9.72$

79. $(3.759+11.61)x^2-4.71x+6.59x$
$\quad +2.85-3.716$
$= 15.369x^2+1.88x-0.866$

81. $\frac{2}{4}x+\frac{4}{2}+\frac{1}{4}x-\frac{1}{2}=\frac{3}{4}x+\frac{3}{2}$

83. $\frac{1}{2}x^2+\frac{1}{3}x-\frac{1}{5}-\frac{2}{2}x^2+\frac{2}{3}x+\frac{1}{5}$
$= -\frac{1}{2}x^2+x$

85. Combine like terms within the first set of brackets:
$x^2-3-x^2-5x+4=-5x+1$
Combine like terms within the second set of brackets:
$x-3x^2+15x=-3x^2+16x$
Now subtract these results:
$[-5x+1]-[-3x^2+16x]$
$= -5x+1+3x^2-16x$
$= 3x^2-21x+1$

87. $[5x-4(x-3)][3x-7(x+2)]$
$= [x+12][-4x-14]$
$= [x+12](-4x)+[x+12](-14)$
$= -4x^2-48x-14x-168$
$= -4x^2-62x-168$

89. $[x^2-(m+2)][x^2+(m+2)]$
$= [x^2-m-2][x^2+m+2]$
$= [x^2-m-2]x^2+[x^2-m-2]m$
$\quad + [x^2-m-2]2$
$= x^4-mx^2-2x^2+mx^2-m^2-2m$
$\quad + 2x^2-2m-4$
$= x^4-m^2-4m-4$

91. $(a^{2m}+3a^m-3)+(-5a^{2m}-7a^m+8)$
$= -4a^{2m}-4a^m+5$

93. $(x^n-1)(x^n+3)$
$= (x^n-1)x^n+(x^n-1)3$
$= x^{2n}-x^n+3x^n-3$
$= x^{2n}+2x^n-3$

95. $z^{3w}-z^{2w}(z^{1-w}-4z^w)$
$= z^{3w}-z^{1+w}+4z^{3w}=5z^{3w}-z^{1+w}$

97. $(x^{2r}+y)x^{4r}+(x^{2r}+y)(-x^{2r}y)$
$\quad + (x^{2r}+y)y^2$
$= x^{6r}+x^{4r}y-x^{4r}y-x^{2r}y^2+x^{2r}y^2+y^3$
$= x^{6r}+y^3$

99. $C(3)=20(3)+15=\$75$

101. $C(3)=50\cdot 3-0.01(3)^4=149.19$
$C(2)=50\cdot 2-0.01(2)^4=99.84$
$C(3)-C(2)=49.35$
Marginal cost of 3rd window is \$49.35.
$C(10)-C(9)=400-384.39$
Marginal cost of 10th window is \$15.61.

103. a) $F(1950) - M(1950) \approx 6.2$ years
b) The lines are not parallel because the slopes are different.
c) If y is the year in which female life expectancy is 8 years larger then male life expectancy, then y satisfies the following equation.
$0.18268y - 284.98$
$= 0.16252y - 251.91 + 8$
$0.02016y = 41.07$
$y \approx 2037$
105. The binomial $x^4 + 1$ has degree 4.
107. There are 4 terms in the product $(a + b)(c + d)$. There are 8 terms in the product $(a + b)(c + d)(e + f)$. There are 16 terms in the product of four binomials with no like terms. There are 2^n terms in the product of n binomials with no like terms.

5.4 WARM-UPS

1. True, because the binomials are multiplied correctly.
2. True, because the binomials are multiplied correctly.
3. False, $(2 + 3)^2 = 5^2 = 25$ and $2^2 + 3^2 = 4 + 9 = 13$.
4. True, because the binomial is squared correctly.
5. False, $(8 - 3)^2 = 5^2 = 25$ and $64 - 9 = 55$.
6. True, because $(a + b)(a - b) = a^2 - b^2$.
7. True, $(60 - 1)(60 + 1) = 60^2 - 1^2$
$= 3600 - 1$.
8. True, because the binomial is squared correctly.
9. False, $(x - 3)^2 = x^2 - 6x + 9$ for any value of x.
10. False, it is a product of two monomials.

5.4 EXERCISES

1. The distributive property is used in multiplying binomials.
3. The purpose of FOIL is to provide a fast way to find the product of two binomials.
5. The square of a difference is the square of the first term minus twice the product of the two terms plus the square of the last term.
7. In general, $(a + b)^2 = a^2 + 2ab + b^2$.
9. $(x + 3)(x + 5) = x^2 + 3x + 5x + 15$
$= x^2 + 8x + 15$
11. $(x - 2)(x + 4) = x^2 + 4x - 2x - 8$
$= x^2 + 2x - 8$
13. $(1 + 2x)(3 + x) = 3 + x + 6x + 2x^2$
$= 2x^2 + 7x + 3$
15. $(-2a - 3)(-a + 5)$
$= 2a^2 - 10a + 3a - 15$
$= 2a^2 - 7a - 15$
17. $(2x^2 - 7)(2x^2 + 7)$
$= 4x^2 + 14x^2 - 14x^2 - 49$
$= 4x^4 - 49$
19. $(2x^3 - 1)(x^3 + 4) = 2x^6 + 8x^3 - x^3 - 4$
$= 2x^6 + 7x^3 - 4$
21. $(6z + w)(w - z)$
$= 6zw - 6z^2 + w^2 - wz$
$= w^2 + 5wz - 6z^2$
23. $(3k - 2t)(4t + 3k)$
$= 12kt + 9k^2 - 8t^2 - 6tk$
$= 9k^2 + 6kt - 8t^2$
25. $(x - 3)(y + w) = xy - 3y + xw - 3w$
27. $(m + 3)^2 = m^2 + 2(3)(m) + 3^2$
$= m^2 + 6m + 9$
29. $(4 - a)^2 = 4^2 - 2(a)(4) + a^2$
$= a^2 - 8a + 16$
31. $(2w + 1)^2 = (2w)^2 + 2(2w)(1) + 1^2$
$= 4w^2 + 4w + 1$
33. $(3t - 5u)^2 = (3t)^2 - 2(3t)(5u) + (5u)^2$
$= 9t^2 - 30tu + 25u^2$
35. $(-x - 1)^2 = (-x)^2 - 2(-x)(1) + 1^2$
$= x^2 + 2x + 1$
37. $(a - 3y^3)^2 = a^2 - 2(a)(3y^3) + (3y^3)^2$
$= a^2 - 6ay^3 + 9y^6$
39. $(w - 9)(w + 9) = w^2 - 9^2 = w^2 - 81$
41. $(w^3 - y)(w^3 + y) = (w^3)^2 - y^2$
$= w^6 - y^2$
43. $(7 - 2x)(7 + 2x) = (7)^2 - (2x)^2$
$= 49 - 4x^2$
45. $(3x^2 - 2)(3x^2 + 2) = (3x^2)^2 - 2^2$
$= 9x^4 - 4$
47. $(5a^3 - 2b)(5a^3 + 2b) = 25a^6 - 4b^2$
49. $[(m + t) + 5][(m + t) - 5]$
$= (m + t)^2 - 25$
$= m^2 + 2mt + t^2 - 25$

51. $[y-(r+5)][y+(r+5)]$
$= y^2-(r+5)^2$
$= y^2-r^2-10r-25$

53. $[(2y-t)+3]^2$
$= (2y-t)^2+2\cdot 3(2y-t)+3^2$
$= 4y^2-4yt+t^2+12y-6t+9$

55. $[3h+(k-1)]^2$
$= 9h^2+2\cdot 3h(k-1)+(k-1)^2$
$= 9h^2+6hk-6h+k^2-2k+1$

57. $(x+1)^3=(x+1)^2(x+1)$
$= (x^2+2x+1)(x+1)$
$= (x^2+2x+1)x+(x^2+2x+1)1$
$= x^3+2x^2+x+x^2+2x+1$
$= x^3+3x^2+3x+1$

59. $(w-2)^3=(w-2)^2(w-2)$
$= (w^2-4w+4)(w-2)$
$= (w^2-4w+4)w+(w^2-4w+4)(-2)$
$= w^3-4w^2+4w-2w^2+8w-8$
$= w^3-6w^2+12w-8$

61. $(2x+1)^3=(2x+1)^2(2x+1)$
$= (4x^2+4x+1)(2x+1)$
$= (4x^2+4x+1)2x+(4x^2+4x+1)(1)$
$= 8x^3+8x^2+2x+4x^2+4x+1$
$= 8x^3+12x^2+6x+1$

63. $(3x-1)^3=(3x-1)^2(3x-1)$
$= (9x^2-6x+1)(3x-1)$
$= 3x(9x^2-6x+1)-1(9x^2-6x+1)$
$= 27x^3-18x^2+3x-9x^2+6x-1$
$= 27x^3-27x^2+9x-1$

65. $(x+1)^4=(x+1)^2(x+1)^2$
$= (x^2+2x+1)(x^2+2x+1)$
$= x^2(x^2+2x+1)+2x(x^2+2x+1)$
$+1(x^2+2x+1)$
$= x^4+2x^3+x^2+2x^3+4x^2+2x$
$+x^2+2x+1$
$= x^4+4x^3+6x^2+4x+1$

67. $(h-3)^4=(h-3)^2(h-3)^2$
$= (h^2-6h+9)(h^2-6h+9)$
$= h^2(h^2-6h+9)-6h(h^2-6h+9)$
$+9(h^2-6h+9)$
$= h^4-6h^3+9h^2-6h^3+36h^2-54h$
$+9h^2-54h+81$
$= h^4-12h^3+54h^2-108h+81$

69. $(x-6)(x+9)=x^2+9x-6x-54$
$= x^2+3x-54$

71. $(5-x)(5+x)=5^2-x^2=25-x^2$

73. $(3x-4a)(2x+5a)$
$= 6x^2+15ax-8ax-20a^2$
$= 6x^2+7ax-20a^2$

75. $(2t-3)(t+w)=2t^2+2tw-3t-3w$

77. $(3x^2+2y^3)^2$
$= (3x^2)^2+2(3x^2)(2y^3)+(2y^3)^2$
$= 9x^4+12x^2y^3+4y^6$

79. $(2+2y)(3y-5)=6y-10+6y^2-10y$
$= 6y^2-4y-10$

81. $(2m-7)^2=(2m)^2-2\cdot 2m\cdot 7+7^2$
$= 4m^2-28m+49$

83. $(3+7x)^2=3^2+2(3)(7x)+(7x)^2$
$= 49x^2+42x+9$

85. $4y\left(3y+\frac{1}{2}\right)^2$
$= 4y\left((3y)^2+2(3y)\frac{1}{2}+\left(\frac{1}{2}\right)^2\right)$
$= 4y\left(9y^2+3y+\frac{1}{4}\right)$
$= 36y^3+12y^2+y$

87. $(a+h)^2-a^2=a^2+2ah+h^2-a^2$
$= 2ah+h^2$

89. $(x+2)(x+2)^2=(x+2)(x^2+4x+4)$
$= x(x^2+4x+4)+2(x^2+4x+4)$
$= x^3+6x^2+12x+8$

91. $(y+3)^3=(y+3)^2(y+3)$
$= (y^2+6y+9)(y+3)$
$= (y^2+6y+9)y+(y^2+6y+9)3$
$= y^3+6y^2+9y+3y^2+18y+27$
$= y^3+9y^2+27y+27$

93. $4x-3(x-5)^2$
$= 4x-3(x^2-10x+25)$
$= 4x-3x^2+30x-75$
$= -3x^2+34x-75$

95. $(3.2x-4.5)(5.1x+3.9)$
$= (3.2)(5.1)x^2+(3.9)(3.2)x-(4.5)(5.1)x$
$-(4.5)(3.9)$
$= 16.32x^2-10.47x-17.55$

97. $(3.6y+4.4)^2$
$= (3.6y)^2+2(3.6)(4.4)y+(4.4)^2$
$= 12.96y^2+31.68y+19.36$

99. $(x^m+2)(x^{2m}+3)$
$= x^mx^{2m}+3x^m+2x^{2m}+6$
$= x^{3m}+2x^{2m}+3x^m+6$

101. $a^{n+1}(a^{2n}+a^n-3)$
$= a^{n+1+2n}+a^{n+1+n}-3a^{n+1}$
$= a^{3n+1}+a^{2n+1}-3a^{n+1}$

103. $(a^m+a^n)^2=(a^m)^2+2a^ma^n+(a^n)^2$
$= a^{2m}+2a^{m+n}+a^{2n}$

105. $(5y^m + 8z^k)(3y^{2m} + 4z^{3-k})$
$= 5y^m 3y^{2m} + 3 \cdot 8y^{2m}z^k + 4 \cdot 5y^m z^{3-k} + 32z^{k+3-k}$
$= 15y^{3m} + 24y^{2m}z^k + 20\, y^m z^{3-k} + 32z^3$

107. Use $L = x + 3$ and $W = x + 1$ in the formula for the area of a rectangle, $A = LW$. So a function for the area is
$A(x) = (x + 3)(x + 1)$ or
$A(x) = x^2 + 4x + 3$

109. a) $A(x) = (8 - 2x)(10 - 2x)$
$A(x) = 4x^2 - 36x + 80$
b) If $x = 0.4$, then
$A(0.4) = 4(0.4)^2 - 36(0.4) + 80$
$= 66.24 \text{ km}^2$

111. a) Since x is taken off of both sides of the rectangle, the sides are $4 - 2x$ and $6 - 2x$
$V(x) = x(4 - 2x)(6 - 2x)$
$= x(4x^2 - 20x + 24)$
$V(x) = 4x^3 - 20x^2 + 24x$
b) If $x = 4$ in. $= \frac{1}{3}$ ft, then
$V(1/3) = 4\left(\frac{1}{3}\right)^3 - 20\left(\frac{1}{3}\right)^2 + 24\left(\frac{1}{3}\right)$
$= 5\frac{25}{27} \text{ ft}^3 \approx 5.9 \text{ ft}^3$

113. a) $(a + b)^3 = a^3 + 3a^2b + 3ab^2 + b^3$
b) $(a + b)^4 = a^4 + 4a^3b + 6a^2b^2 + 4ab^3 + b^4$
$(a + b)^5 = a^5 + 5a^4b + 10a^3b^2 + 10a^2b^3 + 5ab^4 + b^5$
c) There are 4, 5, and 6 terms, respectively.
d) The number of terms in $(a + b)^n$ after combining like terms is $n + 1$.

115. The area of the four regions is a^2, ab, ab, and b^2. If you add these you get $a^2 + 2ab + b^2$. So we see geometrically that $(a + b)^2 = a^2 + 2ab + b^2$.

5.5 WARM-UPS

1. True, $3xy(x - 2y) = -3xy(-x + 2y)$.
2. True.
3. True, because $-2(2 - x)$ $= -4 + 2x = 2x - 4$ for any value of x.
4. True, $a^2 - b^2 = (a - b)(a + b)$.
5. False, because $x^2 + 12x + 36$ is a perfect square trinomial.
6. False, because $y^2 + 8y + 16$ is a perfect square trinomial.
7. False, because $(3x + 7)^2 = 9x^2 + 42x + 49$.
8. True, $x^3 + 1 = (x + 1)(x^2 - x + 1)$.
9. False, $x^3 - 27 = (x - 3)(x^2 + 3x + 9)$.
10. False, $x^3 - 8 = (x - 2)(x^2 + 2x + 4)$.

5.5 EXERCISES

1. A prime number is a natural number greater than 1 that has no factors other than itself and 1.

3. The greatest common factor for the terms of a polynomial is a monomial that includes every number or variable that is a factor of all terms of the polynomial.

5. A linear polynomial is a polynomial of the form $ax + b$ with $a \neq 0$.

7. A prime polynomial is a polynomial that cannot be factored.

9. Since $48 = 2 \cdot 2 \cdot 2 \cdot 2 \cdot 3$ and $36x = 2 \cdot 2 \cdot 3 \cdot 3x$, the greatest common factor is $2 \cdot 2 \cdot 3$ or 12.

11. The gcf for 9, 21, and 15 is 3. Since there are no variables in common to all three terms, the gcf is 3.

13. Since $24 = 2 \cdot 2 \cdot 2 \cdot 3$, $42 = 2 \cdot 3 \cdot 7$, and $66 = 2 \cdot 3 \cdot 11$, the GCF of 24, 42, and 66 is $2 \cdot 3$ or 6. Using each common variable with the lowest power, we get the GCF $6xy$.

15. $x^3 - 5x = x \cdot x^2 - x \cdot 5 = x(x^2 - 5)$

17. $48wx + 36wy = 12w \cdot 4x + 12w \cdot 3y$
$= 12w(4x + 3y)$

19. $2x^3 - 4x^2 + 6x$
$= 2x \cdot x^2 - 2x \cdot 2x + 2x \cdot 3$
$= 2x(x^2 - 2x + 3)$

21. $36a^3b^6 - 24a^4b^2 + 60a^5b^3$
$= 12a^3b^2(3b^4 - 2a + 5a^2b)$

23. $2x - 2y = 2 \cdot x - 2 \cdot y = 2(x - y)$
$2x - 2y = -2(-x) - 2 \cdot y = -2(-x + y)$

25. $6x^2 - 3x = 3x(2x - 1)$
$6x^2 - 3x = -3x(-2x + 1)$

27. $-w^3 + 3w^2 = w^2 \cdot (-w) + w^2(3)$
$= w^2(-w + 3)$
$-w^3 + 3w^2 = -w^2 \cdot w - (-w^2) \cdot 3$
$= -w^2(w - 3)$

29. $-a^3 + a^2 - 7a = a(-a^2 + a - 7)$
$-a^3 + a^2 - 7a = -a(a^2 - a + 7)$

31. Factor out the quantity $x - 6$:
$(x - 6)a + (x - 6)b = (x - 6)(a + b)$

33. $(y - 4)x - (y - 4)3 = (y - 4)(x - 3)$

35. Factor out $(y - 1)^2$:
$(y - 1)^2 y + (y - 1)^2 z = (y - 1)^2(y + z)$

37. $a(a - b)^2 - b(a - b)^2$
$= (a - b)(a - b)^2$
$= (a - b)^3$

39. $ax + ay + 3x + 3y$
$= a(x + y) + 3(x + y)$
$= (x + y)(a + 3)$

41. $xy - 3y + x - 3$
$= y(x - 3) + 1(x - 3)$
$= (x - 3)(y + 1)$

43. $4a - 4b - ca + cb$
$= 4(a - b) - c(a - b)$
$= (a - b)(4 - c)$

45. $xy - y - 6x + 6$
$= y(x - 1) - 6(x - 1)$
$= (x - 1)(y - 6)$

47. The polynomial is a difference of two squares and so it factors as the product of a sum and a difference: $(x - 10)(x + 10)$

49. $4y^2 - 49 = (2y)^2 - 7^2$
$= (2y - 7)(2y + 7)$

51. $9x^2 - 25a^2 = (3x)^2 - (5a)^2$
$= (3x - 5a)(3x + 5a)$

53. $144w^2z^2 - h^2 = (12wz)^2 - h^2$
$= (12wz - h)(12wz + h)$

55. This polynomial is a perfect square trinomial. The middle term is twice the product of the square roots of the first term and last term. $x^2 - 2(x)(10) + 10^2 = (x - 10)^2$

57. $4m^2 - 4m + 1 = (2m)^2 - 2 \cdot 2m + 1^2$
$= (2m - 1)^2$

59. $w^2 - 2wt + t^2 = (w - t)^2$

61. $a^3 - 1 = a^3 - 1^3 = (a - 1)(a^2 + a + 1)$

63. $w^3 + 27 = w^3 + 3^3$
$= (w + 3)(w^2 - 3w + 9)$

65. $8x^3 - 1 = (2x)^3 - 1^3$
$= (2x - 1)(4x^2 + 2x + 1)$

67. $64x^3 + 125 = (4x)^3 + 5^3$
$= (4x + 5)(16x^2 - 20x + 25)$

69. $8a^3 - 27b^3 = (2a)^3 - (3b)^3$
$= (2a - 3b)(4a^2 + 6ab + 9b^2)$

71. $2x^2 - 8 = 2(x^2 - 4) = 2(x + 2)(x - 2)$

73. $x^3 + 10x^2 + 25x = x(x^2 + 10x + 25)$
$= x(x + 5)^2$

75. $4x^2 + 4x + 1 = (2x)^2 + 2(2x)(1) + 1^2$
$= (2x + 1)^2$

77. $(x + 3)x + (x + 3)7 = (x + 3)(x + 7)$

79. $6y^2 + 3y = 3y \cdot 2y + 3y \cdot 1 = 3y(2y + 1)$

81. $4x^2 - 20x + 25 = (2x)^2 - 2(2x)(5) + 5^2$
$= (2x - 5)^2$

83. $2m^4 - 2mn^3 = 2m(m^3 - n^3)$
$= 2m(m - n)(m^2 + mn + n^2)$

85. $(2x - 3)x - (2x - 3)2$
$= (2x - 3)(x - 2)$

87. $9a^3 - aw^2 = a(9a^2 - w^2)$
$= a(3a + w)(3a - w)$

89. $-5a^2 + 30a - 45 = -5(a^2 - 6a + 9)$
$= -5(a - 3)^2$

91. $16 - 54x^3 = 2(8 - 27x^3)$
$= 2(2 - 3x)(4 + 6x + 9x^2)$

93. $-3y^3 - 18y^2 - 27y = -3y(y^2 + 6y + 9)$
$= -3y(y + 3)^2$

95. $-7a^2b^2 + 7 = -7(a^2b^2 - 1)$
$= -7(ab + 1)(ab - 1)$

97. $7x - 7h - hx + h^2$
$= 7(x - h) - h(x - h)$
$= (x - h)(7 - h)$

99. $a^2x + 3a^2 - 4x - 12$
$= a^2(x + 3) - 4(x + 3)$
$= (x + 3)(a^2 - 4)(x + 3)$
$= (x + 3)(a - 2)(a + 2)$

101. If we choose $k = 9$, then $x^2 + 6x + k$ will be the perfect square $x^2 + 6x + 9$.

103. If we choose $k = 20$, then $4a^2 - ka + 25$ will be the perfect square $4a^2 - 20a + 25$.

105. If we choose $k = 16$, then $km^2 - 24m + 9$ will be the perfect square $16m^2 - 24m + 9$.

107. a) $b^3 - 6b^2 + 9b$
$= b(b^2 - 6b + 9)$
$= b(b - 3)^2$

Since $V = LWH$ and the height is b, the sides of the square bottom are each $b - 3$ in.

b) $V = 18^3 - 6 \cdot 18^2 + 9 \cdot 18 = 4050 \text{ in.}^3$

c) From the graph it appears that the height of a cage with a volume of 20,000 in.^{3} is 30 in.

5.6 WARM-UPS

1. True. Check by multiplying $(x + 3)(x + 6)$.
2. False, $y^2 + 2y - 35 = (y + 7)(y - 5)$.
3. False, $x^2 + 4$ cannot be factored. It is a sum of two squares.
4. False, $x^2 - 5x - 6 = (x - 6)(x + 1)$.
5. True. Check by multiplying $(x - 6)(x + 2)$.
6. False, $x^2 + 15x + 36 = (x + 12)(x + 3)$.
7. False, $3x^2 + 4x - 15 = (3x - 5)(x + 3)$.
8. False, $4x^2 + 4x - 3 = (2x + 3)(2x - 1)$.
9. True. Check by multiplying.
10. True. Check by multiplying.

5.6 EXERCISES

1. To factor $x^2 + bx + c$, find two integers whose sum is b and whose product is c.
3. Trial and error means to simply write down possible factors and then use FOIL to check until you get the correct factors.
5. Two numbers that have a product of 3 and a sum of 4 are 1 and 3:
$x^2 + 4x + 3 = (x + 1)(x + 3)$
7. Two numbers that have a product of 50 and a sum of 15 are 5 and 10:
$a^2 + 15a + 50 = (a + 10)(a + 5)$
9. Two numbers that have a product of -14 and a sum of -5 are -7 and 2:
$y^2 - 5y - 14 = (y - 7)(y + 2)$
11. Two numbers that have a product of 8 and a sum of -6 are -2 and -4:
$x^2 - 6x + 8 = (x - 2)(x - 4)$
13. Two numbers with product 27 and sum -12 are -9 and -3:
$a^2 - 12a + 27 = (a - 9)(a - 3)$
15. Two numbers that have a product of -30 and a sum of 7 are -3 and 10:
$a^2 + 7a - 30 = (a + 10)(a - 3)$
17. Two numbers with a product of $6 \cdot 1 = 6$ and a sum of 5 are 2 and 3:
$$\begin{aligned} 6w^2 + 5w + 1 &= 6w^2 + 3w + 2w + 1 \\ &= 3w(2w + 1) + 1(2w + 1) \\ &= (3w + 1)(2w + 1) \end{aligned}$$
19. Using the ac method we find two numbers that have a product of $2(-3) = -6$ and a sum of -5. These numbers are -6 and 1:
$$\begin{aligned} 2x^2 - 5x - 3 &= 2x^2 - 6x + 1x - 3 \\ &= 2x(x - 3) + 1(x - 3) \\ &= (2x + 1)(x - 3) \end{aligned}$$
21. Two numbers with a product of $4(15) = 60$ and a sum of 16 are 6 and 10:
$$\begin{aligned} 4x^2 + 16x + 15 &= 4x^2 + 6x + 10x + 15 \\ &= 2x(2x + 3) + 5(2x + 3) \\ &= (2x + 5)(2x + 3) \end{aligned}$$
23. Two numbers with a product of $6 \cdot 1 = 6$ and a sum of -5 are -2 and -3:
$$\begin{aligned} 6x^2 - 5x + 1 &= 6x^2 - 2x - 3x + 1 \\ &= 2x(3x - 1) - 1(3x - 1) \\ &= (2x - 1)(3x - 1) \end{aligned}$$
25. Two numbers with a product of $12(-1) =$
-12 and a sum of 1 are 4 and -3:
$$\begin{aligned} 12y^2 + y - 1 &= 12y^2 - 3y + 4y - 1 \\ &= 3y(4y - 1) + 1(4y - 1) \\ &= (3y + 1)(4y - 1) \end{aligned}$$
27. Two numbers with a product of $-5(6) = -30$ and a sum of 1 are 6 and -5:
$$\begin{aligned} 6a^2 + a - 5 &= 6a^2 + 6a - 5a - 5 \\ &= 6a(a + 1) - 5(a + 1) \\ &= (6a - 5)(a + 1) \end{aligned}$$
29. $2x^2 + 15x - 8 = (2x - 1)(x + 8)$
31. $3b^2 - 16b - 35 = (3b + 5)(b - 7)$
33. $6w^2 + w - 12 = (3w - 4)(2w + 3)$
35. $4x^2 - 5x + 1 = (4x - 1)(x - 1)$
37. $5m^2 + 13m - 6 = (5m - 2)(m + 3)$
39. $6y^2 - 7y - 20 = (3y + 4)(2y - 5)$
41. If $a = x^5$ then
$$\begin{aligned} x^{10} - 9 &= (x^5)^2 - 3^2 = a^2 - 3^2 \\ &= (a - 3)(a + 3) = (x^5 - 3)(x^5 + 3) \end{aligned}$$
43. $z^{12} - 6z^6 + 9 = (z^6 - 3)^2$
45. $$\begin{aligned} 2x^7 + 8x^4 + 8x &= 2x(x^6 + 4x^3 + 4) \\ &= 2x(x^3 + 2)^2 \end{aligned}$$
47. $$\begin{aligned} 4x^5 + 4x^3 + x &= x(4x^4 + 4x^2 + 1) \\ &= x(2x^2 + 1)^2 \end{aligned}$$
49. $$\begin{aligned} x^6 - 8 &= (x^2)^3 - 2^3 \\ &= (x^2 - 2)(x^4 + 2x^2 + 4) \end{aligned}$$
51. $a^{2n} - 1 = (a^n)^2 - 1^2 = (a^n - 1)(a^n + 1)$
53. $$\begin{aligned} a^{2r} + 6a^r + 9 &= (a^r)^2 + 6a^r + 3^2 \\ &= (a^r + 3)^2 \end{aligned}$$

55. $x^{3m} - 8 = (x^m)^3 - 2^3$
$= (x^m - 2)(x^{2m} + 2x^m + 4)$

57. $a^{3m} - b^3 = (a^m)^3 - b^3$
$= (a^m - b)(a^{2m} + a^m b + b^2)$

59. $k^{2w+1} - 10k^{w+1} + 25k$
$= k(k^{2w} - 10k^w + 25) = k(k^w - 5)^2$

61. $x^6 - 2x^3 - 35 = (x^3 + 5)(x^3 - 7)$

63. $a^{20} - 20a^{10} + 100 = (a^{10} - 10)^2$

65. $-12a^5 - 10a^3 - 2a$
$= -2a(6a^4 + 5a^2 + 1)$
$= -2a(3a^2 + 1)(2a^2 + 1)$

67. Two numbers with a product of -15 and a sum of 2 are 5 and -3:
$x^{2a} + 2x^a - 15 = (x^a + 5)(x^a - 3)$

69. $x^{2a} - y^{2b} = (x^a)^2 - (y^b)^2$
$= (x^a - y^b)(x^a + y^b)$

71. $x^8 - x^4 - 6 = (x^4 - 3)(x^4 + 2)$

73. $x^{a+2} - x^a = x^a(x^2 - 1)$
$= x^a(x - 1)(x + 1)$

75. $x^{2a} + 6x^a + 9 = (x^a)^2 + 2 \cdot x^a \cdot 3 + 3^2$
$= (x^a + 3)^2$

77. $2x^2 + 20x + 50 = 2(x^2 + 10x + 25)$
$= 2(x + 5)^2$

79. $a^3 - 36a = a(a^2 - 36)$
$= a(a - 6)(a + 6)$

81. $10a^2 + 55a - 30 = 5(2a^2 + 11a - 6)$
$= 5(2a - 1)(a + 6)$

83. $2x^2 - 128y^2 = 2(x^2 - 64y^2)$
$= 2(x + 8y)(x - 8y)$

85. $-9x^2 + 33x + 12 = -3(3x^2 - 11x - 4)$
$= -3(3x + 1)(x - 4)$

87. $m^5 + 20m^4 + 100m^3$
$= m^3(m^2 + 20m + 100)$
$= m^3(m + 10)^2$

89. Two numbers with a product of $6 \cdot 20 = 120$ and a sum of 23 are 8 and 15:
$6x^2 + 23x + 20 = 6x^2 + 15x + 8x + 20$
$= 3x(2x + 5) + 4(2x + 5)$
$= (3x + 4)(2x + 5)$

91. $9y^3 - 24y^2 + 16y = y(9y^2 - 24y + 16)$
$= y(3y - 4)^2$

93. $r^2 - 6rs + 8s^2 = (r - 4s)(r - 2s)$

95. $m^3 + 2m + 3m^2 = m(m^2 + 3m + 2)$
$= m(m + 1)(m + 2)$

97. $6m^3 - m^2n - 2mn^2$
$= m(6m^2 - mn - 2n^2)$
$= m(2m + n)(3m - 2n)$

99. $9m^2 - 25n^2 = (3m)^2 - (5n)^2$
$= (3m - 5n)(3m + 5n)$

101. $5a^2 + 20a - 60 = 5(a^2 + 4a - 12)$
$= 5(a + 6)(a - 2)$

103. $-2w^2 + 18w + 20$
$= -2(w^2 - 9w - 10)$
$= -2(w - 10)(w + 1)$

105. $w^2x^2 - 100x^2 = x^2(w^2 - 100)$
$= x^2(w + 10)(w - 10)$

107. $81x^2 - 9 = 9(9x^2 - 1) =$
$= 9(3x + 1)(3x - 1)$

109. Two numbers with a product of $-15 \cdot 8 = -120$ and a sum of -2 are -12 and 10:
$8x^2 - 2x - 15 = 8x^2 - 12x + 10x - 15$
$= 4x(2x - 3) + 5(2x - 3)$
$= (4x + 5)(2x - 3)$

111. $3m^4 - 24m = 3m(m^3 - 8)$
$= 3m(m - 2)(m^2 + 2m + 4)$

113. The polynomials in (a) and (b) are not perfect square trinomials.

115. a) $x^2 + 10x + 25 = (x + 5)^2$
b) $x^2 - 10x + 25 = (x - 5)^2$
c) $x^2 + 26x + 25 = (x + 25)(x + 1)$
d) $x^2 - 25 = (x + 5)(x - 5)$
e) $x^2 + 25$ is not factorable.

117. $ax + 3x + 4a + 12$
$= x(a + 3) + 4(a + 3)$
$= (a + 3)(x + 4)$

119. $ax - 2a + 4x - 8$
$= a(x - 2) + 4(x - 2)$
$= (x - 2)(a + 4)$

121. $bm - 4b - 5m + 20$
$= b(m - 4) - 5(m - 4)$
$= (m - 4)(b - 5)$

123. $nx - ax - ac + nc$
$= x(n - a) + c(n - a)$
$= (n - a)(x + c)$

125. $xr - yr - xw + yw$
$= r(x - y) - w(x - y)$
$= (x - y)(r - w)$

127. $xt - t^2 - ax + at$
$= t(x - t) - a(x - t)$
$= (x - t)(t - a)$

129. $2qh - h + 8q - 4$
$h(2a - 1) + 4(2q - 1)$
$= (2q - 1)(h + 4)$

131. $6t - 3ty + awy - 2aw$
$= -3t(y-2) + aw(y-2)$
$= (-3t + aw)(y-2)$
$= (aw - 3t)(y-2)$
133. $x^3 + 7x - 7a - ax^2$
$= x(x^2+7) - a(x^2+7)$
$= (x^2+7)(x-a)$
135. $m^4 - 5m^2 + m^2p - 5p$
$= m^2(m^2-5) + p(m^2-5)$
$= (m^2-5)(m^2+p)$
137. $y^2 + 3y + 2 = (y+1)(y+2)$
139. $x^2 + 10x + 21 = (x+3)(x+7)$
141. $a^2 + 15a + 54 = (a+9)(a+6)$
143. $y^2 + 3y - 10 = (y-2)(y+5)$
145. $w^2 - 2w - 15 = (w-5)(w+3)$
147. $b^2 + 6b - 16 = (b-2)(b+8)$
149. $a^2 - 8a - 33 = (a-11)(a+3)$
151. $a^2 - 9a + 18 = (a-3)(a-6)$
153. $x^2 - 11x + 24 = (x-3)(x-8)$
155. $y^2 - 23y + 130 = (y-10)(y-13)$
157. $2w^2 + 7w + 3 = (2w+1)(w+3)$
159. $2x^2 - 9x - 5 = (2x+1)(x-5)$
161. $3x^2 + 25x + 8 = (3x+1)(x+8)$
163. $3x^2 + 26x - 9 = (3x-1)(x+9)$
165. $5y^2 + 16y + 3 = (5y+1)(y+3)$
167. $5y^2 - 21y + 4 = (5y-1)(y-4)$
169. $7a^2 + 6a - 1 = (7a-1)(a+1)$
171. $7a^2 - 8a + 1 = (7a-1)(a-1)$
173. $2w^2 + 23w + 11 = (2w+1)(w+11)$
175. $2w^2 + 21w - 11 = (2w-1)(w+11)$

5.7 WARM-UPS

1. False, $x^2 - 9 = (x-3)(x+3)$ for any value of x.
2. True, because $4x^2 + 12x + 9 = (2x+3)^2$.
3. True.
4. False, because $x^2 - 4$ is not a prime polynomial.
5. False, because
$y^3 - 27 = (y-3)(y^2+3y+9)$
for any value of y.
6. True, because $y^6 - 1 = (y^3)^2 - 1^2$.
7. False, because $2x - 4$ is not prime.
8. True, because it cannot be factored.
9. True, because $a^6 - 1 = (a^2)^3 - 1^3$.
10. False, because if we factor x out of the first two terms and a out of the last two terms we get
$x^2 + 3x - ax + 3a = x(x+3) + a(-x+3)$
and we cannot finish. If we factor x out of the first two terms and $-a$ out of the last two terms we get $x(x+3) - a(x-3)$ and again we cannot finish the factoring.

5.7 EXERCISES

1. Always factor out the greatest common factor first.
3. In factoring a trinomial look for the perfect square trinomials.
5. Prime, because it is a sum of two squares.
7. Not prime, because
$-9w^2 - 9 = -9(w^2+1)$.
9. Not prime, $x^2 - 2x - 3 = (x-3)(x+1)$.
11. Prime, because no two integers have a product of 3 and a sum of 2.
13. Prime, because no two numbers have a product of -3 and a sum of -4.
15. Prime, because no two numbers have a product of $6 \cdot (-4) = -24$ and a sum of 3.
17. Let $y = a^2$ in the polynomial:
$a^4 - 10a^2 + 25 = y^2 - 10y + 25$
$= (y-5)^2 = (a^2-5)^2$
19. Let $y = x^2$ in the polynomial:
$x^4 - 6x^2 + 8 = y^2 - 6y + 8 = (y-2)(y-4)$
$= (x^2-2)(x^2-4)$
$= (x^2-2)(x-2)(x+2)$
21. If we let $a = y^3$, then $a^2 = (y^3)^2 = y^6$.
$2y^6 - 128 = 2(y^6 - 64) = 2(a^2 - 64)$
$= 2(a-8)(a+8) = 2(y^3-8)(y^3+8)$
$= 2(y-2)(y^2+2y+4)(y+2)(y^2-2y+4)$

23. $32a^4 - 18 = 2(16a^4 - 9)$
$= 2(4a^2+3)(4a^2-3)$

25. Let $y = 3x - 5$ in the polynomial:
$(3x-5)^2 - 1 = y^2 - 1 = (y-1)(y+1)$
$= (3x-5-1)(3x-5+1)$
$= (3x-6)(3x-4)$
$= 3(x-2)(3x-4)$

27. Let $a = x^2$ and $b = x - 6$:
$x^4 - (x-6)^2 = a^2 - b^2 = (a+b)(a-b)$
$= (x^2 + x - 6)(x^2 - [x-6])$
$= (x+3)(x-2)(x^2 - x + 6)$

29. Let $y = m + 2$ in the polynomial:
$(m+2)^2 + 2(m+2) - 3 = y^2 + 2y - 3$
$= (y+3)(y-1) = (m+2+3)(m+2-1)$
$= (m+5)(m+1)$

31. Let $w = y - 1$ in the polynomial:
$3(y-1)^2 + 11(y-1) - 20$
$= 3w^2 + 11w - 20$
$= (3w-4)(w+5)$
$= [3(y-1) - 4][y - 1 + 5]$
$= (3y-7)(y+4)$

33. Let $a = y^2 - 3$:
$(y^2-3)^2 - 4(y^2-3) - 12 = a^2 - 4a - 12$
$= (a-6)(a+2)$
$= (y^2 - 3 - 6)(y^2 - 3 + 2)$
$= (y^2 - 9)(y^2 - 1)$
$= (y-3)(y+3)(y-1)(y+1)$

35. $x^2 - 2b - 2x + bx = x^2 - 2x + bx - 2b$
$= x(x-2) + b(x-2)$
$= (x-2)(x+b)$

37. $x^2 - ay - xy + ax = x^2 - xy + ax - ay$
$= x(x-y) + a(x-y)$
$= (x-y)(x+a)$

39. $x^2 + 2x + 1 - a^2 = (x+1)^2 - a^2$
$= (x+1-a)(x+1+a)$

41. $x^2 - 4x + 4 - w^2 = (x-2)^2 - w^2$
$= (x-2-w)(x-2+w)$

43. $x^2 - z^2 + 4x + 4 = x^2 + 4x + 4 - z^2$
$= (x+2)^2 - z^2$
$= (x+2-z)(x+2+z)$

45. This is a perfect square trinomial.
$9x^2 - 24x + 16 = (3x-4)^2$

47. Two numbers whose product is 36 and whose sum is -13 are -4 and -9.
$12x^2 - 13x + 3 = 12x^2 - 4x - 9x + 3$
$= 4x(3x-1) - 3(3x-1)$
$= (3x-1)(4x-3)$

49. $3a^4 + 81a = 3a(a^3 + 27)$
$= 3a(a+3)(a^2 - 3a + 9)$

51. $32 + 2x^2 = 2(x^2 + 16)$

53. Prime, because there are no two integers that have a product of 72 and a sum of -5.

55. Let $a = x + y$: $(x+y)^2 - 1 = a^2 - 1$
$= (a-1)(a+1) = (x+y-1)(x+y+1)$

57. $a^3b - ab^3 = ab(a^2 - b^2)$
$= ab(a-b)(a+b)$

59. $x^4 - 16 = (x^2 - 4)(x^2 + 4)$
$= (x-2)(x+2)(x^2+4)$

61. $x^4 + 2x^3 - 8x - 16$
$= x^3(x+2) - 8(x+2)$
$= (x+2)(x^3 - 8)$
$= (x+2)(x-2)(x^2 + 2x + 4)$

63. $m^2n + 2mn^2 + n^3$
$= n(m^2 + 2mn + n^2)$
$= n(m+n)^2$

65. $2m + wn + 2n + wm =$
$= 2m + 2n + wm + wn$
$= 2(m+n) + w(m+n)$
$= (m+n)(2+w)$

67. Two numbers with a product of -12 and a sum of 4 are 6 and -2:
$4w^2 + 4w - 3 = 4w^2 - 2w + 6w - 3$
$= 2w(2w-1) + 3(2w-1)$
$= (2w+3)(2w-1)$

69. Let $a = t^2$:
$t^4 + 4t^2 - 21 = a^2 + 4a - 21$
$= (a+7)(a-3)$
$= (t^2+7)(t^2-3)$

71. $-a^3 - 7a^2 + 30a = -a(a^2 + 7a - 30)$
$= -a(a+10)(a-3)$

73. $a^4 - w^4 = (a^2 - w^2)(a^2 + w^2)$
$= (a-w)(a+w)(a^2+w^2)$

75. Let $a = y + 5$: $(y+5)^2 - 2(y+5) - 3$
$= a^2 - 2a - 3 = (a-3)(a+1)$
$= (y+5-3)(y+5+1)$
$= (y+2)(y+6)$

77. $-2w^4 + 1250 = -2(w^4 - 625)$
$= -2(w^2 - 25)(w^2 + 25)$
$= -2(w-5)(w+5)(w^2+25)$

79. $4a^2 + 16 = 4(a^2 + 4)$

81. $8a^3 + 8a = 8a(a^2 + 1)$

83. Let $y = w + 5$:
$(w+5)^2 - 9 = y^2 - 9$
$= (y-3)(y+3)$
$= (w+5-3)(w+5+3)$
$= (w+2)(w+8)$

85. $4aw^2 - 12aw + 9a = a(4w^2 - 12w + 9)$
$= a(2w-3)^2$

87. $x^2 - 6xy + 9y^2 = (x-3y)^2$

89. $3x^4 - 75x^2 = 3x^2(x^2 - 25)$
$= 3x^2(x-5)(x+5)$

91. $m^3n - n = n(m^3 - 1)$
$= n(m-1)(m^2 + m + 1)$

93. $12x^2 + 2x - 30 = 2(6x^2 + x - 15)$
$= 2[6x^2 - 9x + 10x - 15]$
$= 2[3x(2x-3) + 5(2x-3)]$
$= 2(3x+5)(2x-3)$

95. $2a^3 - 32 = 2(a^3 - 16)$

97. $x^6 - y^6 = (x^3 - y^3)(x^3 + y^3)$
$= (x-y)(x^2 + xy + y^2)(x+y)(x^2 - xy + y^2)$

99. $a^{3m} - 1 = (a^m)^3 - 1^3$
$= (a^m - 1)(a^{2m} + a^m + 1)$

101. $a^{3w} - b^{6n} = (a^w)^3 - (b^{2n})^3$
$= (a^w - b^{2n})(a^{2w} + a^w b^{2n} + b^{4n})$

103. $t^{4n} - 16 = (t^{2n})^2 - 4^2$
$= (t^{2n} - 4)(t^{2n} + 4)$
$= (t^n - 2)(t^n + 2)(t^{2n} + 4)$

105. $a^{2n+1} - 2a^{n+1} - 15a$
$= a(a^{2n} - 2a^n - 15)$
$= a(a^n - 5)(a^n + 3)$

107. $a^{2n} - 3a^n + a^n b - 3b$
$= a^n(a^n - 3) + b(a^n - 3)$
$= (a^n - 3)(a^n + \mathrm{b})$

109. If you randomly write down trinomials, then most of them will be prime.

111. $a^2 + 8a + 16 = (a+4)^2$

113. $36 - y^2 = (6-y)(6+y)$

115. $a^2 - 16 = (a-4)(a+4)$

117. $w^2 + 18w + 81 = (w+9)^2$

119. $a^2 - 2a = a(a-2)$

121. $4w^2 + 36w + 81 = (2w+9)^2$

123. $x^2 - 1 = (x-1)(x+1)$

125. $w^3 + 27 = (w+3)(w^2 - 3w + 9)$

127. $aw - 4w - ab + 4b = (a-4)(w-b)$

129. $zw - 3w - 5z + 15 = (z-3)(w-5)$

131. $4b^2 - 4ab + a^2 = (2b-a)^2$

133. $a^3 - 27 = (a-3)(a^2 + 3a + 9)$

135. $3z^2 - 30z + 75$
$= 3(z^2 - 10z + 25)$
$= 3(z-5)^2$

137. $2b^3 - 16$
$= 2(b^3 - 8)$
$= 2(b-2)(b^2 + 2b + 4)$

139. $-2a^3 + 36a^2 - 162a$
$= -2a(a^2 - 18a + 81)$
$= -2a(a-9)^2$

141. $z^4 - 16 = (z^2 - 4)(z^2 + 4)$
$= (z-2)(z+2)(z^2 + 4)$

143. $a^3 + ab^2 = a(a^2 + b^2)$

145. $w^2 - 2w + 3aw - 6a$
$= w(w-2) + 3a(w-2)$
$= (w-2)(w+3a)$

147. $a^2 + 10a + 25 = (a+5)^2$

149. $25b^2 + 30b + 9 = (5b+3)^2$

151. $144 - y^2 = (12-y)(12+y)$

153. $9a^2 - z^2 = (3a-z)(3a+z)$

155. $3w^2 - 38w - 13 = (w-13)(3w+1)$

157. $m^2 + 4m - 21 = (m-3)(m+7)$

159. $b^4 - y^2 = (b^2 - y)(b^2 + y)$

161. $z^6 - 49 = (z^3 - 7)(z^3 + 7)$

163. $a^5 + 4a^3 = a^3(a^2 + 4)$

165. $75w^2 + 120w + 48$
$= 3(25w^2 + 40w + 16)$
$= 3(5w+4)^2$

167. $a^2x^2 - b^2 = (ax-b)(ax+b)$

169. $bx - xy + bz - zy$
$= x(b-y) + z(b-y)$
$= (b-y)(x+z)$

171. $z^3 - 125 = (z-5)(z^2 + 5z + 25)$

173. $27x^2 - 6x - 1 = (3x-1)(9x+1)$

175. $b^2n^2 - y^4 = (bn - y^2)(bn + y^2)$

177. $-6a^2 + 33a - 15$
$= -3(2a^2 - 11a + 5)$
$= -3(2a-1)(a-5)$

179. $4a^2b^2 - 4abw + w^2 = (2ab - w)^2$

181. $t^3 - 27 = (t-3)(t^2 + 3t + 9)$

183. $3z^6 - 30z^3 + 75$
$= 3(z^6 - 10z^3 + 25)$
$= 3(z^3 - 5)^2$

185. $5b^3 - 40 = 5(b^3 - 8)$
$= 5(b-2)(b^2 + 2b + 4)$

187. $-2a^4 + 20a^3 - 50a^2$
$= -2a^2(a^2 - 10a + 25)$
$= -2a^2(a-5)^2$

189. $a^4 - 16 = (a^2 - 4)(a^2 + 4)$
$= (a-2)(a+2)(a^2 + 4)$

191. $a^2b^2 + b^4 = b^2(a^2 + b^2)$

193. $x^4 + 2x^2 - 15 = (x^2 + 5)(x^2 - 3)$

195. $w^4 - 2w^2 - 6a + 3aw^2$
$= w^2(w^2 - 2) + 3a(w^2 - 2)$
$= (w^2 - 2)(w^2 + 3a)$

5.8 WARM-UPS

1. False, because 4 is a solution to $x-1=3$, but $(4-1)(4+3) \neq 12$.
2. True, both 2 and 3 satisfy $(x-2)(x-3)=0$.
3. True, because of the zero factor property.
4. False, $|x^2+4|=5$ is equivalent to $x^2+4=5$ or $x^2+4=-5$.
5. True, because of the zero factor property.
6. False, the Pythagorean theorem applies only to right triangles.
7. True, because the sum of the length and width of a rectangle is one-half of the perimeter.
8. True, $x+(8-x)=8$ for any number x.
9. False, the solution set also includes 0 because of the factor x.
10. False, because 3 is not a solution.

5.8 EXERCISES

1. The zero factor property says that if $a \cdot b = 0$ then either $a=0$ or $b=0$.
3. The hypotenuse of a right triangle is the side opposite the right angle.
5. The Pythagorean theorem says that a triangle is a right triangle if and only if the sum of the squares of the legs is equal to the square of the hypotenuse.
7. $(x-5)(x+4)=0$
$x-5=0$ or $x+4=0$
$x=5$ or $x=-4$
Solution set: $\{-4, 5\}$
9. $(2x-5)(3x+4)=0$
$2x-5=0$ or $3x+4=0$
$2x=5$ or $3x=-4$
$x=\frac{5}{2}$ or $x=-\frac{4}{3}$
Solution set: $\left\{\frac{5}{2}, -\frac{4}{3}\right\}$
11. $4(x-2)(x+5)=0$
Note that 4 is a factor, but 4 cannot cause the product to be zero. Only the factors containing variables could make the product zero.
$x-2=0$ or $x+5=0$
$x=2$ or $x=-5$
Solution set: $\{-5, 2\}$
13. $x(x-5)(x+5)=0$
$x=0$ or $x-5=0$ or $x+5=0$
$x=0$ or $x=5$ or $x=-5$
Solution set: $\{-5, 0, 5\}$
15. $w^2+5w-14=0$
$(w+7)(w-2)=0$
$w+7=0$ or $w-2=0$
$w=-7$ or $w=2$
Solution set: $\{-7, 2\}$
17. $m^2-7m=0$
$m(m-7)=0$
$m=0$ or $m-7=0$
$m=0$ or $m=7$
Solution set: $\{0, 7\}$
19. $a^2-a=20$
$a^2-a-20=0$
$(a-5)(a+4)=0$
$a-5=0$ or $a+4=0$
$a=5$ or $a=-4$
Solution set: $\{-4, 5\}$
21. $x^2-6x+9=0$
$(x-3)^2=0$
$x-3=0$
$x=3$
Solution set: $\{3\}$
23. $2a^2+7a=15$
$2a^2+7a-15=0$
$(2a-3)(a+5)=0$
$2a-3=0$ or $a+5=0$
$2a=3$ or $a=-5$
$a=\frac{3}{2}$ or $a=-5$
Solution set: $\left\{-5, \frac{3}{2}\right\}$
25. $(x-3)(x+2)=14$
$x^2-x-6=14$
$x^2-x-20=0$
$(x+4)(x-5)=0$
$x+4=0$ or $x-5=0$
$x=-4$ or $x=5$
Solution set: $\{-4, 5\}$
27. $(x-8)(x-2)=-5$
$x^2-10x+16=-5$
$x^2-10x+21=0$
$(x-7)(x-3)=0$
$x-7=0$ or $x-3=0$
$x=7$ or $x=3$
Solution set: $\{3, 7\}$

29. $(x-6)(x-2)=-4$
$x^2-8x+12=-4$
$x^2-8x+16=0$
$(x-4)^2=0$
$x-4=0$
$x=4$
Solution set: $\{4\}$

31. $10a^2+38a-8=0$
$2(5a^2+19a-4)=0$
$2(5a-1)(a+4)=0$
$5a-1=0$ or $a+4=0$
$5a=1$ or $a=-4$
$a=\frac{1}{5}$ or $a=-4$
Solution set: $\left\{-4, \frac{1}{5}\right\}$

33. $3x^2-3x-36=0$
$3(x^2-x-12)=0$
$3(x-4)(x+3)=0$
$x-4=0$ or $x+3=0$
$x=4$ or $x=-3$
Solution set: $\{-3, 4\}$

35. $z^2+\frac{3}{2}z=10$
$2\left(z^2+\frac{3}{2}z\right)=2(10)$
$2z^2+3z=20$
$2z^2+3z-20=0$
$(2z-5)(z+4)=0$
$2z-5=0$ or $z+4=0$
$2z=5$ or $z=-4$
$z=\frac{5}{2}$
Solution set: $\left\{-4, \frac{5}{2}\right\}$

37. $x^3-4x=0$
$x(x^2-4)=0$
$x(x-2)(x+2)=0$
$x=0$ or $x-2=0$ or $x+2=0$
$x=0$ or $x=2$ or $x=-2$
Solution set: $\{-2, 0, 2\}$

39. $-4x^3+x=3x^2$
$-4x^3-3x^2+x=0$
$-x(4x^2+3x-1)=0$
$-x(4x-1)(x+1)=0$
$-x=0$ or $4x-1=0$ or $x+1=0$
$x=0$ or $x=\frac{1}{4}$ or $x=-1$
The solution set is $\left\{-1, 0, \frac{1}{4}\right\}$.

41. $w^3+4w^2-25w-100=0$
$w^2(w+4)-25(w+4)=0$
$(w^2-25)(w+4)=0$
$(w-5)(w+5)(w+4)=0$
$w-5=0$ or $w+5=0$ or $w+4=0$
$w=5$ or $w=-5$ or $w=-4$
Solution set: $\{-5, -4, 5\}$

43. $n^3-2n^2-n+2=0$
$n^2(n-2)-1(n-2)=0$
$(n^2-1)(n-2)=0$
$(n-1)(n+1)(n-2)=0$
$n-1=0$ or $n+1=0$ or $n-2=0$
$n=1$ or $n=-1$ or $n=2$
Solution set: $\{-1, 1, 2\}$

45. $|x^2-5|=4$
$x^2-5=4$ or $x^2-5=-4$
$x^2-9=0$ or $x^2-1=0$
$(x-3)(x+3)=0$ or $(x-1)(x+1)=0$
$x-3=0$ or $x+3=0$ or $x-1=0$
or $x+1=0$
$x=3$ or $x=-3$ or $x=1$ or $x=-1$
Solution set: $\{-3, -1, 1, 3\}$

47. $|x^2+2x-36|=12$
$x^2+2x-36=12$ or $x^2+2x-36=-12$
$x^2+2x-48=0$ or $x^2+2x-24=0$
$(x+8)(x-6)=0$ or $(x+6)(x-4)=0$
$x=-8$ or $x=6$ or $x=-6$ or $x=4$
Solution set: $\{-8, -6, 4, 6\}$

49. $|x^2+4x+2|=2$
$x^2+4x+2=2$ or $x^2+4x+2=-2$
$x^2+4x=0$ or $x^2+4x+4=0$
$x(x+4)=0$ or $(x+2)^2=0$
$x=0$ or $x+4=0$ or $x+2=0$
$x=0$ or $x=-4$ or $x=-2$
Solution set: $\{-4, -2, 0\}$

51. $|x^2+6x+1|=8$
$x^2+6x+1=8$ or $x^2+6x+1=-8$
$x^2+6x-7=0$ or $x^2+6x+9=0$
$(x+7)(x-1)=0$ or $(x+3)^2=0$
$x+7=0$ or $x-1=0$ or $x+3=0$
$x=-7$ or $x=1$ or $x=-3$
Solution set: $\{-7, -3, 1\}$

53. $2x^2-x=6$
$2x^2-x-6=0$
$(2x+3)(x-2)=0$
$2x+3=0$ or $x-2=0$
$x=-\frac{3}{2}$ or $x=2$
Solution set: $\left\{-\frac{3}{2}, 2\right\}$

55. $|x^2 + 5x| = 6$
$x^2 + 5x = 6$ or $x^2 + 5x = -6$
$x^2 + 5x - 6 = 0$ or $x^2 + 5x + 6 = 0$
$(x + 6)(x - 1) = 0$ or $(x + 2)(x + 3) = 0$
$x + 6 = 0$ or $x - 1 = 0$ or $x + 2 = 0$ or $x + 3 = 0$
$x = -6$ or $x = 1$ or $x = -2$ or $x = -3$
Solution set: $\{-6, -3, -2, 1\}$

57. $x^2 + 5x = 6$
$x^2 + 5x - 6 = 0$
$(x + 6)(x - 1) = 0$
$x + 6 = 0$ or $x - 1 = 0$
$x = -6$ or $x = 1$
Solution set: $\{-6, 1\}$

59. $(x + 2)(x + 1) = 12$
$x^2 + 3x + 2 = 12$
$x^2 + 3x - 10 = 0$
$(x + 5)(x - 2) = 0$
$x + 5 = 0$ or $x - 2 = 0$
$x = -5$ or $x = 2$
The solution set is $\{-5, 2\}$.

61. $y^3 + 9y^2 + 20y = 0$
$y(y^2 + 9y + 20) = 0$
$y(y + 4)(y + 5) = 0$
$y = 0$ or $y + 4 = 0$ or $y + 5 = 0$
$y = 0$ or $y = -4$ or $y = -5$

Solution set: $\{-5, -4, 0\}$

63. $5a^3 = 45a$
$5a^3 - 45a = 0$
$5a(a^2 - 9) = 0$
$5a(a - 3)(a + 3) = 0$
$a = 0$ or $a - 3 = 0$ or $a + 3 = 0$
$a = 0$ or $a = 3$ or $a = -3$
Solution set: $\{-3, 0, 3\}$

65. $(2x - 1)(x^2 - 9) = 0$
$(2x - 1)(x - 3)(x + 3) = 0$
$2x - 1 = 0$ or $x - 3 = 0$ or $x + 3 = 0$
$x = \frac{1}{2}$ or $x = 3$ or $x = -3$
The solution set is $\left\{-3, \frac{1}{2}, 3\right\}$.

67. $(2x - 1)(3x + 1)(4x - 1) = 0$
$2x - 1 = 0$ or $3x + 1 = 0$ or $4x - 1 = 0$
$x = \frac{1}{2}$ or $x = -\frac{1}{3}$ or $x = \frac{1}{4}$
The solution set is $\left\{-\frac{1}{3}, \frac{1}{4}, \frac{1}{2}\right\}$.

69. $4x^2 - 12x + 9 = 0$
$(2x - 3)^2 = 0$
$2x - 3 = 0$
$x = \frac{3}{2}$
The solution set is $\left\{\frac{3}{2}\right\}$.

71. $y^2 + by = 0$
$y(y + b) = 0$
$y = 0$ or $y + b = 0$
$y = 0$ or $y = -b$
Solution set: $\{0, -b\}$

73. $a^2y^2 - b^2 = 0$
$(ay - b)(ay + b) = 0$
$ay - b = 0$ or $ay + b = 0$
$ay = b$ or $ay = -b$
$y = \frac{b}{a}$ or $y = -\frac{b}{a}$
Solution set: $\{-\frac{b}{a}, \frac{b}{a}\}$

75. $4y^2 + 4by + b^2 = 0$
$(2y + b)^2 = 0$
$2y + b = 0$
$2y = -b$
$y = -\frac{b}{2}$

Solution set: $\left\{-\frac{b}{2}\right\}$

77. $ay^2 + 3y - ay = 3$
$ay^2 + 3y - ay - 3 = 0$
$y(ay + 3) - 1(ay + 3) = 0$
$(y - 1)(ay + 3) = 0$
$y - 1 = 0$ or $ay + 3 = 0$
$y = 1$ or $y = -\frac{3}{a}$
Solution set: $\left\{-\frac{3}{a}, 1\right\}$

79. Let x = the width and $x + 2$ = the length. Since $A = LW$, we can write the equation
$x(x + 2) = 24$
$x^2 + 2x = 24$
$x^2 + 2x - 24 = 0$
$(x + 6)(x - 4) = 0$
$x + 6 = 0$ or $x - 4 = 0$
$x = -6$ or $x = 4$
$x + 2 = 6$
Since the width cannot be -6, we have a width of 4 inches and a length of 6 inches.

81. Let x = one number and $13 - x$ = the other. Their product is 36:
$x(13 - x) = 36$
$-x^2 + 13x - 36 = 0$
$x^2 - 13x + 36 = 0$
$(x - 4)(x - 9) = 0$

$$x-4=0 \text{ or } x-9=0$$
$$x=4 \text{ or } x=9$$
$$13-x=9 \text{ or } 13-x=4$$

The numbers are 4 and 9.

83. Let $x =$ the width and $x+21 =$ the length.

$$x(x+21)=946$$
$$x^2+21x-946=0$$
$$(x+43)(x-22)=0$$
$$x+43=0 \text{ or } x-22=0$$
$$x=-43 \text{ or } x=22$$
$$x+21=43$$

The length is 43 in. and the width is 22 in.

85. Let $x =$ the width and $2x+2 =$ the length. Using the Pythagorean theorem we can write

$$x^2+(2x+2)^2=13^2$$
$$x^2+4x^2+8x+4=169$$
$$5x^2+8x-165=0$$
$$(5x+33)(x-5)=0$$
$$5x+33=0 \text{ or } x-5=0$$
$$x=-\frac{33}{5} \text{ or } x=5$$
$$2x+2=12$$

Since the dimensions must be positive numbers, the width is 5 feet and the length is 12 feet.

87. a) From the graph it appears the ball is in the air for 4 seconds.

b) To find the time the ball is in the air we need the values of t that gives $h(t)$ a value of 0. We must solve the equation

$$-16t^2+64t=0$$
$$-16t(t-4)=0$$
$$t=0 \text{ or } t-4=0$$
$$t=0 \text{ or } t=4$$

The ball is in the air for 4 seconds.

c) From the graph it appears that the maximum height is 64 feet.

d) The arrow reaches its maximum height at time $t=2$ seconds.

89. a) $h(5)=-16\cdot 5^2+325\cdot 5$
$=1225 \text{ ft}$

b) $-16t^2+325t=0$

$$t(-16t+325)=0$$
$$t=0 \text{ or } -16t+325=0$$
$$\text{or } t=\frac{-325}{-16}=20.3125$$

It takes 20.3125 sec to return to earth.

91. $-16t^2+32t+48=0$

$$-16(t^2-2t-3)=0$$
$$-16(t-3)(t+1)=0$$
$$t-3=0 \text{ or } t+1=0$$
$$t=3 \text{ or } t=-1$$

The ball strikes the earth after 3 sec.

93. Since the perimeter is 34 feet the sum of the length and width is 17 feet. Let $x =$ the width and $17-x =$ the length. Using the Pythagorean theorem we can write

$$x^2+(17-x)^2=13^2$$
$$x^2+289-34x+x^2=169$$
$$2x^2-34x+120=0$$
$$x^2-17x+60=0$$
$$(x-5)(x-12)=0$$
$$x-5=0 \text{ or } x-12=0$$
$$x=5 \text{ or } x=12$$
$$17-x=12 \text{ or } 17-x=5$$

The width is 5 feet and the length is 12 feet.

95. Let $x =$ the distance from the bottom of the ladder to the wall and $x+3 =$ the distance from the top of the ladder to the ground.

$$x^2+(x+3)^2=15^2$$
$$x^2+x^2+6x+9=225$$
$$2x^2+6x-216=0$$
$$x^2+3x-108=0$$
$$(x+12)(x-9)=0$$
$$x+12=0 \text{ or } x-9=0$$
$$x=-12 \text{ or } x=9$$
$$x+3=12$$

The distance from the top of the ladder to the ground is 12 feet.

97. Let $x =$ the number.

$$x^2-x=12$$
$$x^2-x-12=0$$
$$(x-4)(x+3)=0$$
$$x-4=0 \text{ or } x+3=0$$
$$x=4 \text{ or } x=-3$$

The number is either -3 or 4.

99. Let $x =$ the first integer and $x+1 =$ the second integer. The sum of their squares is 25:

$$x^2+(x+1)^2=25$$
$$x^2+x^2+2x+1=25$$
$$2x^2+2x-24=0$$
$$x^2+x-12=0$$
$$(x+4)(x-3)=0$$
$$x+4=0 \text{ or } x-3=0$$
$$x=-4 \text{ or } x=3$$
$$x+1=-3 \text{ or } x+1=4$$

The integers are 3 and 4, or -4 and -3.

101. Let $x =$ the original length. Since the length times the width is 240 square feet, the width is $240/x$. The new length is $x - 4$ and the new width is $(240/x) + 3$. Since the area is still 240 square feet, we can write the following equation.

$$(x-4)\left(\frac{240}{x}+3\right) = 240$$
$$x \cdot \frac{240}{x} - 4 \cdot \frac{240}{x} + 3x - 12 = 240$$
$$240 - \frac{960}{x} + 3x - 252 = 0$$
$$-\frac{960}{x} + 3x - 12 = 0$$
$$x\left(-\frac{960}{x} + 3x - 12\right) = x \cdot 0$$
$$-960 + 3x^2 - 12x = 0$$
$$3x^2 - 12x - 960 = 0$$
$$x^2 - 4x - 320 = 0$$
$$(x-20)(x+16) = 0$$
$$x - 20 = 0 \text{ or } x + 16 = 0$$
$$x = 20 \text{ or } \quad x = -16$$
$$\frac{240}{x} = 12$$

The length is 20 feet and the width is 12 feet.

103. Dividing each side of an equation by a variable can cause you to lose a solution to the equation. Both 0 and 1 satisfy $x^2 = x$.

105. The 5 can be disregarded. If a product is equal to zero, it is because one of the factors is equal to zero and 5 is not 0.

Chapter 5 Wrap-Up

Enriching Your Mathematical Word Power

1. d **2.** b **3.** c **4.** d **5.** b **6.** a
7. c **8.** a **9.** a **10.** d **11.** c **12.** a
13. c **14.** b **15.** c **16.** a **17.** c

CHAPTER 5 REVIEW

1. $2 \cdot 2 \cdot 2^{-1} = 4 \cdot \frac{1}{2} = 2$

3. $2^2 \cdot 3^2 = 4 \cdot 9 = 36$

5. $(-3)^{-3} = -\frac{1}{3^3} = -\frac{1}{27}$

7. $-(-1)^{-3} = -\frac{1}{(-1)^3} = -\frac{1}{-1} = 1$

9. $2x^3 \cdot 4x^{-6} = 8x^{-3} = \frac{8}{x^3}$

11. $\frac{y^{-5}}{y^{-3}} = y^{-2} = \frac{1}{y^2}$

13. $\frac{a^5a^{-2}}{a^{-4}} = \frac{a^3}{a^{-4}} = a^7$

15. $\frac{6x^{-2}}{3x^2} = 2x^{-4} = \frac{2}{x^4}$

17. Move the decimal 6 places to the right: $8.36 \times 10^6 = 8{,}360{,}000$

19. Move the decimal point 4 places to the left: $5.7 \times 10^{-4} = 0.00057$

21. Move the decimal point 6 places to the left: $8{,}070{,}000 = 8.07 \times 10^6$

23. Move the decimal point 4 places to the right: $0.000709 = 7.09 \times 10^{-4}$

25. $(5(2 \times 10^4))^3 = 5^3(2 \times 10^4)^3$
$= 125 \cdot 8 \times 10^{12} = 1000 \times 10^{12}$
$= 1 \times 10^3 \times 10^{12}$
$= 1 \times 10^{15}$

27. $\frac{(2 \times 10^{-9})(3 \times 10^7)}{5(6 \times 10^{-4})} = \frac{6 \times 10^{-2}}{30 \times 10^{-4}}$
$= 0.2 \times 10^2 = 2 \times 10^1$

29. $\frac{(4 \times 10^9)(6 \times 10^{-7})}{(1.2 \times 10^{-5})(2 \times 10^6)} = \frac{24 \times 10^2}{2.4 \times 10^1}$
$= 10 \times 10^1 = 1 \times 10^2$

31. $(a^{-3})^{-2} \cdot a^{-7} = a^6 \cdot a^{-7} = a^{-1} = \frac{1}{a}$

33. $(m^2n^3)^{-2}(m^{-3}n^2)^4 = m^{-4}n^{-6}m^{-12}n^8$
$= m^{-16}n^2 = \frac{n^2}{m^{16}}$

35. $\left(\frac{2}{3}\right)^{-4} = \left(\frac{3}{2}\right)^4 = \frac{3^4}{2^4} = \frac{81}{16}$

37. $\left(\frac{1}{2} + \frac{1}{3}\right)^2 = \left(\frac{5}{6}\right)^2 = \frac{25}{36}$

39. $\left(-\frac{3a}{4b^{-1}}\right)^{-1} = -\frac{4b^{-1}}{3a} = -\frac{4}{3ab}$

41. $\frac{(a^{-3}b)^4}{(ab^2)^{-5}} = \frac{a^{-12}b^4}{a^{-5}b^{-10}} = \frac{b^{14}}{a^7}$

43. $5^{2w} \cdot 5^{4w} \cdot 5^{-1} = 5^{6w-1}$

45. $\left(\frac{7^{3a}}{7^8}\right)^5 = \frac{7^{15a}}{7^{40}} = 7^{15a-40}$

47. $(2w - 3) + (6w + 5) = 8w + 2$

49. $(x^2 - 3x - 4) - (x^2 + 3x - 7)$
$= x^2 - 3x - 4 - x^2 - 3x + 7 = -6x + 3$

51. $(x^2 - 2x + 4)(x - 2)$
$= (x^2 - 2x + 4)(x) - (x^2 - 2x + 4)(2)$
$= x^3 - 2x^2 + 4x - 2x^2 + 4x - 8$
$= x^3 - 4x^2 + 8x - 8$

53. $xy + 7z - 5(xy - 3z)$
$= xy + 7z - 5xy + 15z = -4xy + 22z$

55. $m^2(5m^3 - m + 2) = 5m^5 - m^3 + 2m^2$

57. $(x - 3)(x + 7) = x^2 - 3x + 7x - 21$
$= x^2 + 4x - 21$

59. $(z - 5y)(z + 5y) = z^2 - (5y)^2$
$= z^2 - 25y^2$

61. $(m + 8)^2 = m^2 + 2 \cdot 8m + 8^2$
$= m^2 + 16m + 64$

63. $(w - 6x)(w - 4x)$
$= w^2 - 6xw - 4xw + 24x^2$
$= w^2 - 10xw + 24x^2$

65. $(k - 3)^2 = k^2 - 2 \cdot 3 \cdot k + 9$
$= k^2 - 6k + 9$

67. $(m^2 - 5)(m^2 + 5)$
$= m^4 - 5m^2 + 5m^2 - 25$
$= m^4 - 25$

69. $3x - 6 = 3(x - 2)$

71. $4a - 20 = -4(-a + 5)$

73. $3w - w^2 = -w(-3 + w) = -w(w - 3)$

75. $y^2 - 81 = (y - 9)(y + 9)$

77. $4x^2 + 28x + 49$
$= (2x)^2 + 2 \cdot 2x \cdot 7 + 7^2$
$= (2x + 7)^2$

79. $t^2 - 18t + 81$
$= t^2 - 2 \cdot 9t + 9^2 = (t - 9)^2$

81. $t^3 - 125 = (t - 5)(t^2 + 5t + 25)$

83. Two numbers with a product of 40 and a sum of 14 are 10 and 4.
$x^2 + 14x + 40 = (x + 4)(x + 10)$

85. Two numbers with a product of -30 and a sum of -7 are -10 and 3.
$x^2 - 7x - 30 = (x - 10)(x + 3)$

87. Two numbers with a product of -28 and a sum of -3 are -7 and 4.
$w^2 - 3w - 28 = (w - 7)(w + 4)$

89. Two numbers with a product of -14 and a sum of 5 are -2 and 7.
$2m^2 + 5m - 7 = 2m^2 - 2m + 7m - 7$
$= 2m(m - 1) + 7(m - 1)$
$= (2m + 7)(m - 1)$

91. $m^7 - 3m^4 - 10m = m(m^6 - 3m^3 - 10)$
$= m(m^3 - 5)(m^3 + 2)$

93. $5x^3 + 40 = 5(x^3 + 8)$
$= 5(x + 2)(x^2 - 2x + 4)$

95. Two numbers with a product of 18 and a sum of 9 are 6 and 3.
$9x^2 + 9x + 2 = 9x^2 + 3x + 6x + 2$
$= 3x(3x + 1) + 2(3x + 1)$
$= (3x + 2)(3x + 1)$

97. $x^3 + x^2 - x - 1$
$= x^2(x + 1) - 1(x + 1)$
$= (x + 1)(x^2 - 1)$
$= (x + 1)(x + 1)(x - 1)$
$= (x + 1)^2(x - 1)$

99. $-x^2y + 16y = -y(x^2 - 16)$
$= -y(x - 4)(x + 4)$

101. $-a^3b^2 + 2a^2b^2 - ab^2$
$= -ab^2(a^2 - 2a + 1)$
$= -ab^2(a - 1)^2$

103. $x^3 - x^2 + 9x - 9$
$= x^2(x - 1) + 9(x - 1)$
$= (x - 1)(x^2 + 9)$

105. $x^4 - x^2 - 12 = (x^2 - 4)(x^2 + 3)$
$= (x - 2)(x + 2)(x^2 + 3)$

107. $a^6 - a^3 = a^3(a^3 - 1)$
$= a^3(a - 1)(a^2 + a + 1)$

109. $-8m^2 - 24m - 18$
$= -2(4m^2 + 12m + 9)$
$= -2(2m + 3)^2$

111. Let $y = 2x - 3$:
$(2x - 3)^2 - 16 = y^2 - 16$
$= (y - 4)(y + 4)$
$= (2x - 3 - 4)(2x - 3 + 4)$
$= (2x - 7)(2x + 1)$

113. $x^6 + 7x^3 - 8 = (x^3 + 8)(x^3 - 1)$
$= (x + 2)(x^2 - 2x + 4)(x - 1)(x^2 + x + 1)$

115. Let $y = a^2 - 9$:
$(a^2 - 9)^2 - 5(a^2 - 9) + 6$
$= y^2 - 5y + 6$
$= (y - 2)(y - 3)$
$= (a^2 - 9 - 2)(a^2 - 9 - 3)$
$= (a^2 - 11)(a^2 - 12)$

117. $x^{2k} - 49 = (x^k - 7)(x^k + 7)$

119. $m^{2a} - 2m^a - 3 = (m^a - 3)(m^a + 1)$

121. $9z^{2k} - 12z^k + 4 = (3z^k - 2)^2$

123. $y^{2a} - by^a + cy^a - bc$
$= y^a(y^a - b) + c(y^a - b) = (y^a - b)(y^a + c)$

125. $x^3 - 5x^2 = 0$
$x^2(x - 5) = 0$
$x^2 = 0$ or $x - 5 = 0$
$x = 0$ or $x = 5$
Solution set: $\{0, 5\}$

127. $(a - 2)(a - 3) = 6$
$a^2 - 5a + 6 = 6$
$a^2 - 5a = 0$
$a(a - 5) = 0$
$a = 0$ or $a - 5 = 0$
$a = 0$ or $a = 5$
Solution set: $\{0, 5\}$

129.
$$2m^2 - 9m - 5 = 0$$
$$(2m+1)(m-5) = 0$$
$$2m + 1 = 0 \text{ or } m - 5 = 0$$
$$m = -1/2 \text{ or } m = 5$$
Solution set: $\{-1/2, 5\}$

131.
$$w^3 + 5w^2 - w - 5 = 0$$
$$w^2(w+5) - 1(w+5) = 0$$
$$(w^2 - 1)(w+5) = 0$$
$$(w-1)(w+1)(w+5) = 0$$
$$w - 1 = 0 \text{ or } w + 1 = 0 \text{ or } w + 5 = 0$$
$$w = 1 \text{ or } w = -1 \text{ or } w = -5$$
Solution set: $\{-5, -1, 1\}$

133.
$$|x^2 - 5| = 4$$
$$x^2 - 5 = 4 \text{ or } x^2 - 5 = -4$$
$$x^2 - 9 = 0 \text{ or } x^2 - 1 = 0$$
$$(x-3)(x+3) = 0 \text{ or } (x-1)(x+1) = 0$$
$$x = 3 \text{ or } x = -3 \text{ or } x = 1 \text{ or } x = -1$$
Solution set: $\{-3, -1, 1, 3\}$

135. Let $x =$ the distance from the bottom of the ladder and the cactus and $x + 2 =$ the distance from the top of the ladder to the ground. Since the ladder is the hypotenuse of a right triangle, we can use the Pythagorean theorem:
$$x^2 + (x+2)^2 = 10^2$$
$$x^2 + x^2 + 4x + 4 = 100$$
$$2x^2 + 4x - 96 = 0$$
$$x^2 + 2x - 48 = 0$$
$$(x+8)(x-6) = 0$$
$$x + 8 = 0 \text{ or } x - 6 = 0$$
$$x = -8 \text{ or } x = 6$$
The ladder should be placed 6 feet from the cactus.

137. Since the perimeter is 32 meters, the sum of the length and width is 16 meters. So if x represents the length, then $16 - x$ represents the width. Now use the fact that the area is 63 square meters:
$$x(16 - x) = 63$$
$$-x^2 + 16x - 63 = 0$$
$$x^2 - 16x + 63 = 0$$
$$(x-7)(x-9) = 0$$
$$x - 7 = 0 \text{ or } x - 9 = 0$$
$$x = 7 \text{ or } x = 9$$
$$16 - x = 9 \text{ or } 16 - x = 7$$
The width is 7 meters and the length is 9 meters.

139. Let x represent the width and $x + 17$ represent the length. The diagonal is 25.
$$x^2 + (x+17)^2 = 25^2$$
$$x^2 + x^2 + 34x + 289 = 625$$
$$2x^2 + 34x - 336 = 0$$
$$x^2 + 17x - 168 = 0$$
$$(x+24)(x-7) = 0$$
$$x + 24 = 0 \text{ or } x - 7 = 0$$
$$x = -24 \text{ or } x = 7$$
$$x + 17 = 24$$
The screen is 7 inches by 24 inches.

141. a) $L = 64.3(1.0033)^{20} \approx 68.7$ yr
b) From Exercise 95 Section 5.2, a 20-yr old white male is expected to live 75.1 yr. The White male is expected to live 6.5 years longer. To get this answer subtract before rounding each life expectancy.

143. a) The interest rate is approximately 15%.
b) $500{,}000 = R \cdot \dfrac{(1+0.07)^{20} - 1}{0.07}$
$$R = 500{,}000 \div \frac{(1+0.07)^{20} - 1}{0.07}$$
$$\approx \$12,196.46$$

CHAPTER 5 TEST

1. $3^{-2} = \dfrac{1}{3^2} = \dfrac{1}{9}$

2. $\dfrac{1}{6^{-2}} = 6^2 = 36$

3. $\left(\dfrac{1}{2}\right)^{-3} = 2^3 = 8$

4. $3x^4 \cdot 4x^3 = 3 \cdot 4x^4x^3 = 12x^7$

5. $\dfrac{8y^9}{2y^{-3}} = 4y^{9-(-3)} = 4y^{12}$

6. $(4a^2b)^3 = 4^3(a^2)^3b^3 = 64a^6b^3$

7. $\left(\dfrac{x^2}{3}\right)^{-3} = \left(\dfrac{3}{x^2}\right)^3 = \dfrac{27}{x^6}$

8. $\dfrac{(2^{-1})^{-3}(a^2)^{-3}b^{-3}}{4a^{-9}} = \dfrac{2^3a^{-6}b^{-3}}{4a^{-9}} = \dfrac{8a^3}{4b^3}$
$$= \frac{2a^3}{b^3}$$

9. $3.24 \times 10^9 = 3{,}240{,}000{,}000$

10. $8.673 \times 10^{-4} = 0.0008673$

11. $\dfrac{(8 \times 10^4)(6 \times 10^{-4})}{2 \times 10^6} = 24 \times 10^{-6}$
$$= 2.4 \times 10^1 \times 10^{-6} = 2.4 \times 10^{-5}$$

12. $\dfrac{(6 \times 10^{-5})^2(5 \times 10^2)}{(3 \times 10^4)^2\,(1 \times 10^{-2})}$
$$= \frac{36 \times 10^{-10} \cdot 5 \times 10^2}{9 \times 10^8 \cdot 1 \times 10^{-2}}$$
$$= 20 \times 10^{-14}$$
$$= 2 \times 10^{-13}$$

13. Add like terms to get $3x^3 + 3x^2 - 2x + 3$.

14. $(x^2-6x-7)-(3x^2+2x-4)$
$=x^2-6x-7-3x^2-2x+4$
$=-2x^2-8x-3$

15. $(x^2-3x+7)(x-2)$
$=(x^2-3x+7)(x)-(x^2-3x+7)(2)$
$=x^3-3x^2+7x-2x^2+6x-14$
$=x^3-5x^2+13x-14$

16. $(x-2)^3=(x-2)^2(x-2)$
$=(x^2-4x+4)(x-2)$
$=(x^2-4x+4)(x)-(x^2-4x+4)(2)$
$=x^3-4x^2+4x-2x^2+8x-8$
$=x^3-6x^2+12x-8$

17. $(x-7)(2x+3)=2x^2-14x+3x-21$
$=2x^2-11x-21$

18. $(x-6)^2=x^2-2\cdot 6x+36$
$=x^2-12x+36$

19. $(2x+5)^2=4x^2+20x+25$

20. $(3y^2-5)(3y^2+5)=9y^4-25$

21. Two numbers with a product of -24 and a sum of -2 are -6 and 4.
$a^2-2a-24=(a-6)(a+4)$

22. $4x^2+28x+49$
$=(2x)^2+2\cdot 2x\cdot 7+7^2=(2x+7)^2$

23. $3m^3-24=3(m^3-8)$
$=3(m-2)(m^2+2m+4)$

24. $2x^2y-32y=2y(x^2-16)$
$=2y(x-4)(x+4)$

25. $12m^2+28m+15=(2m+3)(6m+5)$

26. $3x^{10}+5x^5-12=(2x^5-3)(x^5+4)$

27. $2xa+3a-10x-15$
$=a(2x+3)-5(2x+3)$
$=(2x+3)(a-5)$

28. $x^4+3x^2-4=(x^2-1)(x^2+4)$
$=(x-1)(x+1)(x^2+4)$

29. $a^4-1=(a^2-1)(a^2+1)$
$=(a-1)(a+1)(a^2+1)$

30. $2m^2+7m-15=0$
$(2m-3)(m+5)=0$
$2m-3=0$ or $m+5=0$
$m=\frac{3}{2}$ or $m=-5$
Solution set: $\left\{-5, \frac{3}{2}\right\}$

31. $a^2+10a+25=0$
$(a+5)^2=0$
$a+5=0$
$a=-5$
The solution set is $\{-5\}$.

32. $x^3-4x=0$
$x(x^2-4)=0$
$x(x-2)(x+2)=0$
$x=0$ or $x-2=0$ or $x+2=0$
$x=0$ or $x=2$ or $x=-2$
Solution set: $\{-2, 0, 2\}$

33. $|x^2+x-9|=3$
$x^2+x-9=3$ or $x^2+x-9=-3$
$x^2+x-12=0$ or $x^2+x-6=0$
$(x+4)(x-3)=0$ or $(x+3)(x-2)=0$
$x=-4$ or $x=3$ or $x=-3$ or $x=2$
Solution set: $\{-4, -3, 2, 3\}$

34. Let $x=$ the height and $x+2=$ the width. Using the Pythagorean theorem we can write
$x^2+(x+2)^2=10^2$
$x^2+x^2+4x+4=100$
$2x^2+4x-96=0$
$x^2+2x-48=0$
$(x+8)(x-6)=0$
$x=-8$ or $x=6$
$x+2=-6$ or $x+2=8$
The width is 8 inches and the height is 6 inches.

35. $d=(1.8\times 10^{28})(1.032)^{-1950}\approx 38.0$
$d=(1.8\times 10^{28})(1.032)^{-1990}\approx 10.8$
$d=(1.8\times 10^{28})(1.032)^{-2000}\approx 7.9$

36. a) $h(2)=-16\cdot 2^2+80\cdot 2=96$ ft
$h(3)=-16\cdot 3^2+80\cdot 3=96$ ft
b) $-16t^2+80t=0$
$-16t(t-5)=0$
$-16t=0$ or $t-5=0$
$t=0$ or $t=5$
The stone is in the air for 5 seconds.

37. The sum of the length and width is 44 ft. So if x represents the width, then $44-x$ represents the length.
$x(44-x)=480$
$-x^2+44x-480=0$
$x^2-44x+480=0$
$(x-20)(x-24)=0$
$x-20=0$ or $x-24=0$
$x=20$ or $x=24$
$44-x=24$ or $44-x=20$

So the width is 20 ft and the length is 24 ft.

Making Connections

Chapters 1 - 5

1. $4^2 = 4 \cdot 4 = 16$

2. $4(-2) = -8$

3. $4^{-2} = \frac{1}{4^2} = \frac{1}{16}$

4. $2^3 \cdot 4^{-1} = 8 \cdot \frac{1}{4} = 2$

5. $2^{-1} + 2^{-1} = \frac{1}{2} + \frac{1}{2} = 1$

6. $2^{-1} \cdot 3^{-1} = \frac{1}{2} \cdot \frac{1}{3} = \frac{1}{6}$

7. $2^{-2} \cdot 3^2 = \frac{1}{4} \cdot 9 = \frac{9}{4}$

8. $-3^4 \cdot 6^{-2} = -81 \cdot \frac{1}{36} = -\frac{9}{4}$

9. $(-2)^3 \cdot 6^{-1} = -8 \cdot \frac{1}{6} = -\frac{4}{3}$

10. $8^{-3} \cdot 8^3 = 8^0 = 1$

11. $\left(\frac{2^{-2}}{2} + \frac{1}{2}\right)^2 = \left(\frac{1}{8} + \frac{1}{2}\right)^2$
$= \left(\frac{5}{8}\right)^2 = \frac{25}{64}$

12. $\left(\frac{2^2+1}{2^{-2}+1}\right)^{-3} = \left(\frac{5}{\frac{1}{4}+1}\right)^{-3}$
$= \left(\frac{5}{\frac{5}{4}}\right)^{-3} = 4^{-3} = \frac{1}{64}$

13. $\left(\frac{3-6}{8-4}\right)^{-2} = \left(-\frac{3}{4}\right)^{-2}$
$= \left(-\frac{4}{3}\right)^2 = \frac{16}{9}$

14. $\left(\frac{6-9}{14-20}\right)^{-3} = \left(\frac{1}{2}\right)^{-3}$
$= (2)^3 = 8$

15. $\left(\frac{1}{2^{-3}+1}\right)^{-1} = 2^{-3} + 1$
$= \frac{1}{8} + 1 = \frac{9}{8}$

16. $(2^{-1}+1)^2 = \left(\frac{3}{2}\right)^2 = \frac{9}{4}$

17. $3^{-1} - 2^{-2} = \frac{1}{3} - \frac{1}{4} = \frac{4}{12} - \frac{3}{12} = \frac{1}{12}$

18. $3^2 - 4(5)(-2) = 9 - (-40) = 49$

19. $2^7 - 2^6 = 128 - 64 = 64$

20. $0.08(32) + 0.08(68) = 0.08(32 + 68)$
$= 0.08(100) = 8$

21. $3 - 2\,|\,5 - 7 \cdot 3\,| = 3 - 2\,|-16\,|$
$= 3 - 2 \cdot 16 = 3 - 32 = -29$

22. $5^{-1} + 6^{-1} = \frac{1}{5} + \frac{1}{6} = \frac{6}{30} + \frac{5}{30} = \frac{11}{30}$

23.
$$0.05a - 0.04(a - 50) = 4$$
$$0.05a - 0.04a + 2 = 4$$
$$0.01a = 2$$
$$a = 200$$
Solution set: $\{200\}$

24.
$$15b - 27 = 0$$
$$15b = 27$$
$$b = \frac{27}{15} = \frac{9}{5}$$
Solution set: $\left\{\frac{9}{5}\right\}$

25.
$$2c^2 + 15c - 27 = 0$$
$$(2c - 3)(c + 9) = 0$$
$$2c - 3 = 0 \text{ or } c + 9 = 0$$
$$c = \frac{3}{2} \text{ or } c = -9$$
Solution set: $\left\{-9, \frac{3}{2}\right\}$

26.
$$2t^2 + 15t = 0$$
$$t(2t + 15) = 0$$
$$t = 0 \text{ or } 2t + 15 = 0$$
$$t = 0 \text{ or } t = -\frac{15}{2}$$
Solution set: $\left\{-\frac{15}{2}, 0\right\}$

27.
$$|\,15u - 27\,| = 3$$
$$15u - 27 = 3 \text{ or } 15u - 27 = -3$$
$$15u = 30 \text{ or } 15u = 24$$
$$u = 2 \text{ or } u = \frac{8}{5}$$
Solution set: $\left\{2, \frac{8}{5}\right\}$

28.
$$\begin{aligned} |15v - 27| &= 0 \\ 15v - 27 &= 0 \\ 15v &= 27 \\ v &= \frac{27}{15} = \frac{9}{5} \end{aligned}$$
Solution set: $\left\{\frac{9}{5}\right\}$

29. Absolute value of any quantity is greater than or equal to 0. So the solution set is $\emptyset$.

30.
$$|x^2 + x - 4| = 2$$
$x^2 + x - 4 = 2$ or $x^2 + x - 4 = -2$
$x^2 + x - 6 = 0$ or $x^2 + x - 2 = 0$
$(x + 3)(x - 2) = 0$ or $(x + 2)(x - 1) = 0$
$x = -3$ or $x = 2$ or $x = -2$ or $x = 1$
Solution set: $\{-3, -2, 1, 2\}$

31. $(2x - 1)(x + 5) = 0$
$2x - 1 = 0$ or $x + 5 = 0$
$x = \frac{1}{2}$ or $x = -5$
Solution set: $\left\{-5, \frac{1}{2}\right\}$

32.
$$\begin{aligned} |3x - 1| + 6 &= 9 \\ |3x - 1| &= 3 \end{aligned}$$
$3x - 1 = 3$ or $3x - 1 = -3$
$3x = 4$ or $3x = -2$
$x = \frac{4}{3}$ or $x = -\frac{2}{3}$
Solution set: $\left\{-\frac{2}{3}, \frac{4}{3}\right\}$

33. $(1.5 \times 10^{-4})w = 7 \times 10^6 + 5 \times 10^5$
$$\begin{aligned} w &= \frac{7 \times 10^6 + 5 \times 10^5}{1.5 \times 10^{-4}} \\ &= 5 \times 10^{10} \end{aligned}$$

34. $(3 \times 10^7)(y - 5 \times 10^3) = 6 \times 10^{12}$
$$\begin{aligned} y - 5 \times 10^3 &= \frac{6 \times 10^{12}}{3 \times 10^7} \\ y &= \frac{6 \times 10^{12}}{3 \times 10^7} + 5 \times 10^3 \\ y &= 2.05 \times 10^5 \end{aligned}$$

35. a) If $E = \$100{,}000$, then
$$\begin{aligned} D &= 0.75(100{,}000) + 5000 \\ &= \$80{,}000 \end{aligned}$$
$\$100{,}000 - \$80{,}000 = \$20{,}000$
So the tax is \$20,000.

b) If $E = \$10{,}000$, then
$$D = 0.75(10{,}000) + 5000 = 12{,}500$$
So Uncle Sam pays the person \$2500.

c) If $D = E$, then $E = 0.75E + 5000$ or $0.25E = 5000$, or $E = 20{,}000$.
So the lines intersect at an earned income of \$20,000 or at the point $(20{,}000, 20{,}000)$

d) At the intersection the earned income is equal to the disposable income, which means that if you earn \$20,000 then you pay no tax.

6.1 WARM-UPS

1. True, because a number is a monomial.
2. True, because it is a ratio of two binomials.
3. False, because the domain is any number except 2.
4. True, because 9 and $-\frac{1}{2}$ both give a denominator of 0.
5. False, because the numerator may be zero in a rational expression.
6. False, because 5 is not a factor of the numerator.
7. False, because $x \cdot x = x^2$, but $(x-1)x = x^2 - x$.
8. True, because we can multiply the numerator and denominator by -1.
9. True, because it is a correctly reduced rational expression.
10. False, it reduces to $x + y$.

6.1 EXERCISES

1. A rational expression is a ratio of two polynomials with the denominator not zero.
3. The basic principle of rational numbers says that $(ab)/(ac) = b/c$ provided a and c are not 0.
5. We build up the denominator by multiplying the numerator and denominator by the same expression.
7. If $x = 1$, then $x - 1 = 0$. So the domain is $\{x \mid x \neq 1\}$ or $(-\infty, 1) \cup (1, \infty)$.
9. If $z = 0$ then $7z = 0$. So the domain is $\{z \mid z \neq 0\}$ or $(-\infty, 0) \cup (0, \infty)$.
11. If $y = \pm 2$, then $y^2 - 4 = 0$. So the domain is $\{y \mid y \neq -2 \text{ and } y \neq 2\}$ or $(-\infty, -2) \cup (-2, 2) \cup (2, \infty)$.
13. To find the domain solve

$$x^2 + 4 = 0$$
$$x^2 = -4$$

Since $x^2 = -4$ has no solution there are no numbers that cause the denominator to be zero. Any real number can be used in place of x. The domain is $(-\infty, \infty)$.
15. Numbers that cannot be used in place of a are numbers for which the denominator is zero.

$$a^2 + 5a + 6 = 0$$
$$(a+2)(a+3) = 0$$
$$a = -2 \text{ or } a = -3$$

So -3 and -2 cannot be used for a.
17. Numbers that cannot be used in place of x are numbers for which the denominator is zero.

$$x^3 + x^2 - 6x = 0$$
$$x(x+3)(x-2) = 0$$
$$x = 0 \text{ or } x = -3 \text{ or } x = 2$$

So $-3, 0,$ and 2 cannot be use in place of x.

19. $\frac{6}{57} = \frac{3 \cdot 2}{3 \cdot 19} = \frac{2}{19}$
21. $\frac{42}{210} = \frac{42 \cdot 1}{42 \cdot 5} = \frac{1}{5}$
23. $\frac{2x+2}{4} = \frac{2(x+1)}{2 \cdot 2} = \frac{x+1}{2}$
25. $\frac{3x-6y}{10y-5x} = \frac{3(x-2y)}{-5(x-2y)} = -\frac{3}{5}$
27. $\frac{ab^2}{a^3b} = \frac{b}{a^2}$
29. $\frac{-2w^2x^3y}{6wx^5y^2} = \frac{-w}{3x^2y}$
31. $\frac{a^3b^2}{a^3+a^4} = \frac{a^3 \cdot b^2}{a^3(1+a)} = \frac{b^2}{1+a}$
33. $\frac{a-b}{2b-2a} = \frac{1(a-b)}{-2(a-b)} = -\frac{1}{2}$
35. $\frac{3x+6}{3x} = \frac{3(x+2)}{3x} = \frac{x+2}{x}$
37. $\frac{a^3-b^3}{a-b} = \frac{(a-b)(a^2+ab+b^2)}{a-b}$
$= a^2 + ab + b^2$
39. $\frac{4x^2-4}{4x^2+4} = \frac{4(x^2-1)}{4(x^2+1)} = \frac{x^2-1}{x^2+1}$
41. $\frac{12x^2-26x-10}{4x^2-25}$
$= \frac{2(3x+1)(2x-5)}{(2x-5)(2x+5)} = \frac{6x+2}{2x+5}$
43. $\frac{x^3+7x^2-4x}{x^3-16x} = \frac{x(x^2+7x-4)}{x(x-4)(x+4)}$
$= \frac{x^2+7x-4}{(x-4)(x+4)}$
45. $\frac{2ab+2by+3a+3y}{2b^2-7b-15} = \frac{(a+y)(2b+3)}{(2b+3)(b-5)}$
$= \frac{a+y}{b-5}$
47. $\frac{4x^2-10x-6}{2x^2+11x+5} = \frac{2(2x+1)(x-3)}{(2x+1)(x+5)}$
$= \frac{2x-6}{x+5}$
49. $\frac{1}{5} = \frac{1 \cdot 10}{5 \cdot 10} = \frac{10}{50}$
51. $\frac{1}{xy} = \frac{1 \cdot 3xy^2}{xy \cdot 3xy^2} = \frac{3xy^2}{3x^2y^3}$
53. $\frac{5}{x-1} = \frac{5(x-1)}{(x-1)(x-1)} = \frac{5x-5}{x^2-2x+1}$
55. $\frac{3}{2x-5} = \frac{3(2x+5)}{(2x-5)(2x+5)}$
$= \frac{6x+15}{4x^2-25}$

57. $\frac{1}{2x+2} = \frac{1(-3)}{(2x+2)(-3)} = \frac{-3}{-6x-6}$

59. $5 = \frac{5 \cdot a}{1 \cdot a} = \frac{5a}{a}$

61. $\frac{x+2}{x+3} = \frac{(x+2)(x-1)}{(x+3)(x-1)} = \frac{x^2+x-2}{x^2+2x-3}$

63. $\frac{7}{x-1} = \frac{7(-1)}{(x-1)(-1)} = \frac{-7}{1-x}$

65. $\frac{3}{x+2} = \frac{3(x^2-2x+4)}{(x+2)(x^2-2x+4)}$
$= \frac{3x^2-6x+12}{x^3+8}$

67. $\frac{x+2}{3x-1} = \frac{(x+2)(2x+5)}{(3x-1)(2x+5)}$
$= \frac{2x^2+9x+10}{6x^2+13x-5}$

69. $R(3) = \frac{3 \cdot 3 - 5}{3+4} = \frac{4}{7}$

71. $H(-2) = \frac{(-2)^2-5}{3(-2)-4} = \frac{-1}{-10} = \frac{1}{10}$

73. $W(-2) = \frac{4(-2)^3-1}{(-2)^2-(-2)-6} = \frac{-33}{0}$
$W(-2)$ is undefined.

75. $\frac{1}{3} = \frac{1 \cdot 7}{3 \cdot 7} = \frac{7}{21}$

77. $5 = \frac{5 \cdot 2}{1 \cdot 2} = \frac{10}{2}$

79. $\frac{3}{a} = \frac{3a}{a \cdot a} = \frac{3a}{a^2}$

81. $\frac{2}{a-b} = \frac{2(-1)}{(a-b)(-1)} = \frac{-2}{b-a}$

83. $\frac{2}{x-1} = \frac{2(x+1)}{(x-1)(x+1)} = \frac{2x+2}{x^2-1}$

85. $\frac{2}{w-3} = \frac{2(-1)}{(w-3)(-1)} = \frac{-2}{3-w}$

87. $\frac{2x+4}{6} = \frac{2(x+2)}{2 \cdot 3} = \frac{x+2}{3}$

89. $\frac{3a+3}{3a} = \frac{3(a+1)}{3a} = \frac{a+1}{a}$

91. $\frac{1}{x-1} = \frac{1(x^2+x+1)}{(x-1)(x^2+x+1)}$
$= \frac{x^2+x+1}{x^3-1}$

93. Since $D = RT$ we have $R = \frac{D}{T}$. So a rational function for his average speed is $S(x) = \frac{500}{x}$.

95. a) The average cost is the total cost divided by the number of items. So $C(x) = \frac{150}{x}$.
b) $C(5) = \frac{150}{5} = \$30$
$C(10) = \frac{150}{10} = \$15$
$C(30) = \frac{150}{30} = \$5$

97. a) Average cost per invitation is $A(n) = \frac{0.50n+45}{n}$ dollars.
b) $A(200) = \frac{0.50(200)+45}{200} = \$0.725$
$A(300) = \frac{0.50(300)+45}{300} = \$0.65$
It costs 7.5 cents less per invitation to print 300 rather than 200 invitations.
c) As the number of invitations increases, the average cost per invitation decreases.
d) As the number of invitations increases, the total cost of the invitations increases.

99. a) $p(n) = \frac{0.053n^2-0.64n+6.71}{3.43n+87.24}$

b) $p(0) = \frac{0.053(0)^2-0.64(0)+6.71}{3.43(0)+87.24}$
$= 7.7\%$

$p(30) = \frac{0.053(30)^2-0.64(30)+6.71}{3.43(30)+87.24}$
$= 18.5\%$

$p(50) = \frac{0.053(50)^2-0.64(50)+6.71}{3.43(50)+87.24}$
$= 41.4\%$

101. The value of $R(x)$ gets closer and closer to $1/2$.

6.2 WARM-UPS

1. False, because we can multiply any two fractions.

2. False, $\frac{2}{7} \cdot \frac{3}{7} = \frac{6}{49}$.

3. True.

4. False, $a \div b = a \cdot \frac{1}{b}$ for $b \neq 0$.

5. True, because the expressions are correctly multiplied.

6. True, $\frac{1}{2} \cdot \frac{1}{3} = \frac{1}{6}$.

7. True, $\frac{1}{3} \div \frac{1}{2} = \frac{1}{3} \cdot \frac{2}{1} = \frac{2}{3}$.

8. True, $\frac{w-z}{z-w} = -1$.

9. True, $\frac{x}{3} \div 2 = \frac{x}{3} \cdot \frac{1}{2} = \frac{x}{6}$.

10. False, $\frac{a}{b} \div \frac{b}{a} = \frac{a}{b} \cdot \frac{a}{b} = \frac{a^2}{b^2}$.

6.2 EXERCISES

1. To multiply rational numbers, multiply the numerators and the denominators.

3. The expressions $a - b$ and $b - a$ are opposites.

5. $\frac{12}{42} \cdot \frac{35}{22} = \frac{2 \cdot 2 \cdot 3}{2 \cdot 3 \cdot 7} \cdot \frac{5 \cdot 7}{2 \cdot 11} = \frac{5}{11}$

7. $\frac{2x}{3} \cdot \frac{6}{5x} = \frac{2x}{3} \cdot \frac{2 \cdot 3}{5x} = \frac{4}{5}$

9. $\frac{3a}{10b} \cdot \frac{5b^2}{6} = \frac{3a}{2 \cdot 5b} \cdot \frac{5b^2}{2 \cdot 3} = \frac{ab}{4}$

11. $\frac{3x-3}{6} \cdot \frac{x}{x^2 - x}$
$= \frac{3(x-1)}{2 \cdot 3} \cdot \frac{x}{x(x-1)} = \frac{1}{2}$

13. $\frac{10x+5}{5x^2+5} \cdot \frac{2x^2+x-1}{4x^2-1}$
$= \frac{5(2x+1)}{5(x^2+1)} \cdot \frac{(2x-1)(x+1)}{(2x-1)(2x+1)} = \frac{x+1}{x^2+1}$

15. $\frac{ax+aw+bx+bw}{x^2-w^2} \cdot \frac{x-w}{a^2-b^2}$
$= \frac{(a+b)(x+w)}{(x-w)(x+w)} \cdot \frac{x-w}{(a-b)(a+b)} = \frac{1}{a-b}$

17. $\frac{a^2-2a+4}{a^3+8} \cdot \frac{(a+2)^3}{2a+4}$
$= \frac{a^2-2a+4}{(a+2)(a^2-2a+4)} \cdot \frac{(a+2)^3}{2(a+2)}$
$= \frac{a+2}{2}$

19. $\frac{x-9}{12y} \cdot \frac{8y}{9-x}$
$= \frac{x-9}{4 \cdot 3y} \cdot \frac{4 \cdot 2y}{-1(x-9)} = \frac{2}{-3} = -\frac{2}{3}$

21. $(a^2-4) \cdot \frac{7}{2-a}$
$= (a-2)(a+2) \cdot \frac{7}{-1(a-2)}$
$= -7(a+2) = -7a - 14$

23. $\frac{(3x+1)^3}{2x-1} \cdot \frac{4x^2-4x+1}{9x^2+6x+1}$
$= \frac{(3x+1)^3}{2x-1} \cdot \frac{(2x-1)^2}{(3x+1)^2}$
$= (3x+1)(2x-1)$
$= 6x^2 - x - 1$

25. $\frac{15}{17} \div \frac{10}{17} = \frac{3 \cdot 5}{17} \cdot \frac{17}{2 \cdot 5} = \frac{3}{2}$

27. $\frac{36x}{5y} \div \frac{20x}{35y} = \frac{2^2 3^2 x}{5y} \cdot \frac{5 \cdot 7y}{2^2 5x} = \frac{3^2 7}{5}$
$= \frac{63}{5}$

29. $\frac{24a^5b^2}{5c^3} \div (4a^5bc^5)$
$= \frac{2^3 \cdot 3a^5b^2}{5c^3} \cdot \frac{1}{2^2a^5bc^5} = \frac{6b}{5c^8}$

31. $(w+1) \div \frac{w^2-1}{w} = (w+1) \cdot \frac{w}{w^2-1}$
$= (w+1) \cdot \frac{w}{(w-1)(w+1)} = \frac{w}{w-1}$

33. $\frac{x-y}{5} \div \frac{x^2-2xy+y^2}{10}$
$= \frac{x-y}{5} \cdot \frac{2 \cdot 5}{(x-y)^2}$
$= \frac{2}{x-y}$

35. $\frac{4x-2}{x^2-5x} \div \frac{2x^2+9x-5}{x^2-25}$

$= \frac{2(2x-1)}{x(x-5)} \cdot \frac{(x-5)(x+5)}{(2x-1)(x+5)} = \frac{2}{x}$

37. $\frac{x-y}{3} \div \frac{1}{6} = \frac{x-y}{3} \cdot \frac{6}{1} = 2x - 2y$

39. $\frac{x^2-25}{3} \div \frac{x-5}{6}$
$= \frac{(x-5)(x+5)}{3} \cdot \frac{2 \cdot 3}{x-5} = 2x + 10$

41. $\frac{a-b}{2} \div 3 = \frac{a-b}{2} \cdot \frac{1}{3} = \frac{a-b}{6}$

43. $(a^2-b^2) \div \frac{a+b}{3}$
$(a-b)(a+b) \cdot \frac{3}{a+b} = 3a - 3b$

45. $\frac{5x}{2} \div 3 = \frac{5x}{2} \cdot \frac{1}{3} = \frac{5x}{6}$

47. $\frac{3}{4} \div \frac{1}{4} = \frac{3}{4} \cdot \frac{4}{1} = 3$

49. $\frac{1}{2} \cdot \frac{1}{6} = \frac{1}{12}$

51. $\frac{1}{2} \cdot \frac{4x}{3} = \frac{2x}{3}$

53. $\frac{a-b}{b-a} = \frac{-1(b-a)}{b-a} = -1$

55. $(a-b) \div (-1) = (a-b) \cdot (-1) = b - a$

57. $\frac{x-y}{3} \cdot \frac{6}{y-x} = \frac{-1(y-x)}{3} \cdot \frac{2 \cdot 3}{y-x}$
$= -2$

59. $\frac{2a+2b}{a} \cdot \frac{1}{2} = \frac{2(a+b)}{a} \cdot \frac{1}{2} = \frac{a+b}{a}$

61. $-1\left(\frac{9-x}{2}\right) = \frac{-1(9-x)}{2} = \frac{x-9}{2}$

63. $\frac{4}{y-7} \div \frac{2}{7-y} = \frac{2 \cdot 2}{y-7} \cdot \frac{-1(y-7)}{2}$
$= -2$

65. $(a+b) \div \frac{1}{2} = (a+b) \cdot 2 = 2a + 2b$

67. $\frac{3x}{5} \div y = \frac{3x}{5} \cdot \frac{1}{y} = \frac{3x}{5y}$

69. $\frac{3a}{5b} \div 2 = \frac{3a}{5b} \cdot \frac{1}{2} = \frac{3a}{10b}$

71. $\frac{3x^2+13x-10}{x} \cdot \frac{x^3}{9x^2-4} \cdot \frac{7x-35}{x^2-25}$
$= \frac{(3x-2)(x+5)}{x} \cdot \frac{x^3}{(3x-2)(3x+2)} \cdot \frac{7(x-5)}{(x-5)(x+5)}$
$= \frac{7x^2}{3x+2}$

73. $\frac{(a^2b^3c)^2}{(-2ab^2c)^3} \cdot \frac{(a^3b^2c)^3}{(abc)^4} = \frac{a^4b^6c^2a^9b^6c^3}{-8a^3b^6c^3a^4b^4c^4}$
$= -\frac{a^6b^2}{8c^2}$

75. $\frac{(2mn)^3}{6mn^2} \div \frac{2m^2n^3}{(m^2n)^4} = \frac{(2mn)^3}{6mn^2} \cdot \frac{(m^2n)^4}{2m^2n^3}$
$= \frac{8m^3n^3m^8n^4}{12m^3n^5} = \frac{2m^8n^2}{3}$

77. $\frac{2x^2+7x-15}{4x^2-100} \cdot \frac{2x^2-9x-5}{4x^2-1}$
$= \frac{(2x-3)(x+5)}{4(x-5)(x+5)} \cdot \frac{(2x+1)(x-5)}{(2x-1)(2x+1)}$
$= \frac{2x-3}{4(2x-1)}$

79. $\frac{2h^2-5h-3}{5h^2-4h-1} \div \frac{2h^2+7h+3}{h^2+2h-3}$
$= \frac{(2h+1)(h-3)}{(5h+1)(h-1)} \cdot \frac{(h+3)(h-1)}{(2h+1)(h+3)}$
$= \frac{h-3}{5h+1}$

81. $\frac{9a-3}{1-9a^2} \cdot \frac{9a^2+6a+1}{6}$
$= \frac{-3(1-3a)}{(1-3a)(1+3a)} \cdot \frac{(3a+1)^2}{6}$
$= \frac{-3a-1}{2}$

83. We can factor $k^2-2km+m^2$ as $(k-m)^2$ or $(m-k)^2$.
$\frac{k^2+2km+m^2}{k^2-2km+m^2} \cdot \frac{m^2+3m-mk-3k}{m^2+mk+3m+3k}$
$= \frac{(k+m)^2}{(m-k)^2} \cdot \frac{(m-k)(m+3)}{(m+3)(m+k)} = \frac{k+m}{m-k}$

85. $\frac{x^a}{y^2} \cdot \frac{y^{b+2}}{x^{2a}} = \frac{y^b}{x^a}$

87. $\frac{x^{2a}+x^a-6}{x^{2a}+6x^a+9} \div \frac{x^{2a}-4}{x^{2a}+2x^a-3}$
$= \frac{(x^a+3)(x^a-2)}{(x^a+3)^2} \cdot \frac{(x^a+3)(x^a-1)}{(x^a-2)(x^a+2)}$
$= \frac{x^a-1}{x^a+2}$

89. $\frac{m^kv^k+3v^k-2m^k-6}{m^{2k}-9} \cdot \frac{m^{2k}-2m^k-3}{v^km^k-2m^k+2v^k-4}$
$= \frac{(m^k+3)(v^k-2)}{(m^k-3)(m^k+3)} \cdot \frac{(m^k-3)(m^k+1)}{(v^k-2)(m^k+2)}$
$= \frac{m^k+1}{m^k+2}$

91. $\frac{1}{50} \div \frac{7}{25} = \frac{1}{50} \cdot \frac{25}{7} = \frac{1}{14} \approx 7.1\%$

93. Her rate is $\frac{100}{x}$ miles/hour. Since $D = RT$, in $\frac{3}{4}$ hr she traveled $\frac{100}{x} \cdot \frac{3}{4}$ or $\frac{75}{x}$ miles.

95. Because $x \div \frac{y}{z} = x \cdot \frac{z}{y} = \frac{xz}{y}$, only (e) is not equivalent to the given expression.

6.3 WARM-UPS

1. False, the LCM is 30.
2. False, the LCM is $24a^2b^3$.
3. True, because $x-1$ is a factor of x^2-1.
4. False, the LCD is $x(x+1)$.
5. False, $\frac{1}{2}+\frac{2}{3}=\frac{7}{6}$.
6. False, $5+\frac{1}{x}=\frac{5x}{x}+\frac{1}{x}=\frac{5x+1}{x}$ for any nonzero x.
7. True, because the expressions are added correctly.
8. True, because the expressions are subtracted correctly.
9. True, because $\frac{8}{12}+\frac{9}{12}=\frac{17}{12}$.
10. False, he uses x reams per day

6.3 EXERCISES

1. The sum of a/b and c/b is $(a+c)/b$.
3. The least common multiple of some numbers is the smallest number that is a multiple of all of the numbers.
5. To add or subtract rational expressions with different denominators, you must build-up the expressions to equivalent expressions with the same denominator.
7. $\frac{3x}{2}+\frac{5x}{2}=\frac{8x}{2}=4x$
9. $\frac{7x}{2}-\frac{9x}{2}=\frac{-2x}{2}=-x$
11. $\frac{x-3}{2x}-\frac{3x-5}{2x}=\frac{x-3-3x+5}{2x}$
$= \frac{-2x+2}{2x} = \frac{2(-x+1)}{2x} = \frac{-x+1}{x}$

13.
$$\frac{3x-4}{2x-4}+\frac{2x-6}{2x-4}=\frac{5x-10}{2x-4}=\frac{5(x-2)}{2(x-2)}=\frac{5}{2}$$

15. $$\frac{x^2+4x-6}{x^2-9}-\frac{x^2+2x-12}{x^2-9}=\frac{2x+6}{x^2-9}=\frac{2(x+3)}{(x-3)(x+3)}=\frac{2}{x-3}$$

17. $$\frac{2x^2-8x-4}{2x^2+7x+3}+\frac{4x^2+x-1}{2x^2+7x+3}=\frac{6x^2-7x-5}{2x^2+7x+3}=\frac{(3x-5)(2x+1)}{(2x+1)(x+3)}=\frac{3x-5}{x+3}$$

19. Since $24=2^3\cdot 3$ and $20=2^2\cdot 5$, the $\text{LCM}=2^3\cdot 3\cdot 5=120$.

21. Since $12=2^2\cdot 3$, $18=2\cdot 3^2$, and $22=2\cdot 11$, the $\text{LCM}=2^2\cdot 3^2\cdot 11=396$.

23. The LCM for 10 and 15 is 30. So the LCM is $30x^3y$.

25. The highest power of a is 3, the highest power of b is 5, and the highest power of c is 2. So the $\text{LCM}=a^3b^5c^2$.

27. Since the three polynomials have no common factors, the LCM is the product of the polynomials: $x(x+2)(x-2)$

29. $4a+8=4(a+2)$, $6a+12=6(a+2)$
$\text{LCM}=12(a+2)$

31. $x^2-1=(x-1)(x+1)$
$x^2+2x+1=(x+1)^2$
$\text{LCM}=(x-1)(x+1)^2$

33. $x^2-4x=x(x-4)$
$x^2-16=(x-4)(x+4)$
$x^2+6x+8=(x+4)(x+2)$
$\text{LCM}=x(x-4)(x+4)(x+2)$

35. $6x^2+17x+12=(2x+3)(3x+4)$
$9x^2-16=(3x-4)(3x+4)$
$\text{LCM}=(2x+3)(3x+4)(3x-4)$

37. $$\frac{1}{28}+\frac{3}{35}=\frac{1\cdot 5}{28\cdot 5}+\frac{3\cdot 4}{35\cdot 4}=\frac{5}{140}+\frac{12}{140}=\frac{17}{140}$$

39. $$\frac{7}{48}-\frac{5}{36}=\frac{7\cdot 3}{48\cdot 3}-\frac{5\cdot 4}{36\cdot 4}=\frac{21}{144}-\frac{20}{144}=\frac{1}{144}$$

41. $$\frac{3}{wz^2}+\frac{5}{w^2z}=\frac{3\cdot w}{wz^2\cdot w}+\frac{5\cdot z}{w^2z\cdot z}=\frac{3w}{w^2z^2}+\frac{5z}{w^2z^2}=\frac{3w+5z}{w^2z^2}$$

43. $$\frac{2x-3}{8}-\frac{x-2}{6}=\frac{(2x-3)3}{8\cdot 3}-\frac{(x-2)4}{6\cdot 4}=\frac{6x-9}{24}-\frac{4x-8}{24}=\frac{2x-1}{24}$$

45. $$\frac{xa^3}{2a^4}+\frac{21x^2}{35ax}=\frac{x}{2a}+\frac{3x}{5a}=\frac{5x}{10a}+\frac{6x}{10a}=\frac{11x}{10a}$$

47. $$\frac{9}{4y}-x=\frac{9}{4y}-\frac{x\cdot 4y}{4y}=\frac{9}{4y}-\frac{4xy}{4y}=\frac{9-4xy}{4y}$$

49. $$\frac{5}{a+2}-\frac{7}{a}=\frac{5\cdot a}{(a+2)a}-\frac{7(a+2)}{a(a+2)}=\frac{5a}{a(a+2)}-\frac{7a+14}{a(a+2)}=\frac{5a-(7a+14)}{a(a+2)}=\frac{-2a-14}{a(a+2)}$$

51. $$\frac{1}{a-b}+\frac{2}{a+b}=\frac{1(a+b)}{(a-b)(a+b)}+\frac{2(a-b)}{(a+b)(a-b)}=\frac{a+b}{(a-b)(a+b)}+\frac{2a-2b}{(a-b)(a+b)}=\frac{3a-b}{(a-b)(a+b)}$$

53. $$\frac{1}{a-b}+\frac{1(-1)}{(b-a)(-1)}=\frac{1}{a-b}+\frac{-1}{a-b}=\frac{0}{a-b}=0$$

55. $$\frac{1-2x}{x^2+3x+2}+\frac{5}{x+2}=\frac{1-2x}{(x+1)(x+2)}+\frac{5(x+1)}{(x+2)(x+1)}=\frac{3x+6}{(x+1)(x+2)}=\frac{3(x+2)}{(x+1)(x+2)}=\frac{3}{x+1}$$

57. $$\frac{2x^2}{2x^3-18x}+\frac{15}{5x-15}=\frac{2x^2}{2x(x+3)(x-3)}+\frac{3\cdot 5}{5(x-3)}=\frac{x}{(x+3)(x-3)}+\frac{3}{x-3}=\frac{x}{(x+3)(x-3)}+\frac{3(x+3)}{(x-3)(x+3)}=\frac{4x+9}{(x+3)(x-3)}$$

59. $\dfrac{6x-11}{x^2+x-12}-\dfrac{5}{x+4}$
$=\dfrac{6x-11}{(x-3)(x+4)}-\dfrac{5(x-3)}{(x+4)(x-3)}$
$=\dfrac{x+4}{(x-3)(x+4)}=\dfrac{1}{x-3}$

61. $\dfrac{10}{4x-8}-\dfrac{15}{10-5x}$
$=\dfrac{10}{4(x-2)}-\dfrac{15}{-5(x-2)}$
$=\dfrac{5}{2(x-2)}+\dfrac{3}{(x-2)}$
$=\dfrac{5}{2(x-2)}+\dfrac{3\cdot 2}{(x-2)2}$
$=\dfrac{5}{2x-4}+\dfrac{6}{2x-4}$
$=\dfrac{11}{2(x-2)}$

63. $\dfrac{5}{x^2+x+2}-\dfrac{6}{x^2+2x-3}$
$=\dfrac{5(x+3)}{(x+2)(x-1)(x+3)}$
$-\dfrac{6(x+2)}{(x+3)(x-1)(x+2)}$
$=\dfrac{5x+15}{(x-1)(x+2)(x+3)}$
$-\dfrac{6x+12}{(x-1)(x+2)(x+3)}$
$=\dfrac{-x+3}{(x-1)(x+2)(x+3)}$

65. $\dfrac{x}{2x^2+x-1}+\dfrac{3}{3x^2+2x-1}$
$=\dfrac{x(3x-1)}{(2x-1)(x+1)(3x-1)}$
$+\dfrac{3(2x-1)}{(3x-1)(x+1)(2x-1)}$
$=\dfrac{3x^2-x}{(x+1)(2x-1)(3x-1)}$
$+\dfrac{6x-3}{(x+1)(2x-1)(3x-1)}$
$=\dfrac{3x^2+5x-3}{(x+1)(2x-1)(3x-1)}$

67. $\dfrac{1}{x}+\dfrac{2}{x-1}-\dfrac{3}{x+2}$
$=\dfrac{1(x-1)(x+2)}{x(x-1)(x+2)}+\dfrac{2x(x+2)}{(x-1)x(x+2)}$
$-\dfrac{3x(x-1)}{(x+2)x(x-1)}$
$=\dfrac{x^2+x-2}{x(x-1)(x+2)}+\dfrac{2x^2+4x}{x(x-1)(x+2)}$
$-\dfrac{3x^2-3x}{x(x-1)(x+2)}$
$=\dfrac{8x-2}{x(x-1)(x+2)}$

69. $\frac{1}{3}+\frac{1}{4}=\dfrac{1\cdot 4+1\cdot 3}{12}=\dfrac{7}{12}$

71. $\frac{1}{8}-\frac{3}{5}=\dfrac{1\cdot 5-8\cdot 3}{40}=-\dfrac{19}{40}$

73. $\frac{x}{3}+\frac{x}{2}=\dfrac{x\cdot 2+x\cdot 3}{6}=\dfrac{5x}{6}$

75. $\frac{a}{b}-\frac{2}{3}=\dfrac{3\cdot a-2\cdot b}{3b}=\dfrac{3a-2b}{3b}$

77. $a+\frac{2}{3}=\dfrac{a\cdot 3+1\cdot 2}{3\cdot 1}=\dfrac{3a+2}{3}$

79. $\frac{3}{a}+1=\dfrac{a\cdot 1+3\cdot 1}{a\cdot 1}=\dfrac{a+3}{a}$

81. $\dfrac{3+x}{x}-1=\dfrac{3+x-x}{x}=\dfrac{3}{x}$

83. $\frac{2}{3}+\dfrac{1}{4x}=\dfrac{2\cdot 4x+1\cdot 3}{12x}=\dfrac{8x+3}{12x}$

85. $\dfrac{w^2-3w+6}{w-5}+\dfrac{9-w^2}{w-5}=\dfrac{-3w+15}{w-5}$
$=\dfrac{-3(w-5)}{w-5}=-3$

87. $\dfrac{1}{3x-6}-\dfrac{6}{5x-10}$
$=\dfrac{1}{3(x-2)}-\dfrac{6}{5(x-2)}$
$=\dfrac{5}{15(x-2)}-\dfrac{18}{15(x-2)}$
$=\dfrac{-13}{15(x-2)}$

89. $\dfrac{x-1}{2x^2+3x+1}-\dfrac{x+1}{2x^2-x-1}$
$=\dfrac{x-1}{(2x+1)(x+1)}-\dfrac{x+1}{(2x+1)(x-1)}$
$=\dfrac{(x-1)(x-1)}{(2x+1)(x+1)(x-1)}$
$-\dfrac{(x+1)(x+1)}{(2x+1)(x-1)(x+1)}$
$=\dfrac{x^2-2x+1}{(x+1)(x-1)(2x+1)}$
$-\dfrac{x^2+2x+1}{(x+1)(x-1)(2x+1)}$
$=\dfrac{-4x}{(x+1)(x-1)(2x+1)}$

91. $\dfrac{(a^2b^3)^4}{(ab^4)^3}\cdot\dfrac{(ab)^3}{(a^4b)^2}$
$=\dfrac{a^8b^{12}}{a^3b^{12}}\cdot\dfrac{a^3b^3}{a^8b^2}=\dfrac{a^{11}b^{15}}{a^{11}b^{14}}=b$

93. $\dfrac{x^2+4}{25x^2-20x+4}+\dfrac{10x+4}{25x^2-4}$
$=\dfrac{x^2+4}{(5x-2)^2}+\dfrac{10x+4}{(5x-2)(5x+2)}$
$=\dfrac{x^2+4}{(5x-2)^2}+\dfrac{2(5x+2)}{(5x-2)(5x+2)}$
$=\dfrac{x^2+4}{(5x-2)^2}+\dfrac{2}{5x-2}$
$=\dfrac{x^2+4}{(5x-2)^2}+\dfrac{2(5x-2)}{(5x-2)(5x-2)}$
$=\dfrac{x^2+4}{(5x-2)^2}+\dfrac{10x-4}{(5x-2)^2}$
$=\dfrac{x^2+10x}{(5x-2)^2}$

95. $\frac{4x^2+9}{4x^2-9} \cdot \frac{4x^2+12x+9}{2x^2+3x}$

$= \frac{4x^2+9}{(2x-3)(2x+3)} \cdot \frac{(2x+3)^2}{x(2x+3)}$

$= \frac{4x^2+9}{x(2x-3)}$

97. $\frac{w^2-3}{3w^3+81} - \frac{2}{6w+18} - \frac{w-4}{w^2-3w+9}$

$= \frac{w^2-3}{3w^3+81} - \frac{1}{3(w+3)} - \frac{w-4}{w^2-3w+9}$

$= \frac{w^2-3}{3(w+3)(w^2-3w+9)}$

$- \frac{1(w^2-3w+9)}{3(w+3)(w^2-3w+9)}$

$- \frac{3(w-4)(w+3)}{3(w+3)(w^2-3w+9)}$

$= \frac{w^2-3-w^2+3w-9-3(w^2-w-12)}{3(w+3)(w^2-3w+9)}$

$= \frac{-3w^2+6w+24}{3(w+3)(w^2-3w+9)}$

$= \frac{-w^2+2w+8}{(w+3)(w^2-3w+9)}$

99. $\frac{a^2-6a+9}{a^3+8} \div \frac{a^2-a-6}{a^2-4}$

$= \frac{(a-3)^2}{(a-2)(a^2+2a+4)} \cdot \frac{(a-2)(a+2)}{(a-3)(a+2)}$

$= \frac{a-3}{a^2+2a+4}$

101. $\frac{w^2+3}{w^3-8} - \frac{2w}{w^2-4}$

$= \frac{(w^2+3)(w+2)}{(w-2)(w^2+2w+4)(w+2)}$

$- \frac{2w(w^2+2w+4)}{(w-2)(w+2)(w^2+2w+4)}$

$= \frac{w^3+3w+2w^2+6-(2w^3+4w^2+8w)}{(w-2)(w+2)(w^2+2w+4)}$

$= \frac{-w^3-2w^2-5w+6}{(w-2)(w+2)(w^2+2w+4)}$

103. $\frac{1}{x^3-1} - \frac{1}{x^2-1} + \frac{1}{x-1}$

$= \frac{x+1}{(x-1)(x^2+x+1)(x+1)}$

$- \frac{x^2+x+1}{(x-1)(x+1)(x^2+x+1)}$

$+ \frac{(x+1)(x^2+x+1)}{(x-1)(x+1)(x^2+x+1)}$

$= \frac{x+1-x^2-x-1+x^3+2x^2+2x+1}{(x-1)(x+1)(x^2+x+1)}$

$= \frac{x^3+x^2+2x+1}{(x-1)(x+1)(x^2+x+1)}$

105. a) Joe processes $\frac{1}{x}$ claims/hr while Ellen processes $\frac{1}{x+1}$ claims/hr. Together they process $\frac{1}{x} + \frac{1}{x+1}$ or $\frac{2x+1}{x(x+1)}$ claims/hr. In 8 hours they process $8 \cdot \frac{(2x+1)}{x(x+1)}$ or $\frac{16x+8}{x(x+1)}$ claims. So a rational function for the number of claims in 8 hours is $C(x) = \frac{16x+8}{x(x+1)}$

b) $C(2) = \frac{16(2)+8}{2(2+1)} = \frac{40}{6} = 6\frac{2}{3}$ claims

107. a) George's rate is $\frac{1}{20}$ mag/min while Theresa's rate is $\frac{1}{x}$ mag/min. Their rate together is $\frac{1}{x} + \frac{1}{20}$ or $\frac{x+20}{20x}$ mag/min. In 60 min they will sell $60 \cdot \frac{x+20}{20x}$ or $\frac{3x+60}{x}$ magazines. So a rational function for the number they will sell in one hour is $M(x) = \frac{3x+60}{x}$.

b) $M(5) = \frac{3(5)+60}{5} = 15$ subscriptions

109. a) Let x be Joan's original speed and $x+5$ be her increased speed. Since $T = \frac{D}{R}$, her total travel time is $\frac{100}{x} + \frac{200}{x+5}$ or $\frac{300x+500}{x(x+5)}$ hours. So a rational function for her total travel time is $T(x) = \frac{300x+500}{x(x+5)}$.

b) $T(65) = \frac{300(65)+500}{65(65+5)} \approx 4.3956$ hours or 4 hours 24 minutes.

111. By definition, $a/b + c/b = (a+c)/b$ provided $b \neq 0$. By definition of fraction multiplication, common denominators are not needed.

6.4 WARM-UPS

1. False, the LCM is $6x^2$.
2. True, since $2b - 2a = -2(a - b)$.
3. True.
4. True, because $\frac{1}{2} + \frac{1}{3} = \frac{5}{6}$ and $1 + \frac{1}{2} = \frac{3}{2}$.
5. False, because $2^{-1} + 3^{-1} = \frac{1}{2} + \frac{1}{3} = \frac{5}{6}$ and $(2+3)^{-1} = \frac{1}{5}$.
6. False, because the left side is the reciprocal of $\frac{5}{6}$ and the right side is 5.
7. False, because $2 + 3^{-1} = \frac{7}{3}$ and $5^{-1} = \frac{1}{5}$.

8. False, because $x + 2^{-1} = x + \frac{1}{2} = \frac{2x+1}{2}$ for any value of x.

9. True, because ab is the LCD.

10. True, because multiplying by a^5b^2 will eliminate all negative exponents.

6.4 EXERCISES

1. A complex fraction is a fraction that contains fractions in the numerator, denominator, or both.

3. $$\frac{\frac{1}{2}+\frac{1}{4}}{\frac{1}{2}+\frac{1}{8}} = \frac{\left(\frac{1}{2}+\frac{1}{4}\right)8}{\left(\frac{1}{2}+\frac{1}{8}\right)8} = \frac{4+2}{4+1} = \frac{6}{5}$$

5. $$\frac{\frac{1}{2}-\frac{1}{3}}{\frac{1}{4}-\frac{1}{5}} = \frac{\left(\frac{1}{2}-\frac{1}{3}\right)60}{\left(\frac{1}{4}-\frac{1}{5}\right)60} = \frac{30-20}{15-12} = \frac{10}{3}$$

7. $$\frac{\frac{2}{3}+\frac{5}{6}-\frac{1}{2}}{\frac{1}{8}-\frac{1}{3}+\frac{1}{12}} = \frac{24\cdot\frac{2}{3}+24\cdot\frac{5}{6}-24\cdot\frac{1}{2}}{24\cdot\frac{1}{8}-24\cdot\frac{1}{3}+24\cdot\frac{1}{12}} = \frac{16+20-12}{3-8+2} = -8$$

9. $$\frac{\frac{x}{2}-\frac{1}{3}}{\frac{x}{2}+\frac{3}{4}} = \frac{\left(\frac{x}{2}-\frac{1}{3}\right)12}{\left(\frac{x}{2}+\frac{3}{4}\right)12} = \frac{6x-4}{6x+9}$$

11. $$\frac{a+\frac{3}{b}}{\frac{b}{a}+\frac{1}{b}} = \frac{ab\cdot a + ab\cdot\frac{3}{b}}{ab\cdot\frac{b}{a}+ab\cdot\frac{1}{b}} = \frac{a^2b+3a}{a+b^2}$$

13. $$\frac{\frac{a+b}{b}}{\frac{a-b}{ab}} = \frac{ab\left(\frac{a+b}{b}\right)}{ab\left(\frac{a-b}{ab}\right)} = \frac{a^2+ab}{a-b}$$

15. $$\frac{\frac{x-3y}{xy}}{\frac{1}{x}+\frac{1}{y}} = \frac{xy\left(\frac{x-3y}{xy}\right)}{xy\cdot\frac{1}{x}+xy\cdot\frac{1}{y}} = \frac{x-3y}{x+y}$$

17. $$\frac{3-\frac{m-2}{6}}{\frac{4}{9}+\frac{2}{m}} = \frac{18m\cdot 3 - 18m\cdot\frac{m-2}{6}}{18m\cdot\frac{4}{9}+18m\cdot\frac{2}{m}} = \frac{54m-3m^2+6m}{8m+36} = \frac{60m-3m^2}{4(2m+9)}$$

19. $$\frac{\frac{a^2-b^2}{a^2b^3}}{\frac{a+b}{a^3b}} = \frac{a^3b^3\left(\frac{a^2-b^2}{a^2b^3}\right)}{a^3b^3\left(\frac{a+b}{a^3b}\right)} = \frac{a(a^2-b^2)}{b^2(a+b)} = \frac{a(a-b)(a+b)}{b^2(a+b)} = \frac{a^2-ab}{b^2}$$

21. $$\frac{\frac{1}{x^2y^2}+\frac{1}{xy^3}}{\frac{1}{x^3y}-\frac{1}{xy}} = \frac{\left(\frac{1}{x^2y^2}+\frac{1}{xy^3}\right)x^3y^3}{\left(\frac{1}{x^3y}-\frac{1}{xy}\right)x^3y^3} = \frac{xy+x^2}{y^2-x^2y^2} = \frac{xy+x^2}{y^2(1-x^2)} = \frac{xy+x^2}{y^2(1-x)(1+x)}$$

23. $$\frac{x+\frac{4}{x+4}}{x-\frac{4x+4}{x+4}} = \frac{(x+4)x+(x+4)\cdot\frac{4}{x+4}}{(x+4)x-(x+4)\cdot\frac{4x+4}{x+4}} = \frac{x^2+4x+4}{x^2+4x-4x-4} = \frac{(x+2)^2}{(x-2)(x+2)} = \frac{x+2}{x-2}$$

25. $\dfrac{1-\dfrac{1}{y-1}}{3+\dfrac{1}{y+1}}$

$$= \frac{(y-1)(y+1)\left(1-\frac{1}{y-1}\right)}{(y-1)(y+1)\left(3+\frac{1}{y+1}\right)}$$

$$= \frac{y^2-1-(y+1)}{3(y^2-1)+y-1} = \frac{y^2-y-2}{3y^2+y-4}$$

$$= \frac{y^2-y-2}{(y-1)(3y+4)}$$

27. $\dfrac{\dfrac{2}{3-x}-4}{\dfrac{1}{x-3}-1} = \dfrac{(x-3)\cdot\dfrac{2}{3-x}-(x-3)4}{(x-3)\cdot\dfrac{1}{x-3}-(x-3)1}$

$$= \frac{-2-4x+12}{1-x+3} = \frac{-4x+10}{-x+4} = \frac{4x-10}{x-4}$$

29. $\dfrac{\dfrac{w+2}{w-1}-\dfrac{w-3}{w}}{\dfrac{w+4}{w}+\dfrac{w-2}{w-1}}$

$$= \frac{w(w-1)\left(\frac{w+2}{w-1}-\frac{w-3}{w}\right)}{w(w-1)\left(\frac{w+4}{w}+\frac{w-2}{w-1}\right)}$$

$$= \frac{w^2+2w-(w-1)(w-3)}{(w-1)(w+4)+w^2-2w}$$

$$= \frac{w^2+2w-w^2+4w-3}{w^2+3w-4+w^2-2w}$$

$$= \frac{6w-3}{2w^2+w-4}$$

31. $\dfrac{\dfrac{1}{a-b}-\dfrac{3}{a+b}}{\dfrac{2}{b-a}+\dfrac{4}{a+b}}$

$$= \frac{(a-b)(a+b)\left(\frac{1}{a-b}-\frac{3}{a+b}\right)}{(a-b)(a+b)\left(\frac{2}{b-a}+\frac{4}{a+b}\right)}$$

$$= \frac{a+b-3(a-b)}{-2(a+b)+4(a-b)}$$

$$= \frac{-2a+4b}{2a-6b}$$

$$= \frac{2(2b-a)}{2(a-3b)} = \frac{2b-a}{a-3b}$$

33. $\dfrac{\dfrac{4}{y}-\dfrac{y+4}{y-3}}{\dfrac{2}{y-3}+\dfrac{y+1}{y}}$

$$= \frac{\left(\frac{4}{y}-\frac{y+4}{y-3}\right)(y)(y-3)}{\left(\frac{2}{y-3}+\frac{y+1}{y}\right)(y)(y-3)}$$

$$= \frac{4(y-3)-y(y+4)}{2y+(y+1)(y-3)}$$

$$= \frac{4y-12-y^2-4y}{2y+y^2-2y-3} = \frac{-y^2-12}{y^2-3}$$

35. $\dfrac{3-\dfrac{4}{a-1}}{5-\dfrac{3}{1-a}} = \dfrac{(a-1)3-(a-1)\cdot\dfrac{4}{a-1}}{(a-1)5-(a-1)\cdot\dfrac{3}{1-a}}$

$$= \frac{3a-3-4}{5a-5+3} = \frac{3a-7}{5a-2}$$

37. $\dfrac{\dfrac{2}{m-3}+\dfrac{4}{m}}{\dfrac{3}{m-2}+\dfrac{1}{m}}$

$$= \frac{m(m-3)(m-2)\left(\frac{2}{m-3}+\frac{4}{m}\right)}{m(m-3)(m-2)\left(\frac{3}{m-2}+\frac{1}{m}\right)}$$

$$= \frac{2m(m-2)+4(m-3)(m-2)}{3m(m-3)+(m-3)(m-2)}$$

$$= \frac{2m^2-4m+4m^2-20m+24}{3m^2-9m+m^2-5m+6}$$

$$= \frac{6m^2-24m+24}{4m^2-14m+6} = \frac{3m^2-12m+12}{2m^2-7m+3}$$

$$= \frac{3m^2-12m+12}{(m-3)(2m-1)}$$

39. $\dfrac{\dfrac{3}{x^2-1}-\dfrac{x-2}{x^3-1}}{\dfrac{3}{x^2+x+1}+\dfrac{x-3}{x^3-1}}$

$$= \frac{(x-1)(x+1)(x^2+x+1)\left(\frac{3}{x^2-1}-\frac{x-2}{x^3-1}\right)}{(x-1)(x+1)(x^2+x+1)\left(\frac{3}{x^2+x+1}+\frac{x-3}{x^3-1}\right)}$$

$$= \frac{3x^2+3x+3-(x+1)(x-2)}{3x^2-3+(x+1)(x-3)}$$

$$= \frac{3x^2+3x+3-x^2+x+2}{3x^2-3+x^2-2x-3}$$

$$= \frac{2x^2+4x+5}{4x^2-2x-6} = \frac{2x^2+4x+5}{2(2x-3)(x+1)}$$

41. $\dfrac{w^{-1}+y^{-1}}{z^{-1}+y^{-1}} = \dfrac{wyz(w^{-1}+y^{-1})}{wyz(z^{-1}+y^{-1})}$

$$= \frac{yz+wz}{wy+wz}$$

$$= \frac{yz+wz}{w(y+z)}$$

43. $\dfrac{1-x^{-1}}{1-x^{-2}} = \dfrac{x^2(1-x^{-1})}{x^2(1-x^{-2})} = \dfrac{x^2-x}{x^2-1}$

$$= \frac{x(x-1)}{(x-1)(x+1)} = \frac{x}{x+1}$$

45. $\dfrac{a^{-2}+b^{-2}}{a^{-1}b} = \dfrac{a^2b^2(a^{-2}+b^{-2})}{a^2b^2(a^{-1}b)}$

$$= \frac{a^2+b^2}{ab^3}$$

47. $1 - a^{-1} = \dfrac{a(1 - a^{-1})}{a} = \dfrac{a-1}{a}$

49. $\dfrac{x^{-1} + x^{-2}}{x + x^{-2}} = \dfrac{x^2(x^{-1} + x^{-2})}{x^2(x + x^{-2})}$
$= \dfrac{x+1}{x^3+1} = \dfrac{x+1}{(x+1)(x^2 - x + 1)}$
$= \dfrac{1}{x^2 - x + 1}$

51. $\dfrac{2m^{-1} - 3m^{-2}}{m^{-2}} = \dfrac{m^2(2m^{-1} - 3m^{-2})}{m^2(m^{-2})}$
$= 2m - 3$

53. $\dfrac{a^{-1} - b^{-1}}{a - b} = \dfrac{ab(a^{-1} - b^{-1})}{ab(a - b)}$
$= \dfrac{b - a}{-ab(b - a)} = -\dfrac{1}{ab}$

55. $\dfrac{x^3 - y^3}{x^{-3} - y^{-3}} = \dfrac{(x^3 - y^3)x^3y^3}{(x^{-3} - y^{-3})x^3y^3}$
$= \dfrac{(x^3 - y^3)x^3y^3}{y^3 - x^3} = (-1)x^3y^3 = -x^3y^3$

57. $\dfrac{1 - 8x^{-3}}{x^{-1} + 2x^{-2} + 4x^{-3}}$
$= \dfrac{x^3(1 - 8x^{-3})}{x^3(x^{-1} + 2x^{-2} + 4x^{-3})}$
$= \dfrac{x^3 - 8}{x^2 + 2x + 4}$
$= \dfrac{(x-2)(x^2 + 2x + 4)}{x^2 + 2x + 4}$
$= x - 2$

59. $(x^{-1} + y^{-1})^{-1} = \dfrac{1}{x^{-1} + y^{-1}}$
$= \dfrac{xy \cdot 1}{xy(x^{-1} + y^{-1})} = \dfrac{xy}{x + y}$

61. $\dfrac{\frac{5}{3} - \frac{4}{5}}{\frac{1}{3} - \frac{5}{6}} \approx -1.7333$ or $-26/15$

63. $\dfrac{4^{-1} - 9^{-1}}{2^{-1} + 3^{-1}} \approx 0.1667$ or $1/6$

65. If $x =$ the number of students at Central, then $\frac{1}{2}x =$ the number at Northside and $\frac{2}{3}x =$ the number at Southside. To find the percentage of black students among the city's elementary students we divide the number of black students by the total number of students:

$\dfrac{\frac{1}{3} \cdot \frac{1}{2}x + \frac{3}{4} \cdot x + \frac{1}{6} \cdot \frac{2}{3}x}{\frac{1}{2}x + x + \frac{2}{3}x} = \dfrac{\frac{1}{6}x + \frac{3}{4}x + \frac{1}{9}x}{\frac{1}{2}x + x + \frac{2}{3}x}$
$= \dfrac{36\left(\frac{1}{6}x + \frac{3}{4}x + \frac{1}{9}x\right)}{36\left(\frac{1}{2}x + x + \frac{2}{3}x\right)} = \dfrac{6x + 27x + 4x}{18x + 36x + 24x}$
$= \dfrac{37x}{78x} = \dfrac{37}{78} = 47.4\%$

67. Let $x =$ the distance and $\frac{x}{45} =$ the time from Clarksville to Leesville. We can also say that $x =$ the distance and $\frac{x}{55} =$ the time for the return trip. To find the average speed for any trip we divide the total distance by the total time. In this case the total distance is $2x$ and the total time is $\frac{x}{45} + \frac{x}{55}$.

$\dfrac{2x}{\frac{x}{45} + \frac{x}{55}} = \dfrac{2x \cdot 5 \cdot 9 \cdot 11}{\left(\frac{x}{45} + \frac{x}{55}\right)5 \cdot 9 \cdot 11} = \dfrac{990x}{11x + 9x}$
$= \dfrac{990x}{20x} = \dfrac{99}{2} = 49.5$

Her average speed for the trip is 49.5 mph.

69. $\dfrac{2}{\frac{1}{45} + \frac{1}{55}} = 49.5$ mph

73. The complex fraction simplifies to $\frac{2x+1}{3x+2}$. The complex fraction is undefined for $x = 0, -1, -1/2, -2/3$.

6.5 WARM-UPS

1. False, c is the quotient.
2. False, (quotient)(divisor) + remainder = dividend.
3. True, because
$(x + 2)(x + 3) = x^2 + 5x + 6$ and
$(x + 2)(x + 3) + 1 = x^2 + 5x + 7$.
4. True. From warm-up number 3 we can see that if $x + 3$ is the divisor then $x + 2$ is the quotient.
5. True. From warm-up number 3 we can see that if $x + 2$ is the divisor then $x + 3$ is the quotient and 1 is the remainder.
6. False, to divide by $x - c$ with synthetic division we use c.
7. False, synthetic division is used only for dividing by $x - c$.
8. True, when dividing by $x - c$ the degree of the quotient is 1 less than the degree of the dividend.
9. True, because if remainder is zero, then (quotient)(divisor) = dividend.
10. True, because if remainder is zero, then (quotient)(divisor) = dividend.

6.5 EXERCISES

1. If $a \div b = c$, then the dividend is a, the divisor is b, and the quotient is c.

3. If the term x^n is missing in the dividend, insert the term $0 \cdot x^n$ for the missing term.

5. Synthetic division is used only for dividing by a binomial of the form $x - c$.

7. $\dfrac{36x^7}{3x^3} = 12x^4$

9. $\dfrac{16x^2}{-8x^2} = -2$

11. $\dfrac{6b-9}{3} = \dfrac{6b}{3} - \dfrac{9}{3} = 2b - 3$

13. $\dfrac{3x^2+3x}{3x} = \dfrac{3x^2}{3x} + \dfrac{3x}{3x} = x + 1$

15. $\dfrac{10x^4 - 8x^3 + 6x^2}{-2x^2}$
$= \dfrac{10x^4}{-2x^2} - \dfrac{8x^3}{-2x^2} + \dfrac{6x^2}{-2x^2}$
$= -5x^2 + 4x - 3$

17. $\dfrac{7x^3 - 4x^2}{2x} = \dfrac{7x^3}{2x} - \dfrac{4x^2}{2x} = \dfrac{7}{2}x^2 - 2x$

19. $(8x - 4) \div (2) = \dfrac{8x}{2} - \dfrac{4}{2}$
$= 4x - 2$

The quotient is $4x - 2$ and the remainder is 0.

21. $(8x - 3) \div (4x) = \dfrac{8x}{4x} + \dfrac{-3}{4x}$
$= 2 + \dfrac{-3}{4x}$

The quotient is 2 and the remainder is -3.

23. $(2x^3 + x^2 + 4x - 3) \div (3x^2)$
$= \dfrac{2x^3}{3x^2} + \dfrac{x^2}{3x^2} + \dfrac{4x}{3x^2} + \dfrac{-3}{3x^2}$
$= \dfrac{2}{3}x + \dfrac{1}{3} + \dfrac{4x-3}{3x^2}$

The quotient is $\dfrac{2}{3}x + \dfrac{1}{3}$ and the remainder is $4x - 3$.

25. $(-10x^4 - 5x^3 - 6x - 7) \div (-5x^2)$
$= \dfrac{-10x^4}{-5x^2} - \dfrac{5x^3}{-5x^2} + \dfrac{-6x-7}{-5x^2}$
$= 2x^2 + x + \dfrac{-6x-7}{-5x^2}$

The quotient is $2x^2 + x$ and the remainder is $-6x - 7$.

27.
$$\begin{array}{r} x+5 \\ x+3\overline{)\,x^2+8x+13} \\ \underline{x^2+3x} \\ 5x+13 \\ \underline{5x+15} \\ -2 \end{array}$$

Quotient: $x + 5$ Remainder: -2

29.
$$\begin{array}{r} x-4 \\ x+2\overline{)\,x^2-2x} \\ \underline{x^2+2x} \\ -4x \\ \underline{-4x-8} \\ 8 \end{array}$$

Quotient: $x - 4$ Remainder: 8

31.
$$\begin{array}{r} x^2-2x+4 \\ x+2\overline{)\,x^3+0x^2+0x+8} \\ \underline{x^3+2x^2} \\ -2x^2+0x \\ \underline{-2x^2-4x} \\ 4x+8 \\ \underline{4x+8} \\ 0 \end{array}$$

Quotient: $x^2 - 2x + 4$ Remainder: 0

33.
$$\begin{array}{r} a^2+2a+8 \\ a-2\overline{)\,a^3+0a^2+4a-5} \\ \underline{a^3-2a^2} \\ 2a^2+4a \\ \underline{2a^2-4a} \\ 8a-5 \\ \underline{8a-16} \\ 11 \end{array}$$

Quotient: $a^2 + 2a + 8$ Remainder: 11

35.
$$\begin{array}{r} x^2-2x+3 \\ x+1\overline{)\,x^3-x^2+x-3} \\ \underline{x^3+x^2} \\ -2x^2+x \\ \underline{-2x^2-2x} \\ 3x-3 \\ \underline{3x+3} \\ -6 \end{array}$$

Quotient: $x^2 - 2x + 3$ Remainder: -6

37.
$$\begin{array}{r} x^3+3x^2+6x+11 \\ x-2\overline{)\,x^4+x^3+0x^2-x-1} \\ \underline{x^4-2x^3} \\ 3x^3+0x^2 \\ \underline{3x^3-6x^2} \\ 6x^2-x \\ \underline{6x^2-12x} \\ 11x-1 \\ \underline{11x-22} \\ 21 \end{array}$$

Quotient: $x^3 + 3x^2 + 6x + 11$ Remainder: 21

39.
$$\begin{array}{r|l} & -3x^2 \qquad\qquad -1 \\ \hline x^2-2 & -3x^4 + 0x^3 + 5x^2 + x - 2 \\ & \underline{-3x^4 \qquad\quad + 6x^2} \\ & \qquad\qquad -x^2 + x - 2 \\ & \qquad\qquad \underline{-x^2 \qquad + 2} \\ & \qquad\qquad\qquad x - 4 \end{array}$$

Quotient: $-3x^2 - 1$ Remainder: $x - 4$

41.
$$\begin{array}{r|l} & 3x + 5 \\ \hline 2x-3 & 6x^2 + x - 16 \\ & \underline{6x^2 - 9x} \\ & \qquad 10x - 16 \\ & \qquad \underline{10x - 15} \\ & \qquad\qquad -1 \end{array}$$

Quotient: $3x + 5$ Remainder: -1

43.
$$\begin{array}{r|l} & 2b - 5 \\ \hline 5b+4 & 10b^2 - 17b - 22 \\ & \underline{10b^2 + 8b} \\ & \qquad -25b - 22 \\ & \qquad \underline{-25b - 20} \\ & \qquad\qquad -2 \end{array}$$

Quotient: $2b - 5$ Remainder: -2

45.
$$\begin{array}{r|l} & x^2 + x - 2 \\ \hline 2x+1 & 2x^3 + 3x^2 - 3x - 2 \\ & \underline{2x^3 + x^2} \\ & \qquad 2x^2 - 3x \\ & \qquad \underline{2x^2 + x} \\ & \qquad\qquad -4x - 2 \\ & \qquad\qquad \underline{-4x - 2} \\ & \qquad\qquad\qquad 0 \end{array}$$

Quotient: $x^2 + x - 2$ Remainder: 0

47.
$$\begin{array}{r|l} & x - 5 \\ \hline x^2+x+2 & x^3 - 4x^2 - 3x - 10 \\ & \underline{x^3 + x^2 + 2x} \\ & \quad -5x^2 - 5x - 10 \\ & \quad \underline{-5x^2 - 5x - 10} \\ & \qquad\qquad\qquad 0 \end{array}$$

Quotient: $x - 5$ Remainder: 0

49.
$$\begin{array}{r|l} & 2 \\ \hline x-5 & 2x + 0 \\ & \underline{2x - 10} \\ & \qquad 10 \end{array}$$

$$\frac{2x}{x-5} = 2 + \frac{10}{x-5}$$

51.
$$\begin{array}{r|l} & x - 1 \\ \hline 2x+1 & 2x^2 - x + 0 \\ & \underline{2x^2 + x} \\ & \qquad -2x + 0 \\ & \qquad \underline{-2x - 1} \\ & \qquad\qquad 1 \end{array}$$

$$\frac{2x^2 - x}{2x+1} = x - 1 + \frac{1}{2x+1}$$

53.
$$\begin{array}{r|l} & x^2 - 2x + 4 \\ \hline x+2 & x^3 + 0x^2 + 0x + 0 \\ & \underline{x^3 + 2x^2} \\ & \quad -2x^2 + 0x \\ & \quad \underline{-2x^2 - 4x} \\ & \qquad\qquad 4x + 0 \\ & \qquad\qquad \underline{4x + 8} \\ & \qquad\qquad\quad -8 \end{array}$$

$$\frac{x^3}{x+2} = x^2 - 2x + 4 + \frac{-8}{x+2}$$

55. $\dfrac{x^3 + 2x}{x^2} = \dfrac{x^3}{x^2} + \dfrac{2x}{x^2} = x + \dfrac{2}{x}$

57.
$$\begin{array}{r|l} & x - 6 \\ \hline 2x+1 & 2x^2 - 11x - 4 \\ & \underline{2x^2 + x} \\ & \qquad -12x - 4 \\ & \qquad \underline{-12x - 6} \\ & \qquad\qquad 2 \end{array}$$

$$\frac{2x^2 - 11x - 4}{2x+1} = x - 6 + \frac{2}{2x+1}$$

59.
$$\begin{array}{r|l} & 3x^2 - x - 1 \\ \hline x-1 & 3x^3 - 4x^2 + 0x + 7 \\ & \underline{3x^3 - 3x^2} \\ & \quad -x^2 + 0x \\ & \quad \underline{-x^2 + x} \\ & \qquad\qquad -x + 7 \\ & \qquad\qquad \underline{-x + 1} \\ & \qquad\qquad\quad 6 \end{array}$$

$$\frac{3x^3-4x^2+7}{x-1} = 3x^2 - x - 1 + \frac{6}{x-1}$$

61.
$$\begin{array}{r|l} & 6x^2+12x+20 \\ \hline x-2 & 6x^3+0x^2-4x+5 \\ & \underline{6x^3-12x^2} \\ & \quad 12x^2-4x \\ & \quad \underline{12x^2-24x} \\ & \qquad\quad 20x+5 \\ & \qquad\quad \underline{20x-40} \\ & \qquad\qquad\quad 45 \end{array}$$

$$\frac{6x^3-4x+5}{x-2} = 6x^2 + 12x + 20 + \frac{45}{x-2}$$

63.
$$\begin{array}{r|l} & x^2 - x + 2 \\ \hline x+1 & x^3+0x^2+x+0 \\ & \underline{x^3+x^2} \\ & \quad -x^2+x \\ & \quad \underline{-x^2-x} \\ & \qquad\quad 2x+0 \\ & \qquad\quad \underline{2x+2} \\ & \qquad\qquad -2 \end{array}$$

$$\frac{x^3+x}{x+1} = x^2 - x + 2 + \frac{-2}{x+1}$$

65.
$$\begin{array}{r|rrr} 1 & 1 & 1 & 7 \\ & & 1 & 2 \\ \hline & 1 & 2 & 9 \end{array}$$

The quotient is $x+2$ and the remainder is 9.

67.
$$\begin{array}{r|rrr} -1 & 2 & -4 & 5 \\ & & -2 & 6 \\ \hline & 2 & -6 & 11 \end{array}$$

The quotient is $2x-6$ and the remainder is 11.

69.
$$\begin{array}{r|rrrr} 2 & 1 & -5 & 6 & -3 \\ & & 2 & -6 & 0 \\ \hline & 1 & -3 & 0 & -3 \end{array}$$

The quotient is x^2-3x and the remainder is -3.

71.
$$\begin{array}{r|rrrrr} 3 & 3 & 0 & -15 & 7 & -9 \\ & & 9 & 27 & 36 & 129 \\ \hline & 3 & 9 & 12 & 43 & 120 \end{array}$$

The quotient is $3x^3+9x^2+12x+43$ and the remainder is 120.

73.
$$\begin{array}{r|rrrrrr} 1 & 1 & 0 & 0 & 0 & 0 & -1 \\ & & 1 & 1 & 1 & 1 & 1 \\ \hline & 1 & 1 & 1 & 1 & 1 & 0 \end{array}$$

The quotient is $x^4+x^3+x^2+x+1$ and the remainder is 0.

75.
$$\begin{array}{r|rrrr} -2 & 1 & 0 & -5 & 6 \\ & & -2 & 4 & 2 \\ \hline & 1 & -2 & -1 & 8 \end{array}$$

Quotient is x^2-2x-1 and remainder is 8.

77.
$$\begin{array}{r|rrrr} 5 & 1 & -3 & 0 & 5 \\ & & 5 & 10 & 50 \\ \hline & 1 & 2 & 10 & 55 \end{array}$$

The quotient is $x^2+2x+10$ and the remainder is 55.

79. Divide $x^3+x^2-11x+8$ by $x+4$:
$$\begin{array}{r|rrrr} -4 & 1 & 1 & -11 & 8 \\ & & -4 & 12 & -4 \\ \hline & 1 & -3 & 1 & 4 \end{array}$$

Since the remainder is not 0, $x+4$ is not a factor of $x^3+x^2-11x+8$.

81. Divide x^2-6x+8 by $x-4$:
$$\begin{array}{r|rrr} 4 & 1 & -6 & 8 \\ & & 4 & -8 \\ \hline & 1 & -2 & 0 \end{array}$$

Since the remainder is 0, $x-4$ is a factor of x^2-6x+8: $x^2-6x+8=(x-4)(x-2)$.

83. Divide w^3-27 by $w-3$:
$$\begin{array}{r|rrrr} 3 & 1 & 0 & 0 & -27 \\ & & 3 & 9 & 27 \\ \hline & 1 & 3 & 9 & 0 \end{array}$$

Since the remainder is zero, $w-3$ is a factor of w^2-27: $w^3-27=(w-3)(w^2+3w+9)$.

85. Divide $2x^3-3x^2-4x+7$ by $2x-3$ using long division to get a quotient of x^2-2 and a remainder of 1. By the remainder theorem, $2x-3$ is not a factor of $2x^3-3x^2-4x+7$.

87. Divide y^3-4y^2+6y-4 by $y-2$:
$$\begin{array}{r|rrrr} 2 & 1 & -4 & 6 & -4 \\ & & 2 & -4 & 4 \\ \hline & 1 & -2 & 2 & 0 \end{array}$$

Since the remainder is 0, $y-2$ is a factor of y^3-4y^2+6y-4 and $y^3-4y^2+6y-4=(y-2)(y^2-2y+2)$.

89. Divide x^2-5x-9 by $x-3$:
$$\begin{array}{r|rrr} 3 & 1 & -5 & -9 \\ & & 3 & -6 \\ \hline & 1 & -2 & -15 \end{array}$$

By the remainder theorem $P(3)=-15$.

91. Divide $4y^3 - 6y + 7$ by $y + 1$:

$$\begin{array}{r|rrrr} -1 & 4 & 0 & -6 & 7 \\ & & -4 & 4 & 2 \\ \hline & 4 & -4 & -2 & 9 \end{array}$$

By the remainder theorem $P(-1) = 9$.

93. Divide $-w^3 + 5w^2 + 3w$ by $w - 4$:

$$\begin{array}{r|rrrr} 4 & -1 & 5 & 3 & 0 \\ & & -4 & 4 & 28 \\ \hline & -1 & 1 & 7 & 28 \end{array}$$

By the remainder theorem $P(4) = 28$.

95. a) Divide $0.03x^2 + 300x$ by x to get $AC(x) = 0.03x + 300$.

b) $AC(x)$ is not constant, it is a linear function.

c) Because $AC(x)$ is very close to 300 for x less than 15, the graph looks horizontal.

97. Divide $x^2 - 1$ by $x + 1$ to get $x - 1$. The width is $x - 1$ ft.

99. $V = \dfrac{10(30^3 - 20^3)}{3(30 - 20)} \approx 6{,}333.3 \text{ m}^3$

101. The answer is wrong because the degree of the remainder $9x + 7$ is not less than the degree of the divisor $x - 3$.

6.6 WARM-UPS

1. True, because that will eliminate all denominators.

2. False, we should multiply each side by $6x$.

3. False, extraneous roots are real numbers that are not roots to the equation.

4. True, $x^2 - 4$ is the LCD.

5. False, $-\frac{1}{2}$ cannot be a solution because it causes $2x + 1$ to be 0.

6. True, because the equation is equivalent to $2x = 15$.

7. False, because the extremes-means property is only applied to equations of the type $\frac{a}{b} = \frac{c}{d}$.

8. False, because $x^2 = x$ has two solutions, 0 and 1.

9. False, the solution set is $\left\{\frac{3}{2}, -\frac{4}{3}\right\}$.

10. True, because of the extremes-means property.

6.6 EXERCISES

1. The first step is to multiply each side of the equation by the LCD.

3. A proportion is an equation expressing equality of two rational expressions.

5. In $a/b = c/d$ the extremes are a and d.

7.

$$\begin{aligned} \frac{1}{x} + \frac{1}{6} &= \frac{1}{8} \\ 24x\left(\frac{1}{x} + \frac{1}{6}\right) &= 24x\left(\frac{1}{8}\right) \\ 24 + 4x &= 3x \\ x &= -24 \end{aligned}$$

Solution set: $\{-24\}$

9.

$$\begin{aligned} \frac{2}{3x} + \frac{1}{15x} &= \frac{1}{2} \\ 30x\left(\frac{2}{3x} + \frac{1}{15x}\right) &= 30x\left(\frac{1}{2}\right) \\ 20 + 2 &= 15x \\ 22 &= 15x \\ \frac{22}{15} &= x \end{aligned}$$

Solution set: $\left\{\frac{22}{15}\right\}$

11.

$$\begin{aligned} \frac{3}{x-2} + \frac{5}{x} &= \frac{10}{x} \\ x(x-2)\left(\frac{3}{x-2} + \frac{5}{x}\right) &= x(x-2)\left(\frac{10}{x}\right) \\ 3x + 5x - 10 &= 10x - 20 \\ 8x - 10 &= 10x - 20 \\ 10 &= 2x \\ 5 &= x \end{aligned}$$

Solution set: $\{5\}$

13.

$$\begin{aligned} \frac{x}{x-2} + \frac{3}{x} &= 2 \\ x(x-2)\left(\frac{x}{x-2} + \frac{3}{x}\right) &= x(x-2)2 \\ x^2 + 3x - 6 &= 2x^2 - 4x \\ -x^2 + 7x - 6 &= 0 \\ x^2 - 7x + 6 &= 0 \\ (x-6)(x-1) &= 0 \end{aligned}$$

$$x - 6 = 0 \quad \text{or} \quad x - 1 = 0$$
$$x = 6 \quad \text{or} \quad x = 1$$

Solution set: $\{1, 6\}$

15.

$$\begin{aligned} \frac{100}{x} &= \frac{150}{x+5} - 1 \\ x(x+5)\left(\frac{100}{x}\right) &= x(x+5)\left(\frac{150}{x+5} - 1\right) \\ 100x + 500 &= 150x - x(x+5) \\ 100x + 500 &= 150x - x^2 - 5x \\ x^2 - 45x + 500 &= 0 \\ (x-25)(x-20) &= 0 \\ x = 25 \text{ or } x &= 20 \end{aligned}$$

Solution set: $\{20, 25\}$

17.
$$\frac{3x-5}{x-1} = 2 - \frac{2x}{x-1}$$
$$(x-1)\left(\frac{3x-5}{x-1}\right) = (x-1)\left(2 - \frac{2x}{x-1}\right)$$
$$3x - 5 = 2x - 2 - 2x$$
$$3x - 5 = -2$$
$$3x = 3$$
$$x = 1$$
Since replacing x by 1 gives 0 in the denominator, 1 is an extraneous root. The solution set is $\emptyset$.

19.
$$x + 1 + \frac{2x-5}{x-5} = \frac{x}{x-5}$$
$$(x-5)\left(x + 1 + \frac{2x-5}{x-5}\right) = (x-5)\frac{x}{x-5}$$
$$x^2 - 4x - 5 + 2x - 5 = x$$
$$x^2 - 3x - 10 = 0$$
$$(x-5)(x+2) = 0$$
$$x = 5 \quad \text{or} \quad x = -2$$
Since $x - 5$ is in the denominator, 5 is an extraneous root. The solution set is $\{-2\}$

21.
$$5 + \frac{9}{x-2} = 2 + \frac{x+7}{x-2}$$
$$(x-2)\left(5 + \frac{9}{x-2}\right) = (x-2)\left(2 + \frac{x+7}{x-2}\right)$$
$$5x - 10 + 9 = 2x - 4 + x + 7$$
$$2x = 4$$
$$x = 2$$
Since replacing x by 2 gives 0 in the denominator, 2 is an extraneous root. The solution set is $\emptyset$.

23. $\frac{2}{x+2} + \frac{x}{x-3} + \frac{1}{x^2 - x - 6} = 0$

Multiply each side by $(x+2)(x-3)$.
$$(x-3)2 + (x+2)x + 1 = 0$$
$$2x - 6 + x^2 + 2x + 1 = 0$$
$$x^2 + 4x - 5 = 0$$
$$(x+5)(x-1) = 0$$
$$x = -5 \text{ or } x = 1$$
The solution set is $\{-5, 1\}$.

25.
$$\frac{2}{x} = \frac{3}{4}$$
$$3x = 8$$
$$x = \frac{8}{3}$$
Solution set: $\left\{\frac{8}{3}\right\}$

27.
$$\frac{a}{3} = \frac{-1}{4}$$
$$4a = -3$$
$$a = -\frac{3}{4}$$
Solution set: $\left\{-\frac{3}{4}\right\}$

29.
$$\frac{-5}{7} = \frac{2}{x}$$
$$-5x = 14$$
$$x = -\frac{14}{5}$$
Solution set: $\left\{-\frac{14}{5}\right\}$

31.
$$\frac{10}{x} = \frac{20}{x+20}$$
$$10x + 200 = 20x$$
$$-10x = -200$$
$$x = 20$$
Solution set: $\{20\}$

33.
$$\frac{2}{x+1} = \frac{x-1}{4}$$
$$x^2 - 1 = 8$$
$$x^2 - 9 = 0$$
$$(x-3)(x+3) = 0$$
$$x = 3 \quad \text{or} \quad x = -3$$
Solution set: $\{-3, 3\}$

35.
$$\frac{x}{6} = \frac{5}{x-1}$$
$$x^2 - x = 30$$
$$x^2 - x - 30 = 0$$
$$(x-6)(x+5) = 0$$
$$x = 6 \quad \text{or} \quad x = -5$$
Solution set: $\{-5, 6\}$

37.
$$\frac{x+7}{x+4} = \frac{x+1}{x-2}$$
$$x^2 + 5x - 14 = x^2 + 5x + 4$$
$$-14 = 4$$
Solution set: $\emptyset$

39.
$$\frac{x-2}{x-3} = \frac{x+5}{x+2}$$
$$(x-2)(x+2) = (x-3)(x+5)$$
$$x^2 - 4 = x^2 + 2x - 15$$
$$-2x = -11$$
$$x = \frac{11}{2}$$
Solution set: $\left\{\frac{11}{2}\right\}$

41.
$$\frac{3w}{3w-5} = \frac{w}{w+2}$$
$$3w^2 + 6w = 3w^2 - 5w$$
$$6w = 5w$$
$$w = 0$$
The solution set is $\{0\}$.

43.
$$\frac{a}{9} = \frac{4}{a}$$
$$a^2 = 36$$
$$a^2 - 36 = 0$$
$$(a-6)(a+6) = 0$$
$$a = 6 \quad \text{or} \quad a = -6$$
Solution set: $\{-6, 6\}$

45.
$$\frac{x}{9} = \frac{-20}{9x} + 1$$
$$9x\left(\frac{x}{9}\right) = 9x\left(\frac{-20}{9x} + 1\right)$$

$$x^2 = -20 + 9x$$
$$x^2 - 9x + 20 = 0$$
$$(x-4)(x-5) = 0$$
$$x - 4 = 0 \text{ or } x - 5 = 0$$
$$x = 4 \text{ or } x = 5$$

The solution set is $\{4, 5\}$.

47. $\frac{1}{2x-4} + \frac{1}{x-2} = \frac{1}{4}$

$$4(x-2)\left(\frac{1}{2(x-2)} + \frac{1}{x-2}\right) = 4(x-2)\cdot\frac{1}{4}$$
$$2 + 4 = x - 2$$
$$8 = x$$

Solution set: $\{8\}$

49. $\frac{x-2}{4} = \frac{x-2}{x}$

$$x^2 - 2x = 4x - 8$$
$$x^2 - 6x + 8 = 0$$
$$(x-2)(x-4) = 0$$
$$x = 2 \text{ or } x = 4$$

Solution set: $\{2, 4\}$

51. $\frac{5}{2x+4} - \frac{1}{x-1} = \frac{3}{x+2}$

$$2(x+2)(x-1)\left(\frac{5}{2(x+2)} - \frac{1}{x-1}\right) = 2(x+2)(x-1)\left(\frac{3}{x+2}\right)$$
$$5x - 5 - 2(x+2) = 6(x-1)$$
$$5x - 5 - 2x - 4 = 6x - 6$$
$$3x - 9 = 6x - 6$$
$$-3x = 3$$
$$x = -1$$

Solution set: $\{-1\}$

53. $\frac{5}{x-3} = \frac{x}{x-3}$

$$5x - 15 = x^2 - 3x$$
$$-x^2 + 8x - 15 = 0$$
$$x^2 - 8x + 15 = 0$$
$$(x-3)(x-5) = 0$$
$$x = 3 \text{ or } x = 5$$

Since replacing x by 3 causes 0 to appear in a denominator, 3 is an extraneous root. The solution set is $\{5\}$.

55. $\frac{w}{6} = \frac{3}{2w}$

$$2w^2 = 18$$
$$w^2 = 9$$
$$w^2 - 9 = 0$$
$$(w-3)(w+3) = 0$$
$$w = 3 \text{ or } w = -3$$

Solution set: $\{-3, 3\}$

57. $\frac{5}{4x-2} - \frac{1}{1-2x} = \frac{7}{3x+6}$

$$6(2x-1)(x+2)\left(\frac{5}{2(2x-1)} - \frac{-1}{2x-1}\right) = 6(2x-1)(x+2)\left(\frac{7}{3(x+2)}\right)$$
$$15(x+2) + 6(x+2) = 14(2x-1)$$
$$15x + 30 + 6x + 12 = 28x - 14$$
$$21x + 42 = 28x - 14$$
$$-7x = -56$$
$$x = 8$$

Solution set: $\{8\}$

59. $\frac{5}{x} = \frac{2}{5}$

$$2x = 25$$
$$x = \frac{25}{2}$$

Solution set: $\left\{\frac{25}{2}\right\}$

61. $\frac{x}{x-2} - \frac{x+2}{x^2-2x} = \frac{1}{x}$

$$x(x-2)\left(\frac{x}{x-2} - \frac{x+2}{x(x-2)}\right) = x(x-2)\cdot\frac{1}{x}$$
$$x^2 - x - 2 = x - 2$$
$$x^2 - 2x = 0$$
$$x(x-2) = 0$$
$$x = 0 \text{ or } x = 2$$

Both solutions are extraneous. The solution set is $\emptyset$.

63. $\frac{5}{x^2-9} + \frac{2}{x+3} = \frac{1}{x-3}$

$$(x-3)(x+3)\left(\frac{5}{x^2-9} + \frac{2}{x+3}\right) = (x-3)(x+3)\left(\frac{1}{x-3}\right)$$
$$5 + 2x - 6 = x + 3$$
$$x = 4$$

Solution set: $\{4\}$

65. $\frac{9}{x^3-1} - \frac{1}{x-1} = \frac{2}{x^2+x+1}$

$$(x-1)(x^2+x+1)\left(\frac{9}{x^3-1} - \frac{1}{x-1}\right) = (x-1)(x^2+x+1)\left(\frac{2}{x^2+x+1}\right)$$
$$9 - (x^2 + x + 1) = 2(x-1)$$
$$9 - x^2 - x - 1 = 2x - 2$$
$$-x^2 - 3x + 10 = 0$$
$$x^2 + 3x - 10 = 0$$
$$(x+5)(x-2) = 0$$
$$x = -5 \text{ or } x = 2$$

Solution set: $\{-5, 2\}$

67. $\frac{4}{x} = \frac{3}{4}$
$$3x = 16$$
$$x = \frac{16}{3}$$
The solution set is $\left\{\frac{16}{3}\right\}$.

69. $\frac{4}{x} + \frac{3}{4} = \frac{4 \cdot 4}{x \cdot 4} + \frac{3x}{4x} = \frac{3x + 16}{4x}$

71. $\frac{2}{x} - \frac{3}{4} = \frac{1}{2}$
$$4x \cdot \frac{2}{x} - 4x \cdot \frac{3}{4} = 4x \cdot \frac{1}{2}$$
$$8 - 3x = 2x$$
$$8 = 5x$$
$$x = \frac{8}{5}$$
The solution set is $\left\{\frac{8}{5}\right\}$.

73. $\frac{2}{x} - \frac{3}{4} - \frac{1}{2} = \frac{2 \cdot 4}{x \cdot 4} - \frac{3x}{4x} - \frac{1 \cdot 2x}{2 \cdot 2x}$
$$= \frac{8 - 3x - 2x}{4x} = \frac{-5x + 8}{4x}$$

75. $\frac{1}{x} + \frac{1}{2x} = \frac{3}{2x}$
$$2x \cdot \frac{1}{x} + 2x \cdot \frac{1}{2x} = 2x \cdot \frac{3}{2x}$$
$$2 + 1 = 3$$
The equation is an identity. Zero cannot be used for x. So the solution set is $(-\infty, 0) \cup (0, \infty)$.

77. $\frac{1}{x} + \frac{1}{2x} = \frac{5}{4x}$
$$4x \cdot \frac{1}{x} + 4x \cdot \frac{1}{2x} = 4x \cdot \frac{5}{4x}$$
$$4 + 2 = 5$$
$$6 = 5$$
The equation is inconsistent. The solution set it the empty set $\emptyset$.

79. $\frac{1}{x} + \frac{1}{2x} = \frac{x+2}{2x}$
$$2x \cdot \frac{1}{x} + 2x \cdot \frac{1}{2x} = 2x \cdot \frac{x+2}{2x}$$
$$2 + 1 = x + 2$$
$$1 = x$$
The equation is conditional equation and the solution set is $\{1\}$.

81. $\frac{1}{x} + \frac{1}{x-1} = \frac{2x-1}{x^2 - x}$
$$x(x-1) \cdot \frac{1}{x} + x(x-1) \cdot \frac{1}{x-1} = x(x-1) \cdot \frac{2x-1}{x(x-1)}$$
$$x - 1 + x = 2x - 1$$
$$2x - 1 = 2x - 1$$
The equation is an identity. Since 0 and 1 cannot be used for x, the solution set is $(-\infty, 0) \cup (0, 1) \cup (1, \infty)$.

83. $\frac{1}{x} + \frac{1}{x-1} = \frac{2x}{x^2 - x}$
$$x(x-1) \cdot \frac{1}{x} + x(x-1) \cdot \frac{1}{x-1} = x(x-1) \cdot \frac{2x}{x(x-1)}$$
$$x - 1 + x = 2x$$
$$2x - 1 = 2x$$
$$-1 = 0$$
The equation is an inconsistent equation. The solution set is the empty set $\emptyset$.

85. Let x represent the height of the screen. The ratio of the width to height is 16 to 9.
$$\frac{16}{9} = \frac{48}{x}$$
$$16x = 9 \cdot 48$$
$$x = \frac{9 \cdot 48}{16} = 27 \text{ inches}$$

87. $\frac{300{,}000}{250{,}000} = \frac{200{,}000}{a}$
$$300{,}000a = 200{,}000 \cdot 250{,}000$$
$$a = \frac{200{,}000 \cdot 250{,}000}{300{,}000}$$
$$a = \$166{,}666.67$$

89. $\frac{2}{23} = \frac{12}{p}$
$$2p = 12 \cdot 23$$
$$p = \frac{12 \cdot 23}{2} = 138$$

91. Let $w =$ the width and $w + 22 =$ the length.
$$\frac{7}{6} = \frac{w + 22}{w}$$
$$7w = 6w + 132$$
$$w = 132$$
$$w + 22 = 154$$
The width is 132 cm and the length is 154 cm.

93. $1{,}000{,}000 = \frac{4{,}000{,}000p}{100 - p}$
$$100{,}000{,}000 - 1{,}000{,}000p = 4{,}000{,}000p$$
$$100{,}000{,}000 = 5{,}000{,}000p$$
$$p = 20$$
For \$1 million, 20% of the pollution can be cleaned up. For \$100 million, 96% of the pollution can be cleaned up.

95. a) Let $s =$ the amount invested in stocks and $s - 20{,}000 =$ the amount invested in bonds. For a wealth-building portfolio, the ratio of stocks to bonds should be 65 to 30:

$$\frac{65}{30} = \frac{s}{s - 20{,}000}$$
$$65s - 1{,}300{,}000 = 30s$$
$$35s = 1{,}300{,}000$$
$$s = \$37{,}142.86$$
$$s - 20{,}000 = \$17{,}142.86$$

She invested \$17,142.86 in bonds.

b) Since the stock investment is 65% of her bonus, b, we have $0.65b = 37{,}142.86$ or $b = 37{,}142.86/0.65 = \$57{,}142.86$.

97. To solve the equation, multiply each side by the LCD. To find the sum, build up each rational expression so that its denominator is the LCD.

6.7 WARM-UPS

1. False, t must not appear on both sides of the formula.
2. True, pqs is the LCD.
3. False, the price is $\frac{x}{50}$ dollars per pound.
4. True, rate is distance divided by time.
5. True, because this is his rate.
6. True, this is the rate at which he is working.
7. False, because $y - 1 = x$.
8. True, multiply by B and divide by m.
9. False, if $a = \frac{x}{y}$ then $y = \frac{x}{a}$.
10. False, the correct equation is $x - 3 = y$.

6.7 EXERCISES

1.
$$\frac{y-5}{x+3} = -\frac{4}{3}$$
$$3(y-5) = -4(x+3)$$
$$3y - 15 = -4x - 12$$
$$3y = -4x + 3$$
$$y = -\frac{4}{3}x + 1$$

3.
$$M = \frac{1}{y}$$
$$My = 1$$
$$y = \frac{1}{M}$$

5.
$$\frac{1}{y} = \frac{a}{w} + \frac{w}{a}$$
$$awy\frac{1}{y} = awy\frac{a}{w} + awy\frac{w}{a}$$
$$aw = a^2y + w^2y$$
$$aw = y(a^2 + w^2)$$
$$y = \frac{aw}{a^2 + w^2}$$

7.
$$h = \frac{b}{y} + 3$$
$$y \cdot h = y\frac{b}{y} + y \cdot 3$$
$$hy = b + 3y$$
$$hy - 3y = b$$
$$y(h-3) = b$$
$$y = \frac{b}{h-3}$$

9.
$$M = \frac{F}{f}$$
$$f \cdot M = f \cdot \frac{F}{f}$$
$$fM = F$$
$$f = \frac{F}{M}$$

11.
$$\frac{1}{a} + \frac{1}{b} = \frac{1}{2}$$
$$2ab\left(\frac{1}{a} + \frac{1}{b}\right) = 2ab \cdot \frac{1}{2}$$
$$2b + 2a = ab$$
$$2a - ab = -2b$$
$$a(2-b) = -2b$$
$$a = \frac{-2b}{2-b}$$
$$a = \frac{2b}{b-2}$$

13.
$$\frac{1}{2x} + \frac{1}{2} - \frac{2}{y} = 0$$
$$2xy\left(\frac{1}{2x} + \frac{1}{2} - \frac{2}{y}\right) = 2xy \cdot 0$$
$$y + xy - 4x = 0$$
$$xy - 4x = -y$$
$$x(y-4) = -y$$
$$x = \frac{-y}{y-4}$$
$$x = \frac{y}{4-y}$$

15.
$$\frac{1}{R} = \frac{1}{R_1} + \frac{1}{R_2}$$
$$RR_1R_2 \cdot \frac{1}{R} = RR_1R_2 \cdot \frac{1}{R_1} + RR_1R_2 \cdot \frac{1}{R_2}$$
$$R_1R_2 = RR_2 + RR_1$$
$$R_1R_2 - RR_1 = RR_2$$
$$R_1(R_2 - R) = RR_2$$
$$R_1 = \frac{RR_2}{R_2 - R}$$

17.
$$\frac{P_1V_1}{T_1} = \frac{P_2V_2}{T_2}$$
$$T_1T_2\left(\frac{P_1V_1}{T_1}\right) = T_1T_2\left(\frac{P_2V_2}{T_2}\right)$$
$$T_2P_1V_1 = T_1P_2V_2$$

$$\frac{T_2P_1V_1}{P_2V_2} = \frac{T_1P_2V_2}{P_2V_2}$$

$$T_1 = \frac{P_1V_1T_2}{P_2V_2}$$

19.

$$V = \frac{4}{3}\pi r^2 h$$
$$3V = 3 \cdot \frac{4}{3}\pi r^2 h$$
$$3V = 4\pi r^2 h$$
$$\frac{3V}{4\pi r^2} = \frac{4\pi r^2 h}{4\pi r^2}$$
$$h = \frac{3V}{4\pi r^2}$$

21.

$$M = \frac{F}{f}$$
$$10 = \frac{5}{f}$$
$$10f = 5$$
$$f = \frac{1}{2}$$

23.

$$r^2 = \frac{V}{\pi h}$$
$$3^2 = \frac{12\pi}{\pi h}$$
$$12\pi = 9\pi h$$
$$\frac{12\pi}{9\pi} = h$$
$$h = \frac{12\pi}{9\pi} = \frac{12}{9} = \frac{4}{3}$$

25.

$$F = \frac{mv^2}{r}$$
$$10 = \frac{8(6)^2}{r}$$
$$10r = 8 \cdot 36$$
$$r = \frac{8 \cdot 36}{10} = \frac{144}{5}$$

27.

$$\frac{1}{R} = \frac{1}{R_1} + \frac{1}{R_2}$$
$$\frac{1}{1.29} = \frac{1}{0.045} + \frac{1}{R_2}$$
$$\frac{1}{R_2} = \frac{1}{1.29} - \frac{1}{0.045} \approx -21.447$$
$$R_2 \approx -0.047$$

29. Let $x =$ her walking speed and $x + 10 =$ her riding speed. Since $T = D/R$, her time to school is $7/(x + 10)$ and her time to the post office is $2/x$. Since the times are equal, we can write

$$\frac{2}{x} = \frac{7}{x + 10}$$
$$7x = 2x + 20$$
$$5x = 20$$
$$x = 4$$

She walks 4 mph.

31. Let $x =$ the speed of John and $x - 5 =$ the speed of George. John's time is $240/x$ and George's time is $220/(x - 5)$. Their times are equal:

$$\frac{240}{x} = \frac{220}{x - 5}$$
$$240x - 1200 = 220x$$
$$20x = 1200$$
$$x = 60$$

So John's speed is 60 mph.

33. Let $x =$ the speed of each. Since $T = D/R$, Patrick's time is $\frac{40}{x}$ and Guy's time is $\frac{60}{x}$. Since Guy's time is $\frac{1}{5}$ hr longer than Patrick's we can write the equation

$$\frac{60}{x} - \frac{1}{5} = \frac{40}{x}$$
$$5x\left(\frac{60}{x} - \frac{1}{5}\right) = 5x \cdot \frac{40}{x}$$
$$300 - x = 200$$
$$100 = x$$

They are both driving 100 mph. Patrick takes $\frac{40}{100} = \frac{2}{5}$ hr $= 24$ minutes to get to work and Guy takes 12 minutes longer or 36 minutes. If Guy's claim is correct, then they are both driving 100 mph. So it is not likely that Guy's claim is correct.

35. Let $x =$ his walking speed and $x + 6 =$ his running speed. His time walking was $\frac{1}{x}$ hours and his time running was $\frac{5}{x + 6}$ hours. Since his total time was $3/4$ of an hour, we can write the equation

$$\frac{1}{x} + \frac{5}{x + 6} = \frac{3}{4}$$
$$4x(x + 6)\left(\frac{1}{x} + \frac{5}{x + 6}\right) = 4x(x + 6)\frac{3}{4}$$
$$4x + 24 + 20x = 3x^2 + 18x$$
$$-3x^2 + 6x + 24 = 0$$
$$x^2 - 2x - 8 = 0$$
$$(x - 4)(x + 2) = 0$$
$$x = 4 \text{ or } x = -2$$

Disregard the negative solution. Use $x = 4$ to find that his running speed was 10 mph.

37. Let $x =$ the number of hours for the smaller pump to drain the pool working alone. In one hour of operation, the larger pump drains $\frac{1}{3}$ of the pool, the smaller pump drains $\frac{1}{x}$ of the pool, and together $\frac{1}{2}$ of the pool is emptied.

$$\frac{1}{3}+\frac{1}{x}=\frac{1}{2}$$
$$6x\left(\frac{1}{3}+\frac{1}{x}\right)=6x\left(\frac{1}{2}\right)$$
$$2x+6=3x$$
$$6=x$$

The smaller pump would take 6 hours to drain the pool by itself.

39. Let $x =$ the number of minutes to fill the tub with the drain left open. In one minute, the faucet fills $\frac{1}{10}$ of the tub, the drain takes $\frac{1}{12}$ of the tub, but the tub is $\frac{1}{x}$ full.

$$\frac{1}{10}-\frac{1}{12}=\frac{1}{x}$$
$$60x\left(\frac{1}{10}-\frac{1}{12}\right)=60x\cdot\frac{1}{x}$$
$$6x-5x=60$$
$$x=60$$

The tub is filled in 60 minutes.

41. Let $x =$ their time working together. Since Gina takes 90 minutes and Hilda works twice as fast, Hilda can do the job alone in 45 minutes. Hilda does $\frac{1}{45}$ of the job per minute and Gina does $\frac{1}{90}$ or the job per minute. Since together they do $\frac{1}{x}$ of the job per minute, we can write the following equation.

$$\frac{1}{45}+\frac{1}{90}=\frac{1}{x}$$
$$90x\left(\frac{1}{45}+\frac{1}{90}\right)=90x\cdot\frac{1}{x}$$
$$2x+x=90$$
$$3x=90$$
$$x=30$$

It will take them 30 minutes working together.

43. Let $x =$ the number of pounds of apples and $x+2 =$ the number of pounds of oranges. The apples sell for $\frac{8.80}{x}$ dollars per pound and the oranges sell for $\frac{5.28}{x+2}$. The apples cost twice as much per pound is expressed as

$$\frac{8.80}{x}=2\left(\frac{5.28}{x+2}\right)$$
$$\frac{8.80}{x}=\frac{10.56}{x+2}$$
$$8.80x+17.60=10.56x$$
$$17.60=1.76x$$
$$10=x$$

She bought 10 pounds of apples and 12 pounds of oranges.

45.
$$\frac{1}{2}=\frac{1}{3}+\frac{1}{R_2}$$
$$6R_2\left(\frac{1}{2}\right)=6R_2\left(\frac{1}{3}+\frac{1}{R_2}\right)$$
$$3R_2=2R_2+6$$
$$R_2=6\text{ ohms}$$

47.
$$\frac{1}{S_o}+\frac{1}{S_i}=\frac{1}{F}$$
$$\frac{1}{500}+\frac{1}{S_i}=\frac{1}{100}$$
$$500S_i\left(\frac{1}{500}+\frac{1}{S_i}\right)=500S_i\cdot\frac{1}{100}$$
$$S_i+500=5S_i$$
$$-4S_i=-500$$
$$S_i=125\text{ mm}$$

49. Let $x =$ the number in the original group. Their cost is $\frac{1000}{x}$ per person. For 3 more people, the cost is $\frac{1000}{x+3}$. We have

$$\frac{1000}{x}=\frac{1000}{x+3}+75$$

$$x(x+3)\frac{1000}{x}=x(x+3)\left(\frac{1000}{x+3}+75\right)$$

$$1000x+3000=1000x+75x^2+225x$$
$$75x^2+225x-3000=0$$
$$x^2+3x-40=0$$
$$(x+8)(x-5)=0$$
$$x=-8 \quad\text{or}\quad x=5$$

There are 5 people in the initial group.

51. a) Let $x =$ the number in the original group. Their cost is $\frac{24{,}000}{x}$ per person. For 40 more people, the cost is $\frac{24{,}000}{x+40}$. We have

$$\frac{24{,}000}{x}=\frac{24{,}000}{x+40}+100$$
$$x(x+40)\frac{24{,}000}{x}=x(x+40)\left(\frac{24{,}000}{x+40}+100\right)$$
$$24{,}000x+960{,}000$$
$$=24{,}000x+100x^2+4000x$$
$$100x^2+4000x-960{,}000=0$$
$$x^2+40x-9600=0$$
$$(x-80)(x+120)=0$$
$$x=80 \quad\text{or}\quad x=-120$$

There are 80 people in the initial group.

b) $C(n)=\frac{24{,}000}{n}$ where n is the number of people sharing the cost and $C(n)$ is the cost per person in dollars.

53. Let $x =$ the number of days for both dogs to eat a 50 pound bag together. Muffy eats at a rate of $\frac{25}{28}$ pounds per day, Missy eats at the rate of $\frac{25}{23}$ pounds per day, and together they eat at a rate of $\frac{50}{x}$ pounds per day.

$$\frac{25}{28} + \frac{25}{23} = \frac{50}{x}$$
$$\frac{1}{28} + \frac{1}{23} = \frac{2}{x}$$
$$28 \cdot 23x\left(\frac{1}{28} + \frac{1}{23}\right) = 28 \cdot 23x\left(\frac{2}{x}\right)$$
$$23x + 28x = 1288$$
$$51x = 1288$$
$$x = \frac{1288}{51} \approx 25.255$$

It would take 25.255 days for them to eat 50 pounds of dog food together.

Chapter 6 Wrap-Up

Enriching Your Mathematical Word Power

1. b **2.** d **3.** b **4.** d **5.** b
6. a **7.** a **8.** d **9.** a **10.** b
11. d **12.** d **13.** a **14.** c **15.** d

CHAPTER 6 REVIEW

1. If $x = 1$, then $3x - 3 = 0$. So the domain is $\{x \mid x \neq 1\}$ or $(-\infty, 1) \cup (1, \infty)$.

3. Solve $x^2 - x - 2 = 0$
$$(x - 2)(x + 1) = 0$$
$$x - 2 = 0 \text{ or } x + 1 = 0$$
$$x = 2 \text{ or } \quad x = -1$$
The domain is $\{x \mid x \neq -1 \text{ and } x \neq 2\}$ or $(-\infty, -1) \cup (-1, 2) \cup (2, \infty)$.

5. $\frac{a^3bc^3}{a^5b^2c} = \frac{c^2}{a^2b}$

7. $\frac{68x^3}{51xy} = \frac{2 \cdot 2 \cdot 17x^3}{3 \cdot 17xy} = \frac{4x^2}{3y}$

9. $\frac{a^3b^2}{b^3a} \cdot \frac{ab - b^2}{ab - a^2} = \frac{a^3b^2b(a - b)}{b^3a(-a)(a - b)} = -a$

11. $\frac{w - 4}{3w} \div \frac{2w - 8}{9w} = \frac{w - 4}{3w} \cdot \frac{3 \cdot 3w}{2(w - 4)}$
$$= \frac{3}{2}$$

13. Since $6x = 2 \cdot 3x$, $3x - 6 = 3(x - 2)$, and $x^2 - 2x = x(x - 2)$, the LCM $= 6x(x - 2)$.

15. Since $6ab^3 = 2 \cdot 3ab^3$ and $4a^5b^2 = 2 \cdot 2a^5b^2$, the LCM $= 2 \cdot 2 \cdot 3a^5b^3 = 12a^5b^3$.

17. $\frac{3}{2x - 6} + \frac{1}{x^2 - 9}$
$$= \frac{3(x + 3)}{2(x - 3)(x + 3)} + \frac{1 \cdot 2}{(x - 3)(x + 3)2}$$
$$= \frac{3x + 9}{2(x - 3)(x + 3)} + \frac{2}{2(x - 3)(x + 3)}$$
$$= \frac{3x + 11}{2(x - 3)(x + 3)}$$

19. $\frac{w}{ab^2} - \frac{5}{a^2b} = \frac{w \cdot a}{ab^2 \cdot a} - \frac{5 \cdot b}{a^2b \cdot b}$
$$= \frac{aw - 5b}{a^2b^2}$$

21. $\frac{-x - 17}{x^2 + 6x + 5} + \frac{4}{x + 1}$
$$= \frac{-x - 17}{(x + 5)(x + 1)} + \frac{4(x + 5)}{(x + 1)(x + 5)}$$
$$= \frac{3x + 3}{(x + 5)(x + 1)} = \frac{3(x + 1)}{(x + 5)(x + 1)}$$
$$= \frac{3}{x + 5}$$

23. $\frac{\frac{3}{2x} - \frac{4}{5x}}{\frac{1}{3} - \frac{2}{x}} = \frac{30x\left(\frac{3}{2x} - \frac{4}{5x}\right)}{30x\left(\frac{1}{3} - \frac{2}{x}\right)}$
$$= \frac{45 - 24}{10x - 60} = \frac{21}{10(x - 6)}$$

25. $\frac{\frac{1}{y - 2} - 3}{\frac{5}{y - 2} + 4} = \frac{(y - 2)\left(\frac{1}{y - 2} - 3\right)}{(y - 2)\left(\frac{5}{y - 2} + 4\right)}$
$$= \frac{1 - 3(y - 2)}{5 + 4(y - 2)} = \frac{7 - 3y}{4y - 3}$$

27. $\frac{a^{-2} - b^{-3}}{a^{-1}b^{-2}} = \frac{a^2b^3(a^{-2} - b^{-3})}{a^2b^3(a^{-1}b^{-2})}$
$$= \frac{b^3 - a^2}{ab}$$

29.
$$\begin{array}{r} x^2 + 3x - 5 \\ x - 2 \overline{\smash{\big)}\, x^3 + x^2 - 11x + 10} \\ \underline{x^3 - 2x^2} \\ 3x^2 - 11x \\ \underline{3x^2 - 6x} \\ -5x + 10 \\ \underline{-5x + 10} \\ 0 \end{array}$$

Quotient: $x^2 + 3x - 5$ Remainder: 0

31.
$$\begin{array}{r|l} & m^3 - m^2 + m - 1 \\ \hline m+1 & m^4 + 0m^3 + 0m^2 + 0m - 1 \\ & \underline{m^4 + m^3} \\ & -m^3 + 0m^2 \\ & \underline{-m^3 - m^2} \\ & m^2 + 0m \\ & \underline{m^2 + m} \\ & -m - 1 \\ & \underline{-m - 1} \\ & 0 \end{array}$$

Quotient: $m^3 - m^2 + m - 1$ Remainder: 0

33.
$$\begin{array}{r|l} & a^6 + 2a^3 + 4 \\ \hline a^3 - 2 & a^9 + 0a^6 + 0a^3 - 8 \\ & \underline{a^9 - 2a^6} \\ & 2a^6 + 0a^3 \\ & \underline{2a^6 - 4a^3} \\ & 4a^3 - 8 \\ & \underline{4a^3 - 8} \\ & 0 \end{array}$$

Quotient: $a^6 + 2a^3 + 4$ Remainder: 0

35. $\dfrac{3m^3 + 6m^2 - 18m}{3m}$
$$= \frac{3m^3}{3m} + \frac{6m^2}{3m} - \frac{18m}{3m}$$
$$= m^2 + 2m - 6$$

Quotient: $m^2 + 2m - 6$ Remainder: 0

37.
$$\begin{array}{r|rrr} 1 & 1 & 0 & -5 \\ & & 1 & 1 \\ \hline & 1 & 1 & -4 \end{array}$$

Quotient is $x + 1$ and remainder is -4.

$$\frac{x^2 - 5}{x - 1} = x + 1 + \frac{-4}{x - 1}$$

39.
$$\begin{array}{r|rr} 2 & 3 & 0 \\ & & 6 \\ \hline & 3 & 6 \end{array}$$

Quotient is 3 and remainder is 6.

$$\frac{3x}{x - 2} = 3 + \frac{6}{x - 2}$$

41. When $x^3 - 2x^2 + 3x + 22$ is divided by $x + 2$ the quotient is $x^2 - 4x + 11$ and the remainder is 0. So the first polynomial is a factor of the second.

43. If $x^3 - x - 120$ is divided by $x - 5$ the quotient is $x^2 + 5x + 24$ and remainder is 0. So the first polynomial is a factor of the second.

45. When $x^3 + x^2 - 3$ is divided by $x - 1$ the quotient is $x^2 + 2x + 2$ and the remainder is -1. So the first polynomial is not a factor of the second.

47. When $x^4 + x^3 + 5x^2 + 2x + 6$ is divided by $x^2 + 2$ the quotient is $x^2 + x + 3$ and the remainder is 0. So the first polynomial is a factor of the second.

49.
$$\frac{-3}{8} = \frac{2}{x}$$
$$-3x = 16$$
$$x = -\frac{16}{3}$$

Solution set: $\left\{-\frac{16}{3}\right\}$

51.
$$5 + \frac{x + 1}{x - 1} = 3 + \frac{5x - 3}{x - 1}$$
$$(x - 1)\left(5 + \frac{x + 1}{x - 1}\right) = (x - 1)\left(3 + \frac{5x - 3}{x - 1}\right)$$
$$5x - 5 + x + 1 = 3x - 3 + 5x - 3$$
$$6x - 4 = 8x - 6$$
$$2 = 2x$$
$$1 = x$$

The solution set is the empty set, $\emptyset$.

53.
$$\frac{15}{a^2 - 25} + \frac{1}{a - 5} = \frac{6}{a + 5}$$
$$(a - 5)(a + 5)\left(\frac{15}{a^2 - 25} + \frac{1}{a - 5}\right) = (a - 5)(a + 5)\left(\frac{6}{a + 5}\right)$$
$$15 + a + 5 = 6a - 30$$
$$-5a = -50$$
$$a = 10$$

Solution set: $\{10\}$

55.
$$\frac{y - b}{m} = x$$
$$m\left(\frac{y - b}{m}\right) = mx$$
$$y - b = mx$$
$$y = mx + b$$

57.
$$\frac{1}{x} + \frac{1}{2} = w$$
$$2x\left(\frac{1}{x} + \frac{1}{2}\right) = 2x \cdot w$$
$$2 + x = 2xw$$
$$x - 2xw = -2$$
$$x(1 - 2w) = -2$$
$$x = \frac{-2}{1 - 2w}$$
$$x = \frac{2}{2w - 1}$$

59.
$$F=\frac{mv^2}{r}$$
$$r\cdot F=r\left(\frac{mv^2}{r}\right)$$
$$rF=mv^2$$
$$m=\frac{Fr}{v^2}$$

61.
$$A=\frac{2}{3}\pi rh$$
$$3\cdot A=3\cdot\frac{2}{3}\pi rh$$
$$3A=2\pi rh$$
$$\frac{3A}{2\pi h}=\frac{2\pi rh}{2\pi h}$$
$$r=\frac{3A}{2\pi h}$$

63.
$$\frac{y+3}{x-7}=2$$
$$(x-7)\left(\frac{y+3}{x-7}\right)=(x-7)2$$
$$y+3=2x-14$$
$$y=2x-17$$

65.
$$\frac{1}{p}+\frac{1}{q}=\frac{1}{f}$$
$$pqf\left(\frac{1}{p}+\frac{1}{q}\right)=pqf\cdot\frac{1}{f}$$
$$qf+pf=pq$$
$$qf-pq=-pf$$
$$q(f-p)=-pf$$
$$q=\frac{-pf}{f-p}$$
$$q=\frac{pf}{p-f}$$

67. Let $x=$ the number of reported female AIDs cases and $x+16{,}000=$ the number of reported male AIDs cases.
$$\frac{15}{7}=\frac{x+16{,}000}{x}$$
$$15x=7x+112{,}000$$
$$8x=112{,}000$$
$$x\approx 14{,}000$$
$$x+16{,}000=30{,}000$$
There were 30,000 reported male AIDs cases.

69. If $x=$ the time east of Louisville, then we also have $x=$ the time west of Louisville. His speed east of Louisville was $\frac{310}{x}$ and his speed west of Louisville was $\frac{360}{x}$. We can write an equation expressing the fact that his speed west of Louisville was 10 mph greater than east of Louisville:
$$\frac{360}{x}-10=\frac{310}{x}$$
$$x\left(\frac{360}{x}-10\right)=x\left(\frac{310}{x}\right)$$
$$360-10x=310$$
$$-10x=-50$$
$$x=5$$
His journey took a total of 10 hours.

71. Let $x=$ the time for all three to make the quilt working together. Debbie makes $\frac{1}{2000}$ of the quilt per hour, Pat makes $\frac{1}{1000}$ of the quilt per hour, and Cheryl makes $\frac{1}{1000}$ of the quilt per hour. Working together, $\frac{1}{x}$ of the quilt gets done per hour:
$$\frac{1}{1000}+\frac{1}{1000}+\frac{1}{2000}=\frac{1}{x}$$
$$2000x\left(\frac{1}{1000}+\frac{1}{1000}+\frac{1}{2000}\right)=2000x\cdot\frac{1}{x}$$
$$2x+2x+x=2000$$
$$5x=2000$$
$$x=400$$
It will take them 400 hours to make the quilt together.

73. a) Let $x=$ Trine's time in minutes to pick a bushel. Let $x-6=$ Thud's time per bushel. working together. In two hours, Trine picks $\frac{120}{x}$ bushel and Thud picks $\frac{120}{x-6}$. So a function for the total bushels picked by them is $B(x)=\frac{120}{x}+\frac{120}{x-6}$ or $B(x)=\frac{240x-720}{x(x-6)}$

b) $B(12)=\dfrac{240(12)-720}{12(12-6)}=30$ bushels

75.
$$\frac{5x}{3x^2y}+\frac{7a^2}{6a^2x}=\frac{5}{3xy}+\frac{7}{6x}$$
$$=\frac{5\cdot 2}{3xy\cdot 2}+\frac{7\cdot y}{6x\cdot y}=\frac{10+7y}{6xy}$$

77.
$$\frac{5}{a-5}-\frac{3}{-a-5}$$
$$=\frac{5}{a-5}-\frac{3}{-1(a+5)}$$
$$=\frac{5}{a-5}+\frac{3}{a+5}$$
$$=\frac{5(a+5)}{(a-5)(a+5)}+\frac{3(a-5)}{(a+5)(a-5)}$$
$$=\frac{8a+10}{(a-5)(a+5)}$$

79. $\frac{1}{x-2} - \frac{1}{x+2} = \frac{1}{15}$

$$15(x-2)(x+2)\left(\frac{1}{x-2} - \frac{1}{x+2}\right) = 15(x-2)(x+2)\left(\frac{1}{15}\right)$$

$$15x + 30 - 15(x-2) = x^2 - 4$$
$$60 = x^2 - 4$$
$$x^2 - 64 = 0$$
$$(x-8)(x+8) = 0$$
$$x = 8 \quad \text{or } x = -8$$

Solution set: $\{-8, 8\}$

81. $\frac{-3}{x+2} \cdot \frac{5x+10}{10} = \frac{-3}{x+2} \cdot \frac{5(x+2)}{2 \cdot 5} = -\frac{3}{2}$

83. $\frac{x}{-3} = \frac{-27}{x}$

$$x^2 = 81$$
$$x^2 - 81 = 0$$
$$(x-9)(x+9) = 0$$
$$x = 9 \text{ or } x = -9$$

Solution set: $\{-9, 9\}$

85. $\frac{wx + wm + 3x + 3m}{w^2 - 9} \div \frac{x^2 - m^2}{w-3}$

$$= \frac{(w+3)(x+m)}{(w-3)(w+3)} \cdot \frac{w-3}{(x-m)(x+m)} = \frac{1}{x-m}$$

87. $\frac{5}{a^2 - 25} + \frac{3}{a^2 - 4a - 5}$

$$= \frac{5(a+1)}{(a-5)(a+5)(a+1)} + \frac{3(a+5)}{(a-5)(a+1)(a+5)}$$

$$= \frac{8a+20}{(a-5)(a+5)(a+1)}$$

89. $\frac{-7}{2a^2 - 18} - \frac{4}{a^2 + 5a + 6}$

$$= \frac{-7(a+2)}{2(a-3)(a+3)(a+2)} - \frac{4 \cdot 2(a-3)}{(a+2)(a+3)2(a-3)}$$

$$= \frac{-7a - 14 - (8a - 24)}{2(a+2)(a+3)(a-3)}$$

$$= \frac{-15a + 10}{2(a+2)(a+3)(a-3)}$$

91. $\frac{7}{a^2 - 1} + \frac{2}{1-a} = \frac{1}{a+1}$

$$= (a-1)(a+1)\left(\frac{7}{a^2-1} + \frac{2}{1-a}\right) = (a-1)(a+1)\left(\frac{1}{a+1}\right)$$

$$7 - 2(a+1) = a - 1$$
$$-2a + 5 = a - 1$$
$$-3a = -6$$
$$a = 2$$

Solution set: $\{2\}$

93. $\frac{2x}{x-3} + \frac{3}{x-2} = \frac{6}{(x-2)(x-3)}$

$$(x-2)(x-3)\left(\frac{2x}{x-3} + \frac{3}{x-2}\right) = (x-2)(x-3)\frac{6}{(x-2)(x-3)}$$

$$2x(x-2) + 3(x-3) = 6$$
$$2x^2 - x - 15 = 0$$
$$(2x+5)(x-3) = 0$$
$$2x + 5 = 0 \text{ or } x - 3 = 0$$
$$x = -\frac{5}{2} \text{ or } x = 3$$

Since 3 is an extraneous root, the solution set is $\left\{-\frac{5}{2}\right\}$.

95. $\frac{x-2}{6} \div \frac{2-x}{2} = \frac{x-2}{2 \cdot 3} \cdot \frac{2}{2-x} = -\frac{1}{3}$

97. $\frac{x-3}{x^2 + 3x + 2} \cdot \frac{x^2 - 4}{3x - 9}$

$$= \frac{x-3}{(x+2)(x+1)} \cdot \frac{(x-2)(x+2)}{3(x-3)} = \frac{x-2}{3x+3} = \frac{x-2}{3(x+1)}$$

99. $\frac{a+4}{a^3 - 8} - \frac{3}{2-a}$

$$= \frac{a+4}{(a-2)(a^2 + 2a + 4)} - \frac{-3}{a-2}$$

$$= \frac{a+4}{(a-2)(a^2 + 2a + 4)} - \frac{-3(a^2 + 2a + 4)}{(a-2)(a^2 + 2a + 4)}$$

$$= \frac{a + 4 + 3a^2 + 6a + 12}{a^3 - 8}$$

$$= \frac{3a^2 + 7a + 16}{(a-2)(a^2 + 2a + 4)}$$

101. $\frac{x^3 - 9x}{1 - x^2} \div \frac{x^3 + 6x^2 + 9x}{x-1}$

$$= \frac{x(x-3)(x+3)}{(1-x)(1+x)} \cdot \frac{x-1}{x(x+3)^2}$$

$$= \frac{3-x}{(x+1)(x+3)}$$

103.
$$\frac{a^2+3a+3w+aw}{a^2+6a+8}\cdot\frac{a^2-aw-2w+2a}{a^2+3a-3w-aw}$$
$$=\frac{(a+w)(a+3)}{(a+2)(a+4)}\cdot\frac{(a+2)(a-w)}{(a-w)(a+3)}$$
$$=\frac{a+w}{a+4}$$

105. $\frac{5}{x}-\frac{4}{x+2}=\frac{1}{5}+\frac{1}{5x}$
$$5x(x+2)\left(\frac{5}{x}-\frac{4}{x+2}\right)$$
$$=5x(x+2)\left(\frac{1}{5}+\frac{1}{5x}\right)$$
$$25x+50-20x=x^2+2x+x+2$$
$$5x+50=x^2+3x+2$$
$$-x^2+2x+48=0$$
$$x^2-2x-48=0$$
$$(x-8)(x+6)=0$$
$$x=8 \text{ or } x=-6$$
Solution set: $\{-6, 8\}$

107. $\frac{6}{x}=\frac{6\cdot 3}{x\cdot 3}=\frac{18}{3x}$

109. $\frac{3}{a-b}=\frac{3(-1)}{(a-b)(-1)}=\frac{-3}{b-a}$

111. $4=4\cdot\frac{x}{x}=\frac{4x}{x}$

113. $5x\div\frac{1}{2}=5x\cdot 2=10x$

115. $4a\div\frac{1}{3}=4a\cdot 3=12a$

117. $\frac{a-3}{a^2-9}=\frac{1(a-3)}{(a-3)(a+3)}=\frac{1}{a+3}$

119. $\frac{1}{2}-\frac{1}{5}=\frac{5}{10}-\frac{2}{10}=\frac{3}{10}$

121. $\frac{a}{3}+\frac{a}{2}=\frac{2a}{6}+\frac{3a}{6}=\frac{5a}{6}$

CHAPTER 6 TEST

1. The solution to $4-3x=0$ is $\{4/3\}$. So the domain of the rational expression is $\left\{x \mid x\neq\frac{4}{3}\right\}$ or $(-\infty,4/3)\cup(4/3,\infty)$

.

2. The solution to $x^2-9=0$ is $\{-3,3\}$. So the domain of the rational expression is $\{x \mid x\neq 3 \text{ and } x\neq -3\}$ or $(-\infty,-3)\cup(-3,3)\cup(3,\infty)$.

3. There is no solution to $x^2+9=0$ in the real numbers. So the domain is the set of all real numbers, $(-\infty,\infty)$.

4. $\frac{12a^9b^8}{(2a^2b^3)^3}=\frac{12a^9b^8}{8a^6b^9}=\frac{3a^3}{2b}$

5. $\frac{y^2-x^2}{2x^2-4xy+2y^2}=\frac{(y-x)(y+x)}{2(x-y)^2}$
$$=\frac{-1(y+x)}{2(x-y)}=\frac{-x-y}{2(x-y)}$$

6. $\frac{5y}{12y}-\frac{4x}{9x}=\frac{5}{12}-\frac{4}{9}=\frac{15}{36}-\frac{16}{36}=-\frac{1}{36}$

7. $\frac{3}{y}+7y=\frac{3}{y}+\frac{7y^2}{y}=\frac{7y^2+3}{y}$

8. $\frac{4}{a-9}-\frac{1}{(9-a}=\frac{4}{a-9}-\frac{1(-1)}{(9-a)(-1)}$
$$=\frac{4}{a-9}+\frac{1}{a-9}=\frac{5}{a-9}$$

9. $\frac{1}{6ab^2}+\frac{1}{8a^2b}=\frac{1\cdot 4a}{6ab^2\cdot 4a}+\frac{1\cdot 3b}{8a^2b\cdot 3b}$
$$=\frac{4a}{24a^2b^2}+\frac{3b}{24a^2b^2}=\frac{4a+3b}{24a^2b^2}$$

10. $\frac{3a^3b}{20ab}\cdot\frac{2a^2b}{9ab^3}=\frac{3a^3b}{2\cdot 2\cdot 5ab}\cdot\frac{2a^2b}{3\cdot 3ab^3}$
$$=\frac{a^3}{30b^2}$$

11. $\frac{a-b}{7}\div\frac{b^2-a^2}{21}$
$$=\frac{a-b}{7}\cdot\frac{3\cdot 7}{(b-a)(b+a)}=-\frac{3}{a+b}$$

12. $\frac{x-3}{x-1}\div(x^2-2x-3)$
$$=\frac{x-3}{x-1}\cdot\frac{1}{(x-3)(x+1)}$$
$$=\frac{1}{(x-1)(x+1)}$$

13. $\frac{2}{x^2-4}-\frac{6}{x^2-3x-10}$
$$=\frac{2}{(x-2)(x+2)}-\frac{6}{(x-5)(x+2)}$$
$$=\frac{2(x-5)}{(x-2)(x+2)(x-5)}$$
$$-\frac{6(x-2)}{(x-5)(x+2)(x-2)}$$
$$=\frac{2x-10-(6x-12)}{(x+2)(x-2)(x-5)}$$
$$=\frac{-4x+2}{(x+2)(x-2)(x-5)}$$

14. $\dfrac{m^3-1}{(m-1)^2}\cdot\dfrac{m^2-1}{3m^2+3m+3}$

$=\dfrac{(m-1)(m^2+m+1)}{(m-1)^2}\cdot\dfrac{(m-1)(m+1)}{3(m^2+m+1)}$

$=\dfrac{m+1}{3}$

15.
$$\frac{3}{x}=\frac{7}{4}$$
$$7x=12$$
$$x=\frac{12}{7}$$
The solution set is $\left\{\frac{12}{7}\right\}$.

16. $\frac{x}{x-2}-\frac{5}{x}=\frac{3}{4}$
$$4x(x-2)\left(\frac{x}{x-2}-\frac{5}{x}\right)=4x(x-2)\left(\frac{3}{4}\right)$$
$$4x^2-5(4x-8)=3x^2-6x$$
$$x^2-14x+40=0$$
$$(x-4)(x-10)=0$$
$$x=4\quad\text{or}\quad x=10$$
Solution set: $\{4, 10\}$

17.
$$\frac{3m}{2}=\frac{6}{m}$$
$$3m^2=12$$
$$m^2=4$$
$$m^2-4=0$$
$$(m-4)(m+4)=0$$
$$m=2\ \text{or}\ m=-2$$
Solution set: $\{-2, 2\}$

18.
$$W=\frac{a^2}{t}$$
$$tW=a^2$$
$$t=\frac{a^2}{W}$$

19.
$$\frac{1}{a}+\frac{1}{b}=\frac{1}{2}$$
$$2ab\left(\frac{1}{a}+\frac{1}{b}\right)=2ab\cdot\frac{1}{2}$$
$$2b+2a=ab$$
$$2b-ab=-2a$$
$$b(2-a)=-2a$$
$$b=\frac{-2a}{2-a}$$
$$b=\frac{2a}{a-2}$$

20. $\dfrac{\frac{1}{x}+\frac{1}{3x}}{\frac{3}{4x}-\frac{1}{2}}=\dfrac{12x\left(\frac{1}{x}+\frac{1}{3x}\right)}{12x\left(\frac{3}{4x}-\frac{1}{2}\right)}$

$=\dfrac{12+4}{9-6x}=\dfrac{16}{3(3-2x)}$

21. $\dfrac{m^{-2}-w^{-2}}{m^{-2}w^{-1}+m^{-1}w^{-2}}$

$=\dfrac{m^2w^2(m^{-2}-w^{-2})}{m^2w^2(m^{-2}w^{-1}+m^{-1}w^{-2})}$

$=\dfrac{w^2-m^2}{w+m}=\dfrac{(w-m)(w+m)}{w+m}=w-m$

22. $\dfrac{a^2b^3}{4a}\div\dfrac{ab^3}{6a^2}=\dfrac{a^2b^3}{2\cdot 2a}\cdot\dfrac{2\cdot 3a^2}{ab^3}=\dfrac{3a^2}{2}$

23.
```
             3x + 2
2x + 1 ) 6x^2 + 7x - 6
         6x^2 + 3x
                4x - 6
                4x + 2
                    -8
```

Quotient: $3x+2$ Remainder: -8

24. $(x-3)\div(3-x)=\dfrac{x-3}{3-x}$

$=\dfrac{-1(3-x)}{3-x}=-1$

Quotient: -1 Remainder: 0

25.
```
          5
x + 3 ) 5x + 0
        5x + 15
             -15
```

$\dfrac{5x}{x+3}=5+\dfrac{-15}{x+3}$

26.
```
         x + 5
x - 2 ) x^2 + 3x - 6
        x^2 - 2x
              5x - 6
              5x - 10
                    4
```

$\dfrac{x^2+3x-6}{x-2}=x+5+\dfrac{4}{x-2}$

27. Let $x =$ the number of minutes to fill the leaky pool. In one minute, the hose is supposed to fill $\frac{1}{6}$ of the pool, the leak removes $\frac{1}{8}$ of the pool, but together $\frac{1}{x}$ of the pool actually gets filled.

$$\frac{1}{6} - \frac{1}{8} = \frac{1}{x}$$
$$24x\left(\frac{1}{6} - \frac{1}{8}\right) = 24x \cdot \frac{1}{x}$$
$$4x - 3x = 24$$
$$x = 24$$

The leaky pool will be filled in 24 minutes.

28. Let $x =$ the number of miles hiked in one day. Milton's time is $\frac{x}{4}$ and Bonnie's time is $\frac{x}{3}$ hours. Since Milton's time is 2.5 hours less than Bonnie's, we can write the equation

$$\frac{x}{4} + \frac{5}{2} = \frac{x}{3}$$
$$12\left(\frac{x}{4} + \frac{5}{2}\right) = 12 \cdot \frac{x}{3}$$
$$3x + 30 = 4x$$
$$30 = x$$

They hiked 30 miles that day.

29. Let $x =$ the number of sailors in the original group and $x + 3 =$ the number in the larger group. The cost is $\frac{72{,}000}{x}$ dollars per person for x people and $\frac{72{,}000}{x+3}$ dollars per person for $x + 3$ people.

$$\frac{72{,}000}{x} = \frac{72{,}000}{x+3} + 2000$$
$$x(x+3)\frac{72{,}000}{x} = x(x+3)\left(\frac{72{,}000}{x+3} + 2000\right)$$
$$72{,}000x + 216{,}000$$
$$= 72{,}000x + 2000x^2 + 6000x$$
$$2000x^2 + 6000x - 216{,}000 = 0$$
$$x^2 + 3x - 108 = 0$$
$$(x-9)(x+12) = 0$$
$$x = 9 \quad \text{or} \quad x = -12$$

There are 9 sailors in the original group.

30. a) Let $x =$ her rate before lunch and $x - 3 =$ her rate after lunch. Her time before lunch is $\frac{20}{x}$ hours and her time after lunch is $\frac{30}{x-3}$ hours. So a rational function for her total time is $T(x) = \frac{20}{x} + \frac{30}{x-3}$ or $T(x) = \frac{50x - 60}{x(x-3)}$.

b) $T(12) = \frac{50(12) - 60}{12(12-3)} = 5$ hours

Making Connections

Chapters 1 - 6

1.
$$\frac{3}{x} = \frac{4}{5}$$
$$4x = 15$$
$$x = \frac{15}{4}$$

Solution set: $\left\{\frac{15}{4}\right\}$

2.
$$\frac{2}{x} = \frac{x}{8}$$
$$x^2 = 16$$
$$x^2 - 16 = 0$$
$$(x-4)(x+4) = 0$$
$$x = 4 \text{ or } x = -4$$

Solution set: $\{-4, 4\}$

3.
$$\frac{x}{3} = \frac{4}{5}$$
$$5x = 12$$
$$x = \frac{12}{5}$$

Solution set: $\left\{\frac{12}{5}\right\}$

4.
$$\frac{3}{x} = \frac{x+3}{6}$$
$$x^2 + 3x = 18$$
$$x^2 + 3x - 18 = 0$$
$$(x+6)(x-3) = 0$$
$$x = -6 \text{ or } x = 3$$

Solution set: $\{-6, 3\}$

5.
$$\frac{1}{x} = 4$$
$$4x = 1$$
$$x = \frac{1}{4}$$

Solution set: $\left\{\frac{1}{4}\right\}$

6.
$$\frac{2}{3}x = 4$$
$$2x = 12$$
$$x = 6$$

Solution set: $\{6\}$

7.
$$2x + 3 = 4$$
$$2x = 1$$
$$x = \frac{1}{2}$$

Solution set: $\left\{\frac{1}{2}\right\}$

8. $$\begin{aligned} 2x + 3 &= 4x \\ 3 &= 2x \\ \frac{3}{2} &= x \end{aligned}$$
Solution set: $\left\{\frac{3}{2}\right\}$

9. $$\begin{aligned} \frac{2a}{3} &= \frac{6}{a} \\ 2a^2 &= 18 \\ a^2 &= 9 \\ a^2 - 9 &= 0 \\ (a-3)(a+3) &= 0 \\ a = 3 \quad &\text{or} \quad a = -3 \end{aligned}$$
Solution set: $\{-3, 3\}$

10. $$\begin{aligned} \frac{12}{x} - \frac{14}{x+1} &= \frac{1}{2} \\ 2x(x+1)\left(\frac{12}{x} - \frac{14}{x+1}\right) &= 2x(x+1) \cdot \frac{1}{2} \\ 24x + 24 - 28x &= x^2 + x \\ -x^2 - 5x + 24 &= 0 \\ x^2 + 5x - 24 &= 0 \\ (x+8)(x-3) &= 0 \\ x = -8 \quad &\text{or} \quad x = 3 \end{aligned}$$
Solution set: $\{-8, 3\}$

11. $$\begin{aligned} |\,6x - 3\,| &= 1 \\ 6x - 3 = 1 \quad \text{or} \quad 6x - 3 &= -1 \\ 6x = 4 \quad \text{or} \quad 6x &= 2 \\ x = \frac{2}{3} \quad \text{or} \quad x &= \frac{1}{3} \end{aligned}$$
Solution set: $\left\{\frac{1}{3}, \frac{2}{3}\right\}$

12. $$\begin{aligned} \frac{x}{2x+9} &= \frac{3}{x} \\ x^2 &= 6x + 27 \\ x^2 - 6x - 27 &= 0 \\ (x-9)(x+3) &= 0 \\ x = 9 \quad &\text{or} \quad x = -3 \end{aligned}$$
Solution set: $\{-3, 9\}$

13. $$\begin{aligned} 4(6x-3)(2x+9) &= 0 \\ 6x - 3 = 0 \quad \text{or} \quad 2x + 9 &= 0 \\ 6x = 3 \quad \text{or} \quad 2x &= -9 \\ x = \frac{1}{2} \quad \text{or} \quad x &= -\frac{9}{2} \end{aligned}$$
Solution set: $\left\{-\frac{9}{2}, \frac{1}{2}\right\}$

14. $$\begin{aligned} \frac{x-1}{x+2} - \frac{1}{5(x+2)} &= 1 \\ 5(x+2)\left(\frac{x-1}{x+2} - \frac{1}{5(x+2)}\right) &= 5(x+2) \cdot 1 \\ 5x - 5 - 1 &= 5x + 10 \\ -6 &= 10 \end{aligned}$$
Solution set: $\emptyset$

15. $$\begin{aligned} Ax + By &= C \\ By &= C - Ax \\ y &= \frac{C - Ax}{B} \end{aligned}$$

16. $$\begin{aligned} \frac{y-3}{x+5} &= -\frac{1}{3} \\ y - 3 &= -\frac{1}{3}(x+5) \\ y - 3 &= -\frac{1}{3}x - \frac{5}{3} \\ y &= -\frac{1}{3}x - \frac{5}{3} + \frac{9}{3} \\ y &= -\frac{1}{3}x + \frac{4}{3} \end{aligned}$$

17. $$\begin{aligned} Ay &= By + C \\ Ay - By &= C \\ y(A - B) &= C \\ y &= \frac{C}{A - B} \end{aligned}$$

18. $$\begin{aligned} \frac{A}{y} &= \frac{y}{A} \\ y^2 &= A^2 \\ y^2 - A^2 &= 0 \\ (y - A)(y + A) &= 0 \\ y - A = 0 \quad \text{or} \quad y + A &= 0 \\ y = A \quad \text{or} \quad y &= -A \end{aligned}$$

19. $$\begin{aligned} \frac{A}{y} - \frac{1}{2} &= \frac{B}{y} \\ 2y\left(\frac{A}{y} - \frac{1}{2}\right) &= 2y \cdot \frac{B}{y} \\ 2A - y &= 2B \\ 2A - 2B &= y \\ y &= 2A - 2B \end{aligned}$$

20. $$\begin{aligned} \frac{A}{y} - \frac{1}{2} &= \frac{B}{C} \\ 2Cy\left(\frac{A}{y} - \frac{1}{2}\right) &= 2Cy \cdot \frac{B}{C} \\ 2AC - Cy &= 2yB \\ 2AC &= Cy + 2yB \\ 2AC &= y(C + 2B) \\ \frac{2AC}{C + 2B} &= y \\ y &= \frac{2AC}{2B + C} \end{aligned}$$

21.
$$3x - 4y = 6$$
$$-4y = -3x + 6$$
$$y = \frac{-3x + 6}{-4}$$
$$y = \frac{3}{4}x - \frac{3}{2}$$

22.
$$y^2 - 2y - Ay + 2A = 0$$
$$y(y-2) - A(y-2) = 0$$
$$(y - A)(y - 2) = 0$$
$$y - A = 0 \quad \text{or} \quad y - 2 = 0$$
$$y = A \quad \text{or} \quad y = 2$$

23.
$$A = \frac{1}{2}B(C + y)$$
$$2A = B(C + y)$$
$$2A = BC + By$$
$$2A - BC = By$$
$$\frac{2A - BC}{B} = y$$
$$y = \frac{2A - BC}{B}$$

24.
$$y^2 + Cy = BC + By$$
$$y^2 + Cy - BC - By = 0$$
$$y(y + C) - B(y + C) = 0$$
$$(y - B)(y + C) = 0$$
$$y - B = 0 \quad \text{or} \quad y + C = 0$$
$$y = B \quad \text{or} \quad y = -C$$

25. $3x^5 \cdot 4x^8 = 12x^{13}$

26. $3x^2(x^3 + 5x^6) = 3x^5 + 15x^8$

27. $(5x^6)^2 = 25x^{12}$

28. $(3a^3b^2)^3 = 27a^9b^6$

29. $\dfrac{12a^9b^4}{-3a^3b^{-2}} = -4a^6b^6$

30. $\left(\dfrac{x^{-2}}{2}\right)^5 = \dfrac{x^{-10}}{2^5} = \dfrac{1}{32x^{10}}$

31. $\left(\dfrac{2x^{-4}}{3y^5}\right)^{-3} = \dfrac{2^{-3}x^{12}}{3^{-3}y^{-15}} = \dfrac{27x^{12}y^{15}}{8}$

32.
$$(-2a^{-1}b^3c)^{-2} = (-2)^{-2}a^2b^{-6}c^{-2}$$
$$= \frac{a^2}{4b^6c^2}$$

33.
$$\frac{a^{-1} + b^3}{a^{-2} + b^{-1}} = \frac{a^2b(a^{-1} + b^3)}{a^2b(a^{-2} + b^{-1})}$$
$$= \frac{ab + a^2b^4}{b + a^2}$$

34. $\dfrac{(a+b)^{-1}}{(a+b)^{-2}} = a + b$

35. a)
$$B = 655 + 9.56(292/2.2) + 1.85(86 \cdot 2.54) - 4.68(30)$$
$$= 2188 \text{ calories}$$

b) The graph shows the energy requirement increasing as the weight increases.

c)
$$B = 655 + 9.56W + 1.85(86 \cdot 2.54) - 4.68(30)$$
$$B = 9.56W + 918.714$$

d)
$$B = 655 + 9.56(292/2.2) + 1.85(86 \cdot 2.54) - 4.68(A)$$
$$B = 2328 - 4.68A$$

7.1 WARM-UPS

1. True, because of the definition of square root.
2. False, because $\sqrt[3]{2}\cdot\sqrt[3]{2}\cdot\sqrt[3]{2}=2$.
3. True, because $(-3)^3=-27$.
4. False, because $(-5)^2=25$.
5. True, because 2^4=16.
6. False, because $\sqrt{9}=3$.
7. False, because $(2^3)^2=2^6$.
8. False, because $\dfrac{\sqrt{10}}{\sqrt{2}}=\sqrt{5}$.
9. True, because $\left(\frac{1}{2}\right)^2=\frac{1}{4}$.
10. True, because $\dfrac{\sqrt{6}}{\sqrt{3}}=\sqrt{\frac{6}{3}}=\sqrt{2}$

7.1 EXERCISES

1. If $b^n=a$ then b is an nth root of a.
3. If $b^n=a$ then b is an even root of a provided n is even or an odd root provided n is odd.
5. The product rule for radicals says that $\sqrt[n]{a}\cdot\sqrt[n]{b}=\sqrt[n]{ab}$ provided all of these roots are real.
7. Because $6^2=36,\ \sqrt{36}=6$.
9. Because $10^2=100,\ \sqrt{100}=10$.
11. Because $3^2=9,\ \sqrt{9}=3$ and $-\sqrt{9}=-3$.
13. Because $2^3=8,\ \sqrt[3]{8}=2$.
15. Because $(-2)^3=-8,\ \sqrt[3]{-8}=-2$.
17. Because $2^5=32,\ \sqrt[5]{32}=2$.
19. Because $10^3=1000,\ \sqrt[3]{1000}=10$.
21. The expression $\sqrt[4]{-16}$ is not a real number since it is an even root of a negative number.
23. Since $m\cdot m=m^2,\ \sqrt{m^2}=m$.
25. Since $(x^8)^2=x^{16},\ \sqrt{x^{16}}=x^8$.
27. Since $(y^3)^5=y^{15},\ \sqrt[5]{y^{15}}=y^3$.
29. Since $(y^5)^3=y^{15},\ \sqrt[3]{y^{15}}=y^5$.
31. Since $(m)^3=m^3,\ \sqrt[3]{m^3}=m$.
33. Since $(w^3)^4=w^{12},\ \sqrt[4]{w^{12}}=w^3$.
35. $\sqrt{9y}=\sqrt{9}\sqrt{y}=3\sqrt{y}$
37. $\sqrt{4a^2}=\sqrt{4}\sqrt{a^2}=2a$
39. $\sqrt{x^4y^2}=\sqrt{x^4}\sqrt{y^2}=x^2y$
41. $\sqrt{5m^{12}}=\sqrt{m^{12}}\sqrt{5}=m^6\sqrt{5}$
43. $\sqrt[3]{8y}=\sqrt[3]{8}\sqrt[3]{y}=2\sqrt[3]{y}$
45. $\sqrt[3]{3a^6}=\sqrt[3]{a^6}\sqrt[3]{3}=a^2\sqrt[3]{3}$
47. $\sqrt{20}=\sqrt{4}\cdot\sqrt{5}=2\sqrt{5}$
49. $\sqrt{50}=\sqrt{25}\cdot\sqrt{2}=5\sqrt{2}$
51. $\sqrt{72}=\sqrt{36}\cdot\sqrt{2}=6\sqrt{2}$
53. $\sqrt[3]{40}=\sqrt[3]{8}\cdot\sqrt[3]{5}=2\sqrt[3]{5}$
55. $\sqrt[3]{81}=\sqrt[3]{27}\cdot\sqrt[3]{3}=3\sqrt[3]{3}$
57. $\sqrt[4]{48}=\sqrt[4]{16}\cdot\sqrt[4]{3}=2\sqrt[4]{3}$
59. $\sqrt[5]{96}=\sqrt[5]{32}\cdot\sqrt[5]{3}=2\sqrt[5]{3}$
61. $\sqrt{a^3}=\sqrt{a^2}\sqrt{a}=a\sqrt{a}$
63. $\sqrt{18a^6}=\sqrt{9a^6}\sqrt{2}=3a^3\sqrt{2}$
65. $\sqrt{20x^5y}=\sqrt{4x^4}\sqrt{5xy}=2x^2\sqrt{5xy}$
67. $\sqrt[3]{24m^4}=\sqrt[3]{8m^3}\cdot\sqrt[3]{3m}=2m\sqrt[3]{3m}$
69. $\sqrt[4]{32a^5}=\sqrt[4]{16a^4}\cdot\sqrt[4]{2a}=2a\sqrt[4]{2a}$
71. $\sqrt[5]{64x^6}=\sqrt[5]{32x^5}\cdot\sqrt[5]{2x}=2x\sqrt[5]{2x}$
73. $\sqrt{48x^3y^8z^7}=\sqrt{16x^2y^8z^6}\cdot\sqrt{3xz}$
$=4xy^4z^3\sqrt{3xz}$
75. $\sqrt{\frac{t}{4}}=\dfrac{\sqrt{t}}{\sqrt{4}}=\dfrac{\sqrt{t}}{2}$
77. $\sqrt{\frac{625}{16}}=\dfrac{\sqrt{625}}{\sqrt{16}}=\dfrac{25}{4}$
79. $\dfrac{\sqrt{30}}{\sqrt{3}}=\sqrt{\frac{30}{3}}=\sqrt{10}$
81. $\sqrt[3]{\frac{t}{8}}=\dfrac{\sqrt[3]{t}}{\sqrt[3]{8}}=\dfrac{\sqrt[3]{t}}{2}$
83. $\sqrt[3]{\frac{-8x^6}{y^3}}=\dfrac{\sqrt[3]{-8x^6}}{\sqrt[3]{y^3}}=\dfrac{-2x^2}{y}$
85. $\sqrt{\frac{4a^6}{9}}=\dfrac{\sqrt{4a^6}}{\sqrt{9}}=\dfrac{2a^3}{3}$
87. $\sqrt{\frac{12}{25}}=\dfrac{\sqrt{12}}{\sqrt{25}}=\dfrac{\sqrt{4}\sqrt{3}}{5}=\dfrac{2\sqrt{3}}{5}$
89. $\sqrt{\frac{27}{16}}=\dfrac{\sqrt{27}}{\sqrt{16}}=\dfrac{\sqrt{9}\sqrt{3}}{4}=\dfrac{3\sqrt{3}}{4}$
91. $\sqrt[3]{\frac{a^4}{125}}=\dfrac{\sqrt[3]{a^4}}{\sqrt[3]{125}}=\dfrac{\sqrt[3]{a^3}\cdot\sqrt[3]{a}}{5}$
$=\dfrac{a\cdot\sqrt[3]{a}}{5}$
93. $\sqrt[3]{\frac{81}{8b^3}}=\dfrac{\sqrt[3]{81}}{\sqrt[3]{8b^3}}=\dfrac{\sqrt[3]{27}\cdot\sqrt[3]{3}}{2b}=\dfrac{3\sqrt[3]{3}}{2b}$
95. $\sqrt[4]{\frac{x^7}{y^8}}=\dfrac{\sqrt[4]{x^7}}{\sqrt[4]{y^8}}=\dfrac{\sqrt[4]{x^4}\cdot\sqrt[4]{x^3}}{y^2}=\dfrac{x\sqrt[4]{x^3}}{y^2}$

97. $\sqrt[4]{\dfrac{a^5}{16b^{12}}} = \dfrac{\sqrt[4]{a^5}}{\sqrt[4]{16b^{12}}} = \dfrac{\sqrt[4]{a^4} \cdot \sqrt[4]{a}}{2b^3}$
$= \dfrac{a\sqrt[4]{a}}{2b^3}$

99. In $f(x) = \sqrt{x-2}$ we must have $x - 2 \geq 0$ or $x \geq 2$. So the domain of f is the interval $[2, \infty)$.

101. In $f(x) = \sqrt[3]{3x-7}$, $3x - 7$ can be any real number. So any real number can be used for x. So the domain of f is the interval $(-\infty, \infty)$.

103. In $f(x) = \sqrt{9-3x}$ we must have $9 - 3x \geq 0$, $-3x \geq -9$, or $x \leq 3$. So the domain of f is the interval $(-\infty, 3]$.

105. In $f(x) = \sqrt{2x+1}$ we must have $2x + 1 \geq 0$, $2x \geq -1$, or $x \geq -1/2$. So the domain of f is the interval $[-1/2, \infty)$.

107. a) $w =$
$91.4 - \dfrac{(10.5 + 6.7\sqrt{20} - 0.45 \cdot 20)(457 - 5 \cdot 25)}{110}$
$w \approx -4°\text{F}$

b) If the air temp is 25°F and wind is 30 mph, then from the graph w is approximately -10°F.

109. a) $t = \sqrt{\dfrac{h}{16}} = \dfrac{\sqrt{h}}{4}$

b) $t = \dfrac{\sqrt{40}}{4} = \dfrac{2\sqrt{10}}{4} = \dfrac{\sqrt{10}}{2}$ sec

c) From the graph it appears that if a diver takes 2.5 seconds then the height is 100 feet.

111. $M = 1.3\sqrt{20} \approx 5.8$ knots

113. a) $V = \sqrt{\dfrac{841 \cdot 8700}{2.81 \cdot 200}} \approx 114.1 \text{ ft/sec}$

b) $114.1 \dfrac{\text{ft}}{\text{sec}} \cdot \dfrac{1 \text{ mi}}{5280 \text{ ft}} \cdot \dfrac{3600 \text{ sec}}{1 \text{ hr}} \approx 77.8$ mph

115. The equations in (a), (c), and (d) are identities because the radical symbol always indicates a positive root. In (b), $\sqrt[3]{x^3}$ is negative if x is negative, but $|x|$ is positive if x is negative. So (b) is not an identity.

117. The arithmetic mean of 80 and 100 is 90. Solve $80/h = h/100$ to get $h^2 = 8000$ or $h = \sqrt{8000} \approx 89.4$. So you are better of with the arithmetic mean.

7.2 WARM-UPS

1. True, by definition of exponent $1/3$.
2. False, because $8^{5/3} = \sqrt[3]{8^5}$.
3. False, because $-16^{1/2} = -4$, while $(-16)^{1/2}$ is not a real number.
4. True, $9^{-3/2} = \dfrac{1}{(\sqrt{9})^3} = \dfrac{1}{3^3} = \dfrac{1}{27}$.
5. True, $\dfrac{\sqrt{6}}{6} = \dfrac{6^{1/2}}{6^1} = 6^{1/2-1} = 6^{-1/2}$.
6. True, because $\dfrac{2^1}{2^{1/2}} = 2^{1-1/2} = 2^{1/2}$.
7. True, because $2^{1/2} \cdot 2^{1/2} = 2^1 = 2$ and $4^{1/2} = 2$.
8. False, because $16^{-1/4} = \dfrac{1}{16^{1/4}} = \dfrac{1}{2}$.
9. True, $6^{1/6} \cdot 6^{1/6} = 6^{1/6+1/6} = 6^{2/6} = 6^{1/3}$.
10. True, $(2^8)^{3/4} = 2^{24/4} = 2^6$.

7.2 EXERCISES

1. The nth root of a is $a^{1/n}$.

3. The expression $a^{-m/n}$ means $\dfrac{1}{a^{m/n}}$.

5. The operations can be performed in any order, but the easiest is usually root, power, and then reciprocal.

7. $\sqrt[4]{7} = 7^{1/4}$
9. $\sqrt{5x} = (5x)^{1/2}$
11. $9^{1/5} = \sqrt[5]{9}$
13. $a^{1/2} = \sqrt{a}$
15. $25^{1/2} = \sqrt{25} = 5$
17. $(-125)^{1/3} = \sqrt[3]{-125} = -5$
19. $16^{1/4} = \sqrt[4]{16} = 2$
21. $(-4)^{1/2}$ is not a real number because it is $\sqrt{-4}$.
23. $\sqrt[3]{w^7} = w^{7/3}$
25. $\dfrac{1}{\sqrt[3]{2^{10}}} = 2^{-10/3}$
27. $w^{-3/4} = \sqrt[4]{\dfrac{1}{w^3}}$
29. $(ab)^{3/2} = \sqrt{(ab)^3}$
31. $125^{2/3} = (\sqrt[3]{125})^2 = 5^2 = 25$
33. $25^{3/2} = (\sqrt{25})^3 = 5^3 = 125$
35. $27^{-4/3} = \dfrac{1}{(\sqrt[3]{27})^4} = \dfrac{1}{3^4} = \dfrac{1}{81}$
37. $16^{-3/2} = \dfrac{1}{(\sqrt{16})^3} = \dfrac{1}{4^3} = \dfrac{1}{64}$
39. $(-27)^{-1/3} = \dfrac{1}{(-27)^{1/3}} = \dfrac{1}{-3} = -\dfrac{1}{3}$
41. $(-16)^{-1/4}$ is not a real number because it is a fourth root of a negative number.

43. $3^{1/3} \cdot 3^{1/4} = 3^{\frac{1}{3}+\frac{1}{4}} = 3^{7/12}$

45. $3^{1/3} \cdot 3^{-1/3} = 3^0 = 1$

47. $\frac{8^{1/3}}{8^{2/3}} = 8^{1/3-2/3} = 8^{-1/3} = \frac{1}{8^{1/3}} = \frac{1}{2}$

49. $4^{3/4} \div 4^{1/4} = 4^{3/4-1/4} = 4^{1/2} = 2$

51. $18^{1/2} \cdot 2^{1/2} = 36^{1/2} = 6$

53. $(2^6)^{1/3} = 2^{6/3} = 2^2 = 4$

55. $(3^8)^{1/2} = 3^{8/2} = 3^4 = 81$

57. $(2^{-4})^{1/2} = 2^{-2} = \frac{1}{4}$

59. $\left(\frac{3^4}{2^6}\right)^{1/2} = \frac{3^2}{2^3} = \frac{9}{8}$

61. $(x^4)^{1/4} = |\, x \,|$

63. $(a^8)^{1/2} = a^4$

65. $(y^3)^{1/3} = y$

67. $(9x^6y^2)^{1/2} = |\, 3x^3y \,|$

69. $\left(\frac{81x^{12}}{y^{20}}\right)^{1/4} = \left|\frac{3x^3}{y^5}\right|$

71. $x^{1/2}x^{1/4} = x^{2/4+1/4} = x^{3/4}$

73. $(x^{1/2}y)(x^{-3/4}y^{1/2}) = x^{-1/4}y^{3/2} = \frac{y^{3/2}}{x^{1/4}}$

75. $\frac{w^{1/3}}{w^3} = w^{\frac{1}{3}-3} = w^{-8/3} = \frac{1}{w^{8/3}}$

77. $(144x^{16})^{1/2} = 12x^8$

79. $\left(\frac{a^{-1/2}}{b^{-1/4}}\right)^{-4} = \frac{a^{4/2}}{b^{4/4}} = \frac{a^2}{b}$

81. $\left(\frac{2w^{1/3}}{w^{-3/4}}\right)^{3} = \frac{8w}{w^{-9/4}} = 8w^{1-(-9/4)}$
$= 8w^{13/4}$

83. $(9^2)^{1/2} = 9$

85. $-16^{-3/4} = -\frac{1}{2^3} = -\frac{1}{8}$

87. $125^{-4/3} = \frac{1}{5^4} = \frac{1}{625}$

89. $2^{1/2} \cdot 2^{-1/4} = 2^{2/4-1/4} = 2^{1/4}$

91. $3^{0.26}3^{0.74} = 3^{0.26+0.74} = 3^1 = 3$

93. $3^{1/4} \cdot 27^{1/4} = (3 \cdot 27)^{1/4} = 81^{1/4} = 3$

95. $\left(-\frac{8}{27}\right)^{2/3} = \frac{(-8)^{2/3}}{27^{2/3}} = \frac{4}{9}$

97. Not a real number, because the fourth root of $-1/16$ is not real.

99. $\left(\frac{9}{16}\right)^{-1/2} = \left(\frac{16}{9}\right)^{1/2} = \frac{4}{3}$

101. $-\left(\frac{25}{36}\right)^{-3/2} = -\left(\frac{36}{25}\right)^{3/2} = -\frac{216}{125}$

103. $(9x^9)^{1/2} = 9^{1/2}x^{9/2} = 3x^{9/2}$

105. $(3a^{-2/3})^{-3} = 3^{-3}a^2 = \frac{a^2}{27}$

107. $(a^{1/2}b)^{1/2}(ab^{1/2}) = a^{1/4}b^{1/2}a^1b^{1/2}$
$= a^{5/4}b$

109. $(km^{\,1/2})^3(k^3m^5)^{1/2} = k^3m^{3/2}k^{3/2}m^{5/2}$
$= k^{9/2}m^4$

111. $2^{1/3} \approx 2^{0.33333333} \approx 1.2599$

113. $-2^{1/2} = -(2^{0.5}) \approx -1.4142$

115. $1024^{1/10} = 2$

117. $\left(\frac{64}{15{,}625}\right)^{-1/6} = 2.5$

119. $a^{m/2} \cdot a^{m/4} = a^{m/2+m/4} = a^{3m/4}$

121. $\frac{a^{-m/5}}{a^{-m/3}} = a^{-m/5+m/3} = a^{2m/15}$

123. $\left(a^{-1/m}b^{-1/n}\right)^{-mn} = a^nb^m$

125. $\left(\frac{a^{-3m}b^{-6n}}{a^{9m}}\right)^{-1/3} = \frac{a^mb^{2n}}{a^{-3m}} = a^{4m}b^{2n}$

127. **a)** $D = (12^2 + 4^2 + 3^2)^{1/2}$
$= (144 + 16 + 9)^{1/2}$
$= 169^{1/2} =$ 13 inches

b) $D = (1^2 + 1^2 + 1^2)^{1/2} = \sqrt{3}$ or approximately 1.73 inches.

129. $S = (13.0368 + 7.84(18.42)^{1/3}$
$- 0.8(21.45))^2$
$S \approx 274.96$ m^2

131. $r = \left(\frac{62{,}760}{10{,}000}\right)^{1/5} - 1 \approx 0.4439$
$= 44.39\%$

133. $r = \left(\frac{141{,}600{,}000{,}000}{450{,}000}\right)^{1/213} - 1$
$\approx 0.061 = 6.12\%$

135. The expression $(-1)^{1/2}$ is not a real number. The rules of exponents hold only for expressions that represent real numbers. So we should not be using either rule of exponents to perform these computations. Using complex numbers from Section 7.6 we can compute $(-1)^{1/2} \cdot (-1)^{1/2} = i \cdot i = -1$.

7.3 WARM-UPS

1. False, because $\sqrt{3} + \sqrt{3} = 2\sqrt{3}$.

2. True, because
$\sqrt{8} + \sqrt{2} = 2\sqrt{2} + \sqrt{2} = 3\sqrt{2}$.

3. False, because $2\sqrt{3} \cdot 3\sqrt{3} = 6 \cdot 3 = 18$.

4. False, because $\sqrt[3]{2} \cdot \sqrt[3]{2} = \sqrt[3]{4}$.

5. True, because $\sqrt{5} \cdot \sqrt{2} = \sqrt{10}$.

6. False, because $2\sqrt{5} + 3\sqrt{5} = 5\sqrt{5}$.

7. True, because $\sqrt{2}\sqrt{3} = \sqrt{6}$ and $\sqrt{2}\sqrt{2} = 2$.

8. False, because $\sqrt{12} = \sqrt{4}\sqrt{3} = 2\sqrt{3}$.

9. False, because $(\sqrt{2} + \sqrt{3})^2$
$= 2 + 2\sqrt{2}\sqrt{3} + 3 = 5 + 2\sqrt{6}$.

10. True,
$(\sqrt{3} - \sqrt{2})(\sqrt{3} + \sqrt{2}) = 3 - 2 = 1$.

7.3 EXERCISES

1. Like radicals are radicals with the same index and same radicand.

3. In the product rule the radicals must have the same index, but do not have to have the same radicand.

5. $\sqrt{3} - 2\sqrt{3} = 1\sqrt{3} - 2\sqrt{3}$
$= -1\sqrt{3} = -\sqrt{3}$

7. $5\sqrt{7x} + 4\sqrt{7x} = (5 + 4)\sqrt{7x} = 9\sqrt{7x}$

9. $2 \cdot \sqrt[3]{2} + 3 \cdot \sqrt[3]{2} = (2 + 3)\sqrt[3]{2} = 5\sqrt[3]{2}$

11. $\sqrt{3} - \sqrt{5} + 3\sqrt{3} - \sqrt{5}$
$= \sqrt{3} + 3\sqrt{3} - \sqrt{5} - \sqrt{5} = 4\sqrt{3} - 2\sqrt{5}$

13. $\sqrt[3]{2} + \sqrt[3]{x} - \sqrt[3]{2} + 4\sqrt[3]{x}$
$= \sqrt[3]{2} - \sqrt[3]{2} + \sqrt[3]{x} + 4\sqrt[3]{x} = 5\sqrt[3]{x}$

15. $\sqrt[3]{x} - \sqrt{2x} + \sqrt[3]{x} = \sqrt[3]{x} + \sqrt[3]{x} - \sqrt{2x}$
$= 2\sqrt[3]{x} - \sqrt{2x}$

17. $\sqrt{8} + \sqrt{28} = \sqrt{4}\sqrt{2} + \sqrt{4}\sqrt{7}$
$= 2\sqrt{2} + 2\sqrt{7}$

19. $\sqrt{8} + \sqrt{18} = \sqrt{4}\sqrt{2} + \sqrt{9}\sqrt{2}$
$= 2\sqrt{2} + 3\sqrt{2} = 5\sqrt{2}$

21. $2\sqrt{45} - 3\sqrt{20} = 2\sqrt{9}\sqrt{5} - 3\sqrt{4}\sqrt{5}$
$= 6\sqrt{5} - 6\sqrt{5} = 0$

23. $\sqrt{2} - \sqrt{8} = \sqrt{2} - 2\sqrt{2} = -\sqrt{2}$

25. $\sqrt{45x^3} - \sqrt{18x^2} + \sqrt{50x^2} - \sqrt{20x^3}$
$= 3x\sqrt{5x} - 3x\sqrt{2} + 5x\sqrt{2} - 2x\sqrt{5x}$
$= x\sqrt{5x} + 2x\sqrt{2}$

27. $2\sqrt[3]{24} + \sqrt[3]{81} = 2\sqrt[3]{8} \cdot \sqrt[3]{3} + \sqrt[3]{27} \cdot \sqrt[3]{3}$
$= 4\sqrt[3]{3} + 3 \cdot \sqrt[3]{3} = 7\sqrt[3]{3}$

29. $\sqrt[4]{48} - 2\sqrt[4]{243} = \sqrt[4]{16 \cdot 3} - 2\sqrt[4]{81 \cdot 3}$
$= 2\sqrt[4]{3} - 6\sqrt[4]{3}$
$= -4\sqrt[4]{3}$

31. $\sqrt[3]{54t^4y^3} - \sqrt[3]{16t^4y^3}$
$= \sqrt[3]{27t^3y^3} \cdot \sqrt[3]{2t} - \sqrt[3]{8t^3y^3} \cdot \sqrt[3]{2t}$
$= 3ty \cdot \sqrt[3]{2t} - 2ty \cdot \sqrt[3]{2t} = ty\sqrt[3]{2t}$

33. $\sqrt{3}\sqrt{5} = \sqrt{3 \cdot 5} = \sqrt{15}$

35. $(2\sqrt{5})(3\sqrt{10}) = 6\sqrt{50} = 6\sqrt{25}\sqrt{2}$
$= 6 \cdot 5\sqrt{2} = 30\sqrt{2}$

37. $(2\sqrt{7a})(3\sqrt{2a}) = 6\sqrt{14a^2} = 6\sqrt{a^2}\sqrt{14}$
$= 6a\sqrt{14}$

39. $(\sqrt[4]{9})(\sqrt[4]{27}) = \sqrt[4]{243} = \sqrt[4]{81} \cdot \sqrt[4]{3}$
$= 3\sqrt[4]{3}$

41. $(2\sqrt{3})^2 = 4 \cdot 3 = 12$

43. $\sqrt{5x^3}\sqrt{8x^4} = \sqrt{40x^7}$
$= \sqrt{4x^6}\sqrt{10x} = 2x^3\sqrt{10x}$

45. $\sqrt[4]{\frac{x^5}{3}} \cdot \sqrt[4]{\frac{x^2}{27}} = \sqrt[4]{\frac{x^7}{81}}$
$= \sqrt[4]{\frac{x^4}{81}} \cdot \sqrt[4]{x^3} = \frac{x}{3}\sqrt[4]{x^3} = \frac{x\sqrt[4]{x^3}}{3}$

47. $2\sqrt{3}(\sqrt{6} + 3\sqrt{3}) = 2\sqrt{18} + 18$
$= 2\sqrt{9}\sqrt{2} + 18 = 6\sqrt{2} + 18$

49. $\sqrt{5}(\sqrt{10} - 2) = \sqrt{50} - 2\sqrt{5}$
$= \sqrt{25}\sqrt{2} - 2\sqrt{5} = 5\sqrt{2} - 2\sqrt{5}$

51. $\sqrt[3]{3t}\left(\sqrt[3]{9t} - \sqrt[3]{t^2}\right) = \sqrt[3]{27t^2} - \sqrt[3]{3t^3}$
$= 3\sqrt[3]{t^2} - t\sqrt[3]{3}$

53. $\left(\sqrt{3} + 2\right)\left(\sqrt{3} - 5\right)$
$= 3 + 2\sqrt{3} - 5\sqrt{3} - 10 = -7 - 3\sqrt{3}$

55. $(\sqrt{11} - 3)(\sqrt{11} + 3) = 11 - 9 = 2$

57. $(2\sqrt{5} - 7)(2\sqrt{5} + 4)$
$= 20 - 14\sqrt{5} + 8\sqrt{5} - 28$
$= -8 - 6\sqrt{5}$

59. $(2\sqrt{3} - \sqrt{6})(\sqrt{3} + 2\sqrt{6})$
$= 6 - \sqrt{18} + 4\sqrt{18} - 12$
$= -6 - 3\sqrt{2} + 4 \cdot 3\sqrt{2} = -6 + 9\sqrt{2}$

61. $(\sqrt{3} - 2)(\sqrt{3} + 2) = 3 - 4 = -1$

63. $(\sqrt{5} + \sqrt{2})(\sqrt{5} - \sqrt{2}) = 5 - 2 = 3$

65. $(2\sqrt{5} + 1)(2\sqrt{5} - 1) = 4 \cdot 5 - 1 = 19$

67. $(3\sqrt{2} + \sqrt{5})(3\sqrt{2} - \sqrt{5}) = 9 \cdot 2 - 5$
$= 13$

69. $(5 - 3\sqrt{x})(5 + 3\sqrt{x}) = 25 - 9x$

71. $\sqrt[3]{3} \cdot \sqrt{3} = 3^{1/3}3^{1/2} = 3^{5/6} = \sqrt[6]{3^5}$
$= \sqrt[6]{243}$

73. $\sqrt[3]{5} \cdot \sqrt[4]{5} = 5^{1/3}5^{1/4} = 5^{7/12} = \sqrt[12]{5^7}$

75. $\sqrt[3]{2} \cdot \sqrt{5} = 2^{1/3}5^{1/2} = 2^{2/6}5^{3/6}$
$= \sqrt[6]{2^2 5^3} = \sqrt[6]{500}$

77. $\sqrt[3]{2} \cdot \sqrt[4]{3} = 2^{1/3}3^{1/4} = 2^{4/12}3^{3/12}$
$= \sqrt[12]{2^4 3^3} = \sqrt[12]{432}$

79. $\sqrt{300} + \sqrt{3} = 10\sqrt{3} + \sqrt{3} = 11\sqrt{3}$

81. $2\sqrt{5} \cdot 5\sqrt{6} = 10\sqrt{30}$

83. $(3+2\sqrt{7})(\sqrt{7}-2)$
$= 3\sqrt{7}-6+2\cdot 7-4\sqrt{7}$
$= 8-\sqrt{7}$

85. $4\sqrt{w}\cdot 4\sqrt{w} = 16(\sqrt{w})^2 = 16w$

87. $\sqrt{3x^3}\cdot\sqrt{6x^2} = \sqrt{18x^5} = \sqrt{9x^4}\cdot\sqrt{2x}$
$= 3x^2\sqrt{2x}$

89. $(2\sqrt{5}+\sqrt{2})(3\sqrt{5}-\sqrt{2})$
$= 30+3\sqrt{10}-2\sqrt{10}-2 = 28+\sqrt{10}$

91. $\frac{\sqrt{2}}{3}+\frac{\sqrt{2}}{5} = \frac{5\sqrt{2}}{5\cdot 3}+\frac{3\sqrt{2}}{3\cdot 5} = \frac{8\sqrt{2}}{15}$

93. $(5+2\sqrt{2})(5-2\sqrt{2}) = 25-4\cdot 2 = 17$

95. $(3+\sqrt{x})^2 = 9+2\cdot 3\sqrt{x}+x$
$= 9+6\sqrt{x}+x$

97. $(5\sqrt{x}-3)^2 = 25x-2\cdot 3\cdot 5\sqrt{x}+9$
$= 25x-30\sqrt{x}+9$

99. $(1+\sqrt{x+2})^2 = 1+2\sqrt{x+2}+x+2$
$= x+3+2\sqrt{x+2}$

101. $\sqrt{4w}-\sqrt{9w} = 2\sqrt{w}-3\sqrt{w}$
$= -\sqrt{w}$

103. $2\sqrt{a^3}+3\sqrt{a^3}-2a\sqrt{4a}$
$= 2a\sqrt{a}+3a\sqrt{a}-4a\sqrt{a} = a\sqrt{a}$

105. $\sqrt{x^5}+2x\sqrt{x^3}$
$= \sqrt{x^4}\sqrt{x}+2x\sqrt{x^2}\sqrt{x}$
$= x^2\sqrt{x}+2x^2\sqrt{x} = 3x^2\sqrt{x}$

107. $\sqrt[3]{-16x^4}+5x\sqrt[3]{54x}$
$= -2x\sqrt[3]{2x}+5x\cdot 3\cdot\sqrt[3]{2x} = 13x\sqrt[3]{2x}$

109. $\sqrt[3]{2x}\cdot\sqrt{2x} = (2x)^{1/3}(2x)^{1/2} = (2x)^{5/6}$
$= \sqrt[6]{32x^5}$

111. $A = LW = \sqrt{6}\cdot\sqrt{3} = \sqrt{18}$
$= 3\sqrt{2}$ ft^2

113. $A = \frac{1}{2}h(b_1+b_2)$
$= \frac{1}{2}\sqrt{6}\left(\sqrt{3}+\sqrt{12}\right)$
$= \frac{1}{2}\sqrt{6}\left(\sqrt{3}+2\sqrt{3}\right)$
$= \frac{1}{2}\sqrt{6}\cdot 3\sqrt{3} = \frac{3\sqrt{18}}{2}$
$= \frac{9\sqrt{2}}{2}$ ft^2

115. No because $\sqrt{9}+\sqrt{16} = 3+4 = 7$ and $\sqrt{9+16} = \sqrt{25} = 5$.

117. a) $y^2-3 = \left(y+\sqrt{3}\right)\left(y-\sqrt{3}\right)$
$2a^2-7 = \left(\sqrt{2}a+\sqrt{7}\right)\left(\sqrt{2}a-\sqrt{7}\right)$

b) $x^2-8 = 0$
$\left(x+\sqrt{8}\right)\left(x-\sqrt{8}\right) = 0$
$x+\sqrt{8} = 0$ or $x-\sqrt{8} = 0$
$x = -\sqrt{8}$ or $x = \sqrt{8}$
$x = -2\sqrt{2}$ or $x = 2\sqrt{2}$
The solution set is $\left\{\pm 2\sqrt{2}\right\}$.

c) $x^2-a = 0$
$(x+\sqrt{a})(x-\sqrt{a}) = 0$
$x+\sqrt{a} = 0$ or $x-\sqrt{a} = 0$
$x = -\sqrt{a}$ or $x = \sqrt{a}$
The solution set is $\{\pm\sqrt{a}\}$.

7.4 WARM-UPS

1. True, because $\sqrt{3}\sqrt{2} = \sqrt{6}$.

2. True, because
$\frac{2}{\sqrt{2}} = \frac{2\sqrt{2}}{\sqrt{2}\sqrt{2}} = \frac{2\sqrt{2}}{2} = \sqrt{2}.$

3. False, because $\frac{4-\sqrt{10}}{2} = 2-\frac{\sqrt{10}}{2}$.

4. True, because $\frac{1}{\sqrt{3}} = \frac{1\cdot\sqrt{3}}{\sqrt{3}\sqrt{3}} = \frac{\sqrt{3}}{3}$.

5. False, because $\frac{8\sqrt{7}}{2\sqrt{7}} = 4$.

6. True, because
$(2-\sqrt{3})(2+\sqrt{3}) = 4-3 = 1.$

7. False, because $\frac{\sqrt{12}}{3} = \frac{2\sqrt{3}}{3}$.

8. True, because
$\frac{\sqrt{20}}{\sqrt{5}} = \frac{\sqrt{4}\sqrt{5}}{\sqrt{5}} = \sqrt{4} = 2.$

9. True, because $(2\sqrt{4})^2 = 4\cdot 4 = 16$.

10. True, because
$(3\sqrt{5})^3 = 3^3\cdot\sqrt{5^3} = 27\sqrt{125}.$

7.4 EXERCISES

1. $\frac{2}{\sqrt{5}} = \frac{2\sqrt{5}}{\sqrt{5}\sqrt{5}} = \frac{2\sqrt{5}}{5}$

3. $\frac{\sqrt{3}}{\sqrt{7}} = \frac{\sqrt{3}\sqrt{7}}{\sqrt{7}\sqrt{7}} = \frac{\sqrt{21}}{\sqrt{49}} = \frac{\sqrt{21}}{7}$

5. $\frac{1}{\sqrt[3]{4}} = \frac{1\cdot\sqrt[3]{2}}{\sqrt[3]{4}\cdot\sqrt[3]{2}} = \frac{\sqrt[3]{2}}{\sqrt[3]{8}} = \frac{\sqrt[3]{2}}{2}$

7. $\frac{\sqrt[3]{6}}{\sqrt[3]{5}} = \frac{\sqrt[3]{6}\cdot\sqrt[3]{25}}{\sqrt[3]{5}\cdot\sqrt[3]{25}} = \frac{\sqrt[3]{150}}{\sqrt[3]{125}} = \frac{\sqrt[3]{150}}{5}$

9. $\frac{\sqrt{5}}{\sqrt{12}} = \frac{\sqrt{5}\sqrt{3}}{\sqrt{12}\sqrt{3}} = \frac{\sqrt{15}}{\sqrt{36}} = \frac{\sqrt{15}}{6}$

11. $\frac{\sqrt{3}}{\sqrt{12}} = \frac{\sqrt{3}}{\sqrt{4}\sqrt{3}} = \frac{1}{\sqrt{4}} = \frac{1}{2}$

13. $\sqrt{\frac{1}{2}} = \frac{1}{\sqrt{2}} = \frac{1 \cdot \sqrt{2}}{\sqrt{2}\sqrt{2}} = \frac{\sqrt{2}}{2}$

15. $\sqrt[3]{\frac{2}{3}} = \frac{\sqrt[3]{2} \cdot \sqrt[3]{9}}{\sqrt[3]{3} \cdot \sqrt[3]{9}} = \frac{\sqrt[3]{18}}{3}$

17. $\sqrt[3]{\frac{7}{4}} = \frac{\sqrt[3]{7} \cdot \sqrt[3]{2}}{\sqrt[3]{4} \cdot \sqrt[3]{2}} = \frac{\sqrt[3]{14}}{2}$

19. $\sqrt{\frac{x}{y}} = \frac{\sqrt{x}\sqrt{y}}{\sqrt{y}\sqrt{y}} = \frac{\sqrt{xy}}{y}$

21. $\frac{\sqrt{a^3}}{\sqrt{b^7}} = \frac{\sqrt{a^2}\sqrt{a}}{\sqrt{b^6}\sqrt{b}} = \frac{a\sqrt{a}\sqrt{b}}{b^3\sqrt{b}\sqrt{b}} = \frac{a\sqrt{ab}}{b^4}$

23. $\sqrt{\frac{a}{3b}} = \frac{\sqrt{a}\sqrt{3b}}{\sqrt{3b}\sqrt{3b}} = \frac{\sqrt{3ab}}{3b}$

25. $\sqrt[3]{\frac{a}{b}} = \frac{\sqrt[3]{a} \cdot \sqrt[3]{b^2}}{\sqrt[3]{b} \cdot \sqrt[3]{b^2}} = \frac{\sqrt[3]{ab^2}}{b}$

27. $\sqrt[3]{\frac{5}{2b^2}} = \frac{\sqrt[3]{5} \cdot \sqrt[3]{4b}}{\sqrt[3]{2b^2} \cdot \sqrt[3]{4b}} = \frac{\sqrt[3]{20b}}{2b}$

29. $\sqrt{15} \div \sqrt{5} = \sqrt{\frac{15}{5}} = \sqrt{3}$

31. $\sqrt{3} \div \sqrt{5} = \frac{\sqrt{3}}{\sqrt{5}} = \frac{\sqrt{3}\sqrt{5}}{\sqrt{5}\sqrt{5}} = \frac{\sqrt{15}}{5}$

33. $(3\sqrt{3}) \div (5\sqrt{6}) = \frac{3\sqrt{3}}{5\sqrt{6}} = \frac{3\sqrt{3}}{5\sqrt{2}\sqrt{3}}$
$= \frac{3}{5\sqrt{2}} = \frac{3\sqrt{2}}{5\sqrt{2}\sqrt{2}} = \frac{3\sqrt{2}}{10}$

35. $(2\sqrt{3}) \div (3\sqrt{6}) = \frac{2\sqrt{3}}{3\sqrt{6}} = \frac{2}{3\sqrt{2}}$
$= \frac{2\sqrt{2}}{3\sqrt{2}\sqrt{2}} = \frac{2\sqrt{2}}{3 \cdot 2} = \frac{\sqrt{2}}{3}$

37. $(\sqrt{24a^2}) \div (\sqrt{72a}) = \frac{\sqrt{24a^2}}{\sqrt{72a}} = \frac{2a\sqrt{6}}{6\sqrt{2a}}$
$= \frac{a\sqrt{3}}{3\sqrt{a}} = \frac{a\sqrt{3}\sqrt{a}}{3\sqrt{a}\sqrt{a}} = = \frac{a\sqrt{3a}}{3a} = \frac{\sqrt{3a}}{3}$

39. $\sqrt[3]{20} \div \sqrt[3]{2} = \frac{\sqrt[3]{20}}{\sqrt[3]{2}} = \sqrt[3]{\frac{20}{2}} = \sqrt[3]{10}$

41. $\sqrt[4]{48} \div \sqrt[4]{3} = \frac{\sqrt[4]{48}}{\sqrt[4]{3}} = \frac{\sqrt[4]{16} \cdot \sqrt[4]{3}}{\sqrt[4]{3}} = 2$

43. $\sqrt[4]{16w} \div \sqrt[4]{w^5} = \frac{\sqrt[4]{16w}}{\sqrt[4]{w^5}} = \frac{2\sqrt[4]{w}}{w\sqrt[4]{w}}$
$= \frac{2}{w}$

45. $\frac{6+\sqrt{45}}{3} = \frac{6+3\sqrt{5}}{3} = 2+\sqrt{5}$

47. $\frac{-2+\sqrt{12}}{-2} = \frac{-2+2\sqrt{3}}{-2} = 1-\sqrt{3}$

49. $\frac{4}{2+\sqrt{8}} = \frac{4(2-\sqrt{8})}{(2+\sqrt{8})(2-\sqrt{8})}$
$= \frac{8-4\sqrt{8}}{4-8} = \frac{8-8\sqrt{2}}{-4} = -2+2\sqrt{2}$
$= 2\sqrt{2}-2$

51. $\frac{3}{\sqrt{11}-\sqrt{5}}$
$= \frac{3(\sqrt{11}+\sqrt{5})}{(\sqrt{11}-\sqrt{5})(\sqrt{11}+\sqrt{5})}$
$= \frac{3\sqrt{11}+3\sqrt{5}}{11-5} = \frac{3\sqrt{11}+3\sqrt{5}}{6}$
$= \frac{\sqrt{11}+\sqrt{5}}{2}$

53. $\frac{1+\sqrt{2}}{\sqrt{3}-1} = \frac{(1+\sqrt{2})(\sqrt{3}+1)}{(\sqrt{3}-1)(\sqrt{3}+1)}$
$= \frac{1+\sqrt{6}+\sqrt{2}+\sqrt{3}}{2}$

55. $\frac{\sqrt{2}}{\sqrt{6}+\sqrt{3}} = \frac{\sqrt{2}(\sqrt{6}-\sqrt{3})}{(\sqrt{6}+\sqrt{3})(\sqrt{6}-\sqrt{3})}$
$= \frac{\sqrt{12}-\sqrt{6}}{6-3} = \frac{2\sqrt{3}-\sqrt{6}}{3}$

57. $\frac{2\sqrt{3}}{3\sqrt{2}-\sqrt{5}}$
$= \frac{2\sqrt{3}(3\sqrt{2}+\sqrt{5})}{(3\sqrt{2}-\sqrt{5})(3\sqrt{2}+\sqrt{5})}$
$= \frac{6\sqrt{6}+2\sqrt{15}}{18-5} = \frac{6\sqrt{6}+2\sqrt{15}}{13}$

59. $(2\sqrt{2})^5 = 2^5 \cdot \sqrt{2^5} = 32\sqrt{16}\sqrt{2}$
$= 128\sqrt{2}$

61. $(\sqrt{x})^5 = \sqrt{x^5} = \sqrt{x^4}\sqrt{x} = x^2\sqrt{x}$

63. $(-3\sqrt{x^3})^3 = -27\sqrt{x^9} = -27\sqrt{x^8}\sqrt{x}$
$= -27x^4\sqrt{x}$

65. $(2x\sqrt[3]{x^2})^3 = 8x^3\sqrt[3]{x^6} = 8x^3 \cdot x^2 = 8x^5$

67. $(-2\sqrt[3]{5})^2 = (-2)^2 \cdot \sqrt[3]{5^2} = 4\sqrt[3]{25}$

69. $(\sqrt[3]{x^2})^6 = (x^2)^{6/3} = (x^2)^2 = x^4$

71. $\frac{\sqrt{3}}{\sqrt{2}} + \frac{2}{\sqrt{2}} = \frac{\sqrt{3}\sqrt{2}}{\sqrt{2}\sqrt{2}} + \frac{2\sqrt{2}}{\sqrt{2}\sqrt{2}}$
$= \frac{\sqrt{6}}{2} + \frac{2\sqrt{2}}{2} = \frac{\sqrt{6}+2\sqrt{2}}{2}$

73. $\dfrac{\sqrt{3}}{\sqrt{2}} + \dfrac{3\sqrt{6}}{2} = \dfrac{\sqrt{3}\sqrt{2}}{\sqrt{2}\sqrt{2}} + \dfrac{3\sqrt{6}}{2}$
$= \dfrac{\sqrt{6}}{2} + \dfrac{3\sqrt{6}}{2} = \dfrac{4\sqrt{6}}{2} = 2\sqrt{6}$

75. $\dfrac{\sqrt{6}}{2} \cdot \dfrac{1}{\sqrt{3}} = \dfrac{\sqrt{2}\sqrt{3}}{2\sqrt{3}} = \dfrac{\sqrt{2}}{2}$

77. $\dfrac{8 - \sqrt{32}}{20} = \dfrac{8 - 4\sqrt{2}}{20} = \dfrac{4(2 - \sqrt{2})}{4 \cdot 5}$
$= \dfrac{2 - \sqrt{2}}{5}$

79. $\dfrac{5 + \sqrt{75}}{10} = \dfrac{5 + 5\sqrt{3}}{10} = \dfrac{5(1 + \sqrt{3})}{5 \cdot 2}$
$= \dfrac{1 + \sqrt{3}}{2}$

81. $\sqrt{a}(\sqrt{a} - 3) = a - 3\sqrt{a}$

83. $4\sqrt{a}(a + \sqrt{a}) = 4a\sqrt{a} + 4a$

85. $(2\sqrt{3m})^2 = 4 \cdot 3m = 12m$

87. $\left(-2\sqrt{xy^2z}\right)^2 = (-2)^2xy^2z = 4xy^2z$

89. $\sqrt[3]{m}\left(\sqrt[3]{m^2} - \sqrt[3]{m^5}\right) = \sqrt[3]{m^3} - \sqrt[3]{m^6}$
$= m - m^2$

91. $\sqrt[3]{8x^4} + \sqrt[3]{27x^4}$
$= \sqrt[3]{8x^3} \cdot \sqrt[3]{x} + \sqrt[3]{27x^3} \cdot \sqrt[3]{x}$
$= 2x \cdot \sqrt[3]{x} + 3x \cdot \sqrt[3]{x} = 5x\sqrt[3]{x}$

93. $\left(2m\sqrt[4]{2m^2}\right)^3 = 8m^3\sqrt[4]{8m^6}$
$= 8m^3\sqrt[4]{m^4}\sqrt[4]{8m^2}$
$= 8m^3 \cdot m \cdot \sqrt[4]{8m^2} = 8m^4\sqrt[4]{8m^2}$

95. $\dfrac{x - 9}{\sqrt{x} - 3} = \dfrac{(x - 9)(\sqrt{x} + 3)}{(\sqrt{x} - 3)(\sqrt{x} + 3)}$
$= \dfrac{(x - 9)(\sqrt{x} + 3)}{x - 9} = \sqrt{x} + 3$

97. $\dfrac{3\sqrt{k}}{\sqrt{k} + \sqrt{7}} = \dfrac{3\sqrt{k}(\sqrt{k} - \sqrt{7})}{(\sqrt{k} + \sqrt{7})(\sqrt{k} - \sqrt{7})}$
$= \dfrac{3k - 3\sqrt{7k}}{k - 7}$

99. $\dfrac{5}{\sqrt{2} - 1} + \dfrac{3}{\sqrt{2} + 1}$
$= \dfrac{5(\sqrt{2} + 1)}{(\sqrt{2} - 1)(\sqrt{2} + 1)} + \dfrac{3(\sqrt{2} - 1)}{(\sqrt{2} + 1)(\sqrt{2} - 1)}$
$= \dfrac{5\sqrt{2} + 5}{2 - 1} + \dfrac{3\sqrt{2} - 3}{2 - 1} = 2 + 8\sqrt{2}$

101. $\dfrac{1}{\sqrt{2}} + \dfrac{1}{\sqrt{3}} = \dfrac{\sqrt{2}}{2} + \dfrac{\sqrt{3}}{3}$
$= \dfrac{3\sqrt{2}}{3 \cdot 2} + \dfrac{2\sqrt{3}}{2 \cdot 3} = \dfrac{3\sqrt{2} + 2\sqrt{3}}{6}$

103. $\dfrac{3}{\sqrt{2} - 1} + \dfrac{4}{\sqrt{2} + 1}$
$= \dfrac{3(\sqrt{2} + 1)}{(\sqrt{2} - 1)(\sqrt{2} + 1)} + \dfrac{4(\sqrt{2} - 1)}{(\sqrt{2} + 1)(\sqrt{2} - 1)}$
$= \dfrac{3\sqrt{2} + 3}{1} + \dfrac{4\sqrt{2} - 4}{1} = 7\sqrt{2} - 1$

105. $\dfrac{\sqrt{x}}{\sqrt{x} + 2} + \dfrac{3\sqrt{x}}{\sqrt{x} - 2}$
$= \dfrac{\sqrt{x}(\sqrt{x} - 2)}{(\sqrt{x} + 2)(\sqrt{x} - 2)}$
$+ \dfrac{3\sqrt{x}(\sqrt{x} + 2)}{(\sqrt{x} - 2)(\sqrt{x} + 2)}$
$= \dfrac{x - 2\sqrt{x}}{x - 4} + \dfrac{3x + 6\sqrt{x}}{x - 4} = \dfrac{4x + 4\sqrt{x}}{x - 4}$

107. $\dfrac{1}{\sqrt{x}} + \dfrac{1}{1 - \sqrt{x}}$
$= \dfrac{1(1 - \sqrt{x})}{\sqrt{x}(1 - \sqrt{x})} + \dfrac{1\sqrt{x}}{(1 - \sqrt{x})\sqrt{x}}$
$= \dfrac{1 - \sqrt{x}}{\sqrt{x} - x} + \dfrac{\sqrt{x}}{\sqrt{x} - x} = \dfrac{1}{\sqrt{x} - x}$
$= \dfrac{1(\sqrt{x} + x)}{(\sqrt{x} - x)(\sqrt{x} + x)} = \dfrac{x + \sqrt{x}}{x - x^2}$
$= \dfrac{x + \sqrt{x}}{x(1 - x)}$

109. a) Use $(a - b)(a^2 + ab + b^2) = a^3 - b^3$ with $a = x$ and $b = \sqrt[3]{2}$ to get $x^3 - 2$.
b) Use $a^3 + b^3 = (a + b)(a^2 - ab + b^2)$ to get
$x^3 + 5 = x^3 + \left(\sqrt[3]{5}\right)^3$
$= (x + \sqrt[3]{5})(x^2 - \sqrt[3]{5}\,x + \sqrt[3]{25})$

c) Use $(a - b)(a^2 + ab + b^2) = a^3 - b^3$ with $a = \sqrt[3]{5}$ and $b = \sqrt[3]{2}$ to get
$\left(\sqrt[3]{5} - \sqrt[3]{2}\right)\left(\sqrt[3]{25} + \sqrt[3]{10} + \sqrt[3]{4}\right)$
$= 5 - 2 = 3$

d)
$a + b = \left(\sqrt[3]{a} + \sqrt[3]{b}\right)\left(\sqrt[3]{a^2} - \sqrt[3]{ab} + \sqrt[3]{b^2}\right)$
$a - b = \left(\sqrt[3]{a} - \sqrt[3]{b}\right)\left(\sqrt[3]{a^2} + \sqrt[3]{ab} + \sqrt[3]{b^2}\right)$

7.5 WARM-UPS

1. False, because $x^2 = 4$ is equivalent to the compound equation $x = 2$ or $x = -2$.
2. True, because the square of any real number is nonnegative.
3. False, 0 is a solution.
4. False, because -2 is not a solution to $x^3 = 8$.
5. True, because if x is positive the left side of the equation is negative and the right side is positive.
6. False, we should square each side.
7. False, extraneous roots are found but they do not satisfy the equation.
8. True, because we get $x = 49$, and 49 does not satisfy $\sqrt{x} = -7$.
9. True, because both square roots of 6 satisfy $x^2 - 6 = 0$.
10. True, we get extraneous roots only by raising each side to an even power.

7.5 EXERCISES

1. The odd-root property says that if n is an odd positive integer then $x^n = k$ is equivalent to $x = \sqrt[n]{k}$ for any real number k.
3. An extraneous solution is a solution that appears when solving an equation, but does not satisfy the original equation.
5. $x^3 = -1000$
$x = \sqrt[3]{-1000} = -10$
Solution set: $\{-10\}$
7. $32m^5 - 1 = 0$
$32m^5 = 1$
$m^5 = \frac{1}{32}$
$m = \sqrt[5]{\frac{1}{32}} = \frac{1}{2}$
Solution set: $\left\{\frac{1}{2}\right\}$
9. $(y-3)^3 = -8$
$y - 3 = \sqrt[3]{-8}$
$y - 3 = -2$
$y = 1$
Solution set: $\{1\}$
11. $\frac{1}{2}x^3 + 4 = 0$
$\frac{1}{2}x^3 = -4$
$x^3 = -8$
$x = \sqrt[3]{-8} = -2$
Solution set: $\{-2\}$
13. $x^2 = 25$
$x = \pm 5$
Solution set: $\{-5, 5\}$
15. $x^2 - 20 = 0$
$x^2 = 20$
$x = \pm\sqrt{20} = \pm 2\sqrt{5}$
Solution set: $\{-2\sqrt{5}, 2\sqrt{5}\}$
17. $x^2 = -9$
Since an even root of a negative number is not a real number, there is no real solution.
19. $(x-3)^2 = 16$
$x - 3 = \pm 4$
$x = 3 \pm 4$
$x = 3 + 4$ or $x = 3 - 4$
$x = 7$ or $x = -1$
Solution set: $\{-1, 7\}$
21. $(x+1)^2 - 8 = 0$
$(x+1)^2 = 8$
$x + 1 = \pm\sqrt{8}$
$x = -1 \pm 2\sqrt{2}$
Solution set: $\left\{-1 - 2\sqrt{2}, -1 + 2\sqrt{2}\right\}$
23. $\frac{1}{2}x^2 = 5$
$x^2 = 10$
$x = \pm\sqrt{10}$
Solution set: $\left\{-\sqrt{10}, \sqrt{10}\right\}$
25. $(y-3)^4 = 0$
$y - 3 = \sqrt[4]{0} = 0$
$y = 3$
Solution set: $\{3\}$
27. $2x^6 = 128$
$x^6 = 64$
$x = \pm\sqrt[6]{64}$
$x = \pm 2$
Solution set: $\{-2, 2\}$
29. $\sqrt{x-3} - 3 = 4$
$\sqrt{x-3} = 7$
$(\sqrt{x-3})^2 = 7^2$
$x - 3 = 49$
$x = 52$
Check 52 in the original equation.
Solution set: $\{52\}$

31.
$$\begin{aligned} 2\sqrt{w+4} &= 5 \\ (2\sqrt{w+4})^2 &= 5^2 \\ 4(w+4) &= 25 \\ 4w+16 &= 25 \\ 4w &= 9 \\ w &= \frac{9}{4} \end{aligned}$$

Check $\frac{9}{4}$ in the original equation.

Solution set: $\left\{\frac{9}{4}\right\}$

33.
$$\begin{aligned} \sqrt[3]{2x+3} &= \sqrt[3]{x+12} \\ \left(\sqrt[3]{2x+3}\right)^3 &= \left(\sqrt[3]{x+12}\right)^3 \\ 2x+3 &= x+12 \\ x &= 9 \end{aligned}$$

Check 9 in the original equation.
Solution set: $\{9\}$

35.
$$\begin{aligned} \sqrt{2t-4} &= \sqrt{t-1} \\ (\sqrt{2t-4})^2 &= (\sqrt{t-1})^2 \\ 2t-4 &= t-1 \\ t &= 3 \end{aligned}$$

Check: $\sqrt{2(3)-4} = \sqrt{3-1}$

Since each side of the equation is $\sqrt{2}$, the solution set is $\{3\}$.

37.
$$\begin{aligned} \sqrt{4x^2+x-3} &= 2x \\ \left(\sqrt{4x^2+x-3}\right)^2 &= (2x)^2 \\ 4x^2+x-3 &= 4x^2 \\ x-3 &= 0 \\ x &= 3 \end{aligned}$$

Check 3 in the original equation.
Solution set: $\{3\}$

39.
$$\begin{aligned} \sqrt{x^2+2x-6} &= 3 \\ \left(\sqrt{x^2+2x-6}\right)^2 &= 3^2 \\ x^2+2x-6 &= 9 \\ x^2+2x-15 &= 0 \\ (x+5)(x-3) &= 0 \\ x+5=0 \text{ or } x-3 &= 0 \\ x=-5 \text{ or } x &= 3 \end{aligned}$$

Check 3 and -5 in the original equation.
Solution set: $\{-5, 3\}$

41.
$$\begin{aligned} \sqrt{2x^2-1} &= x \\ \left(\sqrt{2x^2-1}\right)^2 &= x^2 \\ 2x^2-1 &= x^2 \\ x^2 &= 1 \\ x &= \pm 1 \end{aligned}$$

Checking in the original we find that if $x=-1$ we get $\sqrt{1}=-1$, which is incorrect. So the solution set is $\{1\}$.

43.
$$\begin{aligned} \sqrt{2x^2+5x+6} &= x \\ \left(\sqrt{2x^2+5x+6}\right)^2 &= x^2 \\ 2x^2+5x+6 &= x^2 \\ x^2+5x+6 &= 0 \\ (x+2)(x+3) &= 0 \\ x=-2 \text{ or } x &= -3 \end{aligned}$$

If we use $x=-2$ in the original equation we get $\sqrt{4}=-2$, and if $x=-3$ in the original we get $\sqrt{9}=-3$. Since both of these equations are incorrect, the solution set is $\emptyset$.

45.
$$\begin{aligned} \sqrt{x-1} &= x-1 \\ \left(\sqrt{x-1}\right)^2 &= (x-1)^2 \\ x-1 &= x^2-2x+1 \\ x^2-3x+2 &= 0 \\ (x-2)(x-1) &= 0 \\ x-2=0 \text{ or } x-1 &= 0 \\ x=2 \text{ or } x &= 1 \end{aligned}$$

Both 1 and 2 satisfy the original equation. So the solution set is $\{1, 2\}$.

47.
$$\begin{aligned} x+\sqrt{x-9} &= 9 \\ \left(\sqrt{x-9}\right)^2 &= (9-x)^2 \\ x-9 &= 81-18x+x^2 \\ x^2-19x+90 &= 0 \\ (x-10)(x-9) &= 0 \\ x-10=0 \text{ or } x-9 &= 0 \\ x=10 \text{ or } x &= 9 \end{aligned}$$

Only 9 satisfies the original equation. So the solution set is $\{9\}$.

49.
$$\begin{aligned} \sqrt{x}+\sqrt{x-3} &= 3 \\ \sqrt{x} &= 3-\sqrt{x-3} \\ \left(\sqrt{x}\right)^2 &= \left(3-\sqrt{x-3}\right)^2 \\ x &= 9-6\sqrt{x-3}+x-3 \\ -6 &= -6\sqrt{x-3} \\ 1 &= \sqrt{x-3} \\ 1^2 &= \left(\sqrt{x-3}\right)^2 \\ 1 &= x-3 \\ 4 &= x \end{aligned}$$

Check 4 in the original equation. The solution set is $\{4\}$.

51. $\sqrt{x+2}+\sqrt{x-1}=3$

$$\sqrt{x+2}=3-\sqrt{x-1}$$
$$\left(\sqrt{x+2}\right)^2=\left(3-\sqrt{x-1}\right)^2$$
$$x+2=9-6\sqrt{x-1}+x-1$$
$$-6=-6\sqrt{x-1}$$
$$1=\sqrt{x-1}$$
$$1^2=\left(\sqrt{x-1}\right)^2$$
$$1=x-1$$
$$2=x$$

Check 2 in the original equation. The solution set is $\{2\}$.

53. $\sqrt{x+3}-\sqrt{x-2}=1$

$$\sqrt{x+3}=1+\sqrt{x-2}$$
$$\left(\sqrt{x+3}\right)^2=\left(1+\sqrt{x-2}\right)^2$$
$$x+3=1+2\sqrt{x-2}+x-2$$
$$4=2\sqrt{x-2}$$
$$2=\sqrt{x-2}$$
$$2^2=\left(\sqrt{x-2}\right)^2$$
$$4=x-2$$
$$6=x$$

Check 6 in the original equation. The solution set is $\{6\}$.

55. $\sqrt{3x+1}-\sqrt{2x-1}=1$

$$\sqrt{3x+1}=1+\sqrt{2x-1}$$
$$\left(\sqrt{3x+1}\right)^2=(1+\sqrt{2x-1})^2$$
$$3x+1=1+2\sqrt{2x-1}+2x-1$$
$$x+1=2\sqrt{2x-1}$$
$$(x+1)^2=(2\sqrt{2x-1})^2$$
$$x^2+2x+1=4(2x-1)$$
$$x^2-6x+5=0$$
$$(x-1)(x-5)=0$$
$$x-1=0 \text{ or } x-5=0$$
$$x=1 \text{ or } \quad x=5$$

The solution set is $\{1,5\}$.

57. $\sqrt{2x+2}-\sqrt{x-3}=2$

$$\sqrt{2x+2}=2+\sqrt{x-3}$$
$$\left(\sqrt{2x+2}\right)^2=\left(2+\sqrt{x-3}\right)^2$$
$$2x+2=4+4\sqrt{x-3}+x-3$$
$$x+1=4\sqrt{x-3}$$
$$(x+1)^2=\left(4\sqrt{x-3}\right)^2$$
$$x^2+2x+1=16(x-3)$$
$$x^2-14x+49=0$$
$$(x-7)^2=0$$
$$x=7$$

Check 7 in the original equation.
Solution set: $\{7\}$

59. $\sqrt{4-x}-\sqrt{x+6}=2$

$$\sqrt{4-x}=2+\sqrt{x+6}$$
$$\left(\sqrt{4-x}\right)^2=\left(2+\sqrt{x+6}\right)^2$$
$$4-x=4+4\sqrt{x+6}+x+6$$
$$-6-2x=4\sqrt{x+6}$$
$$-3-x=2\sqrt{x+6}$$
$$(-3-x)^2=\left(2\sqrt{x+6}\right)^2$$
$$9+6x+x^2=4(x+6)$$
$$x^2+2x-15=0$$
$$(x+5)(x-3)=0$$
$$x=-5 \text{ or } x=3$$

If $x=3$ in the original equation we get $1-3=2$, which is incorrect. If $x=-5$ we get $3-1=2$. So the solution set is $\{-5\}$.

61. $\sqrt{x-5}-\sqrt{x}=3$

$$\sqrt{x-5}=3+\sqrt{x}$$
$$\left(\sqrt{x-5}\right)^2=(3+\sqrt{x})^2$$
$$x-5=9+6\sqrt{x}+x$$
$$-14=6\sqrt{x}$$
$$-\frac{7}{3}=\sqrt{x}$$
$$x=\left(-\frac{7}{3}\right)^2=\frac{49}{9}$$

Since $49/9$ does not satisfy the original equation, the solution set is $\emptyset$.

63. $\sqrt{3x+1}+\sqrt{2x+4}=3$

$$\sqrt{3x+1}=3-\sqrt{2x+4}$$
$$\left(\sqrt{3x+1}\right)^2=(3-\sqrt{2x+4})^2$$
$$3x+1=9-6\sqrt{2x+4}+2x+4$$
$$x-12=-6\sqrt{2x+4}$$
$$(x-12)^2=(-6\sqrt{2x+4})^2$$
$$x^2-24x+144=36(2x+4)$$
$$x^2-96x=0$$
$$(x)(x-96)=0$$
$$x=0 \text{ or } x-96=0$$
$$x=6 \text{ or } \quad x=96$$

Since 96 does not satisfy the original equation, the solution set is $\{0\}$.

65. $$x^{2/3} = 3$$
$$\left(x^{2/3}\right)^3 = 3^3$$
$$x^2 = 27$$
$$x = \pm\sqrt{27} = \pm 3\sqrt{3}$$
Solution set: $\left\{-3\sqrt{3}, 3\sqrt{3}\right\}$

67. $$y^{-2/3} = 9$$
$$\left(y^{-2/3}\right)^{-3} = (9)^{-3}$$
$$y^2 = \frac{1}{729}$$
$$y = \pm\sqrt{\frac{1}{729}} = \pm\frac{1}{27}$$
Solution set: $\left\{-\frac{1}{27}, \frac{1}{27}\right\}$

69. $$w^{1/3} = 8$$
$$\left(w^{1/3}\right)^3 = 8^3$$
$$w = 512$$
Solution set: $\{512\}$

71. $$t^{-1/2} = 9$$
$$\left(t^{-1/2}\right)^{-2} = (9)^{-2}$$
$$t = \frac{1}{81}$$
Solution set: $\left\{\frac{1}{81}\right\}$

73. $$\left((3a-1)^{-2/5}\right)^{-5} = 1^{-5}$$
$$(3a-1)^2 = 1$$
$$3a - 1 = \pm 1$$
$$3a - 1 = 1 \quad \text{or} \quad 3a - 1 = -1$$
$$3a = 2 \quad \text{or} \quad 3a = 0$$
$$a = \frac{2}{3} \quad \text{or} \quad a = 0$$
Solution set: $\left\{0, \frac{2}{3}\right\}$

75. $(t-1)^{-2/3} = 2$
$$\left((t-1)^{-2/3}\right)^{-3} = 2^{-3}$$
$$(t-1)^2 = \frac{1}{8}$$
$$t - 1 = \pm\sqrt{\frac{1}{8}} = \pm\frac{\sqrt{2}}{4}$$
$$t = 1 \pm \frac{\sqrt{2}}{4} = \frac{4}{4} \pm \frac{\sqrt{2}}{4} = \frac{4 \pm \sqrt{2}}{4}$$
Solution set: $\left\{\frac{4-\sqrt{2}}{4}, \frac{4+\sqrt{2}}{4}\right\}$

77. $$(x-3)^{2/3} = -4$$
$$\left((x-3)^{2/3}\right)^3 = (-4)^3$$
$$(x-3)^2 = -64$$
Because the square of any real number is nonnegative, the solution set is $\emptyset$. There is no real solution.

79. $$2x^2 + 3 = 7$$
$$2x^2 = 4$$
$$x^2 = 2$$
$$x = \pm\sqrt{2}$$
Solution set: $\left\{-\sqrt{2}, \sqrt{2}\right\}$

81. $$\sqrt[3]{2w+3} = \sqrt[3]{w-2}$$
$$\left(\sqrt[3]{2w+3}\right)^3 = \left(\sqrt[3]{w-2}\right)^3$$
$$2w + 3 = w - 2$$
$$w = -5$$
Solution set: $\{-5\}$

83. $(w+1)^{2/3} = -3$
$$\left((w+1)^{2/3}\right)^3 = (-3)^3$$
$$(w+1)^2 = -27$$
This equation has no real solution by the even root property. The solution set is $\emptyset$.

85. $(a+1)^{1/3} = -2$
$$\left((a+1)^{1/3}\right)^3 = (-2)^3$$
$$a + 1 = -8$$
$$a = -9$$
Solution set: $\{-9\}$

87. $(4y-5)^7 = 0$
$$4y - 5 = \sqrt[7]{0} = 0$$
$$4y = 5$$
$$y = \frac{5}{4}$$
Solution set: $\left\{\frac{5}{4}\right\}$

89. $$\sqrt{5x^2 + 4x + 1} - x = 0$$
$$\sqrt{5x^2 + 4x + 1} = x$$
$$5x^2 + 4x + 1 = x^2$$
$$4x^2 + 4x + 1 = 0$$
$$(2x+1)^2 = 0$$
$$2x + 1 = 0$$
$$x = -\frac{1}{2}$$
Since $-1/2$ does not satisfy the original equation, the solution set is $\emptyset$.

91. $$\sqrt{4x^2} = x + 2$$
$$\left(\sqrt{4x^2}\right)^2 = (x+2)^2$$
$$4x^2 = x^2 + 4x + 4$$
$$3x^2 - 4x - 4 = 0$$
$$(3x+2)(x-2) = 0$$
$$3x + 2 = 0 \quad \text{or} \quad x - 2 = 0$$

$$x = -\frac{2}{3} \quad \text{or} \quad x = 2$$

Solution set: $\left\{-\frac{2}{3}, 2\right\}$

93. $(t+2)^4 = 32$

$$t + 2 = \pm\sqrt[4]{32} = \pm 2 \cdot \sqrt[4]{2}$$
$$t = -2 \pm 2 \cdot \sqrt[4]{2}$$

Solution set: $\left\{-2 - 2\sqrt[4]{2}, -2 + 2\sqrt[4]{2}\right\}$

95. $\sqrt{x^2 - 3x} = x$

$$\left(\sqrt{x^2 - 3x}\right)^2 = x^2$$
$$x^2 - 3x = x^2$$
$$-3x = 0$$
$$x = 0$$

Solution set: $\{0\}$

97. $x^{-3} = 8$

$$\left(x^{-3}\right)^{-1} = 8^{-1}$$
$$x^3 = \frac{1}{8}$$
$$x = \sqrt[3]{\frac{1}{8}} = \frac{1}{2}$$

Solution set: $\left\{\frac{1}{2}\right\}$

99. Let $x =$ the length of a side. Two sides and the diagonal of a square form a right triangle. By the Pythagorean theorem we can write the equation

$$x^2 + x^2 = 8^2$$
$$2x^2 = 64$$
$$x^2 = 32$$
$$x = \pm\sqrt{32} = \pm 4\sqrt{2}$$

The length of the side is not a negative number. So the side is $4\sqrt{2}$ feet in length.

101. Let $s =$ the length of the side of the square. Since $A = s^2$ for a square, we can write the equation

$$s^2 = 50$$
$$s = \pm\sqrt{50} = \pm 5\sqrt{2}$$

Since the side of a square is not negative, the length of the side is $5\sqrt{2}$ feet.

103. Let $d =$ the length of a diagonal of the rectangle whose sides are 30 and 40 feet. By the Pythagorean theorem we can write the equation

$$d^2 = 30^2 + 40^2$$
$$d^2 = 2500$$
$$d = \pm\sqrt{2500} = \pm 50$$

Since the diagonal is not negative, the length of the diagonal is 50 feet.

105. a) If the side opposite 30° is 1, then the hypotenuse is 2.

b) If the side opposite 30° is 1, then the hypotenuse is 2. If x is the length of the side opposite 60°, then $x^2 + 1^2 = 2^2$ or $x^2 = 3$. So the side opposite 60° is $\sqrt{3}$.

c) If the hypotenuse is 1, then the side opposite 30° is $\frac{1}{2}$. If x is the length of the side opposite 60°, then $x^2 + \left(\frac{1}{2}\right)^2 = 1^2$ or $x^2 = \frac{3}{4}$. So the side opposite 60° is $\frac{\sqrt{3}}{2}$.

107. a) $C = 4(23{,}245)^{-1/3}(13.5) \approx 1.89$

b) $C = 4d^{-1/3}b$

$$d^{-1/3} = \frac{C}{4b}$$
$$(d^{-1/3})^{-3} = \left(\frac{C}{4b}\right)^{-3}$$
$$d = \frac{64b^3}{C^3}$$

c) The capsize screening value is less than 2 when $d > 19{,}683$ pounds.

109. If the volume of the cube is 2, each side of the cube has length $\sqrt[3]{2}$, because $V = s^3$. Let $d =$ the length of the diagonal of a side. The diagonal of a side is the diagonal of a square with sides of length $\sqrt[3]{2}$. By the Pythagorean theorem we can write

$$d^2 = \left(\sqrt[3]{2}\right)^2 + \left(\sqrt[3]{2}\right)^2$$
$$d^2 = \sqrt[3]{4} + \sqrt[3]{4}$$
$$d^2 = 2 \cdot \sqrt[3]{4} = \sqrt[3]{8} \cdot \sqrt[3]{4} = \sqrt[3]{32}$$
$$d = \left(\sqrt[3]{32}\right)^{1/2} = \left(32^{1/3}\right)^{1/2} = 32^{1/6}$$

The length of the diagonal is $\sqrt[6]{32}$ meters.

111. Let $x =$ the third side to the triangle whose given sides are 3 and 5. By the Pythagorean theorem we can find x:

$$x^2 + 3^2 = 5^2$$
$$x^2 = 16$$
$$x = 4$$

Since $x = 4$, the base of length 12 is divided into 2 parts, one of length 4 and the other of length 8. The side marked a is the hypotenuse of a right triangle with legs 3 and 8:

$$a^2 = 3^2 + 8^2$$
$$a^2 = 73$$
$$a = \sqrt{73}\text{ km}$$

113. a) $r = \left(\frac{S}{P}\right)^{1/n} - 1$

$1 + r = \left(\frac{S}{P}\right)^{1/n}$

$\frac{S}{P} = (1+r)^n$

$S = P(1+r)^n$

b) Solve for P:

$\frac{S}{P} = (1+r)^n$

$P = \frac{S}{(1+r)^n}$

$P = S(1+r)^{-n}$

115. $\frac{11.86^2}{5.2^3} = \frac{29.46^2}{R^3}$

$11.86^2 R^3 = 5.2^3 \cdot 29.46^2$

$R^3 = \frac{5.2^3 \cdot 29.46^2}{11.86^2}$

$R = \sqrt[3]{\frac{5.2^3 \cdot 29.46^2}{11.86^2}} \approx 9.5$ AU

117. $x^2 = 3.24$

$x = \pm\sqrt{3.24} = \pm 1.8$

Solution set: $\{-1.8, 1.8\}$

119. $\sqrt{x-2} = 1.73$

$x - 2 = (1.73)^2$

$x = 2 + (1.73)^2 \approx 4.993$

Solution set: $\{4.993\}$

121. $x^{2/3} = 8.86$

$\left(x^{2/3}\right)^3 = (8.86)^3$

$x^2 = 695.506$

$x = \pm\sqrt{695.506} \approx \pm 26.372$

Solution set: $\{-26.372, 26.372\}$

123. Since $V = s^3$, $s = \sqrt[3]{V}$.

7.6 WARM-UPS

1. True, because every real number is a complex number.

2. False, because $2 - \sqrt{-6} = 2 - i\sqrt{6}$.

3. False, because $\sqrt{-9} = 3i$.

4. True, because $(\pm 3i)^2 = -9$.

5. True, because we subtract complex numbers just like we subtract binomials.

6. True, because $i^4 = i^2 \cdot i^2 = (-1)(-1) = 1$.

7. True, because $(2 - i)(2 + i) = 4 - i^2 = 4 - (-1) = 5$.

8. False, $i^3 = i^2 \cdot i = -1 \cdot i = -i$.

9. True, $i^{48} = (i^4)^{12} = 1^{12} = 1$.

10. False, $x^2 = 0$ has only one solution.

7.6 EXERCISES

1. A complex number is a number of the form $a + bi$ where a and b are real numbers.

3. The union of the real numbers and the imaginary numbers is the set of complex numbers.

5. The conjugate of $a + bi$ is $a - bi$.

7. $(2 + 3i) + (-4 + 5i) = -2 + 8i$

9. $(2 - 3i) - (6 - 7i) = 2 - 3i - 6 + 7i$
$= -4 + 4i$

11. $(-1 + i) + (-1 - i) = -2$

13. $(-2 - 3i) - (6 - i) = -2 - 3i - 6 + i$
$= -8 - 2i$

15. $3(2 + 5i) = 3 \cdot 2 + 3 \cdot 5i = 6 + 15i$

17. $2i(i - 5) = 2i^2 - 10i = 2(-1) - 10i$
$= -2 - 10i$

19. $-4i(3 - i) = -12i + 4i^2$
$= -12i + 4(-1) = -4 - 12i$

21. $(2 + 3i)(4 + 6i) = 8 + 24i + 18i^2$
$= 8 + 24i + 18(-1) = -10 + 24i$

23. $(-1 + i)(2 - i) = -2 + 3i - i^2$
$= -2 + 3i - (-1) = -1 + 3i$

25. $(-1 - 2i)(2 + i) = -2 - 5i - 2i^2$
$= -2 - 5i - 2(-1) = -5i$

27. $(5 - 2i)(5 + 2i) = 25 - 4i^2$
$= 25 - 4(-1) = 29$

29. $(1 - i)(1 + i) = 1 - i^2 = 1 - (-1) = 2$

31. $(4 + 2i)(4 - 2i) = 16 - 4i^2$
$= 16 - 4(-1) = 20$

33. $(3i)^2 = 9i^2 = 9(-1) = -9$

35. $(-5i)^2 = (-5)^2 i^2 = 25(-1) = -25$

37. $(2i)^4 = 2^4 i^4 = 16(1) = 16$

39. $i^9 = (i^4)^2 \cdot i = 1^2 \cdot i = i$

41. $i^{18} = (i^4)^4 \cdot i^2 = 1^4(-1) = -1$

43. $i^{25} = (i^4)^6 \cdot i = 1^6(i) = i$

45. $(3 + 5i)(3 - 5i) = 9 - 25i^2 = 9 + 25$
$= 34$

47. $(1 - 2i)(1 + 2i) = 1 - 4i^2 = 1 - 4(-1)$
$= 5$

49. $(-2 + i)(-2 - i) = 4 - i^2 = 4 - (-1)$
$= 5$

51. $(2 - i\sqrt{3})(2 + i\sqrt{3}) = 4 - 3i^2$
$= 4 - 3(-1) = 7$

53. $\frac{3}{4+i} = \frac{3(4-i)}{(4+i)(4-i)} = \frac{12-3i}{16-i^2}$
$= \frac{12-3i}{17} = \frac{12}{17} - \frac{3}{17}i$

55. $\frac{2+i}{3-2i} = \frac{(2+i)(3+2i)}{(3-2i)(3+2i)}$
$= \frac{6+7i+2i^2}{9-4i^2} = \frac{4+7i}{13} = \frac{4}{13} + \frac{7}{13}i$

57. $\frac{4+3i}{i} = \frac{(4+3i)(-i)}{(i)(-i)} = \frac{-4i-3i^2}{-i^2}$
$= \frac{-4i+3}{1} = 3 - 4i$

59. $\frac{2+6i}{2} = \frac{2}{2} + \frac{6i}{2} = 1 + 3i$

61. $\frac{1+i}{3i-2} = \frac{(1+i)(3i+2)}{(3i-2)(3i+2)}$
$= \frac{2+5i+3i^2}{9i^2-4} = \frac{-1+5i}{-13} = \frac{1}{13} - \frac{5}{13}i$

63. $\frac{6}{3i} = \frac{6\cdot(-3i)}{(3i)(-3i)} = \frac{-18i}{-9i^2} = \frac{-18i}{9} = -2i$

65. $\sqrt{-25} = i\sqrt{25} = 5i$

67. $2 + \sqrt{-4} = 2 + i\sqrt{4} = 2 + 2i$

69. $2\sqrt{-9} + 5 = 2i\sqrt{9} + 5 = 2i \cdot 3 + 5$
$= 5 + 6i$

71. $7 - \sqrt{-6} = 7 - i\sqrt{6}$

73. $\sqrt{-8} + \sqrt{-18} = i\sqrt{4}\sqrt{2} + i\sqrt{9}\sqrt{2}$
$= 2i\sqrt{2} + 3i\sqrt{2} = 5i\sqrt{2}$

75. $\frac{2+\sqrt{-12}}{2} = \frac{2+i\sqrt{4}\sqrt{3}}{2} = 1 + i\sqrt{3}$

77. $\frac{-4-\sqrt{-24}}{4} = \frac{-4-i\sqrt{4}\sqrt{6}}{4}$
$= \frac{-4}{4} - \frac{2i\sqrt{6}}{4} = -1 - \frac{1}{2}i\sqrt{6}$

79. $\sqrt{-2}\cdot\sqrt{-6} = i\sqrt{2}\cdot i\sqrt{6} = i^2\sqrt{12}$
$= -1 \cdot 2\sqrt{3} = -2\sqrt{3}$

81. $\sqrt{-3}\cdot\sqrt{-27} = i\sqrt{3}\cdot i\sqrt{27} = i^2\sqrt{81}$
$= -1 \cdot 9 = -9$

83. $\frac{\sqrt{8}}{\sqrt{-4}} = \frac{\sqrt{8}}{i\sqrt{4}} = \frac{\sqrt{2}}{i} = \frac{\sqrt{2}\cdot(-i)}{i(-i)}$
$= \frac{-i\sqrt{2}}{-i^2} = \frac{-i\sqrt{2}}{1} = -i\sqrt{2}$

85. $x^2 = -36$
$x = \pm\sqrt{-36} = \pm 6i$
Solution set: $\{\pm 6i\}$

87. $x^2 = -12$
$x = \pm\sqrt{-12} = \pm i\sqrt{4}\sqrt{3} = \pm 2i\sqrt{3}$
Solution set: $\{\pm 2i\sqrt{3}\}$

89. $2x^2 + 5 = 0$
$x^2 = -\frac{5}{2}$
$x = \pm\sqrt{-\frac{5}{2}} = \pm\frac{i\sqrt{5}}{\sqrt{2}} = \pm\frac{i\sqrt{10}}{2}$
Solution set: $\left\{\pm\frac{i\sqrt{10}}{2}\right\}$

91. $3x^2 + 6 = 0$
$x^2 = -2$
$x = \pm\sqrt{-2} = \pm i\sqrt{2}$
Solution set: $\{\pm i\sqrt{2}\}$

93. $(2 - 3i)(3 + 4i) = 6 - i - 12i^2$
$= 6 - i - 12(-1) = 18 - i$

95. $(2 - 3i) + (3 + 4i) = 5 + i$

97. $\frac{2-3i}{3+4i} = \frac{(2-3i)(3-4i)}{(3+4i)(3-4i)}$
$= \frac{6-17i+12i^2}{9-16i^2} = \frac{6-17i+12(-1)}{9-16(-1)}$
$= \frac{-6-17i}{25} = -\frac{6}{25} - \frac{17}{25}i$

99. $i(2 - 3i) = 2i - 3i^2 = 2i + 3 = 3 + 2i$

101. $(-3i)^2 = 9i^2 = -9$

103. $\sqrt{-12} + \sqrt{-3} = 2i\sqrt{3} + i\sqrt{3}$
$= 3i\sqrt{3}$

105. $(2 - 3i)^2 = 4 - 12i + 9i^2 = -5 - 12i$

107. $\frac{-4+\sqrt{-32}}{2} = \frac{-4+4i\sqrt{2}}{2}$
$= -2 + 2i\sqrt{2}$

109. If $x = 2 - i$ then
$x^2 - 4x + 5 = (2 - i)^2 - 4(2 - i) + 5$
$= 4 - 4i + i^2 - 8 + 4i + 5$
$= 1 + i^2 = 0$
So $2 - i$ is a solution to $x^2 - 4x + 5 = 0$.

111. The product rule for radicals is valid only for radicals that represent real numbers. So the product of 4 is not correct. The correct product is found as follows:
$\sqrt{-4}\cdot\sqrt{-4} = 2i \cdot 2i = 4i^2 = -4$

Chapter 7 Wrap-Up

Enriching Your Mathematical Word Power

1. d **2.** b **3.** b **4.** b **5.** d
6. b **7.** c **8.** a **9.** a **10.** d
11. c **12.** a **13.** c **14.** d **15.** b

CHAPTER 7 REVIEW

1. $\sqrt[5]{32} = 2$ because $2^5 = 32$.

3. $\sqrt[3]{1000} = 10$ because $10^3 = 1000$.

5. $\sqrt{72} = \sqrt{36}\sqrt{2} = 6\sqrt{2}$

7. $\sqrt{x^{12}} = x^6$ because $(x^6)^2 = x^{12}$.

9. $\sqrt[3]{x^6} = x^2$ because $(x^2)^3 = x^6$.

11. $\sqrt{2x^9} = \sqrt{x^8}\sqrt{2x} = x^4\sqrt{2x}$

13. $\sqrt{8w^5} = \sqrt{4w^4}\sqrt{2w} = 2w^2\sqrt{2w}$

15. $\sqrt[3]{16x^4} = \sqrt[3]{8x^3}\sqrt[3]{2x} = 2x\sqrt[3]{2x}$

17. $\sqrt[4]{a^9b^5} = \sqrt[4]{a^8b^4}\sqrt[4]{ab} = a^2b\sqrt[4]{ab}$

19. $\sqrt{\frac{x^3}{16}} = \frac{\sqrt{x^2}\sqrt{x}}{\sqrt{16}} = \frac{x\sqrt{x}}{4}$

21. In $f(x) = \sqrt{2x - 5}$ we must have $2x - 5 \geq 0$ or $x \geq 2.5$. So the domain of f is the interval $[2.5, \infty)$.

23. In $f(x) = \sqrt[3]{7x - 1}$, $7x - 1$ can be any real number. So any real number can be used for x. So the domain of f is the interval $(-\infty, \infty)$.

25. In $f(x) = \sqrt[4]{-3x + 1}$ we must have $-3x + 1 \geq 0$, $-3x \geq -1$, or $x \leq 1/3$. So the domain of f is the interval $(-\infty, 1/3]$.

27. In $f(x) = \sqrt{\frac{1}{2}x + 1}$ we must have $\frac{1}{2}x + 1 \geq 0$, $\frac{1}{2}x \geq -1$, or $x \geq -2$. So the domain of f is the interval $[-2, \infty)$.

29. $(-27)^{-2/3} = \frac{1}{(-3)^2} = \frac{1}{9}$

31. $(2^6)^{1/3} = 2^{6/3} = 2^2 = 4$

33. $100^{-3/2} = \frac{1}{10^3} = \frac{1}{1000}$

35. $\frac{3x^{-1/2}}{3^{-2}x^{-1}} = 3^3x^{-1/2+1} = 27x^{1/2}$

37. $(a^{1/2}b)^3(ab^{1/4})^2 = a^{3/2}b^3a^2b^{2/4}$
$= a^{7/2}b^{7/2}$

39. $(x^{1/2}y^{1/4})(x^{1/4}y) = x^{1/2+1/4}y^{1/4+1}$
$= x^{3/4}y^{5/4}$

41. $\sqrt{13}\sqrt{13} = 13$

43. $\sqrt{27} + \sqrt{45} - \sqrt{75}$
$= 3\sqrt{3} + 3\sqrt{5} - 5\sqrt{3} = 3\sqrt{5} - 2\sqrt{3}$

45. $3\sqrt{2}(5\sqrt{2} - 7\sqrt{3}) = 15 \cdot 2 - 21\sqrt{6}$
$= 30 - 21\sqrt{6}$

47. $(2 - \sqrt{3})(3 + \sqrt{2})$
$= 6 - 3\sqrt{3} + 2\sqrt{2} - \sqrt{6}$

49. $5 \div \sqrt{2} = \frac{5}{\sqrt{2}} = \frac{5\sqrt{2}}{\sqrt{2}\sqrt{2}} = \frac{5\sqrt{2}}{2}$

51. $\sqrt{\frac{2}{5}} = \frac{\sqrt{2}\sqrt{5}}{\sqrt{5}\sqrt{5}} = \frac{\sqrt{10}}{5}$

53. $\sqrt[3]{\frac{2}{3}} = \frac{\sqrt[3]{2} \cdot \sqrt[3]{9}}{\sqrt[3]{3} \cdot \sqrt[3]{9}} = \frac{\sqrt[3]{18}}{\sqrt[3]{27}} = \frac{\sqrt[3]{18}}{3}$

55. $\frac{2}{\sqrt{3x}} = \frac{2\sqrt{3x}}{\sqrt{3x}\sqrt{3x}} = \frac{2\sqrt{3x}}{3x}$

57. $\frac{\sqrt{10y^3}}{\sqrt{6}} = \frac{\sqrt{2}\sqrt{5y}\sqrt{y^2}}{\sqrt{2}\sqrt{3}} = \frac{y\sqrt{5y}}{\sqrt{3}}$
$= \frac{y\sqrt{5y}\sqrt{3}}{\sqrt{3}\sqrt{3}} = \frac{y\sqrt{15y}}{3}$

59. $\frac{3}{\sqrt[3]{2a}} = \frac{3\sqrt[3]{4a^2}}{\sqrt[3]{2a}\sqrt[3]{4a^2}} = \frac{3\sqrt[3]{4a^2}}{2a}$

61. $\frac{5}{\sqrt[4]{3x^2}} = \frac{5\sqrt[4]{27x^2}}{\sqrt[4]{3x^2} \cdot \sqrt[4]{27x^2}}$
$= \frac{5\sqrt[4]{27x^2}}{\sqrt[4]{81x^4}} = \frac{5\sqrt[4]{27x^2}}{3x}$

63. $(\sqrt{3})^4 = 3^{4/2} = 3^2 = 9$

65. $\frac{2 - \sqrt{8}}{2} = \frac{2 - 2\sqrt{2}}{2} = \frac{2(1 - \sqrt{2})}{2}$
$= 1 - \sqrt{2}$

67. $\frac{\sqrt{6}}{1 - \sqrt{3}} = \frac{\sqrt{6}(1 + \sqrt{3})}{(1 - \sqrt{3})(1 + \sqrt{3})}$
$= \frac{\sqrt{6} + \sqrt{18}}{1 - 3} = \frac{\sqrt{6} + 3\sqrt{2}}{-2} = \frac{-\sqrt{6} - 3\sqrt{2}}{2}$

69. $\frac{2\sqrt{3}}{3\sqrt{6} - \sqrt{12}} = \frac{2\sqrt{3}}{3\sqrt{6} - 2\sqrt{3}}$
$= \frac{2\sqrt{3}(3\sqrt{6} + 2\sqrt{3})}{(3\sqrt{6} - 2\sqrt{3})(3\sqrt{6} + 2\sqrt{3})}$

$= \frac{6\sqrt{18} + 12}{54 - 12} = \frac{6 \cdot 3\sqrt{2} + 12}{42} = \frac{6 \cdot 3\sqrt{2} + 6 \cdot 2}{6 \cdot 7}$

$= \frac{3\sqrt{2} + 2}{7}$

71. $\left(2w\sqrt[3]{2w^2}\right)^6 = 2^6w^6\sqrt[3]{2^6w^{12}}$
$= 2^6 2^2 w^6 w^4 = 256w^{10}$

73. $x^2 = 16$
$x = \pm 4$
Solution set: $\{-4, 4\}$

75. $(a-5)^2 = 4$
$a - 5 = \pm 2$
$a = 5 \pm 2$
$a = 5 + 2$ or $a = 5 - 2$
$a = 7$ or $a = 3$
Solution set: $\{3, 7\}$

77. $(a+1)^2 = 5$
$a + 1 = \pm\sqrt{5}$
$a = -1 \pm \sqrt{5}$
Solution set: $\{-1-\sqrt{5}, -1+\sqrt{5}\}$

79. $(m+1)^2 = -8$
Since the square root of -8 is not a real number, there is no real solution. The solution set is $\emptyset$.

81. $\sqrt{m-1} = 3$
$(\sqrt{m-1})^2 = 3^2$
$m - 1 = 9$
$m = 10$
Solution set: $\{10\}$

83. $\sqrt[3]{2x+9} = 3$
$(\sqrt[3]{2x+9})^3 = 3^3$
$2x + 9 = 27$
$2x = 18$
$x = 9$
Solution set: $\{9\}$

85. $w^{2/3} = 4$
$(w^{2/3})^3 = 4^3$
$w^2 = 64$
$w = \pm 8$
Solution set: $\{-8, 8\}$

87. $(m+1)^{1/3} = 5$
$((m+1)^{1/3})^3 = 5^3$
$m + 1 = 125$
$m = 124$
Solution set: $\{124\}$

89. $\sqrt{x-3} = \sqrt{x+2} - 1$
$\left(\sqrt{x-3}\right)^2 = \left(\sqrt{x+2}-1\right)^2$
$x - 3 = x + 2 - 2\sqrt{x+2} + 1$
$-6 = -2\sqrt{x+2}$
$3 = \sqrt{x+2}$
$3^2 = (\sqrt{x+2})^2$
$9 = x + 2$
$7 = x$
Solution set: $\{7\}$

91. $\sqrt{5x-x^2} = \sqrt{6}$
$\left(\sqrt{5x-x^2}\right)^2 = (\sqrt{6})^2$
$5x - x^2 = 6$
$-x^2 + 5x - 6 = 0$
$x^2 - 5x + 6 = 0$
$(x-2)(x-3) = 0$
$x = 2$ or $x = 3$
Solution set: $\{2, 3\}$

93. $\sqrt{x+7} - 2\sqrt{x} = -2$
$\sqrt{x+7} = 2\sqrt{x} - 2$
$\left(\sqrt{x+7}\right)^2 = \left(2\sqrt{x}-2\right)^2$
$x + 7 = 4x - 8\sqrt{x} + 4$
$8\sqrt{x} = 3x - 3$
$\left(8\sqrt{x}\right)^2 = \left(3x-3\right)^2$
$64x = 9x^2 - 18x + 9$
$0 = 9x^2 - 82x + 9$
$0 = (9x-1)(x-9)$
$x = \frac{1}{9}$ or $x = 9$
Since $\sqrt{\frac{1}{9}+7} - 2\sqrt{\frac{1}{9}} = \frac{8}{3} - \frac{2}{3} = 2$, 1/9 is an extraneous root. The solution set is $\{9\}$.

95. $2\sqrt{x} - \sqrt{x-3} = 3$
$2\sqrt{x} = \sqrt{x-3} + 3$
$\left(2\sqrt{x}\right)^2 = \left(\sqrt{x-3}+3\right)^2$
$4x = x - 3 + 6\sqrt{x-3} + 9$
$3x - 6 = 6\sqrt{x-3}$
$x - 2 = 2\sqrt{x-3}$
$(x-2)^2 = (2\sqrt{x-3})^2$
$x^2 - 4x + 4 = 4(x-3)$
$x^2 - 8x + 16 = 0$
$(x-4)^2 = 0$
$x = 4$
Solution set: $\{4\}$

97. $(2-3i)(-5+5i) = -10 + 25i - 15i^2$
$= -10 + 25i + 15 = 5 + 25i$

99. $(2+i) + (5-4i) = 7 - 3i$

101. $(1-i) - (2-3i) = 1 - i - 2 + 3i$
$= -1 + 2i$

103. $\frac{6+3i}{3} = \frac{6}{3} + \frac{3i}{3} = 2 + i$

105. $\frac{4-\sqrt{-12}}{2} = \frac{4-2i\sqrt{3}}{2} = 2 - i\sqrt{3}$

107. $\frac{2-3i}{4+i} = \frac{(2-3i)(4-i)}{(4+i)(4-i)}$
$= \frac{5-14i}{17} = \frac{5}{17} - \frac{14}{17}i$

109. $(-2i)^4 = (-2)^4 i^4 = 16 \cdot 1 = 16$

111. $i^{14} = (i^4)^3 i^2 = 1^3 \cdot (-1) = -1$

113. $x^2 + 100 = 0$
$x^2 = -100$
$x = \pm\sqrt{-100} = \pm 10i$
The solution set is $\{\pm 10i\}$.

115. $2b^2 + 9 = 0$
$2b^2 = -9$
$b^2 = -\frac{9}{2}$
$b = \sqrt{-\frac{9}{2}} = \pm\frac{3i}{\sqrt{2}} = \pm\frac{3i\sqrt{2}}{2}$
The solution set is $\left\{\pm\frac{3i\sqrt{2}}{2}\right\}$.

117. False, because $2^3 \cdot 3^2 = 8 \cdot 9 = 72$.

119. True, because $(\sqrt{2})^3 = \sqrt{8} = 2\sqrt{2}$.

121. True, because
$8^{200}8^{200} = (8 \cdot 8)^{200} = 64^{200}$.

123. False, because $4^{1/2} = \sqrt{4} = 2$.

125. False, because $5^2 \cdot 5^2 = 5^4 = 625$.

127. False, because $\sqrt{w^{10}} = |w^5|$.

129. False, because $\sqrt{x^6} = |x^3|$.

131. True, $\sqrt{x^8} = x^{8/2} = x^4$.

133. False, $\sqrt{16} = 4$.

135. True, $2^{600} = (2^2)^{300} = 4^{300}$.

137. False, $\frac{2+\sqrt{6}}{2} = 1 + \frac{\sqrt{6}}{2}$.

139. False, $\sqrt{\frac{4}{6}} = \sqrt{\frac{2}{3}} = \frac{\sqrt{2}\sqrt{3}}{\sqrt{3}\sqrt{3}} = \frac{\sqrt{6}}{3}$.

141. True, $81^{2/4} = 3^2 = 9 = \sqrt{81}$.

143. True, because $(a^4b^2)^{1/2}$ is nonnegative and a^2b could be negative, absolute value symbols are necessary.

145. To find the time for which $s = 12{,}000$, solve the equation
$16t^2 = 12{,}000$
$t^2 = 750$
$t = \sqrt{750} = 5\sqrt{30}$
The time is $5\sqrt{30}$ seconds or approximately 27.4 seconds.

147. The guy wire of length 40 is the hypotenuse of a right triangle where one leg is length 30 and the other is length x. By the Pythagorean theorem we can write
$x^2 + 30^2 = 40^2$
$x^2 = 700$
$x = \sqrt{700} = 10\sqrt{7}$
The wire is attached to the ground $10\sqrt{7}$ feet from the base of the antenna.

149. Let $x =$ the length of the guy wire. The height of the antenna is 200 feet and the distance from the base of the antenna to the point on the ground where the guy wire is attached is 200 feet. The guy wire is the hypotenuse of a right triangle whose legs each have length 200. By the Pythagorean theorem we can write
$x^2 = 200^2 + 200^2$
$x^2 = 80{,}000$
$x = \sqrt{80{,}000} = \sqrt{40{,}000}\sqrt{2} = 200\sqrt{2}$
The length of the guy wires should be $200\sqrt{2}$ feet.

151. If the volume is 40 ft^3, then each side is $\sqrt[3]{40}$ ft in length. The surface area of the six square sides is $6(\sqrt[3]{40})^2$ ft^2. Multiply by 1.1 to get $1.1 \cdot 6(\sqrt[3]{40})^2$ ft^2 as the amount of cardboard needed to make the box. Simplify:
$$6.6\left(2\sqrt[3]{5}\right)^2 = 26.4\sqrt[3]{25}\ \text{ft}^2$$
This is the exact answer. You can find an approximate answer with a calculator.

153. a) $1821.5 = 993.3(1+r)^{11}$
$(1+r)^{11} = \frac{1821.5}{993.3}$
$1 + r = \left(\frac{1821.5}{993.3}\right)^{1/11}$
$r = \left(\frac{1821.5}{993.3}\right)^{1/11} - 1$
$r \approx 0.057 = 5.7\%$

b) From the graph it appears that the annual cost of health care in 2015 will about \$3000 billion or \$3 trillion.

155. $V = \sqrt{\frac{841L}{CS}} = \frac{\sqrt{841} \cdot \sqrt{L} \cdot \sqrt{CS}}{\sqrt{CS} \cdot \sqrt{CS}}$
$V = \frac{29\sqrt{LCS}}{CS}$

CHAPTER 7 TEST

1. $8^{2/3} = 2^2 = 4$

2. $4^{-3/2} = \frac{1}{2^3} = \frac{1}{8}$

3. $\sqrt{21} \div \sqrt{7} = \frac{\sqrt{21}}{\sqrt{7}} = \sqrt{\frac{21}{7}} = \sqrt{3}$

4. $2\sqrt{5}\cdot 3\sqrt{5} = 6\cdot 5 = 30$

5. $\sqrt{20}+\sqrt{5} = 2\sqrt{5}+\sqrt{5} = 3\sqrt{5}$

6. $\sqrt{5}+\dfrac{1\sqrt{5}}{\sqrt{5}\sqrt{5}} = \sqrt{5}+\dfrac{\sqrt{5}}{5}$

$= \dfrac{5\sqrt{5}}{5}+\dfrac{\sqrt{5}}{5} = \dfrac{6\sqrt{5}}{5}$

7. $2^{1/2}\cdot 2^{1/2} = 2^1 = 2$

8. $\sqrt{72} = \sqrt{36}\sqrt{2} = 6\sqrt{2}$

9. $\sqrt{\dfrac{5}{12}} = \dfrac{\sqrt{5}\sqrt{3}}{\sqrt{12}\sqrt{3}} = \dfrac{\sqrt{15}}{\sqrt{36}} = \dfrac{\sqrt{15}}{6}$

10. $\dfrac{6+\sqrt{18}}{6} = \dfrac{6+3\sqrt{2}}{6} = \dfrac{3(2+\sqrt{2})}{3\cdot 2}$

$= \dfrac{2+\sqrt{2}}{2}$

11. $(2\sqrt{3}+1)(\sqrt{3}-2)$

$= 6+\sqrt{3}-4\sqrt{3}-2 = 4-3\sqrt{3}$

12. $\sqrt[4]{32a^5y^8} = \sqrt[4]{16a^4y^8}\cdot\sqrt[4]{2a}$

$= 2ay^2\sqrt[4]{2a}$

13. $\dfrac{1}{\sqrt[3]{2x^2}} = \dfrac{1\cdot\sqrt[3]{4x}}{\sqrt[3]{2x^2}\cdot\sqrt[3]{4x}} = \dfrac{\sqrt[3]{4x}}{\sqrt[3]{8x^3}}$

$= \dfrac{\sqrt[3]{4x}}{2x}$

14. $\sqrt{\dfrac{8a^9}{b^3}} = \dfrac{2a^4\sqrt{2a}}{b\sqrt{b}} = \dfrac{2a^4\sqrt{2a}\sqrt{b}}{b\sqrt{b}\sqrt{b}}$

$= \dfrac{2a^4\sqrt{2ab}}{b^2}$

15. $\sqrt[3]{-27x^9} = -3x^{9/3} = -3x^3$

16. $\sqrt{20m^3} = \sqrt{4m^2}\sqrt{5m} = 2m\sqrt{5m}$

17. $x^{1/2}x^{1/4} = x^{1/2+1/4} = x^{3/4}$

18. $(9y^4x^{1/2})^{1/2} = 3y^2x^{1/4}$

19. $\sqrt[3]{40x^7} = \sqrt[3]{8x^6}\cdot\sqrt[3]{5x} = 2x^2\sqrt[3]{5x}$

20. $(4+\sqrt{3})^2 = 16+8\sqrt{3}+3 = 19+8\sqrt{3}$

21. In $f(x) = \sqrt{4-x}$ we must have $4-x\geq 0, -x\geq -4,$ or $x\leq 4$. So the domain of f is the interval $(-\infty, 4]$.

22. In $f(x) = \sqrt[3]{5x-3}$, $5x-3$ can be any real number. So any real number can be used for x. So the domain of f is the interval $(-\infty,\infty)$.

23. $\dfrac{2}{5-\sqrt{3}} = \dfrac{2(5+\sqrt{3})}{(5-\sqrt{3})(5+\sqrt{3})}$

$= \dfrac{2(5+\sqrt{3})}{25-3} = \dfrac{2(5+\sqrt{3})}{22} = \dfrac{5+\sqrt{3}}{11}$

24. $\dfrac{\sqrt{6}}{4\sqrt{3}+\sqrt{2}} = \dfrac{\sqrt{6}(4\sqrt{3}-\sqrt{2})}{(4\sqrt{3}+\sqrt{2})(4\sqrt{3}-\sqrt{2})}$

$= \dfrac{4\sqrt{18}-\sqrt{12}}{48-2} = \dfrac{12\sqrt{2}-2\sqrt{3}}{46}$

$= \dfrac{6\sqrt{2}-\sqrt{3}}{23}$

25. $(3-2i)(4+5i) = 12+7i-10i^2$

$= 22+7i$

26. $i^4-i^5 = 1-i^4\cdot i = 1-i$

27. $\dfrac{3-i}{1+2i} = \dfrac{(3-i)(1-2i)}{(1+2i)(1-2i)}$

$= \dfrac{3-7i+2i^2}{1-4i^2} = \dfrac{1-7i}{5} = \dfrac{1}{5}-\dfrac{7}{5}i$

28. $\dfrac{-6+\sqrt{-12}}{8} = \dfrac{-6+2i\sqrt{3}}{8}$

$= \dfrac{-3+i\sqrt{3}}{4} = -\dfrac{3}{4}+\dfrac{1}{4}i\sqrt{3}$

29. $(x-2)^2 = 49$

$x-2 = \pm 7$

$x = 2\pm 7$

$x = 2+7$ or $x = 2-7$

$x = 9$ or $x = -5$

Solution set: $\{-5, 9\}$

30. $2\sqrt{x+4} = 3$

$\left(2\sqrt{x+4}\right)^2 = (3)^2$

$4(x+4) = 9$

$4x = -7$

$x = -\dfrac{7}{4}$

Solution set: $\left\{-\dfrac{7}{4}\right\}$

31. $w^{2/3} = 4$

$(w^{2/3})^3 = 4^3$

$w^2 = 64$

$w = \pm 8$

Solution set: $\{-8, 8\}$

32. $9y^2+16 = 0$

$9y^2 = -16$

$y^2 = -\dfrac{16}{9}$

$y = \pm\sqrt{-\dfrac{16}{9}} = \pm\dfrac{4}{3}i$

The solution set is $\left\{\pm\dfrac{4}{3}i\right\}$.

33. $\sqrt{2x^2+x-12} = x$

$\left(\sqrt{2x^2+x-12}\right)^2 = x^2$

$2x^2+x-12 = x^2$

$x^2+x-12 = 0$

$(x+4)(x-3) = 0$

$x+4 = 0$ or $x-3 = 0$

$x = -4$ or $x = 3$

Since -4 does not check in the original equation, the solution set is $\{3\}$.

34. $\sqrt{x-1}+\sqrt{x+4} = 5$

$\sqrt{x-1} = 5-\sqrt{x+4}$

$$\left(\sqrt{x-1}\right)^2 = \left(5-\sqrt{x+4}\right)^2$$
$$x-1 = 25-10\sqrt{x+4}+x+4$$
$$10\sqrt{x+4} = 30$$
$$\left(\sqrt{x+4}\right)^2 = (3)^2$$
$$x+4 = 9$$
$$x = 5$$

Solution set: $\{5\}$

35. Let $x =$ the length of the side. Since the diagonal is the hypotenuse of a right triangle, we can write the equation

$$x^2+x^2 = 3^2$$
$$2x^2 = 9$$
$$x^2 = \frac{9}{2}$$
$$x = \sqrt{\frac{9}{2}} = \frac{3\sqrt{2}}{\sqrt{2}\sqrt{2}} = \frac{3\sqrt{2}}{2}$$

The length of each side is $\frac{3\sqrt{2}}{2}$ feet.

36. Let $x =$ one number and $x+11 =$ the other. Since their square roots differ by 1, we can write

$$\sqrt{x+11}-\sqrt{x} = 1$$
$$\sqrt{x+11} = \sqrt{x}+1$$
$$\left(\sqrt{x+11}\right)^2 = \left(\sqrt{x}+1\right)^2$$
$$x+11 = x+2\sqrt{x}+1$$
$$10 = 2\sqrt{x}$$
$$10^2 = (2\sqrt{x})^2$$
$$4x = 100$$
$$x = 25$$
$$x+11 = 36$$

The numbers are 25 and 36.

37. If the perimeter is 20, the sum of the length and width is 10. Let $x =$ the length and $10-x =$ the width. Use the Pythagorean theorem to write the equation

$$x^2+(10-x)^2 = \left(2\sqrt{13}\right)^2$$
$$x^2+100-20x+x^2 = 52$$
$$2x^2-20x+48 = 0$$
$$x^2-10x+24 = 0$$
$$(x-4)(x-6) = 0$$
$$x = 4 \quad \text{or} \quad x = 6$$
$$10-x = 6 \quad \text{or} \quad 10-x = 4$$

The length and width are 4 feet and 6 feet.

38. $R = (248.530)^{2/3} \approx 39.53$ AU

$$30.08 = T^{2/3}$$
$$T = 30.08^{3/2} = 164.97 \text{ years}$$

Making Connections

Chapters 1 - 7

1. $3+2\sqrt{14-2\cdot 5} = 3+2\sqrt{4}$
$= 3+2(2) = 7$

2. $4-3|5-2\cdot 4| = 4-3|-3|$
$= 4-3(3) = -5$

3. $5-2(6-2\cdot 4^2) = 5-2(6-32)$
$= 5-2(-26) = 57$

4. $\sqrt{13^2-12^2+6} = \sqrt{169-144}+6$
$= \sqrt{25}+6 = 5+6 = 11$

5. $\sqrt[3]{6^2-3^2}-2^5 = \sqrt[3]{27}-32$
$= 3-32 = -29$

6. $\sqrt[3]{4(7+3^2)}-2^3 = \sqrt[3]{64}-8$
$= 4-8 = -4$

7. $(4+3^2) \div |5-2\cdot 9|$
$= 13 \div |-13| = 13 \div 13 = 1$

8. $\sqrt{9+16}-|9-16| = 5-|-7|$
$= 5-7 = -2$

9. $\sqrt{(-30)^2-4\cdot 9\cdot 25} = \sqrt{900-900}$
$= \sqrt{0} = 0$

10. $\sqrt{(-23)^2-4\cdot 12\cdot 5} = \sqrt{529-240}$
$= \sqrt{289} = 17$

11. $3(x-2)+5 = 7-4(x+3)$

$$3x-6+5 = 7-4x-12$$
$$3x-1 = -4x-5$$
$$7x = -4$$
$$x = -\frac{4}{7}$$

Solution set: $\left\{-\frac{4}{7}\right\}$

12.

$$\sqrt{6x+7} = 4$$
$$\left(\sqrt{6x+7}\right)^2 = (4)^2$$
$$6x+7 = 16$$
$$6x = 9$$
$$x = \frac{3}{2}$$

Solution set: $\left\{\frac{3}{2}\right\}$

13.

$$|\,2x+5\,| > 1$$
$$2x+5 > 1 \quad \text{or} \quad 2x+5 < -1$$
$$2x > -4 \quad \text{or} \quad 2x < -6$$
$$x > -2 \quad \text{or} \quad x < -3$$

$(-\infty, -3) \cup (-2, \infty)$

−5 −4 −3 −2 −1 0

14.

$$8x^3-27 = 0$$
$$x^3 = \frac{27}{8}$$

$$x = \sqrt[3]{\frac{27}{8}} = \frac{3}{2}$$

Solution set: $\left\{\frac{3}{2}\right\}$

15. $2x - 3 > 3x - 4$

$1 > x$

$(-\infty, 1)$

16. $\sqrt{2x-3} - \sqrt{3x+4} = 0$

$\left(\sqrt{2x-3}\right)^2 = \left(\sqrt{3x+4}\right)^2$

$2x - 3 = 3x + 4$

$-7 = x$

Checking -7 gives us a square root of a negative number. So the solution set is $\emptyset$.

17. $\frac{w}{3} + \frac{w-4}{2} = \frac{11}{2}$

$6\left(\frac{w}{3} + \frac{w-4}{2}\right) = 6\left(\frac{11}{2}\right)$

$2w + 3w - 12 = 33$

$5w = 45$

$w = 9$

Solution set: $\{9\}$

18. $2(x+7) - 4 = x - (10 - x)$

$2x + 14 - 4 = x - 10 + x$

$2x + 10 = 2x - 10$

$10 = -10$

Solution set: $\emptyset$

19. $(x+7)^2 = 25$

$x + 7 = \pm 5$

$x = -7 \pm 5$

Solution set: $\{-12, -2\}$

20. $a^{-1/2} = 4$

$\left(a^{-1/2}\right)^{-2} = (4)^{-2}$

$a = \frac{1}{16}$

Solution set: $\left\{\frac{1}{16}\right\}$

21. $x - 3 > 2$ or $x < 2x + 6$

$x > 5$ or $-x < 6$

$x > 5$ or $x > -6$

$(-6, \infty)$

22. $a^{-2/3} = 16$

$\left(a^{-2/3}\right)^{-3} = (16)^{-3}$

$a^2 = 2^{-12}$

$a = \pm\sqrt{2^{-12}} = \pm 2^{-6} = \pm\frac{1}{64}$

Solution set: $\left\{-\frac{1}{64}, \frac{1}{64}\right\}$

23. $3x^2 - 1 = 0$

$x^2 = \frac{1}{3}$

$x = \pm\sqrt{\frac{1}{3}} = \pm\frac{\sqrt{3}}{3}$

Solution set: $\left\{-\frac{\sqrt{3}}{3}, \frac{\sqrt{3}}{3}\right\}$

24. $5 - 2(x-2) = 3x - 5(x-2) - 1$

$5 - 2x + 4 = 3x - 5x + 10 - 1$

$-2x + 9 = -2x + 9$

Solution set is all real numbers R.

25. $|\,3x - 4\,| < 5$

$-5 < 3x - 4 < 5$

$-1 < 3x < 9$

$-\frac{1}{3} < x < 3$

$\left(-\frac{1}{3}, 3\right)$

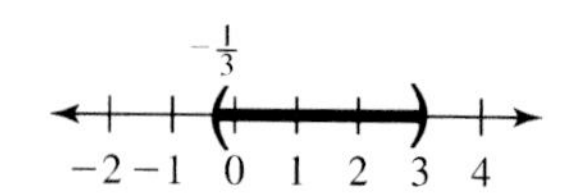

26. $3x - 1 = 0$

$x = \frac{1}{3}$

Solution set: $\left\{\frac{1}{3}\right\}$

27. $\sqrt{y-1} = 9$

$\left(\sqrt{y-1}\right)^2 = 9^2$

$y - 1 = 81$

$y = 82$

Solution set: $\{82\}$

28. $|\,5(x-2) + 1\,| = 3$

$|\,5x - 10 + 1\,| = 3$

$|\,5x - 9\,| = 3$

$5x - 9 = 3$ or $5x - 9 = -3$

$5x = 12 \qquad 5x = 6$

$x = \frac{12}{5}$ or $x = \frac{6}{5}$

Solution set: $\left\{\frac{6}{5}, \frac{12}{5}\right\}$

29. $0.06x - 0.04(x - 20) = 2.8$

$0.06x - 0.04x + 0.8 = 2.8$

$0.02x = 2.0$

$x = 100$

Solution set: $\{100\}$

30. $|\,3x - 1\,| > -2$

Since absolute value of any quantity is greater than or equal to zero, any real number satisfies this inequality. Solution set: R

31. $\frac{3\sqrt{2}}{x} = \frac{\sqrt{3}}{4\sqrt{5}}$

$\sqrt{3}x = 12\sqrt{10}$

$$x = \frac{12\sqrt{10}}{\sqrt{3}} = \frac{12\sqrt{10}\sqrt{3}}{\sqrt{3}\sqrt{3}} = 4\sqrt{30}$$

Solution set: $\left\{4\sqrt{30}\right\}$

32. $\frac{\sqrt{x}-4}{x} = \frac{1}{\sqrt{x}+5}$

$$x = (\sqrt{x}-4)(\sqrt{x}+5)$$
$$x = x + \sqrt{x} - 20$$
$$20 = \sqrt{x}$$
$$400 = x$$

Solution set: $\{400\}$

33. $\frac{3\sqrt{2}+4}{\sqrt{2}} = \frac{x\sqrt{18}}{3\sqrt{2}+2}$

$$6x = (3\sqrt{2}+4)(3\sqrt{2}+2)$$
$$6x = 18 + 18\sqrt{2} + 8$$
$$x = \frac{26+18\sqrt{2}}{6} = \frac{13+9\sqrt{2}}{3}$$

Solution set: $\left\{\frac{13+9\sqrt{2}}{3}\right\}$

34. $\frac{x}{2\sqrt{5}-\sqrt{2}} = \frac{2\sqrt{5}+\sqrt{2}}{x}$

$$x^2 = 20 - 2$$
$$x = \pm\sqrt{18} = \pm 3\sqrt{2}$$

Solution set: $\{-3\sqrt{2}, 3\sqrt{2}\}$

35. $\frac{\sqrt{2x}-5}{x} = \frac{-3}{\sqrt{2x}+5}$

$$-3x = 2x - 25$$
$$-5x = -25$$
$$x = 5$$

Solution set: $\{5\}$

36. $\frac{\sqrt{6}+2}{x} = \frac{2}{\sqrt{6}+4}$

$$2x = (\sqrt{6}+2)(\sqrt{6}+4)$$
$$2x = 14 + 6\sqrt{6}$$
$$x = 7 + 3\sqrt{6}$$

Solution set: $\{7 + 3\sqrt{6}\}$

37. $\frac{x-1}{\sqrt{6}} = \frac{\sqrt{6}}{x}$

$$x^2 - x = 6$$
$$x^2 - x - 6 = 0$$
$$(x-3)(x+2) = 0$$
$$x = 3 \text{ or } x = -2$$

Solution set: $\{-2, 3\}$

38. $\frac{x+3}{\sqrt{10}} = \frac{\sqrt{10}}{x}$

$$x^2 + 3x = 10$$
$$x^2 + 3x - 10 = 0$$
$$(x+5)(x-2) = 0$$
$$x = -5 \text{ or } x = 2$$

Solution set: $\{-5, 2\}$

39. $\frac{1}{x} - \frac{1}{x-1} = -\frac{1}{6}$

$$6x(x-1)\left(\frac{1}{x} - \frac{1}{x-1}\right) = 6x(x-1)\left(-\frac{1}{6}\right)$$
$$6x - 6 - 6x = -x^2 + x$$
$$x^2 - x - 6 = 0$$
$$(x-3)(x+2) = 0$$
$$x - 3 = 0 \text{ or } x + 2 = 0$$
$$x = 3 \text{ or } x = -2$$

Solution set: $\{-2, 3\}$

40. $\frac{1}{x^2-2x} + \frac{1}{x} = \frac{2}{3}$

$$3x(x-2)\left(\frac{1}{x^2-2x} + \frac{1}{x}\right) = 3x(x-2)\frac{2}{3}$$
$$3 + 3x - 6 = 2x^2 - 4x$$
$$-2x^2 + 7x - 3 = 0$$
$$2x^2 - 7x + 3 = 0$$
$$(2x-1)(x-3) = 0$$
$$2x - 1 = 0 \text{ or } x - 3 = 0$$
$$x = \frac{1}{2} \text{ or } x = 3$$

Solution set: $\left\{\frac{1}{2}, 3\right\}$

41. $\frac{-2+\sqrt{2^2-4(1)(-15)}}{2(1)}$

$$= \frac{-2+\sqrt{64}}{2} = 3$$

42. $\frac{-8+\sqrt{8^2-4(1)(12)}}{2(1)}$

$$= \frac{-8+\sqrt{16}}{2} = -2$$

43. $\frac{-5+\sqrt{5^2-4(2)(-3)}}{2(2)}$

$$= \frac{-5+\sqrt{49}}{4} = \frac{1}{2}$$

44. $\frac{-7+\sqrt{7^2-4(6)(-3)}}{2(6)}$

$$= \frac{-7+\sqrt{121}}{12} = \frac{1}{3}$$

45. a) $v = -94.8 + 21.4x - 0.761x^2$

$$v = -94.8 + 21.4(11) - 0.761(11)^2$$
$$\approx 48.5 \text{ cm}^3$$

b) From the graph it appears that a moisture content of 14% will produce the maximum volume of popped corn.

c) From the graph it appears that the maximum volume is about 56 cm^3.

8.1 WARM-UPS

1. False, completing the square involves perfect square trinomials.
2. False, it is equivalent to $x - 3 = \pm 2\sqrt{3}$.
3. False, because some quadratic polynomials cannot be factored.
4. False, because one-half of $4/3$ is $2/3$ and $2/3$ squared is $4/9$.
5. True.
6. False, we must first divide each side of the equation by 2.
7. False, $x = 3/2$ or $x = -5/3$.
8. True, one-half of 3 is $3/2$ and $3/2$ squared is $9/4$.
9. False, $x = \pm 2i\sqrt{2}$.
10. False, $(x-3)^2 = 0$ is a quadratic equation with only one solution.

8.1 EXERCISES

1. In this section, quadratic equations are solved by factoring, the even root property, and completing the square.
3. The last term is the square of one-half the coefficient of the middle term.

5.
$$x^2 - x - 6 = 0$$
$$(x-3)(x+2) = 0$$
$$x - 3 = 0 \text{ or } x + 2 = 0$$
$$x = 3 \text{ or } x = -2$$
Solution set: $\{-2, 3\}$

7.
$$a^2 + 2a = 15$$
$$a^2 + 2a - 15 = 0$$
$$(a+5)(a-3) = 0$$
$$a + 5 = 0 \text{ or } a - 3 = 0$$
$$a = -5 \text{ or } a = 3$$
Solution set: $\{-5, 3\}$

9.
$$2x^2 - x - 3 = 0$$
$$(2x-3)(x+1) = 0$$
$$2x - 3 = 0 \text{ or } x + 1 = 0$$
$$x = \frac{3}{2} \text{ or } x = -1$$
Solution set: $\left\{-1, \frac{3}{2}\right\}$

11.
$$y^2 + 14y + 49 = 0$$
$$(y+7)^2 = 0$$
$$y + 7 = 0$$
$$y = -7$$
Solution set: $\{-7\}$

13.
$$a^2 - 16 = 0$$
$$(a-4)(a+4) = 0$$
$$a - 4 = 0 \text{ or } a + 4 = 0$$
$$a = 4 \text{ or } a = -4$$
Solution set: $\{-4, 4\}$

15.
$$x^2 = 81$$
$$x = \pm\sqrt{81} = \pm 9$$
Solution set: $\{-9, 9\}$

17.
$$x^2 = \frac{16}{9}$$
$$x = \pm\sqrt{\frac{16}{9}} = \pm\frac{4}{3}$$
Solution set: $\left\{-\frac{4}{3}, \frac{4}{3}\right\}$

19.
$$(x-3)^2 = 16$$
$$x - 3 = \pm\sqrt{16}$$
$$x = 3 \pm 4$$
$$x = 3 + 4 \text{ or } x = 3 - 4$$
$$x = 7 \text{ or } x = -1$$
Solution set: $\{-1, 7\}$

21.
$$(z+1)^2 = 5$$
$$z + 1 = \pm\sqrt{5}$$
$$z = -1 \pm \sqrt{5}$$
Solution set: $\{-1 - \sqrt{5}, -1 + \sqrt{5}\}$

23.
$$\left(w - \frac{3}{2}\right)^2 = \frac{7}{4}$$
$$w - \frac{3}{2} = \pm\sqrt{\frac{7}{4}}$$
$$w = \frac{3}{2} \pm \frac{\sqrt{7}}{2} = \frac{3 \pm \sqrt{7}}{2}$$
Solution set: $\left\{\frac{3 - \sqrt{7}}{2}, \frac{3 + \sqrt{7}}{2}\right\}$

25. One-half of 2 is 1, and 1 squared is 1: $x^2 + 2x + 1$
27. One-half of -3 is $-3/2$, and $-3/2$ squared is $9/4$: $x^2 - 3x + \frac{9}{4}$
29. One-half of $1/4$ is $1/8$, and $1/8$ squared is $1/64$: $y^2 + \frac{1}{4}y + \frac{1}{64}$
31. One-half of $2/3$ is $1/3$, and $1/3$ squared is $1/9$: $x^2 + \frac{2}{3}x + \frac{1}{9}$
33. $x^2 + 8x + 16 = (x+4)^2$
35. $y^2 - 5y + \frac{25}{4} = \left(y - \frac{5}{2}\right)^2$
37. $z^2 - \frac{4}{7}z + \frac{4}{49} = \left(z - \frac{2}{7}\right)^2$
39. $t^2 + \frac{3}{5}t + \frac{9}{100} = \left(t + \frac{3}{10}\right)^2$

41.
$$x^2 - 2x - 15 = 0$$
$$x^2 - 2x = 15$$
$$x^2 - 2x + 1 = 15 + 1$$

$$(x-1)^2 = 16$$
$$x - 1 = \pm 4$$
$$x = 1 \pm 4$$
$$x = 5 \text{ or } x = -3$$

Solution set: $\{-3, 5\}$

43. $2x^2 - 4x = 70$
$$x^2 - 2x = 35$$
$$x^2 - 2x + 1 = 35 + 1$$
$$(x-1)^2 = 36$$
$$x - 1 = \pm 6$$
$$x = 1 \pm 6$$
$$x = 7 \text{ or } x = -5$$

Solution set: $\{-5, 7\}$

45. $w^2 - w - 20 = 0$
$$w^2 - w = 20$$
$$w^2 - w + \frac{1}{4} = 20 + \frac{1}{4}$$
$$\left(w - \frac{1}{2}\right)^2 = \frac{81}{4}$$
$$w - \frac{1}{2} = \pm \frac{9}{2}$$
$$w = \frac{1}{2} \pm \frac{9}{2}$$
$$x = \frac{10}{2} = 5 \quad \text{or} \quad x = \frac{-8}{2} = -4$$

Solution set: $\{-4, 5\}$

47. $q^2 + 5q = 14$
$$q^2 + 5q + \frac{25}{4} = 14 + \frac{25}{4}$$
$$\left(q + \frac{5}{2}\right)^2 = \frac{81}{4}$$
$$q + \frac{5}{2} = \pm \frac{9}{2}$$
$$q = -\frac{5}{2} \pm \frac{9}{2}$$
$$q = \frac{4}{2} = 2 \quad \text{or} \quad q = \frac{-14}{2} = -7$$

Solution set: $\{-7, 2\}$

49. $2h^2 - h - 3 = 0$
$$2h^2 - h = 3$$
$$h^2 - \frac{1}{2}h = \frac{3}{2}$$
$$h^2 - \frac{1}{2}h + \frac{1}{16} = \frac{3}{2} + \frac{1}{16}$$
$$\left(h - \frac{1}{4}\right)^2 = \frac{25}{16}$$
$$h - \frac{1}{4} = \pm \frac{5}{4}$$
$$h = \frac{1}{4} \pm \frac{5}{4}$$

Solution set: $\left\{-1, \frac{3}{2}\right\}$

51. $x^2 + 4x = 6$
$$x^2 + 4x + 4 = 6 + 4$$
$$(x+2)^2 = 10$$
$$x + 2 = \pm\sqrt{10}$$
$$x = -2 \pm \sqrt{10}$$

Solution set: $\{-2 - \sqrt{10}, -2 + \sqrt{10}\}$

53. $x^2 + 8x - 4 = 0$
$$x^2 + 8x = 4$$
$$x^2 + 8x + 16 = 4 + 16$$
$$(x+4)^2 = 20$$
$$x + 4 = \pm\sqrt{20}$$
$$x = -4 \pm 2\sqrt{5}\tfrac{1}{2}$$

Solution set: $\{-4 - 2\sqrt{5}, -4 + 2\sqrt{5}\}$

55. $x^2 + 5x + 5 = 0$
$$x^2 + 5x = -5$$
$$x^2 + 5x + \frac{25}{4} = -5 + \frac{25}{4}$$
$$x^2 + 5x + \frac{25}{4} = \frac{5}{4}$$
$$\left(x + \frac{5}{2}\right)^2 = \frac{5}{4}$$
$$x + \frac{5}{2} = \pm\sqrt{\frac{5}{4}}$$
$$x = -\frac{5}{2} \pm \frac{\sqrt{5}}{2}$$
$$x = \frac{-5 \pm \sqrt{5}}{2}$$

Solution set: $\left\{\frac{-5 \pm \sqrt{5}}{2}\right\}$

57. $4x^2 - 4x - 1 = 0$
$$x^2 - x - \frac{1}{4} = 0$$
$$x^2 - x = \frac{1}{4}$$
$$x^2 - x + \frac{1}{4} = \frac{1}{4} + \frac{1}{4}$$
$$\left(x - \frac{1}{2}\right)^2 = \frac{1}{2}$$
$$x - \frac{1}{2} = \pm\sqrt{\frac{1}{2}}$$
$$x = \frac{1}{2} \pm \frac{\sqrt{2}}{2}$$
$$x = \frac{1 \pm \sqrt{2}}{2}$$

Solution set: $\left\{\frac{1 \pm \sqrt{2}}{2}\right\}$

59. $2x^2 + 3x - 4 = 0$
$$2x^2 + 3x = 4$$
$$x^2 + \frac{3}{2}x = 2$$
$$x^2 + \frac{3}{2}x + \frac{9}{16} = 2 + \frac{9}{16}$$
$$\left(x + \frac{3}{4}\right)^2 = \frac{41}{16}$$
$$x + \frac{3}{4} = \pm\frac{\sqrt{41}}{4}$$
$$x = -\frac{3}{4} \pm \frac{\sqrt{41}}{4}$$

Solution set: $\left\{\frac{-3 - \sqrt{41}}{4}, \frac{-3 + \sqrt{41}}{4}\right\}$

61. $\sqrt{2x+1} = x - 1$

$(\sqrt{2x+1})^2 = (x-1)^2$

$2x + 1 = x^2 - 2x + 1$

$0 = x^2 - 4x$

$x(x-4) = 0$

$x = 0$ or $x = 4$

Since 0 is an extraneous root, the solution set is $\{4\}$.

63. $w = \dfrac{\sqrt{w+1}}{2}$

$2w = \sqrt{w+1}$

$(2w)^2 = (\sqrt{w+1})^2$

$4w^2 = w + 1$

$4w^2 - w = 1$

$w^2 - \frac{1}{4}w = \frac{1}{4}$

$w^2 - \frac{1}{4}w + \frac{1}{64} = \frac{1}{4} + \frac{1}{64}$

$\left(w - \frac{1}{8}\right)^2 = \frac{17}{64}$

$w - \frac{1}{8} = \pm\frac{\sqrt{17}}{8}$

$w = \frac{1}{8} \pm \frac{\sqrt{17}}{8}$

The number $\dfrac{1-\sqrt{17}}{8}$ is a negative number. No negative number can be a solution to the original equation because the left side would be negative and the right side is a principal square root. The solution set is $\left\{\dfrac{1+\sqrt{17}}{8}\right\}$.

65. $\dfrac{t}{t-2} = \dfrac{2t-3}{t}$

$t^2 = (t-2)(2t-3)$

$t^2 = 2t^2 - 7t + 6$

$-t^2 + 7t - 6 = 0$

$t^2 - 7t + 6 = 0$

$(t-6)(t-1) = 0$

$t = 6$ or $t = 1$

Solution set: $\{1, 6\}$

67. $\dfrac{2}{x^2} + \dfrac{4}{x} + 1 = 0$

$x^2\left(\dfrac{2}{x^2} + \dfrac{4}{x} + 1\right) = x^2(0)$

$2 + 4x + x^2 = 0$

$x^2 + 4x = -2$

$x^2 + 4x + 4 = -2 + 4$

$(x+2)^2 = 2$

$x + 2 = \pm\sqrt{2}$

$x = -2 \pm \sqrt{2}$

Solution set: $\{-2 - \sqrt{2}, -2 + \sqrt{2}\}$

69. $x^2 + 2x + 5 = 0$

$x^2 + 2x + 1 = -5 + 1$

$(x+1)^2 = -4$

$x + 1 = \pm\sqrt{-4}$

$x = -1 \pm 2i$

The solution set is $\{-1 - 2i, -1 + 2i\}$.

71. $x^2 - 6x + 11 = 0$

$x^2 - 6x + 9 = -11 + 9$

$(x-3)^2 = -2$

$x - 3 = \pm\sqrt{-2}$

$x = 3 \pm i\sqrt{2}$

The solution set is $\left\{3 - i\sqrt{2}, 3 + i\sqrt{2}\right\}$.

73. $x^2 = -\frac{1}{2}$

$x = \pm\sqrt{-\frac{1}{2}} = \pm i\frac{\sqrt{2}}{2}$

Solution set: $\left\{\pm i\frac{\sqrt{2}}{2}\right\}$

75. $x^2 + 12 = 0$

$x^2 = -12$

$x = \pm\sqrt{-12}$

$x = \pm 2i\sqrt{3}$

The solution set is $\{-2i\sqrt{3}, 2i\sqrt{3}\}$

77. $5z^2 - 4z + 1 = 0$

$z^2 - \frac{4}{5}z + \frac{4}{25} = -\frac{1}{5} + \frac{4}{25}$

$\left(z - \frac{2}{5}\right)^2 = \frac{-1}{25}$

$z - \frac{2}{5} = \pm\sqrt{\frac{-1}{25}}$

$z = \frac{2}{5} \pm \sqrt{\frac{-1}{25}}$

$z = \frac{2}{5} \pm \frac{i}{5} = \frac{2 \pm i}{5}$

The solution set is $\left\{\frac{2 \pm i}{5}\right\}$.

79. $x^2 = -121$

$x = \pm\sqrt{-121} = \pm 11i$

Solution set: $\{\pm 11i\}$

81. $4x^2 + 25 = 0$

$4x^2 = -25$

$x^2 = -\frac{25}{4}$

$x = \pm\sqrt{-\frac{25}{4}} = \pm\frac{5}{2}i$

Solution set: $\left\{-\frac{5}{2}i, \frac{5}{2}i\right\}$

83. $\left(p + \frac{1}{2}\right)^2 = \frac{9}{4}$

$p + \frac{1}{2} = \pm\frac{3}{2}$

$p = -\frac{1}{2} \pm \frac{3}{2}$

The solution set is $\{-2, 1\}$

85. $5t^2 + 4t - 3 = 0$

$$5t^2 + 4t = 3$$
$$t^2 + \frac{4}{5}t = \frac{3}{5}$$
$$t^2 + \frac{4}{5}t + \frac{4}{25} = \frac{3}{5} + \frac{4}{25}$$
$$\left(t + \frac{2}{5}\right)^2 = \frac{19}{25}$$
$$t + \frac{2}{5} = \pm\frac{\sqrt{19}}{5}$$
$$t = -\frac{2}{5} \pm \frac{\sqrt{19}}{5}$$

Solution set: $\left\{\frac{-2-\sqrt{19}}{5}, \frac{-2+\sqrt{19}}{5}\right\}$

87. $m^2 + 2m - 24 = 0$

$$m^2 + 2m = 24$$
$$m^2 + 2m + 1 = 24 + 1$$
$$(m+1)^2 = 25$$
$$m + 1 = \pm 5$$
$$m = -1 \pm 5$$
$$m = -1 - 5 = -6 \text{ or } m = -1 + 5 = 4$$

The solution set is $\{-6, 4\}$.

89. $\left(x - 2\right)^2 = -9$

$$x - 2 = \pm\sqrt{-9}$$
$$x = 2 \pm 3i$$

Solution set: $\{2 \pm 3i\}$

91. $-x^2 + x + 6 = 0$

$$-x^2 + x = -6$$
$$x^2 - x = 6$$
$$x^2 - x + \frac{1}{4} = 6 + \frac{1}{4}$$
$$\left(x - \frac{1}{2}\right)^2 = \frac{25}{4}$$
$$x - \frac{1}{2} = \pm\frac{5}{2}$$
$$x = \frac{1}{2} \pm \frac{5}{2}$$

Solution set: $\{-2, 3\}$

93. $x^2 - 6x + 10 = 0$

$$x^2 - 6x + 9 = -10 + 9$$
$$(x-3)^2 = -1$$
$$x - 3 = \pm\sqrt{-1} = \pm i$$
$$x = 3 \pm i$$

The solution set is $\{3 - i, 3 + i\}$.

95. $2x - 5 = \sqrt{7x + 7}$

$$(2x-5)^2 = (\sqrt{7x+7})^2$$
$$4x^2 - 20x + 25 = 7x + 7$$
$$4x^2 - 27x + 18 = 0$$
$$(4x - 3)(x - 6) = 0$$
$$4x - 3 = 0 \text{ or } x - 6 = 0$$
$$x = \frac{3}{4} \text{ or } x = 6$$

Since $3/4$ does not check, the solution set is $\{6\}$.

97. $\frac{1}{x} + \frac{1}{x-1} = \frac{1}{4}$

$$4x(x-1)\left(\frac{1}{x} + \frac{1}{x-1}\right) = 4x(x-1)\left(\frac{1}{4}\right)$$
$$4x - 4 + 4x = x^2 - x$$
$$-x^2 + 9x - 4 = 0$$
$$x^2 - 9x + 4 = 0$$
$$x^2 - 9x + \frac{81}{4} = -4 + \frac{81}{4}$$
$$\left(x - \frac{9}{2}\right)^2 = \frac{65}{4}$$
$$x - \frac{9}{2} = \pm\frac{\sqrt{65}}{2}$$
$$x = \frac{9}{2} \pm \frac{\sqrt{65}}{2}$$

Solution set: $\left\{\frac{9-\sqrt{65}}{2}, \frac{9+\sqrt{65}}{2}\right\}$

99. Since the graph of $y = x^2 + 2x - 15$ has x-intercepts at $(-5, 0)$ and $(3, 0)$, the solution set to the equation $x^2 + 2x - 15 = 0$ is $\{-5, 3\}$.

101. Since the graph of $y = x^2 + 4x + 15$ has no x-intercepts, the solution set to the equation $x^2 + 4x + 15 = 0$ is the empty set, $\emptyset$.

103. $1211.1 \cdot 8700 = 2.81A^2 \cdot 200$

$$A^2 = \frac{1211.1 \cdot 8700}{2.81 \cdot 200}$$
$$A = \sqrt{\frac{1211.1 \cdot 8700}{2.81 \cdot 200}} \approx 136.9 \text{ ft/sec}$$

105. $17{,}568 = 1500x - 3x^2$

$$3x^2 - 1500x = -17{,}568$$
$$x^2 - 500x = -5856$$
$$x^2 - 500x + 62500 = -5856 + 62500$$
$$(x - 250)^2 = 56644$$
$$x - 250 = \pm 238$$
$$x = 250 \pm 238$$
$$x = 12 \quad \text{or} \quad x = 488$$

Since x is less than 25, the answer is 12.

107. Equation (c) is not quadratic because there is no x^2 term.

111. $\{4.56, 2.74\}$

113. $\{3.53\}$

8.2 WARM-UPS

1. True.
2. False, before identifying a, b, and c, we must write the equation as $3x^2 - 4x + 7 = 0$.
3. True, this is just the quadratic formula with $a = d$, $b = e$, and $c = f$.
4. False, the quadratic formula works to solve any quadratic equation.
5. True, because $(-3)^2 - 4(2)(-4) = 9 + 32 = 41$.
6. True, if the discriminant is 0, then the quadratic equation has one real solution.
7. True.
8. True, because we can write the equation in the form $-1x^2 + 2x + 0 = 0$.
9. False, x and $6 - x$ have a sum of 6.
10. False, there can be one real solution, 2 real solutions, or 2 imaginary solutions.

8.2 EXERCISES

1. The quadratic formula can be used to solve any quadratic equation.
3. Factoring is used when the quadratic polynomial is simple enough to factor.
5. The discriminant is $b^2 - 4ac$.
7. $x^2 - 3x + 2 = 0$
$a = 1, b = -3, c = 2$
$$x = \frac{3 \pm \sqrt{(-3)^2 - 4(1)(2)}}{2(1)} = \frac{3 \pm \sqrt{1}}{2}$$
$$= \frac{3 \pm 1}{2} = \frac{4}{2}, \frac{2}{2}$$
Solution set: $\{1, 2\}$
9. $x^2 + 5x + 6 = 0$
$a = 1, b = 5, c = 6$
$$x = \frac{-5 \pm \sqrt{5^2 - 4(1)(6)}}{2(1)} = \frac{-5 \pm \sqrt{1}}{2}$$
$$= \frac{-5 \pm 1}{2} = \frac{-4}{2}, \frac{-6}{2}$$
Solution set: $\{-3, -2\}$
11. $y^2 + y - 6 = 0$
$a = 1, b = 1, c = -6$
$$y = \frac{-1 \pm \sqrt{1^2 - 4(1)(-6)}}{2(1)} = \frac{-1 \pm \sqrt{25}}{2}$$
$$= \frac{-1 \pm 5}{2} = \frac{4}{2}, \frac{-6}{2}$$
Solution set: $\{-3, 2\}$
13. $-6z^2 + 7z + 3 = 0$
$a = -6, b = 7, c = 3$
$$z = \frac{-7 \pm \sqrt{(7)^2 - 4(-6)(3)}}{2(-6)} = \frac{-7 \pm \sqrt{121}}{-12}$$
$$= \frac{-7 \pm 11}{-12} = \frac{-18}{-12}, \frac{4}{-12}$$
Solution set: $\left\{-\frac{1}{3}, \frac{3}{2}\right\}$
15. $4x^2 - 4x + 1 = 0$
$a = 4,\ b = -4,\ c = 1$
$$x = \frac{4 \pm \sqrt{16 - 4(4)(1)}}{2(4)} = \frac{4 \pm 0}{8} = \frac{1}{2}$$
Solution set: $\left\{\frac{1}{2}\right\}$
17. $-9x^2 + 6x - 1 = 0$
$a = -9, b = 6, c = -1$
$$x = \frac{-6 \pm \sqrt{36 - 4(-9)(-1)}}{2(-9)} = \frac{-6 \pm \sqrt{0}}{-18}$$
$$= \frac{1}{3}$$
Solution set: $\left\{\frac{1}{3}\right\}$
19. $16x^2 + 24x + 9 = 0$
$a = 16, b = 24, c = 9$
$$x = \frac{-24 \pm \sqrt{576 - 4(16)(9)}}{2(16)} = \frac{-24 \pm \sqrt{0}}{32}$$
$$= -\frac{3}{4}$$
Solution set: $\left\{-\frac{3}{4}\right\}$
21. $v^2 + 8v + 6 = 0$
$a = 1, b = 8, c = 6$
$$v = \frac{-8 \pm \sqrt{8^2 - 4(1)(6)}}{2(1)} = \frac{-8 \pm \sqrt{40}}{2}$$
$$= \frac{-8 \pm 2\sqrt{10}}{2} = -4 \pm \sqrt{10}$$
Solution set: $\{-4 \pm \sqrt{10}\}$
23. $x^2 + 5x - 1 = 0$: $a = 1, b = 5, c = -1$
$$x = \frac{-5 \pm \sqrt{5^2 - 4(1)(-1)}}{2(1)} = \frac{-5 \pm \sqrt{29}}{2}$$
Solution set: $\left\{\frac{-5 \pm \sqrt{29}}{2}\right\}$
25. $2t^2 - 6t + 1 = 0$
$a = 2, b = -6, c = 1$
$$t = \frac{6 \pm \sqrt{36 - 4(2)(1)}}{2(2)} = \frac{6 \pm \sqrt{28}}{4} = \frac{6 \pm 2\sqrt{7}}{4}$$
$$= \frac{2(3 \pm \sqrt{7})}{2(2)} = \frac{3 \pm \sqrt{7}}{2}$$
Solution set: $\left\{\frac{3 \pm \sqrt{7}}{2}\right\}$

27. $2t^2 - 6t + 5 = 0$

$$t = \frac{6 \pm \sqrt{36 - 4(2)(5)}}{2(2)} = \frac{6 \pm \sqrt{-4}}{4} = \frac{6 \pm 2i}{4}$$

$= \frac{3 \pm i}{2}$ Solution set: $\left\{\frac{3 \pm i}{2}\right\}$

29. $-2x^2 + 3x - 6 = 0$

$a = -2, b = 3, c = -6$

$$x = \frac{-3 \pm \sqrt{(3)^2 - 4(-2)(-6)}}{2(-2)} = \frac{-3 \pm \sqrt{-39}}{-4}$$

$$= \frac{-3 \pm i\sqrt{39}}{-4} = \frac{3 \pm i\sqrt{39}}{4}$$

Solution set: $\left\{\frac{3 \pm i\sqrt{39}}{4}\right\}$

31. $\frac{1}{2}x^2 + 13 = 5x, \quad x^2 - 10x + 26 = 0$

$$x = \frac{10 \pm \sqrt{100 - 4(1)(26)}}{2(1)} = \frac{10 \pm \sqrt{-4}}{2}$$

$= \frac{10 \pm 2i}{2} = 5 \pm i$ Solution set: $\{5 \pm i\}$

33. $x^2 - 6x + 2 = 0$

$a = 1, b = -6, c = 2$

$b^2 - 4ac = 36 - 4(1)(2) = 28$

So there are two real solutions to the equation.

35. $-2x^2 + 5x - 6 = 0$

$a = -2, b = 5, c = -6$

$b^2 - 4ac = 25 - 4(-2)(-6) = -23$

So there are no real solutions to the equation.

37. $4m^2 - 20m + 25 = 0$

$a = 4, b = -20, c = 25$

$b^2 - 4ac = 400 - 4(4)(25) = 0$

So there is one real solution to the equation.

39. $y^2 - \frac{1}{2}y + \frac{1}{4} = 0$

$a = 1, b = -\frac{1}{2}, c = \frac{1}{4}$

$b^2 - 4ac = \frac{1}{4} - 4(1)(\frac{1}{4}) = -\frac{3}{4}$

So there are no real solutions to the equation.

41. $-3t^2 + 5t + 6 = 0$

$a = -3, \ b = 5, c = 6$

$b^2 - 4ac = 25 - 4(-3)(6) = 97$

So there are two real solutions to the equation.

43. $16z^2 - 24z + 9 = 0$

$a = 16, b = -24, c = 9$

$b^2 - 4ac = (-24)^2 - 4(16)(9) = 0$

So there is one real solution to the equation.

45. $5x^2 - 7 = 0$

$a = 5, b = 0, c = -7$

$b^2 - 4ac = 0 - 4(5)(-7) = 140$

So there are two real solutions to the equation.

47. $x^2 - x = 0$

$a = 1, b = -1, c = 0$

$b^2 - 4ac = 1 - 4(1)(0) = 1$

So there are two real solutions to the equation.

49. $\frac{1}{4}y^2 + y = 1$

$4\left(\frac{1}{4}y^2 + y\right) = 4 \cdot 1$

$y^2 + 4y = 4$

$y^2 + 4y - 4 = 0$

$a = 1, b = 4, c = -4$

$$y = \frac{-4 \pm \sqrt{16 - 4(1)(-4)}}{2(1)} = \frac{-4 \pm \sqrt{32}}{2}$$

$$= \frac{-4 \pm 4\sqrt{2}}{2} = -2 \pm 2\sqrt{2}$$

Solution set: $\left\{-2 \pm 2\sqrt{2}\right\}$

51. $\frac{1}{3}x^2 + \frac{1}{2}x = \frac{1}{3}$

$6\left(\frac{1}{3}x^2 + \frac{1}{2}x\right) = 6\left(\frac{1}{3}\right)$

$2x^2 + 3x = 2$

$2x^2 + 3x - 2 = 0$

$a = 2, b = 3, c = -2$

$$x = \frac{-3 \pm \sqrt{9 - 4(2)(-2)}}{2(2)} = \frac{-3 \pm 5}{4}$$

$$= \frac{2}{4}, \frac{-8}{4}$$

Solution set: $\left\{-2, \frac{1}{2}\right\}$

53. $3y^2 + 2y - 4 = 0$

$a = 3, b = 2, c = -4$

$$y = \frac{-2 \pm \sqrt{4 - 4(3)(-4)}}{2(3)} = \frac{-2 \pm \sqrt{52}}{6}$$

$$= \frac{-2 \pm 2\sqrt{13}}{6} = \frac{-1 \pm \sqrt{13}}{3}$$

Solution set: $\left\{\frac{-1 \pm \sqrt{13}}{3}\right\}$

55. $\frac{w}{w-2} = \frac{w}{w-3}$

$w^2 - 3w = w^2 - 2w$

$0 = w$

Solution set: $\{0\}$

57. $\frac{9(3x-5)^2}{4} = 1$

$(3x - 5)^2 = \frac{4}{9}$

$3x - 5 = \pm\frac{2}{3}$

$3x = 5 + \frac{2}{3}$ or $3x = 5 - \frac{2}{3}$

$3x = \frac{17}{3}$ or $3x = \frac{13}{3}$

$x = \frac{17}{9}$ or $x = \frac{13}{9}$

Solution set: $\left\{\frac{13}{9}, \frac{17}{9}\right\}$

59. $25 - \frac{1}{3}x^2 = 0$
$$-\frac{1}{3}x^2 = -25$$
$$x^2 = 75$$
$$x = \pm\sqrt{75} = \pm 5\sqrt{3}$$
Solution set: $\left\{\pm 5\sqrt{3}\right\}$

61. $1 + \frac{20}{x^2} = \frac{8}{x}$
$$x^2\left(1 + \frac{20}{x^2}\right) = x^2 \cdot \frac{8}{x}$$
$$x^2 - 8x + 20 = 0$$
$$x^2 - 8x + 16 = -20 + 16$$
$$(x-4)^2 = -4$$
$$x - 4 = \pm 2i$$
$$x = 4 \pm 2i$$
Solution set: $\{4 \pm 2i\}$

63. $(x-8)(x+4) = -42$
$$x^2 - 4x + 10 = 0$$
$$x^2 - 4x + 4 = -10 + 4$$
$$(x-2)^2 = -6$$
$$x - 2 = \pm i\sqrt{6}$$
$$x = 2 \pm i\sqrt{6}$$
Solution set: $\{2 \pm i\sqrt{6}\}$

65. $y = \dfrac{3(2y+5)}{8(y-1)}$
$$8y^2 - 8y = 6y + 15$$
$$8y^2 - 14y - 15 = 0$$
$$(4y+3)(2y-5) = 0$$
$$4y + 3 = 0 \text{ or } 2y - 5 = 0$$
$$y = -\frac{3}{4} \text{ or } y = \frac{5}{2}$$
Solution set: $\left\{-\frac{3}{4}, \frac{5}{2}\right\}$

67. $x = \dfrac{-3.2 \pm \sqrt{(3.2)^2 - 4(1)(-5.7)}}{2(1)}$
$$= \frac{-3.2 \pm \sqrt{33.04}}{2} \quad \{-4.474, 1.274\}$$

69. $x = \dfrac{7.4 \pm \sqrt{(-7.4)^2 - 4(1)(13.69)}}{2(1)}$
$$= \frac{7.4 \pm \sqrt{0}}{2(1)} = 3.7 \quad \{3.7\}$$

71. $x = \dfrac{-6.72 \pm \sqrt{(6.72)^2 - 4(1.85)(3.6)}}{2(1.85)}$
$$= \frac{-6.72 \pm \sqrt{18.5184}}{3.7} \quad \{-2.979, -0.653\}$$

73. $x = \dfrac{-14379 \pm \sqrt{14379^2 - 4(3)(243)}}{2(3)}$
$$= \frac{-14379 \pm 14378.8986}{6}$$
Solution set: $\{-4792.983, -0.017\}$

75. $x = \dfrac{-0.00075 \pm \sqrt{0.00075^2 - 4 \cdot 1(-0.0062)}}{2(1)}$
$$= \frac{-0.00075 \pm 0.1574819}{2} \quad \{-0.079, 0.078\}$$

77. Let x = one number and $x + 1$ = the other. Since their product is 16, we can write
$$x(x+1) = 16$$
$$x^2 + x = 16$$
$$x^2 + x - 16 = 0$$
$$x = \frac{-1 \pm \sqrt{1^2 - 4(1)(-16)}}{2(1)} = \frac{-1 \pm \sqrt{65}}{2}$$
Since the numbers are positive,
$$x = \frac{-1 + \sqrt{65}}{2} \text{ and } x + 1 = \frac{-1 + \sqrt{65}}{2} + \frac{2}{2}$$
$= \dfrac{1 + \sqrt{65}}{2}$. The numbers are $\dfrac{1 + \sqrt{65}}{2}$ and $\dfrac{-1 + \sqrt{65}}{2}$ or approximately 4.5 and 3.5.

79. Let x = one of the numbers. If the numbers are to have a sum of 6, then $6 - x$ = the other number. Since their product is 4, we can write the equation
$$x(6 - x) = 4$$
$$-x^2 + 6x - 4 = 0$$
$$x^2 - 6x + 4 = 0$$
$$x = \frac{6 \pm \sqrt{36 - 4(1)(4)}}{2(1)} = \frac{6 \pm 2\sqrt{5}}{2}$$
$$= 3 \pm \sqrt{5}$$
If $x = 3 + \sqrt{5}$, then
$6 - x = 6 - (3 + \sqrt{5}) = 3 - \sqrt{5}$.
If $x = 3 - \sqrt{5}$, then
$6 - x = 6 - (3 - \sqrt{5}) = 3 + \sqrt{5}$.
So the numbers are $3 + \sqrt{5}$ and $3 - \sqrt{5}$ or approximately 5.2 and 0.8

81. Let x = the width and $x + 1$ = the length. Since the diagonal is the hypotenuse of a right triangle, we can write
$$x^2 + (x+1)^2 = (\sqrt{3})^2$$
$$x^2 + x^2 + 2x + 1 = 3$$
$$2x^2 + 2x - 2 = 0$$
$$x^2 + x - 1 = 0$$
$$x = \frac{-1 \pm \sqrt{1 - 4(1)(-1)}}{2(1)} = \frac{-1 \pm \sqrt{5}}{2}$$
Since the width of a rectangle is a positive number, we have width $= \dfrac{-1 + \sqrt{5}}{2}$ feet and length $= \dfrac{-1 + \sqrt{5}}{2} + \dfrac{2}{2} = \dfrac{1 + \sqrt{5}}{2}$ feet or width approximately 0.6 ft and length approximately 1.6 ft.

83. Let $x =$ the width, and $x + 4 =$ the length. Since the area is 10 square feet, we can write

$$x(x+4) = 10$$
$$x^2 + 4x - 10 = 0$$
$$x = \frac{-4 \pm \sqrt{16 - 4(1)(-10)}}{2(1)} = \frac{-4 \pm 2\sqrt{14}}{2}$$
$$= -2 \pm \sqrt{14}$$

Since $x > 0$, the width is $-2 + \sqrt{14}$ feet, and the length is $-2 + \sqrt{14} + 4 = 2 + \sqrt{14}$ feet or approximately 1.7 ft and 5.7 ft.

85. The time it takes to reach the ground is found by solving the equation

$$-16t^2 + 16t + 96 = 0$$
$$t^2 - t - 6 = 0$$
$$(t-3)(t+2) = 0$$
$$t = 3 \quad \text{or} \quad t = -2$$

The pine cone reaches the earth 3 seconds after it is tossed.

87. The time it takes to reach the ground is found by solving the equation

$$-16t^2 + 10t + 5 = 0$$
$$16t^2 - 10t - 5 = 0$$
$$t = \frac{-(-10) \pm \sqrt{(-10)^2 - 4(16)(-5)}}{2(16)}$$
$$= \frac{10 \pm \sqrt{420}}{32} = \frac{10 \pm 2\sqrt{105}}{32}$$
$$= \frac{5 \pm \sqrt{105}}{16}$$

Since $5 - \sqrt{105}$ is a negative number, the only possibility for t is $\frac{5 + \sqrt{105}}{16}$ or approximately 1.0 seconds.

89. The time it takes to reach the river is found by solving the equation

$$-16t^2 - 30t + 1000 = 0$$
$$8t^2 + 15t - 500 = 0$$
$$t = \frac{-15 \pm \sqrt{15^2 - 4(8)(-500)}}{2(8)}$$
$$= \frac{-15 \pm \sqrt{16225}}{16} = \frac{-15 \pm 127.377}{16}$$

Since the time is positive, the time is $\frac{-15 + 127.377}{16}$, or 7.0 seconds.

91. If x is the width of the border, then

$$(30 - 2x)(40 - 2x) = 704$$
$$1200 - 140x + 4x^2 = 704$$
$$4x^2 - 140x + 496 = 0$$
$$x^2 - 35x + 124 = 0$$
$$x = \frac{35 \pm \sqrt{35^2 - 4 \cdot 1 \cdot 124}}{2} = \frac{35 \pm \sqrt{729}}{2}$$
$$= \frac{35 \pm 27}{2} = 31 \text{ or } 4$$

The width of the border is 4 inches.

93. Let $x =$ the number presently sharing the cost. The cost is $\frac{100}{x}$ dollars each. If 6 more join, the cost is $\frac{100}{x} - 15$. We have

$$(x+6)(\frac{100}{x} - 15) = 100$$
$$100 + \frac{600}{x} - 15x - 90 = 100$$
$$\frac{600}{x} - 15x - 90 = 0$$
$$600 - 15x^2 - 90x = 0$$
$$x^2 + 6x - 40 = 0$$
$$(x-4)(x+10) = 0$$
$$x - 4 = 0 \quad \text{or } x + 10 = 0$$
$$x = 4 \quad \text{or} \quad x = -10$$

There are 4 workers in the original group.

95. Let $x =$ the original number of melons purchased. The cost is $\frac{750}{x}$ dollars each. He sold them for $\frac{750}{x} + 2$ dollars each. When he sold $x - 100$ of them he broke even. So

$$\left(\frac{750}{x} + 2\right)(x - 100) = 750$$
$$750 + \frac{75000}{x} + 2x - 200 = 750$$
$$2x - 200 + \frac{75000}{x} = 0$$
$$2x^2 - 200x + 75000 = 0$$
$$x^2 - 100x + 37500 = 0$$
$$(x - 250)(x + 150) = 0$$
$$x - 250 = 0 \quad \text{or } x + 150 = 0$$
$$x = 250 \quad \text{or} \quad x = -150$$

So he bought 250 melons originally.

97.

$$6x^2 + 5x - 4 = 0$$
$$(3x+4)(2x-1) = 0$$
$$x = -\frac{4}{3} \quad \text{or} \quad x = \frac{1}{2}$$
$$-\frac{4}{3} + \frac{1}{2} = -\frac{8}{6} + \frac{3}{6} = -\frac{5}{6}$$
$$-\frac{b}{a} = -\frac{5}{6}$$

The sum of the solutions is equal to $-\frac{b}{a}$.

99. The solutions to $6x^2 + 5x - 4 = 0$ are $-\frac{4}{3}$ and $\frac{1}{2}$. Their product is $-\frac{2}{3}$, which is the value of $\frac{c}{a}$

101. Since $y = x^2 - 6.33x + 3.7$ crosses the x-axis twice there are 2 real solutions.

103. Since $y = 4x^2 - 67.1x + 344$ does not cross the x-axis there are no real solutions.

105. Since the graph does not cross the x-axis there are no real solutions.

8.3 WARM-UPS

1. True, because if $w = x^2$ the equation becomes $w^2 - 5w + 6 = 0$.
2. False, the equation cannot be solved by substitution.
3. False, we can use factoring.
4. True, because $(x^{1/6})^2 = x^{2/6} = x^{1/3}$.
5. False, we should let $w = \sqrt{x}$.
6. False, because $(2^{1/2})^2 = 2^{2/2} = 2^1 = 2$.
7. False, his rate is $1/x$ of the fence per hour.
8. True, because $R = D/T$.
9. False, against a 5 mph current it will go 5 mph.
10. False, the dimensions of the bottom will be $11 - 2x$ by $14 - 2x$.

8.3 EXERCISES

1. If the coefficients are integers and the discriminant is a perfect square, then the quadratic polynomial can be factored.
3. If the solutions are a and b, then the quadratic equation $(x - a)(x - b) = 0$ has those solutions.
5. $(x - 3)(x + 7) = 0$
$x^2 + 4x - 21 = 0$
7. $(x - 4)(x - 1) = 0$
$x^2 - 5x + 4 = 0$
9. $(x - \sqrt{5})(x + \sqrt{5}) = 0$
$x^2 - 5 = 0$
11. $(x - 4i)(x + 4i) = 0$
$x^2 + 16 = 0$
13. $(x - i\sqrt{2})(x + i\sqrt{2}) = 0$
$x^2 + 2 = 0$
15. $(2x - 1)(3x - 1) = 0$
$6x^2 - 5x + 1 = 0$
17. For $x^2 + 9$, $b^2 - 4ac = (0)^2 - 4(1)9$ $= -36$. So the polynomial is prime.
19. $b^2 - 4ac = (-1)^2 - 4(2)4 = -31$
Polynomial is prime.
21. $b^2 - 4ac = 6^2 - 4(2)(-5) = 76$
Polynomial is prime.
23. $b^2 - 4ac = 19^2 - 4(6)(-36) = 1225$
Since $\sqrt{1225} = 35$, the polynomial is not prime.
$6x^2 + 19x - 36 = (3x - 4)(2x + 9)$
25. $b^2 - 4ac = 25 - 4(4)(-12) = 217$
Polynomial is prime.
27. $b^2 - 4ac = (-18)^2 - 4 \cdot 8(-45) = 1764$
Since $\sqrt{1764} = 42$, the polynomial is not prime.
$8x^2 - 18x - 45 = (4x - 15)(2x + 3)$
29. Let $w = x - 1$ in
$(x - 1)^2 - 2(x - 1) - 8 = 0$:
$$w^2 - 2w - 8 = 0$$
$$(w - 4)(w + 2) = 0$$
$$w - 4 = 0 \quad \text{or} \quad w + 2 = 0$$
$$w = 4 \quad \text{or} \quad w = -2$$
$$x - 1 = 4 \quad \text{or} \quad x - 1 = -2$$
$$x = 5 \quad \text{or} \quad x = -1$$
Solution set: $\{-1, 5\}$
31. Let $w = 2a - 1$ in
$(2a - 1)^2 + 2(2a - 1) - 8 = 0$:
$$w^2 + 2w - 8 = 0$$
$$(w + 4)(w - 2) = 0$$
$$w + 4 = 0 \quad \text{or} \quad w - 2 = 0$$
$$w = -4 \quad \text{or} \quad w = 2$$
$$2a - 1 = -4 \quad \text{or} \quad 2a - 1 = 2$$
$$2a = -3 \quad \text{or} \quad 2a = 3$$
$$a = -\frac{3}{2} \quad \text{or} \quad a = \frac{3}{2}$$
Solution set: $\left\{-\frac{3}{2}, \frac{3}{2}\right\}$
33. Let $y = w - 1$ in
$(w - 1)^2 + 5(w - 1) + 5 = 0$:
$$y^2 + 5y + 5 = 0$$
$$y = \frac{-5 \pm \sqrt{25 - 4(1)(5)}}{2(1)} = \frac{-5 \pm \sqrt{5}}{2}$$
$$w - 1 = \frac{-5 \pm \sqrt{5}}{2}$$
$$w = \frac{-5 \pm \sqrt{5}}{2} + 1 = \frac{-5 \pm \sqrt{5}}{2} + \frac{2}{2} = \frac{-3 \pm \sqrt{5}}{2}$$
Solution set: $\left\{\frac{-3 \pm \sqrt{5}}{2}\right\}$

35. Let $w = x^2$ and $w^2 = x^4$ in $x^4 - 13x^2 + 36 = 0$:

$$w^2 - 13w + 36 = 0$$
$$(w - 4)(w - 9) = 0$$
$$w - 4 = 0 \text{ or } w - 9 = 0$$
$$w = 4 \text{ or } w = 9$$
$$x^2 = 4 \text{ or } x^2 = 9$$
$$x = \pm 2 \text{ or } x = \pm 3$$

Solution set: $\{\pm 2, \pm 3\}$

37. Let $w = x^3$ and $w^2 = x^6$ in $x^6 - 28x^3 + 27 = 0$:

$$w^2 - 28w + 27 = 0$$
$$(w - 27)(w - 1) = 0$$
$$w - 27 = 0 \text{ or } w - 1 = 0$$
$$w = 27 \text{ or } w = 1$$
$$x^3 = 27 \text{ or } x^3 = 1$$
$$x = 3 \text{ or } x = 1$$

Solution set: $\{1, 3\}$

39. Let $w = x^2$ and $w^2 = x^4$ in $x^4 - 14x^2 + 45 = 0$:

$$w^2 - 14w + 45 = 0$$
$$(w - 5)(w - 9) = 0$$
$$w - 5 = 0 \text{ or } w - 9 = 0$$
$$w = 5 \text{ or } w = 9$$
$$x^2 = 5 \text{ or } x^2 = 9$$
$$x = \pm\sqrt{5} \text{ or } x = \pm 3$$

Solution set: $\{\pm\sqrt{5}, \pm 3\}$

41. Let $w = x^3$, and $w^2 = x^6$ in $x^6 + 7x^3 = 8$:

$$w^2 + 7w = 8$$
$$w^2 + 7w - 8 = 0$$
$$(w - 1)(w + 8) = 0$$
$$w - 1 = 0 \text{ or } w + 8 = 0$$
$$w = 1 \text{ or } w = -8$$
$$x^3 = 1 \text{ or } x^3 = -8$$
$$x = 1 \text{ or } x = -2$$

Solution set: $\{-2, 1\}$

43. Let $w = x^2 + 1$ in $(x^2 + 1)^2 - 11(x^2 + 1) = -10$:

$$w^2 - 11w + 10 = 0$$
$$(w - 10)(w - 1) = 0$$
$$w - 10 = 0 \text{ or } w - 1 = 0$$
$$w = 10 \text{ or } w = 1$$
$$x^2 + 1 = 10 \text{ or } x^2 + 1 = 1$$
$$x^2 = 9 \text{ or } x^2 = 0$$
$$x = \pm 3 \text{ or } x = 0$$

Solution set: $\{0, \pm 3\}$

45. Let $w = x^2 + 2x$ in $(x^2 + 2x)^2 - 7(x^2 + 2x) + 12 = 0$:

$$w^2 - 7w + 12 = 0$$
$$(w - 3)(w - 4) = 0$$
$$w - 3 = 0 \text{ or } w - 4 = 0$$
$$w = 3 \text{ or } w = 4$$
$$x^2 + 2x = 3 \text{ or } x^2 + 2x = 4$$
$$x^2 + 2x - 3 = 0 \text{ or } x^2 + 2x - 4 = 0$$
$$(x + 3)(x - 1) = 0 \text{ or}$$
$$x = \frac{-2 \pm \sqrt{4 - 4(1)(-4)}}{2}$$
$$x = -3 \text{ or } x = 1 \text{ or } x = \frac{-2 \pm 2\sqrt{5}}{2}$$

Solution set: $\{-1 \pm \sqrt{5}, -3, 1\}$

47. Let $w = y^2 + y$ in $(y^2 + y)^2 - 8(y^2 + y) + 12 = 0$:

$$w^2 - 8w + 12 = 0$$
$$(w - 6)(w - 2) = 0$$
$$w = 6 \text{ or } w = 2$$
$$y^2 + y = 6 \text{ or } y^2 + y = 2$$
$$y^2 + y - 6 = 0 \text{ or } y^2 + y - 2 = 0$$
$$(y + 3)(y - 2) = 0 \text{ or } (y + 2)(y - 1) = 0$$
$$y = -3 \text{ or } y = 2 \text{ or } y = -2 \text{ or } y = 1$$

Solution set: $\{-3, -2, 1, 2\}$

49. Let $w = x^{1/2}$ and $w^2 = x$ in $x - 3x^{1/2} + 2 = 0$:

$$w^2 - 3w + 2 = 0$$
$$(w - 2)(w - 1) = 0$$
$$w - 2 = 0 \text{ or } w - 1 = 0$$
$$w = 2 \text{ or } w = 1$$
$$x^{1/2} = 2 \text{ or } x^{1/2} = 1$$
$$(x^{1/2})^2 = 2^2 \text{ or } (x^{1/2})^2 = 1^2$$
$$x = 4 \text{ or } x = 1$$

Solution set: $\{1, 4\}$

51. Let $w = x^{1/3}$ and $w^2 = x^{2/3}$ in $x^{2/3} + 4x^{1/3} + 3 = 0$:

$$w^2 + 4w + 3 = 0$$
$$(w + 3)(w + 1) = 0$$
$$w + 3 = 0 \text{ or } w + 1 = 0$$
$$w = -3 \text{ or } w = -1$$
$$x^{1/3} = -3 \text{ or } x^{1/3} = -1$$
$$(x^{1/3})^3 = (-3)^3 \text{ or } (x^{1/3})^3 = (-1)^3$$
$$x = -27 \text{ or } x = -1$$

Solution set: $\{-27, -1\}$

53. Let $w = x^{1/4}$ and $w^2 = x^{1/2}$ in $x^{1/2} - 5x^{1/4} + 6 = 0$:

$$w^2 - 5w + 6 = 0$$
$$(w - 3)(w - 2) = 0$$
$$w - 3 = 0 \text{ or } w - 2 = 0$$
$$w = 3 \text{ or } w = 2$$

$x^{1/4} = 3$ or $x^{1/4} = 2$
$(x^{1/4})^4 = 3^4$ or $(x^{1/4})^4 = 2^4$
$x = 81$ or $x = 16$
Solution set: $\{16, 81\}$

55. Let $w = x^{1/2}$ and $w^2 = x$ in $2x - 5x^{1/2} - 3 = 0$.
$2w^2 - 5w - 3 = 0$
$(2w + 1)(w - 3) = 0$
$2w + 1 = 0$ or $w - 3 = 0$
$w = -\frac{1}{2}$ or $w = 3$
$x^{1/2} = -\frac{1}{2}$ or $x^{1/2} = 3$
$x = \frac{1}{4}$ or $x = 9$
The solution $1/4$ does not check in the original equation. So the solution set is $\{9\}$.

57. Let $t = x^{-1}$ in $x^{-2} + x^{-1} - 6 = 0$:
$t^2 + t - 6 = 0$
$(t + 3)(t - 2) = 0$
$t = -3$ or $t = 2$
$x^{-1} = -3$ or $x^{-1} = 2$
$x = -\frac{1}{3}$ or $x = \frac{1}{2}$
Solution set: $\left\{-\frac{1}{3}, \frac{1}{2}\right\}$

59. Let $w = x^{1/6}$ and $w^2 = x^{1/3}$ in $x^{1/6} - x^{1/3} + 2 = 0$:
$w - w^2 + 2 = 0$
$w^2 - w - 2 = 0$
$(w - 2)(w + 1) = 0$
$w = 2$ or $w = -1$
$x^{1/6} = 2$ or $x^{1/6} = -1$
$(x^{1/6})^6 = 2^6$ or $(x^{1/6})^6 = (-1)^6$
$x = 64$ or $x = 1$

The original equation is not satisfied for $x = 1$. So the solution set is $\{64\}$.

61. Let $w = \frac{1}{y - 1}$ in
$\left(\frac{1}{y-1}\right)^2 + \left(\frac{1}{y-1}\right) = 6$.
$w^2 + w = 6$
$w^2 + w - 6 = 0$
$(w + 3)(w - 2) = 0$
$w = -3$ or $w = 2$
$\frac{1}{y-1} = -3$ or $\frac{1}{y-1} = 2$
$-3y + 3 = 1$ or $2y - 2 = 1$
$-3y = -2$ or $2y = 3$
$y = \frac{2}{3}$ or $y = \frac{3}{2}$
Solution set: $\left\{\frac{2}{3}, \frac{3}{2}\right\}$

63. Let $w = \sqrt{2x^2 - 3}$ and $w^2 = 2x^2 - 3$ in $2x^2 - 3 - 6\sqrt{2x^2 - 3} + 8 = 0$:
$w^2 - 6w + 8 = 0$
$(w - 4)(w - 2) = 0$
$w = 4$ or $w = 2$
$\sqrt{2x^2 - 3} = 4$ or $\sqrt{2x^2 - 3} = 2$
Square each side:
$2x^2 - 3 = 16$ or $2x^2 - 3 = 4$
$2x^2 = 19$ or $2x^2 = 7$
$x^2 = \frac{19}{2}$ or $x^2 = \frac{7}{2}$
$x = \pm\sqrt{\frac{19}{2}}$ or $x = \pm\sqrt{\frac{7}{2}}$
$x = \pm\frac{\sqrt{38}}{2}$ or $x = \pm\frac{\sqrt{14}}{2}$
Solution set: $\left\{\pm\frac{\sqrt{14}}{2}, \pm\frac{\sqrt{38}}{2}\right\}$

65. Let $t = x^{-1}$ and $t^2 = x^{-2}$ in $x^{-2} - 2x^{-1} - 1 = 0$:
$t^2 - 2t - 1 = 0$

$$t = \frac{2 \pm \sqrt{4 - 4(1)(-1)}}{2} = \frac{2 \pm 2\sqrt{2}}{2}$$
$$= 1 \pm \sqrt{2}$$

$x^{-1} = 1 \pm \sqrt{2}$
$x = \frac{1}{1 + \sqrt{2}}$ or $x = \frac{1}{1 - \sqrt{2}}$
$x = \frac{1(1 - \sqrt{2})}{(1 + \sqrt{2})(1 - \sqrt{2})}$ or
$x = \frac{1(1 + \sqrt{2})}{(1 - \sqrt{2})(1 + \sqrt{2})}$

$x = \frac{1 - \sqrt{2}}{-1}$ or $x = \frac{1 + \sqrt{2}}{-1}$
$x = -1 + \sqrt{2}$ or $x = -1 - \sqrt{2}$
Solution set: $\{-1 + \sqrt{2}, -1 - \sqrt{2}\}$

67. $w^2 + 4 = 0$
$w^2 = -4$
$w = \pm\sqrt{-4} = \pm 2i$
The solution set is $\{\pm 2i\}$.

69. $a^4 + 6a^2 + 8 = 0$
$(a^2 + 2)(a^2 + 4) = 0$
$a^2 + 2 = 0$ or $a^2 + 4 = 0$
$a^2 = -2$ or $a^2 = -4$
$a = \pm i\sqrt{2}$ or $a = \pm 2i$
The solution set is $\left\{\pm i\sqrt{2}, \pm 2i\right\}$.

71. $m^4 - 16 = 0$

$$(m^2 - 4)(m^2 + 4) = 0$$
$$b^2 - 4 = 0 \quad \text{or} \quad m^2 + 4 = 0$$
$$b^2 = 4 \quad \text{or} \quad m^2 = -4$$
$$b = \pm 2 \quad \text{or} \quad m = \pm 2i$$

The solution set is $\{\pm 2i, \pm 2\}$.

73. $16b^4 - 1 = 0$

$$(4b^2 - 1)(4b^2 + 1) = 0$$
$$4b^2 - 1 = 0 \quad \text{or} \quad 4b^2 + 1 = 0$$
$$b^2 = \frac{1}{4} \quad \text{or} \quad b^2 = -\frac{1}{4}$$
$$b = \pm\frac{1}{2} \quad \text{or} \quad b = \pm\frac{1}{2}i$$

The solution set is $\left\{\pm\frac{1}{2}, \pm\frac{i}{2}\right\}$.

75. $x^3 + 1 = 0$

$$(x + 1)(x^2 - x + 1) = 0$$
$$x + 1 = 0 \quad \text{or} \quad x^2 - x + 1 = 0$$
$$x = -1 \quad \text{or} \quad x = \frac{1 \pm \sqrt{(-1)^2 - 4(1)(1)}}{2(1)}$$
$$x = -1 \quad \text{or} \quad x = \frac{1 \pm \sqrt{-3}}{2}$$

The solution set is $\left\{\frac{1 \pm i\sqrt{3}}{2}, -1\right\}$.

77. $x^3 + 8 = 0$

$$(x + 2)(x^2 - 2x + 4) = 0$$
$$x + 2 = 0 \quad \text{or} \quad x^2 - 2x + 4 = 0$$
$$x = -2 \quad \text{or} \quad x = \frac{2 \pm \sqrt{(-2)^2 - 4(1)(4)}}{2(1)}$$
$$x = -2 \quad \text{or} \quad x = \frac{2 \pm 2i\sqrt{3}}{2} = 1 \pm i\sqrt{3}$$

The solution set is $\left\{1 \pm i\sqrt{3}, -2\right\}$.

79. $a^{-2} - 2a^{-1} + 5 = 0$

Multiply each side by a^2:

$$1 - 2a + 5a^2 = 0$$
$$5a^2 - 2a + 1 = 0$$
$$a = \frac{2 \pm \sqrt{(-2)^2 - 4(5)(1)}}{2(5)}$$
$$= \frac{2 \pm 4i}{10} = \frac{1 \pm 2i}{5}$$

The solution set is $\left\{\frac{1 \pm 2i}{5}\right\}$.

81. $(2x - 1)^2 - 2(2x - 1) + 5 = 0$

Let $w = 2x - 1$:

$$w^2 - 2w + 5 = 0$$
$$w = \frac{2 \pm \sqrt{(-2)^2 - 4(1)(5)}}{2(1)}$$
$$= \frac{2 \pm 4i}{2} = 1 \pm 2i$$

Now use $w = 2x - 1$:

$$2x - 1 = 1 \pm 2i$$
$$2x = 2 \pm 2i$$
$$x = \frac{2 \pm 2i}{2} = 1 \pm i$$

The solution set is $\{1 \pm i\}$.

83. Let $x =$ Gary's travel time and $x + 1 =$ Harry's travel time. Since $R = D/T$, Gary's speed is $300/x$ and Harry's speed is $300/(x + 1)$. Since Gary travels 10 mph faster, we can write the following equation.

$$\frac{300}{x} = \frac{300}{x + 1} + 10$$
$$x(x + 1)\left(\frac{300}{x}\right) = x(x + 1)\left(\frac{300}{x + 1} + 10\right)$$
$$300x + 300 = 300x + 10x^2 + 10x$$
$$0 = 10x^2 + 10x - 300$$
$$0 = x^2 + x - 30$$
$$0 = (x + 6)(x - 5)$$
$$x + 6 = 0 \quad \text{or} \quad x - 5 = 0$$
$$x = -6 \quad \text{or} \quad x = 5$$

Gary travels 5 hours and they arrive at 2 P.M.

85. Let $x =$ her speed before lunch and $x - 4 =$ her speed after lunch. Since $T = D/R$, her time before lunch was $60/x$ and her time after lunch was $46/(x - 4)$. Since she put in one hour more after lunch, we can write the following equation.

$$\frac{46}{x - 4} - 1 = \frac{60}{x}$$
$$x(x - 4)\left(\frac{46}{x - 4} - 1\right) = x(x - 4)\left(\frac{60}{x}\right)$$
$$46x - x^2 + 4x = 60x - 240$$
$$-x^2 - 10x + 240 = 0$$
$$x^2 + 10x - 240 = 0$$

$$x = \frac{-10 \pm \sqrt{100 - 4(1)(-240)}}{2(1)}$$
$$= \frac{-10 \pm \sqrt{1060}}{2} = \frac{-10 \pm 2\sqrt{265}}{2}$$
$$= -5 \pm \sqrt{265}$$

Since $-5 - \sqrt{265}$ is a negative number, we disregard that answer. Her speed before lunch was $-5 + \sqrt{265} \approx 11.3$ mph, and her speed after lunch was 4 mph slower: $-9 + \sqrt{265} \approx 7.3$ mph.

87. Let $x =$ Andrew's time and $x + 3 =$ John's time. Andrew's rate is $1/x$ job/hr and John's rate is $1/(x + 3)$ job/hr. In 8 hours Andrew does $8/x$ job and John does $8/(x + 3)$ job.

$$\frac{8}{x} + \frac{8}{x + 3} = 1$$

$$x(x+3)\left(\frac{8}{x}+\frac{8}{x+3}\right)=x(x+3)1$$
$$8x+24+8x=x^2+3x$$
$$0=x^2-13x-24$$
$$x=\frac{13\pm\sqrt{169-4(1)(-24)}}{2(1)}=\frac{13\pm\sqrt{265}}{2}$$

Since $\frac{13-\sqrt{265}}{2}$ is a negative number, we disregard that solution. Andrew's time is $\frac{13+\sqrt{265}}{2}\approx 14.6$ hr and John's time is 3 hours more :

$3+\frac{13+\sqrt{265}}{2}=\frac{19+\sqrt{265}}{2}\approx 17.6$ hr

89. Let $x=$ the amount of increase. The new length and width will be $30+x$ and $20+x$. Since the new area is to be 1000, we can write the following equation.

$$(20+x)(30+x)=1000$$
$$x^2+50x-400=0$$
$$x=\frac{-50\pm\sqrt{2500-4(1)(-400)}}{2(1)}$$
$$=\frac{-50\pm\sqrt{4100}}{2}=\frac{-50\pm10\sqrt{41}}{2}$$
$$=-25\pm5\sqrt{41}$$

Disregard the negative solution.
Use $x=-25+5\sqrt{41}$ to get
$30+x=5+5\sqrt{41}\approx 37.02$ feet, and
$20+x=-5+5\sqrt{41}\approx 27.02$ feet.

91. Let $x=$ the number of hours for A to empty the pool and $x+2=$ the number of hours for B to empty the pool. A's rate is $\frac{1}{x}$ pool/hr and B's rate is $\frac{1}{x+2}$ pool/hr. A works for 9 hrs and does $\frac{9}{x}$ of the pool while B works for 6 hours and does $\frac{6}{x+2}$ of the pool. Since the pool is half full we have the following equation.

$$\frac{9}{x}+\frac{6}{x+2}=\frac{1}{2}$$
$$2x(x+2)\left(\frac{9}{x}+\frac{6}{x+2}\right)=2x(x+2)\frac{1}{2}$$
$$18x+36+12x=x^2+2x$$
$$-x^2+28x+36=0$$
$$x^2-28x-36=0$$
$$x=\frac{28\pm\sqrt{28^2-4(1)(-36)}}{2}=\frac{28\pm\sqrt{928}}{2}$$
$$=\frac{28\pm4\sqrt{58}}{2}=14\pm2\sqrt{58}=-1.2\text{ or }29.2$$

A would take 29.2 hours working alone.

93. $\frac{10}{W}=\frac{W}{10-W}$
$$100-10W=W^2$$
$$W^2+10W-100=0$$
$$W=\frac{-10\pm\sqrt{10^2-4(1)(-100)}}{2}$$
$$=\frac{-10\pm10\sqrt{5}}{2}$$
$$W=-5\pm5\sqrt{5}\approx-16.2\text{ or }6.2$$

So the width is $-5+5\sqrt{5}$ or 6.2 meters.

95. a) $P(x)=x^4+6x^2-27$
$P(3i)=(3i)^4+6(3i)^2-27=0$
$P(-3i)=(-3i)^4+6(-3i)^2-27=0$
$P(\sqrt{3})=(\sqrt{3})^4+6(\sqrt{3})^2-27=0$
$P(-\sqrt{3})=(-\sqrt{3})^4+6(-\sqrt{3})^2-27=0$

b) All four numbers are solutions to the equation $x^4+6x^2-27=0$. The solutions occur in conjugate pairs.

97. $\{1, 2\}$

99. $\{-4.25, -3.49, 0.49, 1.25\}$

8.4 WARM-UPS

1. True, because $-1=(-2)^2-5$ is correct.
2. False, the y-intercept is (0, 9).
3. True, because the solution to $x^2-5=0$ is $x=\pm\sqrt{5}$.
4. True, because $a>0$.
5. False, because in $y=x^2+4$, $a>0$.
6. True, because $x=\frac{-2}{2(1)}=-1$ and $(-1)^2+2(-1)=-1$.
7. True, because $x^2+1=0$ has no real solution.
8. True, because if $x=0$, then $y=c$.
9. True, because the parabola opens downward from the vertex (0, 9)
10. False, the minimum value of y occurs when $x=7/6$.

8.4 EXERCISES

1. A quadratic function is a function of the form $f(x) = ax^2 + bx + c$ with $a \neq 0$.

3. If $a > 0$ then the parabola opens upward. If $a < 0$ then the parabola opens downward.

5. The vertex is the highest point on a parabola that opens downward or the lowest point on a parabola the opens upward.

7. If $x = 4$, then $f(4) = 4^2 = 16$.
If $y = 9$, then $x^2 = 9$ or $x = \pm 3$.
So the ordered pairs are $(4, 16)$, $(-3, 9)$, and $(3, 9)$.

9. If $x = 3$, then $f(3) = 3^2 - 3 - 12 = -6$.
If $y = 0$, then $x^2 - x - 12 = 0$.

$$(x - 4)(x + 3) = 0$$
$$x - 4 = 0 \quad \text{or } x + 3 = 0$$
$$x = 4 \quad \text{or} \qquad x = -3$$

So the ordered pairs are $(3, -6)$, $(4, 0)$, and $(-3, 0)$.

11. If $t = 4$, then
$s = -16 \cdot 4^2 + 32(4) = -128$.
It $s = 0$, then $-16t^2 + 32t = 0$.

$$-16t(t - 2) = 0$$
$$-16t = 0 \quad \text{or} \quad t - 2 = 0$$
$$t = 0 \quad \text{or} \qquad t = 2$$

The ordered pairs are $(4, -128)$, $(0, 0)$ and $(2, 0)$.

13. Because $a = 1$ in $f(x) = x^2 + 5$ the graph opens upward.

15. Because $a = -3$ in $y = -3x^2 + 4x + 2$ the graph opens downward.

17. $f(x) = (-2x + 3)^2 = 4x^2 - 12x + 9$
Because $a = 4$ in $f(x) = 4x^2 - 12x + 9$ the graph opens upward.

19. The ordered pairs $(-2, 6)$, $(-1, 3)$, $(0, 2)$, $(1, 3)$, and $(2, 6)$ satisfy $f(x) = x^2 + 2$. The domain is $(-\infty, \infty)$ and the range is $[2, \infty)$.

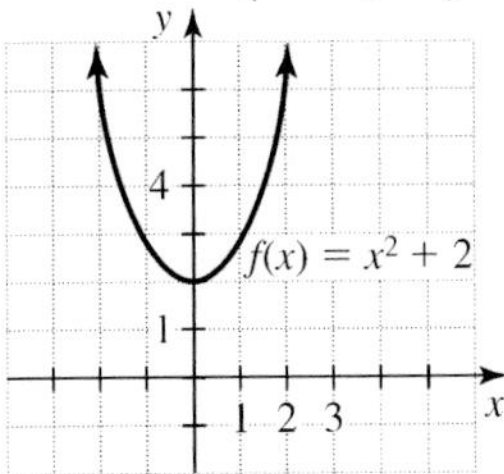

21. The ordered pairs $(-4, 4)$, $(-2, -2)$, $(0, -4)$, $(2, -2)$, and $(4, 4)$ satisfy $y = \frac{1}{2}x^2 - 4$. The domain is $(-\infty, \infty)$ and the range is $[-4, \infty)$.

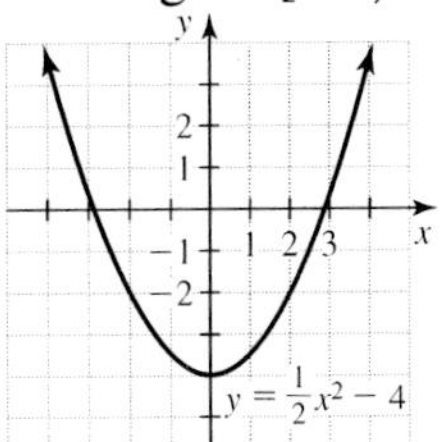

23. The ordered pairs $(-2, -3)$, $(-1, 3)$, $(0, 5)$, $(1, 3)$, and $(2, -3)$ satisfy $f(x) = -2x^2 + 5$. The domain is $(-\infty, \infty)$ and the range is $(-\infty, 5]$.

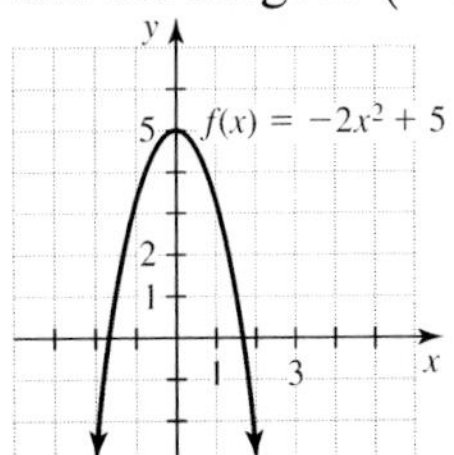

25. The ordered pairs $(-6, -7)$, $(-3, 2)$, $(0, 5)$, $(3, 2)$, and $(6, -7)$ satisfy $y = -\frac{1}{3}x^2 + 5$. The domain is $(-\infty, \infty)$ and the range is $(-\infty, 5]$.

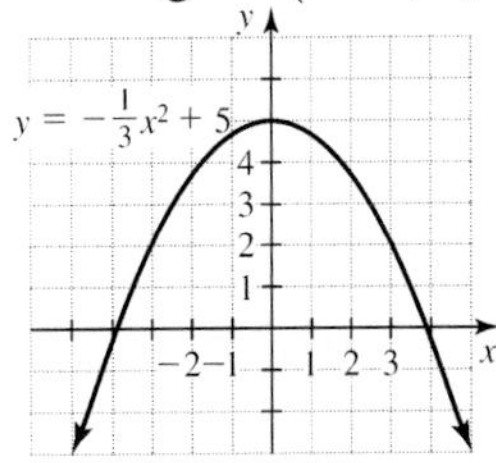

27. The ordered pairs $(0, 4)$, $(1, 1)$, $(2, 0)$, $(3, 1)$, and $(4, 4)$ satisfy $h(x) = (x - 2)^2$. The domain is $(-\infty, \infty)$ and the range is $[0, \infty)$.

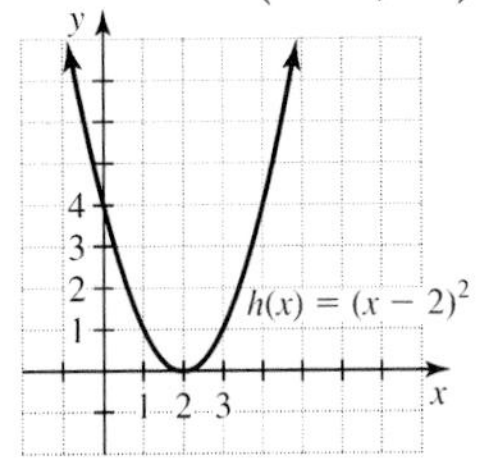

29. For $f(x) = x^2 - 9$ we have $a = 1$ and $b = 0$.

$$x = \frac{-b}{2a} = \frac{-0}{2(1)} = 0$$
$$y = 0^2 - 9 = -9$$

So the vertex is $(0, -9)$.

31. For $y = x^2 - 4x + 1$ we have $a = 1$ and $b = -4$.

$$x = \frac{-b}{2a} = \frac{4}{2(1)} = 2$$
$$y = 2^2 - 4(2) + 1 = -3$$

So the vertex is $(2, -3)$.

33. For $f(x) = -2x^2 + 20x + 1$ we have $a = -2$ and $b = 20$.

$$x = \frac{-b}{2a} = \frac{-20}{2(-2)} = 5$$
$$y = -2(5)^2 + 20(5) + 1 = 51$$

So the vertex is $(5, 51)$.

35. For $y = x^2 - x + 1$ we have $a = 1$ and $b = -1$. $x = \frac{-b}{2a} = \frac{1}{2(1)} = \frac{1}{2}$

$$y = \left(\frac{1}{2}\right)^2 - \left(\frac{1}{2}\right) + 1 = \frac{3}{4}$$

So the vertex is $\left(\frac{1}{2}, \frac{3}{4}\right)$.

37. To find the y-intercept, let $x = 0$:
$f(0) = 16 - 0^2 = 16$.
So the y-intercept is $(0, 16)$.
To find the x-intercepts let $y = 0$:

$$16 - x^2 = 0$$
$$x^2 = 16$$
$$x = \pm 4$$

So the x-intercepts are $(-4, 0)$ and $(4, 0)$.

39. To find the y-intercept, let $x = 0$:

$$y = 0^2 - 2(0) - 8 = -8.$$

So the y-intercept is $(0, -8)$.
To find the x-intercepts let $y = 0$:

$$x^2 - 2x - 8 = 0$$
$$(x - 4)(x + 2) = 0$$
$$x = 4 \quad \text{or} \quad x = -2$$

So the x-intercepts are $(-2, 0)$ and $(4, 0)$.

41. To find the y-intercept, let $x = 0$:

$$f(0) = -4(0)^2 + 12(0) - 9 = -9.$$

So the y-intercept is $(0, -9)$.
To find the x-intercepts let $y = 0$:

$$-4x^2 + 12x - 9 = 0$$
$$4x^2 - 12x + 9 = 0$$
$$(2x - 3)^2 = 0$$
$$2x - 3 = 0$$
$$x = 3/2$$

So the x-intercept is $(3/2, 0)$.

43. $x = \frac{-b}{2a} = \frac{1}{2(1)} = \frac{1}{2}$.

$$f\left(\frac{1}{2}\right) = \left(\frac{1}{2}\right)^2 - \frac{1}{2} - 2 = -\frac{9}{4}$$

The vertex is $\left(\frac{1}{2}, -\frac{9}{4}\right)$. The y-intercept is $(0, -2)$.

$$x^2 - x - 2 = 0$$
$$(x - 2)(x + 1) = 0$$
$$x - 2 = 0 \quad \text{or} \quad x + 1 = 0$$
$$x = 2 \quad \text{or} \quad x = -1$$

The x-intercepts are $(-1, 0)$ and $(2, 0)$.
The domain is $(-\infty, \infty)$ and the range is $\left[-\frac{9}{4}, \infty\right)$.

45. $x = \frac{-b}{2a} = \frac{-2}{2(1)} = -1$

$g(-1) = (-1)^2 + 2(-1) - 8 = -9$
The vertex is $(-1, -9)$. The y-intercept is $(0, -8)$.

$$x^2 + 2x - 8 = 0$$
$$(x - 2)(x + 4) = 0$$
$$x - 2 = 0 \quad \text{or} \quad x + 4 = 0$$
$$x = 2 \quad \text{or} \quad x = -4$$

The x-intercepts are $(-4, 0)$ and $(2, 0)$.
The domain is $(-\infty, \infty)$ and the range is $[-9, \infty)$.

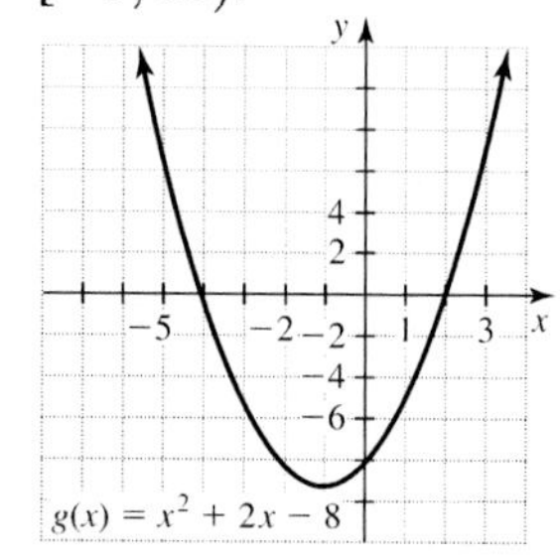

47. $y = -x^2 - 4x - 3$

$$x = \frac{-b}{2a} = \frac{4}{2(-1)} = -2$$
$$y = -(-2)^2 - 4(-2) - 3 = 1$$

Vertex $(-2, 1)$, y-intercept $(0, -3)$.

$$-x^2 - 4x - 3 = 0$$
$$x^2 + 4x + 3 = 0$$
$$(x + 1)(x + 3) = 0$$
$$x + 1 = 0 \quad \text{or} \quad x + 3 = 0$$
$$x = -1 \quad \text{or} \quad x = -3$$

The x-intercepts are $(-1, 0)$ and $(-3, 0)$.
The domain is $(-\infty, \infty)$ and the range is $(-\infty, 1]$.

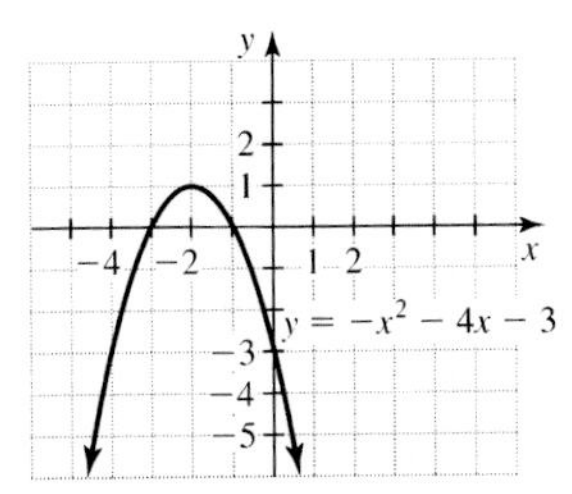

49. $h(x) = -x^2 + 3x + 4$

$$x = \frac{-b}{2a} = \frac{-3}{2(-1)} = \frac{3}{2}$$

$$h\left(\frac{3}{2}\right) = -(\tfrac{3}{2})^2 + 3(\tfrac{3}{2}) + 4 = \frac{25}{4}$$

Vertex $(\frac{3}{2}, \frac{25}{4})$, y-intercept (0, 4).

$$-x^2 + 3x + 4 = 0$$
$$x^2 - 3x - 4 = 0$$
$$(x-4)(x+1) = 0$$
$$x - 4 = 0 \text{ or } x + 1 = 0$$
$$x = 4 \text{ or } x = -1$$

The x-intercepts are (4, 0) and (−1, 0). The domain is $(-\infty, \infty)$ and the range is $\left(-\infty, \frac{25}{4}\right]$.

51. $a = b^2 - 6b - 16$

$$b = \frac{6}{2(1)} = 3$$

$a = 3^2 - 6(3) - 16 = -25$

Vertex (3, −25), b-intercept (0, −16).

$$b^2 - 6b - 16 = 0$$
$$(b-8)(b+2) = 0$$
$$b - 8 = 0 \text{ or } b + 2 = 0$$
$$b = 8 \text{ or } b = -2$$

The b-intercepts are (8, 0) and (−2, 0). The domain is $(-\infty, \infty)$ and the range is $[-25, \infty)$.

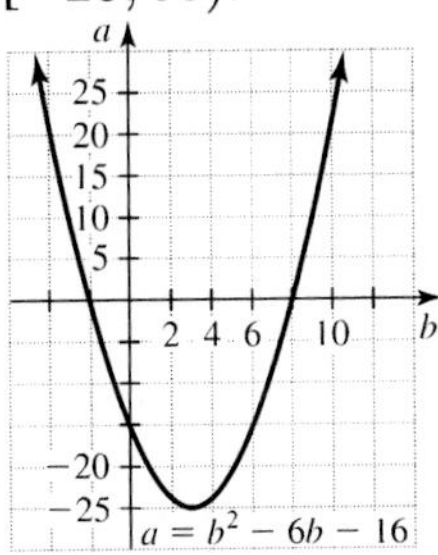

53. $y = x^2 - 8$

$$x = \frac{-b}{2a} = \frac{0}{2(1)} = 0$$

$y = 0^2 - 8 = -8$

The minimum value of y is −8.

55. $y = -3x^2 + 14$

$$x = \frac{-b}{2a} = \frac{0}{2(-3)} = 0$$

$y = -3 \cdot 0^2 + 14 = 14$

The maximum value of y is 14.

57. $y = x^2 + 2x + 3$

$$x = \frac{-b}{2a} = \frac{-2}{2(1)} = -1$$

$y = (-1)^2 + 2(-1) + 3 = 2$

The minimum value of y is 2.

59. $y = -2x^2 - 4x$

$$x = \frac{-b}{2a} = \frac{4}{2(-2)} = -1$$

$y = -2(-1)^2 - 4(-1) = 2$

The maximum value of y is 2.

61. $s(t) = -16t^2 + 64t$

$$t = \frac{-64}{2(-16)} = 2$$

$s(2) = -16 \cdot 2^2 + 64(2) = 64$

Maximum height is 64 feet.

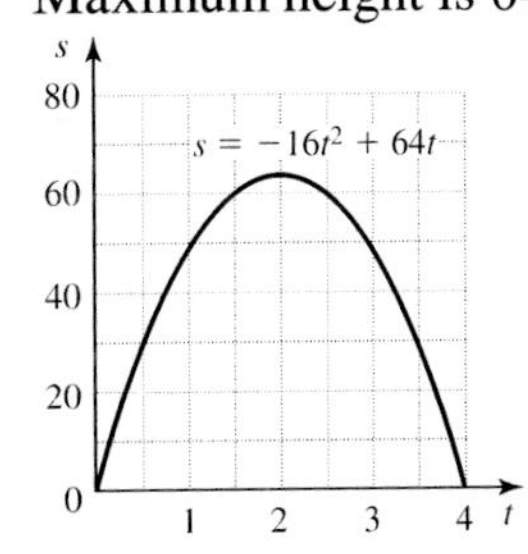

63. The minimum value of C in the formula $C = 0.009x^2 - 1.8x + 100$ is attained when

$$x = \frac{-(-1.8)}{2(0.009)} = 100.$$

The cost per hour will be at a minimum when they produce 100 balls per hour.

65. $A = -w^2 + 50w$

$$w = \frac{-50}{2(-1)} = 25$$

$A = -25^2 + 50(25) = 625$

Maximum area is 625 square meters.

67. $A(t) = -2t^2 + 32t + 12$

$$t = \frac{-32}{2(-2)} = 8$$

The nitrogen dioxide is at its maximum 8 hours after 6 A.M. or at 2 P.M.

69. Since the x-coordinate of the tower on the right is 20, we get $y = 0.0375(20)^2 = 15$. So the height of the towers is 15 meters. Use the Pythagorean theorem to find z:

$$15^2 + 20^2 = z^2$$
$$z^2 = 625$$
$$z = 25$$

So the length of the cable marked z is 25 meters.

71. The graph of $y = ax^2$ gets narrower as a gets larger.

73. The graph of $y = x^2$ has the same shape as $x = y^2$.

75. a)

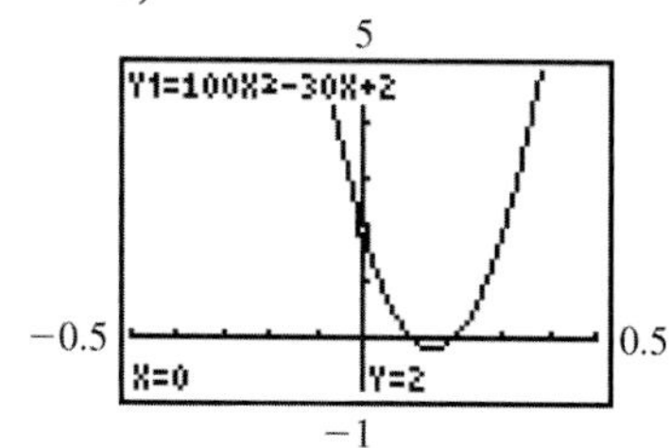

b)

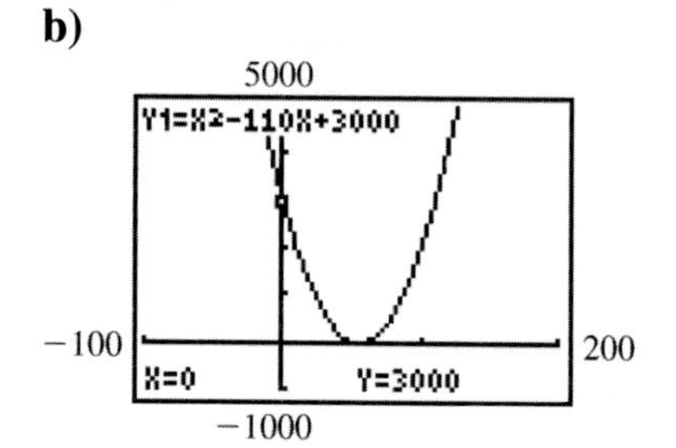

c)

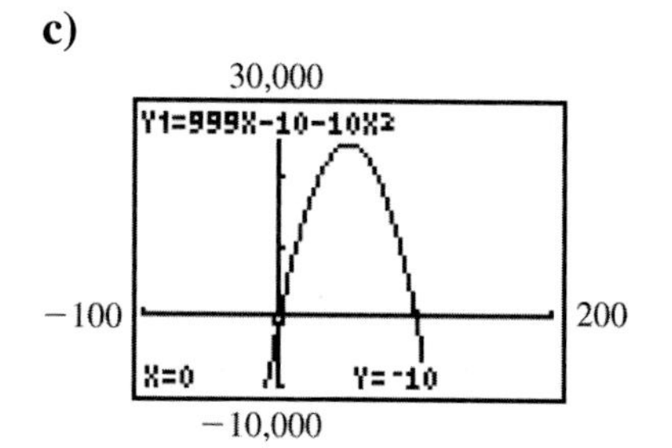

8.5 WARM-UPS

1. False, solution set is $(-\infty, -2) \cup (2, \infty)$.
2. False, we do not multiply each side by a variable.
3. False, that is not how we solve a quadratic inequality.
4. False, we can solve any quadratic inequality.
5. True, that is why we make the sign graph.
6. True, because inequalities change direction when multiplied by a negative number and we do not know if the variable is positive or negative.
7. True, because the solution is based on rules for multiplying or dividing signed numbers.
8. True, multiply each side by 2.
9. True, subtract 1 from each side.
10. False, because 4 causes the denominator to be 0 and so it cannot be in the solution set.

8.5 EXERCISES

1. A quadratic inequality has the form $ax^2 + bx + c > 0$. In place of $>$ we can also use $<$, $\le$, or $\ge$.

3. A rational inequality is an inequality involving a rational expression.

5.
$$x^2 + x - 6 < 0$$
$$(x + 3)(x - 2) < 0$$

$x + 3$ − − − − 0 + + + + + + +
$x - 2$ − − − − − − − − 0 + + + +
−3 2

The product is negative only if the factors have opposite signs. That happens when x is chosen between -3 and 2. Solution set: $(-3, 2)$

−4 −3 −2 −1 0 1 2 3

7.
$$x^2 - 3x - 4 \ge 0$$
$$(x - 4)(x + 1) \ge 0$$

$x - 4$ − − − − − − − 0 + + + + +
$x + 1$ − − 0 + + + + + + + + +
−1 4

The product is positive only if the factors have the same sign. That happens if x is chosen to the left of -1 or to the right of 4. The product is 0 if x is either -1 or 4.
$(-\infty, -1] \cup [4, \infty)$

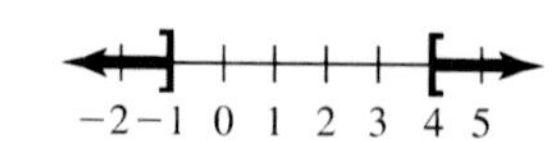

9.
$$x^2 - 2x - 8 \le 0$$
$$(x - 4)(x + 2) \le 0$$

$x - 4$ − − − − − − − − 0 + + +
$x + 2$ − − − 0 + + + + + + +
−2 4

The product is negative only if the factors have opposite signs. That happens if x is chosen between -2 and 4. The product is 0 if x is either -2 or 4.

$[-2, 4]$

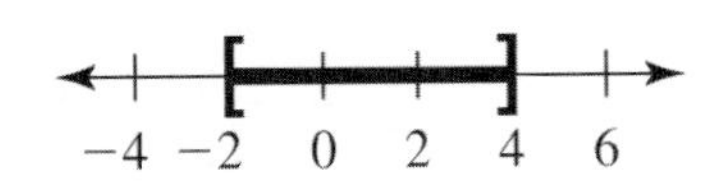

11.
$$2u^2 + 5u \geq 12$$
$$2u^2 + 5u - 12 \geq 0$$
$$(2u - 3)(u + 4) \geq 0$$

$2u - 3$ $- - - - - - - 0 + + +$

$u + 4$ $- - - 0 + + + + + + +$

-4 $\quad 3/2$

The product is positive only if the factors have the same sign. That happens if u is chosen to the left of -4 or to the right of $3/2$. The product is 0 if u is either -4 or $3/2$.

$(-\infty, -4] \cup \left[\frac{3}{2}, \infty\right)$

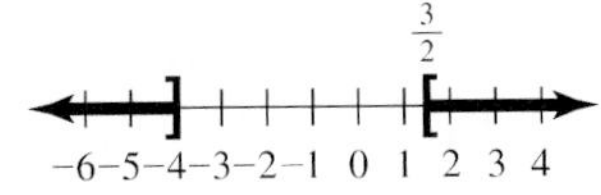

13.
$$4x^2 - 8x \geq 0$$
$$4x(x - 2) \geq 0$$

$4x$ $- - - 0 + + + + + + + +$

$x - 2$ $- - - - - - - 0 + + + +$

$0 \quad 2$

The product is greater than zero when the signs are the same.

$(-\infty, 0] \cup [2, \infty)$

15.
$$5x - 10x^2 < 0$$
$$5x(1 - 2x) < 0$$

$5x$ $- - 0 + + + + + + + +$

$1 - 2x$ $+ + + + + + 0 - - - -$

$0 \quad 1/2$

The product is negative only if the factors have opposite signs: $(-\infty, 0) \cup \left(\frac{1}{2}, \infty\right)$

17.
$$x^2 + 6x + 9 \geq 0$$
$$(x + 3)(x + 3) \geq 0$$

$x + 3$ $- - - - 0 + + + +$

$x + 3$ $- - - - 0 + + + +$

-3

The product is positive only if the factors have the same sign. That happens for every value of x except -3, in which case the product is 0. The solution set is the set of all real numbers, $(-\infty, \infty)$.

19.
$$x^2 + 4 < 4x$$
$$x^2 - 4x + 4 < 0$$
$$(x - 2)^2 < 0$$

$x - 2$ $- - - - - - - 0 + + + +$

$x - 2$ $- - - - - - - 0 + + + +$

2

The product is negative only if the factors have opposite signs. That never happens. So the solution set is the empty set $\emptyset$.

21.
$$4x^2 - 20x + 25 \leq 0$$
$$(2x - 5)^2 \leq 0$$

$2x - 5$ $- - - - - - - 0 + + + +$

$2x - 5$ $- - - - - - - 0 + + + +$

$5/2$

The product is negative only if the factors have opposite signs. That never happens. If $x = 5/2$ then the inequality is satisfied. So the solution set consists of a single number and it is $\left\{\frac{5}{2}\right\}$.

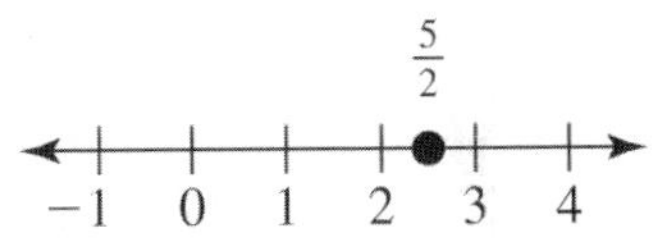

23.
$$25x^2 + 10x + 1 > 0$$
$$(5x + 1)^2 > 0$$

$5x + 1$ $- - - - - - - 0 + + + +$

$5x + 1$ $- - - - - - - 0 + + + +$

$-1/5$

The product is positive only if the factors have the same sign. That happens everywhere except $x = -1/5$. If $x = -1/5$ then the inequality is not satisfied. So the solution set consists of all real numbers except $-1/5$: $(-\infty, -1/5) \cup (-1/5, \infty)$

25. $\frac{1}{x} > 0$

1 + + + + + + + + + + + + +
x − − − − − 0 + + + + + + +
0

The quotient is positive only if the numerator and denominator have the same sign. That happens only when x is greater than 0. So the solution set is $(0, \infty)$.

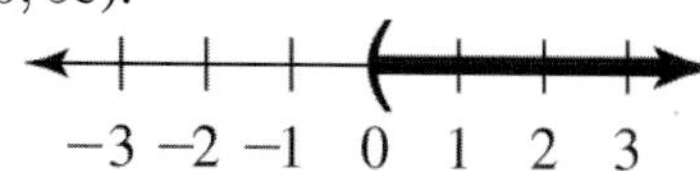

−3 −2 −1 0 1 2 3

27. $\frac{x}{x-3} > 0$

x − − − − 0 + + + + + + +
$x-3$ − − − − − − 0 + + + +
0 3

The quotient is positive only if the numerator and denominator have the same sign.
$(-\infty, 0) \cup (3, \infty)$

−2 −1 0 1 2 3 4 5

29. $\frac{x+2}{x} \leq 0$

$x+2$ − − − − 0 + + + + + + +
x − − − − − − − 0 + + + +
−2 0

The quotient is negative only if the numerator and denominator have opposite signs. Since the denominator must not be 0, we do not include 0 in the solution set: $[-2, 0)$

−4 −3 −2 −1 0 1 2

31. $\frac{t-3}{t+6} > 0$

$t+6$ − − − 0 + + + + + + +
$t-3$ − − − − − − − 0 + + +
−6 3

The quotient is positive only if the numerator and denominator have the same sign.
$(-\infty, -6) \cup (3, \infty)$

−8 −6 −4 −2 0 2 4

33. $\frac{x}{x+2} > -1$

$$\frac{x}{x+2} + 1 > 0$$

$$\frac{x}{x+2} + \frac{x+2}{x+2} > 0$$

$$\frac{2x+2}{x+2} > 0$$

$x+2$ − − − − 0 + + + + + + +
$2x+2$ − − − − − − 0 + + + +
−2 −1

The quotient is positive only if the numerator and denominator have the same sign.
$(-\infty, -2) \cup (-1, \infty)$

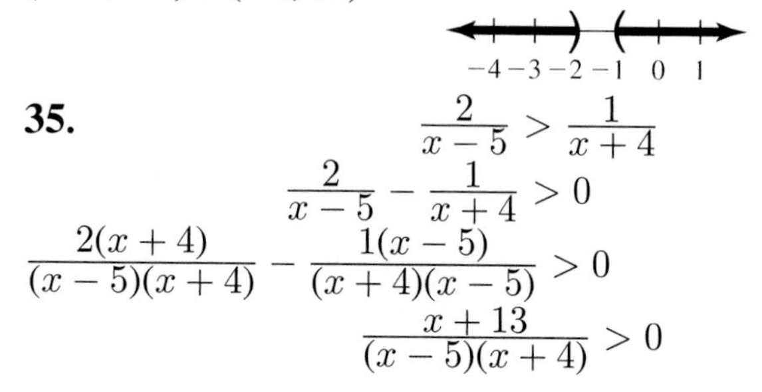

−4 −3 −2 −1 0 1

35. $\frac{2}{x-5} > \frac{1}{x+4}$

$$\frac{2}{x-5} - \frac{1}{x+4} > 0$$

$$\frac{2(x+4)}{(x-5)(x+4)} - \frac{1(x-5)}{(x+4)(x-5)} > 0$$

$$\frac{x+13}{(x-5)(x+4)} > 0$$

$x+4$ − − − − 0 + + + + + + +
$x+13$ − 0 + + + + + + + + +
$x-5$ − − − − − − − − − 0 + +
−13 −4 5

This quotient will be positive only if an even number of the factors have negative values.
$(-13, -4) \cup (5, \infty)$

−4
−13 −9 −5 −1 1 3 5 7

37. $\frac{m}{m-5} + \frac{3}{m-1} > 0$

$$\frac{m(m-1)}{(m-5)(m-1)} + \frac{3(m-5)}{(m-1)(m-5)} > 0$$

$$\frac{m^2+2m-15}{(m-1)(m-5)} > 0$$

$$\frac{(m+5)(m-3)}{(m-1)(m-5)} > 0$$

$m-3$ − − − − − − − − 0 + + + + +
$m-1$ − − − − 0 + + + + + + + +
$m+5$ − − 0 + + + + + + + + + +
$m-5$ − − − − − − − − − − − 0 + +
−5 1 3 5

This quotient is positive only if there is an even number of factors with negative signs.
$(-\infty, -5) \cup (1, 3) \cup (5, \infty)$

−7 −5 −3 −1 1 3 5 7

39. $\frac{x}{x-3} \leq \frac{-8}{x-6}$

$$\frac{x}{x-3} + \frac{8}{x-6} \leq 0$$

$$\frac{x(x-6)}{(x-3)(x-6)} + \frac{8(x-3)}{(x-6)(x-3)} \leq 0$$

$$\frac{x^2+2x-24}{(x-3)(x-6)} \leq 0$$
$$\frac{(x+6)(x-4)}{(x-3)(x-6)} \leq 0$$

$x-4$ − − − − − − − − 0 + + + + +
$x-3$ − − − − 0 + + + + + + + +
$x+6$ − − 0 + + + + + + + + + +
$x-6$ − − − − − − − − − − − − 0 + +
−6 3 4 6

The quotient is negative only if an odd number of factors have negative signs. Note that 3 and 6 cause the denominator to have a value of 0, and so they are excluded from the solution set.
$[-6, 3) \cup [4, 6)$

−6 −4 −2 0 2 4 6

41. First solve $x^2-5=0$:
$$x^2-5=0$$
$$x^2=5$$
$$x=\pm\sqrt{5}$$
The two solutions to the equation divide the number line into three regions. Choose one number in each region, say −10, 0, and 10. Now evaluate x^2-5 for the numbers −10, 0, and 10.
$$(-10)^2-5=95$$
$$0^2-5=-5$$
$$10^2-5=95$$
The signs of these answers indicate the solution to the inequality. The inequality is satisfied to the left of $-\sqrt{5}$ or to the right of $\sqrt{5}$. The solution set is
$(-\infty, -\sqrt{5}) \cup (\sqrt{5}, \infty)$.

$-\sqrt{5}$ $\sqrt{5}$
−3 −2 −1 0 1 2 3

43. To solve $x^2-2x-5 \leq 0$ we first solve $x^2-2x-5=0$:
$$x=\frac{2\pm\sqrt{4-4(1)(-5)}}{2(1)}=\frac{2\pm2\sqrt{6}}{2}=1\pm\sqrt{6}$$
The two solutions to the equation divide the number line into three regions. Choose one number in each region, say −10, 0, and 10. Now evaluate x^2-2x-5 for the numbers −10, 0, and 10.
$$(-10)^2-2(-10)-5=115$$
$$0^2-2(0)-5=-5$$
$$10^2-2(10)-5=75$$
The signs of these answers indicate the solution to the inequality.
$[1-\sqrt{6}, 1+\sqrt{6}]$

$1-\sqrt{6}$ $1+\sqrt{6}$

45. To solve $2x^2-6x+3 \geq 0$, we first solve $2x^2-6x+3=0$.
$$x=\frac{6\pm\sqrt{36-4(2)(3)}}{2(2)}=\frac{3\pm\sqrt{3}}{2}$$
The two solutions to the equation divide the number line into three regions. Choose a test point in each region, say −10, 2, and 10. Evaluate the polynomial $2x^2-6x+3$ at each of these points.
$$2(-10)^2-6(-10)+3=263$$
$$2(2)^2-6(2)+3=-1$$
$$2(10)^2-6(10)+3=143$$
The signs of these answers indicate the regions that satisfy the inequality.
$\left(-\infty, \frac{3-\sqrt{3}}{2}\right] \cup \left[\frac{3+\sqrt{3}}{2}, \infty\right)$

$\frac{3-\sqrt{3}}{2}$ $\frac{3+\sqrt{3}}{2}$

47. To solve $y^2-3y-9 \leq 0$ we first solve $y^2-3y-9=0$:
$$y=\frac{3\pm\sqrt{9-4(1)(-9)}}{2}=\frac{3\pm3\sqrt{5}}{2}$$
Pick test points −10, 4, and 10, and evaluate y^2-3y-9 for these numbers.
$$(-10)^2-3(-10)-9=121$$
$$4^2-3(4)-9=-5$$
$$10^2-3(10)-9=61$$
The signs of these answers indicate which regions satisfy the original inequality.
$$\left[\frac{3-3\sqrt{5}}{2}, \frac{3+3\sqrt{5}}{2}\right]$$

$\frac{3-3\sqrt{5}}{2}$ $\frac{3+3\sqrt{5}}{2}$

49. $x^2+5x+12 \geq 0$
First solve $x^2+5x+12=0$
$$x=\frac{-5\pm\sqrt{25-4(1)(12)}}{2(1)}=\frac{-5\pm\sqrt{-23}}{2}$$
Since b^2-4ac is negative, the equation has no real solutions. So $x^2+5x+12$ does not change sign. It is either always positive or always negative. To see which, test it with a real number, say 0.
$$0^2+5(0)+12=12>0$$

So $x^2 + 5x + 12$ is positive for every real number x. The solution set is the set of all real numbers, $(-\infty, \infty)$.

51. $2x^2 + 5x + 5 < 0$

First solve $2x^2 + 5x + 5 = 0$

$$x = \frac{-5 \pm \sqrt{25 - 4(2)(5)}}{2(1)} = \frac{-5 \pm \sqrt{-15}}{2}$$

Since $b^2 - 4ac$ is negative, the equation has no real solutions. So $2x^2 + 5x + 5$ does not change sign. It is either always positive or always negative. To see which test it with a real number, say 0.

$$2(0)^2 + 5(0) + 5 = 5 > 0$$

So $2x^2 + 5x + 5$ is positive for every real number x. So the inequality $2x^2 + 5x + 5 < 0$ has no solution. The solution set is the empty set, $\emptyset$.

53.

$$-5x^2 + 2x \le 4$$
$$-5x^2 + 2x - 4 \le 0$$

First solve $-5x^2 + 2x - 4 = 0$

$$x = \frac{-2 \pm \sqrt{4 - 4(-5)(-4)}}{2(-5)} = \frac{-2 \pm \sqrt{-76}}{-10}$$

Since $b^2 - 4ac$ is negative, the equation has no real solutions. So $-5x^2 + 2x - 4$ does not change sign. It is either always positive or always negative. To see which, test it with a real number, say 0.

$$-5(0)^2 + 2(0) - 4 = -4 < 0$$

So $-5x^2 + 2x - 4$ is negative for every real number x. The solution set is the set of all real numbers, $(-\infty, \infty)$.

55. Since x^2 is positive except when $x = 0$, all real numbers except zero satisfy $x^2 > 0$. The solution set is $(-\infty, 0) \cup (0, \infty)$.

57. Since x^2 is positive when x is nonzero and x^2 is zero, $x^2 + 4$ is actually greater than 0 for all real numbers. So all real numbers satisfy $x^2 + 4 \ge 0$. The solution set is $(-\infty, \infty)$.

59. Since 1 is positive, $1/x < 0$ is true if and only if x is negative. The solution set is $(-\infty, 0)$.

61.

$$x^2 \le 9$$
$$x^2 - 9 \le 0$$
$$(x - 3)(x + 3) \le 0$$

$$\begin{array}{ll} x + 3 & - - - - 0 + + + + + + + + \\ x - 3 & - - - - - - - - - - 0 + + + \\ \hline & \quad -3 \qquad\qquad 3 \end{array}$$

The product is negative only if the factors have opposite signs. That happens if x is chosen between -3 and 3. At ± 3 the product is 0. The solution set is $[-3, 3]$.

63.

$$16 - x^2 > 0$$
$$-1(16 - x^2) < -1(0)$$
$$x^2 - 16 < 0$$
$$(x - 4)(x + 4) < 0$$

$$\begin{array}{ll} x + 4 & - - - 0 + + + + + + + + + \\ x - 4 & - - - - - - - - - 0 + + + + \\ \hline & \quad -4 \qquad\qquad 4 \end{array}$$

The product is negative only if the value of x is between -4 and 4. The solution set is $(-4, 4)$.

65.

$$x^2 - 4x \ge 0$$
$$x(x - 4) \ge 0$$

$$\begin{array}{ll} x & - - - - 0 + + + + + + + + + \\ x - 4 & - - - - - - - - - 0 + + + + \\ \hline & \quad 0 \qquad\qquad 4 \end{array}$$

The product is positive only if the factors have the same sign. The solution set is $(-\infty, 0] \cup [4, \infty)$.

67.

$$3(2w^2 - 5) < w$$
$$6w^2 - 15 < w$$
$$6w^2 - w - 15 < 0$$
$$(2w + 3)(3w - 5) < 0$$

$$\begin{array}{ll} 2w + 3 & - - - 0 + + + + + + + + \\ 3w - 5 & - - - - - - - - - 0 + + + \\ \hline & \quad -3/2 \qquad\qquad 5/3 \end{array}$$

The product is negative only if the factors have opposite signs. The solution set is $\left(-\frac{3}{2}, \frac{5}{3}\right)$.

69.

$$z^2 \ge 4(z + 3)$$
$$z^2 \ge 4z + 12$$
$$z^2 - 4z - 12 \ge 0$$
$$(z - 6)(z + 2) \ge 0$$

$$\begin{array}{ll} z + 2 & - - - - 0 + + + + + + + + \\ z - 6 & - - - - - - - - - - 0 + + + + \\ \hline & \quad -2 \qquad\qquad 6 \end{array}$$

The product is positive only if the factors have the same sign. The solution set is $(-\infty, -2] \cup [6, \infty)$.

71.

$$(q + 4)^2 > 10q + 31$$
$$q^2 + 8q + 16 > 10q + 31$$

$$\begin{aligned} q^2 - 2q - 15 &> 0 \\ (q+3)(q-5) &> 0 \end{aligned}$$

$$\begin{array}{l} q+3 - - - - 0 + + + + + + + + \\ q-5 - - - - - - - - - 0 + + + + \\ \hline \qquad\qquad -3 \qquad\qquad 5 \end{array}$$

The product is positive only if the factors have the same sign. The solution set is $(-\infty, -3) \cup (5, \infty)$.

73. $$\begin{aligned} \frac{1}{2}x^2 &\geq 4 - x \\ x^2 &\geq 8 - 2x \\ x^2 + 2x - 8 &\geq 0 \\ (x+4)(x-2) &\geq 0 \end{aligned}$$

$$\begin{array}{l} x+4 - - - 0 + + + + + + + + + \\ x-2 - - - - - - - - - 0 + + + + \\ \hline \qquad\qquad -4 \qquad\qquad 2 \end{array}$$

The product is positive only if the factors have the same sign. The solution set is $(-\infty, -4] \cup [2, \infty)$.

75. $\dfrac{x-4}{x+3} \leq 0$

$$\begin{array}{l} x+3 - - - - 0 + + + + + + + + \\ x-4 - - - - - - - - - 0 + + + + \\ \hline \qquad\qquad -3 \qquad\qquad 4 \end{array}$$

The quotient is negative if the factors have opposite signs. The solution set is $(-3, 4]$.

77. $(x-2)(x+1)(x-5) \geq 0$

$$\begin{array}{l} x-2 - - - - 0 + + + + + + + + \\ x+1 - - 0 + + + + + + + + + + \\ x-5 - - - - - - - - - 0 + + + + \\ \hline \qquad -1 \qquad 2 \qquad\qquad 5 \end{array}$$

The product of these three factors is positive only if an even number of the factors have negative signs. This happens between -1 and 2, and also above 5 where no factors have negative signs. The solution set is $[-1, 2] \cup [5, \infty)$.

79. $$\begin{aligned} x^3 + 3x^2 - x - 3 &< 0 \\ x^2(x+3) - 1(x+3) &< 0 \\ (x^2 - 1)(x+3) &< 0 \\ (x-1)(x+1)(x+3) &< 0 \end{aligned}$$

$$\begin{array}{l} x+1 - - - - - 0 + + + + + + + \\ x+3 - - 0 + + + + + + + + + + \\ x-1 - - - - - - - - - - 0 + + + \\ \hline \qquad -3 \qquad -1 \qquad\qquad 1 \end{array}$$

The product of these three factors is negative only if an odd number of them have negative signs. This happens to the left of -3 and also between -1 and 1. The solution set is $(-\infty, -3) \cup (-1, 1)$.

81. To solve $0.23x^2 + 6.5x + 4.3 < 0$, we first solve $0.23x^2 + 6.5x + 4.3 = 0$.

$$x = \frac{-6.5 \pm \sqrt{(6.5)^2 - 4(0.23)(4.3)}}{2(0.23)}$$

$x = -27.58$ or $x = -0.68$

Test the numbers -30, -1, and 0.

$0.23(-30)^2 + 6.5(-30) + 4.3 = 16.3$

$0.23(-1)^2 + 6.5(-1) + 4.3 = -1.97$

$0.23(0)^2 + 6.5(0) + 4.3 = 4.3$

According to the signs of the values of the polynomial at the test points, the value of the polynomial is negative between the two solutions to the equation. The solution set is $(-27.58, -0.68)$.

83. $$\begin{aligned} \frac{x}{x-2} &> \frac{-1}{x+3} \\ \frac{x}{x-2} + \frac{1}{x+3} &> 0 \\ \frac{x(x+3)}{(x-2)(x+3)} + \frac{1(x-2)}{(x+3)(x-2)} &> 0 \\ \frac{x^2 + 4x - 2}{(x+3)(x-2)} &> 0 \end{aligned}$$

Solve $x^2 + 4x - 2 = 0$:

$$\begin{aligned} x &= \frac{-4 \pm \sqrt{16 - 4(1)(-2)}}{2} = \frac{-4 \pm 2\sqrt{6}}{2} \\ &= -2 \pm \sqrt{6} = -4.4, 0.4 \end{aligned}$$

The numbers -4.4, -3, 0.4, and 2 divide the number line into 5 regions. Pick a number in each region and test it in the original inequality. We get the solution set $(-\infty, -2 - \sqrt{6}) \cup (-3, -2 + \sqrt{6}) \cup (2, \infty)$.

85. To solve $x^2 + 5x - 50 > 0$, we first solve $x^2 + 5x - 50 = 0$:

$$\begin{aligned} (x+10)(x-5) &= 0 \\ x = -10 \quad \text{or} \quad x &= 5 \end{aligned}$$

Test a point in each region of the number line, to find that the profit is positive if $x < -10$ or if $x > 5$. He cannot sell negative mobile homes so he must sell more than 5 to have a positive profit. He should sell 6, 7, 8, etc.

87. We must solve the inequality

$$\begin{aligned} -16t^2 + 96t + 6 &> 86 \\ -16t^2 + 96t - 80 &> 0 \\ t^2 - 6t + 5 &< 0 \end{aligned}$$

Solve the equation $t^2 - 6t + 5 = 0$.

$(t-5)(t-1) = 0$

$t = 5$ or $t = 1$

Using a test point we find that the inequality is satisfied for t between 1 and 5 seconds. So the arrow is more than 86 feet high for 4 seconds.

89. **a)** From the graph it appears that the maximum height is 900 feet.

b) The projectile was above 864 feet for approximately 3 seconds.

c) $S = -16t^2 + \frac{240\sqrt{2}}{\sqrt{2}}t + 0$

or $S = -16t^2 + 240t$

Solve

$$-16t^2 + 240t > 864$$
$$-16t^2 + 240t - 864 > 0$$
$$t^2 - 15t + 54 < 0$$

Solve the equation

$(t-6)(t-9) = 0$

$t = 6$ or $t = 9$

Evaluate $t^2 - 15t + 54$ at the test points 0, 7, and 10.

$0^2 - 15(0) + 54 = 54$

$7^2 - 15(7) + 54 = -2$

$10^2 - 15(10) + 54 = 4$

So the inequality is satisfied for t in the interval $(6, 9)$. The projectile was above 864 ft for 3 sec.

91. **a)** (h, k)

b) $(-\infty, h) \cup (k, \infty)$

c) $(-k, -h)$

d) $(-\infty, -k] \cup [-h, \infty)$

e) $(-\infty, h] \cup (k, \infty)$

f) $(-k, -h]$

93. c

95. b

Chapter 8 Wrap-Up

Enriching Your Mathematical Word Power

1. b **2.** a **3.** d **4.** c **5.** b **6.** c

7. c **8.** a **9.** c **10.** a **11.** c

CHAPTER 8 REVIEW

1. $x^2 - 2x - 15 = 0$

$(x-5)(x+3) = 0$

$x - 5 = 0$ or $x + 3 = 0$

$x = 5$ or $x = -3$

Solution set: $\{-3, 5\}$

3. $2x^2 + x = 15$

$2x^2 + x - 15 = 0$

$(2x-5)(x+3) = 0$

$2x - 5 = 0$ or $x + 3 = 0$

$x = \frac{5}{2}$ or $x = -3$

Solution set: $\left\{-3, \frac{5}{2}\right\}$

5. $w^2 - 25 = 0$

$(w-5)(w+5) = 0$

$w - 5 = 0$ or $w + 5 = 0$

$w = 5$ or $w = -5$

Solution set: $\{-5, 5\}$

7. $4x^2 - 12x + 9 = 0$

$(2x-3)^2 = 0$

$2x - 3 = 0$

$x = \frac{3}{2}$

Solution set: $\left\{\frac{3}{2}\right\}$

9. $x^2 = 12$

$x = \pm\sqrt{12} = \pm 2\sqrt{3}$

Solution set: $\left\{\pm 2\sqrt{3}\right\}$

11. $(x-1)^2 = 9$

$x - 1 = \pm 3$

$x = 1 \pm 3$

Solution set: $\{-2, 4\}$

13. $(x-2)^2 = \frac{3}{4}$

$x - 2 = \pm\sqrt{\frac{3}{4}}$

$x = 2 \pm \frac{\sqrt{3}}{2} = \frac{4}{2} \pm \frac{\sqrt{3}}{2}$

Solution set: $\left\{\frac{4 \pm \sqrt{3}}{2}\right\}$

15. $4x^2 = 9$

$x^2 = \frac{9}{4}$

$x = \pm\sqrt{\frac{9}{4}} = \pm\frac{3}{2}$

Solution set: $\left\{\pm\frac{3}{2}\right\}$

17. $x^2 - 6x + 8 = 0$

$x^2 - 6x = -8$

$x^2 - 6x + 9 = -8 + 9$

$(x-3)^2 = 1$

$x - 3 = \pm 1$

$x = 3 \pm 1$

Solution set: $\{2, 4\}$

19. $x^2 - 5x + 6 = 0$

$$x^2 - 5x = -6$$

$$x^2 - 5x + \frac{25}{4} = -\frac{24}{4} + \frac{25}{4}$$

$$(x - \tfrac{5}{2})^2 = \frac{1}{4}$$

$$x - \frac{5}{2} = \pm\frac{1}{2}$$

$$x = \frac{5}{2} \pm \frac{1}{2}$$

Solution set: $\{2, 3\}$

21. $2x^2 - 7x + 3 = 0$

$$2x^2 - 7x = -3$$

$$x^2 - \frac{7}{2}x = -\frac{3}{2}$$

$$x^2 - \frac{7}{2}x + \frac{49}{16} = -\frac{24}{16} + \frac{49}{16}$$

$$(x - \tfrac{7}{4})^2 = \frac{25}{16}$$

$$x - \frac{7}{4} = \pm\frac{5}{4}$$

$$x = \frac{7}{4} \pm \frac{5}{4} \quad \left(\frac{12}{4} \text{ or } \frac{2}{4}\right)$$

Solution set: $\left\{\frac{1}{2}, 3\right\}$

23. $x^2 + 4x + 1 = 0$

$$x^2 + 4x = -1$$

$$x^2 + 4x + 4 = -1 + 4$$

$$(x + 2)^2 = 3$$

$$x + 2 = \pm\sqrt{3}$$

$$x = -2 \pm \sqrt{3}$$

Solution set: $\{-2 \pm \sqrt{3}\}$

25. $x^2 - 3x - 10 = 0$

$$x = \frac{3 \pm \sqrt{9 - 4(1)(-10)}}{2(1)} = \frac{3 \pm \sqrt{49}}{2} = \frac{3 \pm 7}{2}$$

Solution set: $\{-2, 5\}$

27. $6x^2 - 7x = 3$

$$6x^2 - 7x - 3 = 0$$

$$x = \frac{7 \pm \sqrt{49 - 4(6)(-3)}}{2(6)} = \frac{7 \pm \sqrt{121}}{12} = \frac{7 \pm 11}{12}$$

Solution set: $\left\{-\frac{1}{3}, \frac{3}{2}\right\}$

29. $x^2 + 4x + 2 = 0$

$$x = \frac{-4 \pm \sqrt{16 - 4(1)(2)}}{2(1)} = \frac{-4 \pm 2\sqrt{2}}{2} = -2 \pm \sqrt{2}$$

Solution set: $\left\{-2 \pm \sqrt{2}\right\}$

31. $3x^2 + 1 = 5x$

$$3x^2 - 5x + 1 = 0$$

$$x = \frac{5 \pm \sqrt{25 - 4(3)(1)}}{2(3)} = \frac{5 \pm \sqrt{13}}{6}$$

Solution set: $\left\{\frac{5 \pm \sqrt{13}}{6}\right\}$

33. $25x^2 - 20x + 4 = 0$

$b^2 - 4ac = (-20)^2 - 4(25)(4) = 0$

One real solution.

35. $x^2 - 3x + 7 = 0$

$b^2 - 4ac = (-3)^2 - 4(1)(7) = -19$

No real solutions

37. $2x^2 - 5x + 1 = 0$

$b^2 - 4ac = (-5)^2 - 4(2)(1) = 17$

Two real solutions

39. $2x^2 - 4x + 3 = 0$

$$x = \frac{4 \pm \sqrt{16 - 4(2)(3)}}{2(2)} = \frac{4 \pm \sqrt{-8}}{4} = \frac{4 \pm 2i\sqrt{2}}{4} = \frac{2 \pm i\sqrt{2}}{2}$$

Solution set: $\left\{\frac{2 \pm i\sqrt{2}}{2}\right\}$

41. $2x^2 - 3x + 3 = 0$

$$x = \frac{3 \pm \sqrt{9 - 4(2)(3)}}{2(2)} = \frac{3 \pm \sqrt{-15}}{4} = \frac{3 \pm i\sqrt{15}}{4}$$

Solution set: $\left\{\frac{3 \pm i\sqrt{15}}{4}\right\}$

43. $3x^2 + 2x + 2 = 0$

$$x = \frac{-2 \pm \sqrt{4 - 4(3)(2)}}{2(3)} = \frac{-2 \pm \sqrt{-20}}{6} = \frac{-2 \pm 2i\sqrt{5}}{6} = \frac{-1 \pm i\sqrt{5}}{3}$$

Solution set: $\left\{\frac{-1 \pm i\sqrt{5}}{3}\right\}$

45. $\frac{1}{2}x^2 + 3x + 8 = 0$

$$x^2 + 6x + 16 = 0$$

$$x = \frac{-6 \pm \sqrt{36 - 4(1)(16)}}{2(1)} = \frac{-6 \pm \sqrt{-28}}{2} = \frac{-6 \pm 2i\sqrt{7}}{2} = -3 \pm i\sqrt{7}$$

Solution set: $\{-3 \pm i\sqrt{7}\}$

47. $b^2 - 4ac = (-10)^2 - 4(8)(-3) = 196$

Since 196 is a perfect square, the polynomial is not prime.

$8x^2 - 10x - 3 = (4x + 1)(2x - 3)$

49. $b^2 - 4ac = (-5)^2 - 4(4)(2) = -7$

Since -7 is a not a perfect square, the polynomial is prime.

51. $b^2 - 4ac = (10)^2 - 4(8)(-25) = 900$
Since 900 is a perfect square, the polynomial is not prime.
$8y^2 + 10y - 25 = (4y - 5)(2y + 5)$

53. $(x + 3)(x + 6) = 0$
$x^2 + 9x + 18 = 0$

55. $(x + 5\sqrt{2})(x - 5\sqrt{2}) = 0$
$x^2 - 50 = 0$

57. $x^6 + 7x^3 - 8 = 0$
$(x^3 + 8)(x^3 - 1) = 0$
$x^3 + 8 = 0$ or $x^3 - 1 = 0$
$x^3 = -8$ or $x^3 = 1$
$x = -2$ or $x = 1$
Solution set: $\{-2, 1\}$

59. $x^4 - 13x^2 + 36 = 0$
$(x^2 - 4)(x^2 - 9) = 0$
$x^2 - 4 = 0$ or $x^2 - 9 = 0$
$x^2 = 4$ or $x^2 = 9$
$x = \pm 2$ or $x = \pm 3$
Solution set: $\{\pm 2, \pm 3\}$

61. Let $w = x^2 + 3x$ in
$(x^2 + 3x)^2 - 28(x^2 + 3x) + 180 = 0.$
$w^2 - 28w + 180 = 0$
$(w - 10)(w - 18) = 0$
$w = 10$ or $w = 18$
$x^2 + 3x = 10$ or $x^2 + 3x = 18$
$x^2 + 3x - 10 = 0$ or $x^2 + 3x - 18 = 0$
$(x + 5)(x - 2) = 0$ or $(x + 6)(x - 3) = 0$
$x = -5$ or $x = 2$ or $x = -6$ or $x = 3$
Solution set: $\{-6, -5, 2, 3\}$

63. Let $w = \sqrt{x^2 - 6x}$ and $w^2 = x^2 - 6x$ in $x^2 - 6x + 6\sqrt{x^2 - 6x} - 40 = 0.$
$w^2 + 6w - 40 = 0$
$(w + 10)(w - 4) = 0$
$w = -10$ or $w = 4$
$\sqrt{x^2 - 6x} = -10$ or $\sqrt{x^2 - 6x} = 4$
No real solution here. $x^2 - 6x = 16$
$x^2 - 6x - 16 = 0$
$(x - 8)(x + 2) = 0$
$x = 8$ or $x = -2$
Solution set: $\{-2, 8\}$

65. Let $w = t^{-1}$ and $w^2 = t^{-2}$ in
$t^{-2} + 5t^{-1} - 36 = 0.$
$w^2 + 5w - 36 = 0$
$(w + 9)(w - 4) = 0$
$w = -9$ or $w = 4$
$t^{-1} = -9$ or $t^{-1} = 4$
$t = -\frac{1}{9}$ or $t = \frac{1}{4}$
Solution set: $\left\{-\frac{1}{9}, \frac{1}{4}\right\}$

67. Let $y = \sqrt{w}$ and $y^2 = w$ in
$w - 13\sqrt{w} + 36 = 0.$
$y^2 - 13y + 36 = 0$
$(y - 9)(y - 4) = 0$
$y = 9$ or $y = 4$
$\sqrt{w} = 9$ or $\sqrt{w} = 4$
$w = 81$ or $w = 16$
Solution set: $\{16, 81\}$

69. $f(x) = x^2 - 6x$
$x = \frac{-b}{2a} = \frac{6}{2(1)} = 3,\ f(3) = 3^2 - 6(3) = -9$
The vertex is $(3, -9)$. The y-intercept is $(0, 0)$.
$x^2 - 6x = 0$
$x(x - 6) = 0$
$x = 0$ or $x - 6 = 0$
$x = 0$ or $x = 6$
The x-intercepts are $(0, 0)$ and $(6, 0)$.

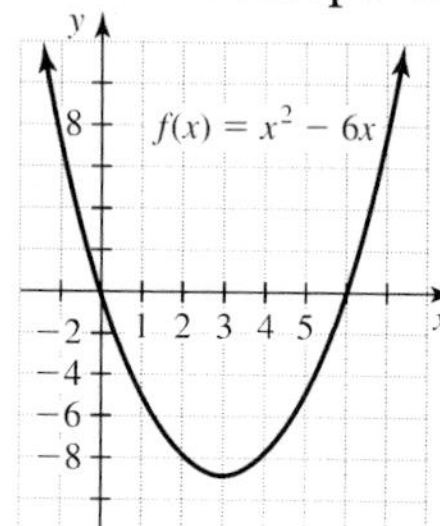

71. $g(x) = x^2 - 4x - 12$
$x = \frac{-b}{2a} = \frac{4}{2(1)} = 2,$
$g(2) = 2^2 - 4(2) - 12 = -16$
The vertex is $(2, -16)$. The y-intercept is $(0, -12)$. The solutions to $x^2 - 4x - 12 = 0$ are 6 and -2. So the x-intercepts are $(-2, 0)$ and $(6, 0)$.

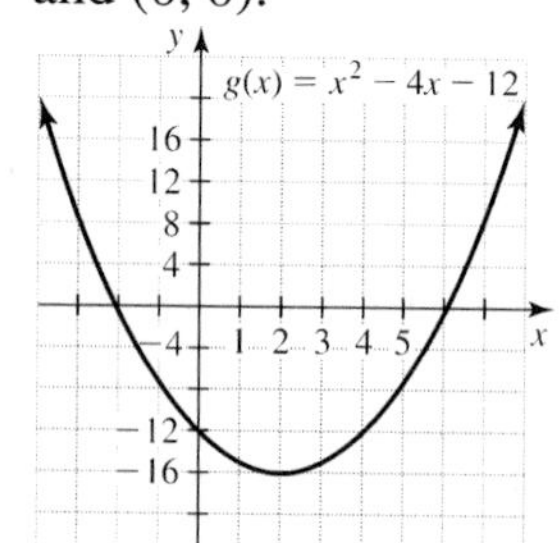

73. $h(x) = -2x^2 + 8x$
$x = \frac{-b}{2a} = \frac{-8}{2(-2)} = 2,$
$h(2) = -2 \cdot 2^2 + 8(2) = 8$
The vertex is $(2, 8)$. The y-intercept is $(0, 0)$. The solutions to $-2x^2 + 8x = 0$ are 0 and 4. So the x-intercepts are $(0, 0)$ and $(4, 0)$.

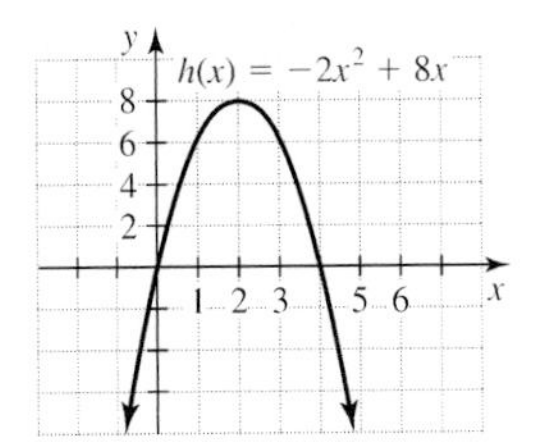

75. $y = -x^2 + 2x + 3$

$x = \frac{-b}{2a} = \frac{-2}{2(-1)} = 1,$

$y = -1^2 + 2(1) + 3 = 4$

The vertex is (1, 4). The y-intercept is (0, 3). The solutions to $-x^2 + 2x + 3 = 0$ are -1 and 3. So the x-intercepts are $(-1, 0)$ and (3, 0).

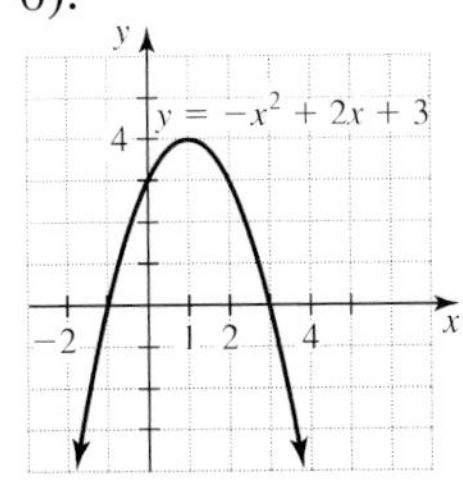

77. $f(x) = x^2 + 4x + 1$

$x = \frac{-b}{2a} = \frac{-4}{2(1)} = -2,$

$f(-2) = (-2)^2 + 4(-2) + 1 = -3$

The domain is $(-\infty, \infty)$ and the range is $[-3, \infty)$.

79. $y = -2x^2 - x + 4$

$x = \frac{-b}{2a} = \frac{1}{2(-2)} = -\frac{1}{4},$

$y = -2(-\frac{1}{4})^2 - (-\frac{1}{4}) + 4 = 4.125$

The domain is $(-\infty, \infty)$ and the range is $(-\infty, 4.125]$.

81.

$$a^2 + a > 6$$
$$a^2 + a - 6 > 0$$
$$(a + 3)(a - 2) > 0$$

$a + 3$ − − − 0 + + + + + + + + +
$a - 2$ − − − − − − − − − 0 + + + +
−3 2

The inequality is satisfied if both factors are the same sign. $(-\infty, -3) \cup (2, \infty)$

83. $x^2 - x - 20 \le 0$

$(x - 5)(x + 4) \le 0$

$x + 4$ − − − 0 + + + + + + + + +
$x - 5$ − − − − − − − − − 0 + + + +
−4 5

The inequality is satisfied if the factors have opposite signs, or if one of the factors is zero. $[-4, 5]$

85.

$$w^2 - w < 0$$
$$w(w - 1) < 0$$

w − − − 0 + + + + + + + + + +
$w - 1$ − − − − − − − − − − 0 + + +
0 1

The inequality is satisfied if the factors have opposite signs. (0, 1)

87. $\frac{x - 4}{x + 2} \ge 0$

$x + 2$ − − − 0 + + + + + + + + +
$x - 4$ − − − − − − − − − 0 + + + +
−2 4

The inequality is satisfied if the factors have the same sign, or if the numerator is equal to zero. $(-\infty, -2) \cup [4, \infty)$

89. $\frac{x - 2}{x + 3} < 1$

$$\frac{x - 2}{x + 3} - 1 < 0$$
$$\frac{x - 2}{x + 3} - \frac{x + 3}{x + 3} < 0$$
$$\frac{-5}{x + 3} < 0$$

Since the numerator is definitely negative, the inequality is satisfied only if the denominator is positive:

$$x + 3 > 0$$
$$x > -3$$

$(-3, \infty)$

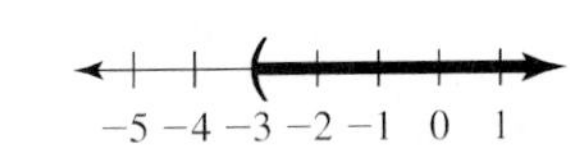

91. $\frac{3}{x + 2} > \frac{1}{x + 1}$

$$\frac{3}{x + 2} - \frac{1}{x + 1} > 0$$
$$\frac{3(x + 1)}{(x + 2)(x + 1)} - \frac{1(x + 2)}{(x + 1)(x + 2)} > 0$$
$$\frac{2x + 1}{(x + 2)(x + 1)} > 0$$

$x + 1$ − − − − 0 + + + + + + + +
$x + 2$ − − 0 + + + + + + + + + +
$2x + 1$ − − − − − − − − − 0 + + +
−2 −1 −1/2

The inequality is satisfied when an even number of the factors have negative signs.

$(-2, -1) \cup \left(-\frac{1}{2}, \infty\right)$

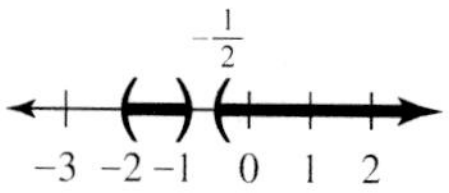

93. $144x^2 - 120x + 25 = 0$

$$(12x - 5)^2 = 0$$
$$12x - 5 = 0$$
$$x = \frac{5}{12}$$

Solution set: $\left\{\frac{5}{12}\right\}$

95. $(2x + 3)^2 + 7 = 12$

$$(2x + 3)^2 = 5$$
$$2x + 3 = \pm\sqrt{5}$$
$$x = \frac{-3 \pm \sqrt{5}}{2}$$

Solution set: $\left\{\frac{-3 \pm \sqrt{5}}{2}\right\}$

97. $1 + \frac{20}{9x^2} = \frac{8}{3x}$

$$9x^2(1 + \frac{20}{9x^2}) = 9x^2 \cdot \frac{8}{3x}$$
$$9x^2 + 20 = 24x$$
$$9x^2 - 24x + 20 = 0$$
$$x = \frac{24 \pm \sqrt{(-24)^2 - 4(9)(20)}}{2(9)} = \frac{24 \pm 12i}{18}$$
$$= \frac{4 \pm 2i}{3}$$

Solution set: $\left\{\frac{4 \pm 2i}{3}\right\}$

99. $\sqrt{3x^2 + 7x - 30} = x$

$$3x^2 + 7x - 30 = x^2$$
$$2x^2 + 7x - 30 = 0$$
$$(2x - 5)(x + 6) = 0$$
$$2x - 5 = 0 \text{ or } x + 6 = 0$$
$$x = \frac{5}{2} \text{ or } x = -6$$

Since -6 does not check, solution set is $\left\{\frac{5}{2}\right\}$.

101. $2(2x + 1)^2 + 5(2x + 1) = 3$

Let $y = 2x + 1$:

$$2y^2 + 5y - 3 = 0$$
$$(2y - 1)(y + 3) = 0$$
$$2y - 1 = 0 \text{ or } y + 3 = 0$$
$$y = \frac{1}{2} \text{ or } y = -3$$
$$2x + 1 = \frac{1}{2} \text{ or } 2x + 1 = -3$$
$$x = -\frac{1}{4} \text{ or } x = -2$$

Solution set: $\left\{-2, -\frac{1}{4}\right\}$

103. $x^{1/2} - 15x^{1/4} + 50 = 0$ Let $y = x^{1/4}$

$$y^2 - 15y + 50 = 0$$
$$(y - 5)(y - 10) = 0$$
$$y = 5 \text{ or } y = 10$$
$$x^{1/4} = 5 \text{ or } x^{1/4} = 10$$
$$x = 5^4 \text{ or } x = 10^4$$
$$x = 625 \text{ or } x = 10{,}000$$

Solution set: $\{625, 10{,}000\}$

105. If $x =$ one of the numbers, then $x + 4 =$ the other number. Since their product is 4, we can write the equation

$$x(x + 4) = 4$$
$$x^2 + 4x = 4$$
$$x^2 + 4x + 4 = 4 + 4$$
$$(x + 2)^2 = 8$$
$$x + 2 = \pm 2\sqrt{2}$$
$$x = -2 \pm 2\sqrt{2}$$

Disregard $-2 - 2\sqrt{2}$ since it is not positive. If $x = -2 + 2\sqrt{2}$ then $x + 4 = 2 + 2\sqrt{2}$. The two numbers are $-2 + 2\sqrt{2} \approx 0.83$ and $2 + 2\sqrt{2} \approx 4.83$.

107. Let $x =$ the width and $x - 4 =$ the height. The diagonal 19 is the hypotenuse of a right triangle with legs x and $x - 4$. By the Pythagorean theorem we can write

$$x^2 + (x - 4)^2 = 19^2$$
$$x^2 + x^2 - 8x + 16 = 361$$
$$2x^2 - 8x - 345 = 0$$
$$x = \frac{8 \pm \sqrt{64 - 4(2)(-345)}}{2(2)}$$
$$= \frac{8 \pm \sqrt{2824}}{4} = \frac{8 \pm 2\sqrt{706}}{4}$$
$$= \frac{4 \pm \sqrt{706}}{2}$$

Since the width must be positive, we have

$$x = \frac{4 + \sqrt{706}}{2} \text{ and}$$
$$x - 4 = \frac{4 + \sqrt{706}}{2} - \frac{8}{2} = \frac{-4 + \sqrt{706}}{2}.$$

So the width is $\frac{4 + \sqrt{706}}{2} \approx 15.3$ inches and the height is $\frac{-4 + \sqrt{706}}{2} \approx 11.3$ inches.

109. Let $x =$ the width of the border. The dimensions of the printed area will be $8 - 2x$ by $10 - 2x$. Since the printed area is to be 24 square inches, we can write the equation

$$(8 - 2x)(10 - 2x) = 24$$

$$80 - 36x + 4x^2 = 24$$
$$4x^2 - 36x + 56 = 0$$
$$x^2 - 9x + 14 = 0$$
$$(x - 7)(x - 2) = 0$$
$$x = 7 \quad \text{or} \quad x = 2$$

Since 7 inches is too wide for a border on an 8 by 10 piece of paper, the border must be 2 inches wide.

111. Let $x =$ width and $x + 4 =$ the length.

$$x(x + 4) = 45$$
$$x^2 + 4x - 45 = 0$$
$$(x + 9)(x - 5) = 0$$
$$x + 9 = 0 \text{ or } x - 5 = 0$$
$$x = -9 \text{ or } x = 5$$
$$x + 4 = 9$$

The table is 5 ft wide and 9 feet long.

113. $C(n) = 0.004n^2 - 3.2n + 660$

If $n = 390$, then

$$C(390) = 0.004(390)^2 - 3.2(390) + 660 = \$20.40$$

$$n = \frac{3.2}{2(0.004)} = 400$$

The unit cost is at a minimum for 400 starters.

115.

$$12 = -16t^2 + 32t$$
$$16t^2 - 32t + 12 = 0$$
$$4t^2 - 8t + 3 = 0$$

$$t = \frac{8 \pm \sqrt{64 - 4(4)(3)}}{2(4)} = \frac{8 \pm \sqrt{16}}{8}$$
$$= 1 \pm \frac{1}{2} = 1.5 \text{ or } 0.5$$

His height was 12 ft for $t = 0.5$ sec and $t = 1.5$ sec.

117.

$$\frac{x}{1} = \frac{1}{x - 1}$$
$$x^2 - x = 1$$
$$x^2 - x - 1 = 0$$
$$x = \frac{-(-1) \pm \sqrt{(-1)^2 - 4(1)(-1)}}{2(1)}$$
$$= \frac{1 \pm \sqrt{5}}{2}$$

Since $\frac{1 - \sqrt{5}}{2}$ is negative, the golden ratio is $\frac{1 + \sqrt{5}}{2}$ or approximately 1.618.

CHAPTER 8 TEST

1. $2x^2 - 3x + 2 = 0$

$(-3)^2 - 4(2)(2) = 9 - 16 = -7$

The equation has no real solutions.

2. $-3x^2 + 5x - 1 = 0$

$(5)^2 - 4(-3)(-1) = 25 - 12 = 13$

The equation has 2 real solutions.

3. $4x^2 - 4x + 1 = 0$

$(-4)^2 - 4(4)(1) = 16 - 16 = 0$

The equation has 1 real solution.

4. $2x^2 + 5x - 3 = 0$

$$x = \frac{-5 \pm \sqrt{25 - 4(2)(-3)}}{2(2)} = \frac{-5 \pm \sqrt{49}}{4}$$
$$= \frac{-5 \pm 7}{4} \quad \left(\frac{2}{4} \text{ or } \frac{-12}{4}\right)$$

Solution set: $\left\{-3, \frac{1}{2}\right\}$

5. $x^2 + 6x + 6 = 0$

$$x = \frac{-6 \pm \sqrt{36 - 4(1)(6)}}{2} = \frac{-6 \pm \sqrt{12}}{2}$$
$$= \frac{-6 \pm 2\sqrt{3}}{2} = -3 \pm \sqrt{3}$$

Solution set: $\{-3 \pm \sqrt{3}\}$

6.

$$x^2 + 10x + 25 = 0$$
$$(x + 5)^2 = 0$$
$$x + 5 = 0$$
$$x = -5$$

Solution set: $\{-5\}$

7.

$$2x^2 + x - 6 = 0$$
$$2x^2 + x = 6$$
$$x^2 + \frac{1}{2}x = 3$$
$$x^2 + \frac{1}{2}x + \frac{1}{16} = 3 + \frac{1}{16}$$
$$(x + \tfrac{1}{4})^2 = \frac{49}{16}$$
$$x + \frac{1}{4} = \pm\frac{7}{4}$$
$$x = -\frac{1}{4} \pm \frac{7}{4} \quad \left(\frac{6}{4} \text{ or } \frac{-8}{4}\right)$$

Solution set: $\left\{-2, \frac{3}{2}\right\}$

8.

$$x(x + 1) = 12$$
$$x^2 + x - 12 = 0$$
$$(x + 4)(x - 3) = 0$$
$$x = -4 \quad \text{or} \quad x = 3$$

Solution set: $\{-4, 3\}$

9.

$$a^4 - 5a^2 + 4 = 0$$
$$(a^2 - 4)(a^2 - 1) = 0$$
$$a^2 - 4 = 0 \text{ or } a^2 - 1 = 0$$
$$a^2 = 4 \text{ or } a^2 = 1$$
$$a = \pm 2 \text{ or } a = \pm 1$$

Solution set: $\{\pm 1, \pm 2\}$

10. Let $w = \sqrt{x - 2}$ and $w^2 = x - 2$ in

$$x - 2 - 8\sqrt{x - 2} + 15 = 0$$
$$w^2 - 8w + 15 = 0$$
$$(w - 3)(w - 5) = 0$$
$$w = 3 \quad \text{or} \quad w = 5$$

$$\sqrt{x-2}=3 \quad \text{or} \quad \sqrt{x-2}=5$$
$$x-2=9 \quad \text{or} \quad x-2=25$$
$$x=11 \quad \text{or} \quad x=27$$

Solution set: $\{11, 27\}$

11. $x^2+36=0$

$$x^2=-36$$
$$x=\pm\sqrt{-36}=\pm 6i$$

Solution set: $\{\pm 6i\}$

12. $x^2+6x+10=0$

$$x=\frac{-6\pm\sqrt{36-4(1)(10)}}{2(1)}=\frac{-6\pm\sqrt{-4}}{2}$$
$$=\frac{-6\pm 2i}{2}=-3\pm i$$

Solution set: $\{-3\pm i\}$

13. $3x^2-x+1=0$

$$x=\frac{1\pm\sqrt{1-4(3)(1)}}{2(3)}=\frac{1\pm\sqrt{-11}}{6}$$

Solution set: $\left\{\frac{1\pm i\sqrt{11}}{6}\right\}$

14. The graph of $f(x)=16-x^2$ goes through $(-2, 12)$, $(-1, 15)$, $(0, 16)$, $(1, 15)$, and $(2, 12)$.

$$x=\frac{-b}{2a}=\frac{-0}{2(-1)}=0$$
$$f(0)=16$$

The vertex is $(0, 16)$. The domain is $(-\infty,\infty)$ and the range is $(-\infty, 16]$. The maximum y-value is 16.

$$16-x^2=0$$
$$x^2=16$$
$$x=\pm 4$$

The intercepts are $(0,16)$, $(-4,0)$, and $(4,0)$.

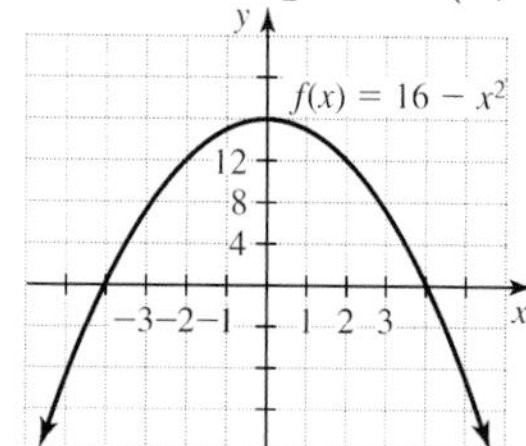

15. The graph of $g(x)=x^2-3x$ goes through $(0, 0)$, $(1, -2)$, $(2, -2)$, and $(3, 0)$.

$$x=\frac{-b}{2a}=\frac{-(-3)}{2(1)}=\frac{3}{2}$$
$$f(3/2)=\left(\frac{3}{2}\right)^2-3\cdot\frac{3}{2}=-\frac{9}{4}$$

The vertex is $(1.5, -2.25)$. The domain is $(-\infty,\infty)$ and the range is $[-2.25,\infty)$. The minimum y-value is $-9/4$ or -2.25.

$$x^2-3x=0$$
$$x(x-3)=0$$
$$x=0 \quad \text{or} \quad x-3=0$$
$$x=0 \quad \text{or} \quad x=3$$

The intercepts are $(0,0)$ and $(3,0)$.

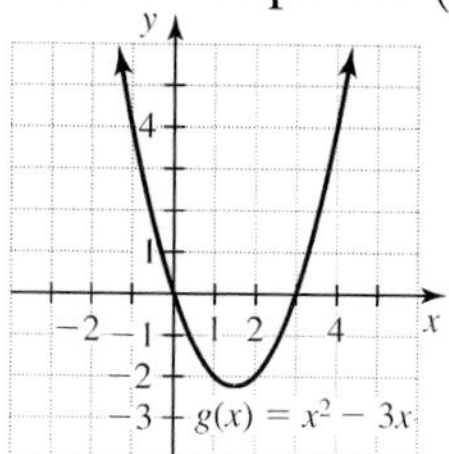

16. $(x+4)(x-6)=0$

$$x^2-2x-24=0$$

17. $(x-5i)(x+5i)=0$

$$x^2+25=0$$

18. $w^2+3w<18$

$$w^2+3w-18<0$$
$$(w+6)(w-3)<0$$

$w+6$ − − − − 0 + + + + + + + +
$w-3$ − − − − − − − − − 0 + + + +
−6 3

The inequality is satisfied when the factors have opposite signs. $(-6, 3)$

−6 −5 −4 −3 −2 −1 0 1 2 3

19.

$$\frac{2}{x-2}<\frac{3}{x+1}$$
$$\frac{2}{x-2}-\frac{3}{x+1}<0$$
$$\frac{2(x+1)}{(x-2)(x+1)}-\frac{3(x-2)}{(x+1)(x-2)}<0$$
$$\frac{-x+8}{(x-2)(x+1)}<0$$

$x-2$ − − − − 0 + + + + + + + +
$x+1$ − − 0 + + + + + + + + + +
$-x+8$ + + + + + + + + 0 − − −
−1 2 8

This quotient will be negative only if an odd number of factors are negative.
$(-1, 2)\cup(8,\infty)$

−1 0 1 2 3 4 5 6 7 8 9 10

20. Let $x=$ the width and $x+2=$ the length. Since the area is 16 square feet, we can write the equation

$$x(x+2)=16$$
$$x^2+2x-16=0$$
$$x=\frac{-2\pm\sqrt{4-4(1)(-16)}}{2}=\frac{-2\pm\sqrt{68}}{2}$$
$$=\frac{-2\pm 2\sqrt{17}}{2}=-1\pm\sqrt{17}$$

Since the width must be positive, we have $x = -1 + \sqrt{17}$ and $x + 2 = 1 + \sqrt{17}$. The width is $-1 + \sqrt{17}$ feet and the length is $1 + \sqrt{17}$ feet.

21. Let x = time for the new computer and $x + 1$ = time for the old computer. New computer's rate is $\frac{1}{x}$ payroll/hr and old computer's rate is $\frac{1}{x+1}$ payroll/hr. In 3 hrs new computer does $\frac{3}{x}$ payroll and old computer does $\frac{3}{x+1}$ payroll.

$$\frac{3}{x} + \frac{3}{x+1} = 1$$
$$x(x+1)\left(\frac{3}{x} + \frac{3}{x+1}\right) = x(x+1)1$$
$$3x + 3 + 3x = x^2 + x$$
$$0 = x^2 - 5x - 3$$
$$x = \frac{5 \pm \sqrt{5^2 - 4(1)(-3)}}{2} = \frac{5 \pm \sqrt{37}}{2}$$
$$\approx 5.5 \text{ or } -0.5$$

It takes the new computer $\frac{5+\sqrt{37}}{2}$ or 5.5 hrs to do the payroll by itself.

22. $s(t) = -16t^2 + 48t$

$t = \frac{-48}{2(-16)} = \frac{3}{2}$, $s = -16(\frac{3}{2})^2 + 48(\frac{3}{2}) = 36$

The maximum height reached by the ball is 36 feet.

Making Connections

Chapters 1 - 8

1. $2x - 15 = 0$
$$2x = 15$$
$$x = \frac{15}{2}$$

Solution set: $\left\{\frac{15}{2}\right\}$

2. $2x^2 - 15 = 0$
$$2x^2 = 15$$
$$x^2 = \frac{15}{2}$$
$$x = \pm\sqrt{\frac{15}{2}} = \pm\frac{\sqrt{15}\sqrt{2}}{\sqrt{2}\sqrt{2}} = \pm\frac{\sqrt{30}}{2}$$

Solution set: $\left\{\pm\frac{\sqrt{30}}{2}\right\}$

3. $2x^2 + x - 15 = 0$
$$(2x - 5)(x + 3) = 0$$
$$2x - 5 = 0 \text{ or } x + 3 = 0$$
$$x = \frac{5}{2} \text{ or } x = -3$$

Solution set: $\left\{-3, \frac{5}{2}\right\}$

4. $2x^2 + 4x - 15 = 0$
$$x = \frac{-4 \pm \sqrt{16 - 4(2)(-15)}}{2(2)} = \frac{-4 \pm \sqrt{136}}{4}$$
$$= \frac{-4 \pm 2\sqrt{34}}{4} = \frac{-2 \pm \sqrt{34}}{2}$$

Solution set: $\left\{\frac{-2 \pm \sqrt{34}}{2}\right\}$

5. $|4x + 11| = 3$
$$4x + 11 = 3 \text{ or } 4x + 11 = -3$$
$$4x = -8 \text{ or } 4x = -14$$
$$x = -2 \text{ or } x = -\frac{7}{2}$$

Solution set: $\left\{-\frac{7}{2}, -2\right\}$

6. $|4x^2 + 11x| = 3$
$$4x^2 + 11x = 3 \text{ or } 4x^2 + 11x = -3$$
$$4x^2 + 11x - 3 = 0 \text{ or } 4x^2 + 11x + 3 = 0$$
$$(4x - 1)(x + 3) = 0 \text{ or } x = \frac{-11 \pm \sqrt{73}}{8}$$
$$4x - 1 = 0 \text{ or } x + 3 = 0$$
$$x = \frac{1}{4} \text{ or } x = -3$$

Solution set: $\left\{-3, \frac{1}{4}, \frac{-11 \pm \sqrt{73}}{8}\right\}$

7. $\sqrt{x} = x - 6$
$$(\sqrt{x})^2 = (x - 6)^2$$
$$x = x^2 - 12x + 36$$
$$0 = x^2 - 13x + 36$$
$$0 = (x - 4)(x - 9)$$
$$x - 4 = 0 \text{ or } x - 9 = 0$$
$$x = 4 \text{ or } x = 9$$

Since $\sqrt{4} = 4 - 6$ is incorrect and $\sqrt{9} = 9 - 6$ is correct, the solution set is $\{9\}$.

8. $(2x - 5)^{2/3} = 4$
$$\left((2x - 5)^{2/3}\right)^3 = 4^3$$
$$(2x - 5)^2 = 64$$
$$2x - 5 = \pm 8$$
$$2x = 5 \pm 8$$
$$x = \frac{5 \pm 8}{2}$$

Solution set: $\left\{-\frac{3}{2}, \frac{13}{2}\right\}$

9. $1 - 2x < 5 - x$
$$-x < 4$$
$$x > -4$$

$(-4, \infty)$

10. $(1 - 2x)(5 - x) \le 0$

$1 - 2x$ + + + 0 − − − − − − − − −

$5 - x$ + + + + + + + + + 0 − − − −

1/2 5

The inequality is satisfied when the factors have opposite signs. The solution set is $\left[\frac{1}{2}, 5\right]$.

11. $\dfrac{1-2x}{5-x} \le 0$ Same as last exercise, but 5 is excluded from the solution set because of the denominator $5 - x$. The solution set is $\left[\frac{1}{2}, 5\right)$.

12.
$$\begin{aligned} |5 - x| &< 3 \\ -3 < 5 - x &< 3 \\ -8 < -x &< -2 \\ 8 > x &> 2 \end{aligned}$$

The solution set is (2, 8).

13. $3x - 1 < 5$ and $-3 \le x$

$x < 2$ and $x \ge -3$

The solution set is $[-3, 2)$.

14. $x - 3 < 1$ or $2x \ge 8$

$x < 4$ or $x \ge 4$

The solution set is $(-\infty, \infty)$.

15.
$$\begin{aligned} 2x - 3y &= 9 \\ -3y &= -2x + 9 \\ y &= \frac{-2x + 9}{-3} \\ y &= \frac{2}{3}x - 3 \end{aligned}$$

16.
$$\begin{aligned} \frac{y-3}{x+2} &= -\frac{1}{2} \\ y - 3 &= -\frac{1}{2}(x + 2) \\ y - 3 &= -\frac{1}{2}x - 1 \\ y &= -\frac{1}{2}x + 2 \end{aligned}$$

17.
$$\begin{aligned} 3y^2 + cy + d &= 0 \\ y &= \frac{-c \pm \sqrt{c^2 - 4(3)(d)}}{2(3)} \\ y &= \frac{-c \pm \sqrt{c^2 - 12d}}{6} \end{aligned}$$

18.
$$\begin{aligned} my^2 - ny - w &= 0 \\ y &= \frac{-(-n) \pm \sqrt{(-n)^2 - 4(m)(-w)}}{2m} \\ y &= \frac{n \pm \sqrt{n^2 + 4mw}}{2m} \end{aligned}$$

19.
$$\begin{aligned} \frac{1}{3}x - \frac{2}{5}y &= \frac{5}{6} \\ 30\left(\frac{1}{3}x - \frac{2}{5}y\right) &= 30\left(\frac{5}{6}\right) \\ 10x - 12y &= 25 \\ -12y &= -10x + 25 \\ y &= \frac{-10}{-12}x + \frac{25}{-12} \\ y &= \frac{5}{6}x - \frac{25}{12} \end{aligned}$$

20.
$$\begin{aligned} y - 3 &= -\frac{2}{3}(x - 4) \\ y - 3 &= -\frac{2}{3}x + \frac{8}{3} \\ y &= -\frac{2}{3}x + \frac{8}{3} + \frac{9}{3} \\ y &= -\frac{2}{3}x + \frac{17}{3} \end{aligned}$$

21. $m = \dfrac{7-3}{5-2} = \dfrac{4}{3}$

22. $m = \dfrac{-6-5}{4-(-3)} = -\dfrac{11}{7}$

23. $m = \dfrac{0.4 - 0.8}{0.5 - 0.3} = \dfrac{-0.4}{0.2} = -2$

24. $m = \dfrac{-\frac{4}{3} - \frac{3}{5}}{\frac{1}{3} - \frac{1}{2}} = \dfrac{-\frac{29}{15}}{-\frac{1}{6}} = \dfrac{29}{15} \cdot \dfrac{6}{1} = \dfrac{58}{5}$

25. At $20 per ticket,

$n = 48{,}000 - 400(20) = 40{,}000$.

At $25 per ticket

$n = 48{,}000 - 400(25) = 38{,}000$

If 35,000 tickets are sold, then the price is $32.50 per ticket.

26. If $p = \$20$, then

$R = 20(48{,}000 - 400 \cdot 20)$

$= \$800{,}000$

If $p = \$25$, then $R = 25(48{,}000 - 400 \cdot 25)$

$= \$950{,}000$

$$\begin{aligned} 1{,}280{,}000 &= p(48{,}000 - 400p) \\ 1{,}280{,}000 &= 48{,}000p - 400p^2 \\ 400p^2 - 48{,}000p + 1{,}280{,}000 &= 0 \\ p^2 - 120p + 3{,}200 &= 0 \\ (p - 80)(p - 40) &= 0 \\ p = 80 \quad \text{or} \quad p &= 40 \end{aligned}$$

A revenue of $1.28 million occurs at a price of $40 and at a price of $80.

The price that determines the maximum revenue is $60 per ticket.

9.1 WARM-UPS

1. True, because the graph of a function consists of all ordered pairs that are in the function.
2. True, that is why they are called linear functions.
3. True, because absolute value functions are generally v-shaped.
4. False, because any real number can be used in place of x in $f(x) = 3$.
5. True, because the graph of $y = ax^2 + bx + c$ for $a \neq 0$ is a parabola.
6. False, the domain of a quadratic function is $(-\infty, \infty)$ or R.
7. True, because $f(x)$ is just another name for the dependent variable y.
8. True, because $y^2 \geq 0$ and x must also be greater than or equal to 0.
9. False, because 1 is in the domain of the function and $(1, \infty)$ does not include 1.
10. True, because any real number can be used for x in $f(x) = ax^2 + bx + c$.

9.1 EXERCISES

1. A linear function is a function of the form $f(x) = mx + b$ where m and b are real numbers with $m \neq 0$.
3. The graph of a constant function is a horizontal line.
5. The graph of a quadratic function is a parabola.
7. The graph of $h(x) = -2$ is the same as the graph of the horizontal line $y = -2$.

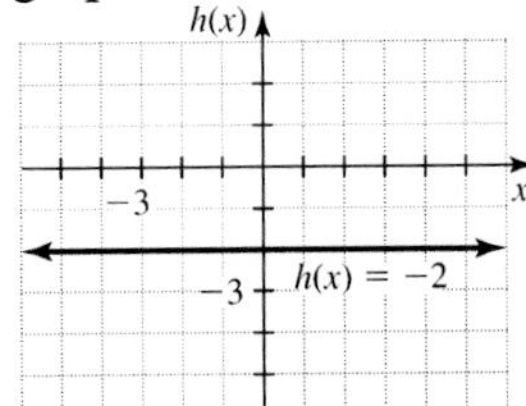

The domain of the function is R or $(-\infty, \infty)$. The only y-coordinate used is -2, and so the range is $\{-2\}$.
9. The graph of $f(x) = 2x - 1$ is the same as the graph of the linear equation $y = 2x - 1$. To draw the graph, start at the y-intercept $(0, -1)$, and use the slope of $2 = 2/1$. Rise 2 and run 1 to locate a second point on the line.

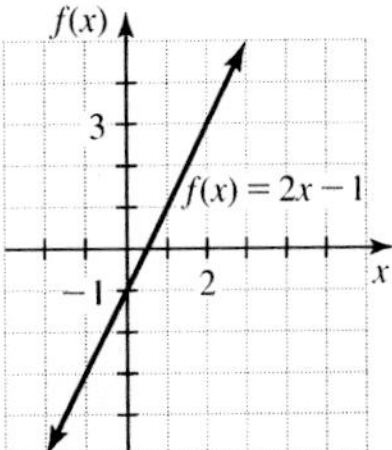

From the graph, we can see that the domain is $(-\infty, \infty)$ and the range is also $(-\infty, \infty)$.
11. Graph the line $y = (1/2)x + 2$ by locating the y-intercept (0, 2) and using a slope of $1/2$. The domain is $(-\infty, \infty)$ and the range is $(-\infty, \infty)$.

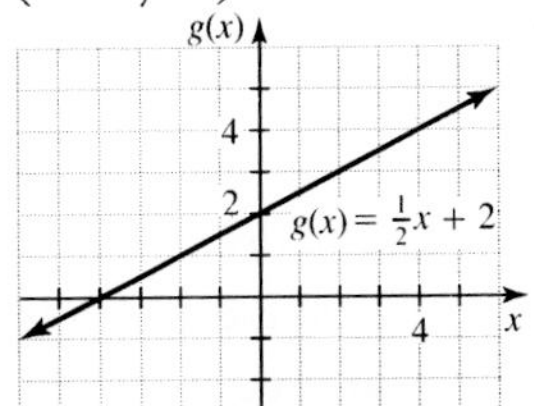

13. The graph of $y = -\frac{2}{3}x + 3$ is a straight line with y-intercept $(0, 3)$ and slope $-2/3$.

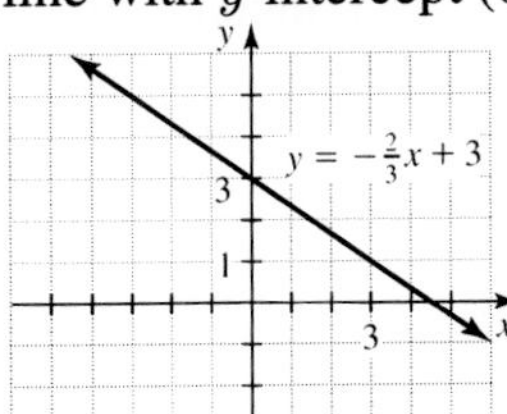

The domain is $(-\infty, \infty)$ and range is $(-\infty, \infty)$.
15. The graph of $y = -0.3x + 6.5$ is a straight line with y-intercept (0, 6.5) and slope $-3/10$.

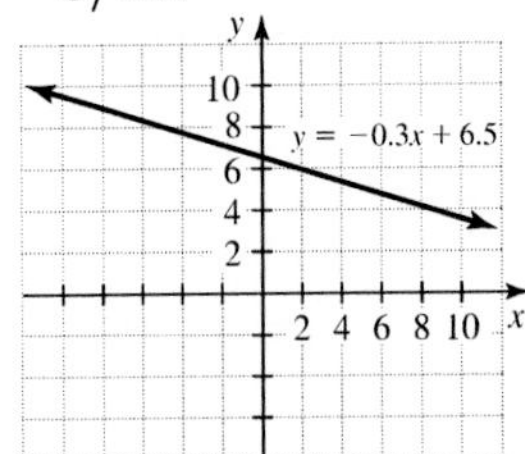

The domain is $(-\infty, \infty)$ and range is $(-\infty, \infty)$.
17. The graph of $f(x) = |\, x \,| + 1$ contains the points (0, 1), (3, 4), and (−3, 4). Plot these points and draw the v-shaped graph.

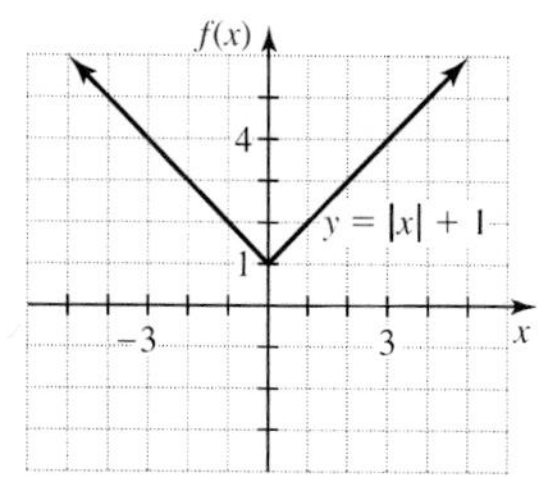

The domain is $(-\infty, \infty)$, and from the graph we can see that the range is $[1, \infty)$.

19. The graph of $h(x) = \mid x + 1 \mid$ includes the points (0, 1), (1, 2), (−1, 0), and (−2, 1).

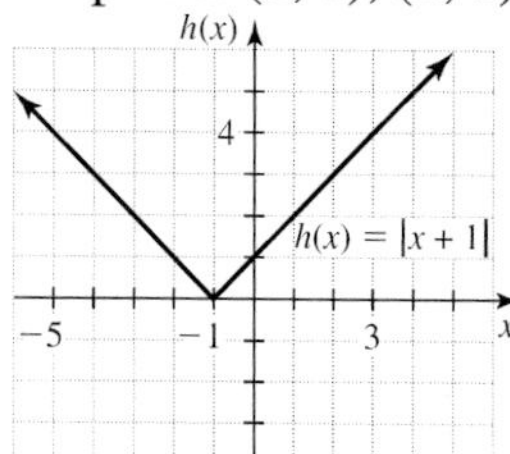

From the graph we can see that the domain is $(-\infty, \infty)$. In the vertical direction the graph is on or above the x-axis. Since all y-coordinates on the graph are greater than or equal to 0, the range is $[0, \infty)$.

21. The graph of $g(x) = \mid 3x \mid$ includes the points (0, 0), (1, 3), and (−1, 3). Plot these points and draw the graph.

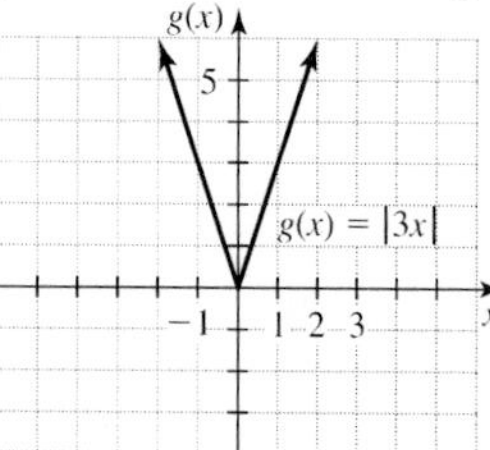

From the graph we can see that the domain is $(-\infty, \infty)$. Since all of the y-coordinates on the graph are greater than or equal to zero, the range is $[0, \infty)$.

23. The graph of $f(x) = \mid 2x - 1 \mid$ includes the points (0, 1), (1/2, 0), (1, 1), and (2, 3). Plot these points and draw a v-shaped graph.

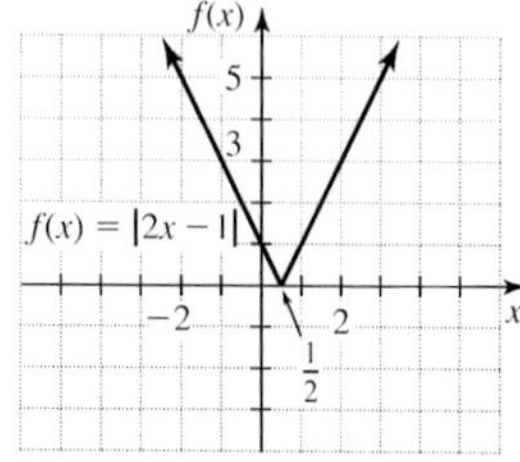

From the graph we can see that the domain is $(-\infty, \infty)$, and the range is $[0, \infty)$.

25. The graph of $f(x) = \mid x - 2 \mid + 1$ includes the points (2, 1), (3, 2), (4, 3), (1, 2), and (0, 3). Plot these points and draw a v-shaped graph. Form the graph we can see that the domain is $(-\infty, \infty)$. Since all y-coordinates are greater than or equal to 1, the range is $[1, \infty)$.

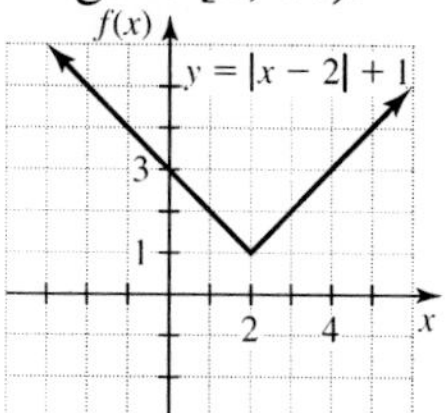

27. The graph of $y = x^2$ includes the points $(0, 0)$, $(-1, 1)$, $(1, 1)$, $(-2, 4)$, and $(2, 4)$. Plot these points and draw the parabola through them. The domain is $(-\infty, \infty)$. Since all y-coordinates are greater than or equal to zero, the range is $[0, \infty)$.

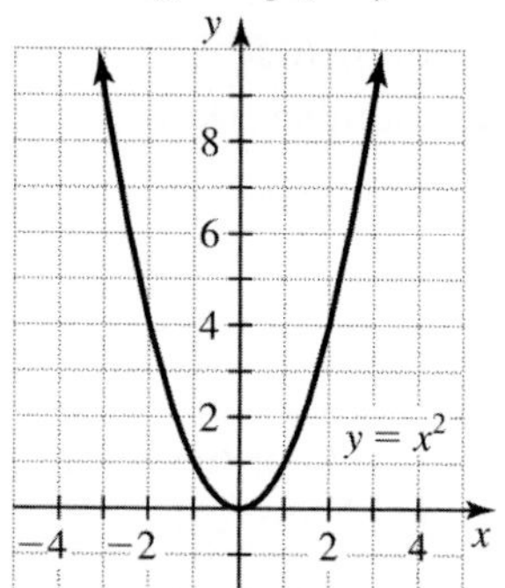

29. The graph of $g(x) = x^2 + 2$ includes the points (−2, 6), (−1, 3), (0, 2), (1, 3), and $(2, 6)$. Plot these points and draw the parabola through them.

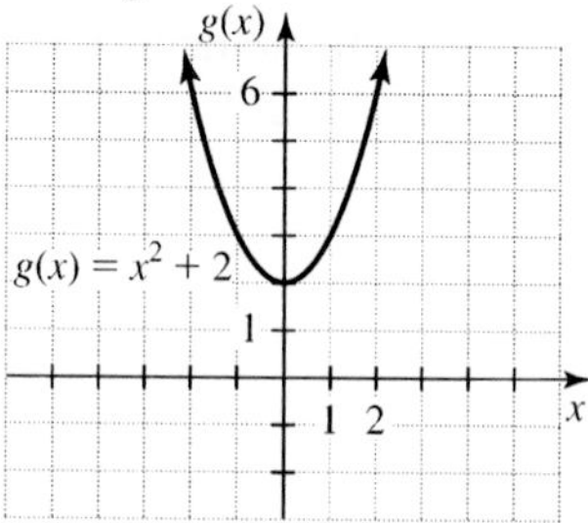

The domain is $(-\infty, \infty)$. Since no y-coordinate on the parabola is below 2, the range is $[2, \infty)$.

31. The graph of $f(x) = 2x^2$ includes the points (0, 0), (1, 2), (−1, 2), (2, 8), and $(-2, 8)$. Plot these points and draw a parabola through them.

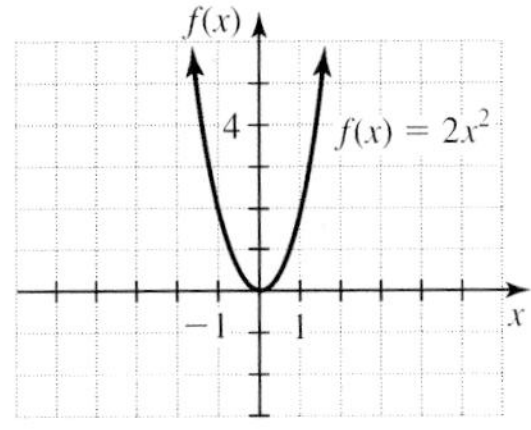

The domain is $(-\infty, \infty)$ and the range is $[0, \infty)$.

33. The graph of $y = 6 - x^2$ includes the points $(-3, -3)$, $(0, 6)$, $(3, -3)$, $(-2, 2)$, and $(2, 2)$. Plot these points and draw a parabola through them.

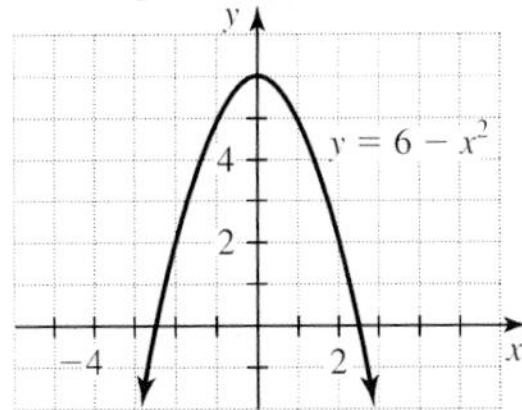

The domain is $(-\infty, \infty)$ and range is $(-\infty, 6]$.

35. The graph of $g(x) = 2\sqrt{x}$ includes the points $(0, 0)$, $(1, 2)$, and $(4, 4)$. Plot these points and draw a curve through them.

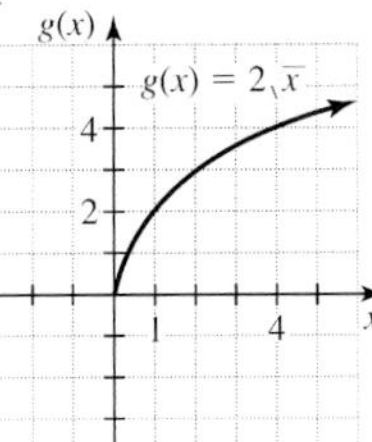

Since we must have $x \geq 0$ in $\sqrt{x}$, the domain is $[0, \infty)$. The range is $[0, \infty)$

37. The graph of $f(x) = \sqrt{x - 1}$ includes the points $(1, 0)$, $(2, 1)$, and $(5, 2)$. Plot these points and draw a smooth curve through them. The curve is half of a parabola, positioned on its side.

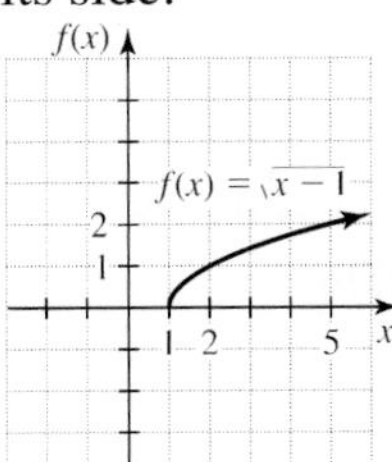

Since we must have $x - 1 \geq 0$, or $x \geq 1$ the domain is $[1, \infty)$. The range is $[0, \infty)$.

39. The graph of $h(x) = -\sqrt{x}$ includes the points $(0, 0)$, $(1, -1)$, and $(4, -2)$. Plot these points and draw a curve through them.

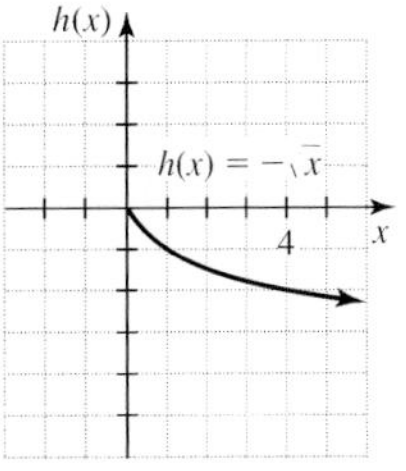

Since we must have $x \geq 0$ in $\sqrt{x}$, the domain is $[0, \infty)$. The range is $(-\infty, 0]$.

41. The graph of $y = \sqrt{x} + 2$ includes the points $(0, 2)$, $(1, 3)$, and $(4, 4)$. Graph these points and draw a curve through them.

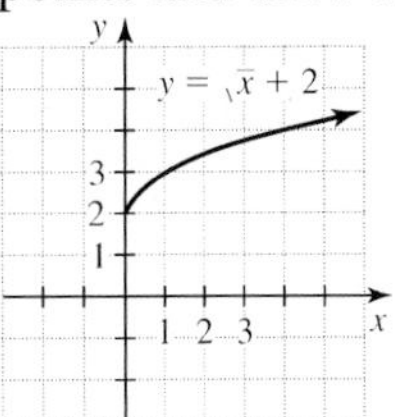

Because $\sqrt{x}$ is a real number only for nonnegative values of x, the domain is $[0, \infty)$. From the graph we see that y-coordinates go no lower than 2, so the range is $[2, \infty)$.

43. $f(x) = \begin{cases} x & \text{for } x \geq 0 \\ -4x & \text{for } x < 0 \end{cases}$

Graph $y = x$ for $x \geq 0$ and $y = -4x$ for $x < 0$.

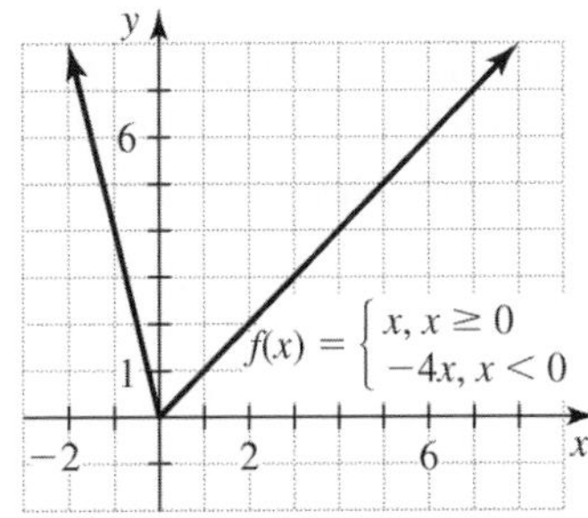

45. $f(x) = \begin{cases} 2 & \text{for } x > 1 \\ -2 & \text{for } x \leq 1 \end{cases}$

Graph $y = 2$ for $x > 1$ and $y = -2$ for $x \leq 1$.

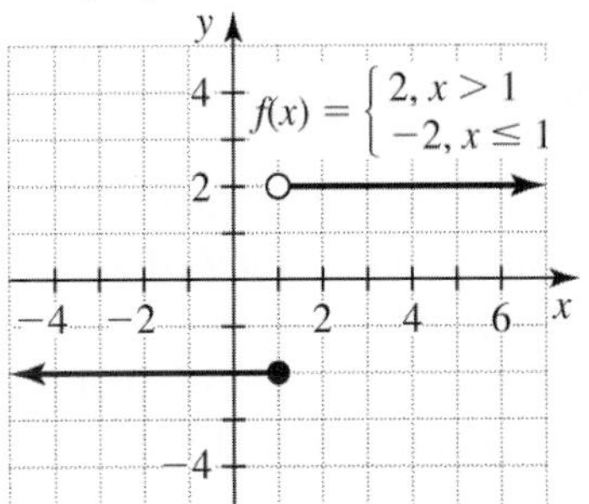

47. $f(x) = \begin{cases} \sqrt{x} & \text{for } x > 1 \\ x+3 & \text{for } x \le 1 \end{cases}$

Graph $y = \sqrt{x}$ for $x > 1$and $y = x + 3$ for $x \le 1$.

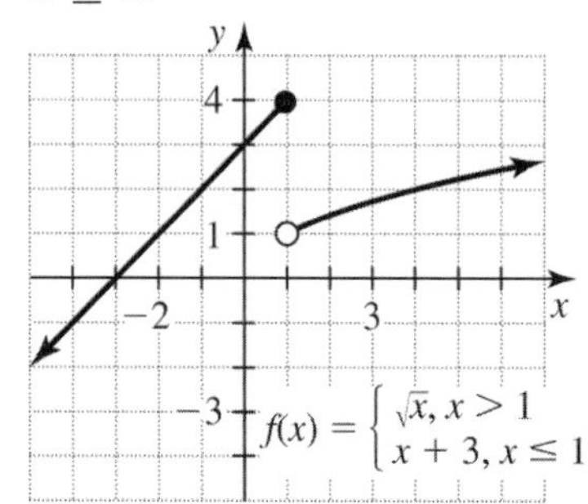

49. $f(x) = \begin{cases} \sqrt{x} & \text{for } 0 \le x \le 4 \\ x-4 & \text{for } x > 4 \end{cases}$

Graph $y = \sqrt{x}$ for $0 \le x \le 4$ and $y = x - 4$ for $x > 4$.

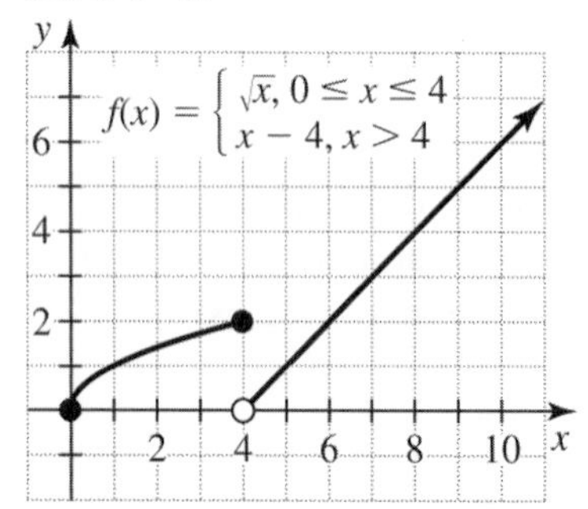

51. The graph of $x = |y|$ includes the points (0, 0), (1, −1), (1, 1), (2, −2), and (2, 2). Note that in this case it is easier to pick the y-coordinate and then find the appropriate x-coordinate using $x = |y|$. Draw the v-shaped graph through these points.

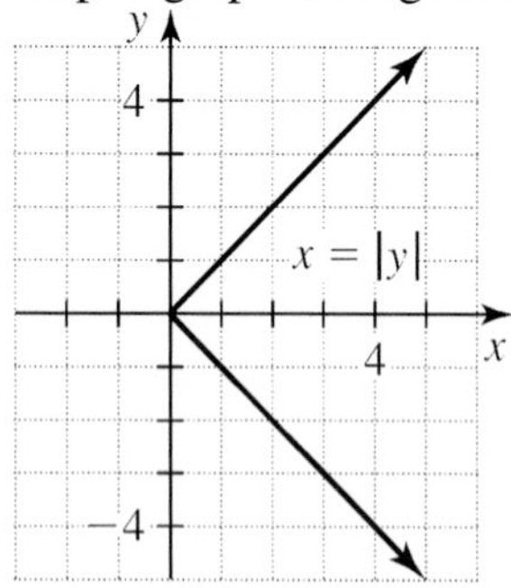

Since the x-coordinates are nonnegative, the domain is $[0, \infty)$. Since we are allowed to select any number for y, the range is $(-\infty, \infty)$.

53. To find pairs that satisfy $x = -y^2$, pick the y-coordinate first and then find the x-coordinate. The points (0, 0), (−1, 1), (1, 1), (−4, 2), and (−4, −2) are on the graph.

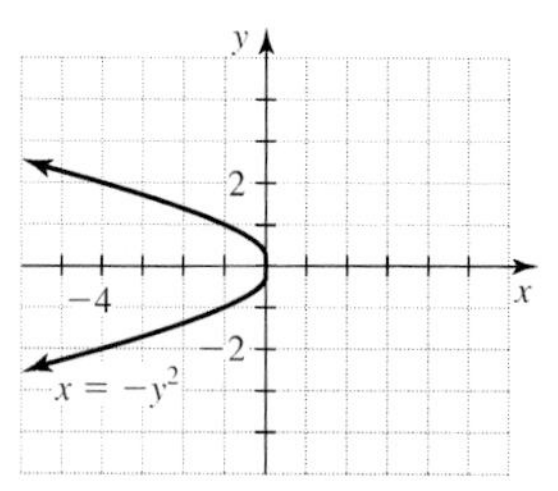

From the graph we see that only nonpositive x-coordinates are used, so the domain is $(-\infty, 0]$. Since any real number is allowable for y, the range is $(-\infty, \infty)$.

55. The equation $x = 5$ is the equation of a vertical line with an x-intercept of (5, 0).

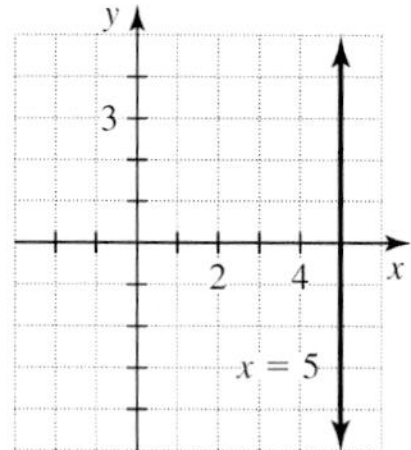

All points on this line have x-coordinate 5, so the domain is $\{5\}$. Every real number occurs as a y-coordinate, so the range is $(-\infty, \infty)$.

57. Rewrite $x + 9 = y^2$ as $x = y^2 - 9$. Select some y-coordinates and calculate the appropriate x-coordinates. The points (−9, 0), (0, 3), and (0, −3) are on the parabola.

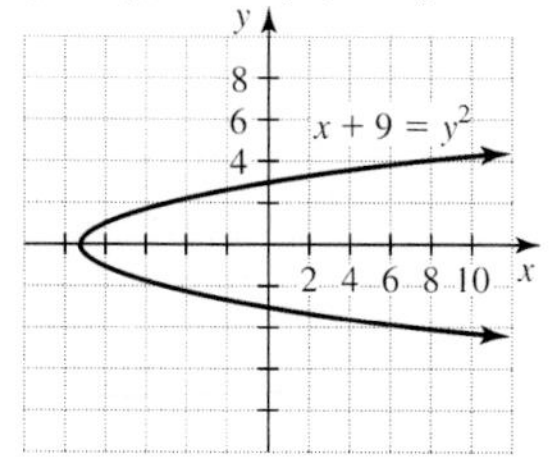

Since no x-coordinate is below −9, the domain is $[-9, \infty)$. Since we could select any value for y, the range is $(-\infty, \infty)$.

59. For the equation $x = \sqrt{y}$, we must select nonnegative values for y and find the corresponding value for x. The points (0, 0), (1, 1), and (2, 4) are on the graph.

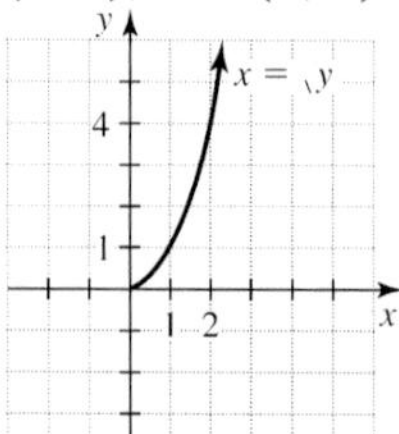

Since the x-coordinates are also nonnegative, the domain is $[0, \infty)$. Since we could only use nonnegative values for y, the range is $[0, \infty)$.

61. The graph of $x = (y - 1)^2$ includes the points (0, 1), (1, 0), (1, 2), (4, 3), and (4, −1). Plot the points and draw a curve through them. From the graph we see that the domain is $[0, \infty)$ and the range is $(-\infty, \infty)$.

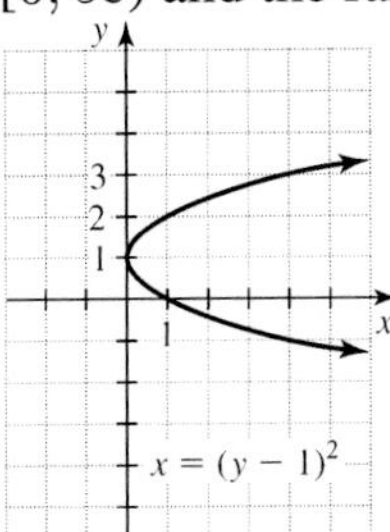

63. To graph $f(x) = 1 - |x|$, we arbitrarily select a value for x and find the corresponding value for y. The points (−2, −1), (−1, 0), (0, 1), (1, 0), and (2, −1) are on the graph.

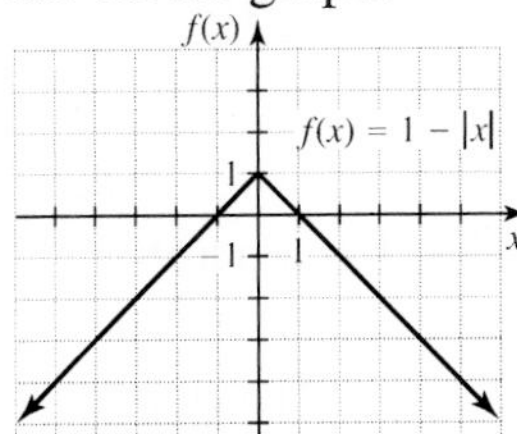

Since any real number could be used for x, the domain is $(-\infty, \infty)$. From the graph we see that the y-coordinates are not higher than 1, so the range is $(-\infty, 1]$.

65. The graph of $y = (x - 3)^2 - 1$ includes the points (1, 3), (2, 0), (3, −1), (4, 0), and (5, 3).

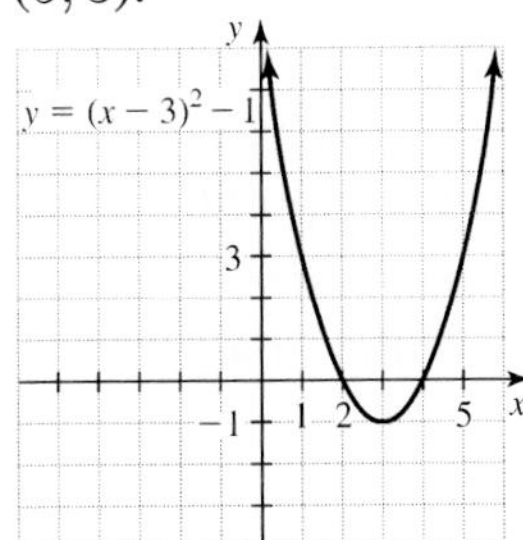

The domain is $(-\infty, \infty)$. Since the y-coordinates are greater than or equal to −1, the range is $[-1, \infty)$.

67. Graph of $y = |x + 3| + 1$ goes through (−4, 2), (−3, 1), (−2, 2), (−1, 3), (0, 4).

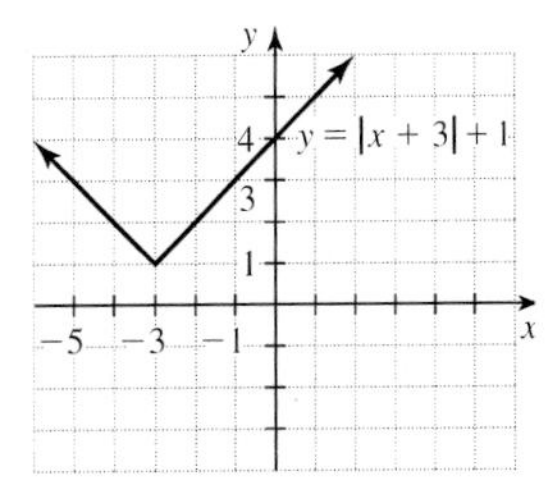

The domain is $(-\infty, \infty)$ and range is $[1, \infty)$.

69. Graph of $y = \sqrt{x} - 3$ goes through (0, −3), (1, −2), and (4, −1). Domain is $[0, \infty)$ and range is $[-3, \infty)$.

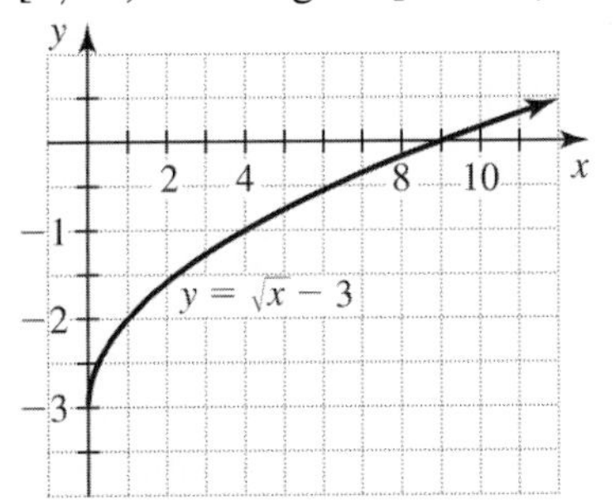

71. The graph of $y = 3x - 5$ is a straight line with y-intercept (0, −5), and a slope of 3.

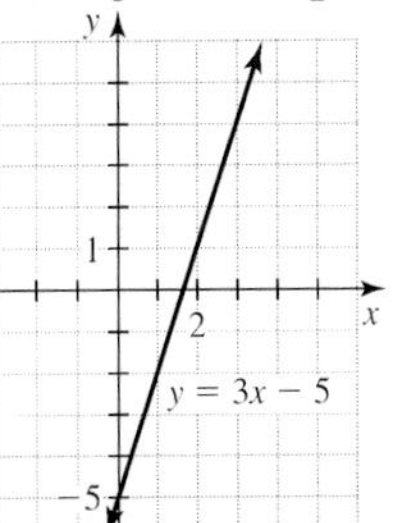

The domain is $(-\infty, \infty)$. From the graph we see that any real number could occur as a y-coordinate, so the range is $(-\infty, \infty)$.

73. The graph of $y = -x^2 + 4x - 4$ goes through $(0, -4)$, $(1, -1)$, $(2, 0)$, $(3, -1)$, and $(4, -4)$. From the graph we can see that the domain is $(-\infty, \infty)$ and range is $(-\infty, 0]$.

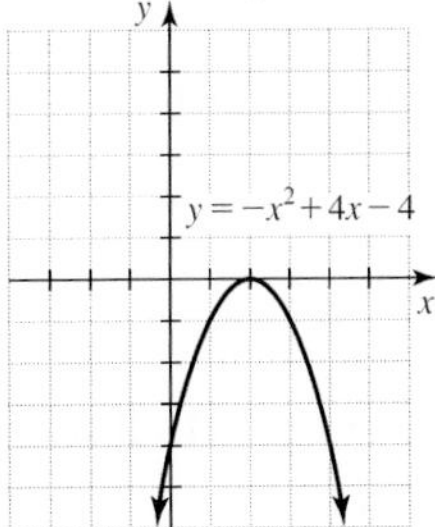

75. The function $f(x) = \sqrt{x - 3}$ is a square root function.

77. The function $f(x) = 4$ is a constant function.

79. The function $f(x) = 4x^2 - 7$ is a quadratic function.
81. The function $f(x) = 5 + \sqrt{x}$ is a square root function.
83. The function $f(x) = 99x - 100$ is a linear function.
85. Graph of $f(x) = \sqrt{x^2}$ is same as graph of $y = |x|$. This graph shows us that $\sqrt{x^2} \neq x$ when $x < 0$.
87. If $a > 0$, then for large values of a the graph gets narrower and for smaller values of a the graph gets broader.
89. The graph of $y = (x - h)^2$ is a shift of the graph of $y = x^2$ to the right for $h > 0$ and to the left for $h < 0$.
91. The graph of $y = f(x - h)$ lies to the right of the graph of $y = f(x)$ when $h > 0$.

9.2 WARM-UPS

1. False, because $(-x)^2 = x^2$ and the graphs of $f(x) = x^2$ and $g(x) = (-x)^2$ are identical.
2. True, because $-2 = -(2)$.
3. True. The graph of $y = x + 3$ is 3 units above $y = x$ or 3 units to the left of $y = x$.
4. False, because $y = |x - 3|$ is 3 units to the right of $y = |x|$.
5. True, because subtracting 3 after the absolute value moves each point on $y = |x|$ down 3 units.
6. True, because the negative sign causes the reflection and the multiplication by 2 causes the stretching.
7. False, because $f(-x) = (-x - 2)^2 = (-1)^2(x + 2)^2 = (x + 2)^2 \neq (x - 2)^2$.
8. True, because $y = \sqrt{\frac{x}{9}} = \frac{1}{3}\sqrt{x}$. The y-coordinates on $y = \sqrt{x}$ are 3 times as large as the y-coordinates on $y = \frac{1}{3}\sqrt{x}$.
9. True, because $y = \sqrt{x - 3} + 5$ is the same as $y = \sqrt{x}$ moved 3 units to the right and 5 units upward.
10. False, because the graph of $y = x^2$ must be moved 2 units to the left, reflected in the x-axis, and then moved 7 units downward to obtain the graph of $y = -(x + 2)^2 - 7$.

9.2 EXERCISES

1. The graph of $y = -f(x)$ is a reflection in the x-axis of the graph of $y = f(x)$.
3. The graph of $y = f(x) + k$ for $k < 0$ is a downward translation of the graph of $y = f(x)$.
5. The graph of $y = f(x - h)$ for $h < 0$ is a translation to the left of the graph of $y = f(x)$.
7. The graph of $f(x) = \sqrt{2x}$ goes through $(0, 0)$, $(2, 2)$, and $(8, 4)$. The graph of $g(x) = -\sqrt{2x}$ is a reflection in the x-axis of the graph of $f(x) = \sqrt{2x}$.

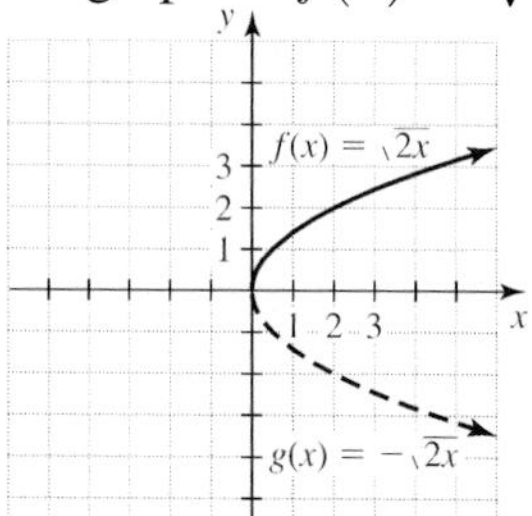

9. The graph of $f(x) = x^2 + 1$ is a parabola through $(0, 0)$, $(\pm 1, 2)$, and $(\pm 2, 5)$. The graph of $g(x) = -(x^2 + 1)$ is a reflection in the x-axis of the graph of $f(x)$.

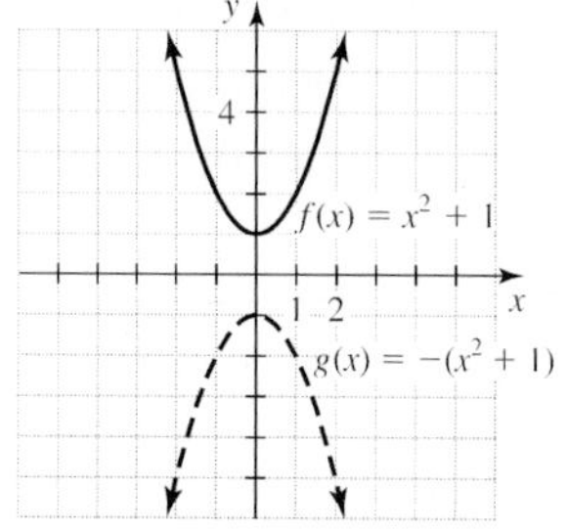

11. The graph of $y = \sqrt{x - 2}$ is half of a parabola. The graph contains the points (2, 0), (3, 1), and (6, 2). The graph of $y = -\sqrt{x - 2}$ is a reflection in the x-axis of the first graph.

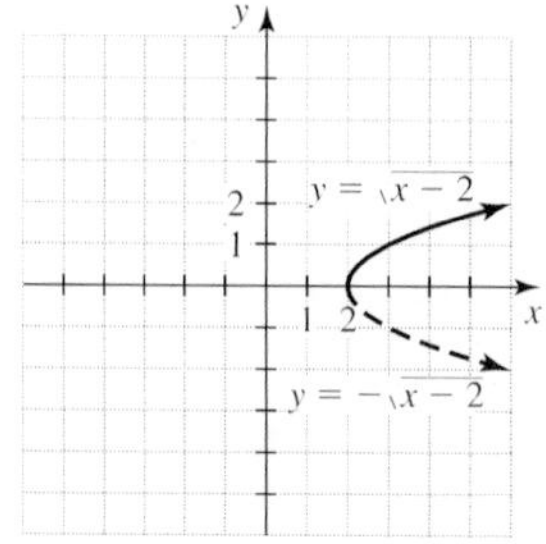

13. The graph of $f(x) = x - 3$ is a straight line through $(0, -3)$ with slope 1. The graph of $g(x) = 3 - x = -(x - 3)$ is a reflection in the x-axis of the graph of $f(x)$.

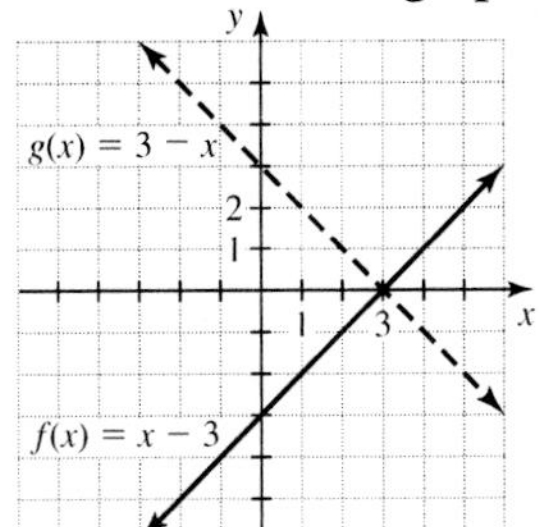

15. The graph of $g(x) = x^2$ is a parabola opening upward with vertex $(0, 0)$. The graph of $f(x) = x^2 - 4$ is a downward translation of 4 units from the graph of g. It includes the points (0,−4), (± 1, −3), and (± 2, 0).

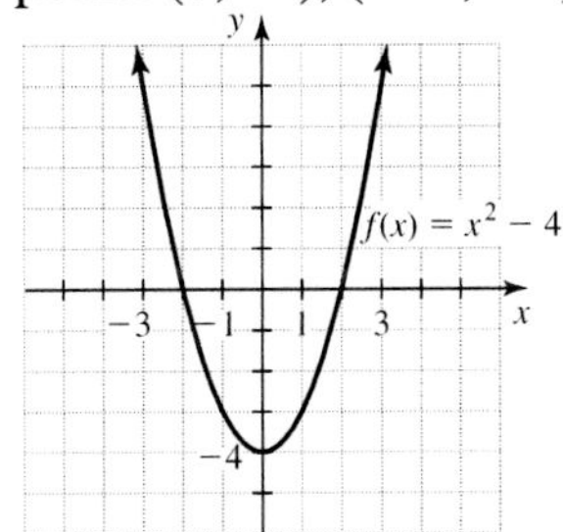

The domain is $(-\infty, \infty)$ and the range is $[-4, \infty)$.

17. The graph of $y = x$ is a straight line with slope 1 and y-intercept $(0, 0)$. The graph of $y = x + 3$ is a translation 3 units upward of the graph of $y = x$.

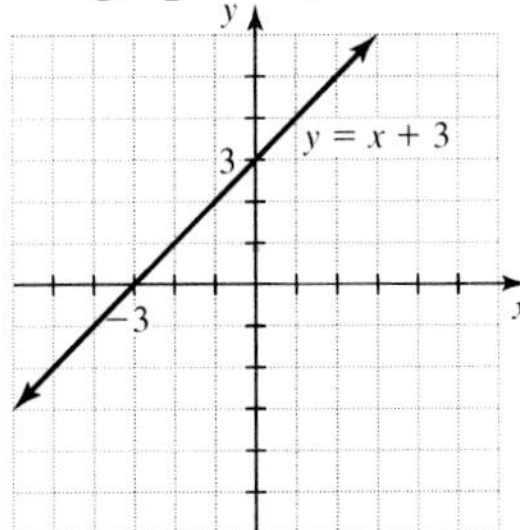

The domain is $(-\infty, \infty)$ and the range is $(-\infty, \infty)$.

19. The graph of $f(x) = (x - 3)^2$ is a translation 3 units to the right of the graph of the parabola $y = x^2$. The graph of $f(x) = (x - 3)^2$ goes through the points (2, 1), (3, 0), and (4, 1).

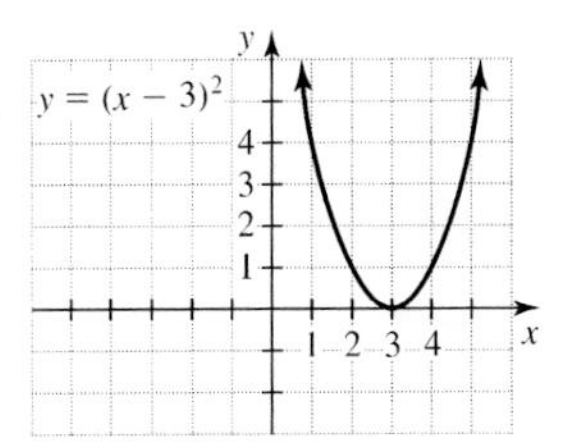

The domain is $(-\infty, \infty)$ and the range is $[0, \infty)$.

21. The graph of $y = \sqrt{x} + 1$ is a translation 1 unit upward of the graph of $y = \sqrt{x}$. The graph of $y = \sqrt{x} + 1$ goes through $(0, 1)$, $(1, 2)$, and $(4, 3)$.

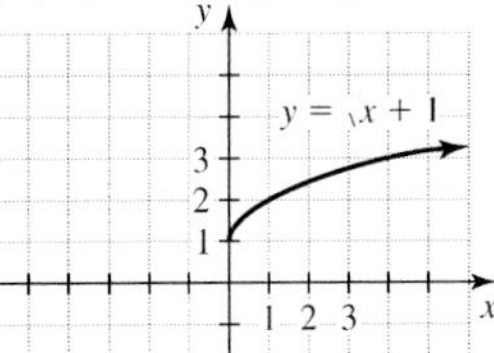

The domain is $[0, \infty)$ and the range is $[1, \infty)$.

23. The graph of $f(x) = |x + 2|$ is a translation 2 units to the left of the graph of $y = |x|$. The graph of $f(x) = |x + 2|$ includes the points (−3, 1), (−2, 0), and $(-1, 1)$.

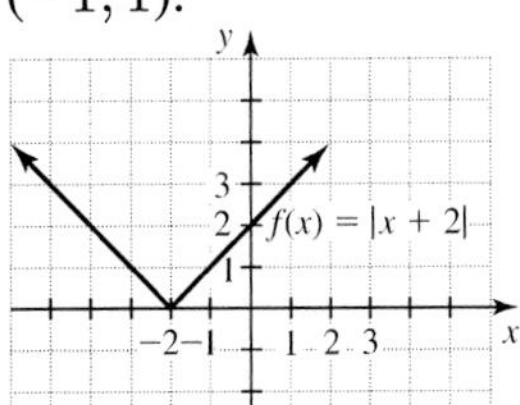

The domain is $(-\infty, \infty)$ and the range is $[0, \infty)$.

25. The graph of $y = |x| + 2$ is a translation 2 units upward of the graph of $y = |x|$. The graph of $y = |x| + 2$ goes through the points (−1, 3), (0, 2), and (1, 3).

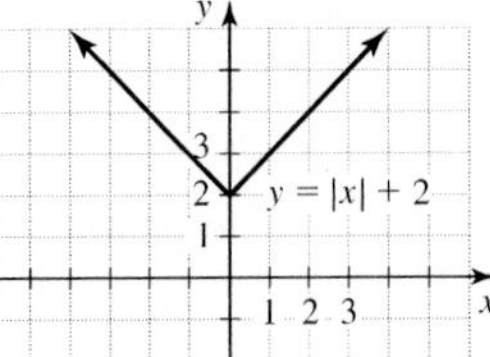

The domain is $(-\infty, \infty)$ and the range is $[2, \infty)$.

27. The graph of $f(x) = \sqrt{x - 1}$ is a translation 1 unit to the right of the graph of $y = \sqrt{x}$. The graph of $f(x) = \sqrt{x - 1}$ goes through the points (1, 0), (2, 1), and (5, 2).

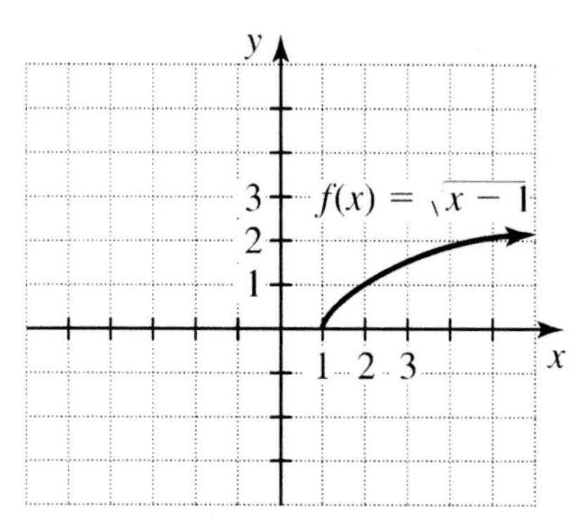

The domain is $[1, \infty)$ and the range is $[0, \infty)$.

29. The graph of $f(x) = 3x^2$ is obtained by stretching the graph of $y = x^2$. The graph of $f(x) = 3x^2$ is a parabola through (−1, 3), (0, 0), and (1, 3).

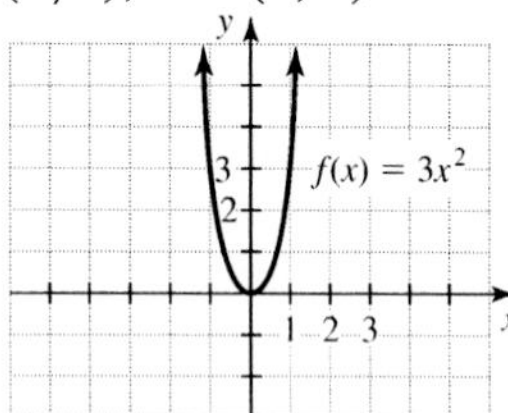

The domain is $(-\infty, \infty)$ and the range is $[0, \infty)$.

31. The graph of $y = \frac{1}{5}x$ is obtained by shrinking the graph of $y = x$. The graph of $y = \frac{1}{5}x$ is a straight line through the points (−5, −1), (0, 0), and (5, 1).

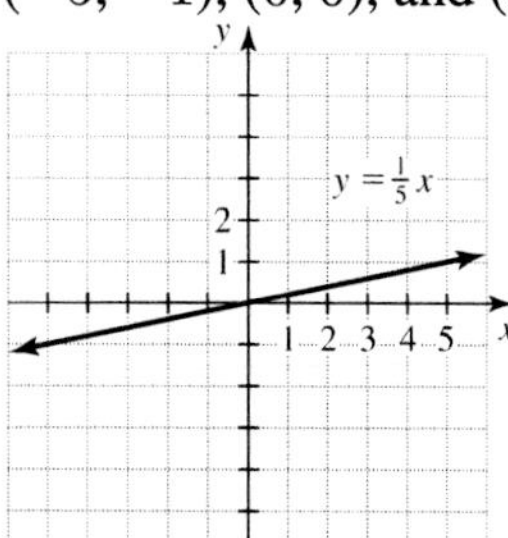

The domain is $(-\infty, \infty)$ and the range is $(-\infty, \infty)$.

33. The graph of $f(x) = 3\sqrt{x}$ is obtained by stretching the graph of $y = \sqrt{x}$. The graph of $f(x) = 3\sqrt{x}$ goes through (0, 0), (1, 3), and (4, 6).

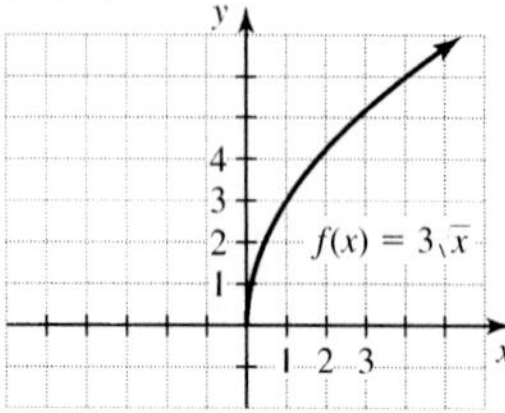

The domain is $[0, \infty)$ and the range is $[0, \infty)$.

35. The graph of $y = \frac{1}{4} \mid x \mid$ is obtained by shrinking the graph of $y = \mid x \mid$. The graph of $y = \frac{1}{4} \mid x \mid$ goes through the points (−4, 1), (0, 0), and (4, 1).

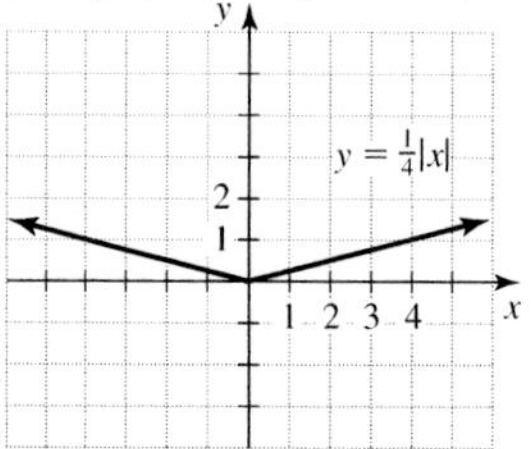

The domain is $(-\infty, \infty)$ and the range is $[0, \infty)$.

37. The graph of $y = \sqrt{x - 2} + 1$ is obtained by translating the graph of $y = \sqrt{x}$ two units to the right and one unit upward. The graph of $y = \sqrt{x - 2} + 1$ goes through the points (2, 1), (3, 2), and (6, 3).

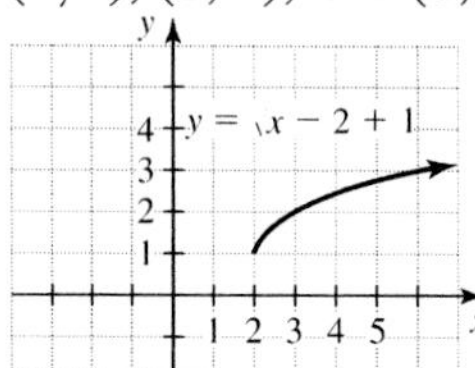

The domain is $[2, \infty)$ and the range is $[1, \infty)$.

39. The graph of $f(x) = (x + 3)^2 - 5$ is a translation of $y = x^2$ three units to the left and five units downward. The graph goes through the points (−4, −4), (−3, −5), (−2, −4), and (0, 4).

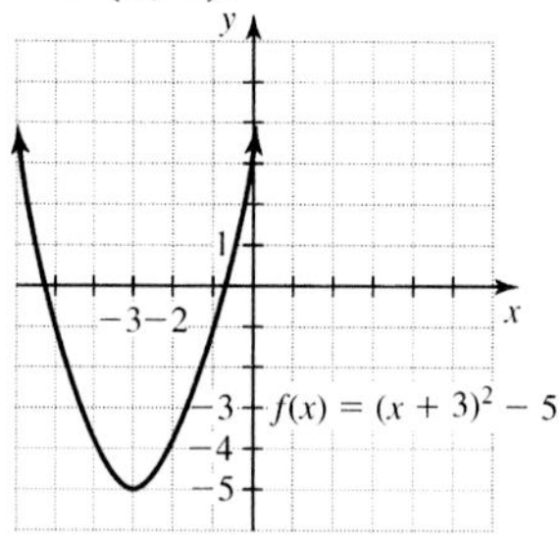

The domain is $(-\infty, \infty)$ and the range is $[-5, \infty)$.

41. The graph of $y = - \mid x + 3 \mid$ is obtained from the graph of $y = \mid x \mid$ by translating it 3 units to the left and then reflecting in the x-axis. The graph of $y = - \mid x + 3 \mid$ contains the points (−4, −1), (−3, 0), and (−2, −1).

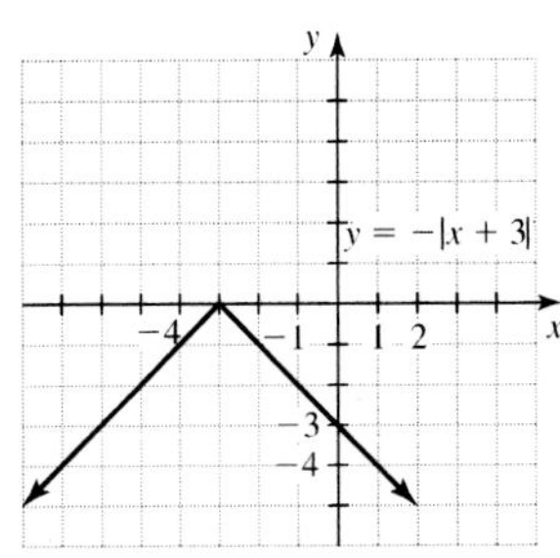

The domain is $(-\infty, \infty)$ and the range is $(-\infty, 0]$.

43. The graph of $y = -\sqrt{x+1} - 2$ is obtained from the graph of $y = \sqrt{x}$ by translating it 1 unit to the left, then reflecting in the x-axis, then translating 2 units downward. The graph of $y = -\sqrt{x+1} - 2$ includes the points (−1, −2), (0, −3), and (3, −4).

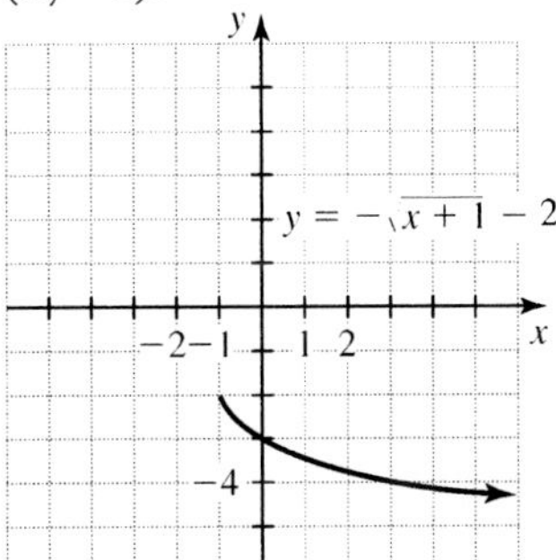

The domain is $[-1, \infty)$ and the range is $(-\infty, -2]$.

45. The graph of $f(x) = -2 \mid x - 3 \mid + 4$ is obtained from the graph of $y = \mid x \mid$ by translating it 3 units to the right, stretching by a factor of 2, reflecting in the x-axis, and then translating 4 units upward. The graph of $f(x) = -2 \mid x - 3 \mid + 4$ includes the points (2, 2), (3, 4), and (4, 2).

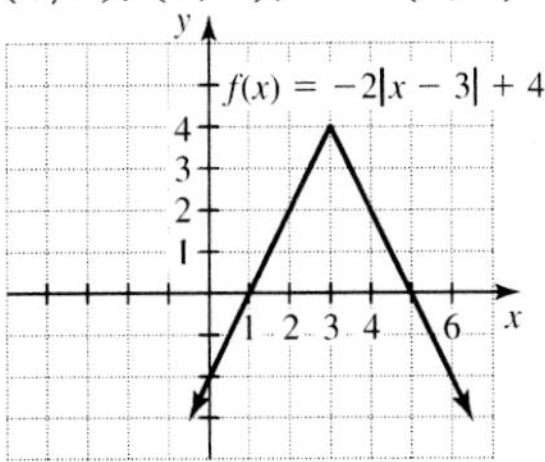

The domain is $(-\infty, \infty)$ and the range is $(-\infty, 4]$.

47. The graph of $y = -2x + 3$ is obtained from the graph of $y = x$ by stretching by a factor of 2, reflecting in the x-axis, and then translating it 3 units upward. The graph of $y = -2x + 3$ includes the points (0, 3), (1, 1), and (2, −1).

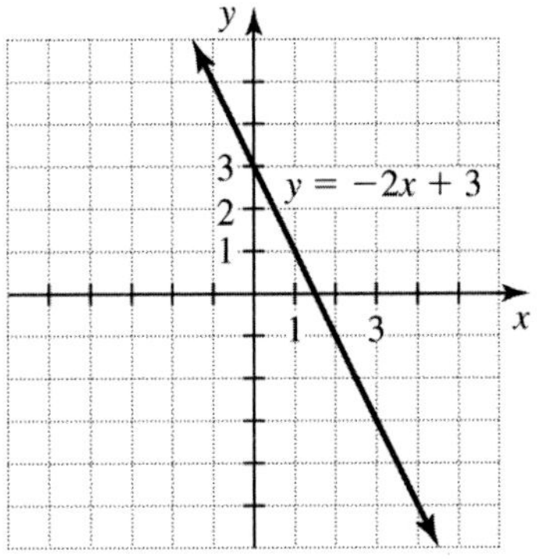

The domain is $(-\infty, \infty)$ and the range is $(-\infty, \infty)$.

49. The graph of $y = 2(x+3)^2 + 1$ is a parabola with vertex at $(-3, 1)$ and it opens upward. The graph includes the points $(-4, 3)$ and $(-2, 3)$.

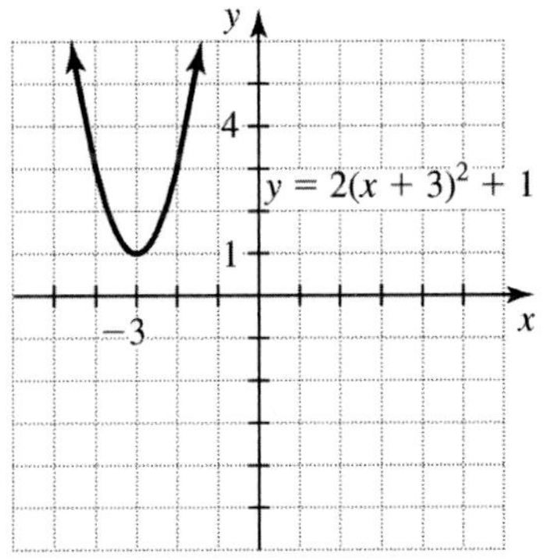

The domain is $(-\infty, \infty)$ and the range is $[1, \infty)$.

51. The graph of $y = -2(x-4)^2 + 2$ is a parabola opening downward with vertex at $(4, 2)$. The graph includes the points $(5, 0)$ and $(3, 0)$.

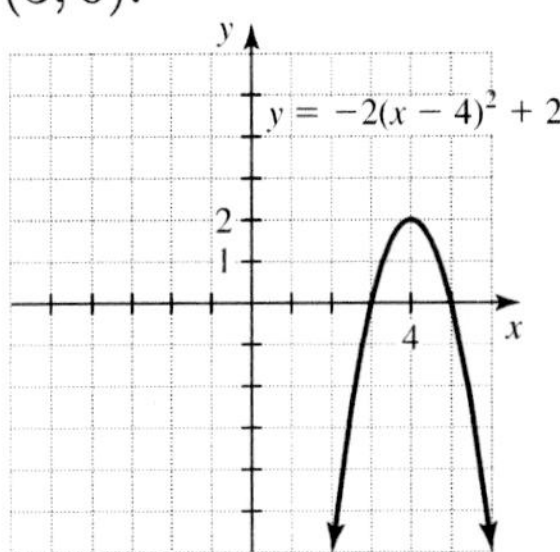

The domain is $(-\infty, \infty)$ and the range is $(-\infty, 2]$.

53. The graph of $y = -3(x-1)^2 + 6$ is a parabola opening downward with vertex at $(1, 6)$. The graph includes the points $(0, 3)$ and $(2, 3)$.

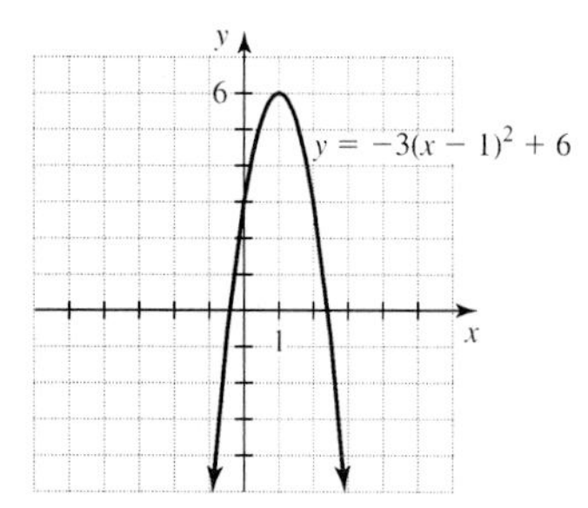

The domain is $(-\infty, \infty)$ and the range is $(-\infty, 6]$.

55. The graph of $y = 2 + \sqrt{x}$ is obtained from $y = \sqrt{x}$ by translating it 2 units upward. The graph of $y = 2 + \sqrt{x}$ includes the points (0, 2), (1, 3), and (4, 4). The graph is (d).

57. The graph of $y = 2\sqrt{x}$ is obtained by stretching the graph of $y = \sqrt{x}$. The graph of $y = 2\sqrt{x}$ goes through (0, 0), (1, 2), and (4, 4). The graph is (e).

59. The graph of $y = \frac{1}{2}\sqrt{x}$ is obtained by shrinking the graph of $y = \sqrt{x}$. The graph of $y = \frac{1}{2}\sqrt{x}$ goes through the points (0, 0), (1, 0.5), and (4, 1). The graph is (h).

61. The graph of $y = -2\sqrt{x}$ is obtained by stretching the graph of $y = \sqrt{x}$ and then reflecting in the x-axis. The graph of $y = -2\sqrt{x}$ contains the points (0, 0), (1, −2), and (4, −4). The graph is (c).

63. If we translate the graph of $y = x^2$ upward 8 units, then the equation is $y = x^2 + 8$.

65. If we translate the graph of $y = \sqrt{x}$ to the left 5 units, then the equation is $y = \sqrt{x + 5}$.

67. If we translate the graph of $y = |\,x\,|$ upward 5 units and the left 3 units, then the equation is $y = |x + 3| + 5$.

69. Translate the graph of $f(x) = |x|$ to the right 20 units and upward 30 units to get the graph of $g(x) = |x - 20| + 30$.

9.3 WARM-UPS

1. True, because $(f - g)(x) = f(x) - g(x)$ $= x - 2 - (x + 3) = -5$.

2. True, because $(f/g)(2) = f(2)/g(2)$ $= (2 + 4)/(3 \cdot 2) = 6/6 = 1$

3. False, because if $f(x) = 3x$ and $g(x) = x + 1$, then $f(g(x)) = 3(x + 1)$ $= 3x + 3$ and $g(f(x)) = 3x + 1$.

4. False, because $(f \circ g)(x) = f(x + 2)$ $= (x + 2)^2 = x^2 + 4x + 4$.

5. False, because if $f(x) = 3x$ and $g(x) = x + 1$, then $(f \circ g)(x) = f(x + 1) = 3x + 3$ and $(f \cdot g)(x) = 3x(x + 1) = 3x^2 + 3x$.

6. False, because $g[f(x)] = \sqrt{x} - 9$ and $f[g(x)] = \sqrt{x - 9}$ which are not equal for all values of x.

7. True, because $(f \circ g)(x) = f(\frac{x}{3}) = 3 \cdot \frac{x}{3} = x$.

8. True, because b determines a and a determines c so b determines the value of c.

9. True, $F(x) = \sqrt{x - 5}$ is a composition of $g(x) = x - 5$ and $h(x) = \sqrt{x}$.

10. True, because $(g \circ h)(x) = g[h(x)]$ $= g[x - 1] = (x - 1)^2 = F(x)$.

9.3 EXERCISES

1. The basic operations of functions are addition, subtraction, multiplication, and division.

3. In the composition of functions the second function is evaluated on the result of the first function.

5.
$$\begin{aligned}(f + g)(x) &= f(x) + g(x)\\ &= 4x - 3 + x^2 - 2x\\ &= x^2 + 2x - 3\end{aligned}$$

7.
$$\begin{aligned}(f \cdot g)(x) &= f(x) \cdot g(x)\\ &= (4x - 3)(x^2 - 2x)\\ &= 4x^3 - 11x^2 + 6x\end{aligned}$$

9. Use $x = 3$ in the formula of Exercise 5:
$$(f + g)(3) = 3^2 + 2(3) - 3 = 12$$

11.
$$\begin{aligned}(f - g)(-3) &= f(-3) - g(-3)\\ &= 4(-3) - 3 - [(-3)^2 - 2(-3)]\\ &= -12 - 3 - [9 + 6]\\ &= -30\end{aligned}$$

13. Use $x = -1$ in the formula of Exercise 7:
$$\begin{aligned}(f \cdot g)(-1) &= 4(-1)^3 - 11(-1)^2 + 6(-1)\\ &= -4 - 11 - 6 = -21\end{aligned}$$

15. $(f/g)(4) = \dfrac{f(4)}{g(4)} = \dfrac{4(4) - 3}{4^2 - 2(4)} = \dfrac{13}{8}$

17. Since $a = 3x$, and $y = 2a$ we have
$$y = 2(3x)$$
$$y = 6x$$

19. Since $a = 2x - 6$, and $y = 3a - 2$ we have
$$y = 3(2x - 6) - 2$$
$$y = 6x - 18 - 2$$
$$y = 6x - 20$$

21. Let $d = (x + 1)/2$ in $y = 2d + 1$.
$$y = 2\left(\frac{x+1}{2}\right) + 1$$
$$y = x + 1 + 1$$
$$y = x + 2$$

23. Let $m = x + 1$ in $y = m^2 - 1$.
$$y = (x + 1)^2 - 1$$
$$y = x^2 + 2x + 1 - 1$$
$$y = x^2 + 2x$$

25.
$$y = \frac{a-3}{a+2}$$
$$y = \frac{\frac{2x+3}{1-x} - 3}{\frac{2x+3}{1-x} + 2}$$
$$y = \frac{\left(\frac{2x+3}{1-x} - 3\right)(1-x)}{\left(\frac{2x+3}{1-x} + 2\right)(1-x)}$$
$$y = \frac{2x + 3 - 3(1-x)}{2x + 3 + 2(1-x)}$$
$$y = \frac{2x + 3 - 3 + 3x}{2x + 3 + 2 - 2x} = \frac{5x}{5}$$
$$y = x$$

27. $(g \circ f)(1) = g[f(1)] = g(-1)$
$= (-1)^2 + 3(-1) = -2$

29. $(f \circ g)(1) = f[g(1)] = f[4]$
$= 2(4) - 3 = 5$

31. $(f \circ f)(4) = f[f(4)] = f[5]$
$= 2(5) - 3 = 7$

33. $(h \circ f)(5) = h[f(5)] = h[7] = \frac{7+3}{2} = 5$

35. $(f \circ h)(5) = f[h(5)] = f[4] = 2(4) - 3$
$= 5$

37. $(g \circ h)(-1) = g[h(-1)] = g[1] = 4$

39. $(f \circ g)(2.36) = f[g(2.36)] = f[12.6496]$
$= 22.2992$

41. $(g \circ f)(x) = g[f(x)] = g[2x - 3]$
$= (2x - 3)^2 + 3(2x - 3)$
$= 4x^2 - 12x + 9 + 6x - 9$
$= 4x^2 - 6x$

43. $(f \circ g)(x) = f[g(x)] = f[x^2 + 3x]$
$= 2(x^2 + 3x) - 3 = 2x^2 + 6x - 3$

45. $(h \circ f)(x) = h[f(x)] = h[2x - 3]$
$$= \frac{2x - 3 + 3}{2} = x$$

47. $(f \circ f)(x) = f[f(x)] = f[2x - 3]$
$= 2(2x - 3) - 3 = 4x - 9$

49. $(h \circ h)(x) = h[h(x)] = h[\frac{x+3}{2}]$
$$= \frac{\frac{x+3}{2} + 3}{2} = \frac{\left(\frac{x+3}{2} + 3\right)2}{2 \cdot 2}$$
$$= \frac{x + 3 + 6}{4} = \frac{x+9}{4}$$

51. $F(x) = \sqrt{x - 3} = \sqrt{h(x)} = f[h(x)]$
Therefore, $F = f \circ h$.

53. $G(x) = x^2 - 6x + 9 = (x - 3)^2$
$= [h(x)]^2 = g[h(x)]$
Therefore $G = g \circ h$.

55. $H(x) = x^2 - 3 = g(x) - 3 = h[g(x)]$
Therefore $H = h \circ g$.

57. $J(x) = x - 6 = x - 3 - 3$
$= h(x) - 3 = h[h(x)]$
Therefore, $J = h \circ h$.

59. $K(x) = x^4 = (x^2)^2 = [g(x)]^2 = g[g(x)]$
Therefore, $K = g \circ g$.

61. $f[g(x)] = f\left(\frac{x-5}{3}\right) = 3\left(\frac{x-5}{3}\right) + 5$
$= x$
$$g[f(x)] = g(3x + 5) = \frac{3x + 5 - 5}{3} = x$$

63. $f[g(x)] = f\left(\sqrt[3]{x + 9}\right)$
$$= \left(\sqrt[3]{x+9}\right)^3 - 9 = x + 9 - 9 = x$$
$$g[f(x)] = g[x^3 - 9] = \sqrt[3]{x^3 - 9 + 9}$$
$$= \sqrt[3]{x^3} = x$$

65. $f[g(x)] = f\left(\frac{x+1}{1-x}\right) = \dfrac{\frac{x+1}{1-x} - 1}{\frac{x+1}{1-x} + 1}$
$$= \frac{\left(\frac{x+1}{1-x} - 1\right)(1-x)}{\left(\frac{x+1}{1-x} + 1\right)(1-x)} = \frac{x + 1 - (1 - x)}{x + 1 + 1 - x}$$
$$= \frac{2x}{2} = x$$
$$g[f(x)] = g\left(\frac{x-1}{x+1}\right) = \frac{\frac{x-1}{x+1} + 1}{1 - \frac{x-1}{x+1}}$$
$$= \frac{\left(\frac{x-1}{x+1} + 1\right)(x+1)}{\left(1 - \frac{x-1}{x+1}\right)(x+1)} = \frac{x - 1 + x + 1}{x + 1 - (x - 1)}$$
$$= \frac{2x}{2} = x$$

67. $f[g(x)] = f\left(\frac{1}{x}\right) = \frac{1}{\frac{1}{x}} = x$

$g[f(x)] = g\left(\frac{1}{x}\right) = \frac{1}{\frac{1}{x}} = x$

69. Since $f(x) = x^2$, $f(3) = 3^2 = 9$ and the statement is true.

71. Since $(f + g)(4) = f(4) + g(4)$
$= 4^2 + 4 + 5$
$= 25$
the statement is false.

73. Since $(f \cdot g)(3) = f(3) \cdot g(3)$
$= 3^2(3 + 5) = 72$
the statement is true.

75. Since $(f \circ g)(2) = f(g(2))$
$= f(7) = 7^2 = 49$
the statement is false.

77. Since $(f(g(x)) = f(g(x))$
$= f(x + 5)$
$= (x + 5)^2$
$= x^2 + 10x + 25$
the statement is false.

79. Since $(f \circ g)(x) = f(x + 5)$
$= (x + 5)^2$
$= x^2 + 10x + 25$
the statement is true.

81. a) Let $s =$ the length of a side of the square.
$10^2 = s^2 + s^2$
$2s^2 = 100$
$s^2 = 50$
$s = \sqrt{50}$
Since $A = s^2$, we have $A = 50$ ft^2

b) In general if d is the diagonal of a square and s is the length of a side, we have $d^2 = s^2 + s^2$ or $s^2 = \frac{d^2}{2}$. Since $A = s^2$, we have $A = \frac{d^2}{2}$.

83. $P(x) = R(x) - C(x)$
$= x^2 - 10x + 30 - (2x^2 - 30x + 200)$
$= -x^2 + 20x - 170$

85. Substitute $F = 0.25I$ into $J = 0.10F$.
$J = 0.10(0.25I)$
$J = 0.025I$

87. a) $x = (30.25/100)^3 \approx 0.0277$
$D = (24665/2240)/0.0277 \approx 397.8$
b) $x = (L/100)^3$
$D = (25000/2240)/x$
$D = (25000/2240)/(L/100)^3$
$D = \frac{25000}{2240} \cdot \frac{100^3}{L^3}$
$D = \frac{1.116 \times 10^7}{L^3}$
c) From the graph it appears that the displacement-length ratio decreases as the length increases.

89. The domain of $f(x) = \sqrt{x} - 4$ is $[0, \infty)$. The domain of $g(x) = \sqrt{x}$ is $[0, \infty)$.
$(g \circ f)(x) = g(\sqrt{x} - 4) = \sqrt{\sqrt{x} - 4}$
So that $\sqrt{x} - 4 \geq 0$ we must have $x \geq 16$. So the domain of $g \circ f$ is $[16, \infty)$.

91. The graph of $y = x + \sqrt{x}$ appears only for $x \geq 0$. So the domain is $[0, \infty)$. The graph starts at $(0, 0)$ and goes up from there. So the range is $[0, \infty)$.

9.4 WARM-UPS

1. False, because the inverse function is obtained by interchanging the coordinates in every ordered pair.
2. False, because the points (0, 3) and (1, 3) both satisfy $f(x) = 3$.
3. False, because the inverse of multiplication by 2 is division by 2. If $g(x) = 2x$, then $g^{-1}(x) = x/2$.
4. True, because if we interchange the coordinates in a function that is not one-to-one, we do not obtain a function.
5. True, because in an inverse function the domain and range of the function are reversed.
6. False, because it is not one-to-one. Both $(2, 16)$ and $(-2, 16)$ satisfy $f(x) = x^4$.
7. True, because taking the opposite of a number twice gives back the original number.
8. True, because the inverse function interchanges the coordinates in all of the ordered pairs of the function.
9. True, the inverse of $k(x) = 3x - 6$ is $k^{-1}(x) = (x + 6)/3 = (1/3)x + 2$.
10. False, because $f^{-1}(x) = (x + 4)/3$.

9.4 EXERCISES

1. The inverse of a function is a function with the same ordered pairs except that the coordinates are reversed.

3. The range of f^{-1} is the domain of f.

5. A function is one-to-one if no two ordered pairs have the same second coordinate with different first coordinates.

7. The switch and solve strategy is used to find a formula for an inverse function.

9. The function is one-to-one. The inverse is $\{(3, 1), (9, 2)\}$.

11. The function is not one-to-one because of the pairs $(-2, 2)$ and $(2, 2)$. Therefore, the function is not invertible.

13. The function is one-to-one. The inverse is $\{(4, 16), (3, 9), (0, 0)\}$.
It is obtained by interchanging the coordinates in each ordered pair.

15. The function is not one-to-one because of the pairs $(5, 0)$ and $(6, 0)$. Therefore, it is not invertible.

17. This function is one-to-one. Its inverse is obtained by interchanging the coordinates in each ordered pair: $\{(0, 0), (2, 2), (9, 9)\}$.

19. This function is not one-to-one because we can draw a horizontal line that crosses the graph twice.

21. This function is one-to-one because we cannot draw a horizontal line that crosses the graph more than once.

23. Since $(f \circ g)(x) = f(g(x)) = f(0.5x)$
$= 2(0.5x) = x$ and $(g \circ f)(x) = g(f(x))$
$= g(2x) = 0.5(2x) = x$, the functions are inverses of each other.

25. $(f \circ g)(x) = f(g(x)) = f(\frac{1}{2}x + 5)$
$= 2(\frac{1}{2}x + 5) - 10 = x + 10 - 10 = x$
$(g \circ f)(x) = g(f(x)) = g(2x - 10)$
$= \frac{1}{2}(2x - 10) + 5 = x - 5 + 5 = x$
Therefore, the functions are inverses of each other.

27. $(f \circ g)(x) = f(g(x)) = f(-x) = -(-x)$
$= x$
$(g \circ f)(x) = g(f(x)) = g(-x) = -(-x) = x$
Therefore, the functions are inverses of each other.

29. $(g \circ f)(x) = g(f(x)) = g(x^4) = (x^4)^{1/4}$
$= |\, x \,|$
For example, $g(f(-2)) = g(16) = 2$. So the functions are not inverses of each other.

31. $y = 5x$, $x = 5y$, $y = \frac{1}{5}x$
So $f^{-1}(x) = \frac{x}{5}$.

33. $y = x - 9$, $x = y - 9$, $y = x + 9$
So $g^{-1}(x) = x + 9$.

35. $y = 5x - 9$, $x = 5y - 9$, $5y = x + 9$,
$y = \frac{x + 9}{5}$
So $k^{-1}(x) = \frac{x + 9}{5}$.

37. $y = \frac{2}{x}$, $x = \frac{2}{y}$, $xy = 2$, $y = \frac{2}{x}$
So $m^{-1}(x) = \frac{2}{x}$.

39. $y = \sqrt[3]{x - 4}$, $x = \sqrt[3]{y - 4}$, $x^3 = y - 4$,
$y = x^3 + 4$ So $f^{-1}(x) = x^3 + 4$.

41. $y = \frac{3}{x - 4}$, $x = \frac{3}{y - 4}$, $x(y - 4) = 3$,
$y - 4 = \frac{3}{x}$, $y = \frac{3}{x} + 4$
So $f^{-1}(x) = \frac{3}{x} + 4$.

43.
$$\begin{aligned} f(x) &= \sqrt[3]{3x + 7} \\ y &= \sqrt[3]{3x + 7} \\ x &= \sqrt[3]{3y + 7} \\ x^3 &= 3y + 7 \\ x^3 - 7 &= 3y \\ \frac{x^3 - 7}{3} &= y \\ f^{-1}(x) &= \frac{x^3 - 7}{3} \end{aligned}$$

45.
$$\begin{aligned} f(x) &= \frac{x + 1}{x - 2} \\ y &= \frac{x + 1}{x - 2} \\ x &= \frac{y + 1}{y - 2} \\ x(y - 2) &= y + 1 \\ xy - 2x &= y + 1 \\ xy - y &= 2x + 1 \\ y(x - 1) &= 2x + 1 \\ y &= \frac{2x + 1}{x - 1} \\ f^{-1}(x) &= \frac{2x + 1}{x - 1} \end{aligned}$$

47.
$$\begin{aligned} f(x) &= \frac{x + 1}{3x - 4} \\ y &= \frac{x + 1}{3x - 4} \\ x &= \frac{y + 1}{3y - 4} \\ x(3y - 4) &= y + 1 \\ 3xy - 4x &= y + 1 \\ 3xy - y &= 1 + 4x \\ y(3x - 1) &= 1 + 4x \end{aligned}$$

$$y = \frac{1+4x}{3x-1}$$
$$f^{-1}(x) = \frac{1+4x}{3x-1}$$

49. The function $p(x) = \sqrt[4]{x}$ finds the fourth root. The inverse must be the fourth power. Since the domain and range of p are both the set of nonnegative real numbers, $p^{-1}(x) = x^4$ for $x \geq 0$.

51. $y = (x-2)^2$, $x = (y-2)^2$, $y - 2 = \pm\sqrt{x}$, $y = 2 \pm \sqrt{x}$
Since $x \geq 2$ in the function, we must have $y \geq 2$ in the inverse function. So $y = 2 - \sqrt{x}$ is not the inverse and $f^{-1}(x) = 2 + \sqrt{x}$.

53. $y = x^2 + 3$, $x = y^2 + 3$, $y^2 = x - 3$, $y = \pm\sqrt{x-3}$. Since $x \geq 0$ in the function, $y \geq 0$ in the inverse function. So $f^{-1}(x) = \sqrt{x-3}$.

55.
$$f(x) = \sqrt{x+2}$$
$$y = \sqrt{x+2}$$
$$x = \sqrt{y+2}$$
$$x^2 = y+2$$
$$x^2 - 2 = y$$

Since $y \geq 0$ in the function, $x \geq 0$ in the inverse. So $f^{-1}(x) = x^2 - 2$ for $x \geq 0$.

57. The inverse of $f(x) = 2x + 3$ is a function that subtracts 3 and then divides by 2, $f^{-1}(x) = \frac{x-3}{2} = \frac{1}{2}x - \frac{3}{2}$. Use y-intercepts and slopes to graph each straight line.

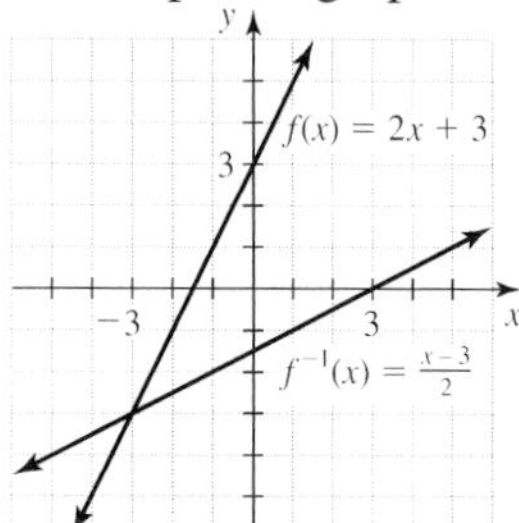

59. The graph of $f(x) = x^2 - 1$ for $x \geq 0$ contains the points $(0, -1)$, $(1, 0)$, and $(2, 3)$. For the inverse function we add 1 and then take the square root, $f^{-1}(x) = \sqrt{x+1}$. The function f^{-1} contains $(-1, 0)$, $(0, 1)$, and $(3, 2)$.

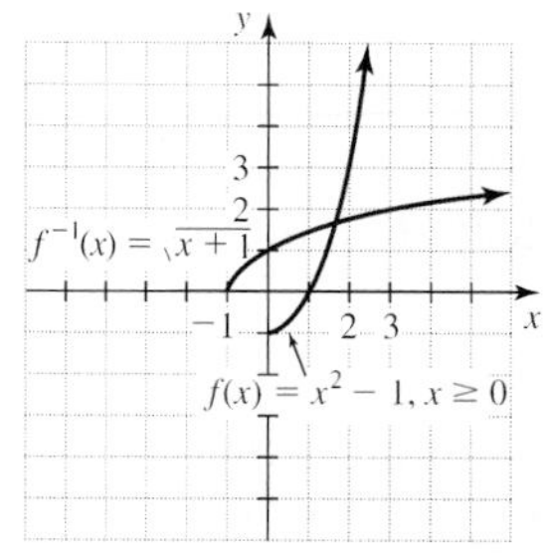

61. The graph of f is a line through (0, 0) with slope 5, and the graph of $f^{-1}(x) = x/5$ is a straight line through (0, 0) with slope $1/5$.

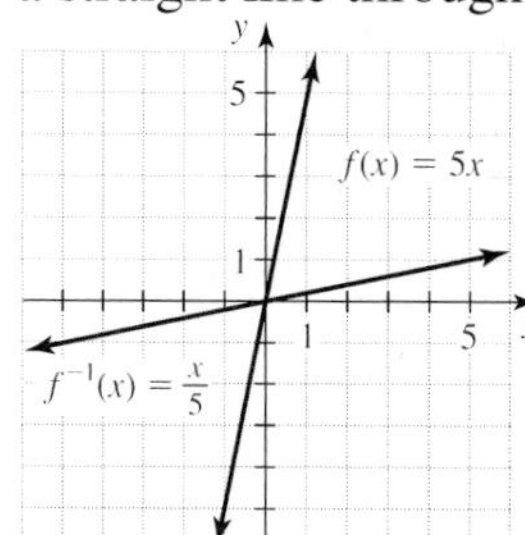

63. The inverse of cubing is the cube root. So $f^{-1}(x) = \sqrt[3]{x}$. The graph of f contains the points $(0, 0)$, $(1, 1)$, $(-1, -1)$, $(2, 8)$, and $(-2, -8)$. The graph of f^{-1} contains the points $(0, 0)$, $(1, 1)$, $(-1, -1)$, $(8, 2)$, and $(-8, -2)$.

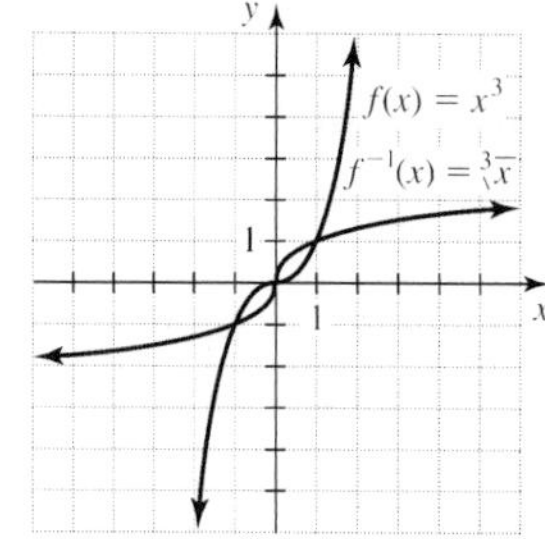

65. The inverse of subtracting 2 and then taking a square root is squaring and then adding 2. So $f^{-1}(x) = x^2 + 2$ for $x \geq 0$. The graph of f contains (2, 0), (3, 1), and (6, 2). The graph of f^{-1} contains (0, 2), (1, 3), and $(2, 6)$.

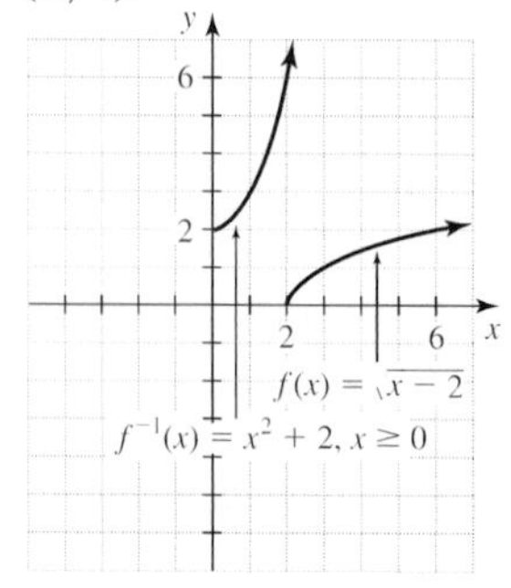

67. The inverse of multiplying by 2 is dividing by 2. So $f^{-1}(x) = \frac{x}{2}$.

69. The inverse of multiplying by 2 and then subtracting 1 is adding 1 and then dividing by 2. So $f^{-1}(x) = \frac{x+1}{2}$.

71. The inverse of cube root is the cubing function. So $f^{-1}(x) = x^3$.

73. The inverse of subtracting 1 and then taking the cube root is cubing and then adding 1. So $f^{-1}(x) = x^3 + 1$.

75. For the function you take the cube root, multiply by 2, and then subtract 1. For the inverse function you add 1, divide by 2, and the cube. So $f^{-1}(x) = \left(\frac{x+1}{2}\right)^3$.

77. $(f^{-1} \circ f)(x) = f^{-1}(f(x)) = f^{-1}(x^3 - 1)$
$= \sqrt[3]{x^3 - 1 + 1} = x$

79. $(f^{-1} \circ f)(x) = f^{-1}(f(x)) = f^{-1}(\frac{1}{2}x - 3)$
$= 2(\frac{1}{2}x - 3) + 6 = x - 6 + 6 = x$

81. $(f^{-1} \circ f)(x) = f^{-1}(f(x)) = f^{-1}(\frac{1}{x} + 2)$
$= \frac{1}{\frac{1}{x} + 2 - 2} = \frac{1}{\frac{1}{x}} = x$

83. $(f^{-1} \circ f)(x) = f^{-1}(f(x)) = f^{-1}(\frac{x+1}{x-2})$
$= \frac{2(\frac{x+1}{x-2}) + 1}{\frac{x+1}{x-2} - 1} = \frac{\left(\frac{2x+2}{x-2} + 1\right)(x-2)}{\left(\frac{x+1}{x-2} - 1\right)(x-2)}$
$= \frac{2x + 2 + x - 2}{x + 1 - x + 2} = \frac{3x}{3} = x$

85. a) $S = \sqrt{30 \cdot 50 \cdot 0.75} \approx 33.5$ mph
b) From the graph you can see that the drag factor decreases when a road gets wet because the skid marks for a given speed will be longer.
c)
$$S = \sqrt{30L \cdot 1}$$
$$S^2 = 30L$$
$$L = \frac{S^2}{30}$$

87. If x is the selling price, then the total cost is given by $T(x) = x + 0.09x + 125$ or $T(x) = 1.09x + 125$. If x is the total cost, then the function $T^{-1}(x) = \frac{x - 125}{1.09}$ gives the selling price as a function of the total cost.

89. Only the odd powers of x are one-to-one functions. So $f(x) = x^n$ is invertible if n is an odd positive integer. If n is an even positive integer, the function $f(x) = x^n$ is not one-to-one and not invertible.

91. The functions f and g are not inverses of each other because the function $y_2 = \sqrt{y_1} = \sqrt{x^2}$ is the absolute value function and not the identity function.

9.5 WARM-UPS

1. True, because direct variation means that a is a constant multiple of b.
2. False, because a inversely proportional to b means $a = k/b$.
3. False, because jointly proportional means $a = kbc$ for some constant k.
4. True, a is a constant multiple of the square root of c.
5. False, b is directly proportional to a means that $b = ka$.
6. True, because a is a multiple of b and b is divided by c.
7. False, a is jointly proportional to c and the square of b means that $a = kcb^2$.
8. False, $a = kc/\sqrt{b}$ is correct.
9. False, b is directly proportional to a and inversely proportional to the square of c means that $b = ka/c^2$.
10. True, b is equal to a constant divided by the square of c.

9.5 EXERCISES

1. If y varies directly as x, then $y = kx$ for some constant k.
3. If y is inversely proportional to x, then $y = k/x$.
5. If y is jointly proportional to x and z, then $y = kxz$ for some constant k.
7. If a varies directly as m, then a is equal to a constant multiple of m, $a = km$.
9. If d varies inversely as e, then d is equal to a constant divided by e, $d = k/e$.

11. If I varies jointly as r and t, then I is a constant multiple of the product of r and t, $I = krt$.

13. If m is directly proportional to the square of p, then m is a constant multiple of the square of p, $m = kp^2$.

15. If B is directly proportional to the cube root of w, then B is a constant multiple of the cube root of w, $B = k\sqrt[3]{w}$.

17. If t is inversely proportional to the square of x, then t is equal to a constant divided by the square of x, $t = \frac{k}{x^2}$.

19. If v varies directly as m and inversely as n, then v is equal to a constant multiple of m divided by n, $v = \frac{km}{n}$.

21. Inverse variation, y varies inversely as x.

23. Direct variation, y varies directly as x.

25. Combined variation, y varies directly as x and inversely as w.

27. Joint variation, y varies jointly as x and z.

29. If y varies directly as x, then $y = kx$. If $y = 6$ when $x = 4$, then $6 = k \cdot 4$, or $k = \frac{6}{4} = \frac{3}{2}$. Therefore, $y = kx$ can be written as $y = \frac{3}{2}x$.

31. If A varies inversely as B, then $A = \frac{k}{B}$. If $A = 10$ when $B = 3$, then $10 = \frac{k}{3}$ or $k = 30$. Therefore, $A = \frac{k}{B}$ can be written as $A = \frac{30}{B}$.

33. If m varies inversely as the square root of p, then $m = \frac{k}{\sqrt{p}}$. If $m = 12$ when $p = 9$, then $12 = \frac{k}{\sqrt{9}}$ or $k = 12\sqrt{9} = 36$.
Therefore, $m = \frac{k}{\sqrt{p}}$ can be written as $m = \frac{36}{\sqrt{p}}$.

35. If A varies jointly as t and u, then $A = ktu$. If $A = 6$ when $t = 5$ and $u = 3$, then $6 = k(5)(3)$ or $k = \frac{2}{5}$. Therefore, $A = \frac{2}{5}tu$.

37. If y varies directly as x and inversely as z, then $y = \frac{kx}{z}$. If $y = 2.37$ when $x = \pi$ and $z = \sqrt{2}$, then $2.37 = \frac{k\pi}{\sqrt{2}}$ or $k = 2.37\sqrt{2}/\pi \approx 1.067$. So, $y = \frac{1.067x}{z}$.

39. If y varies directly as x, then $y = kx$. Since $y = 7$ when $x = 5$, we have $7 = k(5)$ or $k = \frac{7}{5}$. So, $y = kx$ can be written as $y = \frac{7}{5}x$. To find y when $x = -3$, we get $y = \frac{7}{5}(-3) = -\frac{21}{5}$.

41. If w varies inversely as z, then $w = \frac{k}{z}$. Since $w = 6$ when $z = 2$, we have $6 = \frac{k}{2}$ or $k = 12$. Therefore, $w = \frac{12}{z}$. Now when $z = -8$, we get $w = \frac{12}{-8} = -\frac{3}{2}$.

43. If A varies jointly as F and T, then $A = kFT$. Since $A = 6$ when $F = 3\sqrt{2}$ and $T = 4$, we have $6 = k3\sqrt{2}(4)$ or $k = \frac{\sqrt{2}}{4}$. Therefore, $A = \frac{\sqrt{2}}{4}FT$. When $F = 2\sqrt{2}$ and $T = \frac{1}{2}$, we get $A = \frac{\sqrt{2}}{4}(2\sqrt{2})(\frac{1}{2}) = \frac{1}{2}$.

45. If D varies directly with t and inversely with the square of s, then $D = \frac{kt}{s^2}$. Since $D = 12.35$ when $t = 2.8$ and $s = 2.48$, we get $12.35 = \frac{k2.8}{(2.48)^2}$ or

$$k = \frac{12.35(2.48)^2}{2.8} \approx 27.128.$$

When $t = 5.63$ and $s = 6.81$, we get $D = \frac{(27.128)(5.63)}{(6.81)^2} \approx 3.293$.

47. The cost c varies directly with the size of the lawn s means that $c = ks$. Since $c = \$280$ when $s = 4000$, we get $280 = k(4000)$ or $k = 0.07$. The formula is now written as $c = 0.07s$. For a 6000 square foot lawn, the cost is

$$c = 0.07(6000) = \$420.$$

49. The volume v is inversely proportional to the weight w means that $v = k/w$. Since the volume is 6 when the weight is 30, we have $6 = k/30$ or $k = 180$. Therefore, $v = 180/w$. If the weight is 20 kg, then

$$v = 180/20 = 9\,\text{cm}^3.$$

51. The price is jointly proportional to the radius and length means that $P = kRL$. Since a 12 foot culvert with a 6 inch radius costs

\$324, we have $324 = k(6)(12)$ or $k = 4.5$. Therefore, the formula is $P = 4.5RL$. The price of a 10 foot culvert with an 8 inch radius is $P = 4.5(8)(10) = \$360$.

53. The price varies jointly as the length and the square of the diameter means that $P = kLD^2$. Since an 18 foot rod with a diameter of 2 inches has a price of \$12.60, we get $12.60 = k(18)(2)^2$ or $k = 0.175$. Now the formula is written $P = 0.175LD^2$. The cost of a 12 foot rod with a 3 inch diameter is $P = 0.175(12)(3)^2 = \$18.90$.

55. a) The distance varies directly as the square of the time means that $d = kt^2$. If an object falls 0.16 feet in 0.1 seconds, then $0.16 = k(0.1)^2$ or $k = 16$. So the formula is $d = 16t^2$.

b) In 0.5 seconds the object falls $d = 16(0.5)^2 = 4$ feet.

c)
$$16t^2 = 100$$
$$t^2 = 6.25$$
$$t = \pm 2.5$$

It takes 2.5 seconds to reach the ground.

57. The force is inversely proportional to the length means that $F = k/L$. If a force of 2000 pounds is needed at 2 feet, then $2000 = k/2$ or $k = 4000$. The formula is now written as $F = 4000/L$. At 10 feet the force would be $F = 4000/10 = 400$ pounds.

59. Since $t = \frac{kc}{n}$ and $t = 8$ when $n = 3$ and $c = 6$, we have $8 = \frac{k6}{3}$, or $k = 4$. So $t = \frac{4c}{n}$. Now if $c = 9$ and $n = 5$, $t = \frac{4 \cdot 9}{5} = 7.2$ days.

61. a)
$$G = \frac{kNd}{c}$$
$$54 = \frac{k \cdot 52 \cdot 27}{26}$$
$$54 = k \cdot 54$$
$$1 = k$$

So $G = \frac{Nd}{c}$.

b) $G = \frac{42 \cdot 26}{13} = 84$

c)
$$G = \frac{Nd}{c}$$
$$cG = Nd$$
$$c = \frac{Nd}{G}$$

$c = \frac{44 \cdot 27}{52} \approx 23$ $\quad c = \frac{44 \cdot 27}{59} \approx 20$

$c = \frac{44 \cdot 27}{70} \approx 17$ $\quad c = \frac{44 \cdot 27}{79} \approx 15$

$c = \frac{44 \cdot 27}{91} \approx 13$

d) The gear ratio decreases as the number of teeth on the cog increases.

63. The curves cross at (1, 1). The function y_1 is increasing and y_2 is decreasing. The function y_1 represents direct variation and y_2 represents inverse variation.

Chapter 9 Wrap-Up

Enriching Your Mathematical Word Power

1. a **2.** b **3.** c **4.** d **5.** a
6. b **7.** d **8.** d **9.** a **10.** b
11. c **12.** d **13.** a **14.** b **15.** a **16.** d

CHAPTER 9 REVIEW

1. The graph of $f(x) = 3x - 4$ is a straight line with y-intercept $(0, -4)$ and slope 3. Since any number can be used for x, the domain is $(-\infty, \infty)$. From the graph we see that any real number can occur as a y-coordinate, so the range is $(-\infty, \infty)$.

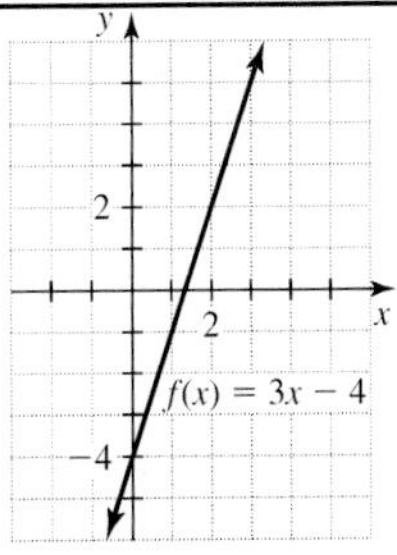

3. The graph of $h(x) = |\,x\,| - 2$ includes the points $(0, -2)$, $(1, -1)$, $(2, 0)$, $(-1, -1)$, and $(-2, 0)$. Draw a v-shaped graph through these points.

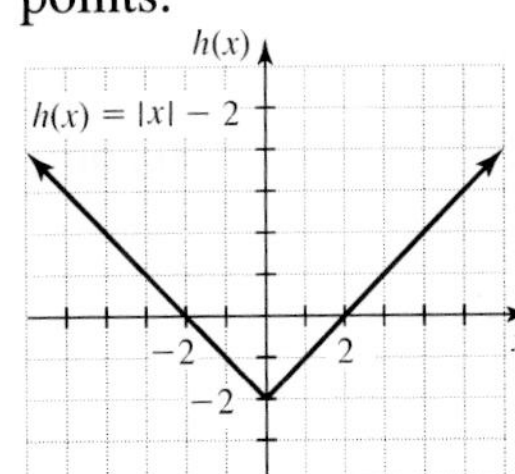

Since any real number can be used for x, the domain is $(-\infty, \infty)$. From the graph we see that all y-coordinates are greater than or equal to -2, and so the range is $[-2, \infty)$.

5. The graph of $y = x^2 - 2x + 1$ includes the points (0, 1), (1, 0), (2, 1), (3, 4), and (−1, 4). Draw a parabola through these points.

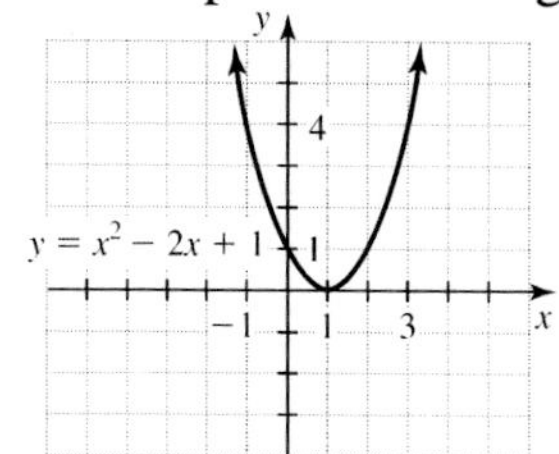

Since any number can be used for x, the domain is $(-\infty, \infty)$. From the graph we see that the y-coordinates are nonnegative, and so the range is $[0, \infty)$.

7. The graph of $k(x) = \sqrt{x} + 2$ includes the points (0, 2), (1, 3), and (4, 4).

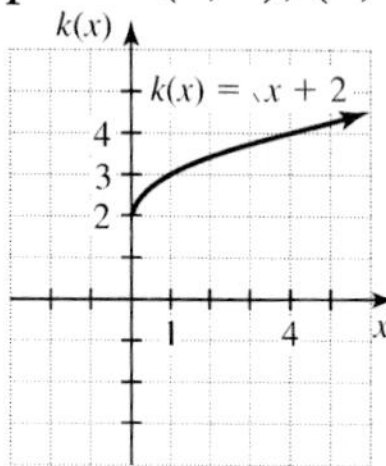

Since we can use only nonnegative numbers for x, the domain is $[0, \infty)$. From the graph we see that no y-coordinate is less than 2, and so the range is $[2, \infty)$.

9. The graph of $y = 30 - x^2$ includes the points $(-5, 5)$, $(0, 30)$, and $(5, 5)$.

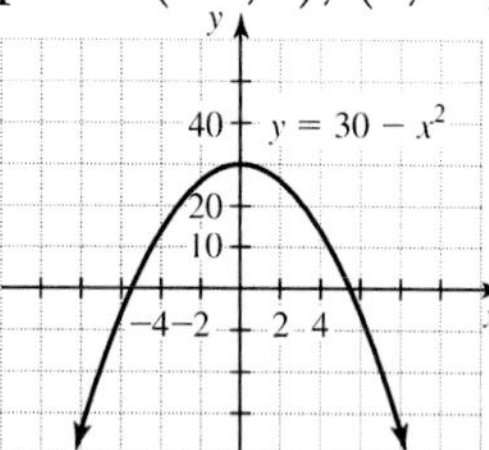

Since x can be any real number, the domain is $(-\infty, \infty)$. From the graph we see that y is at most 30. So the range is $(-\infty, 30]$.

11. Graph $y = \sqrt{x+4}$ for $-4 \leq x \leq 0$ and $y = x + 2$ for $x > 0$.

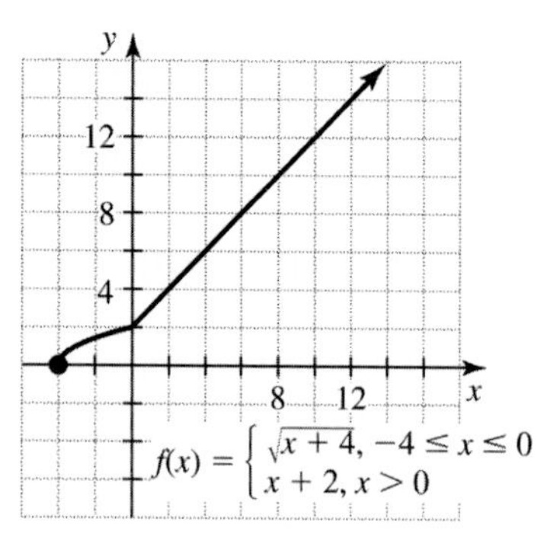

13. The graph of $x = 2$ is the vertical line with x-intercept (2, 0).

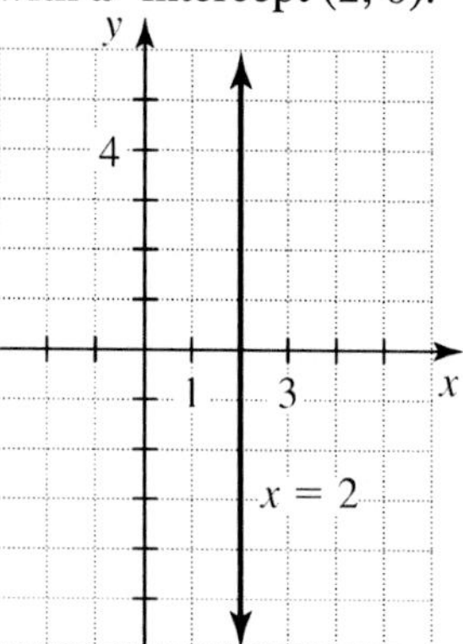

Since the only x-coordinate used on this graph is 2, the domain is $\{2\}$. Since all real numbers occur as y-coordinates, the range is $(-\infty, \infty)$.

15. The graph of $x = |y| + 1$ includes the points (1, 0), (2, 1), (2, −1), (3, 2), and (3, −2). Draw a v-shaped graph through these points.

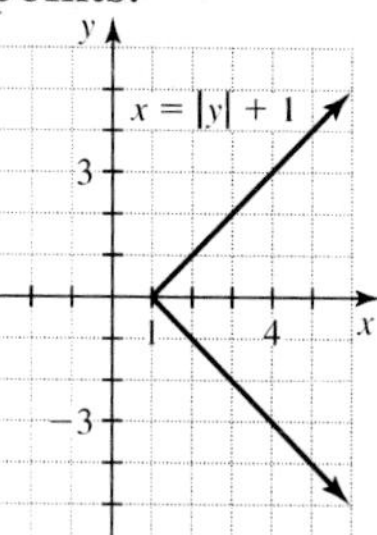

From the graph we see that the x-coordinates are not less than 1, and so the domain is $[1, \infty)$. Any number can be used for y, and so the range is $(-\infty, \infty)$.

17. The graph of $y = \sqrt{x}$ includes the points (0, 0), (1, 1), and (4, 2).

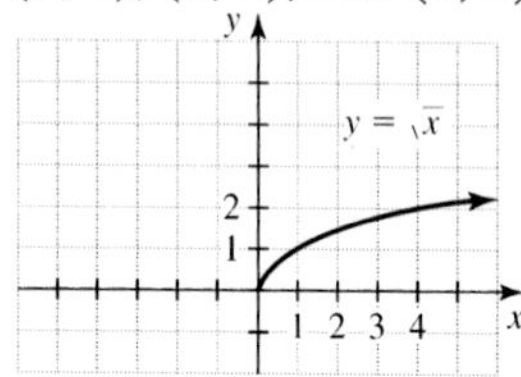

The domain is $[0, \infty)$ and the range is $[0, \infty)$.

19. The graph of $y = -2\sqrt{x}$ includes the points (0, 0), (1, −2), and (4, −4). It can be obtained from the graph of $y = \sqrt{x}$ by stretching and reflecting in the x-axis.

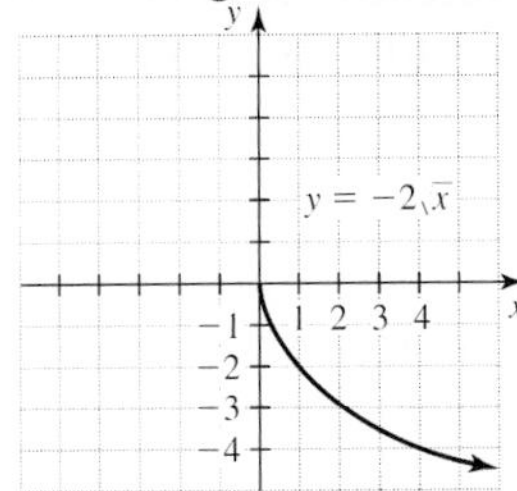

The domain is $[0, \infty)$ and the range is $(-\infty, 0]$.

21. The graph of $y = \sqrt{x - 2}$ can be obtained by shifting the graph of $y = \sqrt{x}$ two spaces to the right. The graph of $y = \sqrt{x - 2}$ includes the points (2, 0), (3, 1), and (6, 2).

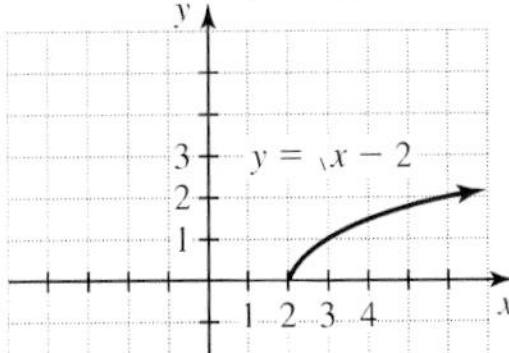

The domain is $[2, \infty)$ and the range is $[0, \infty)$.

23. The graph of $y = \frac{1}{2}\sqrt{x}$ is obtained from the graph of $y = \sqrt{x}$ by shrinking. The graph of $y = \frac{1}{2}\sqrt{x}$ includes the points (0, 0), $(1, 0.5)$, and (4, 1).

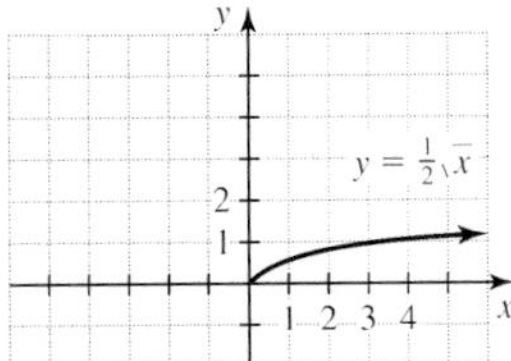

The domain is $[0, \infty)$ and the range is $[0, \infty)$.

25. The graph of $y = -\sqrt{x + 1} + 3$ is obtained from the graph of $y = \sqrt{x}$ by shifting 1 unit to the left, reflecting in the x-axis, and then shifting 3 units upward. The graph of $y = -\sqrt{x + 1} + 3$ includes the points $(-1, 3)$, $(0, 2)$, and $(3, 1)$.

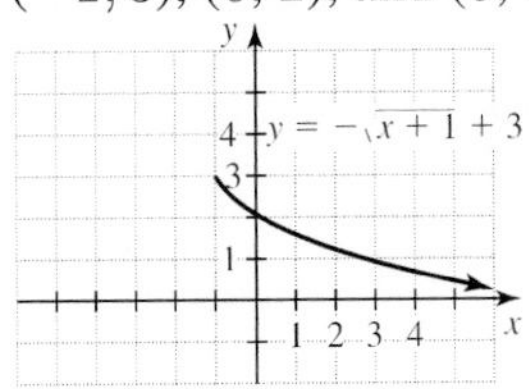

The domain is $[-1, \infty)$ and the range is $(-\infty, 3]$.

27. $f(-3) = 3(-3) + 5 = -4$

29. $(h \circ f)(\sqrt{2}) = h(f(\sqrt{2})) = h(3\sqrt{2} + 5)$
$$= \frac{3\sqrt{2} + 5 - 5}{3} = \frac{3\sqrt{2}}{3} = \sqrt{2}$$

31. $(g \circ f)(2) = g(f(2))$
$$= g(11) = 11^2 - 2(11) = 99$$

33. $(f+g)(3) = f(3) + g(3)$
$$= 3 \cdot 3 + 5 + 3^2 - 2 \cdot 3 = 17$$

35. $(f \cdot g)(x) = f(x) \cdot g(x)$
$$= (3x + 5)(x^2 - 2x)$$
$$= 3x^3 + 5x^2 - 6x^2 - 10x = 3x^3 - x^2 - 10x$$

37. $(f \circ f)(0) = f(f(0)) = f(5) = 3 \cdot 5 + 5$
$$= 20$$

39. $F(x) = |\, x + 2 \,| = |\, g(x) \,| = f(g(x))$
$$= (f \circ g)(x)$$
Therefore, F is the same function as f composite g, and we write $F = f \circ g$.

41. $H(x) = x^2 + 2 = h(x) + 2 = g(h(x))$
$$= (g \circ h)(x)$$
Therefore, H is the same function as g composite h, and we write $H = g \circ h$.

43. $I(x) = x + 4 = x + 2 + 2 = g(x) + 2$
$$= g(g(x)) = (g \circ g)(x)$$
Therefore, I is the same function as g composite g, and we write $I = g \circ g$.

45. This function is not invertible, because it is not one-to-one. The ordered pairs have different first coordinates and the same second coordinate.

47. The function $f(x) = 8x$ is invertible. The inverse of multiplication by 8 is division by 8 and $f^{-1}(x) = x/8$.

49. The function $g(x) = 13x - 6$ is one-to-one and so it is invertible. The inverse of multiplying by 13 and subtracting 6 is adding 6 and then dividing by 13, $g^{-1}(x) = \dfrac{x + 6}{13}$.

51.
$$j(x) = \frac{x + 1}{x - 1}$$
$$y = \frac{x + 1}{x - 1}$$
$$x = \frac{y + 1}{y - 1}$$
$$x(y - 1) = y + 1$$
$$xy - x = y + 1$$
$$xy - y = x + 1$$

$$y(x-1) = x+1$$
$$y = \frac{x+1}{x-1}$$
$$j^{-1}(x) = \frac{x+1}{x-1}$$

53. The function $m(x) = (x-1)^2$ is not invertible because it is not one-to-one. The ordered pairs (2, 1) and (0, 1) are both in the function.

55. The function $f(x) = 3x - 1$ is inverted by adding 1 and then dividing by 3:

$$f^{-1}(x) = \frac{x+1}{3} = \frac{1}{3}x + \frac{1}{3}$$

Use slope and y-intercept to graph both functions on the same coordinate system.

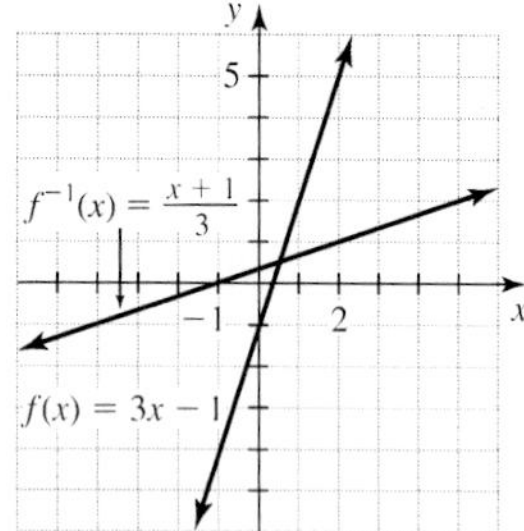

57. The function $f(x) = x^3/2$ is inverted by multiplying by 2 and then taking the cube root:

$$f^{-1}(x) = \sqrt[3]{2x}$$

The graph of f includes the points (0, 0), $(1, 1/2)$, $(-1, -1/2)$, (2, 4), and $(-2, -4)$. The graph of f^{-1} includes the points (0, 0), $(1/2, 1)$, $(-1/2, -1)$, (4, 2), and $(-4, -2)$. Plot these points and graph both functions on the same coordinate system.

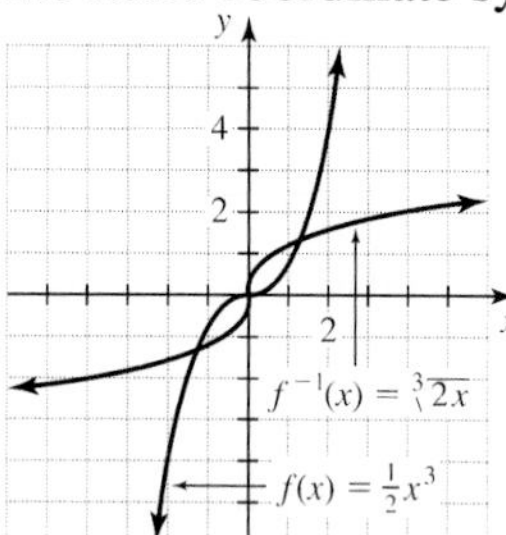

59. If y varies directly as m, then $y = km$. If $y = -3$ when $m = 1/4$, then

$$-3 = k \cdot \frac{1}{4}$$

or $\quad -12 = k.$

Since $y = -12m$, when $m = -2$ we get $y = -12(-2) = 24$.

61. If c varies directly as m and inversely as n, then $c = km/n$. If $c = 20$ when $m = 10$ and $n = 4$, then

$$20 = \frac{k(10)}{4}$$

or $\quad 8 = k.$

Since $c = 8m/n$, when $m = 6$ and $n = -3$, we get

$$c = \frac{8(6)}{-3} = -16.$$

63. If the distance varies directly with the square of the time, then $d = kt^2$. If the ball travels 144 feet in 3 seconds, then $144 = k3^2$, or $k = 16$ and $d = 16t^2$. So in 4 seconds the ball falls $d = 16 \cdot 4^2 = 256$ feet.

65. The area of the circle is $A = \pi r^2$. Let $s =$ the length of the side of the square. Since the square is inscribed in the circle, the length of the diagonal of the square is the same as the diameter of the circle $2r$. Since the diagonal of the square is the hypotenuse of a right triangle whose sides are s and s, we can write the following equation.

$$s^2 + s^2 = (2r)^2$$
$$2s^2 = 4r^2$$
$$s^2 = 2r^2$$

Since the area of the square is s^2, we have $B = s^2$, or $B = 2r^2$. Since $r^2 = \frac{A}{\pi}$, we can write $B = 2 \cdot \frac{A}{\pi}$, or $B = \frac{2A}{\pi}$.

67. Substitute $k = 5w - 6$ into the equation $a = 3k + 2$:

$$a = 3(5w-6) + 2$$
$$a = 15w - 18 + 2$$
$$a = 15w - 16$$

69. The area of a square as a function of the side is $A = s^2$. Solve for s to get the side as a function of the area: $s = \sqrt{A}$.

CHAPTER 9 TEST

1. The line $f(x) = -\frac{2}{3}x + 1$ has a y-intercept of (0, 1) and a slope of $-2/3$. Start at (0, 1) and rise -2 and go 3 units to the right to locate a second point on the line.

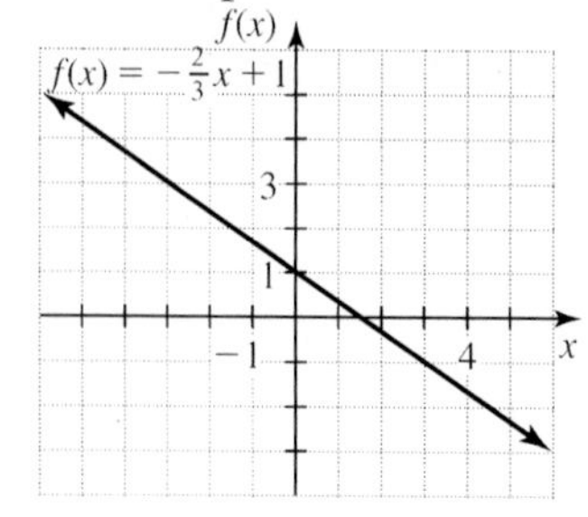

The domain is $(-\infty, \infty)$ and the range is $(-\infty, \infty)$.

2. The graph of $y = |x| - 4$ contains the points $(-2, -2)$, $(-1, -3)$, $(0, -4)$, $(1, -3)$, and $(2, -2)$. Draw a v-shaped graph through these points.

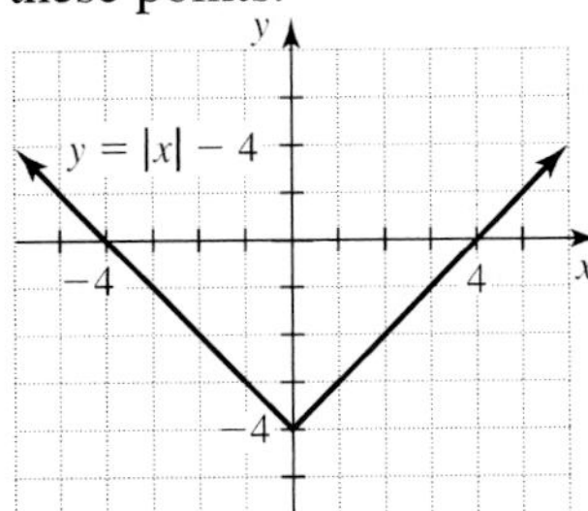

The domain is $(-\infty, \infty)$ and the range is $[-4, \infty)$.

3. The graph of $g(x) = x^2 + 2x - 8$ contains the points $(-2, -8)$, $(-1, -9)$, $(0, -8)$, $(1, -5)$, and $(2, 0)$. Draw a parabola through these points.

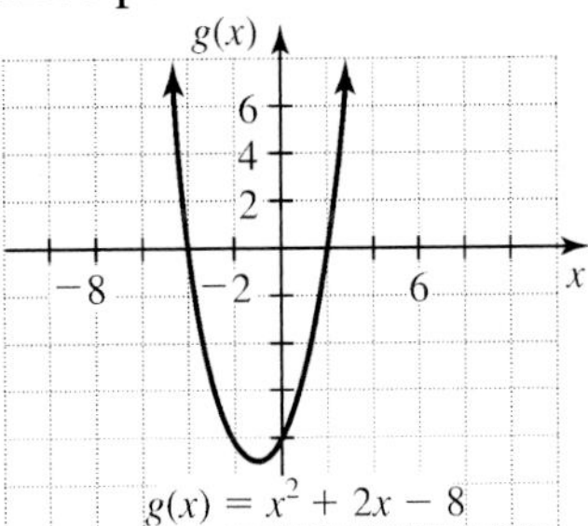

The domain is $(-\infty, \infty)$ and the range is $[-9, \infty)$.

4. The graph of $x = y^2$contains the points $(0, 0)$, $(1, 1)$, $(1, -1)$, $(4, -2)$, and $(4, 2)$. Draw a parabola through these points.

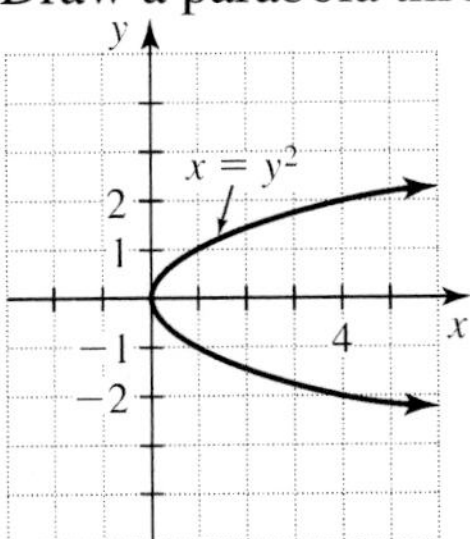

The domain is $[0, \infty)$ and the range is $(-\infty, \infty)$.

5. Graph $y = \sqrt{x}$ for $x \geq 0$ and $y = -x - 3$ for $x < 0$. Note that $(0, 0)$ is on the graph, but $(0, -3)$ is not.

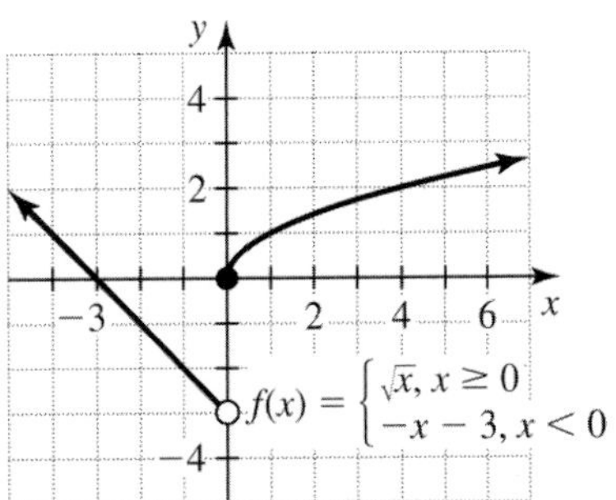

6. The graph of $y = -\mid x - 2 \mid$ is a translation and reflection of the graph of $y = \mid x \mid$. The graph contains (2, 0), (0, −2), and (4, −2).

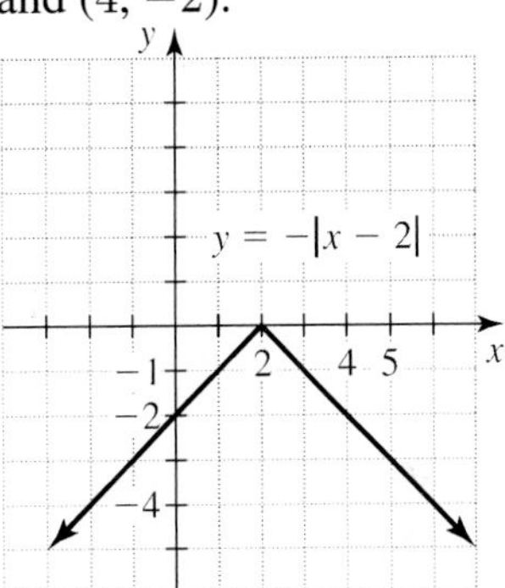

The domain is $(-\infty, \infty)$ and the range is $(-\infty, 0]$.

7. The graph of $y = \sqrt{x + 5} - 2$ is obtained from the graph of $y = \sqrt{x}$ by translating 5 units to the left and 2 units downward. The graph of $y = \sqrt{x + 5} - 2$ goes through (−5, −2), (−4, −1), and (−1, 0).

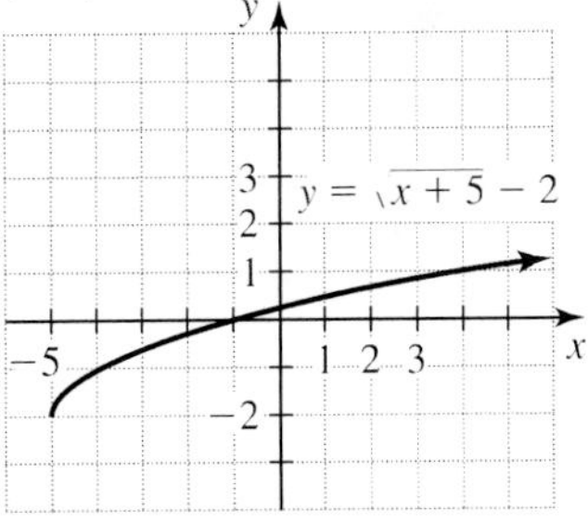

The domain is $[-5, \infty)$ and the range is $[-2, \infty)$.

8. $f(-3) = -2(-3) + 5 = 6 + 5 = 11$

9. $(g \circ f)(-3) = g(f(-3)) = g(11) = 11^2 + 4 = 125$

10. Because $f(-3) = 11$, $f^{-1}(11) = -3$.

11. Because f consists of multiplying by −2 and adding 5, f^{-1} consists of subtracting 5 and dividing by −2:

$f^{-1}(x) = \frac{x - 5}{-2}$ or $f^{-1}(x) = -\frac{1}{2}x + \frac{5}{2}$

12. $(g+f)(x) = g(x) + f(x)$
$= x^2 + 4 + (-2x + 5)$
$= x^2 - 2x + 9.$

13. $(f \cdot g)(1) = f(1) \cdot g(1) = 3 \cdot 5 = 15$

14. Because f^{-1} is the inverse of f we have $(f^{-1} \circ f)(1776) = 1776$

15. $(f/g)(2) = f(2)/g(2) = 1/8$

16. $(f \circ g)(x) = f(g(x)) = f(x^2 + 4)$
$= -2(x^2 + 4) + 5 = -2x^2 - 8 + 5$
$= -2x^2 - 3$

17. $(g \circ f)(x) = g(f(x)) = g(-2x + 5)$
$= (-2x + 5)^2 + 4 = 4x^2 - 20x + 29$

18. $f(g(x)) = f(x^2) = x^2 - 7 = H(x)$
So $H = f \circ g$.

19. $g(f(x)) = g(x - 7) = (x - 7)^2$
$= x^2 - 14x + 49$
So $W = g \circ f$.

20. The function is not invertible because it is not one-to-one. The ordered pairs $(2, 3)$ and $(4, 3)$ have different first coordinates and the same second coordinate.

21. The function is one-to-one and invertible. The inverse is $\{(3, 2), (4, 3), (5, 4)\}$

22. The inverse of subtracting 5 is adding 5. So $f^{-1}(x) = x + 5$.

23. The inverse of multiplying by 3 and then subtracting 5 is adding 5 and then dividing by 3. So $f^{-1}(x) = \frac{x+5}{3}$.

24. The inverse of taking the cube root and then adding 9 is subtracting 9 and then cubing. So the inverse is $f^{-1}(x) = (x - 9)^3$.

25.
$$f(x) = \frac{2x+1}{x-1}$$
$$y = \frac{2x+1}{x-1}$$
$$x = \frac{2y+1}{y-1}$$
$$x(y-1) = 2y + 1$$
$$xy - x = 2y + 1$$
$$xy - 2y = x + 1$$
$$y(x-2) = x + 1$$
$$y = \frac{x+1}{x-2}$$
$$f^{-1}(x) = \frac{x+1}{x-2}$$

26. Since the volume varies directly as the cube of the radius, we have $V = kr^3$. If $V = 36\pi$ when $r = 3$, we have $36\pi = k3^3$ or $k = 4\pi/3$. To find the volume of a sphere with a radius of 2, use $r = 2$and $k = 4\pi/3$ in the formula $V = kr^3$:
$$V = \frac{4\pi}{3}(2)^3 = \frac{32\pi}{3} \text{ cubic feet}$$

27. If y varies directly as x and inversely as the square root of z, then $y = kx/\sqrt{z}$. If $y = 12$ when $x = 7$ and $z = 9$, we have
$$12 = \frac{k(7)}{\sqrt{9}} \text{ or } k = \frac{36}{7}.$$

28. If the cost varies jointly as the length and width, then $C = kLW$.
$$2256 = k(6)(8)$$
$$k = 47$$
So $C = 47LW$. So the cost of a 9 by 12 rug is
$$C = 47(9)(12) = \$5076.$$

Making Connections

Chapters 1 - 9

1. $125^{-2/3} = 5^{-2} = \frac{1}{25}$

2. $\left(\frac{8}{27}\right)^{-1/3} = \left(\frac{27}{8}\right)^{1/3} = \frac{3}{2}$

3. $\sqrt{18} - \sqrt{8} = 3\sqrt{2} - 2\sqrt{2} = \sqrt{2}$

4. $x^5 \cdot x^3 = x^{5+3} = x^8$

5. $16^{1/4} = 2$, because $2^4 = 16$.

6. $\frac{x^{12}}{x^3} = x^{12-3} = x^9$

7. $x^2 = 9$
$x = \pm\sqrt{9} = \pm 3$
The solution set is $\{\pm 3\}$.

8. $x^2 = 8$, $x = \pm\sqrt{8} = \pm 2\sqrt{2}$
The solution set is $\{\pm 2\sqrt{2}\}$.

9. $x^2 - x = 0$
$x(x - 1) = 0$
$x = 0$ or $x - 1 = 0$
The solution set is $\{0, 1\}$.

10. $x^2 - 4x - 6 = 0$
$$x = \frac{4 \pm \sqrt{(-4)^2 - 4(1)(-6)}}{2(1)}$$
$$= \frac{4 \pm \sqrt{40}}{2} = \frac{4 \pm 2\sqrt{10}}{2} = 2 \pm \sqrt{10}$$
The solution set is $\{2 \pm \sqrt{10}\}$.

11. $x^{1/4} = 3$
$(x^{1/4})^4 = 3^4$
$x = 81$
Since we raised each side to an even power we must check. Since the fourth root of 81 is 3, the solution set is $\{81\}$.

12. If we raise each side to the sixth power, we will get $x = 64$. However, the sixth root of 64 is 2and not -2. So 64 does not check. The solution set is $\emptyset$.

13. $$|x| = 8$$
$$x = 8 \text{ or } x = -8$$

The solution set is $\{\pm 8\}$.

14. $$|5x - 4| = 21$$
$$5x - 4 = 21 \text{ or } 5x - 4 = -21$$
$$5x = 25 \quad \text{or} \quad 5x = -17$$
$$x = 5 \quad \text{or} \quad x = -\frac{17}{5}$$

The solution set is $\left\{-\frac{17}{5}, 5\right\}$.

15. $$x^3 = 8$$
$$(x^3)^{1/3} = 8^{1/3}$$
$$x = 2$$
The solution set is $\{2\}$.

16. $$(3x - 2)^3 = 27$$
$$3x - 2 = 3$$
$$3x = 5$$
$$x = \frac{5}{3}$$
The solution set is $\left\{\frac{5}{3}\right\}$.

17. $$\sqrt{2x - 3} = 9$$
$$(\sqrt{2x - 3})^2 = 9^2$$
$$2x - 3 = 81$$
$$2x = 84$$
$$x = 42$$
The solution set is $\{42\}$.

18. $$\sqrt{x - 2} = x - 8$$
$$x - 2 = (x - 8)^2$$
$$x - 2 = x^2 - 16x + 64$$
$$x^2 - 17x + 66 = 0$$
$$(x - 6)(x - 11) = 0$$
$$x = 6 \text{ or } x = 11$$
Only 11 satisfies the original equation. The solution set is $\{11\}$.

19. The graph of the set of points that satisfy the equation $y = 5$ is a horizontal straight line with y-intercept (0, 5).

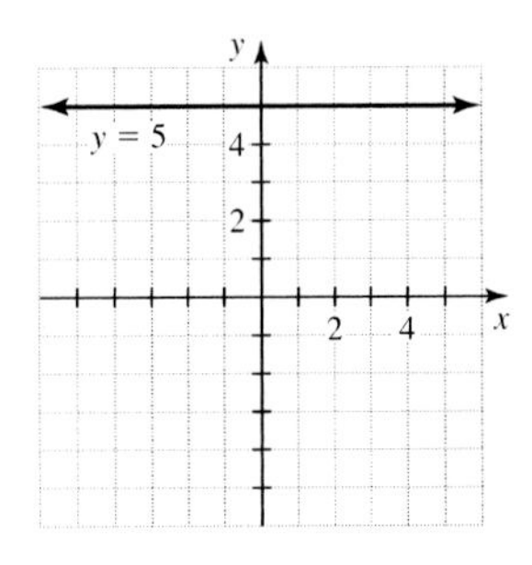

20. The graph of the set of points that satisfy $y = 2x - 5$ is a straight line with y-intercept $(0, -5)$ and slope 2.

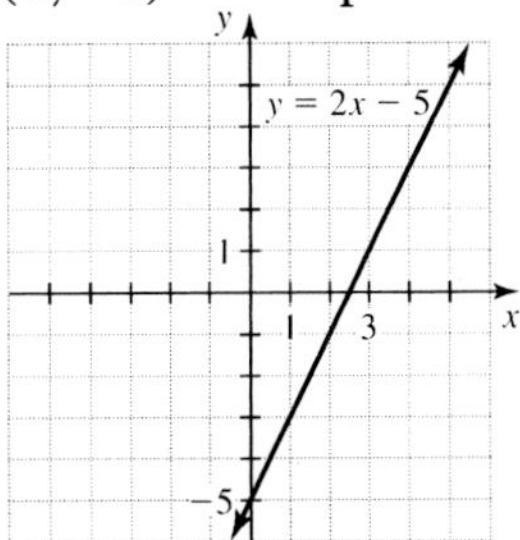

21. The graph of the set of points that satisfy $x = 5$ is a vertical line with x-intercept (5, 0).

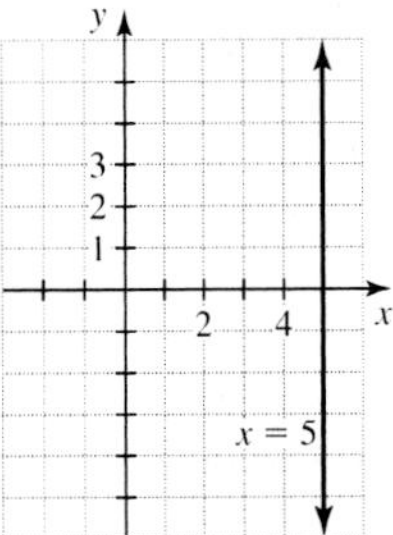

22. The graph of $3y = x$ is the same as the graph of $y = (1/3)x$, a straight line with y-intercept (0, 0) and slope $1/3$.

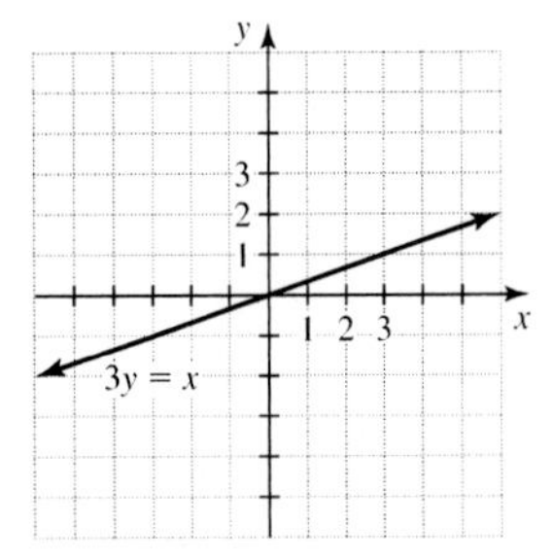

23. The graph of $y = 5x^2$ is a parabola going through (0, 0), (1, 5), and (−1, 5).

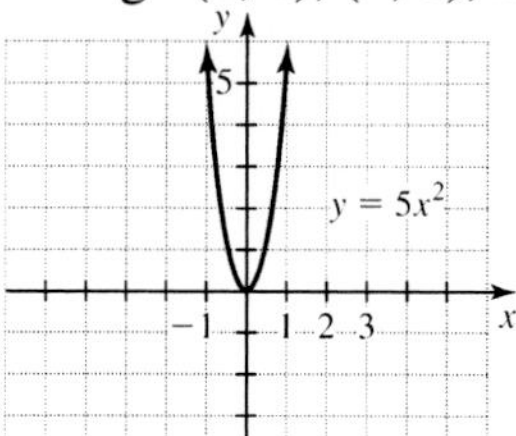

24. The graph of $y = -2x^2$ is a parabola going through (0, 0), (1, −2), and (−1, −2).

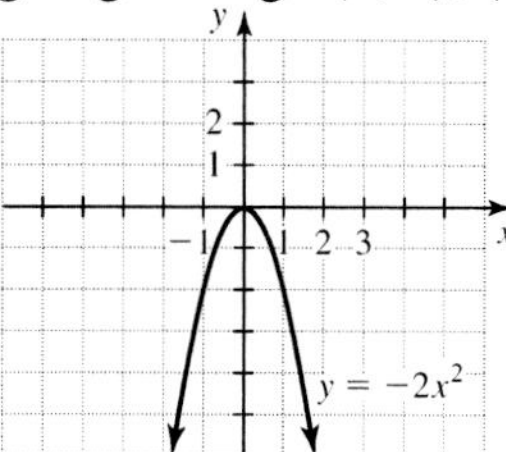

25. Because $2^2 = 4$, and $2^3 = 8$, we have (2, 4) and (3, 8). Because $2^1 = 2$and $2^4 = 16$, we have (1, 2) and (4, 16).

26. Because $4^{1/2} = 2$ and $4^{-1} = 1/4$, we have (1/2, 2) and (−1, 1/4). Because $4^2 = 16$ and $4^0 = 1$, we have (2, 16) and (0, 1).

27. In $\sqrt{x}$ we must have $x \geq 0$. So the domain is $[0, \infty)$.

28. In $\sqrt{6 - 2x}$ we must have $6 - 2x \geq 0$ or $6 \geq 2x$, or $3 \geq x$, or $x \leq 3$. So the domain is $(-\infty, 3]$.

29. Because there are no real solutions to $x^2 + 1 = 0$, we can use any real number for x in this expression. The domain is $(-\infty, \infty)$

30. Solve $x^2 - 10x + 9 = 0$ by factoring.

$$(x - 9)(x - 1) = 0$$
$$x - 9 = 0 \text{ or } x - 1 = 0$$
$$x = 9 \text{ or } \quad x = 1$$

So the domain is all real numbers except 1 and 9, which is written in interval notation as $(-\infty, 1) \cup (1, 9) \cup (9, \infty)$

31. a) $C = 0.12x + 3000$

b) Find the equation of the line through (0, 0.15) and (100,000, 0.25)

$$m = \frac{0.25 - 0.15}{100{,}000 - 0} = 0.000001$$

P-intercept is (0, 0.15)

$$P = 0.000001x + 0.15$$
$$P = 1 \times 10^{-6}x + 0.15$$

32. a) $T = \frac{C}{x} + P$

$$T = \frac{3000 + 0.12x}{x} + 0.000001x + 0.15$$
$$T = \frac{3000}{x} + 0.000001x + 0.27$$

If $x = 20{,}000$, then $T = \$0.44$
If $x = 30{,}000$, then $T = \$0.40$
If $x = 90{,}000$, then $T = \$0.39$

b)

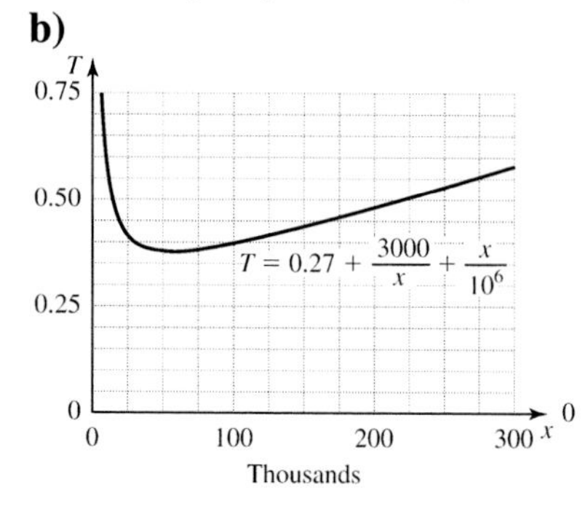

c) $$\frac{3000}{x} + 0.000001x + 0.27 = 0.38$$
$$\frac{3000}{x} + 0.000001x = 0.11$$
$$3000 + 0.000001x^2 = 0.11x$$
$$0.000001x^2 - 0.11x + 3000 = 0$$
$$x = \frac{0.11 \pm \sqrt{0.11^2 - 4(0.000001)(3000)}}{2(0.000001)}$$
$$= 50{,}000 \text{ or } 60{,}000$$

The car will be replaced at 60,000 miles.

d) The total cost is less than or equal to \$0.38 per mile for mileage in the interval [50,000, 60,000]

10.1 WARM-UPS

1. False, because $f(-1/2) = 4^{-1/2} = 1/2$.
2. True, because $(1/3)^{-1} = 3$.
3. False, because the variable is not in the exponent.
4. True, because $(1/2)^x = (2^{-1})^x = 2^{-x}$.
5. True, because it is a one-to-one function.
6. False, because the equation $(1/3)^x = 0$ has no solution.
7. True, because $e^0 = 1$.
8. False, because the domain of the exponential function $f(x) = 2^x$ is the set of all real numbers.
9. True, because $2^{-x} = 1/(2^x)$.
10. False, at the end of 3 years the investment is worth $500(1.005)^{36}$ dollars

10.1 EXERCISES

1. An exponential function has the form $f(x) = a^x$ where $a > 0$ and $a \neq 1$.
3. The two most popular bases are e and 10.
5. The compound interest formula is $A = P(1+i)^n$.
7. $f(2) = 4^2 = 16$
9. $f(1/2) = 4^{1/2} = 2$
11. $g(-2) = (1/3)^{-2+1} = (1/3)^{-1} = 3$
13. $g(0) = (1/3)^{0+1} = (1/3)^1 = \frac{1}{3}$
15. $h(0) = -2^0 = -1$
17. $h(-2) = -2^{-2} = -1/4$
19. $h(0) = 10^0 = 1$
21. $h(2) = 10^2 = 100$
23. $j(1) = e^1 = e \approx 2.718$
25. $j(-2) = e^{-2} \approx 0.135$

27. Evaluate 4^x for $x = -2, -1, 0, 1, 2$ and place the results in the second row of the table:

$4^{-2} = \frac{1}{4^2} = \frac{1}{16}$ $\quad 4^{-1} = \frac{1}{4^1} = \frac{1}{4}$

$4^0 = 1 \quad 4^1 = 4 \quad 4^2 = 16$

x	-2	-1	0	1	2
4^x	1/16	1/4	1	4	16

29. Evaluate $\left(\frac{1}{3}\right)^x$ for $x = -2, -1, 0, 1, 2$ and place the results in the second row of the table:

$\left(\frac{1}{3}\right)^{-2} = 3^2 = 9 \quad \left(\frac{1}{3}\right)^{-1} = 3^1 = 3$

$\left(\frac{1}{3}\right)^0 = 1 \quad \left(\frac{1}{3}\right)^1 = \frac{1}{3} \quad \left(\frac{1}{3}\right)^2 = \frac{1}{9}$

x	-2	-1	0	1	2
$\left(\frac{1}{3}\right)^x$	9	3	1	1/3	1/9

31. The graph of $f(x) = 4^x$ includes the points (0, 1), (1, 4), (2, 16), and $(-1, 1/4)$.

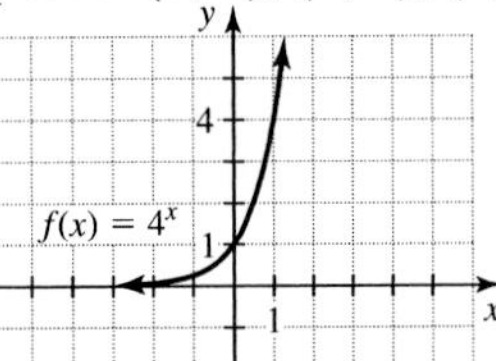

33. The graph of $h(x) = \left(\frac{1}{3}\right)^x$ includes the points (0, 1), (1, 1/3), (2, 1/9), $(-1, 3)$, and $(-2, 9)$.

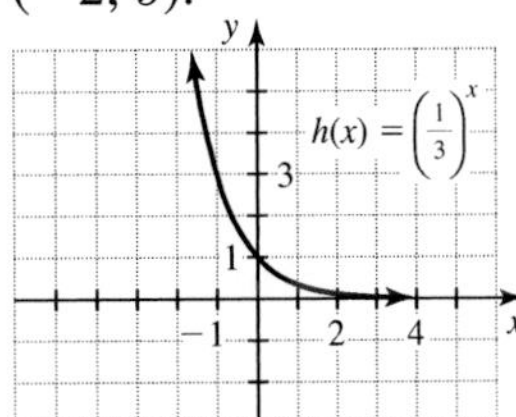

35. The graph of $y = 10^x$ includes the points (0, 1), (1, 10), and $(-1, 1/10)$.

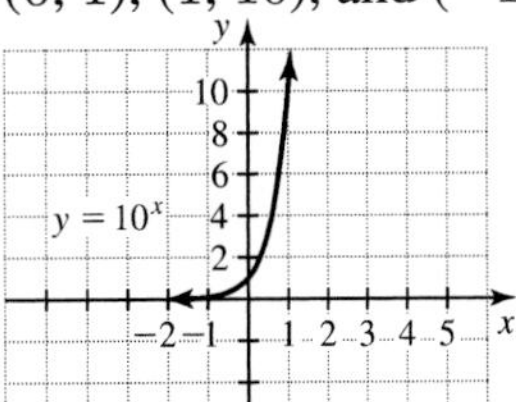

37. Evaluate 10^{x+2} for $x = -4, -3, -2, -1, 0$ and place the results in the second row of the table:

$10^{-4+2} = \frac{1}{10^2} = \frac{1}{100}$

$10^{-3+2} = \frac{1}{10^1} = \frac{1}{10}$

$10^{-2+2} = 1 \quad 10^{-1+2} = 10$

$10^{0+2} = 100$

x	-4	-3	-2	-1	0
4^x	1/100	1/10	1	10	100

39. Evaluate -2^x for $x = -2, -1, 0, 1, 2$ and place the results in the second row of the table:

$-2^{-2} = -\frac{1}{2^2} = -\frac{1}{4}$

$-2^{-1} = -\frac{1}{2^1} = -\frac{1}{2}$

$-2^0 = -1 \qquad -2^1 = -2$

$-2^2 = -4$

x	-2	-1	0	1	2
-2^{2x+1}	$-1/4$	$-1/2$	-1	-2	-4

41. The graph of $f(x) = -3^x$ includes the points $(0, -1), (-1, -1/3), (-2, -1/9)$, and $(1, -3)$. It is a reflection in the x-axis of the graph of $y = 3^x$.

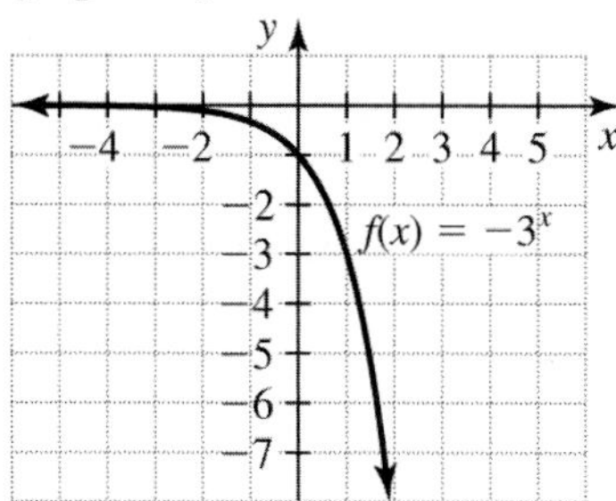

43. The graph of $f(x) = \frac{1}{2} \cdot 3^x$ includes the points $(0, 1/2), (-1, 1/6), (-2, 1/18)$, and $(1, 3/2)$. Shrink the graph of $y = 3^x$ by $1/2$ to obtain the graph.

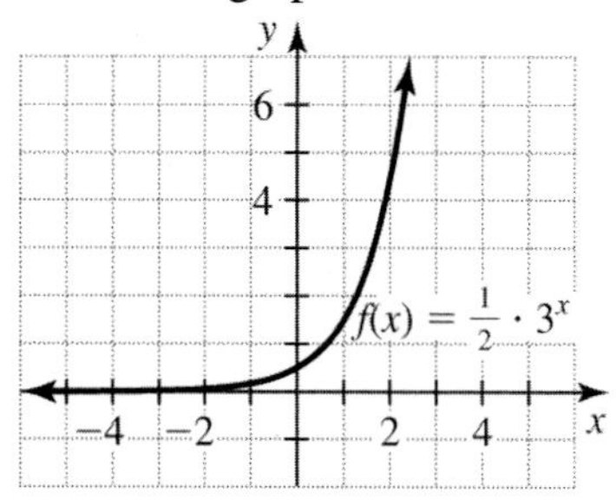

45. The graph of $f(x) = -3^x + 2$ includes the points $(0, 1), (1, -1), (2, -7)$, and $(-1, 5/3)$. Reflect the graph of $y = 3^x$ in the x-axis and then translate it upward 2 units to obtain the graph.

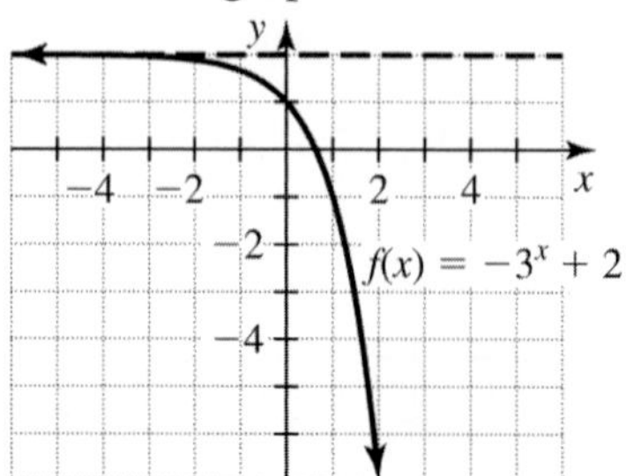

47. The graph of $f(x) = 3^{x-2} + 1$ includes the points $(2, 2), (3, 4), (4, 10)$, and $(1, 4/3)$. Translate the graph of $y = 3^x$ to the right by 2 units and then translate it upward 1 units to obtain the graph.

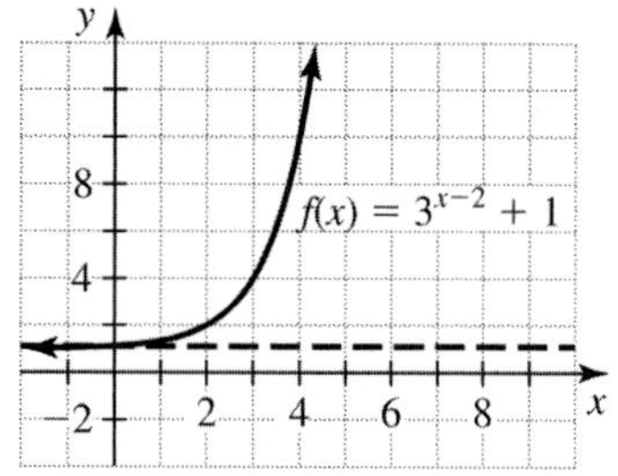

49. The graph of $f(x) = 10^x + 2$ includes the points $(0, 3), (1, 12), (-1, 1.9)$, and $(-2, 1.99)$. Translate the graph of $y = 10^x$ upward 2 units to obtain the graph.

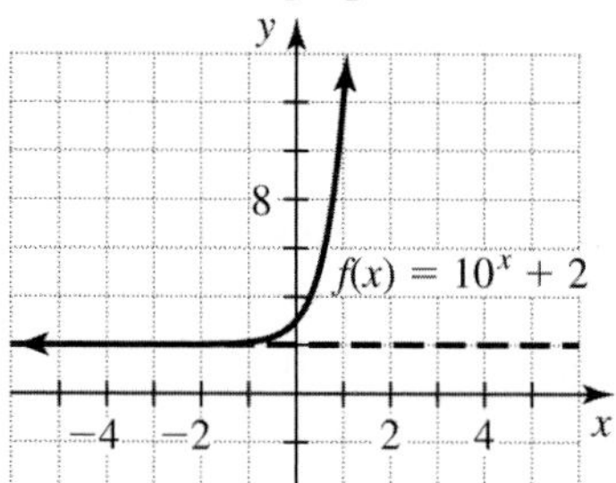

51. The graph of $f(x) = e^{-x} + 2$ includes the points $(0, 3), (1, 2.4), (2, 2.1)$, and $(-1, 4.7)$. Translate the graph of $y = e^{-x}$ upward 2 units to obtain the graph.

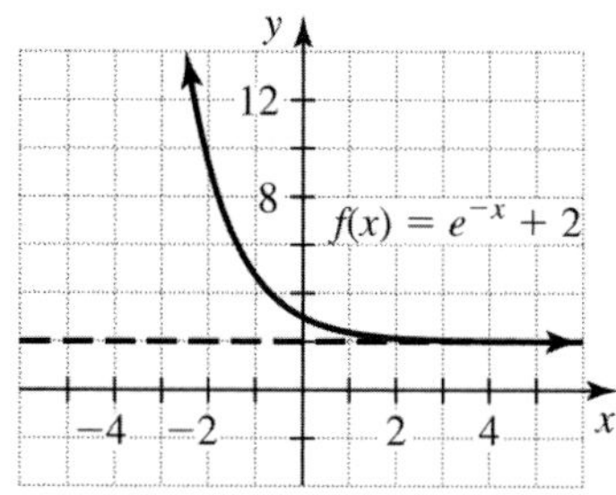

53.

$$2^x = 64$$
$$2^x = 2^6$$
$$x = 6 \quad \text{One-to-one property}$$

The solution set is $\{6\}$.

55.

$$10^x = 0.001$$
$$10^x = 10^{-3}$$
$$x = -3$$

The solution set is $\{-3\}$.

57. $2^x = \frac{1}{4}$
$2^x = 2^{-2}$
$x = -2$
The solution set is $\{-2\}$.

59. $\left(\frac{2}{3}\right)^{x-1} = \frac{9}{4}$
$\left(\frac{2}{3}\right)^{x-1} = \left(\frac{2}{3}\right)^{-2}$
$x - 1 = -2$
$x = -1$
The solution set is $\{-1\}$.

61. $5^{-x} = 25$
$5^{-x} = 5^2$
$-x = 2$ By the one-to-one property
$x = -2$
The solution set is $\{-2\}$.

63. $-2^{1-x} = -8$
$-2^{1-x} = -2^3$
$2^{1-x} = 2^3$
$1 - x = 3$
$x = -2$
The solution set is $\{-2\}$.

65. $10^{|x|} = 1000$
$10^{|x|} = 10^3$
$|\,x\,| = 3$
$x = 3$ or $x = -3$
The solution set is $\{-3, 3\}$.

67. If $f(x) = 2^x$ and $f(x) = 4 = 2^2$, then we must have $x = 2$ by the one-to-one property of exponential functions.

69. Note that $4^{2/3} = (2^2)^{2/3} = 2^{4/3}$. So if $2^x = 2^{4/3}$, then $x = \frac{4}{3}$.

71. Note that $9 = 3^2 = (1/3)^{-2}$. So if $(1/3)^x = (1/3)^{-2}$, then $x = -2$.

73. Note that $1 = (1/3)^0$. So if $(1/3)^x = (1/3)^0$, then $x = 0$.

75. Note that $16 = 4^2$. So if $h(x) = 4^{2x-1} = 4^2$, then
$2x - 1 = 2$
$2x = 3$
$x = \frac{3}{2}$.

77. Since $1 = 4^0$, $h(x) = 4^{2x-1} = 4^0$ implies that $2x - 1 = 0$ or $x = \frac{1}{2}$.

79. Since $2^{-5} = \frac{1}{32}$, $2^{-3} = \frac{1}{8}$, $2^0 = 1$, $2^1 = 2$, and $2^4 = 16$, the table should read as follows.

x	-5	-3	0	1	4
2^x	$\frac{1}{32}$	$\frac{1}{8}$	1	2	16

81. Since $\left(\frac{1}{2}\right)^{-3} = 2^3 = 8$, $\left(\frac{1}{2}\right)^{-2} = 4$, $\left(\frac{1}{2}\right)^0 = 1$, $\left(\frac{1}{2}\right)^1 = \frac{1}{2}$, $\left(\frac{1}{2}\right)^5 = \frac{1}{32}$ the table should read as follows.

x	-3	-2	0	1	5
$\left(\frac{1}{2}\right)^x$	8	4	1	$\frac{1}{2}$	$\frac{1}{32}$

83. Compounded quarterly for 10 years means that interest will be paid 40 times at 1.25%.
$S = 6000\left(1 + \frac{0.05}{4}\right)^{40} = 6000(1.0125)^{40}$
$= \$9861.72$

85. a) $\$10{,}000(1 + 0.155)^{10} \approx \$42{,}249.33$
b) From the graph it appears that the $10,000 will be worth $100,000approximately 16 years after 1996 or around 2012.

87. When the book is new, $t = 0$, and the value is $V = 45 \cdot 2^{-0.9(0)} = \$45$. When $t = 2$, the value is $V = 45 \cdot 2^{-0.9(2)} \approx \$12.92$.

89. A deposit of $500 for 3 years at 7% compounded continuously amounts to $S = 500 \cdot e^{0.07(3)} \approx \$616.84$.

91. A deposit of $80,000 at 7.5% compounded continuously for 1 year amounts to $S = 80{,}000 \cdot e^{0.075(1)} \approx \$86{,}230.73$. The interest earned is $\$86{,}230.73 - \$80{,}000 = \$6230.73$.

93. The amount at time $t = 0$ is $A = 300 \cdot e^{-0.06(0)} = 300$ grams. The amount present after 20 years, $t = 20$, is $A = 300 \cdot e^{-0.06(20)} \approx 90.4$ grams.
One-half of the substance decays in about 12 years. The substance will never completely disappear because the t-axis is the horizontal asymptote.

95. If $t = 0$, then $D = 50e^{-0.03(0)} = 50°$F.
If $t = 15$, then $D = 50e^{-0.03(15)} \approx 31.9°$F.
If $t = 15$, then $D = 50e^{-0.03(15)} \approx 31.9°$F.
The ocean temperature is not changing. So it is 48.6°F at $t = 15$. The human body is 31.9° warmer. So the body is 80.5°F.

97. $1 + 1 + \frac{1}{2} + \frac{1}{6} \approx 2.66666667$
$e - 2.66666667 \approx 0.0516$
$e - \left(1 + 1 + \frac{1}{2} + \frac{1}{6} + \frac{1}{24} + \frac{1}{120} + \frac{1}{720} + \frac{1}{5040}\right)$
$\approx 2.8 \times 10^{-5}$

99. The graph of $y = 3^{x-h}$ lies h units to the right of $y = 3^x$ when $h > 0$ and $|h|$ units to the left of $y = 3^x$ when $h < 0$.

10.2 WARM-UPS

1. True, because 3 is the exponent of a that produces 2.
2. False, because $b = 8^a$ is equivalent to $\log_8(b) = a$.
3. True, because the inverse of the base a exponential function is the base a logarithm function.
4. True, because the inverse of the base e exponential function is the base e logarithm function.
5. False, the domain is $(0, \infty)$.
6. False, $\log_{25}(5) = 1/2$.
7. False, $\log(-10)$ is undefined because -10 is not in the domain of the base 10 logarithm function.
8. False, $\log(0)$ is undefined.
9. True, because $\log_5(125) = 3$ and $5^3 = 125$.
10. True, because $(1/2)^{-5} = 2^5 = 32$.

10.2 EXERCISES

1. If $f(x) = 2^x$ then $f^{-1}(x) = \log_2(x)$.
3. The common logarithm uses the base 10 and the natural logarithm uses base e.
5. The one-to-one property for logarithmic functions says that if $\log_a(m) = \log_a(n)$ then $m = n$.
7. $\log_2(8) = 3$ is equivalent to $2^3 = 8$.
9. $10^2 = 100$ is equivalent to $\log(100) = 2$.
11. $y = \log_5(x)$ is equivalent to $5^y = x$.
13. $2^a = b$ is equivalent to $\log_2(b) = a$.
15. $\log_3(x) = 10$ is equivalent to $3^{10} = x$.
17. $e^3 = x$ is equivalent to $\ln(x) = 3$.
19. Because $2^2 = 4$, $\log_2(4) = 2$.
21. Because $2^4 = 16$, $\log_2(16) = 4$.
23. Because $2^6 = 64$, $\log_2(64) = 6$.
25. Because $4^3 = 64$, $\log_4(64) = 3$.
27. Because $2^{-2} = \frac{1}{4}$, $\log_2(1/4) = -2$.
29. Because $10^2 = 100$, $\log(100) = 2$.
31. Because $10^{-2} = 0.01$, $\log(0.01) = -2$.
33. Because $(1/3)^1 = 1/3$, $\log_{1/3}(1/3) = 1$.
35. Because $(1/3)^{-3} = 27$, $\log_{1/3}(27) = -3$.
37. Because $25^{1/2} = 5$, $\log_{25}(5) = 1/2$.
39. Because $e^2 = e^2$, $\ln(e^2) = 2$.
41. Use a calculator with a base 10 logarithm key to find $\log(5) \approx 0.6990$.
43. Use a calculator with a natural logarithm key to find $\ln(6.238) \approx 1.8307$.
45. Since $3^{-2} = \frac{1}{9}$, $3^{-1} = \frac{1}{3}$, $3^0 = 1$, $3^1 = 3$, and $3^2 = 9$, the table should read as follows. Note that the logarithms are the exponents.

x	$\frac{1}{9}$	$\frac{1}{3}$	1	3	9
$\log_3(x)$	-2	-1	0	1	2

47. Since $\left(\frac{1}{4}\right)^{-2} = 16$,
$\left(\frac{1}{4}\right)^{-1} = 4$, $\left(\frac{1}{4}\right)^{0} = 1$, $\left(\frac{1}{4}\right)^{1} = \frac{1}{4}$,
$\left(\frac{1}{4}\right)^{2} = \frac{1}{16}$ the table should read as follows. Note that the logarithms are the exponents.

x	16	4	1	$\frac{1}{4}$	$\frac{1}{16}$
$\log_{1/4}(x)$	-2	-1	0	1	2

49. The graph of $f(x) = \log_3(x)$ includes the points $(3, 1)$, $(1, 0)$, and $(1/3, -1)$. All graphs of logarithm functions have similar shapes.

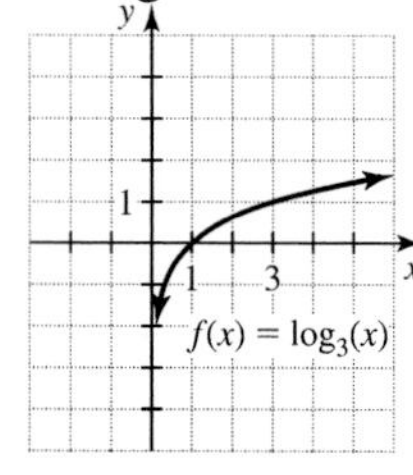

51. The graph of $y = \log_4(x)$ includes the points (4, 1), (1, 0), and (1/4, −1).

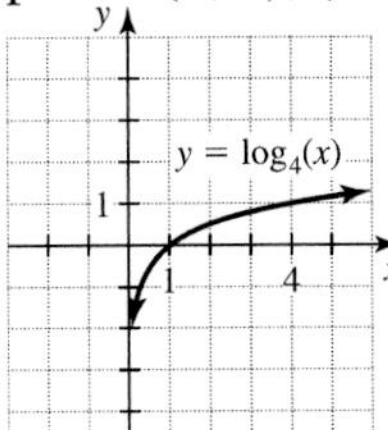

53. The graph of $y = \log_{1/4}(x)$ includes the points (1, 0), (4, −1), and (1/4, 1).

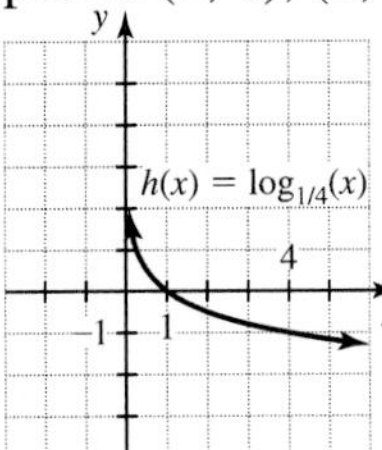

55. The graph of $h(x) = \log_{1/5}(x)$ includes the points (1, 0), (5, −1), and (1/5, 1).

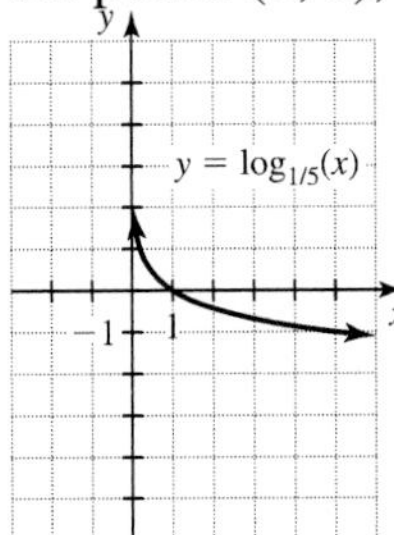

57. The inverse of $f(x) = 6^x$ is $f^{-1}(x) = \log_6(x)$.

59. The inverse of $f(x) = \ln(x)$ is $f^{-1}(x) = e^x$.

61. If $f(x) = \log_{1/2}(x)$ then $f^{-1}(x) = \left(\frac{1}{2}\right)^x$.

63. $x = (1/2)^{-2} = 2^2 = 4$

The solution set is $\{4\}$.

65.
$$5 = 25^x$$
$$x = \log_{25}(5) = 1/2$$

The solution set is $\left\{\frac{1}{2}\right\}$.

67.
$$\log(x) = -3$$
$$x = 10^{-3} = 0.001$$

The solution set is $\{0.001\}$.

69.
$$\log_x(36) = 2$$
$$x^2 = 36$$
$$x = \pm 6$$

Omit −6 because the base of any logarithm function is positive. The solution set is $\{6\}$.

71.
$$\log_x(5) = -1$$
$$x^{-1} = 5$$
$$(x^{-1})^{-1} = 5^{-1}$$
$$x = \frac{1}{5}$$

The solution set is $\left\{\frac{1}{5}\right\}$.

73.
$$\log(x^2) = \log(9)$$
$$x^2 = 9$$
$$x = \pm 3$$

Both 3 and −3 check in the original equation. The solution set is $\{\pm 3\}$.

75.
$$3 = 10^x$$
$$x = \log(3) \approx 0.4771$$

The solution set is $\{0.4771\}$.

77.
$$10^x = \frac{1}{2}$$

$x = \log(1/2) = \log(0.5) \approx -0.3010$

The solution set is $\{-0.3010\}$.

79.
$$e^x = 7.2$$
$$x = \ln(7.2) \approx 1.9741$$

The solution set is $\{1.9741\}$.

81. Since $2^{-2} = \frac{1}{4}$, $2^{-1} = \frac{1}{2}$, $2^0 = 1$, $2^2 = 4$, and $2^4 = 16$, the table should read as follows.

x	$\frac{1}{4}$	$\frac{1}{2}$	1	4	16
$\log_2(x)$	−2	−1	0	2	4

83. Since $\left(\frac{1}{2}\right)^{-4} = 16$, $\left(\frac{1}{2}\right)^{-2} = 4$, $\left(\frac{1}{2}\right)^{0} = 1$, $\left(\frac{1}{2}\right)^{1} = \frac{1}{2}$, $\left(\frac{1}{2}\right)^{2} = \frac{1}{4}$ the table should read as follows.

x	16	4	1	$\frac{1}{2}$	$\frac{1}{4}$
$\log_{1/2}(x)$	−4	−2	0	1	2

85. Use the continuous compounding formula.
$$10{,}000 = 5000 \cdot e^{0.12t}$$
$$2 = e^{0.12t}$$
$$0.12t = \ln(2)$$
$$t = \frac{\ln(2)}{0.12} \approx 5.776 \text{ years}$$

87. To earn \$1000 in interest, the principal must increase from \$6000 to \$7000 in t years.
$$7000 = 6000 \cdot e^{0.08t}$$
$$7/6 = e^{0.08t}$$
$$0.08t = \ln(7/6)$$
$$t = \frac{\ln(7/6)}{0.08} \approx 1.927 \text{ years}$$

89. a) $20{,}733 = 10{,}000 \cdot e^{r5}$
$2.0733 = e^{5r}$
$5r = \ln(2.0733)$
$r = \dfrac{\ln(2.0733)}{5} \approx 0.1458 = 14.58\%$

b) In 2010 the investment grows for 13 years:
$A = 10{,}000e^{0.1458(13)} \approx \$66{,}576.60$

91. $\text{pH} = -\log_{10}(10^{-4.1}) = -(-4.1) = 4.1$

93. $\text{pH} = -\log(1.58 \times 10^{-7}) \approx 6.8$

95. $L = 10 \cdot \log(0.001 \times 10^{12}) = 90$ db

97. $f(x) = 5 + \log_2(x - 3)$

$y = 5 + \log_2(x - 3)$
$x = 5 + \log_2(y - 3)$
$x - 5 = \log_2(y - 3)$
$y - 3 = 2^{x-5}$
$y = 2^{x-5} + 3$
$f^{-1}(x) = 2^{x-5} + 3$

Domain of f^{-1} is $(-\infty, \infty)$ and range is $(3, \infty)$.

99. $y = \ln(e^x) = x$ for $-\infty < x < \infty$
$y = e^{\ln(x)} = x$ for $0 < x < \infty$

10.3 WARM-UPS

1. True, because
$\log_2(x^2/8) = \log_2(x^2) - \log_2(8)$
$= \log_2(x^2) - 3.$

2. False, because
$\log(100) = 2$ and $\log(10) = 1$, and
$2 \div 1 \neq 2 - 1.$

3. True, because
$\ln(\sqrt{2}) = \ln(2^{1/2}) = \frac{1}{2} \cdot \ln(2) = \frac{\ln(2)}{2}.$

4. True, because $\log_3(17)$ is the exponent that we place on 3 to obtain 17.

5. False, because $\log_2(1/8) = -3$ and $\log_2(8) = 3$.

6. True, because $\ln(8) = \ln(2^3) = 3 \cdot \ln(2)$.

7. False, because $\ln(1) = 0$ and $e \neq 0$.

8. False, because $\log(100) = 2, \log(10) = 1$, and $2 \div 10 \neq 1$.

9. False, because $\log_2(8) = 3, \log_2(2) = 1$, $\log_2(4) = 2$, and $3 \div 1 \neq 2$.

10. True, because $\ln(2) + \ln(3) - \ln(7) = \ln(2 \cdot 3) - \ln(7) = \ln(6/7)$.

10.3 EXERCISES

1. The product rule for logarithms says that $\log_a(MN) = \log_a(M) + \log_a(N)$.

3. The power rule for logarithms says that $\log_a(M^N) = N \cdot \log_a(M)$.

5. Since $\log_a(M)$ is the exponent that you would use on a to obtain M, using $\log_a(M)$ as the exponent produces M: $a^{\log_a(M)} = M$.

7. $\log_2(2^{10}) = 10$

9. $5^{\log_5(19)} = 19$

11. $\log(10^8) = 8$

13. $e^{\ln(4.3)} = 4.3$

15. $\log(3) + \log(7) = \log(3 \cdot 7) = \log(21)$

17. $\log_3(\sqrt{5}) + \log_3(\sqrt{x}) = \log_3(\sqrt{5x})$

19. $\log(x^2) + \log(x^3)$
$= \log(x^2 \cdot x^3) = \log(x^5)$

21. $\ln(2) + \ln(3) + \ln(5) = \ln(2 \cdot 3 \cdot 5)$
$= \ln(30)$

23. $\log(x) + \log(x + 3) = \log(x^2 + 3x)$

25. $\log_2(x - 3) + \log_2(x + 2)$
$= \log_2(x^2 - x - 6)$

27. $\log(8) - \log(2) = \log(8/2) = \log(4)$

29. $\log_2(x^6) - \log_2(x^2) = \log_2\left(\frac{x^6}{x^2}\right)$
$= \log_2(x^4)$

31. $\log(\sqrt{10}) - \log(\sqrt{2}) = \log\left(\frac{\sqrt{10}}{\sqrt{2}}\right)$
$= \log\left(\sqrt{5}\right)$

33. $\ln(4h - 8) - \ln(4) = \ln\left(\frac{4h - 8}{4}\right)$
$= \ln(h - 2)$

35. $\log_2(w^2 - 4) - \log_2(w + 2)$
$= \log_2\left(\frac{w^2 - 4}{w + 2}\right) = \log_2\left(\frac{(w - 2)(w + 2)}{w + 2}\right)$
$= \log_2(w - 2)$

37. $\ln(x^2 + x - 6) - \ln(x + 3)$
$= \ln\left(\frac{x^2 + x - 6}{x + 3}\right) = \ln\left(\frac{(x + 3)(x - 2)}{x + 3}\right)$
$= \ln(x - 2)$

39. $\log(27) = \log(3^3) = 3\log(3)$

41. $\log(\sqrt{3}) = \log(3^{1/2}) = \frac{1}{2}\log(3)$

43. $\log(3^x) = x\log(3)$

45. $\log(15) = \log(3 \cdot 5) = \log(3) + \log(5)$

47. $\log(5/3) = \log(5) - \log(3)$

49. $\log(25) = \log(5^2) = 2\log(5)$

51. $\log(75) = \log(5^2 \cdot 3) = 2\log(5) + \log(3)$

53. $\log\left(\frac{1}{3}\right) = \log(1) - \log(3)$
$= 0 - \log(3) = -\log(3)$

55. $\log(0.2) = \log(1/5) = \log(1) - \log(5)$
$= 0 - \log(5) = -\log(5)$

57. $\log(xyz) = \log(x) + \log(y) + \log(z)$

59. $\log_2(8x) = \log_2(8) + \log_2(x)$
$= 3 + \log_2(x)$

61. $\ln(x/y) = \ln(x) - \ln(y)$

63. $\log(10x^2) = \log(10) + \log(x^2)$
$= 1 + 2\log(x)$

65. $\log_5\left[\frac{(x-3)^2}{\sqrt{w}}\right]$
$= \log_5[(x-3)^2] - \log_5(\sqrt{w})$
$= 2\log_5(x-3) - \frac{1}{2}\log_5(w)$

67. $\ln\left[\frac{yz\sqrt{x}}{w}\right] = \ln[yz\sqrt{x}] - \ln(w)$
$= \ln(y) + \ln(z) + \ln(\sqrt{x}) - \ln(w)$
$= \ln(y) + \ln(z) + \frac{1}{2}\ln(x) - \ln(w)$

69. $\log(x) + \log(x-1) = \log(x^2 - x)$

71. $\ln(3x-6) - \ln(x-2) = \ln\left(\frac{3x-6}{x-2}\right)$
$= \ln(3)$

73. $\ln(x) - \ln(w) + \ln(z) = \ln(xz) - \ln(w)$
$= \ln\left(\frac{xz}{w}\right)$

75. $3 \cdot \ln(y) + 2 \cdot \ln(x) - \ln(w)$
$= \ln(y^3) + \ln(x^2) - \ln(w)$
$= \ln(x^2y^3) - \ln(w) = \ln\left(\frac{x^2y^3}{w}\right)$

77. $\frac{1}{2} \cdot \log(x-3) - \frac{2}{3} \cdot \log(x+1)$
$= \log\left((x-3)^{1/2}\right) - \log\left((x+1)^{2/3}\right)$
$= \log\left(\frac{(x-3)^{1/2}}{(x+1)^{2/3}}\right)$

79. $\frac{2}{3} \cdot \log_2(x-1) - \frac{1}{4} \cdot \log_2(x+2)$
$= \log_2\left((x-1)^{2/3}\right) - \log_2\left((x+2)^{1/4}\right)$
$= \log_2\left(\frac{(x-1)^{2/3}}{(x+2)^{1/4}}\right)$

81. False, because
$\log(56) = \log(7 \cdot 8) = \log(7) + \log(8)$.

83. True, because $\log_2(4^2) = \log_2(16) = 4$ and $(\log_2(4))^2 = (2)^2 = 4$.

85. True, because $\ln(25) = \ln(5^2) = 2 \cdot \ln(5)$.

87. False, because $\log_2(64) = 6$ and $\log_2(8) = 3$ and $6 \div 3 \neq 3$.

89. True, because $\log(1/3) = \log(1) - \log(3)$
$= 0 - \log(3) = -\log(3)$.

91. True, because
$\log_2(16^5) = 5 \cdot \log_2(16) = 5 \cdot 4 = 20$.

93. True, $\log(10^3) = 3 \cdot \log(10) = 3 \cdot 1 = 3$.

95. False, because $\log(100+3) = \log(103)$ and
$2 + \log(3) = \log(100) + \log(3) = \log(300)$.

97. $r = \log(I) - \log(I_0)$
$r = \log\left(\frac{I}{I_0}\right)$
If $I = 100 \cdot I_0$, then
$r = \log\left(\frac{100 \cdot I_0}{I_0}\right) = \log(100) = 2$.

99. Only (b) is an identity, because it is the only one that is a correct application of a property of logarithms.

101. The graphs are the same because
$\ln(\sqrt{x}) = \ln(x^{1/2}) = \frac{1}{2}\ln(x) = 0.5 \cdot \ln(x)$.

103. The graph is a straight line because $\log(e^x) = x\log(e) \approx 0.434x$. The slope is $\log(e)$ or approximately 0.434.

10.4 WARM-UPS

1. True, because $\log(x-2) + \log(x+2) = \log[(x-2)(x+2)] = \log(x^2 - 4)$.

2. True, because of the one-to-one property of logarithms.

3. True, because of the one-to-one property of exponential functions.

4. False, because the bases are different and the one-to-one property does not apply.

5. True, because
$\log_2(x^2 - 3x + 5) = 3$ is equivalent to
$x^2 - 3x + 5 = 2^3 = 8$.

6. True, $a^x = y$ is equivalent to $\log_a(y) = x$.

7. True, if $5^x = 23$, then $\ln(5^x) = \ln(23)$, or $x \cdot \ln(5) = \ln(23)$.

8. False, because $\log_3(5) = \frac{\ln(5)}{\ln(3)}$.

9. True, because $\log_6(2) = \frac{\ln(2)}{\ln(6)}$ and
$\log_6(2) = \frac{\log(2)}{\log(6)}$.

10. False, $\log(5) \approx 0.699$ and $\ln(5) \approx 1.609$.

10.4 EXERCISES

1. Equivalent to $\log_a(x) = y$ is $a^y = x$.

3. $\log(x+100) = 3$
$$x + 100 = 10^3$$
$$x + 100 = 1000$$
$$x = 900$$
The solution set is $\{900\}$.

5. $\log_2(x+1) = 3$
$$x + 1 = 2^3$$
$$x + 1 = 8$$
$$x = 7$$
The solution set is $\{7\}$.

7. $3\log_2(x+1) - 2 = 13$
$$3\log_2(x+1) = 15$$
$$\log_2(x+1) = 5$$
$$x + 1 = 2^5$$
$$x = 31$$
The solution set is $\{31\}$.

9. $12 + 2\ln(x) = 14$
$$2\ln(x) = 2$$
$$\ln(x) = 1$$
$$x = e^1$$
The solution set is $\{e\}$.

11. $\log(x) + \log(5) = 1$
$$\log(5x) = 1$$
$$5x = 10^1$$
$$x = 2$$
The solution set is $\{2\}$.

13. $\log_2(x-1) + \log_2(x+1) = 3$
$$\log_2[(x-1)(x+1)] = 3$$
$$\log_2(x^2 - 1) = 3$$
$$x^2 - 1 = 2^3$$
$$x^2 = 9$$
$$x = \pm 3$$
If $x = -3$ in the original equation, $\log_2(-3-1)$ is undefined. The solution set is $\{3\}$.

15. $\log_2(x-1) - \log_2(x+2) = 2$
$$\log_2\left(\frac{x-1}{x+2}\right) = 2$$
$$\frac{x-1}{x+2} = 2^2$$
$$x - 1 = 4(x+2)$$
$$x - 1 = 4x + 8$$
$$-9 = 3x$$
$$-3 = x$$
If $x = -3$ in the original equation, we get the logarithm of a negative number, which is undefined. The solution set is $\emptyset$.

17. $\log_2(x-4) + \log_2(x+2) = 4$
$$\log_2(x^2 - 2x - 8) = 4$$
$$x^2 - 2x - 8 = 2^4$$
$$x^2 - 2x - 8 = 16$$
$$x^2 - 2x - 24 = 0$$
$$(x-6)(x+4) = 0$$
$$x - 6 = 0 \quad \text{or} \quad x + 4 = 0$$
$$x = 6 \quad \text{or} \quad x = -4$$
If $x = -4$ in the original equation, we get $\log_2(-8)$, which is undefined. The solution set is $\{6\}$.

19. $\ln(x) + \ln(x+5)$
$$= \ln(x+1) + \ln(x+3)$$
$$\ln(x^2 + 5x) = \ln(x^2 + 4x + 3)$$
$$x^2 + 5x = x^2 + 4x + 3$$
$$x = 3$$
The solution set is $\{3\}$.

21. $\log(x+3) + \log(x+4)$
$$= \log(x^3 + 13x^2) - \log(x)$$
$$\log(x^2 + 7x + 12) = \log(x^2 + 13x)$$
$$x^2 + 7x + 12 = x^2 + 13x$$
$$12 = 6x$$
$$2 = x$$
The solution set is $\{2\}$.

23. $2 \cdot \log(x) = \log(20 - x)$
$$\log(x^2) = \log(20 - x)$$
$$x^2 = 20 - x$$
$$x^2 + x - 20 = 0$$
$$(x-4)(x+5) = 0$$
$$x - 4 = 0 \quad \text{or} \quad x + 5 = 0$$
$$x = 4 \quad \text{or} \quad x = -5$$
If $x = -5$ in the original equation we get a logarithm of a negative number, which is undefined. The solution set is $\{4\}$.

25. $3^x = 7$
$$x = \log_3(7)$$
The solution set is $\{\log_3(7)\}$.

27. $e^{2x} = 7$
$$2x = \ln(7)$$
$$x = \frac{\ln(7)}{2}$$
The solution set is $\left\{\frac{\ln(7)}{2}\right\}$.

29. $2^{3x+4} = 4^{x-1}$
$$2^{3x+4} = (2^2)^{x-1}$$
$$2^{3x+4} = (2)^{2x-2}$$
$$3x + 4 = 2x - 2$$
$$x = -6$$
The solution set is $\{-6\}$.

31.
$$\begin{aligned} (1/3)^x &= 3^{1+x} \\ (3^{-1})^x &= 3^{1+x} \\ 3^{-x} &= 3^{1+x} \\ -x &= 1+x \\ -2x &= 1 \\ x &= -1/2 \end{aligned}$$
The solution set is $\left\{-\frac{1}{2}\right\}$.

33.
$$\begin{aligned} 2^x &= 3^{x+5} \\ \ln(2^x) &= \ln(3^{x+5}) \\ x\cdot\ln(2) &= (x+5)\ln(3) \\ x\cdot\ln(2) &= x\cdot\ln(3)+5\cdot\ln(3) \\ x\cdot\ln(2)-x\cdot\ln(3) &= 5\cdot\ln(3) \\ x(\ln(2)-\ln(3)) &= 5\cdot\ln(3) \\ x &= \frac{5\cdot\ln(3)}{\ln(2)-\ln(3)} \end{aligned}$$
This is the exact solution.
Use a calculator to find an approximate solution: $\frac{5\ln(3)}{\ln(2)-\ln(3)} \approx -13.548$.

35.
$$\begin{aligned} 5^{x+2} &= 10^{x-4} \\ \log(5^{x+2}) &= \log(10^{x-4}) \\ (x+2)\log(5) &= x-4 \\ x\cdot\log(5)+2\cdot\log(5) &= x-4 \\ x\cdot\log(5)-x &= -4-2\cdot\log(5) \\ x[\log(5)-1] &= -4-2\cdot\log(5) \\ x &= \frac{-4-2\cdot\log(5)}{\log(5)-1} \\ &= \frac{4+2\log(5)}{1-\log(5)} \\ &\approx 17.932 \end{aligned}$$

37.
$$\begin{aligned} 8^x &= 9^{x-1} \\ \ln(8^x) &= \ln(9^{x-1}) \\ x\cdot\ln(8) &= (x-1)\ln(9) \\ x\cdot\ln(8) &= x\cdot\ln(9)-\ln(9) \\ x\cdot\ln(8)-x\cdot\ln(9) &= -\ln(9) \\ x[\ln(8)-\ln(9)] &= -\ln(9) \\ x &= \frac{-\ln(9)}{\ln(8)-\ln(9)} = \frac{\ln(9)}{\ln(9)-\ln(8)} \\ &\approx 18.655 \end{aligned}$$

39. $\log_2(3) = \frac{\ln(3)}{\ln(2)} \approx 1.5850$

41. $\log_3(1/2) = \frac{\ln(0.5)}{\ln(3)} \approx -0.6309$

43. $\log_{1/2}(4.6) = \frac{\ln(4.6)}{\ln(0.5)} \approx -2.2016$

45. $\log_{0.1}(0.03) = \frac{\ln(0.03)}{\ln(0.1)} \approx 1.5229$

47.
$$\begin{aligned} x\cdot\ln(2) &= \ln(7) \\ x &= \frac{\ln(7)}{\ln(2)} \end{aligned}$$
This is the exact solution.
Use a calculator to find the approximate value of x, $x \approx 2.807$.

49.
$$\begin{aligned} 3x - x\cdot\ln(2) &= 1 \\ x(3-\ln(2)) &= 1 \\ x &= \frac{1}{3-\ln(2)} \end{aligned}$$
Exact solution.
Use a calculator to find an approximate value for x: $\frac{1}{3-\ln(2)} \approx 0.433$.

51.
$$\begin{aligned} 3^x &= 5 \\ x &= \log_3(5) = \frac{\ln(5)}{\ln(3)} \end{aligned}$$
This is the exact solution.
Use a calculator to find an approximate value for x that satisfies the equation, $\frac{\ln(5)}{\ln(3)} \approx 1.465$.

53.
$$\begin{aligned} 2^{x-1} &= 9 \\ \ln(2^{x-1}) &= \ln(9) \\ (x-1)\ln(2) &= \ln(9) \\ x-1 &= \frac{\ln(9)}{\ln(2)} \\ x &= 1+\frac{\ln(9)}{\ln(2)} \quad \text{Exact solution} \end{aligned}$$
Use a calculator to find an approximate value for x: $1+\frac{\ln(9)}{\ln(2)} \approx 4.170$.

55.
$$\begin{aligned} 3^x &= 20 \\ x &= \log_3(20) = \frac{\ln(20)}{\ln(3)} \approx 2.727 \end{aligned}$$

57.
$$\begin{aligned} \log_3(x)+\log_3(5) &= 1 \\ \log_3(5x) &= 1 \\ 5x &= 3^1 \\ x &= \frac{3}{5} \end{aligned}$$

59.
$$\begin{aligned} 8^x &= 2^{x+1} \\ (2^3)^x &= 2^{x+1} \\ 2^{3x} &= 2^{x+1} \\ 3x &= x+1 \\ 2x &= 1 \\ x &= \frac{1}{2} \end{aligned}$$

61.
$$\begin{aligned} \log_2(1-x) &= 2 \\ 2^2 &= 1-x \\ x &= 1-4 \\ x &= -3 \end{aligned}$$
The solutions set is $\{-3\}$.

63. $\log_3(1-x)+\log_3(2x+13)=3$

$$\begin{aligned}\log_3((1-x)(2x+13)&=3\\(1-x)(2x+13)&=3^3\\-2x^2-11x+13&=27\\-2x^2-11x-14&=0\\2x^2+11x+14&=0\\(2x+7)(x+2)&=0\\2x+7=0 \quad\text{or}\quad x+2&=0\\x=-\tfrac{7}{2}\quad\text{or}\quad x&=-2\end{aligned}$$

The solution set is $\left\{-\frac{7}{2},-2\right\}$.

65. $\ln(2x-1)-\ln(x+1)=\ln(5)$

$$\begin{aligned}\ln((2x-1)/(x+1))&=\ln(5)\\\frac{2x-1}{x+1}&=5\\2x-1&=5(x+1)\\2x-1&=5x+5\\-3x&=6\\x&=-2\end{aligned}$$

Since $\ln(-2+1)$ is a logarithm of a negative number, -2 does not satisfy the original equation. So the solution set is the empty set, $\emptyset$.

67. $\log_3(x-14)-\log_3(x-6)=2$

$$\begin{aligned}\log_3((x-14)/(x-6))&=2\\\frac{x-14}{x-6}&=3^2\\x-14&=9(x-6)\\x-14&=9x-54\\-8x&=-40\\x&=5\end{aligned}$$

Since $\log_3(5-14)$ is a logarithm of a negative number, 5 does not satisfy the original equation. So the solution set is the empty set, $\emptyset$

69. $\log(x+1)+\log(x-2)=1$

$$\begin{aligned}\log((x+1)(x-2))&=1\\(x+1)(x-2)&=10^1\\x^2-x-2&=10\\x^2-x-12&=0\\(x-4)(x+3)&=0\\x-4=0\quad\text{or}\quad x+3&=0\\x=4\quad\text{or}\quad x&=-3\end{aligned}$$

Since $\log(-3+1)$ is a logarithm of a negative number -3 does not satisfy the original equation. The solution set is $\{4\}$.

71. $2\cdot\ln(x)=\ln(2)+\ln(5x-12)$

$$\begin{aligned}\ln(x^2)&=\ln(10x-24)\\x^2&=10x-24\\x^2-10x+24&=0\\(x-4)(x-6)&=0\\x-4=0\quad\text{or}\quad x-6&=0\\x=4\quad\text{or}\quad x&=6\end{aligned}$$

The solution set is $\{4,6\}$.

73. $\log_3(x^3+16x^2)-\log_3(x)=\log_3(36)$

$$\begin{aligned}\log_3((x^3+16x^2)/x)&=\log_3(36)\\\frac{x^3+16x^2}{x}&=36\\x^3+16x^2&=36x\\x^3+16x^2-36x&=0\\x(x^2+16x-36)&=0\\x(x+18)(x-2)&=0\\x=0\quad\text{or}\quad x=-18\quad&\text{or}\quad x=2\end{aligned}$$

Since $\log_3(x)$ is undefined if $x=0$ or $x=-18$, the solution set is $\{2\}$.

75. $\log(x)+\log(x+5)=2\cdot\log(x+2)$

$$\begin{aligned}\log(x(x+5))&=\log((x+2)^2)\\x^2+5x&=x^2+4x+4\\x&=4\end{aligned}$$

The solution set is $\{4\}$.

77.

$\log_7(x^2+6x+8)-\log_7(x+2)=\log_7(3)$

$$\begin{aligned}\log_7((x^2+6x+8)/(x+2))&=\log_7(3)\\\frac{x^2+6x+8}{x+2}&=3\\x^2+6x+8&=3x+6\\x^2+3x+2&=0\\(x+2)(x+1)&=0\\x+2=0\quad\text{or}\quad x+1&=0\\x=-2\quad\text{or}\quad x&=-1\end{aligned}$$

If $x=-2$, then $\log_7(x+2)$ is undefined. So the solution set is $\{-1\}$.

79. $\ln(6)+2\cdot\ln(x)=\ln(38x-30)-\ln(2)$

$$\begin{aligned}\ln(6x^2)&=\ln((38x-30)/2)\\\ln(6x^2)&=\ln(19x-15)\\6x^2&=19x-15\\6x^2-19x+15&=0\\(2x-3)(3x-5)&=0\\2x-3=0\quad\text{or}\quad 3x-5&=0\\x=\tfrac{3}{2}\quad\text{or}\quad x&=\tfrac{5}{3}\end{aligned}$$

The solution set is $\left\{\frac{3}{2},\frac{5}{3}\right\}$.

81. Use the formula $S=P(1+i)^n$.

$$\begin{aligned}1500&=1000(1+0.01)^n\\1.5&=(1.01)^n\\n=\log_{1.01}(1.5)&=\frac{\ln(1.5)}{\ln(1.01)}\\&\approx 40.749\end{aligned}$$

It takes approximately 41 months.

83. Use the formula $S = P(1+i)^n$.

$$105 = 100(1 + 0.03/365)^n$$
$$1.05 = (1.00008)^n$$
$$n = \log_{1.00008}(1.05) = \frac{\ln(1.05)}{\ln(1.00008)} \approx 593.6$$

It takes approximately 594 days.

85. If $t = 0$, then

$$A = 10e^{-0.0001(0)} = 10 \text{ grams.}$$

If $A = 4$, then

$$4 = 10e^{-0.0001t}$$
$$0.4 = e^{-0.0001t}$$
$$-0.0001t = \ln(0.4)$$
$$t = \frac{\ln(0.4)}{-0.0001} \approx 9162.9$$

The cloth was made about 9163 years ago.

87. $y = 114.308e^{(0.265 \cdot 15.8)} \approx 7524 \text{ ft}^3/\text{sec}$

89.
$$40 = 28e^{0.05t}$$
$$\frac{10}{7} = e^{0.05t}$$
$$0.05t = \ln(10/7)$$
$$t = \frac{\ln(10/7)}{0.05} \approx 7.133$$

There will be 40 million people above the poverty level in approximately 7.1 years.

91.
$$28e^{0.05t} = 20e^{0.07t}$$
$$\frac{e^{0.05t}}{e^{0.07t}} = \frac{20}{28}$$
$$e^{-0.02t} = 5/7$$
$$-0.02t = \ln(5/7)$$
$$t = \frac{\ln(5/7)}{-0.02} \approx 16.824$$

The number of people above the poverty level will equal the number below the poverty level in approximately 16.8 years.

93.
$$\text{pH} = -\log(\text{H}^+)$$
$$3.7 = -\log(\text{H}^+)$$
$$-3.7 = \log(\text{H}^+)$$
$$\text{H}^+ = 10^{-3.7} \approx 2.0 \times 10^{-4}$$

95. $d = \log_2\left(\frac{3\sqrt[3]{2}}{2}\right) = \frac{\ln\left(\frac{3\sqrt[3]{2}}{2}\right)}{\ln(2)}$

$$\approx 0.9183$$

97. Using logarithms:

$$x^3 = 12$$
$$3\ln(x) = \ln(12)$$
$$\ln(x) = \ln(12)/3$$
$$x = e^{\ln(12)/3} \approx 2.2894$$

Using roots:

$$x^3 = 12$$
$$x = \sqrt[3]{12} = 12^{1/3} \approx 2.2894$$

99. (2.71, 6.54)

101. (1.03, 0.04), (4.74, 2.24)

Chapter 10 Wrap-Up

Enriching Your Mathematical Word Power

1. a **2.** d **3.** b **4.** d **5.** d
6. b **7.** a **8.** b **9.** b **10.** c

CHAPTER 10 REVIEW

1. $f(-2) = 5^{-2} = \frac{1}{5^2} = \frac{1}{25}$

3. $f(3) = 5^3 = 125$

5. $g(1) = 10^{1-1} = 10^0 = 1$

7. $g(0) = 10^{0-1} = 10^{-1} = \frac{1}{10}$

9. $h(-1) = (1/4)^{-1} = 4$

11. $h(1/2) = (1/4)^{1/2} = \sqrt{\frac{1}{4}} = \frac{1}{2}$

13. If $f(x) = 25$, then $5^x = 25$, $x = 2$.

15. If $g(x) = 1000$, then $10^{x-1} = 1000$.

$$10^{x-1} = 10^3$$
$$x - 1 = 3$$
$$x = 4$$

17. If $h(x) = 32$, then $(1/4)^x = 32$.

$$(2^{-2})^x = 2^5$$
$$2^{-2x} = 2^5$$
$$-2x = 5$$
$$x = -\frac{5}{2}$$

19. If $h(x) = 1/16$, then $(1/4)^x = 1/16$, or $(1/4)^x = (1/4)^2$, $x = 2$.

21. $f(1.34) = 5^{1.34} \approx 8.6421$

23. $g(3.25) = 10^{3.25-1} = 10^{2.25} \approx 177.828$

25. $h(2.82) = (1/4)^{2.82} = (0.25)^{2.82}$
≈ 0.02005

27. $h(\sqrt{2}) = (1/4)^{\sqrt{2}} = \approx 0.1408$

29. The graph of $f(x) = 5^x$ includes the points (0, 1), (1, 5), and (−1, 1/5).

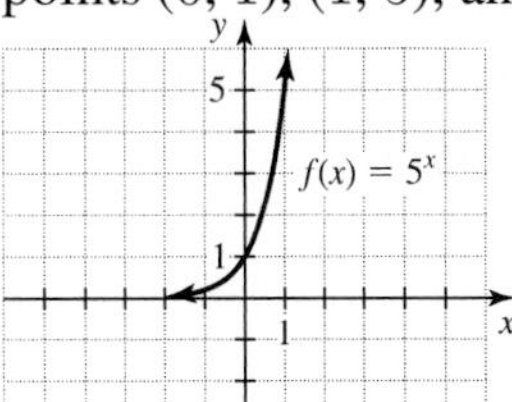

31. The graph of $y = (1/5)^x$ includes the points (1, 1/5), (0, 1) and (−1, 5).

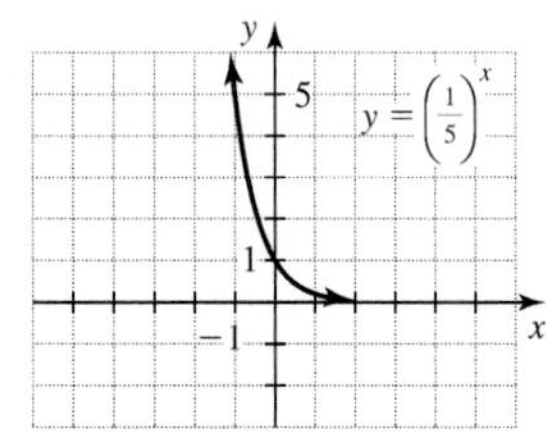

33. The graph of $y = 3^{-x}$ includes the points $(0, 1)$, $(1, 1/3)$, and $(-1, 3)$.

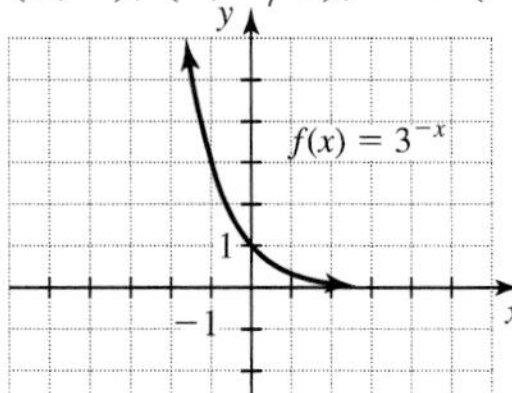

35. The graph of $y = 1 + 2^x$ includes the points $(0, 2)$, $(1, 3)$, $(2, 5)$, and $(-1, 1.5)$.

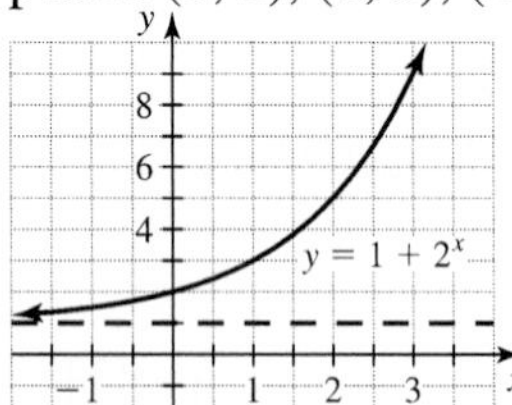

37. $\log(n) = m$

39. $k^h = t$

41. $f(1/8) = \log_2(1/8) = -3$, because $2^{-3} = 1/8$.

43. $g(0.1) = \log(0.1) = -1$, because $10^{-1} = 0.1$.

45. $g(100) = \log(100) = 2$, because $10^2 = 100$.

47. $h(1) = \log_{1/2}(1) = 0$, because $(1/2)^0 = 1$.

49. If $f(x) = 8$, then $\log_2(x) = 8$, $x = 2^8 = 256$.

51. $f(77) = \log_2(77) = \dfrac{\ln(77)}{\ln(2)} \approx 6.267$

53. $h(33.9) = \log_{1/2}(33.9) = \dfrac{\ln(33.9)}{\ln(0.5)}$

≈ -5.083

55. If $f(x) = 2.475$, then $\log_2(x) \approx 2.475$.

$x = 2^{2.475} \approx 5.560$

57. The inverse of the function $f(x) = 10^x$ is the base 10 logarithm function,
$f^{-1}(x) = \log(x)$.
The graph of $f(x)$ includes the points $(1, 10)$, $(0, 1)$, and $(-1, 0.1)$. The graph of $f^{-1}(x)$ includes the points $(10, 1)$, $(1, 0)$, and $(0.1, -1)$.

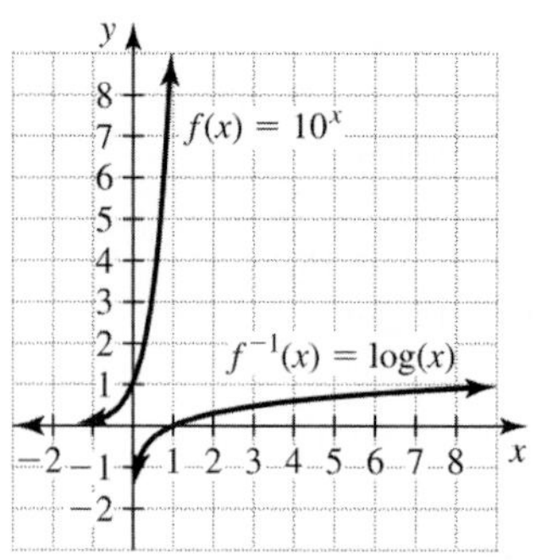

59. The inverse of the function $f(x) = e^x$ is $f^{-1}(x) = \ln(x)$. The graph of $f(x)$ includes the points $(1, e)$, $(0, 1)$, and $(-1, 1/e)$. The graph of $f^{-1}(x)$ includes the points $(e, 1)$, $(1, 0)$, and $(1/e, -1)$.

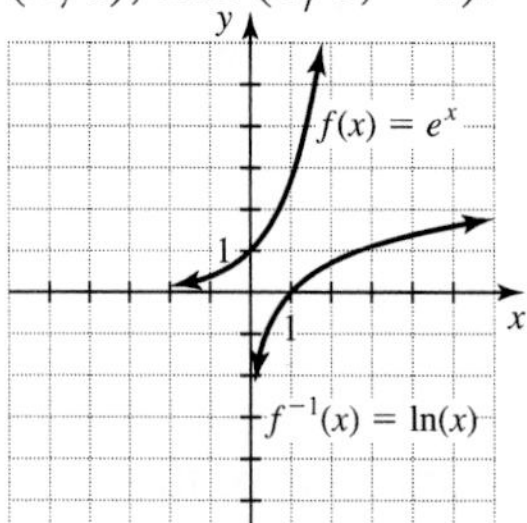

61. $\log(x^2y) = \log(x^2) + \log(y)$
$= 2\log(x) + \log(y)$

63. $\ln(16) = \ln(2^4) = 4\ln(2)$

65. $\log_5(1/x) = \log_5(1) - \log_5(x)$
$= -\log_5(x)$

67. $\frac{1}{2}\log(x + 2) - 2\log(x - 1)$
$= \log\left((x + 2)^{1/2}\right) - \log((x - 1)^2)$
$= \log\left(\dfrac{\sqrt{x + 2}}{(x - 1)^2}\right)$

69. $\log_2(x) = 8$
$x = 2^8 = 256$
The solution set is $\{256\}$.

71. $\log_2(8) = x$
$3 = x$
The solution set is $\{3\}$.

73. $x^3 = 8$
$x = \sqrt[3]{8} = 2$
The solution set is $\{2\}$.

75. $\log_x(27) = 3$
$x^3 = 27$
$x = \sqrt[3]{27} = 3$
The solution set is $\{3\}$.

77. $$x \cdot \ln(3) - x = \ln(7)$$
$$x[\ln(3) - 1] = \ln(7)$$
$$x = \frac{\ln(7)}{\ln(3) - 1}$$
The solution set is $\left\{\frac{\ln(7)}{\ln(3) - 1}\right\}$.

79. $$3^x = 5^{x-1}$$
$$\ln(3^x) = \ln(5^{x-1})$$
$$x \cdot \ln(3) = (x - 1)\ln(5)$$
$$x \cdot \ln(3) = x \cdot \ln(5) - \ln(5)$$
$$x \cdot \ln(3) - x \cdot \ln(5) = -\ln(5)$$
$$x[\ln(3) - \ln(5)] = -\ln(5)$$
$$x = \frac{-\ln(5)}{\ln(3) - \ln(5)} = \frac{\ln(5)}{\ln(5) - \ln(3)}$$
The solution set is $\left\{\frac{\ln(5)}{\ln(5) - \ln(3)}\right\}$.

81. $$4^{2x} = 2^{x+1}$$
$$(2^2)^{2x} = 2^{x+1}$$
$$2^{4x} = 2^{x+1}$$
$$4x = x + 1$$
$$3x = 1$$
$$x = \frac{1}{3}$$
The solution set is $\left\{\frac{1}{3}\right\}$.

83. $$\ln(x + 2) - \ln(x - 10) = \ln(2)$$
$$\ln\left(\frac{x + 2}{x - 10}\right) = \ln(2)$$
$$\frac{x + 2}{x - 10} = 2$$
$$x + 2 = 2(x - 10)$$
$$x + 2 = 2x - 20$$
$$22 = x$$
The solution set is $\{22\}$.

85. $$\log(x) - \log(x - 2) = 2$$
$$\log\left(\frac{x}{x - 2}\right) = 2$$
$$\frac{x}{x - 2} = 10^2$$
$$x = 100(x - 2)$$
$$x = 100x - 200$$
$$-99x = -200$$
$$x = \frac{-200}{-99} = \frac{200}{99}$$
The solution set is $\left\{\frac{200}{99}\right\}$.

87. $$6^x = 12$$
$$x = \log_6(12) = \frac{\ln(12)}{\ln(6)} \approx 1.3869$$
The solution set is $\{1.3869\}$.

89. $$3^{x+1} = 5$$
$$x + 1 = \log_3(5)$$
$$x = -1 + \log_3(5) = -1 + \frac{\ln(5)}{\ln(3)}$$
$$\approx 0.4650$$
The solution set is $\{0.4650\}$.

91. Use the formula $S = P(1 + i)^n$ with $i = 11.5\% = 0.115$, $n = 15$, and $P = \$10{,}000$.
$S = 10{,}000(1.115)^{15} \approx \$51{,}182.68$

93. Use the formula $A = A_0 e^{-0.0003t}$ with $A_0 = 218$ and $t = 1000$.
$$A = 218e^{-0.0003(1000)} = 218e^{-0.3}$$
$$\approx 161.5 \text{ grams}$$

95. The amount in Melissa's account is given by the formula $S = 1000(1.05)^t$ for any number of years t. The amount in Frank's account is given by the formula $S = 900e^{0.07t}$ for any number of years t. To find when they have the same amount, we set the two expressions equal and solve for t.
$$1000(1.05)^t = 900e^{0.07t}$$
$$\ln(1000(1.05)^t) = \ln(900e^{0.07t})$$
$$\ln(1000) + t \cdot \ln(1.05) = \ln(900) + \ln(e^{0.07t})$$
$$\ln(1000) + t \cdot \ln(1.05) = \ln(900) + 0.07t$$
$$t \cdot \ln(1.05) - 0.07t = \ln(900) - \ln(1000)$$
$$t[\ln(1.05) - 0.07] = \ln(900) - \ln(1000)$$
$$t = \frac{\ln(900) - \ln(1000)}{\ln(1.05) - 0.07} \approx 4.9675$$
The amounts will be equal in approximately 5 years.

97. $114.308e^{0.265(20.6-6.87)} \approx 4347.5 \text{ ft}^3/\text{sec}$

CHAPTER 10 TEST

1. $f(2) = 5^2 = 25$

2. $f(-1) = 5^{-1} = \frac{1}{5}$

3. $f(0) = 5^0 = 1$

4. $g(125) = \log_5(125) = 3$, because $5^3 = 125$.

5. $g(1) = \log_5(1) = 0$, because $5^0 = 1$.

6. $g(1/5) = \log_5(1/5) = -1$, because $5^{-1} = 1/5$.

7. The graph of $y = 2^x$ includes the points (0, 1), (1, 2), (2, 4), and (−1, 1/2).

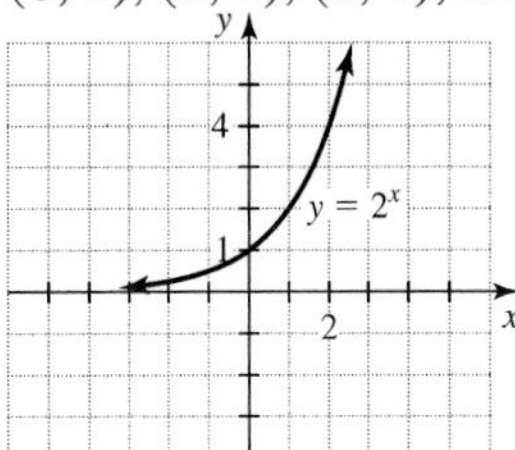

8. The graph of $f(x) = \log_2(x)$ includes the points (1, 0), (2, 1), (4, 2), and (1/2, −1).

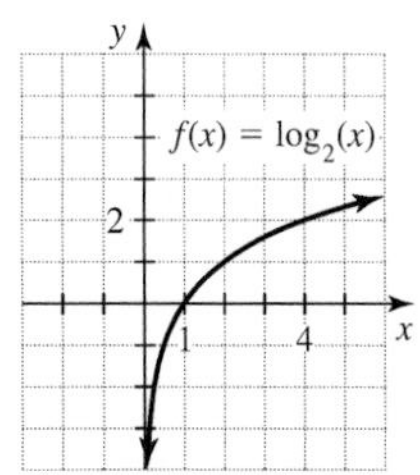

9. The graph of $y = \left(\frac{1}{3}\right)^x$ includes the points (1, 1/3), (0, 1), and (−1, 3).

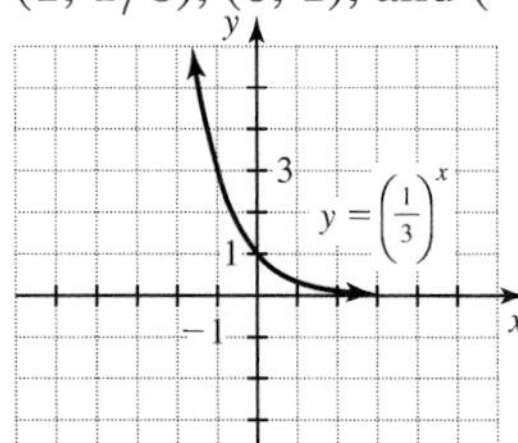

10. The graph of $g(x) = \log_{1/3}(x)$ includes the points (1, 0), (3, −1), and (1/3, 1).

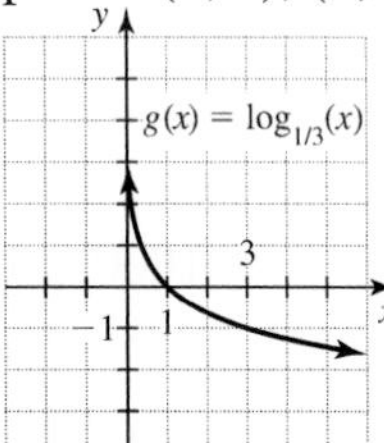

11. The graph of $f(x) = -2^x + 3$ includes the points $(0, 2)$, $(1, 1)$, $(2, -1)$, and $(-1, 2.5)$. Reflect $y = 2^x$ in the x-axis and then translate it upward 3 units to get the graph.

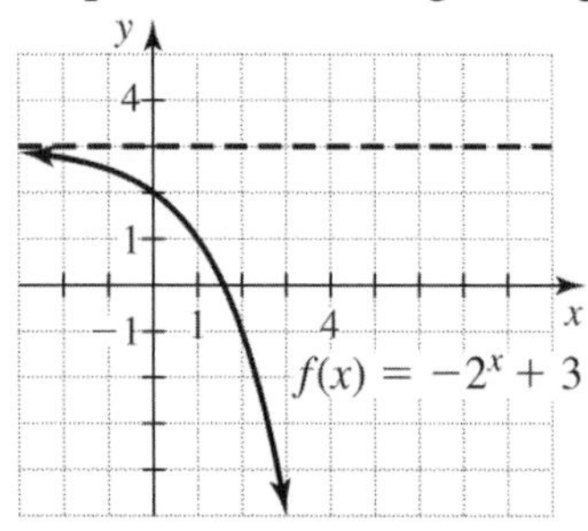

12. The graph of $f(x) = 2^{x-3} - 1$ includes the points $(3, 0)$, $(4, 1)$, $(5, 3)$, and $(2, -0.5)$. Translate $y = 2^x$ to the right 3 units and then downward 1 unit to get the graph.

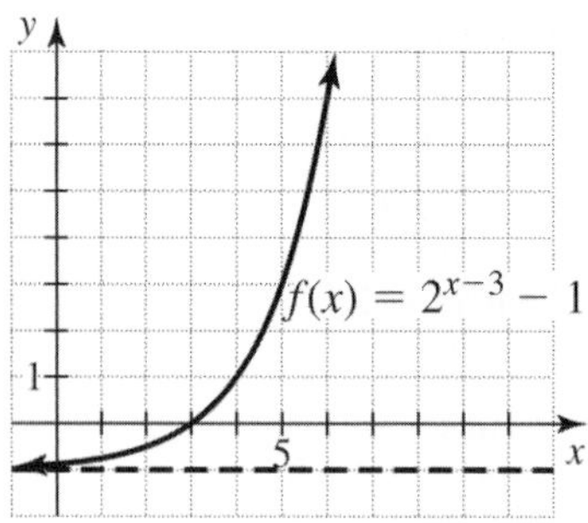

13. $\log_a(MN) = \log_a(M) + \log_a(N) = 6 + 4$
$= 10$

14. $\log_a(M^2/N) = 2 \cdot \log_a(M) - \log_a(N)$
$= 2 \cdot 6 - 4 = 8$

15. $\dfrac{\log_a(M)}{\log_a(N)} = \dfrac{6}{4} = \dfrac{3}{2}$

16. $\log_a(a^3M^2) = 3 \cdot \log_a(a) + 2 \cdot \log_a(M)$
$= 3 \cdot 1 + 2 \cdot 6 = 15$

17. $\log_a(1/N) = \log_a(1) - \log_a(N)$
$= 0 - 4 = -4$

18. $3^x = 12$
$x = \log_3(12)$
The solution set is $\{\log_3(12)\}$ or $\{\ln(12)/\ln(3)\}$.

19. $\log_3(x) = 1/2$
$x = 3^{1/2} = \sqrt{3}$
The solution set is $\{\sqrt{3}\}$.

20.
$$
\begin{aligned}
5^x &= 8^{x-1} \\
\ln(5^x) &= \ln(8^{x-1}) \\
x \cdot \ln(5) &= (x-1)\ln(8) \\
x \cdot \ln(5) &= x \cdot \ln(8) - \ln(8) \\
x \cdot \ln(5) - x \cdot \ln(8) &= -\ln(8) \\
x[\ln(5) - \ln(8)] &= -\ln(8) \\
x &= \frac{-\ln(8)}{\ln(5) - \ln(8)} = \frac{\ln(8)}{-\ln(5) + \ln(8)}
\end{aligned}
$$
The solution set is $\left\{\dfrac{\ln(8)}{\ln(8) - \ln(5)}\right\}$.

21.
$$
\begin{aligned}
\log(x) + \log(x+15) &= 2 \\
\log(x^2 + 15x) &= 2 \\
x^2 + 15x &= 10^2 \\
x^2 + 15x - 100 &= 0 \\
(x+20)(x-5) &= 0 \\
x + 20 = 0 \quad &\text{or} \quad x - 5 = 0 \\
x = -20 \quad &\text{or} \quad x = 5
\end{aligned}
$$
If $x = -20$ in the original equation, we get a logarithm of a negative number, which is undefined. So the solution set is $\{5\}$.

22.
$$2 \cdot \ln(x) = \ln(3) + \ln(6 - x)$$
$$\ln(x^2) = \ln(18 - 3x)$$
$$x^2 = 18 - 3x$$
$$x^2 + 3x - 18 = 0$$
$$(x + 6)(x - 3) = 0$$
$$x + 6 = 0 \quad \text{or} \quad x - 3 = 0$$
$$x = -6 \quad \text{or} \quad x = 3$$
If $x = -6$ in the original equation, we get a logarithm of a negative number, which is undefined. So the solution set is $\{3\}$.

23. $20^x = 5$
$$x = \log_{20}(5) = \frac{\ln(5)}{\ln(20)} \approx 0.5372$$
The solution set is $\{0.5372\}$.

24. $\log_3(x) = 2.75$
$$x = 3^{2.75} \approx 20.5156$$
The solution set is $\{20.5156\}$

25. To find the number present initially, let $t = 0$ in the formula:
$N = 10e^{0.4(0)} = 10e^0 = 10 \cdot 1 = 10$
To find the number present after 24 hours, let $t = 24$ in the formula:
$N = 10e^{0.4(24)} = 10e^{9.6} \approx 147{,}648$

26. To find how long it takes for the population to double, we must find the value of t for which $e^{0.4t} = 2$. Solve for t.
$$0.4t = \ln(2)$$
$$t = \frac{\ln(2)}{0.4} \approx 1.733$$
The bacteria population doubles in 1.733 hours.

Making Connections

Chapters 1 - 10

1. $(x - 3)^2 = 8$
$$x - 3 = \pm\sqrt{8}$$
$$x = 3 \pm 2\sqrt{2}$$
The solution set is $\left\{3 \pm 2\sqrt{2}\right\}$.

2. $\log_2(x - 3) = 8$
$$x - 3 = 2^8$$
$$x = 256 + 3 = 259$$
The solution set is $\{259\}$.

3.
$$2^{x-3} = 8$$
$$2^{x-3} = 2^3$$
$$x - 3 = 3$$
$$x = 6$$
The solution set is $\{6\}$.

4. $2x - 3 = 8$
$$2x = 11$$
$$x = \frac{11}{2}$$
The solution set is $\left\{\frac{11}{2}\right\}$.

5.
$$|x - 3| = 8$$
$$x - 3 = 8 \quad \text{or} \quad x - 3 = -8$$
$$x = 11 \quad \text{or} \quad x = -5$$
The solution set is $\{-5, 11\}$

6.
$$\sqrt{x - 3} = 8$$
$$(\sqrt{x - 3})^2 = 8^2$$
$$x - 3 = 64$$
$$x = 67$$
The solution set is $\{67\}$.

7. $\log_2(x - 3) + \log_2(x) = \log_2(18)$
$$\log_2(x^2 - 3x) = \log_2(18)$$
$$x^2 - 3x = 18$$
$$x^2 - 3x - 18 = 0$$
$$(x - 6)(x + 3) = 0$$
$$x - 6 = 0 \quad \text{or} \quad x + 3 = 0$$
$$x = 6 \quad \text{or} \quad x = -3$$
If $x = -3$ in the original equation, then we get a logarithm of a negative number, which is undefined. The solution set is $\{6\}$.

8.
$$2 \cdot \log_2(x - 3) = \log_2(5 - x)$$
$$\log_2[(x - 3)^2] = \log_2(5 - x)$$
$$(x - 3)^2 = 5 - x$$
$$x^2 - 6x + 9 = 5 - x$$
$$x^2 - 5x + 4 = 0$$
$$(x - 4)(x - 1) = 0$$
$$x - 4 = 0 \quad \text{or} \quad x - 1 = 0$$
$$x = 4 \quad \text{or} \quad x = 1$$
If $x = 1$ in the original equation, we get a logarithm of a negative number. The solution set is $\{4\}$.

9.
$$\frac{1}{2}x - \frac{2}{3} = \frac{3}{4}x + \frac{1}{5}$$
$$60\left(\frac{1}{2}x - \frac{2}{3}\right) = 60\left(\frac{3}{4}x + \frac{1}{5}\right)$$
$$30x - 40 = 45x + 12$$
$$-52 = 15x$$
$$-\frac{52}{15} = x$$
The solution set is $\left\{-\frac{52}{15}\right\}$.

10. To solve $3x^2 - 6x + 2 = 0$ use the quadratic formula.

$$x = \frac{6 \pm \sqrt{(-6)^2 - 4(3)(2)}}{2(3)} = \frac{6 \pm \sqrt{12}}{6}$$
$$= \frac{6 \pm 2\sqrt{3}}{6} = \frac{3 \pm \sqrt{3}}{3}$$

The solution set is $\left\{\frac{3 \pm \sqrt{3}}{3}\right\}$.

11. The inverse of dividing by 3 is multiplying by 3. So $f^{-1}(x) = 3x$.

12. The inverse of the base 3 logarithm function is the base 3 exponential function, $1g^{-1}(x) = 3^x$.

13. The inverse of multiplying by 2 and then subtracting 4 is adding 4 and then dividing by 2, $f^{-1}(x) = \frac{x+4}{2}$.

14. The inverse of the square root function is the squaring function. To keep the domain of one function equal to the range of the inverse function we must restrict the squaring function to the nonnegative numbers:
$h^{-1}(x) = x^2$ for $x \geq 0$

15. The reciprocal function is its own inverse, $j^{-1}(x) = \frac{1}{x}$.

16. The inverse of the base 5 exponential function is the base 5 logarithm function, $k^{-1}(x) = \log_5(x)$

17. We will find the inverse for m by using the technique of interchanging x and y and then solving for y.

$$m(x) = e^{x-1}$$
$$y = e^{x-1}$$
$$x = e^{y-1}$$
$$y - 1 = \ln(x)$$
$$y = 1 + \ln(x)$$
$$m^{-1}(x) = 1 + \ln(x)$$

18. The inverse of the natural logarithm function is the base e exponential function, $n^{-1}(x) = e^x$.

19. The graph of $y = 2x$ is a straight line with slope 2 and y-intercept (0, 0). Start at the origin and rise 2 and go 1 to the right to find a second point on the line.

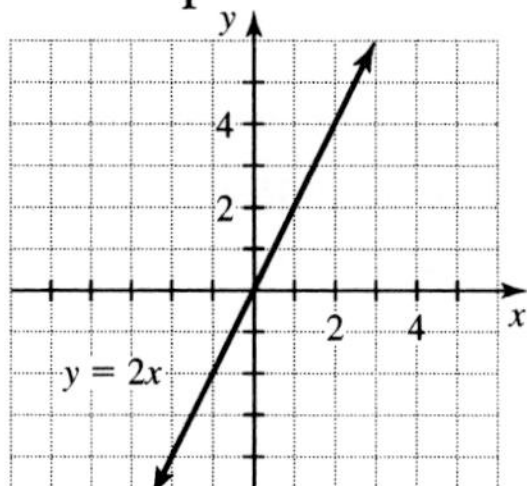

20. The graph of $y = 2^x$ includes the points (0, 1), (1, 2), (2, 2), and (−1, 1/2).

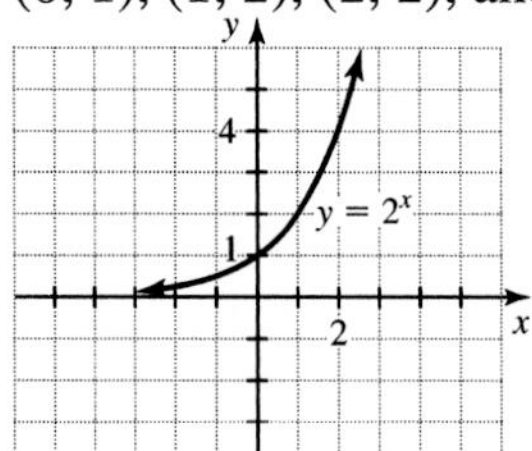

21. The graph of $y = x^2$ is a parabola through (0, 0), (1, 1), (2, 4), (−1, 1), and (−2, 4).

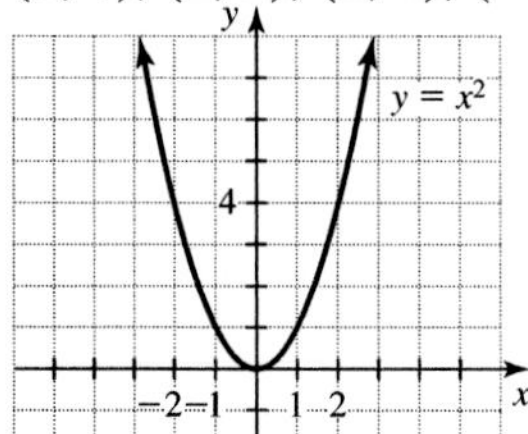

22. The graph of $y = \log_2(x)$ includes the points (1, 0), (2, 1), (1/2, −1) and (4, 2).

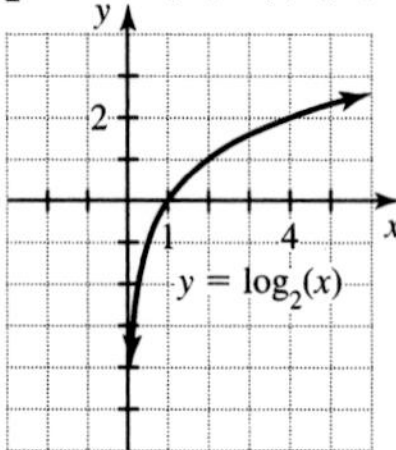

23. The graph of $y = \frac{1}{2}x - 4$ is a straight line with slope 1/2 and y-intercept (0, −4). Start at (0, −4), rise 1 and go 2 to the right to locate a second point on the line.

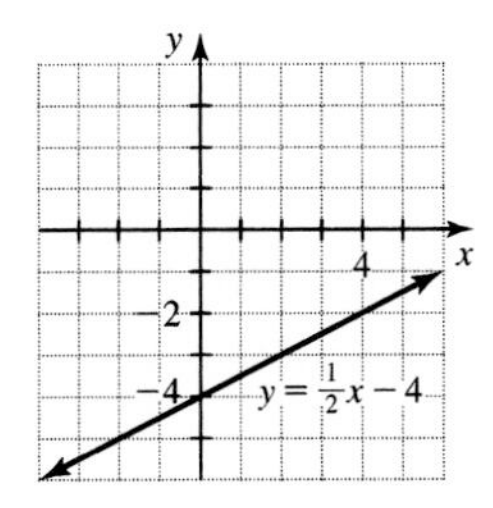

24. The graph of $y = |\,2 - x\,|$ is a v-shaped graph through (−2, 4), (−1, 3), (0, 2), (1, 1), (2, 0), (3, 1), (4, 2), and (5, 3).

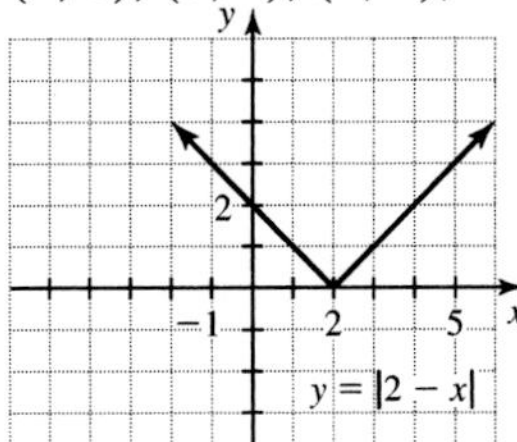

25. The graph of $y = 2 - x^2$ is a parabola with a vertex at (0, 2). It also includes the points (1, 1), (−1, 1), (2, −2), and (−2, −2).

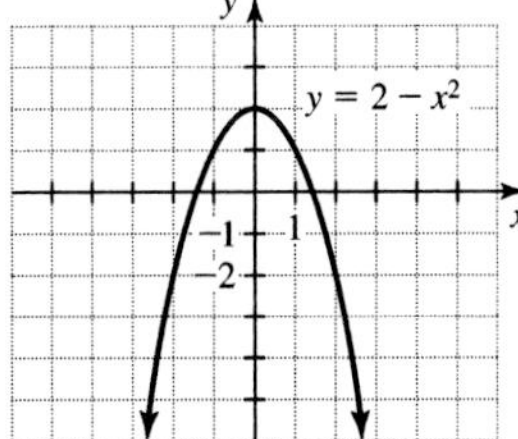

26. Note that e^2 is not a variable. The value of e^2 is approximately 7.389. The graph of $y = e^2$ is a straight line with 0 slope and y-intercept (0, 7.389).

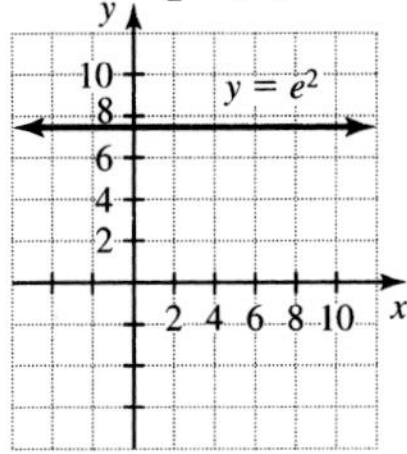

27. a) The graph of $n = 1.51t + 125.5$ is a straight line through (0, 125.5) and (10, 140.6). The graph of $n = 125.6e^{0.011t}$ is an exponential curve through (0, 125.6), (5, 132.7), (10, 140.2), and (15, 148.1).

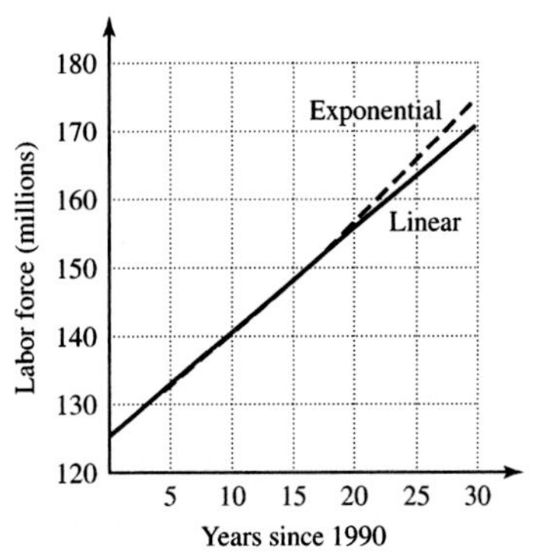

b) In 2010, $t = 20$:
$n = 1.51(20) + 125.5 \approx 155.7$ million
or
$n = 125.6e^{0.011(20)} \approx 156.5$ million

c) Look up current data at www.bls.gov.

28. a) Use $D = RT$ to get $2d_1 = v(0.270)$ or $d_1 = 0.135v$.

b) Using $D = RT$ we get $2d_2 = v(0.432)$ or $d_2 = 0.216v$.

c) Using the Pythagorean theorem we get $(d_2)^2 = (d_1)^2 + 250^2$. Use $d_1 = 0.135v$ and $d_2 = 0.216v$ to get

$$\begin{aligned}(0.216v)^2 &= (0.135v)^2 + 250^2\\ 0.028431v^2 &= 250^2\\ v^2 &= 2198304.667\\ v &= 1482.668091 \text{ m/sec}\end{aligned}$$

$d_1 = 0.135v \approx 200.2$ meters

11.1 WARM-UPS

1. True, because any equation of the form $y = ax^2 + bx + c$ has a graph that is a parabola.
2. False, because absolute value has a v-shaped graph.
3. False, because $-4 \neq \sqrt{5(3)+1}$.
4. True, because $y = \sqrt{x}$ lies entirely in the first quadrant, and $y = -x - 2$ has no points in the first quadrant.
5. False, because we can also use addition.
6. True.
7. True, because a 30-60-90 triangle is half of an equilateral triangle.
8. True, because for any rectangular solid we have $V = LWH$.
9. True, because the surface area consists of 6 rectangles, of which two have area LW, two have area WH, and two have area LH.
10. True, because the area of a triangle is $(bh)/2$.

11.1 EXERCISES

1. If the graph of an equation is not a straight line then it is called nonlinear.
3. Graphing is not an accurate method for solving a system and the graphs might be difficult to draw.
5. The graph of $y = x^2$ is a parabola and the graph of $x + y = 6$ is a straight line.

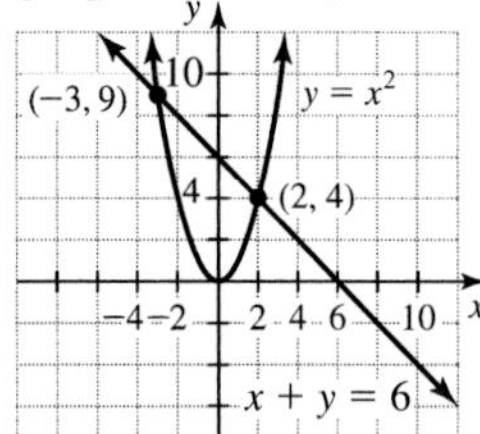

To solve the system, substitute $y = x^2$ into $x + y = 6$.

$$x + x^2 = 6$$
$$x^2 + x - 6 = 0$$
$$(x+3)(x-2) = 0$$
$$x = -3 \text{ or } x = 2$$

If $x = -3$, $y = (-3)^2 = 9$, and if $x = 2$, $y = 2^2 = 4$. The solution set to the system is $\{(2, 4), (-3, 9)\}$.
7. The graph of $y = |x|$ is v-shaped, and the graph of $2y - x = 6$ is a straight line. To solve the system, substitute $x = 2y - 6$ into $y = |x|$.

$$y = |2y - 6|$$
$$y = 2y - 6 \text{ or } y = -(2y - 6)$$
$$6 = y \quad \text{or} \quad y = -2y + 6$$
$$3y = 6$$
$$y = 2$$

Use $y = 6$ in $x = 2y - 6$ to get $x = 6$. Use $y = 2$ in $x = 2y - 6$ to get $x = -2$. The graphs intersect at the points $(-2, 2)$ and $(6, 6)$.

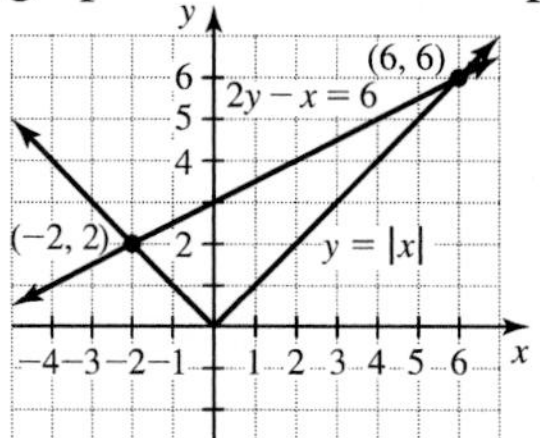

The solution set is $\{(-2, 2), (6, 6)\}$.
9. The graph of $y = \sqrt{2x}$ includes the points $(0, 0)$, $(2, 2)$, and $(4.5, 3)$. The graph of $x - y = 4$ is a straight line with slope 1 and y-intercept $(0, -4)$.

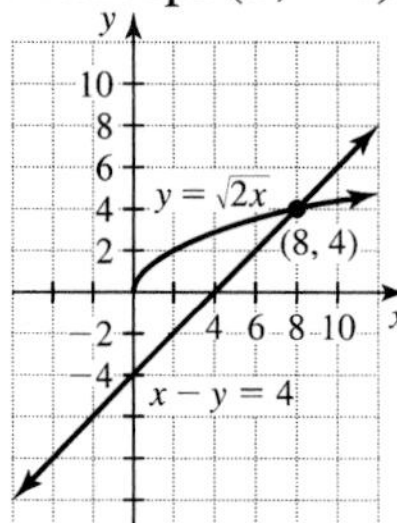

Substitute $y = \sqrt{2x}$ into $y = x - 4$.

$$\sqrt{2x} = x - 4$$
$$(\sqrt{2x})^2 = (x-4)^2$$
$$2x = x^2 - 8x + 16$$
$$0 = x^2 - 10x + 16$$
$$0 = (x-8)(x-2)$$
$$x = 8 \quad \text{or} \quad x = 2$$
$$y = \sqrt{2(8)} \text{ or } y = \sqrt{2(2)}$$
$$= 4 \quad \text{or} \quad = 2$$

Since we squared both sides, we must check. The pair $(8, 4)$ satisfies both equations, but $(2, 2)$ does not. The solution set to the system is $\{(8, 4)\}$.
11. The graph of $4x - 9y = 9$ is a straight line. The equation $xy = 1$ can be written as $y = 1/x$. Substitute $y = 1/x$ into $4x - 9y = 9$.

$$4x - 9\left(\frac{1}{x}\right) = 9$$
$$4x^2 - 9 = 9x$$

$$4x^2 - 9x - 9 = 0$$
$$(4x + 3)(x - 3) = 0$$
$$x = -\frac{3}{4} \quad \text{or} \quad x = 3$$
$$y = \frac{1}{-3/4} = -\frac{4}{3} \quad\quad y = \frac{1}{3}$$

The solution set is $\left\{\left(-\frac{3}{4}, -\frac{4}{3}\right), \left(3, \frac{1}{3}\right)\right\}$.

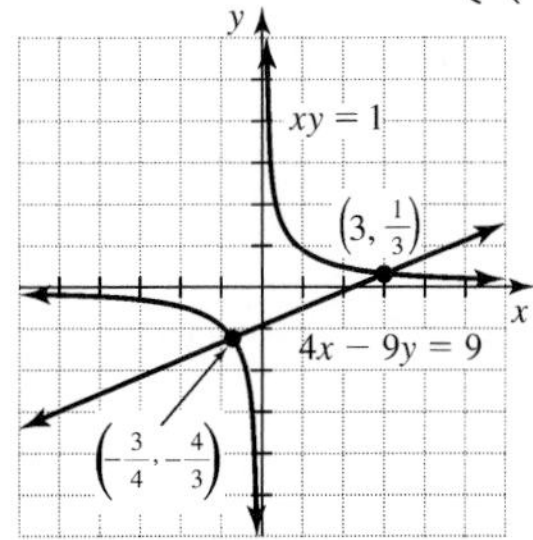

13. The graph of $y = x^2$ is a parabola opening upward, and the graph of $y = -x^2 + 1$ is a parabola opening downward.

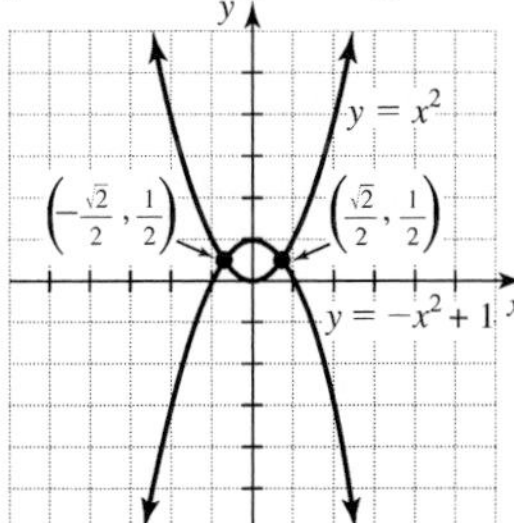

Substitute $y = x^2$ into $y = -x^2 + 1$.

$$x^2 = -x^2 + 1$$
$$2x^2 = 1$$
$$x^2 = \frac{1}{2}$$
$$x = \pm\sqrt{\frac{1}{2}} = \pm\frac{\sqrt{2}}{2}$$

Since $y = x^2$, $y = \frac{1}{2}$ for either value of x.
The

solution set is $\left\{\left(\frac{\sqrt{2}}{2}, \frac{1}{2}\right), \left(-\frac{\sqrt{2}}{2}, \frac{1}{2}\right)\right\}$.

15. Substitute $x = 2$ into $xy = 6$:

$$xy = 6$$
$$2y = 6$$
$$y = 3$$

The solution set is $\{(2, 3)\}$.

17. Substitute $y = x$ into $xy = 1$:

$$xy = 1$$
$$xx = 1$$
$$x^2 = 1$$
$$x = \pm 1$$

Use $y = x$ to get $y = \pm 1$. The solution set is $\{(-1, -1), (1, 1)\}$.

19. Substitute $y = 2$ into $y = x^2$:

$$y = x^2$$
$$2 = x^2$$
$$x = \pm\sqrt{2}$$
$$x = -\sqrt{2} \quad \text{or} \quad x = \sqrt{2}$$
$$y = 2 \quad\quad y = 2$$

The solution set is $\left\{\left(-\sqrt{2}, 2\right), \left(\sqrt{2}, 2\right)\right\}$.

21. Write $y = x^2 - 5$ as $x^2 = y + 5$ and substitute into $x^2 + y^2 = 25$

$$y + 5 + y^2 = 25$$
$$y^2 + y - 20 = 0$$
$$(y + 5)(y - 4) = 0$$
$$y = -5 \quad \text{or} \quad y = 4$$

Use $y = -5$ in $x^2 = y + 5$ to get $x^2 = 0$ or $x = 0$. Use $y = 4$ in $x^2 = y + 5$ to get $x^2 = 9$ or $x = \pm 3$. The solution set to the system is $\{(0, -5), (3, 4), (-3, 4)\}$.

23. Substitute $y = x + 1$ into $xy - 3x = 8$.

$$x(x + 1) - 3x = 8$$
$$x^2 + x - 3x = 8$$
$$x^2 - 2x - 8 = 0$$
$$(x - 4)(x + 2) = 0$$
$$x = 4 \quad \text{or} \quad x = -2$$
$$y = 5 \quad\quad y = -1$$

The solution set is $\{(4, 5), (-2, -1)\}$.

25. Write $xy - x = 8$ as $xy = x + 8$, and substitute for xy in $xy + 3x = -4$.

$$x + 8 + 3x = -4$$
$$4x = -12$$
$$x = -3$$

Use $x = -3$ in $xy = x + 8$.

$$-3y = -3 + 8$$
$$-3y = 5$$
$$y = -5/3$$

The solution set is $\left\{\left(-3, -\frac{5}{3}\right)\right\}$.

27. If we add $x^2 + y^2 = 8$ and $x^2 - y^2 = 2$ we get $2x^2 = 10$.

$$x^2 = 5$$
$$x = \pm\sqrt{5}$$

Use $x = \sqrt{5}$ in $y^2 = 8 - x^2$, to get $y = \pm\sqrt{3}$. Use $x = -\sqrt{5}$ in $y^2 = 8 - x^2$, to get $y = \pm\sqrt{3}$.
The solution set contains four points, $\left\{(\sqrt{5}, \sqrt{3}), (\sqrt{5}, -\sqrt{3}), (-\sqrt{5}, \sqrt{3}), (-\sqrt{5}, -\sqrt{3})\right\}$

29. Write $x^2 + 2y^2 = 8$ as $x^2 = 8 - 2y^2$ and substitute this equation into $2x^2 - y^2 = 1$.

$$2(8 - 2y^2) - y^2 = 1$$

$$16 - 4y^2 - y^2 = 1$$
$$-5y^2 = -15$$
$$y^2 = 3$$
$$y = \pm\sqrt{3}$$

If $y = \sqrt{3}$, then $x^2 = 8 - 2(\sqrt{3})^2 = 2$, or $x = \pm\sqrt{2}$.
If $y = -\sqrt{3}$, then $x^2 = 8 - 2(-\sqrt{3})^2 = 2$, or $x = \pm\sqrt{2}$. The solution set is
$\left\{(\sqrt{2}, \sqrt{3}), (\sqrt{2}, -\sqrt{3}), (-\sqrt{2}, \sqrt{3}), (-\sqrt{2}, -\sqrt{3})\right\}$

31. If we just add the equations as they are given, then y is eliminated.

$$\frac{1}{x} - \frac{1}{y} = 5$$
$$\frac{2}{x} + \frac{1}{y} = -3$$
$$\frac{3}{x} = 2$$
$$3 = 2x$$
$$3/2 = x$$

Use $x = 3/2$ in $\frac{1}{x} - \frac{1}{y} = 5$.

$$\frac{1}{3/2} - \frac{1}{y} = 5$$
$$\frac{2}{3} - \frac{1}{y} = 5$$
$$2y - 3 = 15y$$
$$-3 = 13y$$
$$-3/13 = y$$

The solution set is $\left\{\left(\frac{3}{2}, -\frac{3}{13}\right)\right\}$.

33. Multiply the first equation by -3 and add the result to the second equation.

$$\frac{-6}{x} + \frac{3}{y} = \frac{-15}{12}$$
$$\frac{1}{x} - \frac{3}{y} = -\frac{5}{12}$$
$$\frac{-5}{x} = \frac{-20}{12}$$
$$-20x = -60$$
$$x = 3$$

Use $x = 3$ in the first equation to find y.

$$\frac{2}{3} - \frac{1}{y} = \frac{5}{12}$$
$$12y\left(\frac{2}{3} - \frac{1}{y}\right) = 12y \cdot \frac{5}{12}$$
$$8y - 12 = 5y$$
$$3y = 12$$
$$y = 4$$

The solution set is $\{(3, 4)\}$.

35. Substitute $y = 20/x^2$ into $xy + 2 = 6x$.

$$x \cdot \frac{20}{x^2} + 2 = 6x$$
$$\frac{20}{x} + 2 = 6x$$
$$20 + 2x = 6x^2$$
$$0 = 6x^2 - 2x - 20$$
$$3x^2 - x - 10 = 0$$
$$(3x + 5)(x - 2) = 0$$
$$x = -5/3 \quad \text{or} \quad x = 2$$

Use $y = 20/x^2$ to get

$$y = 36/5 \qquad y = 5.$$

The solution set is $\left\{\left(-\frac{5}{3}, \frac{36}{5}\right), (2, 5)\right\}$.

37. Substitute $y = 7 - x$ in $x^2 + xy - y^2 = -11$.

$$x^2 + x(7 - x) - (7 - x)^2 = -11$$
$$x^2 + 7x - x^2 - 49 + 14x - x^2 = -11$$
$$-x^2 + 21x - 38 = 0$$
$$x^2 - 21x + 38 = 0$$
$$(x - 2)(x - 19) = 0$$
$$x = 2 \quad \text{or} \quad x = 19$$

Since $y = 7 - x$

$$y = 5 \qquad y = -12$$

The solution set is $\{(2, 5), (19, -12)\}$.

39. If $x^2 = y$, then $x^4 = y^2$. Substitute for x^4 in the equation $3y - 2 = x^4$.

$$3y - 2 = y^2$$
$$0 = y^2 - 3y + 2$$
$$0 = (y - 1)(y - 2)$$
$$y = 1 \quad \text{or} \quad y = 2$$

Use $y = 1$ in $x^2 = y$ to get $x^2 = 1$, or $x = \pm 1$. Use $y = 2$ in $x^2 = y$ to get $x^2 = 2$, or $x = \pm\sqrt{2}$. The solution set is $\left\{(\sqrt{2}, 2), (-\sqrt{2}, 2), (1, 1), (-1, 1)\right\}$.

41. Eliminate y by substitution.

$$\log_2(x - 1) = 3 - \log_2(x + 1)$$
$$\log_2(x - 1) + \log_2(x + 1) = 3$$
$$\log_2[x^2 - 1] = 3$$
$$x^2 - 1 = 8$$
$$x^2 = 9$$
$$x = \pm 3$$

If $x = -3$, we get a logarithm of a negative number. If $x = 3$, then $y = \log_2(3 - 1) = 1$. The solution set is $\{(3, 1)\}$.

43. Use substitution to eliminate y.

$$\log_2(x - 1) = 2 + \log_2(x + 2)$$
$$\log_2(x - 1) - \log_2(x + 2) = 2$$
$$\log_2\left(\frac{x - 1}{x + 2}\right) = 2$$
$$\frac{x - 1}{x + 2} = 4$$
$$x - 1 = 4x + 8$$
$$-9 = 3x$$

$$-3 = x$$

If $x = -3$ in either of the original equations, we get a logarithm of a negative number. So the solution set is the empty set, $\emptyset$.

45. Use substitution to eliminate y.

$$2^{3x+4} = 4^{x-1}$$
$$2^{3x+4} = (2^2)^{x-1}$$
$$2^{3x+4} = 2^{2x-2}$$
$$3x + 4 = 2x - 2$$
$$x = -6$$

If $x = -6$, then $y = 4^{-6-1} = 4^{-7}$. So the solution set is $\{(-6, 4^{-7})\}$.

47. Let $x =$ the length of one leg and $y =$ the length of the other. We write one equation for the area: $3 = \frac{1}{2}xy$. We write the other equation from the Pythagorean theorem: $x^2 + y^2 = 15$.

Substitute $y = 6/x$ into $x^2 + y^2 = 15$.

$$x^2 + (\tfrac{6}{x})^2 = 15$$
$$x^2 + \frac{36}{x^2} = 15$$
$$x^4 + 36 = 15x^2$$
$$x^4 - 15x^2 + 36 = 0$$
$$(x^2 - 3)(x^2 - 12) = 0$$
$$x = \sqrt{3} \quad \text{or} \quad x = \sqrt{12} = 2\sqrt{3}$$

If $x = \sqrt{3}$, then $y = 6/\sqrt{3} = 2\sqrt{3}$. If $x = 2\sqrt{3}$, then $y = 6/(2\sqrt{3}) = \sqrt{3}$. So the lengths of the legs are $\sqrt{3}$ feet and $2\sqrt{3}$ feet.

49. Let $h =$ the height of each triangle, and $b =$ the length of the base of each triangle. Since the 7 triangles are to have a total area of 3,500, the area of each must be 500: $\frac{1}{2}hb = 500$. The ratio of the height to base must be 1 to 4 is expressed as $\frac{h}{b} = \frac{1}{4}$. These two equations can be written as $hb = 1000$ and $b = 4h$. Use substitution.

$$h(4h) = 1000$$
$$h^2 = 250$$
$$h = \sqrt{250} = 5\sqrt{10}$$

Since $b = 4h$, $b = 20\sqrt{10}$. So the height is $5\sqrt{10}$ inches and the base is $20\sqrt{10}$ inches.

51. Let $x =$ the number of hours for pump A to fill the tank alone, and $y =$ the number of hours for pump B to fill the tank alone.

$$\frac{1}{x} + \frac{1}{y} = \frac{1}{6}$$
$$\frac{1}{y} - \frac{1}{x} = \frac{1}{12}$$

Adding these two equations eliminates x.

$$\frac{2}{y} = \frac{1}{6} + \frac{1}{12}$$
$$\frac{2}{y} = \frac{1}{4}$$
$$y = 8$$

Use $y = 8$ in the first equation.

$$\frac{1}{x} + \frac{1}{8} = \frac{1}{6}$$
$$\frac{1}{x} = \frac{1}{24}$$
$$x = 24$$

It would take pump A 24 hours to fill the tank alone and pump B 8 hours to fill the tank alone.

53. Let $x =$ the time for Jan to do the job alone. Let $y =$ the time for Beth to do the job alone. Since they do the job together in 24 minutes we have the equation

$$\frac{1}{x} + \frac{1}{y} = \frac{1}{24}.$$

In 50 minutes the job was completed, but not by working together. This implies that the total of their times alone is 100, $x + y = 100$. Substitute $y = 100 - x$ into the first equation.

$$\frac{1}{x} + \frac{1}{100 - x} = \frac{1}{24}$$
$$24(100 - x) + 24x = x(100 - x)$$
$$2400 - 24x + 24x = 100x - x^2$$
$$x^2 - 100x + 2400 = 0$$
$$(x - 40)(x - 60) = 0$$
$$x = 40 \quad \text{or} \quad x = 60$$

Since $y = 100 - x$

$$y = 60 \qquad y = 40$$

Since Jan is the faster worker, it would take her 40 minutes to complete the catfish by herself.

55. Let $x =$ the length and $y =$ the width. The area is 72 is expressed as $xy = 72$. The perimeter is 34 is expressed as $2x + 2y = 34$ or $x + y = 17$. Substitute $y = 17 - x$ into $xy = 72$.

$$x(17 - x) = 72$$
$$-x^2 + 17x - 72 = 0$$
$$x^2 - 17x + 72 = 0$$
$$(x - 9)(x - 8) = 0$$
$$x = 9 \quad \text{or} \quad x = 8$$
$$y = 8 \qquad y = 9$$

The rectangular area is 8 feet by 9 feet.

57. Let x and y represent the numbers.

$$x + y = 8$$
$$xy = 20$$

Substitute $y = 8 - x$ into $xy = 20$.

$$x(8 - x) = 20$$
$$-x^2 + 8x - 20 = 0$$
$$x^2 - 8x + 20 = 0$$
$$x = \frac{8 \pm \sqrt{64 - 4(1)(20)}}{2} = \frac{8 \pm \sqrt{-16}}{2}$$
$$= 4 \pm 2i$$

If $x = 4 + 2i$, then $y = 8 - (4 + 2i) = 4 - 2i$.
If $x = 4 - 2i$, then $y = 8 - (4 - 2i) = 4 + 2i$.
So the numbers are $4 - 2i$ and $4 + 2i$.

59. Let $x =$ the length of the side of the square, and $y =$ the height of the triangle. The total height of 10 feet means that $x + y = 10$. The total area is 72 means that $x^2 + \frac{1}{2}xy = 72$. Substitute $y = 10 - x$ into the second equation.

$$x^2 + \frac{1}{2}x(10 - x) = 72$$
$$x^2 + 5x - \frac{1}{2}x^2 = 72$$
$$\frac{1}{2}x^2 + 5x - 72 = 0$$
$$x^2 + 10x - 144 = 0$$
$$(x - 8)(x + 18) = 0$$
$$x = 8 \text{ or } \quad x = -18$$
$$y = 2$$

The side of the square is 8 feet and the height of the triangle is 2 feet.

61. a) (1.71, 1.55), (−2.98, −3.95)
b) (1, 1) , (0.40, 0.16)
c) (1.17, 1.62), (−1.17, −1.62)

11.2 WARM-UPS

1. False, because if the focus is (2, 3) and the directrix is $y = 1$, then the vertex is (2, 2).
2. True, because the focus is 1 unit above the vertex which is (0, 1).
3. True, because in $y = a(x - h)^2 + k$, the vertex is (h, k).
4. False, because $y = 6x + 3x + 2$ is equivalent to $y = 9x + 2$, which is a straight line.
5. False, because $y = -x^2 + 2x + 9$ opens downward.
6. True, the vertex is (0, 0) and so is the y-intercept.
7. True, because it opens upward from (2, 3) and it cannot intersect the x-axis.
8. True, because the focus is above the directrix in a parabola the opens upward.
9. True, because the vertex is $(2, k)$ and the axis of symmetry is a vertical line through the vertex.
10. True, because $1/(4 \cdot \frac{1}{4}) = 1$.

11.2 EXERCISES

1. A parabola is the set of all points in a plane that are equidistant from a given line and a fixed point not on the line.

3. A parabola can be written in the forms $y = ax^2 + bx + c$ or $y = a(x - h)^2 + k$.

5. We use completing the square to convert $y = ax^2 + bx + c$ into $y = a(x - h)^2 + k$.

7. $\sqrt{(5-2)^2 + (5-1)^2} = \sqrt{9 + 16}$
$= \sqrt{25} = 5$

9. $\sqrt{(5-4)^2 + (-2-(-3))^2} = \sqrt{1 + 1}$
$= \sqrt{2}$

11. $\sqrt{(4-6)^2 + (2-5)^2} = \sqrt{4 + 9} = \sqrt{13}$

13. $\sqrt{(1-3)^2 + (-3-5)^2} = \sqrt{4 + 64}$
$= 2\sqrt{17}$

15. $\sqrt{(-3-4)^2 + (-6-(-2))^2}$
$= \sqrt{49 + 16} = \sqrt{65}$

17. Use the midpoint formula:
$\left(\frac{0+6}{2}, \frac{0+8}{2}\right) = (3, 4)$
Use the distance formula to find the length:
$\sqrt{(0-6)^2 + (0-8)^2} = \sqrt{36 + 64} = \sqrt{100}$
$= 10$

19. Use the midpoint formula:
$\left(\frac{2+5}{2}, \frac{5+1}{2}\right) = \left(\frac{7}{2}, \frac{6}{2}\right) = \left(\frac{7}{2}, 3\right)$
Use the distance formula to find the length:
$\sqrt{(5-2)^2 + (1-5)^2} = \sqrt{9 + 16} = \sqrt{25}$

21. Use the midpoint formula:
$\left(\frac{-2+6}{2}, \frac{4+(-2)}{2}\right) = \left(\frac{4}{2}, \frac{2}{2}\right) = (2, 1)$
Use the distance formula to find the length:
$\sqrt{(6-(-2))^2 + (-2-4)^2} = \sqrt{64 + 36}$
$= \sqrt{100} = 10$

23. Use the midpoint formula:
$\left(\frac{-1+1}{2}, \frac{4+1}{2}\right) = \left(\frac{0}{2}, \frac{5}{2}\right) = \left(0, \frac{5}{2}\right)$

Use the distance formula to find the length:
$\sqrt{(1-(-1))^2+(1-4)^2} = \sqrt{4+9}$
$= \sqrt{13}$

25. The vertex is (0, 0). Since $2 = 1/(4p)$, $p = 1/8$. So the focus is (0, 1/8) and the directrix is $y = -1/8$.

27. The vertex is (0, 0). Since $-1/4 = 1/(4p)$, $4p = -4$, or $p = -1$. So the focus is (0, −1) and the directrix is $y = 1$.

29. The vertex is (3, 2). Since $1/2 = 1/(4p)$, $p = 1/2$. So the focus is (3, 2.5) and the directrix is $y = 1.5$.

31. The vertex is $(-1, 6)$. Since $-1 = 1/(4p)$, $p = -1/4$. So the focus is $(-1, 5.75)$ and the directrix is $y = 6.25$.

33. Since the distance between the focus and directrix is 4, $p = 2$ and $a = \frac{1}{4p} = \frac{1}{4(2)} = \frac{1}{8}$. Since the vertex is half way between the focus and directrix, the vertex is (0, 0). Use the form $y = a(x-h)^2 + k$ to get the equation.

$$y = \tfrac{1}{8}(x-0)^2 + 0$$
$$y = \tfrac{1}{8}x^2$$

35. Since the distance between the focus and directrix is 1 and the focus is below the directrix,
$p = -\frac{1}{2}$ and $a = \frac{1}{4p} = \frac{1}{4(-\frac{1}{2})} = -\frac{1}{2}$. Since the vertex is half way between the focus and directrix, the vertex is (0, 0). Use the form $y = a(x-h)^2 + k$ to get the equation.

$$y = -\tfrac{1}{2}(x-0)^2 + 0$$
$$y = -\tfrac{1}{2}x^2$$

37. Since the distance between the focus and directrix is 1 and the parabola opens upward, $p = \frac{1}{2}$ and $a = \frac{1}{4p} = \frac{1}{4(\frac{1}{2})} = \frac{1}{2}$. Since the vertex is half way between the focus and directrix, the vertex is $\left(3, \frac{3}{2}\right)$. Use the form $y = a(x-h)^2 + k$ to get the equation.

$$y = \tfrac{1}{2}(x-3)^2 + \tfrac{3}{2}$$
$$y = \tfrac{1}{2}x^2 - 3x + \tfrac{9}{2} + \tfrac{3}{2}$$
$$y = \tfrac{1}{2}x^2 - 3x + 6$$

39. Since the distance between the focus and directrix is 4 and the parabola opens downward, $p = -2$ and $a = \frac{1}{4p} = \frac{1}{4(-2)} = -\frac{1}{8}$. Since the vertex is half way between the focus and directrix, the vertex is $(1, 0)$. Use the form $y = a(x-h)^2 + k$ to get the equation.

$$y = -\tfrac{1}{8}(x-1)^2 + 0$$
$$y = -\tfrac{1}{8}x^2 + \tfrac{1}{4}x - \tfrac{1}{8}$$

41. Since the distance between the focus and directrix is $\frac{1}{2}$ and the parabola opens upward, $p = \frac{1}{4}$ and $a = \frac{1}{4p} = \frac{1}{4(0.25)} = 1$. Since the vertex is half way between the focus and directrix, the vertex is $(-3, 1)$. Use the form $y = a(x-h)^2 + k$ to get the equation.

$$y = 1(x+3)^2 + 1$$
$$y = x^2 + 6x + 10$$

43.
$$y = x^2 - 6x + 1$$
$$y = x^2 - 6x + 9 - 9 + 1$$
$$y = (x-3)^2 - 8$$

The vertex is (3, −8). Because $a = 1$, the parabola opens upward, and $p = 1/4$. The focus is (3, −7.75) and the directrix is $y = -8.25$. The axis of symmetry is the vertical line through the vertex, $x = 3$.

45.
$$y = 2x^2 + 12x + 5$$
$$y = 2(x^2 + 6x) + 5$$
$$y = 2(x^2 + 6x + 9) + 5 - 18$$
$$y = 2(x+3)^2 - 13$$

The vertex is (−3, −13) and the parabola opens upward. Because $a = 2$, $p = 1/8 = 0.125$ The focus is (−3, −12.875) and the directrix is $y = -13.125$. The axis of symmetry is $x = -3$.

47.
$$y = -2x^2 + 16x + 1$$
$$y = -2(x^2 - 8x) + 1$$
$$y = -2(x^2 - 8x + 16) + 1 + 32$$
$$y = -2(x-4)^2 + 33$$

The vertex is (4, 33) and the parabola opens downward. Because $a = -2$, $p = -1/8$ $= -0.125$. The focus is (4, $32\frac{7}{8}$) and the directrix is $y = 33\frac{1}{8}$. The axis of symmetry is $x = 4$.

49.
$$y = 5x^2 + 40x$$
$$y = 5(x^2 + 8x)$$
$$y = 5(x^2 + 8x + 16) - 80$$
$$y = 5(x+4)^2 - 80$$

The vertex is $(-4, -80)$ and the parabola opens upward. Because $a = 5$, $p = 1/20$. The focus is $(-4, -79\frac{19}{20})$ and the directrix is $y = -80\frac{1}{20}$. The axis of symmetry is $x = -4$.

51. The x-coordinate of the vertex is $x = \frac{-b}{2a}$.

$$x = \frac{-(-4)}{2(1)} = 2$$
$$y = 2^2 - 4(2) + 1 = -3$$

The vertex is $(2, -3)$ and the parabola opens upward. Because $a = 1$, $p = 1/4$. The focus is $(2, -2\frac{3}{4})$ and the directrix is $y = -3\frac{1}{4}$. Axis of symmetry is $x = 2$.

53. The x-coordinate of the vertex is $x = \frac{-b}{2a}$.

$$x = \frac{-(2)}{2(-1)} = 1$$
$$y = -1^2 + 2(1) - 3 = -2$$

The vertex is $(1, -2)$ and the parabola opens downward. Because $a = -1$, p $= -1/4$. The focus is $(1, -2\frac{1}{4})$ and the directrix is $y = -1\frac{3}{4}$. The axis of symmetry is $x = 1$.

55. The x-coordinate of the vertex is $x = \frac{-b}{2a}$.

$$x = \frac{-(-6)}{2(3)} = 1$$
$$y = 3(1^2) - 6(1) + 1 = -2$$

The vertex is $(1, -2)$ and the parabola opens upward. Because $a = 3$, $p = 1/12$. The focus is $(1, -1\frac{11}{12})$ and the directrix is $y = -2\frac{1}{12}$. The axis of symmetry is $x = 1$.

57. The x-coordinate of the vertex is $x = \frac{-b}{2a}$.

$$x = \frac{-(-3)}{2(-1)} = -\frac{3}{2}$$
$$y = -(-\tfrac{3}{2})^2 - 3(-\tfrac{3}{2}) + 2 = \frac{17}{4}$$

The vertex is $(-\frac{3}{2}, \frac{17}{4})$ and the parabola opens downward. Because $a = -1$, $p = -1/4$. The focus is $(-\frac{3}{2}, 4)$ and the directrix is $y = \frac{9}{2}$. The axis of symmetry is $x = -\frac{3}{2}$.

59. The x-coordinate of the vertex is $x = \frac{-b}{2a}$.

$$x = \frac{-(0)}{2(3)} = 0$$
$$y = 3(0)^2 + 5 = 5$$

The vertex is $(0, 5)$ and the parabola opens upward. Because $a = 3$, $p = 1/12$. The focus is $(0, 5\frac{1}{12})$ and the directrix is $y = 4\frac{11}{12}$. The axis of symmetry is $x = 0$.

61. For $x = (y-2)^2 + 3$ the vertex is $(3, 2)$. Since $a = 1$ we have $p = 1/4$ and the parabola opens to the right. The focus is $\left(\frac{13}{4}, 2\right)$ and the directrix is $x = \frac{11}{4}$.

63. For $x = \frac{1}{4}(y-1)^2 - 2$ the vertex is $(-2, 1)$. Since $a = 1/4$ we have $p = 1$ and the parabola opens to the right. The focus is $(-1, 1)$ and the directrix is $x = -3$.

65. For $x = -\frac{1}{2}(y-2)^2 + 4$ the vertex is $(4, 2)$. Since $a = -1/2$ we have $p = -1/2$ and the parabola opens to the left. The focus is $\left(\frac{7}{2}, 2\right)$ and the directrix is $x = \frac{9}{2}$.

67. For $y = (x-2)^2 + 3$ the vertex is $(2, 3)$ and the parabola opens upward. The graph also goes through $(0, 7)$, $(1, 4)$, $(3, 4)$, and $(4, 7)$.

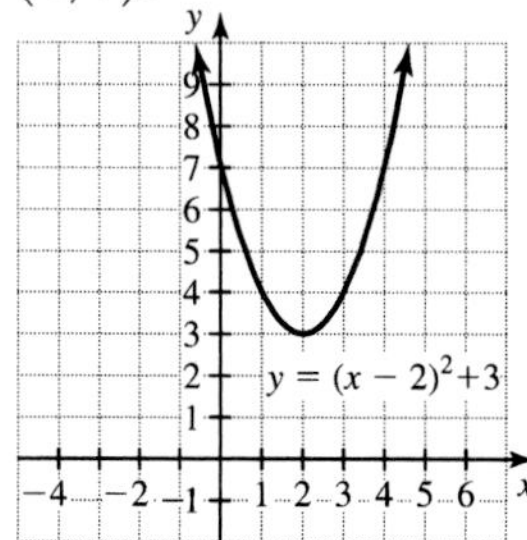

69. For $y = -2(x-1)^2 + 3$ the vertex is $(1, 3)$ and the parabola opens downward. The graph also goes through $(-1, -5)$, $(0, 1)$, $(2, 1)$, and $(3, -5)$.

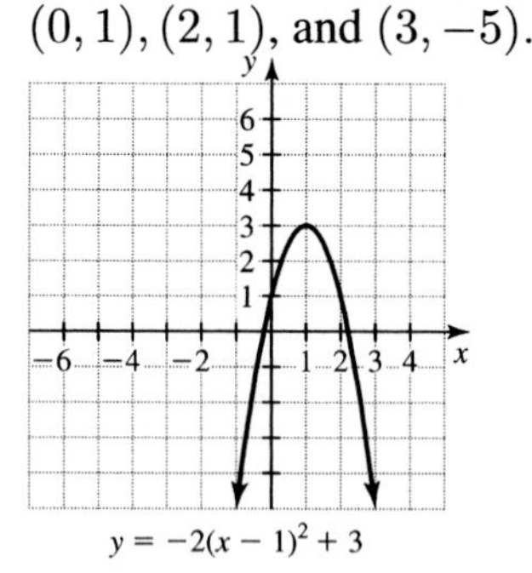

71. For $x = (y-2)^2 + 3$ the vertex is $(3, 2)$ and the parabola opens to the right. The graph also goes through $(4, 3)$, $(4, 1)$, $(7, 4)$, and $(7, 0)$.

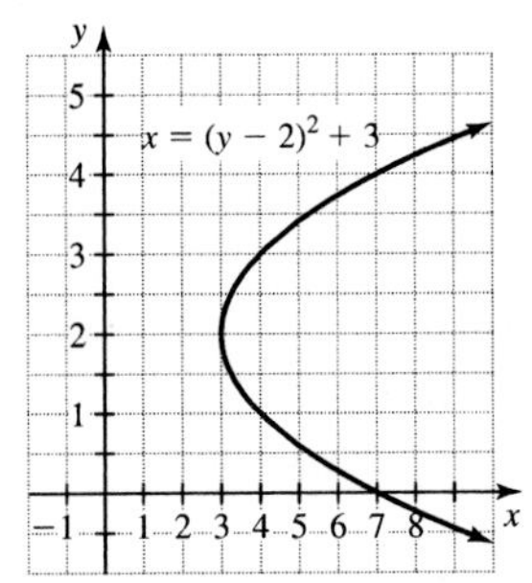

73. For $x = -2(y-1)^2 + 3$ the vertex is $(3, 1)$ and the parabola opens to the left. The graph also goes through $(1, 2)$, $(1, 0)$, $(-5, 3)$, and $(-5, -1)$.

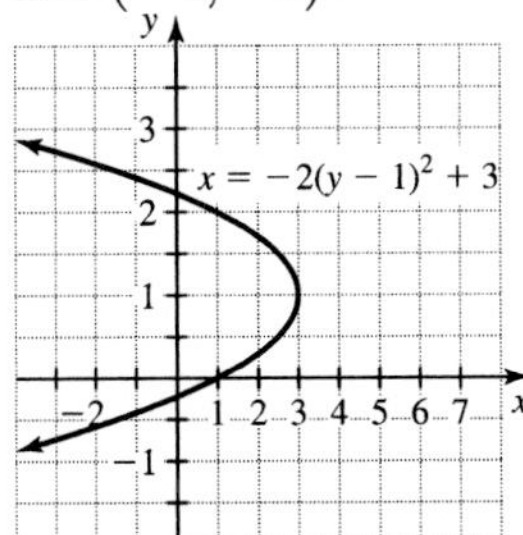

75. The graph of $y = -x^2 + 3$ is a parabola opening downward, and the graph of $y = x^2 + 1$ is a parabola opening upward. To find the points of intersection, eliminate y by substitution.

$$\begin{aligned} x^2 + 1 &= -x^2 + 3 \\ 2x^2 &= 2 \\ x^2 &= 1 \\ x &= \pm 1 \end{aligned}$$

If $x = \pm 1$, then $y = 2$ (since $y = x^2 + 1$). The solution set for the system is $\{(-1, 2), (1, 2)\}$.

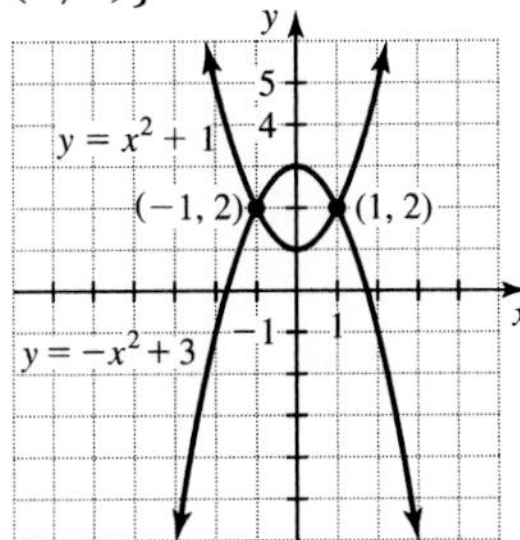

77. The graph of $y = x^2 - 2$ is a parabola opening upward, and the graph of $y = 2x - 3$ is a straight line.

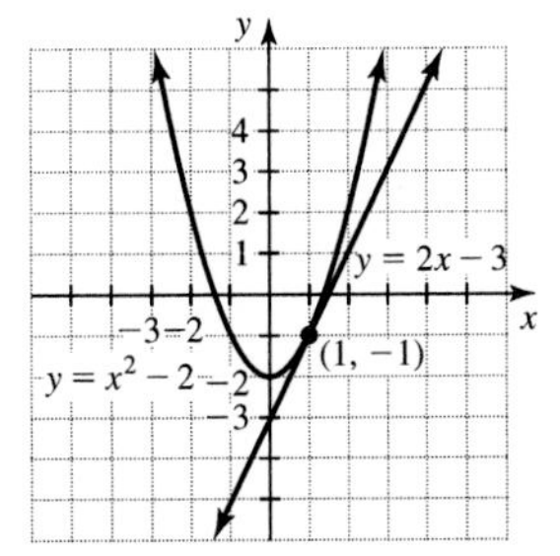

We eliminate y by substitution.

$$\begin{aligned} x^2 - 2 &= 2x - 3 \\ x^2 - 2x + 1 &= 0 \\ (x-1)^2 &= 0 \\ x &= 1 \end{aligned}$$

If $x = 1$, then $y = -1$ (because $y = 2x - 3$). The solution set for the system is $\{(1, -1)\}$.

79. The graph of $y = x^2 + 3x - 4$ is a parabola opening upward, and the graph of $y = -x^2 - 2x + 8$ is a parabola opening downward.

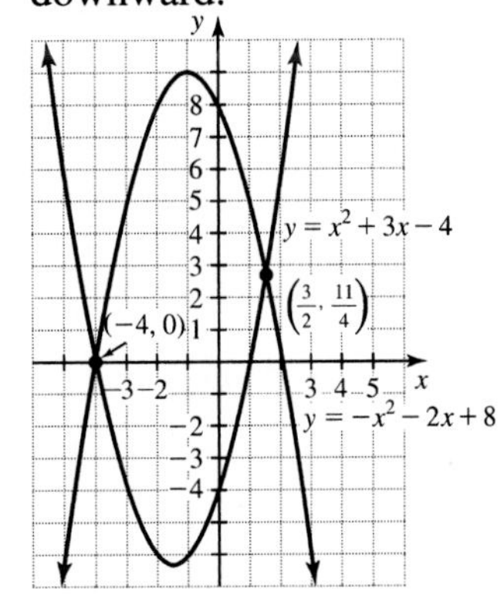

Eliminate y by substitution.

$$\begin{aligned} x^2 + 3x - 4 &= -x^2 - 2x + 8 \\ 2x^2 + 5x - 12 &= 0 \\ (2x - 3)(x + 4) &= 0 \\ x = 3/2 \quad &\text{or} \quad x = -4 \end{aligned}$$

If $x = 3/2$, then

$$y = (3/2)^2 + 3(3/2) - 4 = 11/4.$$

If $x = -4$, then

$$y = (-4)^2 + 3(-4) - 4 = 0.$$

The solution set for the system is

$$\left\{\left(\frac{3}{2}, \frac{11}{4}\right), (-4, 0)\right\}.$$

81. The graph of $y = x^2 + 3x - 4$ is a parabola opening upward. The graph of $y = 2x + 2$ is a straight line with slope 2 and y-intercept (0, 2).

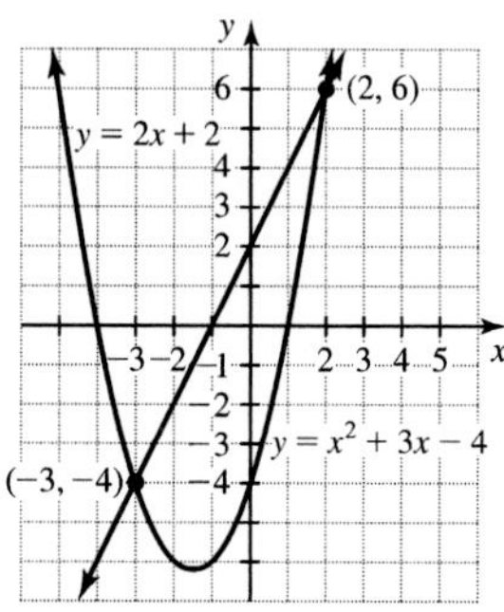

Use substitution to eliminate y.

$x^2 + 3x - 4 = 2x + 2$

$x^2 + x - 6 = 0$

$(x + 3)(x - 2) = 0$

$x = -3$ or $x = 2$

Since $y = 2x + 2$

$y = -4$ $\quad y = 6$

The solution set for the system is $\{(-3, -4), (2, 6)\}$.

83. On the x-axis, the y-coordinate is zero. So we solve $x^2 - 2x - 3 = 0$.

$(x - 3)(x + 1) = 0$

$x - 3 = 0$ or $x + 1 = 0$

$x = 3$ or $x = -1$

The points of intersection are $(3, 0)$ and $(-1, 0)$.

85. Substitute $y = 4$ into $y = 0.01x^2$:

$4 = 0.01x^2$

$400 = x^2$

$\pm 20 = x$

The points of intersection are $(-20, 4)$ and $(20, 4)$

87. Substitute to get $x = (x^2)^2$ or $x = x^4$.

$x = x^4$

$x^4 - x = 0$

$x(x^3 - 1) = 0$

$x(x - 1)(x^2 + x + 1) = 0$

$x = 0$ or $x - 1 = 0$ or $x^2 + x + 1 = 0$

$x = 0$ or $x = 1$

There are no real solutions to $x^2 + x + 1 = 0$. If $x = 0$, then $y = 0^2 = 0$. If $x = 1$, then $y = 1^2 = 1$. So the points of intersection are $(0, 0)$ and $(1, 1)$.

89. **a)** Find the distance;

$\sqrt{(-215 - 185)^2 + (-352 - 234)^2}$

≈ 709.504 yards

At \$84 per yard the cost is \$59,598.

b) $\dfrac{185 + (-215)}{2} = -15$

$\dfrac{234 + (-352)}{2} = -59$

So the midpoint is $(-15, -59)$

91. The distance from the vertex to the focus is 15. So $p = 15$ and $a = 1/(4 \cdot 15) = 1/60$. the equation of the parabola is $y = \frac{1}{60}x^2$.

93. The distance from (x, y) to the focus $(p, 0)$ is $\sqrt{(x - p)^2 + (y - 0)^2}$. The distance from (x, y) to the vertical line $x = -p$ is the distance between the points (x, y) and $(-p, y)$, which is $\sqrt{(x - (-p))^2 + (y - y)^2}$.

Set these equal, square each side, then simplify to get

$x^2 - 2xp + p^2 + y^2 = x^2 + 2xp + p^2$.

Solve for x to get $x = \frac{1}{4p}y^2$ or $x = ay^2$ where $a = 1/(4p)$.

95. The graphs have identical shapes.

11.3 WARM-UPS

1. False, the radius of a circle is a positive real number.

2. False, the coordinates of the center do not satisfy the equation of the circle, only points on the circle satisfy the equation.

3. True, the center is (0, 0).

4. False, the equation of the circle centered at the origin with radius 9 is $x^2 + y^2 = 81$.

5. False, because $(x - 2)^2 + (y - 3)^2 = 4$ has radius 2.

6. False, because $(x - 3)^2 + (y + 5)^2 = 9$ is a circle of radius 3 centered at (3, −5)

7. True, because the distance from (−3, −1) to (0, 0) is the radius.

8. False, the center is (3, 4).

9. True, because there is only an x^2-term and no x-term.

10. False, because if we complete the square for x then the right side will no longer be 4.

11.3 EXERCISES

1. A circle is the set of all points in a plane that lie at a fixed distance from a fixed point.

3. Use $h = 0$, $k = 0$, and $r = 4$ in the standard equation $(x - h)^2 + (y - k)^2 = r^2$ to get the equation $x^2 + y^2 = 16$.

5. Use $h = 0$, $k = 3$, and $r = 5$ in the standard equation $(x - h)^2 + (y - k)^2 = r^2$ to get the equation $x^2 + (y - 3)^2 = 25$.

7. Use $h = 1$, $k = -2$, and $r = 9$ in the standard equation $(x - h)^2 + (y - k)^2 = r^2$ to get the equation $(x - 1)^2 + (y + 2)^2 = 81$.

9. Use $h = 0$, $k = 0$, and $r = \sqrt{3}$ in the standard equation $(x - h)^2 + (y - k)^2 = r^2$ to get the equation $x^2 + y^2 = 3$.

11. Use $h = -6$, $k = -3$, and $r = 1/2$ in the standard equation $(x - h)^2 + (y - k)^2 = r^2$ to get the equation $(x + 6)^2 + (y + 3)^2 = \frac{1}{4}$.

13. Use $h = 1/2$, $k = 1/3$, and $r = 0.1$ in the standard equation $(x - h)^2 + (y - k)^2 = r^2$ to get the equation

$$\left(x - \frac{1}{2}\right)^2 + \left(y - \frac{1}{3}\right)^2 = 0.01.$$

15. Compare $x^2 + y^2 = 1$ with $(x - h)^2 + (y - k)^2 = r^2$ (*where the center is* (h, k) *and radius is* r), to get a center at $(0, 0)$ and radius 1.

17. Compare $(x - 3)^2 + (y - 5)^2 = 2$ with $(x - h)^2 + (y - k)^2 = r^2$ (*where the center is* (h, k) *and radius is* r), to get a center at $(3, 5)$ and radius $\sqrt{2}$.

19. Compare $x^2 + \left(y - \frac{1}{2}\right)^2 = \frac{1}{2}$ with $(x - h)^2 + (y - k)^2 = r^2$ (*where the center is* (h, k) *and radius is* r), to get a center at $\left(0, \frac{1}{2}\right)$ and radius $\sqrt{\frac{1}{2}}$ or $\frac{\sqrt{2}}{2}$.

21. Divide each side by 4 to get $x^2 + y^2 = \frac{9}{4}$, which has center $(0, 0)$ and radius $\sqrt{\frac{9}{4}}$ or $\frac{3}{2}$.

23. Rewrite the equation as $(x - 2)^2 + y^2 = 3$, which has center $(2, 0)$ and radius $\sqrt{3}$.

25. The graph of $x^2 + y^2 = 9$ is a circle with radius 3, centered at $(0, 0)$.

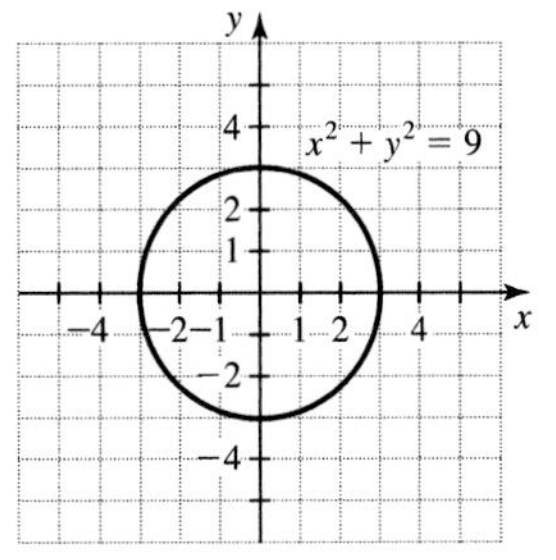

27. The graph of $x^2 + (y - 3)^2 = 9$ is a circle with radius 3, centered at $(0, 3)$.

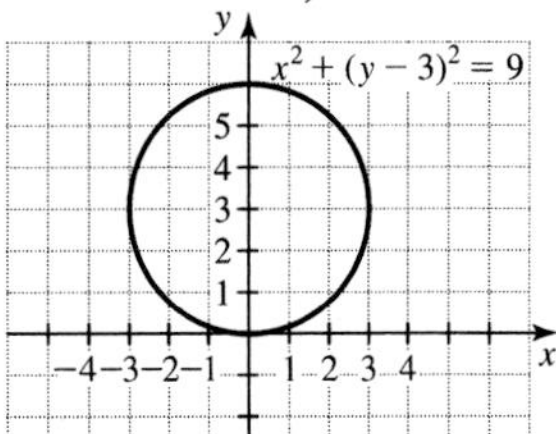

29. The graph of $(x + 1)^2 + (y - 1)^2 = 2$ is a circle with radius $\sqrt{2}$, centered at $(-1, 1)$.

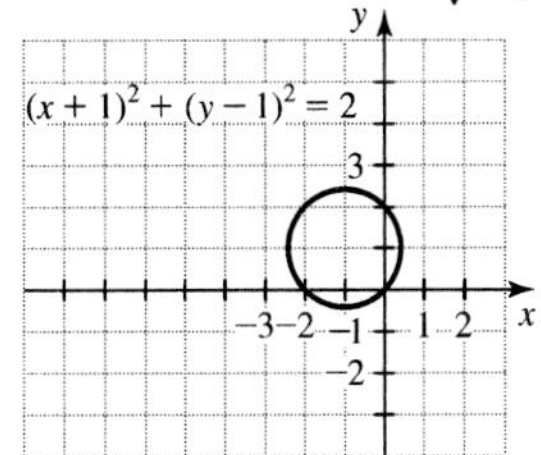

31. The graph of $(x - 4)^2 + (y + 3)^2 = 16$ is a circle of radius 4, centered at $(4, -3)$.

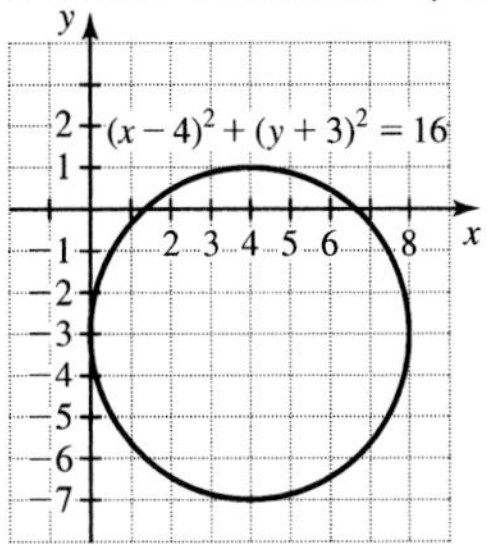

33. The graph of $\left(x - \frac{1}{2}\right)^2 + \left(y + \frac{1}{2}\right)^2 = \frac{1}{4}$ is a circle of radius $1/2$ and center $(1/2, -1/2)$.

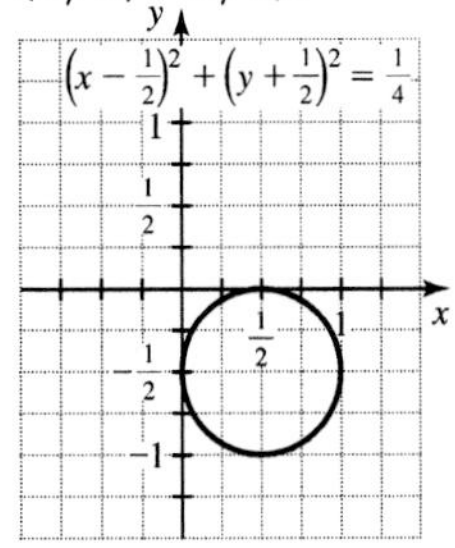

35. $x^2 + 4x + \quad y^2 + 6y \quad = 0$

$$x^2 + 4x + 4 + y^2 + 6y + 9 = 0 + 4 + 9$$
$$(x + 2)^2 + (y + 3)^2 = 13$$

The circle has radius $\sqrt{13}$ and center $(-2, -3)$.

37. $x^2 - 2x + \quad y^2 - 4y - 3 = 0$

$$x^2 - 2x + 1 + y^2 - 4y + 4 = 3 + 1 + 4$$
$$(x - 1)^2 + (y - 2)^2 = 8$$

This circle has radius $\sqrt{8} = 2\sqrt{2}$ and center $(1, 2)$.

39. $x^2 - 10x + y^2 - 8y = -32$

$$x^2 - 10x + 25 + y^2 - 8y + 16$$
$$= -32 + 25 + 16$$
$$(x - 5)^2 + (y - 4)^2 = 9$$

This circle has radius 3 and center (5, 4).

41. $x^2 - x + \quad y^2 + y \quad = 0$

$$x^2 - x + \frac{1}{4} + y^2 + y + \frac{1}{4} = \frac{1}{4} + \frac{1}{4}$$
$$\left(x - \frac{1}{2}\right)^2 + \left(y + \frac{1}{2}\right)^2 = \frac{1}{2}$$

Center is $(1/2, -1/2)$ and radius is $\sqrt{\frac{1}{2}}$ or $\frac{\sqrt{2}}{2}$.

43. $x^2 - 3x + \quad y^2 - y \quad = 1$

$$x^2 - 3x + \frac{9}{4} + y^2 - y + \frac{1}{4} = 1 + \frac{9}{4} + \frac{1}{4}$$
$$\left(x - \frac{3}{2}\right)^2 + \left(y - \frac{1}{2}\right)^2 = \frac{7}{2}$$

Center is $(3/2, 1/2)$ and radius $\sqrt{\frac{7}{2}}$ or $\frac{\sqrt{14}}{2}$.

45. $x^2 - \frac{2}{3}x + \quad y^2 + \frac{3}{2}y \quad = 0$

$$x^2 - \frac{2}{3}x + \frac{1}{9} + y^2 + \frac{3}{2}y + \frac{9}{16} = 0 + \frac{1}{9} + \frac{9}{16}$$
$$\left(x - \frac{1}{3}\right)^2 + \left(y + \frac{3}{4}\right)^2 = \frac{97}{144}$$

Center is $(1/3, -3/4)$ and radius is $\sqrt{\frac{97}{144}}$ or $\frac{\sqrt{97}}{12}$.

47. The graph of $x^2 + y^2 = 10$ is a circle centered at (0, 0) with radius $\sqrt{10}$. The graph of $y = 3x$ is a straight line through (0, 0) with slope 3. Use substitution to eliminate y.

$$x^2 + (3x)^2 = 10$$
$$10x^2 = 10$$
$$x^2 = 1$$
$$x = 1 \quad \text{or} \quad x = -1$$
$$y = 3 \quad \text{or} \quad y = -3 \quad \text{Since } y = 3x$$

The solution set is $\{(1, 3), (-1, -3)\}$.

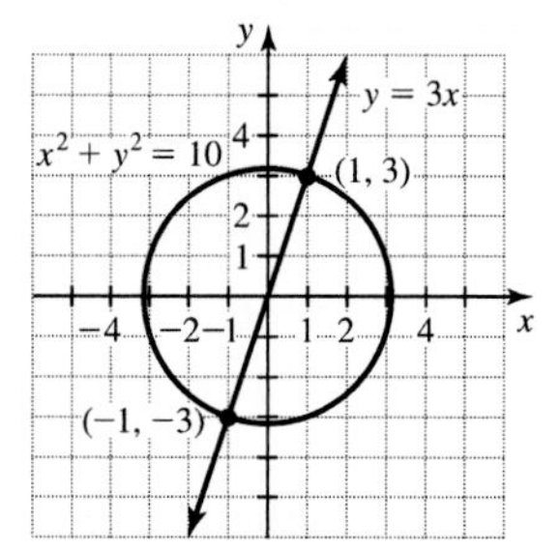

49. The graph of $x^2 + y^2 = 9$ is a circle centered at (0, 0) with radius 3. The graph of $y = x^2 - 3$ is a parabola opening upward.

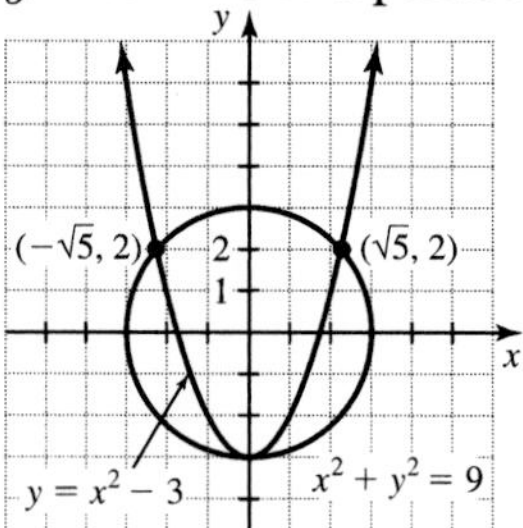

Use $x^2 = y + 3$ to eliminate x.

$$y + 3 + y^2 = 9$$
$$y^2 + y - 6 = 0$$
$$(y + 3)(y - 2) = 0$$
$$y = -3 \quad \text{or} \quad y = 2$$

If $y = -3$, then $x^2 = -3 + 3 = 0$ or $x = 0$. If $y = 2$, then $x^2 = 2 + 3 = 5$ or $x = \pm\sqrt{5}$.

The solution set is $\left\{(0, -3), \left(\sqrt{5}, 2\right), \left(-\sqrt{5}, 2\right)\right\}$.

51. The graph of $(x - 2)^2 + (y + 3)^2 = 4$ is a circle centered at $(2, -3)$ with radius 2. The graph of $y = x - 3$ is a line through $(0, -3)$ with slope 1.

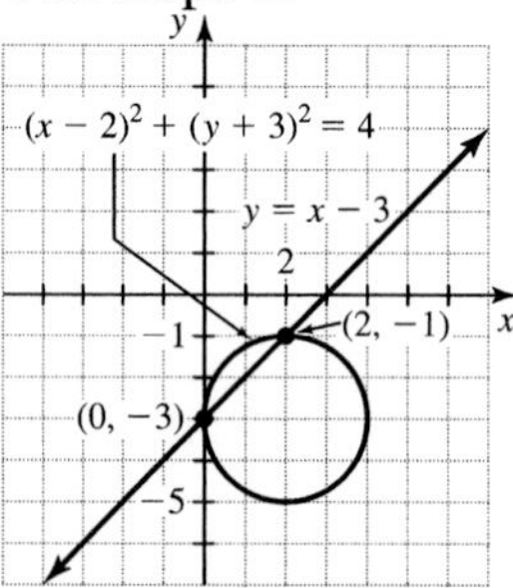

Use substitution to eliminate y.

$$(x - 2)^2 + (x - 3 + 3)^2 = 4$$
$$x^2 - 4x + 4 + x^2 = 4$$
$$2x^2 - 4x = 0$$
$$x^2 - 2x = 0$$
$$x(x - 2) = 0$$
$$x = 0 \quad \text{or} \quad x = 2$$

Since $y = x - 3$

$y = -3 \qquad y = -1$

The solution set is $\{(0, -3), (2, -1)\}$.

53. To find the y-intercepts, use $x = 0$ in the equation of the circle.

$$(0-1)^2 + (y-2)^2 = 4$$
$$1 + y^2 - 4y + 4 = 4$$
$$y^2 - 4y + 1 = 0$$
$$y = \frac{4 \pm \sqrt{16 - 4(1)(1)}}{2(1)} = \frac{4 \pm 2\sqrt{3}}{2}$$
$$= 2 \pm \sqrt{3}$$

The y-intercepts are $(0, 2 + \sqrt{3})$ and $(0, 2 - \sqrt{3})$.

55. The radius is the distance from $(2, -5)$ to $(0, 0)$.

$$r = \sqrt{(2-0)^2 + (-5-0)^2} = \sqrt{4 + 25}$$
$$= \sqrt{29}$$

57. The radius is the distance from $(2, 3)$ to $(-2, -1)$.

$$r = \sqrt{(-2-2)^2 + (-1-3)^2} = \sqrt{16 + 16}$$
$$= \sqrt{32}$$

The equation of a circle centered at $(2, 3)$ with radius $\sqrt{32}$ is $(x - 2)^2 + (y - 3)^2 = 32$.

59. Substitute $y^2 = 9 - x^2$ into $(x - 5)^2 + y^2 = 9$ to eliminate y.

$$(x-5)^2 + 9 - x^2 = 9$$
$$x^2 - 10x + 25 + 9 - x^2 = 9$$
$$-10x = -25$$
$$x = 5/2$$

Use $x = 5/2$ in $y^2 = 9 - x^2$ to find y.

$$y^2 = 9 - \left(\frac{5}{2}\right)^2 = \frac{11}{4}$$
$$y = \pm\sqrt{\frac{11}{4}} = \pm\frac{\sqrt{11}}{2}$$

The circles intersect at $\left(\frac{5}{2}, -\frac{\sqrt{11}}{2}\right)$ and $\left(\frac{5}{2}, \frac{\sqrt{11}}{2}\right)$.

61. From the equation of the bore we see that the radius is $\sqrt{83.72}$. The volume is given by $V = \pi r^2 h = \pi \cdot 83.72 \cdot 2874 \approx 755{,}903 \text{ mm}^3$.

63. Since both x^2 and y^2 are nonnegative for real values of x and y, the only way their sum could be zero is to choose both x and y equal to zero. So the graph consists of $(0, 0)$ only.

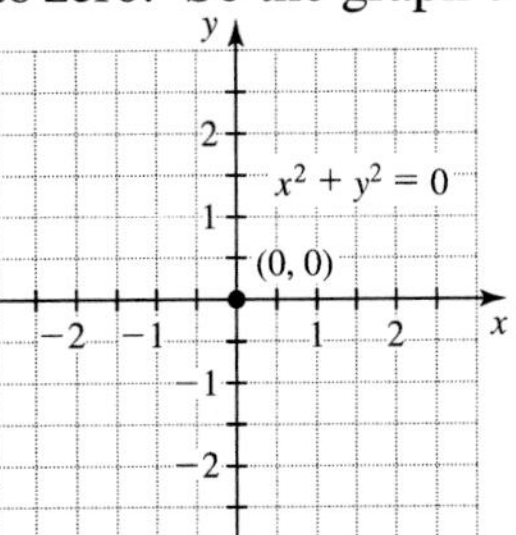

65. If we square both sides of $y = \sqrt{1 - x^2}$ we get $x^2 + y^2 = 1$, which is a circle with center $(0, 0)$ and radius 1. In the equation $y = \sqrt{1 - x^2}$, y must be nonnegative. So the graph of $y = \sqrt{1 - x^2}$ is the top half of the graph of $x^2 + y^2 = 1$.

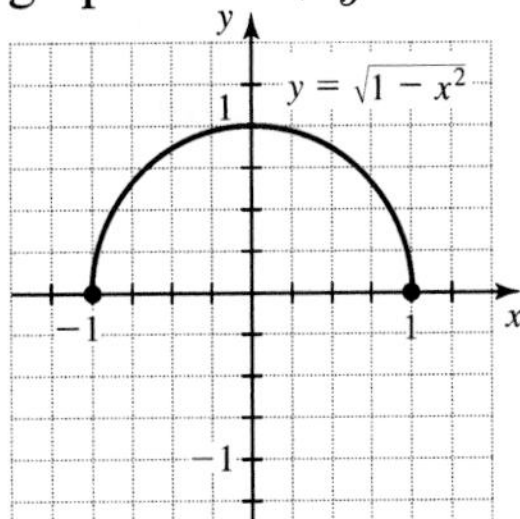

67. B and D can be any real numbers, but A must equal C. So that the radius is positive, we must also have

$\frac{E}{A} + \left(\frac{B}{2A}\right)^2 + \left(\frac{D}{2A}\right)^2 > 0$, or

$4AE + B^2 + D^2 > 0$.

No ordered pairs satisfy $x^2 + y^2 = -9$. So there is no graph.

69. Solve for y to get $y = \pm\sqrt{4 - x^2}$. Then graph $y_1 = \sqrt{4 - x^2}$ and $y_2 = -\sqrt{4 - x^2}$

71. $y = \pm\sqrt{x}$

73.

$$x = y^2 + 2y + 1$$
$$x = (y+1)^2$$
$$y + 1 = \pm\sqrt{x}$$
$$y = -1 \pm \sqrt{x}$$

11.4 WARM-UPS

1. False, the x-intercepts are $(-6, 0)$ and $(6, 0)$.
2. False, because both x^2 and y^2 must appear in the equation for an ellipse.
3. True, because if the foci coincide then every point on the ellipse will be a fixed distance from one fixed point.
4. True, because if we divide each side of the equation by 2, then it fits the standard equation for an ellipse.
5. True, because if we use $x = 0$ in the equation, we get $y = \pm\sqrt{3}$.
6. False, because both x^2 and y^2 must appear in the equation of a hyperbola.
7. False, because in this hyperbola there are no y-intercepts.
8. True, because it has y-intercepts at $(0, -3)$ and $(0, 3)$.
9. True, because if we divide each side of the equation by 4, then it fits the standard form for the equation of a hyperbola.
10. True, the asymptotes are the extended diagonals of the fundamental rectangle.

11.4 EXERCISES

1. An ellipse is the set of all points in a plane such that the sum of their distances from two fixed points is constant.
3. The center of an ellipse is the point that is midway between the foci.
5. The equation of an ellipse centered at (h, k) is $\dfrac{(x-h)^2}{a^2} + \dfrac{(y-k)^2}{b^2} = 1$.
7. The asymptotes of a hyperbola are the extended diagonals of the fundamental rectangle.

9. The graph of $\frac{x^2}{9} + \frac{y^2}{4} = 1$ is an ellipse with x-intercepts $(-3, 0)$ and $(3, 0)$, and y-intercepts $(0, 2)$ and $(0, -2)$.

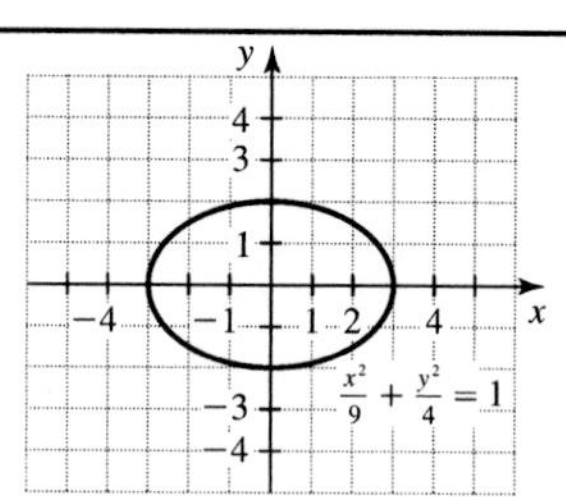

11. The graph of $\frac{x^2}{9} + y^2 = 1$ is an ellipse with x-intercepts $(-3, 0)$ and $(3, 0)$, and y-intercepts $(0, -1)$ and $(0, 1)$.

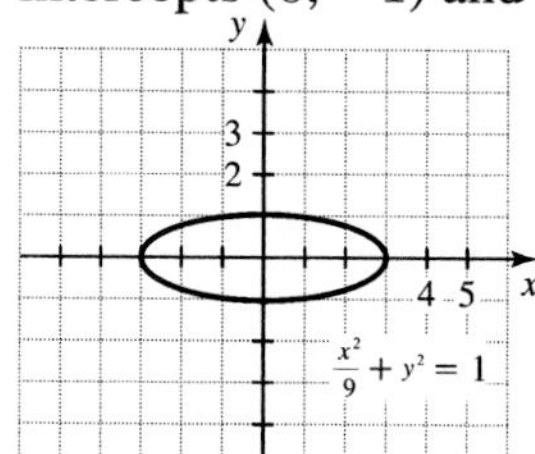

13. The graph of $\frac{x^2}{36} + \frac{y^2}{25} = 1$ is an ellipse with x-intercepts $(-6, 0)$ and $(6, 0)$, and y-intercepts $(0, -5)$ and $(0, 5)$.

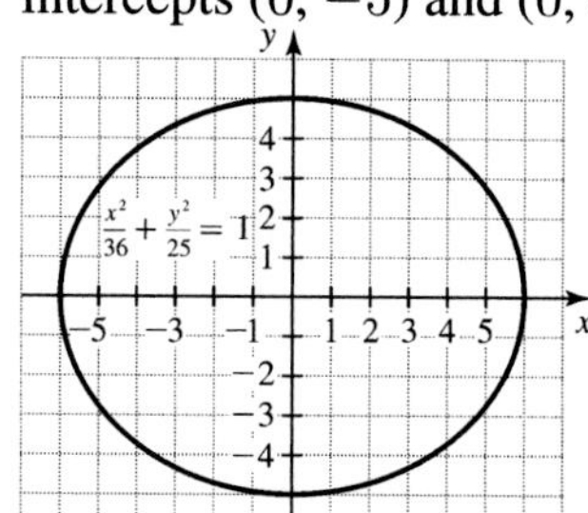

15. The graph of $\frac{x^2}{24} + \frac{y^2}{5} = 1$ is an ellipse with x-intercepts $(-\sqrt{24}, 0)$ and $(\sqrt{24}, 0)$, and y-intercepts $(0, -\sqrt{5})$ and $(0, \sqrt{5})$.

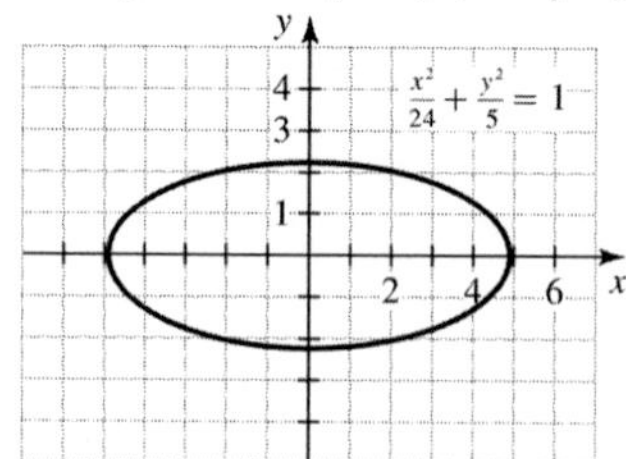

17. The graph $9x^2 + 16y^2 = 144$ is an ellipse with x-intercepts $(-4, 0)$ and $(4, 0)$, and y-intercepts $(0, -3)$ and $(0, 3)$.

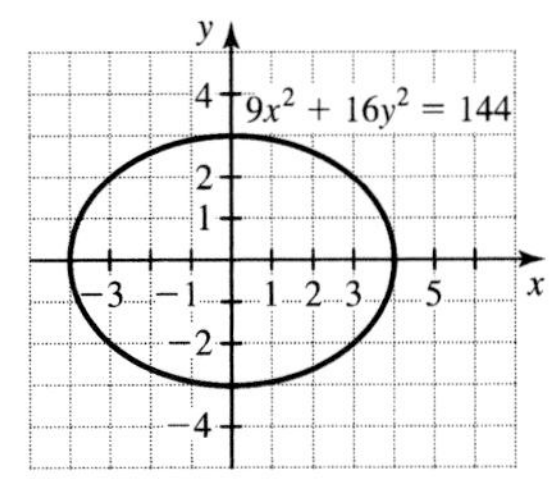

19. The graph of $25x^2 + y^2 = 25$ is an ellipse with x-intercepts $(-1, 0)$ and $(1, 0)$, and y-intercepts $(0, -5)$ and $(0, 5)$.

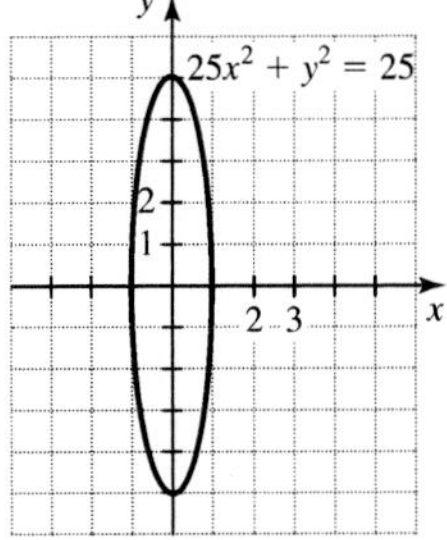

21. The graph of $4x^2 + 9y^2 = 1$ is an ellipse with x-intercepts $(1/2, 0)$ and $(-1/2, 0)$, and y-intercepts $(0, 1/3)$ and $(0, -1/3)$.

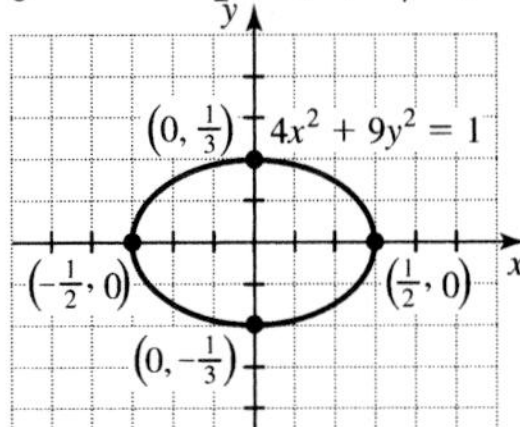

23. The graph of $\dfrac{(x-3)^2}{4} + \dfrac{(y-1)^2}{9} = 1$ is an ellipse centered at $(3, 1)$. The 4 in the denominator indicates that the ellipse passes through points that are 2 units to the right and 2 units to the left of the center: $(5, 1)$ and $(1, 1)$. The 9 in the denominator indicates that the ellipse passes through points that are 3 units above and 3 units below the center: $(3, 4)$ and $(3, -2)$.

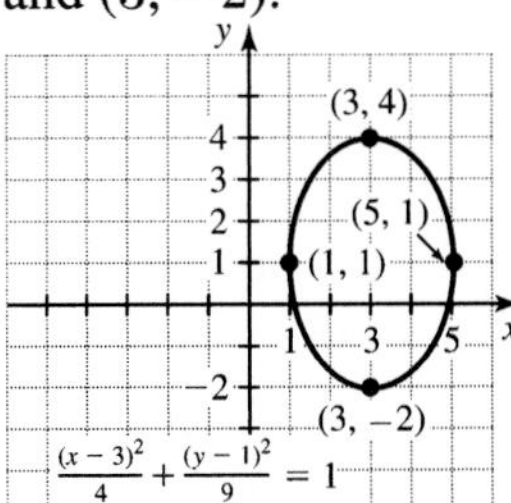

25. The graph of $\dfrac{(x+1)^2}{16} + \dfrac{(y-2)^2}{25} = 1$ is an ellipse centered at $(-1, 2)$. The 16 in the denominator indicates that the ellipse passes through points that are 4 units to the right and 4 units to the left of the center: $(3, 2)$ and $(-5, 2)$. The 25 in the denominator indicates that the ellipse passes through points that are 5 units above and 5 units below the center: $(-1, 7)$ and $(-1, -3)$.

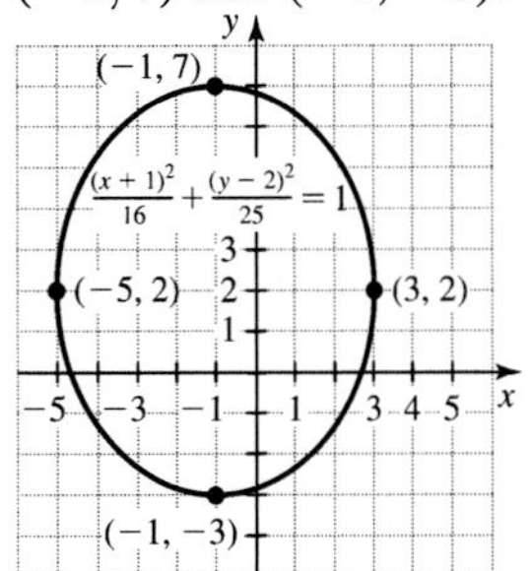

27. The graph of $(x-2)^2 + \dfrac{(y+1)^2}{36} = 1$ is an ellipse centered at $(2, -1)$. The 1 in the denominator indicates that the ellipse passes through points that are 1 unit to the right and 1 unit to the left of the center: $(3, -1)$ and $(1, -1)$. The 36 in the denominator indicates that the ellipse passes through points that are 6 units above and 6 units below the center: $(2, 5)$ and $(2, -7)$.

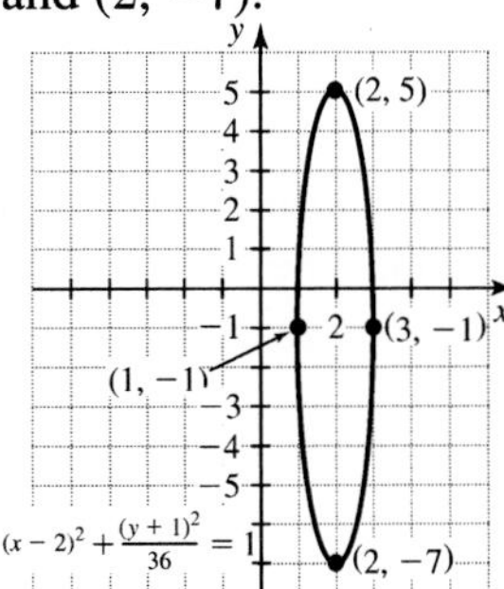

29. The graph of $\dfrac{x^2}{4} - \dfrac{y^2}{9} = 1$ is a hyperbola centered at $(0, 0)$ with x-intercepts at $(-2, 0)$ and $(2, 0)$. There are no y-intercepts. Use 9 to determine the size of the fundamental rectangle. The fundamental rectangle passes through $(0, 3)$ and $(0, -3)$. extend the diagonals of the rectangle to determine the asymptotes. The hyperbola opens to the left and right. The equations of the asymptotes are $y = \pm\frac{3}{2}x$.

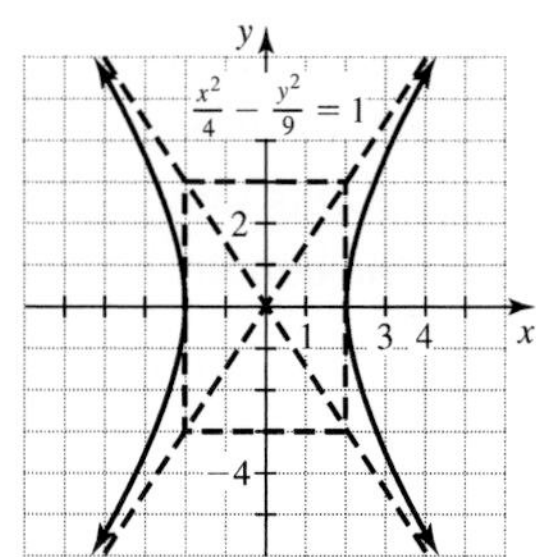

31. The graph of $\frac{y^2}{4} - \frac{x^2}{25} = 1$ is a hyperbola with y-intercepts (0, −2) and (0, 2). The fundamental rectangle passes through the y-intercepts, (−5, 0), and (5, 0). The extended diagonals determine the asymptotes, $y = \pm\frac{2}{5}x$.

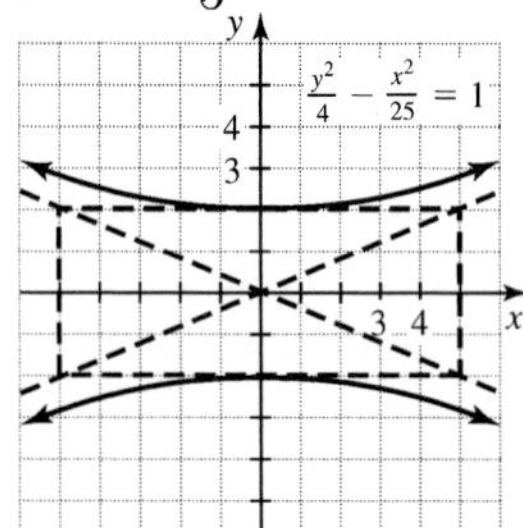

33. The graph of $\frac{x^2}{25} - y^2 = 1$ is a hyperbola with x-intercepts (−5, 0) and (5, 0). The fundamental rectangle passes through the intercepts, (0, −1), and (0, 1). The extended diagonals determine the asymptotes. The equations of the asymptotes are $y = \pm\frac{1}{5}x$.

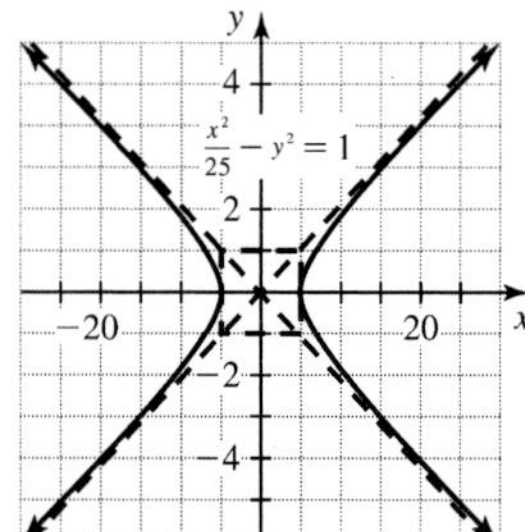

35. The graph of $x^2 - \frac{y^2}{25} = 1$ is a hyperbola with x-intercepts (−1, 0) and (1, 0). Plot (0, 5) and (0, −5) for the fundamental rectangle and extend the diagonals. The equations of the asymptotes are $y = \pm 5x$.

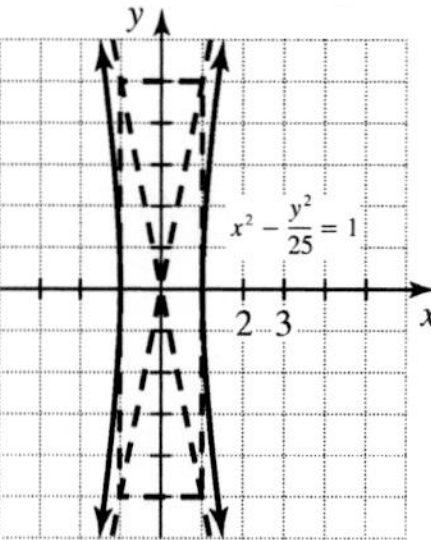

37. Divide each side of $9x^2 - 16y^2 = 144$ by 144 to get $\frac{x^2}{16} - \frac{y^2}{9} = 1$. The graph is a hyperbola with x-intercepts at (−4, 0) and (4, 0). The fundamental rectangle passes through the y-intercepts, (0, −3), and (0, 3). The extended diagonals determine the asymptotes. The equations of the asymptotes are $y = \pm\frac{3}{4}x$.

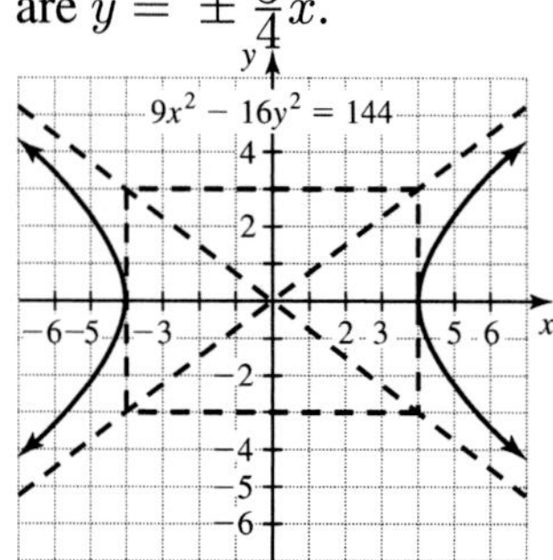

39. The graph of $x^2 - y^2 = 1$ is a hyperbola with x-intercepts (−1, 0) and (1, 0). The fundamental rectangle passes through the y-intercepts, (0, −1), and (0, 1). The extended diagonals determine the asymptotes, $y = \pm x$.

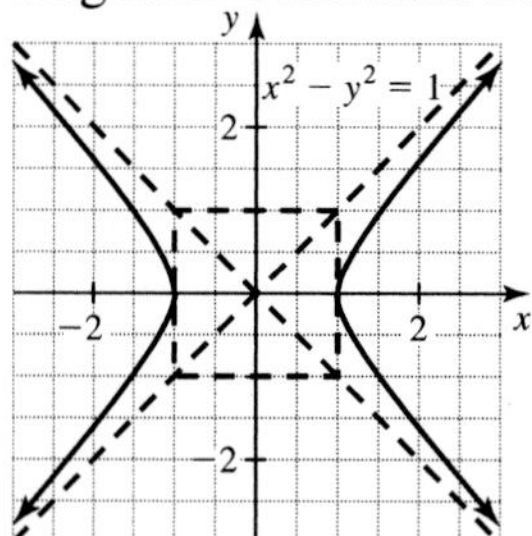

41. The graph of $\frac{(x-2)^2}{4} - (y+1)^2 = 1$ is a hyperbola centered at $(2, -1)$ and opening left and right. It is a transformation of $\frac{x^2}{4} - y^2 = 1$, which has its fundamental rectangle through $(\pm 2, 0)$ and $(0, \pm 1)$. Move the fundamental rectangle 2 to the right and down 1 unit. Then draw the asymptotes and the hyperbola.

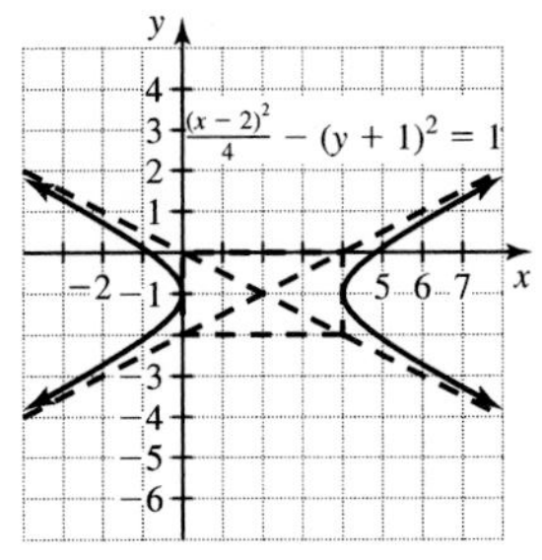

43. The graph of

$$(x+1)^2/16 - (y-1)^2/9 = 1$$

is a hyperbola centered at $(-1, 1)$ and opening left and right. It is a transformation of $x^2/16 - y^2/9 = 1$, which has its fundamental rectangle through $(\pm 4, 0)$ and $(0, \pm 3)$. Move the fundamental rectangle 1 to the left and up 1 unit. Then draw the asymptotes and the hyperbola.

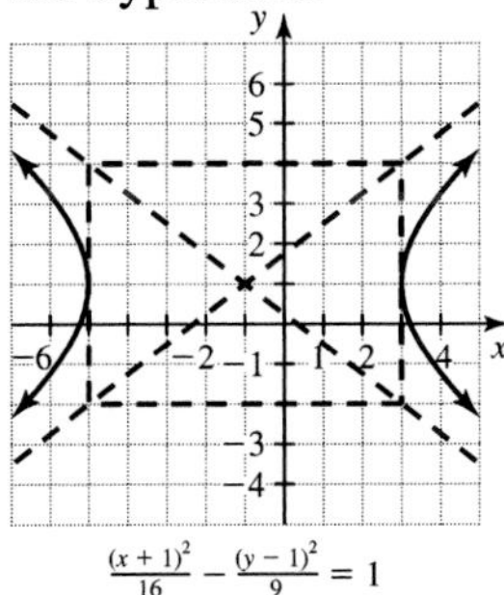

45. The graph of

$$(y-2)^2/9 - (x-4)^2/4 = 1$$

is a hyperbola centered at $(4, 2)$ and opening up and down. It is a transformation of $y^2/9 - x^2/4 = 1$, which has its fundamental rectangle through $(\pm 2, 0)$ and $(0, \pm 3)$. Move the fundamental rectangle 4 to the right and up 2 units. Then draw the asymptotes and the hyperbola.

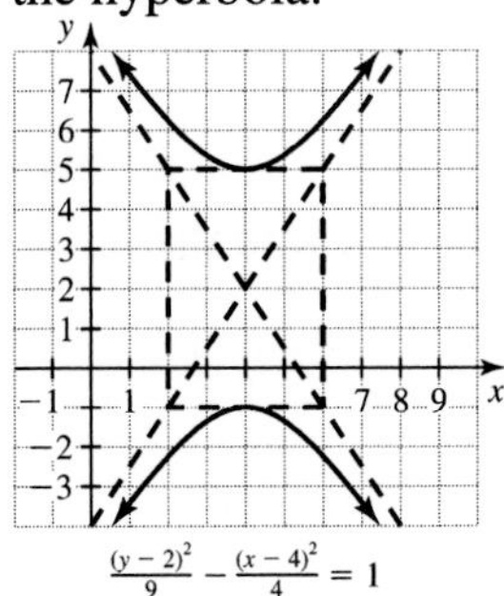

47. The graph of $y = x^2 + 1$ is a parabola because the equation has the form $y = ax^2 + bx + c$.

49. The graph of $x^2 - y^2 = 1$ is a hyperbola because the equation has the form $\frac{x^2}{a^2} - \frac{y^2}{b^2} = 1$.

51. The graph of $\frac{x^2}{2} + y^2 = 1$ is an ellipse because the equation has the form $\frac{x^2}{a^2} + \frac{y^2}{b^2} = 1$.

53. The graph of $(x-2)^2 + (y-4)^2 = 9$ is a circle because the equation has the form $(x-h)^2 + (y-k)^2 = r^2$.

55. First graph the hyperbola and the ellipse.

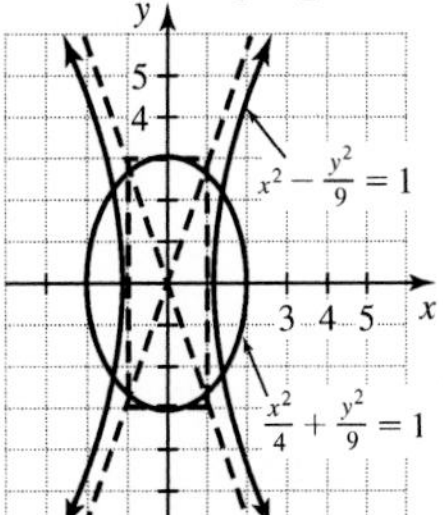

If we add the equations the variable y will be eliminated and we get the following equation.

$$\frac{x^2}{4} + x^2 = 2$$

Now multiply both sides by 4:

$$x^2 + 4x^2 = 8$$
$$5x^2 = 8$$
$$x^2 = \frac{8}{5}$$
$$x = \pm\sqrt{\frac{8}{5}} = \pm\frac{2\sqrt{10}}{5}$$

Use $x^2 = 8/5$ in the second equation.

$$\frac{8}{5} - \frac{y^2}{9} = 1$$
$$-\frac{y^2}{9} = -\frac{3}{5}$$
$$y^2 = \frac{27}{5}$$
$$y = \pm\sqrt{\frac{27}{5}} = \pm\frac{3\sqrt{15}}{5}$$

The graphs intersect at $\left(\frac{2\sqrt{10}}{5}, \frac{3\sqrt{15}}{5}\right)$, $\left(\frac{2\sqrt{10}}{5}, -\frac{3\sqrt{15}}{5}\right)$, $\left(-\frac{2\sqrt{10}}{5}, \frac{3\sqrt{15}}{5}\right)$, and $\left(-\frac{2\sqrt{10}}{5}, -\frac{3\sqrt{15}}{5}\right)$.

57. The graphs are an ellipse and a circle.

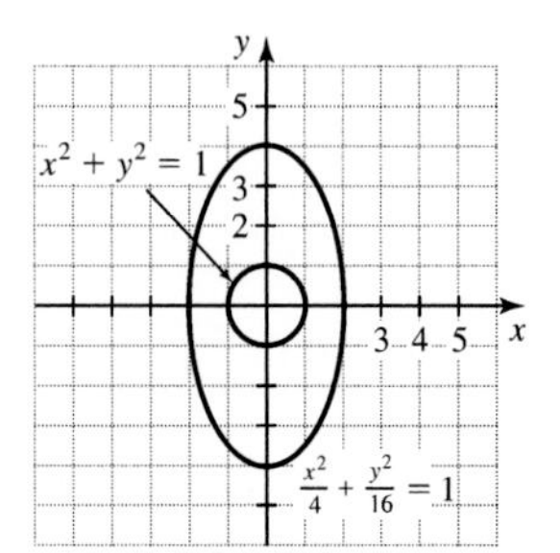

From the graph we can see that there are no points of intersection. If we eliminate y by substituting $y^2 = 1 - x^2$ into the first equation, we get the following equation.

$$\frac{x^2}{4} + \frac{1 - x^2}{16} = 1$$

Now multiply both sides by 16:

$$4x^2 + 1 - x^2 = 16$$
$$3x^2 = 15$$
$$x^2 = 5$$

From $y^2 = 1 - x^2$, we get $y^2 = -4$, which has no real solution. There are no points of intersection.

59. The graphs are a circle and a hyperbola.

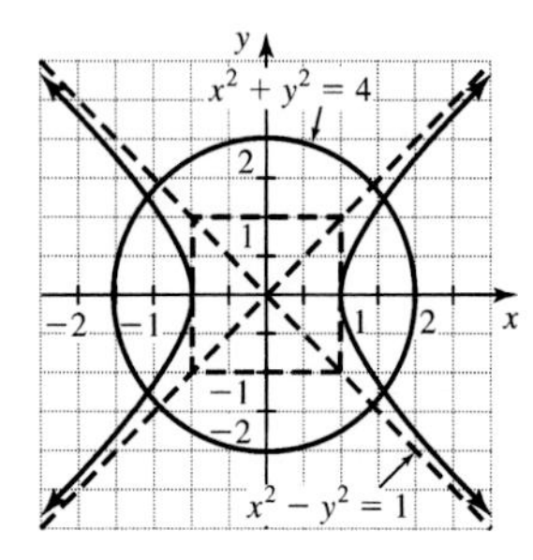

It appears that there are 4 points of intersection. To find them, add the equations to get the following equation.

$$2x^2 = 5$$
$$x^2 = \frac{5}{2}$$
$$x = \pm\sqrt{\frac{5}{2}} = \pm\frac{\sqrt{10}}{2}$$

From $y^2 = 4 - x^2$, we get $y^2 = 4 - \frac{5}{2} = \frac{3}{2}$, or $y = \pm\sqrt{\frac{3}{2}} = \pm\frac{\sqrt{6}}{2}$.

The graphs intersect at $\left(\frac{\sqrt{10}}{2}, \frac{\sqrt{6}}{2}\right)$, $\left(\frac{\sqrt{10}}{2}, -\frac{\sqrt{6}}{2}\right)$, $\left(-\frac{\sqrt{10}}{2}, \frac{\sqrt{6}}{2}\right)$, and $\left(-\frac{\sqrt{10}}{2}, -\frac{\sqrt{6}}{2}\right)$.

61. The graphs are an ellipse and a circle.

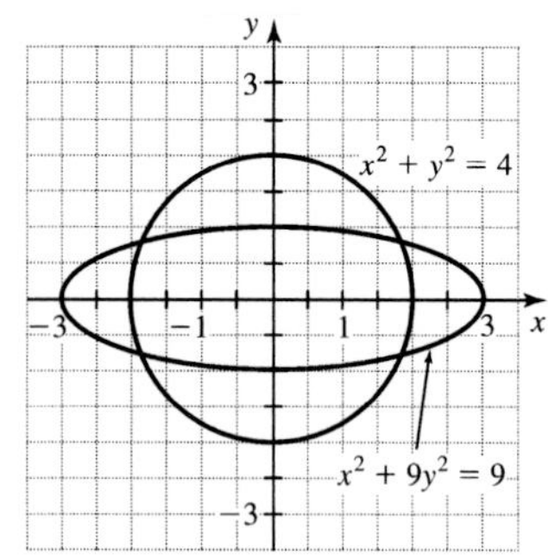

The graphs appear to intersect at 4 points. To find the points, substitute $y^2 = 4 - x^2$ into the first equation.

$$x^2 + 9(4 - x^2) = 9$$
$$-8x^2 + 36 = 9$$
$$-8x^2 = -27$$
$$x^2 = \frac{27}{8}$$
$$x = \pm\sqrt{\frac{27}{8}} = \pm\frac{3\sqrt{6}}{4}$$

Use $x^2 = 27/8$ in $y^2 = 4 - x^2$.

$$y^2 = 4 - \frac{27}{8} = \frac{5}{8}$$
$$y = \pm\sqrt{\frac{5}{8}} = \pm\frac{\sqrt{10}}{4}$$

The graphs intersect at $\left(\frac{3\sqrt{6}}{4}, \frac{\sqrt{10}}{4}\right)$, $\left(\frac{3\sqrt{6}}{4}, -\frac{\sqrt{10}}{4}\right)$, $\left(-\frac{3\sqrt{6}}{4}, \frac{\sqrt{10}}{4}\right)$, and $\left(-\frac{3\sqrt{6}}{4}, -\frac{\sqrt{10}}{4}\right)$.

63. Graph the parabola and the ellipse.

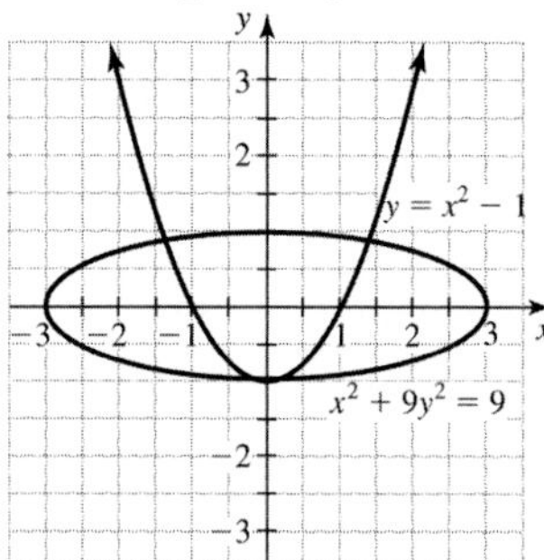

The graphs appear to intersect at 3 points. To find them, substitute $x^2 = y + 1$ into the first equation.

$$y + 1 + 9y^2 = 9$$
$$9y^2 + y - 8 = 0$$
$$(9y - 8)(y + 1) = 0$$
$$y = \frac{8}{9} \quad \text{or} \quad y = -1$$

If $y = -1$, then $x^2 = -1 + 1 = 0$ or $x = 0$. If $y = 8/9$, then

$$x^2 = \frac{8}{9} + 1 = \frac{17}{9} \quad \text{or} \quad x = \pm\frac{\sqrt{17}}{3}.$$

The three points of intersection are $\left(\frac{\sqrt{17}}{3}, \frac{8}{9}\right)$, $\left(-\frac{\sqrt{17}}{3}, \frac{8}{9}\right)$, and $(0, -1)$.

65. Graph the hyperbola and the line.

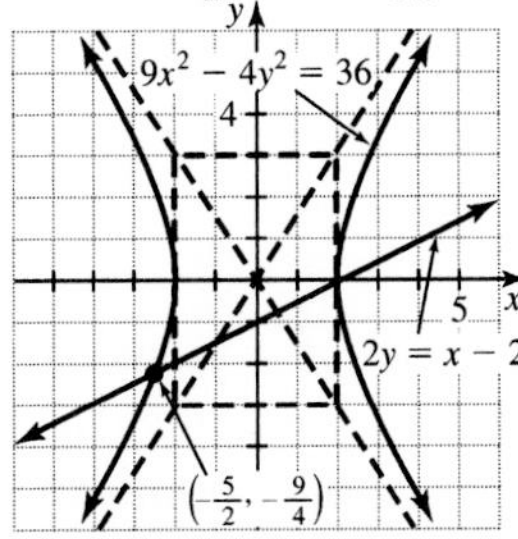

To find the points of intersection, substitute $x = 2y + 2$ into the first equation.

$$9(2y+2)^2 - 4y^2 = 36$$
$$9(4y^2 + 8y + 4) - 4y^2 = 36$$
$$36y^2 + 72y + 36 - 4y^2 = 36$$
$$32y^2 + 72y = 0$$
$$4y^2 + 9y = 0$$
$$y(4y + 9) = 0$$
$$y = 0 \quad \text{or} \quad y = -9/4$$

If $y = 0$, then $x = 2(0) + 2 = 2$. If $y = -9/4$, then $x = 2(-9/4) + 2 = -5/2$. The two points of intersection are $(2, 0)$ and $(-5/2, -9/4)$.

67. a) From the graph it appears that the boat is approximately at (2.5, 1.5)

b) Substitute $x^2 = 1 + 3y^2$ into $4y^2 - x^2 = 1$.

$$4y^2 - (1 + 3y^2) = 1$$
$$y^2 = 2$$
$$y = \pm\sqrt{2}$$
$$x^2 = 1 + 3(\pm\sqrt{2})^2 = 1 \pm 3(2) = 1 \pm 6$$
$$= 7 \text{ or } -5$$

If $x^2 = 7$, then $x = \pm\sqrt{7}$. If $x^2 = -5$, there is no real solution. Since the boat is in the first quadrant, the boat is at $(\sqrt{7}, \sqrt{2})$.

11.5 WARM-UPS

1. False, because $x^2 + y = 4$ is the equation of a parabola.
2. True, because if we divide each side by 9 we get the standard form for an ellipse.
3. True, because it can be written as $y^2 - x^2 = 1$.
4. True, because $2(0)^2 - 0 < 3$ is correct.
5. False, because $0 > 0^2 - 3(0) + 2$ is incorrect.
6. False, because the origin is on the parabola $x^2 = y$.
7. False, because (0, 0) does not satisfy the inequality.
8. True, because $x^2 + y^2 = 4$ is a circle of radius 2 centered at (0, 0), and the points inside this circle satisfy the inequality.
9. True, because both $0^2 - 4^2 < 1$ and $4 > 0^2 - 2(0) + 3$ are correct.
10. True, because both $0^2 + 0^2 < 1$ and $0 < 0^2 + 1$ are correct.

11.5 EXERCISES

1. Graph the parabola $y = x^2$. Since (0, 4) satisfies $y > x^2$, shade the region containing $(0, 4)$.

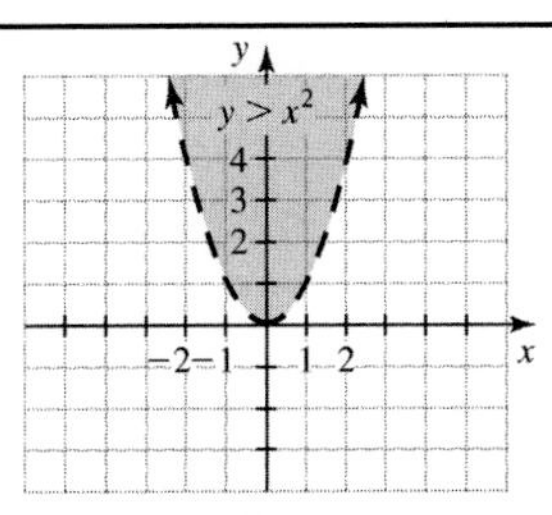

3. First graph the parabola $y = x^2 - x$. Since (5, 0) satisfies $y < x^2 - x$, we shade the region containing (5, 0).

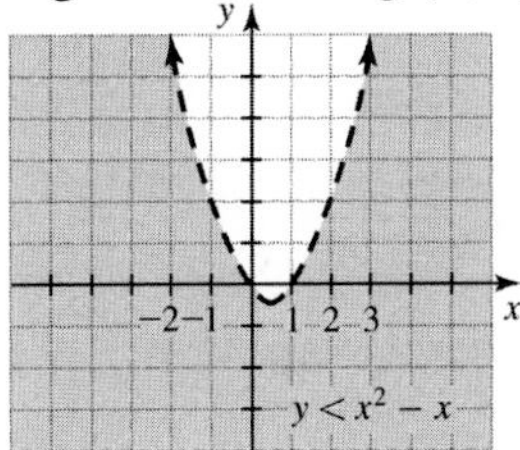

5. First graph the parabola $y = x^2 - x - 2$. Since (0, 5) satisfies $y > x^2 - x - 2$, shade the region containing (0, 5).

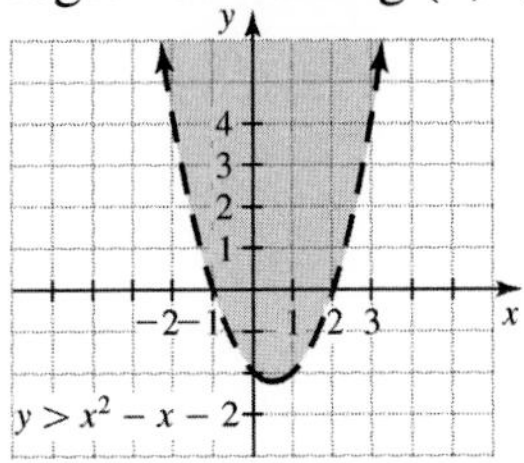

7. First graph the circle $x^2 + y^2 = 9$ using a solid curve. Since (0, 0) satisfies $x^2 + y^2 \leq 9$, shade the inside of the circle.

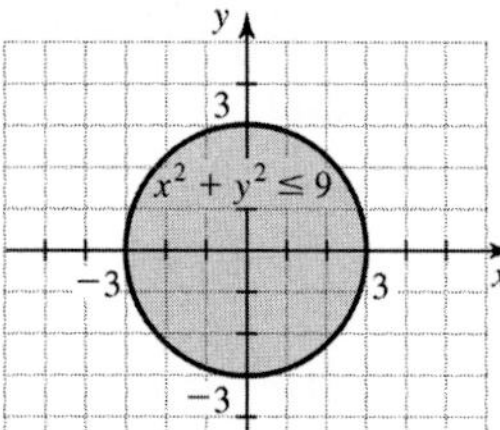

9. First graph the ellipse $x^2 + 4y^2 = 4$ using a dashed curve. Since (0, 0) does not satisfy the inequality $x^2 + 4y^2 > 4$, we shade the region outside the ellipse.

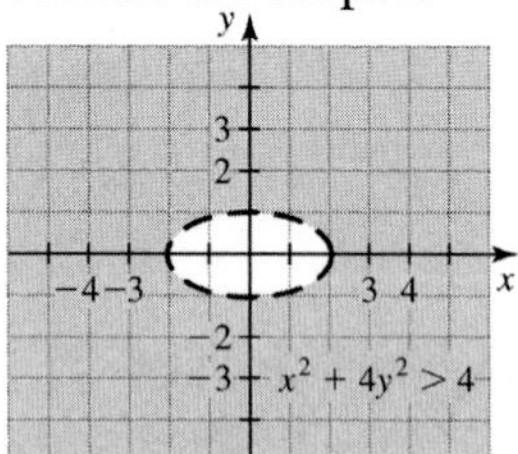

11. First divide both sides of the inequality by 36.

$$\frac{x^2}{9} - \frac{y^2}{4} < 1$$

Now graph the hyperbola $\frac{x^2}{9} - \frac{y^2}{4} = 1$. Testing a point in each of the three regions, we find that only points in the region containing the origin satisfy $4x^2 - 9y^2 < 36$.

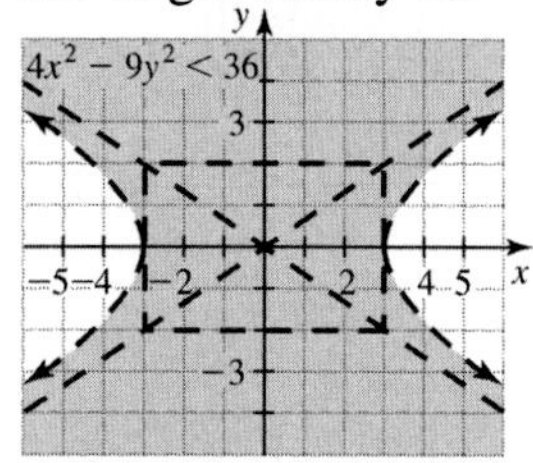

13. First graph the circle centered at (2, 3) with radius 2, using a dashed curve. Since $(2, 3)$ satisfies $(x - 2)^2 + (y - 3)^2 < 4$, shade the region inside the circle.

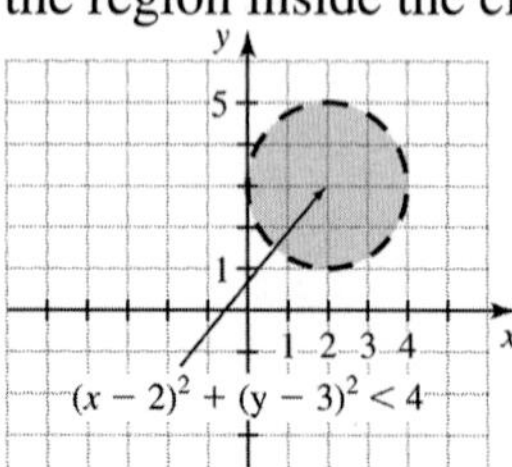

15. First graph the circle centered at (0, 0) with radius 1. Since (0, 0) does not satisfy $x^2 + y^2 > 1$, shade the region outside the circle.

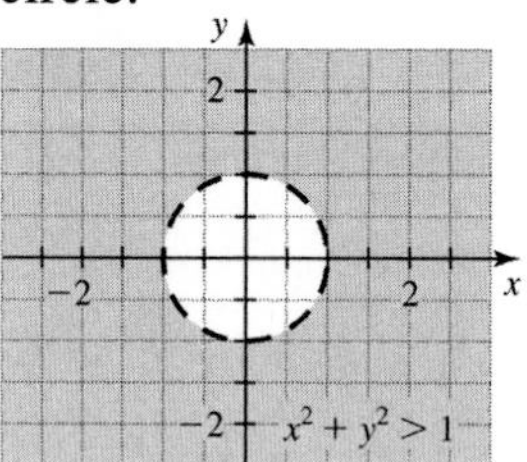

17. Graph the hyperbola $4x^2 - y^2 = 4$, using a dashed curve. After testing a point in each of the three regions, we see that only points in the region containing (0, 0) fail to satisfy $4x^2 - y^2 > 4$.

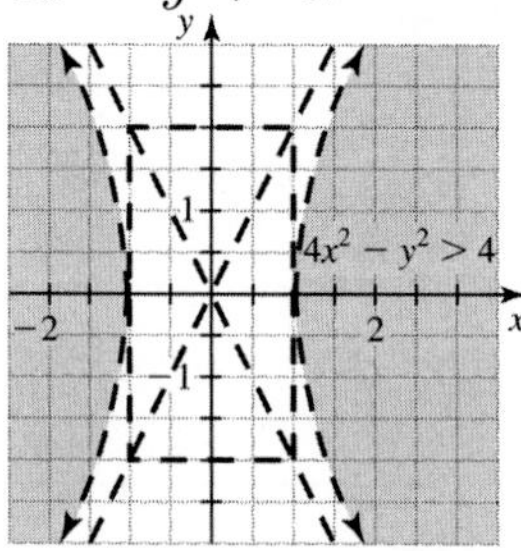

19. Graph the hyperbola $y^2 - x^2 = 1$. After testing a point in each of the three regions, we see that only points in the region containing (0, 0) satisfy $y^2 - x^2 \leq 1$.

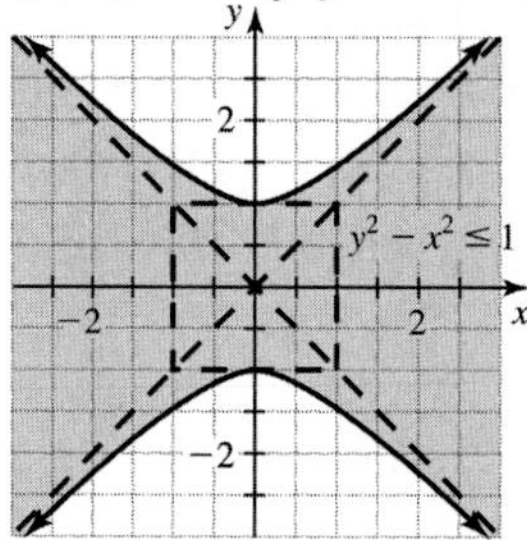

21. The graph of $y = x$ is a line with slope 1 and y-intercept (0, 0). Since (5, −5) satisfies $x > y$, we shade the region below the line.

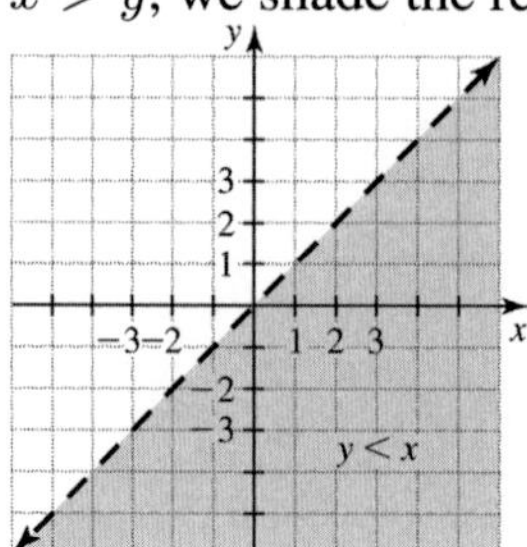

23. Check $(3, -4)$ in each inequality:

$x^2 + y^2 \le 25$	$y \le x^2$
$3^2 + (-4)^2 \le 25$	$-4 \le 3^2$
$25 \le 25$	$-4 \le 9$
True	True

Since $(3, -4)$ satisfies both inequalities it is in the solution set to the system.

25. Check $(3, -4)$ in each inequality:

$x - y > 1$	$y > (x - 2)^2 + 3$
$3 - (-4) > 1$	$-4 > (3 - 2)^2 + 3$
$7 > 1$	$-4 > 4$
True	False

Since $(3, -4)$ does not satisfy both inequalities it is not in the solution set to the system.

27. Graph the circle $x^2 + y^2 = 9$ and the line $y = x$. The graph of $x^2 + y^2 < 9$ is the region inside the circle. the graph of $y > x$ is the region above the line. The intersection of these two regions is shown as follows.

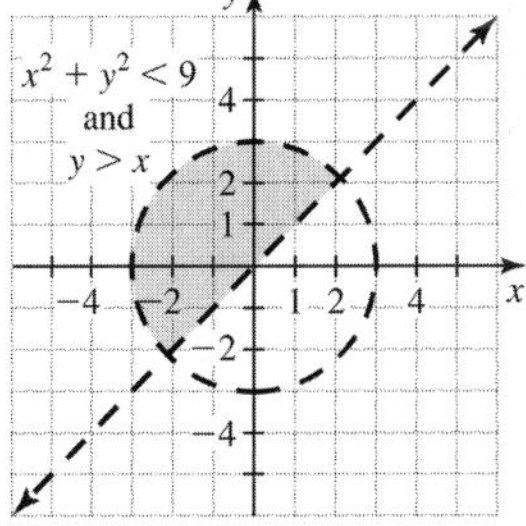

We could have used test points to determine which of the 4 regions determined by the circle and the line satisfies both inequalities.

29. The graph of $x^2 - y^2 = 1$ is a hyperbola. The graph of $x^2 + y^2 = 4$ is a circle of radius 2. Points that satisfy $x^2 + y^2 < 4$ are inside the circle. The hyperbola divides the plane into 3 regions. Points in the 2 regions not containing (0, 0) are the points that satisfy $x^2 - y^2 > 1$. The intersection of these regions is shown in the following graph.

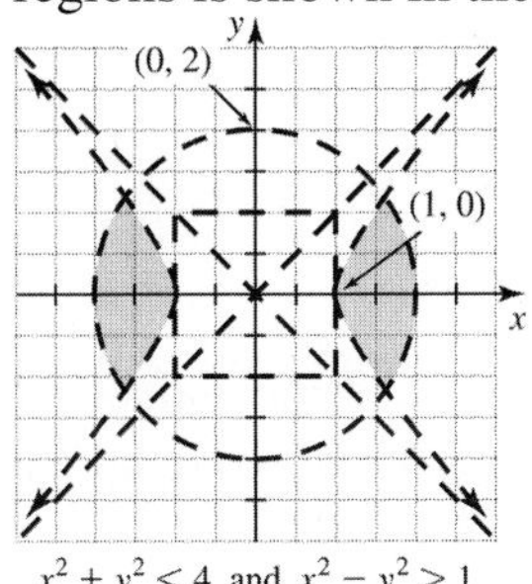

$x^2 + y^2 < 4$ and $x^2 - y^2 > 1$

31. The graph of $y = x^2 + x$ is a parabola opening upward, with x-intercepts at (0, 0) and $(-1, 0)$. The graph of $y > x^2 + x$ is the region inside this parabola. The graph of $y = 5$ is a horizontal line through (0, 5). The graph of $y < 5$ is the region below this horizontal line. The points that satisfy both inequalities are the points inside the parabola and below $y = 5$, as shown in the following graph.

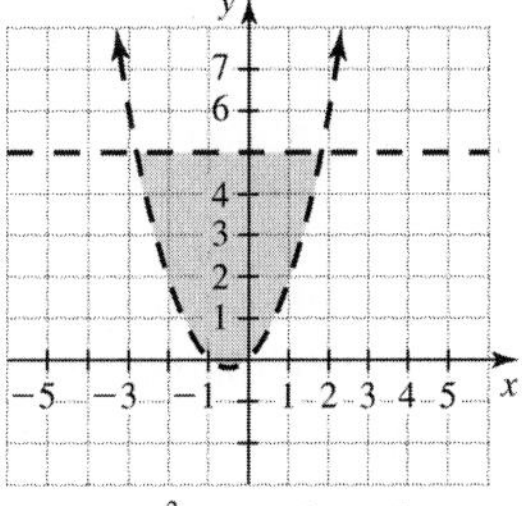

$y > x^2 + x$ and $y < 5$

33. The graph of $y = x + 2$ is a line with y-intercept (0, 2) and slope 1. The graph of $y \ge x + 2$ is the region above and including the line. The graph of $y = 2 - x$ is a line with y-intercept (0, 2) and slope -1. The graph of $y \le 2 - x$ is the region below and including this line. The region above the first line and below the second is shown in the following graph. We could have tested a point in each of the 4 regions determined by the two lines. We would find that only points in the region shown satisfy both inequalities.

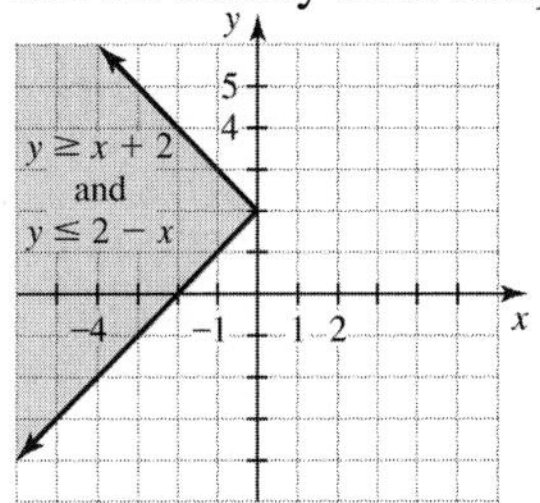

35. The graph of $4x^2 - y^2 = 4$ is a hyperbola opening to the left and right. Points that satisfy $4x^2 - y^2 < 4$ are in the region between the two branches of the hyperbola. The graph of $x^2 + 4y^2 = 4$ is an ellipse passing through (0, 1), $(0, -1)$, (2, 0), and $(-2, 0)$. Points that satisfy $x^2 + 4y^2 > 4$ are outside the ellipse. Points that satisfy both inequalities are outside the ellipse and between the two branches of the hyperbola.

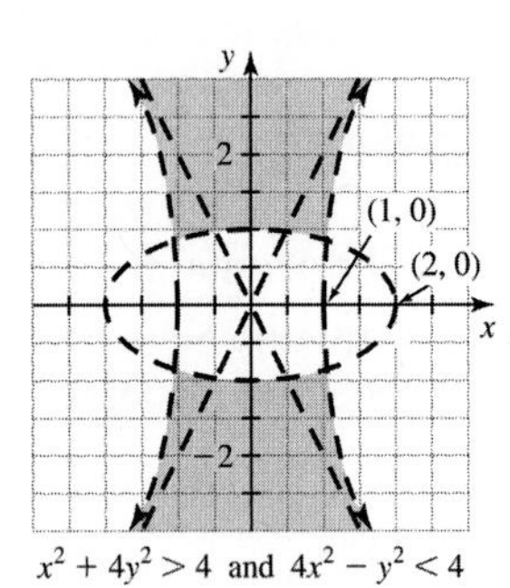

$x^2 + 4y^2 > 4$ and $4x^2 - y^2 < 4$

37. The graph of $x - y = 0$ is the same as the line $y = x$ (through (0, 0) with slope 1). Points that satisfy $x - y < 0$ are above the line $y = x$. The graph of $y + x^2 = 1$ is the same as the parabola $y = -x^2 + 1$, which opens downward. Points that satisfy $y + x^2 < 1$ are below the parabola. The points that satisfy both inequalities, below the parabola and above the line, are shown in the following graph.

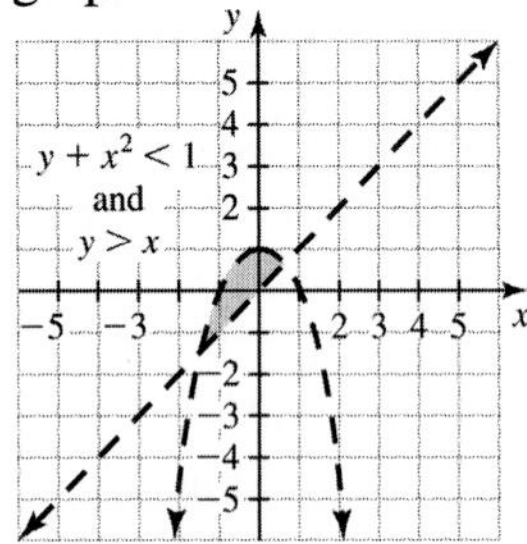

39. The graph of $y = 5x - x^2$ is a parabola opening downward. Points that satisfy $y < 5x - x^2$ are the points below the parabola. The graph of $x^2 + y^2 = 9$ is a circle of radius 3. Points that satisfy $x^2 + y^2 < 9$ are inside the circle. Points that satisfy both inequalities are the points inside the circle and below the parabola.

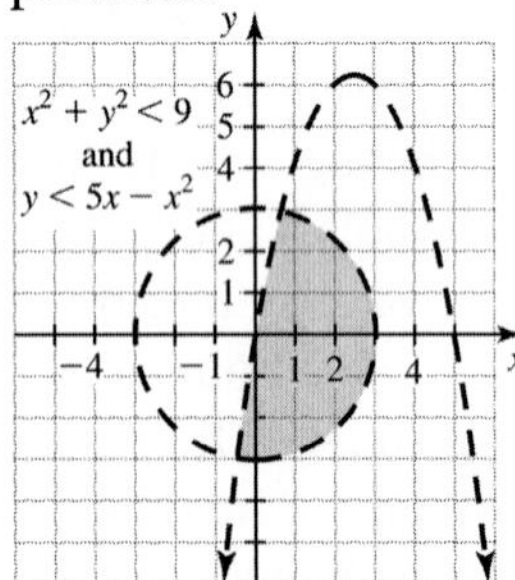

41. Points that satisfy $y \geq 3$ are above or on the horizontal line $y = 3$. Points that satisfy $x \leq 1$ are on or to the left of the vertical line $x = 1$. Points that satisfy both inequalities are shown in the following graph. The points graphed are the points with x-coordinate less than or equal to 1, and y-coordinate greater than or equal to 3.

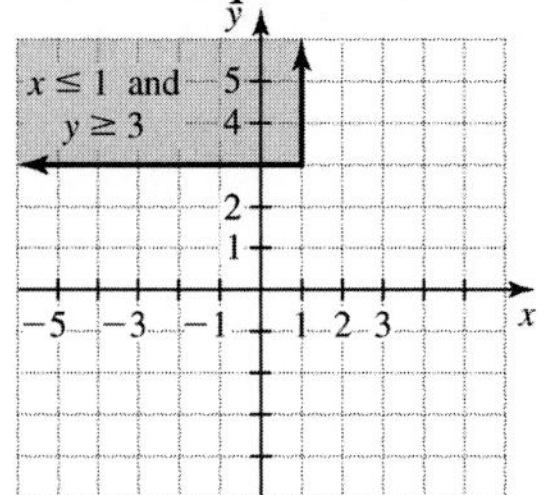

43. The graph of $4y^2 - 9x^2 = 36$ is a hyperbola opening up and down. Points that satisfy $4y^2 - 9x^2 < 36$ are the points between the two branches of the hyperbola. The graph of $x^2 + y^2 = 16$ is a circle of radius 4. Points that satisfy $x^2 + y^2 < 16$ are inside the circle. The points that satisfy both inequalities are indicated as follows.

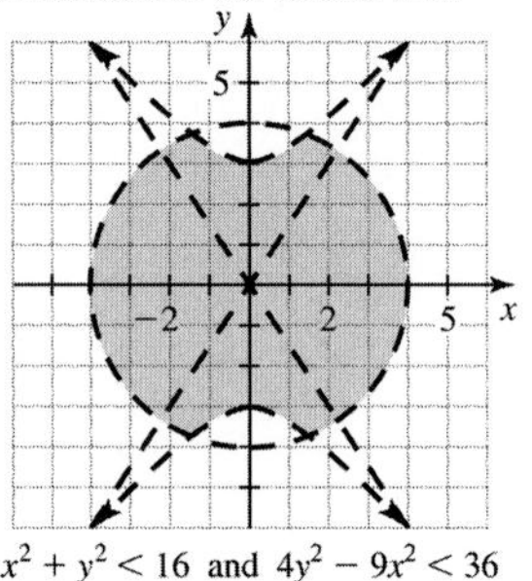

$x^2 + y^2 < 16$ and $4y^2 - 9x^2 < 36$

45. The graph of $y = x^2$ is a parabola opening upward. Points that satisfy $y < x^2$ are below this parabola. The graph of $x^2 + y^2 = 1$ is a circle of radius 1. Points that satisfy $x^2 + y^2 < 1$ are inside the circle. The graph of the system of inequalities consists of points that are below the parabola and inside the circle.

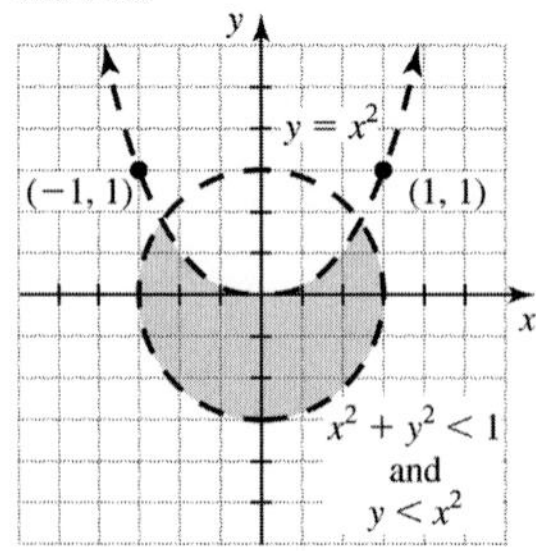

47. Let $x =$ the number of paces he walked to the east and $y =$ the number of paces he walked to the north. So $x + y \geq 50$, $x^2 + y^2 \leq 50^2$, and $y > x$. The graph of this system follows.

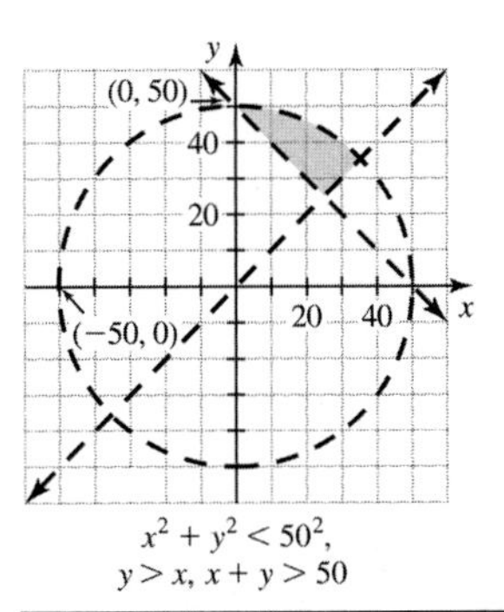

$x^2 + y^2 < 50^2$,
$y > x$, $x + y > 50$

49. From the graphs it appears that there is no intersection of the two regions and there is no solution to the system.

Chapter 11 Wrap-Up

Enriching Your Mathematical Word Power

1. c **2.** a **3.** d **4.** a **5.** c
6. d **7.** b **8.** d **9.** c **10.** a

CHAPTER 11 REVIEW

1. The graph of $y = x^2$ is a parabola and the graph of $y = -2x + 15$ is a straight line.

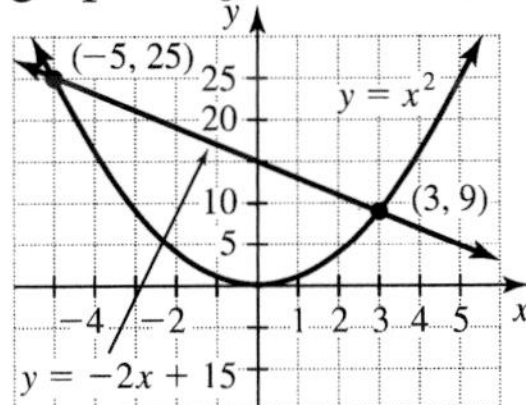

Use substitution to eliminate y.

$$x^2 = -2x + 15$$
$$x^2 + 2x - 15 = 0$$
$$(x + 5)(x - 3) = 0$$
$$x = -5 \text{ or } x = 3$$
$$y = 25 \quad y = 9 \text{ Since } y = x^2$$

The solution set is $\{(3, 9), (-5, 25)\}$.

3. First graph $y = 3x$ and $y = 1/x$.

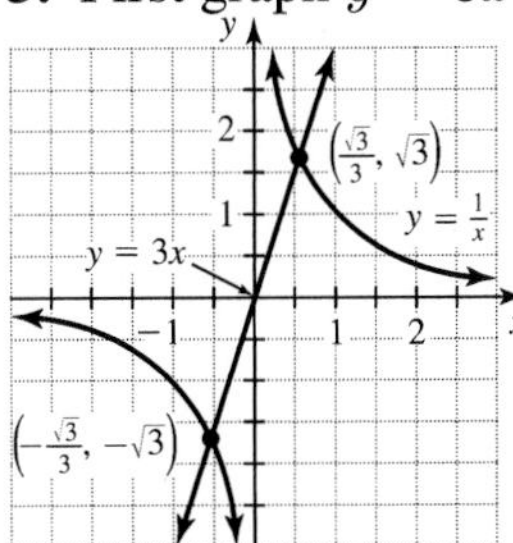

Use substitution to eliminate y.

$$3x = \frac{1}{x}$$
$$3x^2 = 1$$
$$x^2 = \frac{1}{3}$$
$$x = \pm\sqrt{\frac{1}{3}} = \pm\frac{\sqrt{3}}{3}$$

If $x = \frac{\sqrt{3}}{3}$, then $y = \sqrt{3}$. If $x = -\frac{\sqrt{3}}{3}$, then $y = -\sqrt{3}$. The solution set is $\left\{\left(\frac{\sqrt{3}}{3}, \sqrt{3}\right), \left(-\frac{\sqrt{3}}{3}, -\sqrt{3}\right)\right\}$.

5. Substitute $y = x$ into $xy = 9$.

$$x \cdot x = 9$$
$$x^2 = 9$$
$$x = \pm 3$$

If $x = -3$, then $y = -3$. If $x = 3$, then $y = 3$. The solution set is $\{(3, 3), (-3, -3)\}$.

7. The second equation can be written as $x^2 = 3y$. Substitute this equation into the first equation.

$$3y + y^2 = 4$$
$$y^2 + 3y - 4 = 0$$
$$(y + 4)(y - 1) = 0$$
$$y = -4 \text{ or } y = 1$$

If $y = -4$, then $x^2 = 3(-4)$ has no solution.
If $y = 1$, then $x^2 = 3(1)$ gives us $x = \pm\sqrt{3}$.
The solution set is $\left\{(\sqrt{3}, 1), (-\sqrt{3}, 1)\right\}$.

9. Use substitution to eliminate y.

$$x^2 + (x + 2)^2 = 34$$
$$x^2 + x^2 + 4x + 4 = 34$$
$$2x^2 + 4x - 30 = 0$$
$$x^2 + 2x - 15 = 0$$
$$(x + 5)(x - 3) = 0$$
$$x = -5 \text{ or } x = 3$$
$$y = -3 \quad y = 5 \text{ Since } y = x + 2$$

The solution set is $\{(-5, -3), (3, 5)\}$.

11. Use substitution to eliminate y.

$$\log(x - 3) = 1 - \log(x)$$
$$\log(x - 3) + \log(x) = 1$$
$$\log(x^2 - 3x) = 1$$
$$x^2 - 3x = 10$$
$$x^2 - 3x - 10 = 0$$

$$(x-5)(x+2)=0$$
$$x=5 \text{ or } x=-2$$

If $x=5$, then $y=\log(5-3)=\log(2)$. If $x=-2$, we get a logarithm of a negative number. So the solution set is $\{(5, \log(2))\}$.

13. Use substitution to eliminate x.

$$y^2=2(12-y)$$
$$y^2=24-2y$$
$$y^2+2y-24=0$$
$$(y+6)(y-4)=0$$
$$y=-6 \text{ or } y=4$$

If $y=-6$, then $-6=x^2$ has no solution. If $y=4$, then $x^2=4$ and $x=\pm 2$. The solution set is $\{(2, 4), (-2, 4)\}$.

15. $\sqrt{(3-1)^2+(3-1)^2}=\sqrt{4+4}=2\sqrt{2}$

17. $\sqrt{(2-(-4))^2+(-8-6)^2}$
$=\sqrt{36+196}=2\sqrt{58}$

19. Use the midpoint formula:
$\left(\frac{8+2}{2}, \frac{-2+6}{2}\right)=\left(\frac{10}{2}, \frac{4}{2}\right)=(5, 2)$
Use the distance formula to find the length:
$\sqrt{(8-2)^2+(-2-6)^2}=\sqrt{36+64}$
$=\sqrt{100}=10$

21. Use the midpoint formula:
$\left(\frac{2+3}{2}, \frac{-2+1}{2}\right)=\left(\frac{5}{2}, -\frac{1}{2}\right)$
Use the distance formula to find the length:
$\sqrt{(2-3)^2+(-2-1)^2}=\sqrt{1+9}$
$=\sqrt{10}$

23. For $y=x^2+3x-18$, use $x=-b/(2a)$ to find the x-coordinate of the vertex. The vertex is $(-\frac{3}{2}, -\frac{81}{4})$ and the axis of symmetry is $x=-3/2$. Since $a=1$, $p=1/4$. The focus is $(-\frac{3}{2}, -20)$ and the directrix is $y=-\frac{41}{2}$.

25. For $y=x^2+3x+2$, use $x=-b/(2a)$ to find the x-coordinate of the vertex. The vertex is $(-\frac{3}{2}, -\frac{1}{4})$ and the axis of symmetry is $x=-3/2$. Since $a=1$, $p=1/4$. The focus is $(-\frac{3}{2}, 0)$ and the directrix is $y=-\frac{1}{2}$.

27. For $y=-\frac{1}{2}(x-2)^2+3$ the vertex is $(2, 3)$ and the parabola opens down. The axis of symmetry is $x=2$. Since $a=-1/2$, $p=-1/2$. The focus is $(2, \frac{5}{2})$ and the directrix is $y=\frac{7}{2}$.

29.
$$y=2x^2-8x+1$$
$$y=2(x^2-4x)+1$$
$$y=2(x^2-4x+4-4)+1$$
$$y=2(x^2-4x+4)-8+1$$
$$y=2(x-2)^2-7$$

Vertex $(2, -7)$

31.
$$y=-\frac{1}{2}x^2-x+\frac{1}{2}$$
$$y=-\frac{1}{2}(x^2+2x)+\frac{1}{2}$$
$$y=-\frac{1}{2}(x^2+2x+1-1)+\frac{1}{2}$$
$$y=-\frac{1}{2}(x+1)^2+\frac{1}{2}+\frac{1}{2}$$
$$y=-\frac{1}{2}(x+1)^2+1$$

Vertex $(-1, 1)$

33. The graph of $x^2+y^2=100$ is a circle centered at $(0, 0)$ with radius 10.

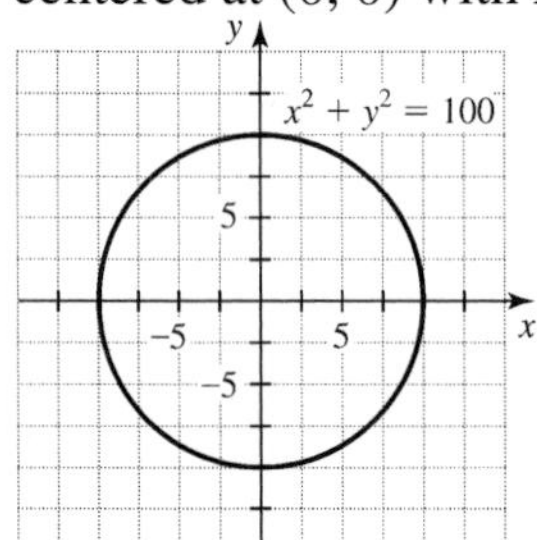

35. The graph of $(x-2)^2+(y+3)^2=81$ is a circle centered at $(2, -3)$ with radius 9.

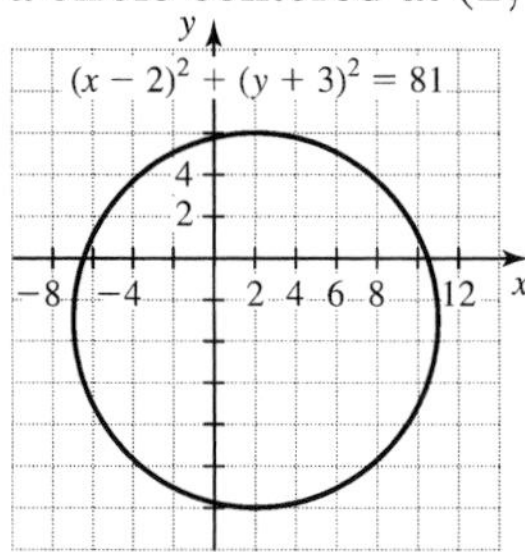

37.
$$9y^2+9x^2=4$$
$$x^2+y^2=\frac{4}{9}$$

The graph is a circle with center $(0, 0)$ and radius $2/3$.

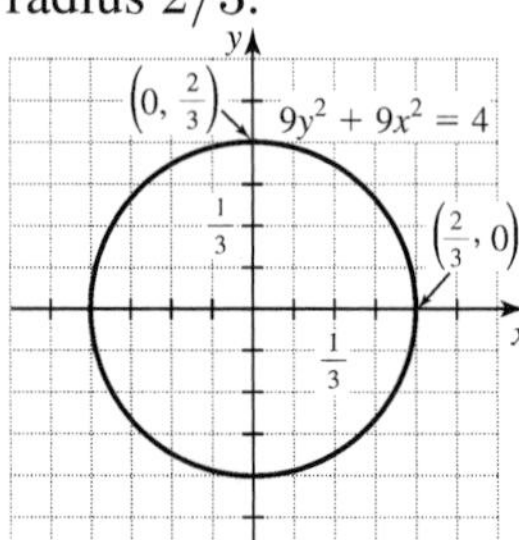

39. The equation of a circle with center (h, k) and radius r is $(x-h)^2+(y-k)^2=r^2$. The equation of a circle with center (0, 3) and radius 6 is $x^2 + (y - 3)^2 = 36$.

41. The equation of a circle with center (2, −7) and radius 5 is $(x - 2)^2 + (y + 7)^2 = 25$.

43. The graph of $\frac{x^2}{36} + \frac{y^2}{49} = 1$ is an ellipse with x-intercepts at (−6, 0) and (6, 0), y-intercepts at (0, −7) and (0, 7), and centered at the origin.

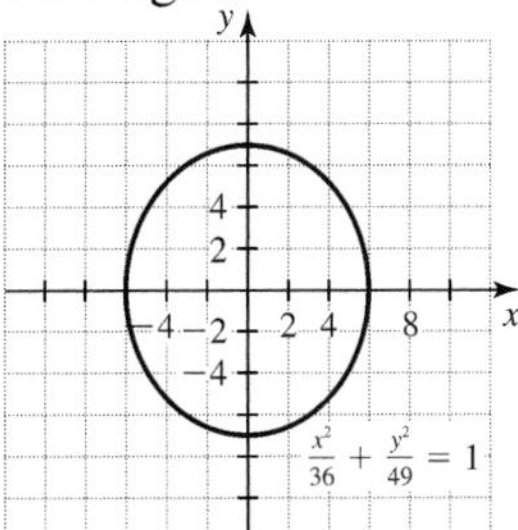

45. The graph of $25x^2 + 4y^2 = 100$ is an ellipse with x-intercepts (−2, 0) and (2, 0), y-intercepts (0, −5) and (0, 5), and centered at the origin.

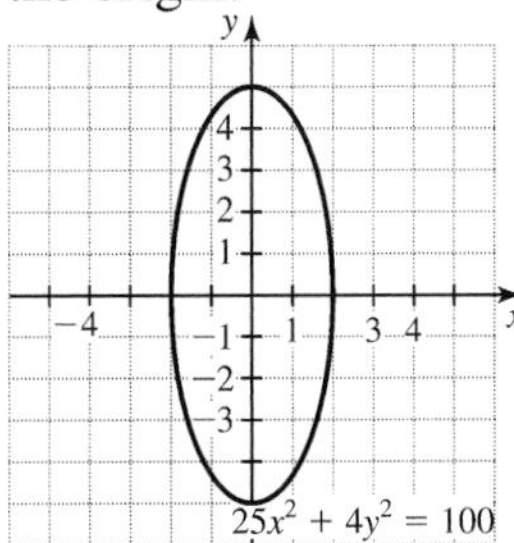

47. The graph of $\frac{x^2}{49} - \frac{y^2}{36} = 1$ is a hyperbola with x-intercepts (−7, 0) and (7, 0). The fundamental rectangle passes through the x-intercepts and the points (0, −6) and (0, 6). Extend the diagonals of the fundamental rectangle to get the asymptotes and draw a hyperbola opening to the left and right.

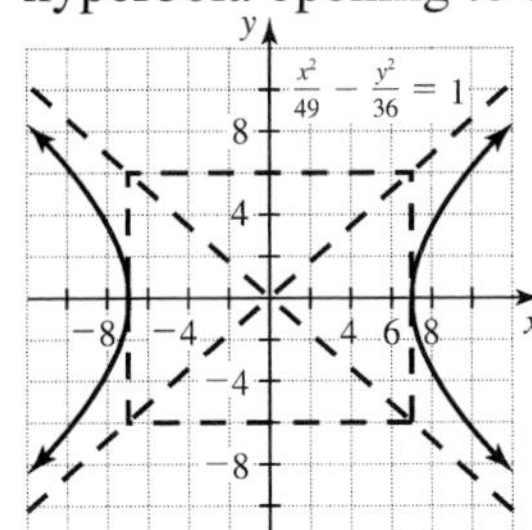

49. Write $4x^2 - 25y^2 = 100$ as $\frac{x^2}{25} - \frac{y^2}{4} = 1$. The graph has x-intercepts at (−5, 0) and $(5, 0)$. The fundamental rectangle passes through the x-intercepts and the points (0, −2) and (0, 2). Extend the diagonals of the rectangle to get the asymptotes and draw a hyperbola opening to the left and right.

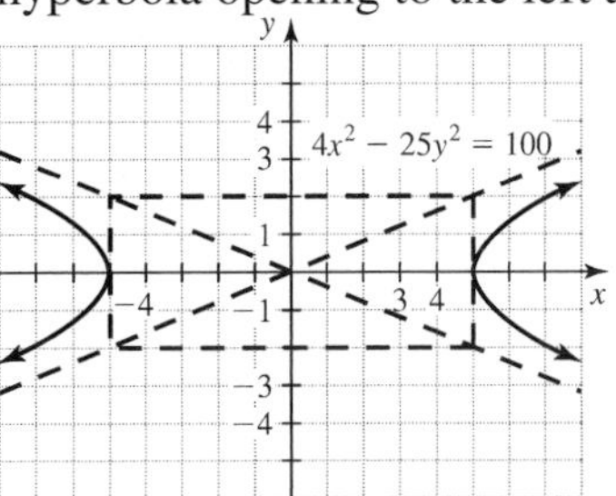

51. First graph the line $4x - 2y = 3$. It goes through (0, −3/2) and (3/4, 0). Since (0, 0) fails to satisfy $4x - 2y > 3$, we shade the region not containing (0, 0).

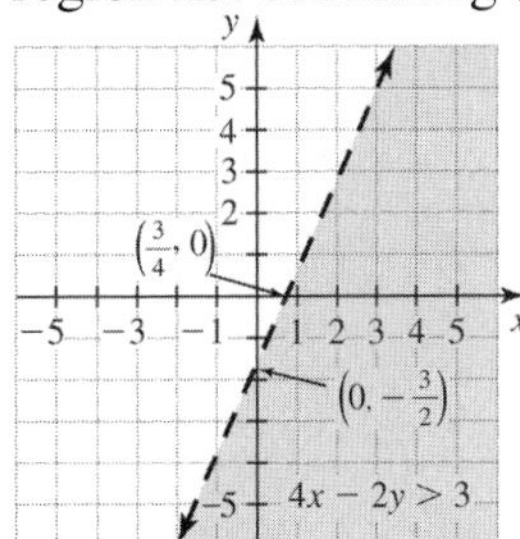

53. Write $y^2 < x^2 - 1$ as $x^2 - y^2 > 1$. Graph the hyperbola $x^2 - y^2 = 1$ and test a point in each of the three regions: (−5, 0), (0, 0), and (5, 0). Since (−5, 0) and (5, 0) satisfy the inequality, we shade the regions containing those points.

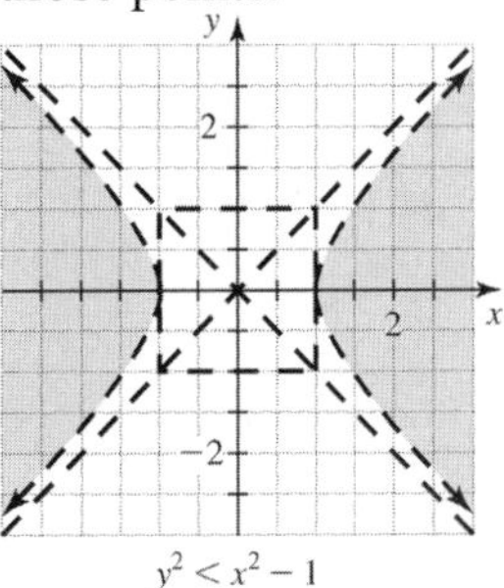

55. Write $4x^2 + 9y^2 > 36$ as $\frac{x^2}{9} + \frac{y^2}{4} > 1$.

First graph the ellipse $\frac{x^2}{9} + \frac{y^2}{4} = 1$ through (−3, 0), (3, 0), (0, −2), and (0, 2). Since (0, 0) fails to satisfy the inequality, we shade the region outside of the ellipse.

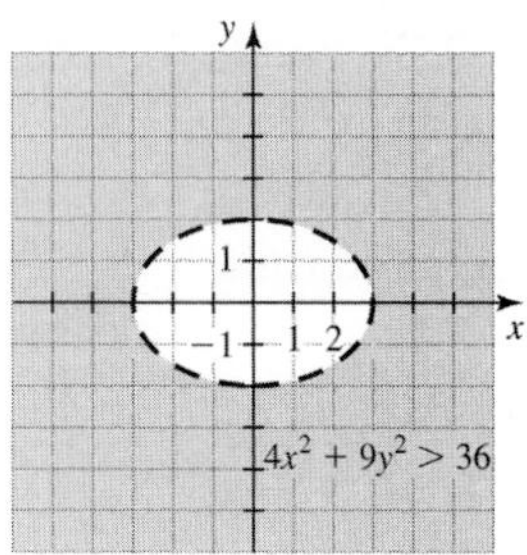

57. The graph of $y < 4x - x^2$ is the region below the parabola $y = 4x - x^2$. The graph of $x^2 + y^2 < 9$ is the region inside the circle $x^2 + y^2 = 9$. Points that are inside the circle and below the parabola are shown in the following graph.

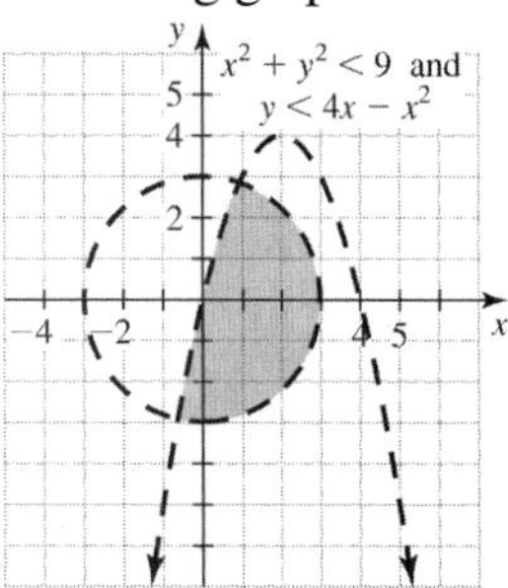

59. The set of points that satisfy $4x^2 + 9y^2 > 36$ is the set of points outside the ellipse $4x^2 + 9y^2 = 36$. The set of points that satisfy $x^2 + y^2 < 9$ is the set of points inside the circle $x^2 + y^2 = 9$. The solution set to the system consists of points that are outside the ellipse and inside the circle, as shown in the following graph.

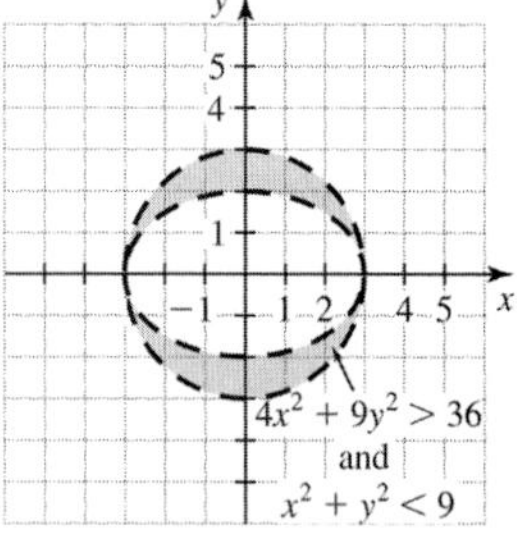

61. The equation $x^2 = y^2 + 1$ is the equation of a hyperbola because it could be written as $x^2 - y^2 = 1$.

63. The equation $x^2 = 1 - y^2$ is the equation of a circle because it could be written as $x^2 + y^2 = 1$.

65. The equation $x^2 + x = 1 - y^2$ is the equation of a circle because we could write $x^2 + x + y^2 = 1$, and then complete the square to get the standard equation for a circle.

67. The equation $x^2 + 4x = 6y - y^2$ is the equation of a circle because we could write it as $x^2 + 4x + y^2 - 6y = 0$, and then complete the squares for both x and y to get the standard equation of a circle.

69. The equation is the equation of a hyperbola in standard form.

71. The equation $4y^2 - x^2 = 8$ is the equation of a hyperbola because we could divide by 8 to get the standard equation $\frac{y^2}{2} - \frac{x^2}{8} = 1$.

73. Write $x^2 = 4 - y^2$ as $x^2 + y^2 = 4$ to see that it is the equation of a circle of radius 2 centered at the origin.

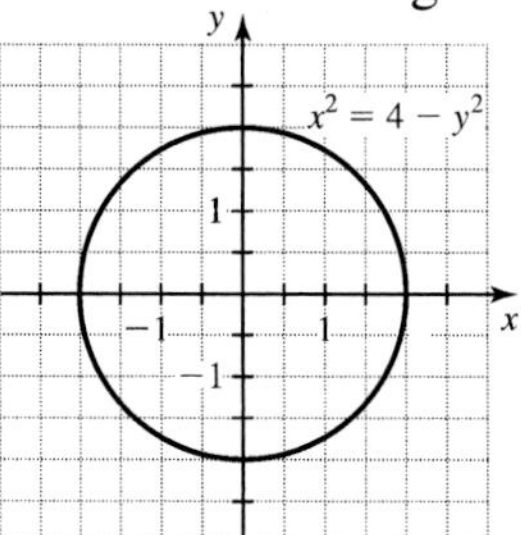

75. Write $x^2 = 4y + 4$ as $y = \frac{1}{4}x^2 - 1$ to see that it is the equation of a parabola opening upward with vertex at (0, −1).

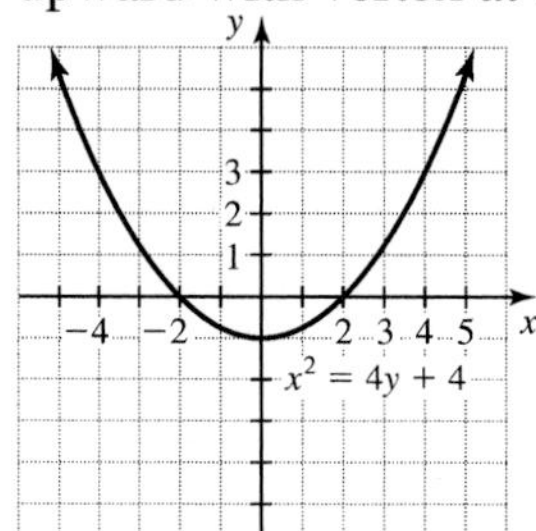

77. Write $x^2 = 4 - 4y^2$ as $\frac{x^2}{4} + y^2 = 1$ to see that it is the equation of an ellipse centered at (0, 0) and passing through (0, −1), (0, 1), $(-2, 0)$, and $(2, 0)$.

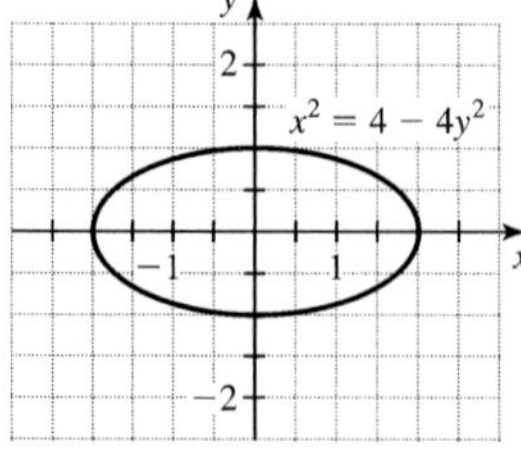

79. Write $x^2 = 4 - (y - 4)^2$ as $x^2 + (y - 4)^2 = 4$ to see that it is the equation of a circle of radius 2 centered at (0, 4).

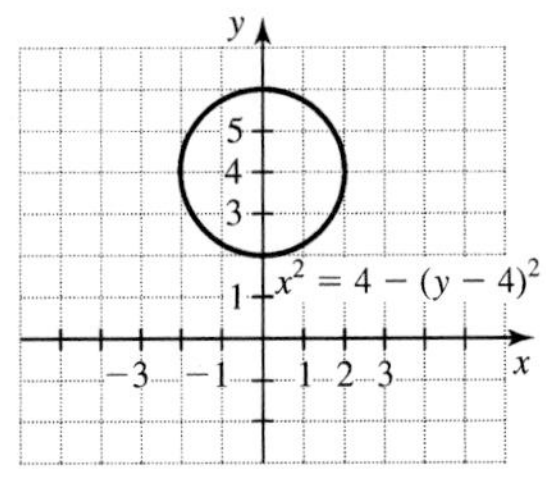

81. The radius of the circle is the distance between (0, 0) and (3, 4).

$$r = \sqrt{(3-0)^2 + (4-0)^2} = \sqrt{9+16}$$
$$= \sqrt{25} = 5$$

The equation of the circle centered at (0, 0) with radius 5 is $x^2 + y^2 = 25$.

83. Use the center $(-1, 5)$ and radius 6 in the form $(x - h)^2 + (y - k)^2 = r^2$ to get the equation $(x + 1)^2 + (y - 5)^2 = 36$.

85. The vertex is half way between the focus and directrix at (1, 3). Since the distance from (1, 3) to (1, 4) is 1, we have $p = 1$, and $a = 1/4$. So the equation is

$$y = \frac{1}{4}(x-1)^2 + 3.$$

87. Since the vertex is below the focus, $p = 1/4$ and $a = 1$. The equation is $y = 1(x - 0)^2 + 0$ or $y = x^2$.

89. Since the vertex is (0, 0) the parabola has equation $y = ax^2$. Since (3, 2) is on the parabola, $2 = a \cdot 3^2$, or $a = 2/9$. The equation is $y = \frac{2}{9}x^2$.

91. Substitute $y = -x + 1$ into $x^2 + y^2 = 25$ to eliminate y.

$$x^2 + (-x+1)^2 = 25$$
$$x^2 + x^2 - 2x + 1 = 25$$
$$2x^2 - 2x - 24 = 0$$
$$x^2 - x - 12 = 0$$
$$(x-4)(x+3) = 0$$
$$x = 4 \quad \text{or} \quad x = -3$$
$$y = -3 \qquad y = 4 \quad \text{Since } y = -x + 1$$

The solution set is $\{(4, -3), (-3, 4)\}$

93. If we add the two equations to eliminate y, we get the equation $5x^2 = 25$. Solving this equation gives $x^2 = 5$ or $x = \pm\sqrt{5}$. Use $x^2 = 5$ in the first equation.

$$4(5) + y^2 = 4$$
$$y^2 = -16$$

There are no real numbers that satisfy the system of equations. The solution set is $\emptyset$.

95. Let $x =$ the length and $y =$ the width. We can write the following 2 equations.

$$2x + 2y = 16$$
$$xy = 12$$

The first equation can be written as $y = 8 - x$. Substitute $y = 8 - x$ into the second equation.

$$x(8-x) = 12$$
$$-x^2 + 8x - 12 = 0$$
$$x^2 - 8x + 12 = 0$$
$$(x-6)(x-2) = 0$$
$$x = 6 \quad \text{or} \quad x = 2$$
$$y = 2 \qquad y = 6 \quad \text{Since } y = 8 - x$$

So the length is 6 feet and the width is 2 feet.

CHAPTER 11 TEST

1. The graph of $x^2 + y^2 = 25$ is a circle of radius 5 centered at (0, 0).

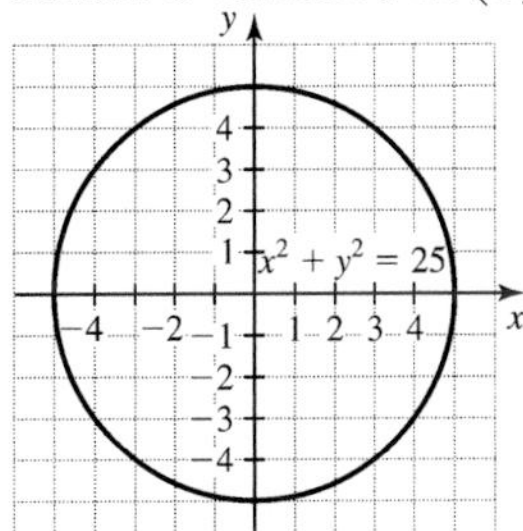

2. The graph of $\frac{x^2}{16} - \frac{y^2}{25} = 1$ is a hyperbola centered at the origin. the x-intercepts are $(-4, 0)$ and $(4, 0)$. The fundamental rectangle passes through the x-intercepts, $(0, -5)$, and $(0, 5)$. Extend the diagonals of the fundamental rectangle to obtain the asymptotes. The hyperbola opens to the left and right.

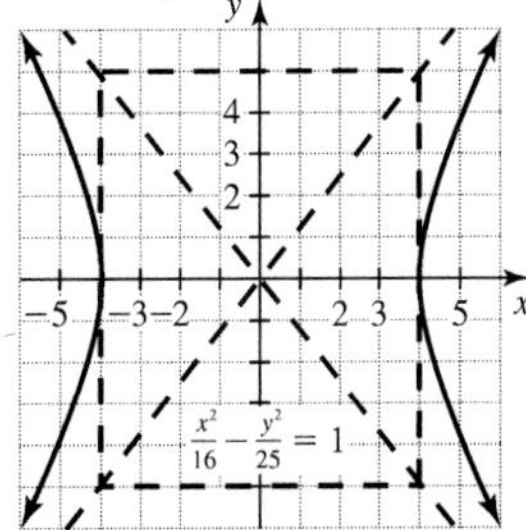

3. The graph of $y^2 + 4x^2 = 4$ is an ellipse with x-intercepts at $(-1, 0)$ and $(1, 0)$. Its y-intercepts are $(0, -2)$ and $(0, 2)$.

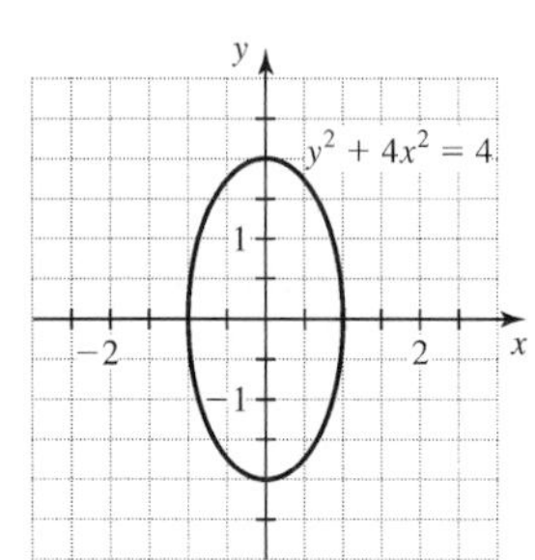

4. The graph of $y = x^2 + 4x + 4$ is a parabola opening upward with y-intercept (0, 4). We can write this equation as $y = (x + 2)^2$. In this form we see that the vertex is $(-2, 0)$.

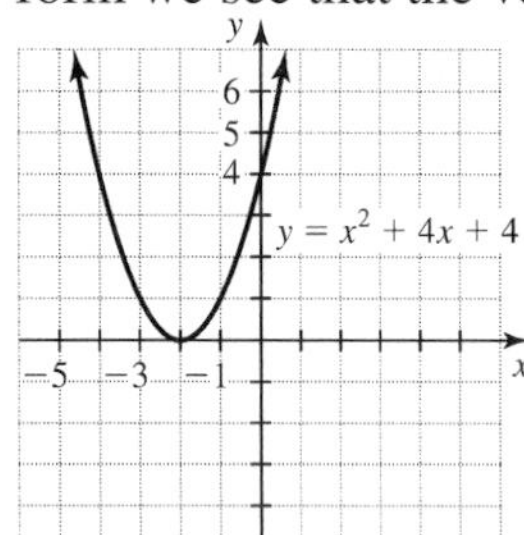

5. Write $y^2 - 4x^2 = 4$ as $\frac{y^2}{4} - x^2 = 1$. The graph is a hyperbola with y-intercepts at $(0, -2)$ and $(0, 2)$. The fundamental rectangle passes through the y-intercepts, $(-1, 0)$, and $(1, 0)$. Extend the diagonals to get the asymptotes. The hyperbola opens up and down.

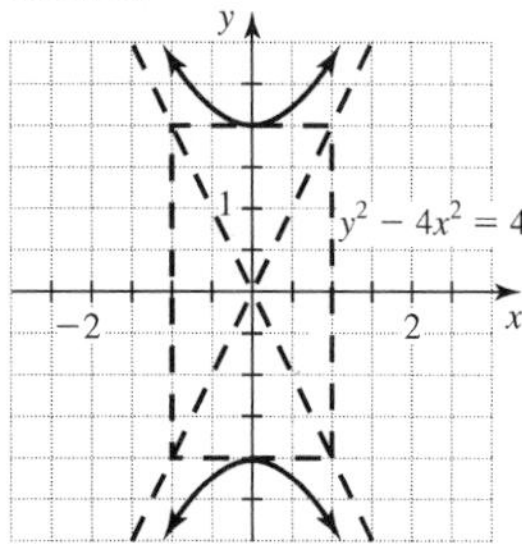

6. The graph of $y = -x^2 - 2x + 3$ is a parabola opening downward. The vertex is at $(-1, 4)$. The y-intercept is (0, 3). The x-intercepts are $(-3, 0)$ and $(1, 0)$.

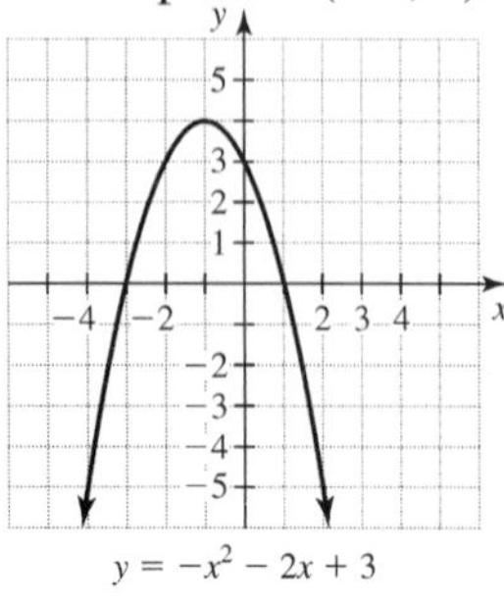

7. First graph the hyperbola $\frac{x^2}{9} - \frac{y^2}{9} = 1$. Since (0, 0) satisfies the inequality, the region containing (0, 0) is shaded.

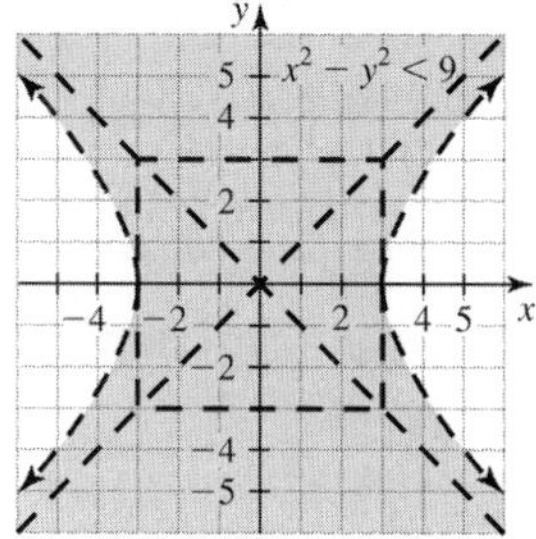

8. First graph the circle $x^2 + y^2 = 9$, centered at (0, 0) with radius 3. Since (0, 0) fails to satisfy $x^2 + y^2 > 9$, we shade the region outside the circle.

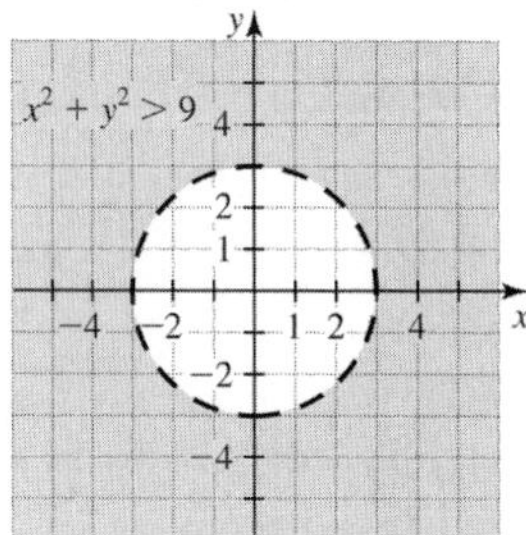

9. The graph of $y = x^2 - 9$ is a parabola opening upward. Its vertex is $(0, -9)$. To find its x-intercepts solve $x^2 - 9 = 0$. The x-intercepts are $(-3, 0)$ and $(3, 0)$. The graph of $y > x^2 - 9$ is the region containing the origin because (0, 0) satisfies the inequality.

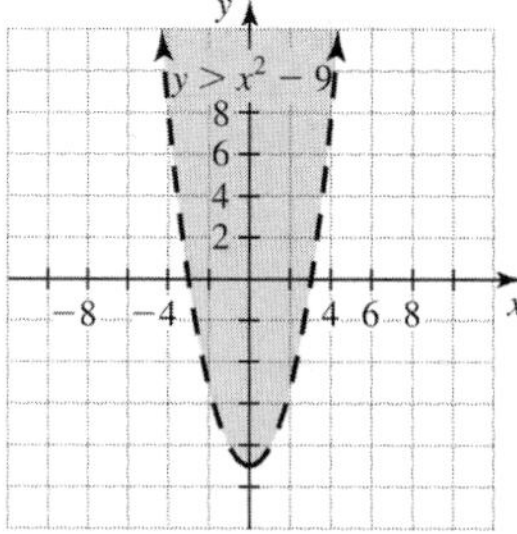

10. The graph of $x^2 + y^2 < 9$ is the region inside the circle $x^2 + y^2 = 9$. To find the graph of $x^2 - y^2 > 1$ first graph the hyperbola $x^2 - y^2 = 1$. By testing points, we can see that the two regions not containing the origin satisfy $x^2 - y^2 > 1$. The points in these two regions that are also inside the circle are the points that satisfy both inequalities of the system.

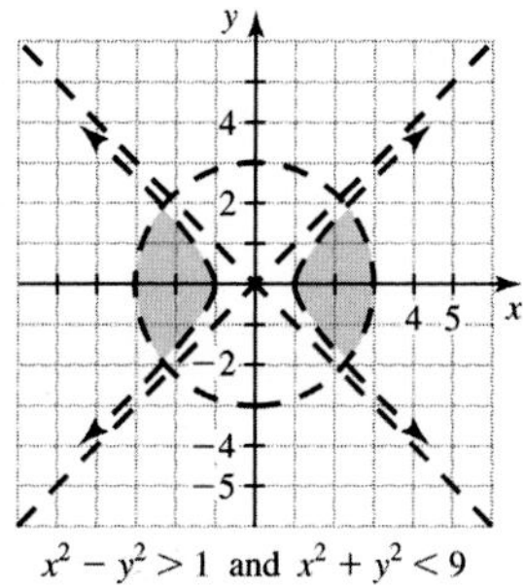

$x^2 - y^2 > 1$ and $x^2 + y^2 < 9$

11. The graph of $y = -x^2 + x$ is a parabola opening downward. Points below this parabola satisfy $y < -x^2 + x$. The graph of $y = x - 4$ is a line through $(0, -4)$ with slope 1. Points below this line satisfy $y < x - 4$. The solution set to the system of inequalities consists of points below the parabola and below the line.

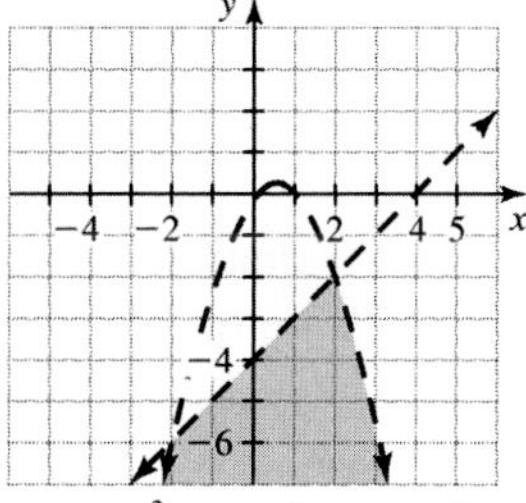

$y < -x^2 + x$ and $y < x - 4$

12. Use substitution to eliminate y.

$$x^2 - 2x - 8 = 7 - 4x$$
$$x^2 + 2x - 15 = 0$$
$$(x + 5)(x - 3) = 0$$
$$x = -5 \text{ or } x = 3$$
$$y = 27 \qquad y = -5 \text{ Since } y = 7 - 4x$$

The solution set is $\{(-5, 27), (3, -5)\}$.

13. Substitute $x^2 = y$ into $x^2 + y^2 = 12$.

$$y + y^2 = 12$$
$$y^2 + y - 12 = 0$$
$$(y + 4)(y - 3) = 0$$
$$y = -4 \text{ or } y = 3$$

If $y = -4$, the $x^2 = -4$ has no solution. If $y = 3$, then $x^2 = 3$ or $x = \pm\sqrt{3}$. The solution set is $\{(\sqrt{3}, 3), (-\sqrt{3}, 3)\}$.

14. $\sqrt{(-1 - 1)^2 + (4 - 6)^2} = \sqrt{4 + 4}$

$$= \sqrt{8} = 2\sqrt{2}$$

15. Use the midpoint formula:

$$\left(\frac{2 + (-3)}{2}, \frac{0 + (-1)}{2}\right) = \left(-\frac{1}{2}, -\frac{1}{2}\right)$$

Use the distance formula to find the length:

$$\sqrt{(2 - (-3))^2 + (0 - (-1))^2} = \sqrt{25 + 1}$$
$$= \sqrt{26}$$

16. Complete the square to get the equation into the standard form.

$$x^2 + 2x + \quad y^2 + 10y \quad = 10$$
$$x^2 + 2x + 1 + y^2 + 10y + 25 = 10 + 1 + 25$$
$$(x + 1)^2 + (y + 5)^2 = 36$$

The center is $(-1, -5)$ and the radius is 6.

17. For $y = x^2 + x + 3$ the x-coordinate of the vertex is

$$x = \frac{-b}{2a} = \frac{-1}{2(1)} = -\frac{1}{2}.$$

Use $x = -1/2$ in $y = x^2 + x + 3$, to get the vertex $(-\frac{1}{2}, \frac{11}{4})$. Since $a = 1$, we have $p = 1/4$. The focus is $(-\frac{1}{2}, 3)$ and the directrix is $y = \frac{5}{2}$. The axis of symmetry is $x = -\frac{1}{2}$ and the parabola opens up.

18.
$$y = \frac{1}{2}x^2 - 3x - \frac{1}{2}$$
$$y = \frac{1}{2}(x^2 - 6x) - \frac{1}{2}$$
$$y = \frac{1}{2}(x^2 - 6x + 9 - 9) - \frac{1}{2}$$
$$y = \frac{1}{2}(x^2 - 6x + 9) - \frac{9}{2} - \frac{1}{2}$$
$$y = \frac{1}{2}(x - 3)^2 - 5$$

19. The radius of the circle is the distance between $(-1, 3)$ and $(2, 5)$.

$$r = \sqrt{(-1 - 2)^2 + (3 - 5)^2} = \sqrt{9 + 4}$$
$$= \sqrt{13}$$

The equation of the circle with center $(-1, 3)$ and radius $\sqrt{13}$ is $(x + 1)^2 + (y - 3)^2 = 13$.

20. Let $x =$ the length and $y =$ the width. We can write the following two equations.

$$2x + 2y = 42$$
$$xy = 108$$

If we solve $2x + 2y = 42$ for y, we get $y = 21 - x$.

$$x(21 - x) = 108$$
$$-x^2 + 21x - 108 = 0$$
$$x^2 - 21x + 108 = 0$$
$$(x - 9)(x - 12) = 0$$
$$x = 9 \text{ or } x = 12$$
$$y = 12 \qquad y = 9 \text{ Since } y = 21 - x$$

The length is 12 feet and the width is 9 feet.

Making Connections

Chapters 1 - 11

1. The graph of $y = 9x - x^2$ is a parabola that opens downward. Solve $9x - x^2 = 0$ to find the x-intercepts.

$$x(9 - x) = 0$$
$$x = 0 \quad \text{or} \quad x = 9$$

The x-intercepts are $(0, 0)$ and $(9, 0)$. The x-coordinate of the vertex is

$x = \frac{-(9)}{2(-1)} = \frac{9}{2}$.

The vertex is $(9/2, 81/4)$.

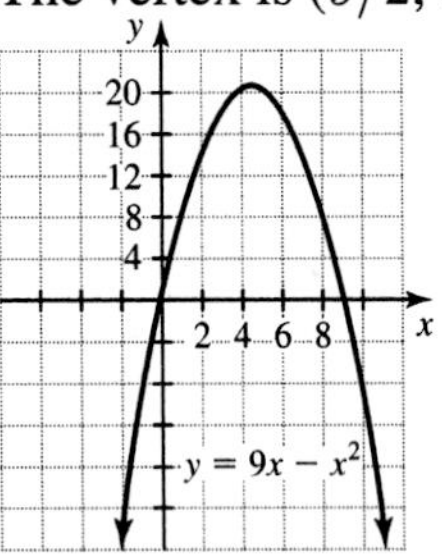

2. The graph of $y = 9x$ is a line through $(0, 0)$ with slope 9.

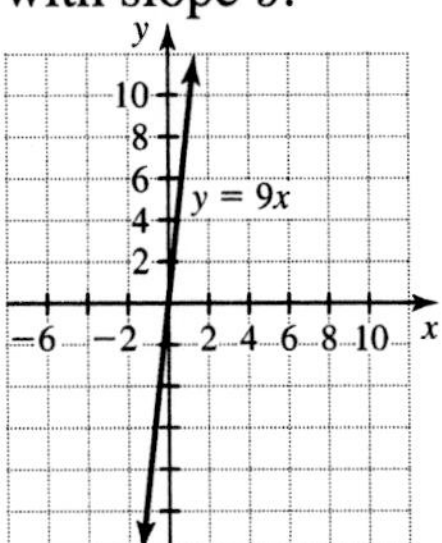

3. The graph of $y = (x - 9)^2$ is a parabola opening upward, with vertex at $(9, 0)$.

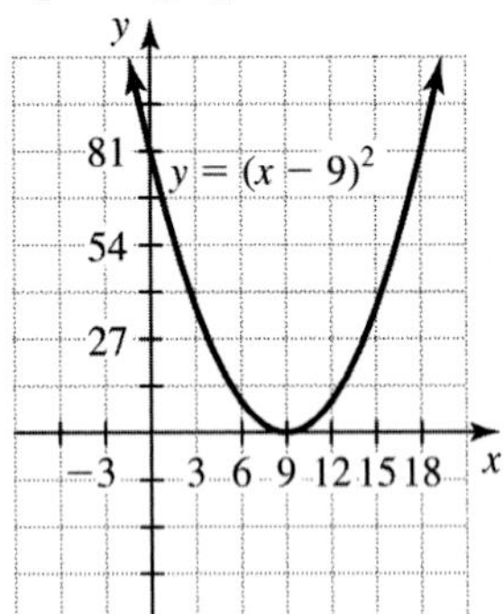

4. Write $y^2 = 9 - x^2$ as $x^2 + y^2 = 9$ to see that it is the equation of a circle of radius 3 centered at the origin.

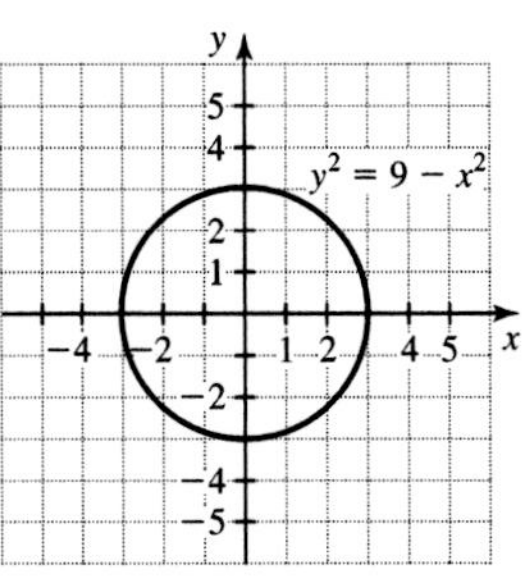

5. The graph of $y = 9x^2$ is a parabola opening upward with vertex at $(0, 0)$.

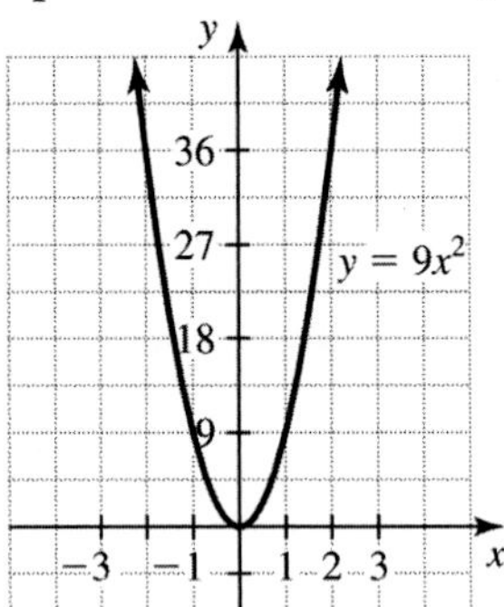

6. The graph of $y = |\, 9x \,|$ is v-shaped. It contains the points $(0, 0)$, $(-1, 9)$, and $(1, 9)$.

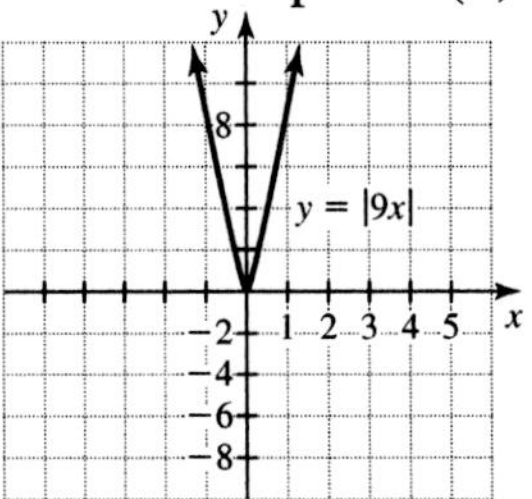

7. Write $4x^2 + 9y^2 = 36$ as $\frac{x^2}{9} + \frac{y^2}{4} = 1$ to see that it is the equation of an ellipse through $(-3, 0)$, $(3, 0)$, $(0, 2)$, and $(0, -2)$.

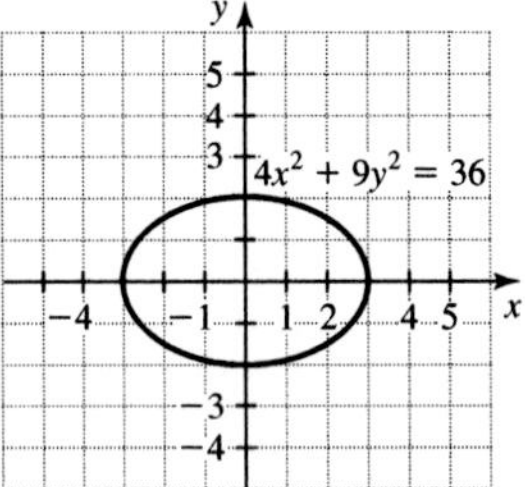

8. Write $4x^2 - 9y^2 = 36$ as $\frac{x^2}{9} - \frac{y^2}{4} = 1$ to see that it is the equation of a hyperbola with x-intercepts $(-3, 0)$ and $(3, 0)$. The fundamental rectangle goes through the x-intercepts, $(0, -2)$, and $(0, 2)$. Extend the diagonals of the rectangle to get the asymptotes.

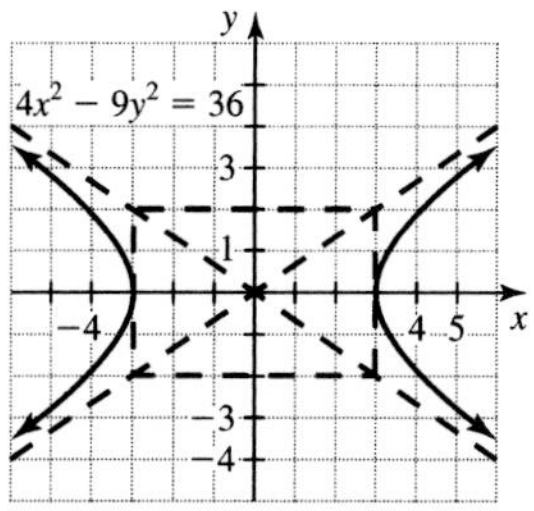

9. The graph of $y = 9 - x$ is a line through (0, 9) with slope -1.

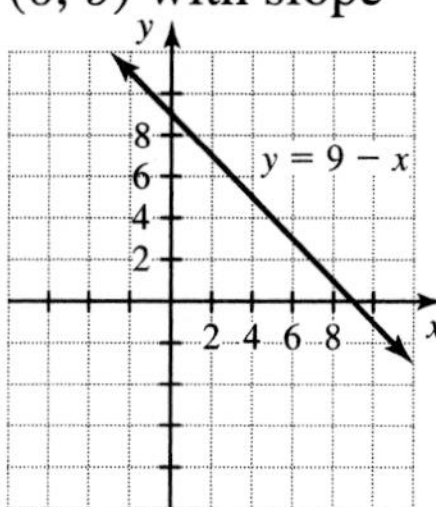

10. The graph of $y = 9^x$ goes through (0, 1), (1, 9), and (-1, 1/9).

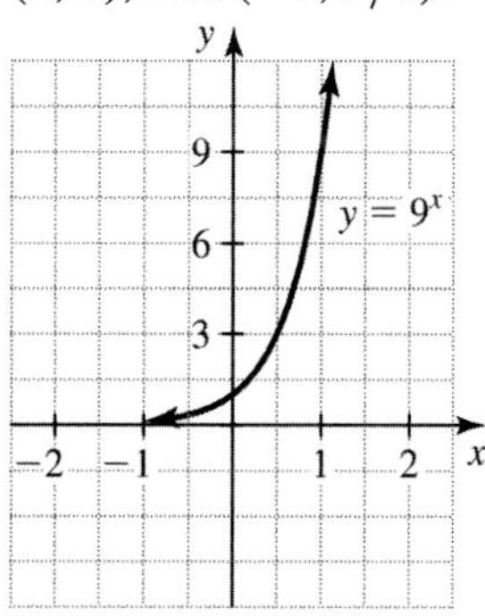

11. $(x + 2y)^2 = x^2 + 2(x)(2y) + (2y)^2$
$= x^2 + 4xy + 4y^2$

12. $(x + y)(x^2 + 2xy + y^2)$
$= x(x^2 + 2xy + y^2) + y(x^2 + 2xy + y^2)$
$= x^3 + 2x^2y + xy^2 + x^2y + 2xy^2 + y^3$
$= x^3 + 3x^2y + 3xy^2 + y^3$

13. $(a + b)^3 = (a + b)(a^2 + 2ab + b^2)$
$= a(a^2 + 2ab + b^2) + b(a^2 + 2ab + b^2)$
$= a^3 + 2a^2b = ab^2 + a^2b + 2ab^2 + b^3$
$= a^3 + 3a^2b + 3ab^2 + b^3$

14. $(a - 3b)^2 = a^2 - 2(a)(3b) + (3b)^2$
$= a^2 - 6ab + 9b^2$

15. $(2a + 1)(3a - 5) = 6a^2 + 3a - 10a - 5$
$= 6a^2 - 7a - 5$

16. $(x - y)(x^2 + xy + y^2)$
$= x(x^2 + xy + y^2) - y(x^2 + xy + y^2)$
$= x^3 + x^2y + xy^2 - x^2y - xy^2 - y^3$
$= x^3 - y^3$

17. Multiply the second equation by -2 and add the result to the first equation.

$$\begin{aligned} 2x - 3y &= -4 \\ -2x - 4y &= -10 \\ \hline -7y &= -14 \\ y &= 2 \end{aligned}$$

Use $y = 2$ in $x + 2y = 5$ to find x.

$$x + 2(2) = 5$$
$$x = 1$$

The solution set is $\{(1, 2)\}$.

18. Substitute $y = 7 - x$ into $x^2 + y^2 = 25$.

$$x^2 + (7 - x)^2 = 25$$
$$x^2 + 49 - 14x + x^2 = 25$$
$$2x^2 - 14x + 24 = 0$$
$$x^2 - 7x + 12 = 0$$
$$(x - 3)(x - 4) = 0$$
$$x = 3 \quad \text{or} \quad x = 4$$
$$y = 4 \qquad y = 3 \quad \text{Since } y = 7 - x$$

The solution set is $\{(3, 4), (4, 3)\}$.

19. Adding the first and second equations to eliminate z, we get $3x - 3y = 9$. Adding the second and third equations to eliminate z, we get $2x - y = 4$. Divide $3x - 3y = 9$ by -3 and add the result to $2x - y = 4$.

$$\begin{aligned} -x + y &= -3 \\ 2x - y &= 4 \\ \hline x &= 1 \end{aligned}$$

Use $x = 1$ in $-x + y = -3$.

$$-1 + y = -3$$
$$y = -2$$

Use $x = 1$ and $y = -2$ in $x + y + z = 2$.

$$1 + (-2) + z = 2$$
$$z = 3$$

The solution set is $\{(1, -2, 3)\}$.

20. Substitute $y = x^2$ into $y - 2x = 3$.

$$x^2 - 2x = 3$$
$$x^2 - 2x - 3 = 0$$
$$(x - 3)(x + 1) = 0$$
$$x = 3 \quad \text{or} \quad x = -1$$
$$y = 9 \qquad y = 1 \quad \text{Since } y = x^2$$

The solution set is $\{(-1, 1), (3, 9)\}$.

21. $ax + b = 0$
$ax = -b$
$x = -\dfrac{b}{a}$

22. Use the quadratic formula with $a = w$, $b = d$, and $c = m$.

$$x = \frac{-d \pm \sqrt{d^2 - 4wm}}{2w}$$

23.
$$A = \tfrac{1}{2}h(B + b)$$
$$2A = h(B + b)$$
$$2A = hB + hb$$

$$2A - bh = hB$$
$$B = \frac{2A - bh}{h}$$

24. $\frac{1}{x} + \frac{1}{y} = \frac{1}{2}$

$$2xy\left(\frac{1}{x} + \frac{1}{y}\right) = 2xy\left(\frac{1}{2}\right)$$
$$2y + 2x = xy$$
$$2y = xy - 2x$$
$$2y = x(y - 2)$$
$$x = \frac{2y}{y - 2}$$

25. $L = m + mxt$

$$L = m(1 + xt)$$
$$m = \frac{L}{1 + xt}$$

26. $y = 3a\sqrt{t}$

$$(y)^2 = (3a\sqrt{t})^2$$
$$y^2 = 9a^2t$$
$$\frac{y^2}{9a^2} = t$$
$$t = \frac{y^2}{9a^2}$$

27. First find the slope.

$$m = \frac{-3 - 1}{2 - (-4)} = \frac{-4}{6} = -\frac{2}{3}$$

Use point-slope form with (2, −3).

$$y - (-3) = -\frac{2}{3}(x - 2)$$
$$y + 3 = -\frac{2}{3}x + \frac{4}{3}$$
$$y = -\frac{2}{3}x - \frac{5}{3}$$

28. Write $2x - 4y = 5$ as $y = \frac{1}{2}x - \frac{5}{4}$. The slope of any line perpendicular to this line is -2. The line through (0, 0) with slope -2 is $y = -2x$.

29. The radius is the distance between (2, 5) and (−1, −1).

$$r = \sqrt{(-1 - 2)^2 + (-1 - 5)^2} = \sqrt{9 + 36}$$
$$= \sqrt{45}$$

The equation of circle with center (2, 5) and radius $\sqrt{45}$ is $(x - 2)^2 + (y - 5)^2 = 45$.

30. Use completing the square to get the equation into standard form for a circle.

$$x^2 + 3x + y^2 - 6y = 0$$
$$x^2 + 3x + \frac{9}{4} + y^2 - 6y + 9 = 0 + \frac{9}{4} + 9$$
$$(x + \tfrac{3}{2})^2 + (y - 3)^2 = \frac{45}{4}$$

The center is $(-\frac{3}{2}, 3)$ and radius is $\sqrt{\frac{45}{4}}$, or $\frac{3\sqrt{5}}{2}$.

31. $2i(3 + 5i) = 6i + 10i^2 = -10 + 6i$

32. $i^6 = i^4 \cdot i^2 = 1(-1) = -1$

33. $(2i - 3) + (6 - 7i) = 3 - 5i$

34. $(3 + i\sqrt{2})^2 = 9 + 6i\sqrt{2} + 2i^2$
$= 7 + 6i\sqrt{2}$

35. $(2 - 3i)(5 - 6i) = 10 - 15i - 12i + 18i^2$
$= -8 - 27i$

36. $(3 - i) + (-6 + 4i) = -3 + 3i$

37. $(5 - 2i)(5 + 2i) = 25 - 4i^2 = 29$

38. $\frac{2 - 3i}{2i} = \frac{(2 - 3i)(-i)}{2i(-i)} = \frac{-2i + 3i^2}{2(1)}$
$= -\frac{3}{2} - i$

39. $\frac{(4 + 5i)(1 + i)}{(1 - i)(1 + i)} = \frac{4 + 9i + 5i^2}{2}$
$= \frac{-1 + 9i}{2} = -\frac{1}{2} + \frac{9}{2}i$

40. $\frac{4 - \sqrt{-8}}{2} = \frac{4 - 2i\sqrt{2}}{2} = 2 - i\sqrt{2}$

41. a) $m = \frac{250 - 200}{0.30 - 0.40} = \frac{50}{-0.1} = -500$

$$q - 250 = -500(x - 0.30)$$
$$q - 250 = -500x + 150$$
$$q = -500x + 400$$

b) $R = qx = (-500x + 400)x$
$R = -500x^2 + 400x$

c) The graph is a parabola through $(0, 0)$,
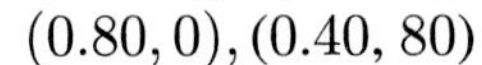
(0.80, 0), (0.40, 80)

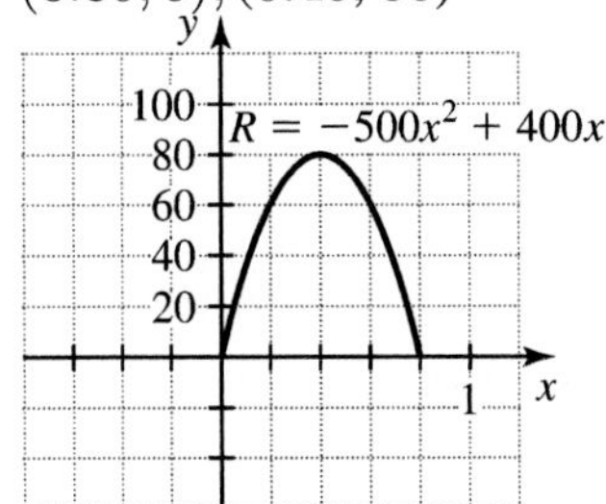

d) The maximum revenue occurs at the vertex of the parabola.

$$x = \frac{-400}{2(-500)} = 0.40$$

The maximum revenue occurs when bananas are \$0.40 per pound.

e) To find the maximum revenue use the formula for revenue with $x = 0.40$.
$R = -500(0.40)^2 + 400(0.40) = 80$
The maximum revenue is \$80 when the bananas are \$0.40 per pound.

12.1 WARM-UPS

1. True, because the formula $a_n = 2n$ will produce even numbers when n is a positive integer.
2. True, because the formula $a_n = 2n - 1$ will produce odd numbers when n is a positive integer.
3. True, by the definition of sequence.
4. False, because the domain of a finite sequence is the set of positive integers less than or equal to some fixed positive integer.
5. False, because if $n = 1$, $a_1 = (-1)^2 \cdot 1^2 = 1$.
6. False, because the independent variable is n.
7. True.
8. False, because the 6th term is $a_6 = (-1)^7 2^6 = -64$.
9. True.
10. True, because the sequence is $a_n = 2^n$ and $a_{10} = 2^{10} = 1024$.

12.1 EXERCISES

1. A sequence is a list of numbers.
3. A finite sequence is a function whose domain is the set of positive integers less than or equal to some fixed positive integer.
5. The terms of the sequence a_n are found by multiplying the integers from 1 through 5 by 2, because of the formula $a_n = 2n$. So the terms are $2, 4, 6, 8,$ and 10.
7. The terms of the sequence a_n are found by squaring the integers from 1 through 8, because of the formula $a_n = n^2$. So the terms are 1, 4, 9, 16, 25, 36, 49, and 64.
9. $b_1 = \frac{(-1)^1}{1} = -1$, $b_2 = \frac{(-1)^2}{2} = \frac{1}{2}$,
$b_3 = \frac{(-1)^3}{3} = -\frac{1}{3}$, $b_4 = \frac{(-1)^4}{4} = \frac{1}{4}$,
$b_5 = \frac{(-1)^5}{5} = -\frac{1}{5}$, etc.
The 10 terms of the sequence are
$-1, \frac{1}{2}, -\frac{1}{3}, \frac{1}{4}, -\frac{1}{5}, \frac{1}{6}, -\frac{1}{7}, \frac{1}{8}, -\frac{1}{9}$, and $\frac{1}{10}$.

11. $c_1 = (-2)^{1-1} = 1$, $c_2 = (-2)^{2-1} = -2$,
$c_3 = (-2)^{3-1} = 4$, $c_4 = (-2)^{4-1} = -8$,
$c_5 = (-2)^{5-1} = 16$
The five terms are 1, -2, 4, -8, and 16.
13. $a_1 = 2^{-1} = \frac{1}{2}$, $a_2 = 2^{-2} = \frac{1}{4}$,
$a_3 = 2^{-3} = \frac{1}{8}$, $a_4 = 2^{-4} = \frac{1}{16}$, etc.
The six terms are $\frac{1}{2}, \frac{1}{4}, \frac{1}{8}, \frac{1}{16}, \frac{1}{32}$, and $\frac{1}{64}$.
15. $b_1 = 2(1) - 3 = -1$, $b_2 = 2(2) - 3 = 1$,
$b_3 = 2(3) - 3 = 3$, $b_4 = 2(4) - 3 = 5$,
$b_5 = 2(5) - 3 = 7$, $b_6 = 2(6) - 3 = 9$,
$b_7 = 2(7) - 3 = 11$
The seven terms are -1, 1, 3, 5, 7, 9, and 11.
17. $c_1 = 1^{-1/2} = 1$, $c_2 = 2^{-1/2} = \frac{1}{\sqrt{2}}$,
$c_3 = 3^{-1/2} = \frac{1}{\sqrt{3}} = \frac{\sqrt{3}}{3}$, $c_4 = 4^{-1/2} = \frac{1}{2}$,
$c_5 = 5^{-1/2} = \frac{1}{\sqrt{5}} = \frac{\sqrt{5}}{5}$
The five terms are 1, $\frac{\sqrt{2}}{2}, \frac{\sqrt{3}}{3}, \frac{1}{2}$, and $\frac{\sqrt{5}}{5}$.
19. $a_1 = \frac{1}{1^2 + 1} = \frac{1}{2}$, $a_2 = \frac{1}{2^2 + 2} = \frac{1}{6}$,
$a_3 = \frac{1}{3^2 + 3} = \frac{1}{12}$, $a_4 = \frac{1}{4^2 + 4} = \frac{1}{20}$
The first four terms are $\frac{1}{2}, \frac{1}{6}, \frac{1}{12}$, and $\frac{1}{20}$.
21. $b_1 = \frac{1}{2(1) - 5} = -\frac{1}{3}$,
$b_2 = \frac{1}{2(2) - 5} = -1$,
$b_3 = \frac{1}{2(3) - 5} = 1$, $b_4 = \frac{1}{2(4) - 5} = \frac{1}{3}$
The first four terms are $-\frac{1}{3}$, -1, 1, and $\frac{1}{3}$.
23. $c_1 = (-1)^1(1 - 2)^2 = -1$,
$c_2 = (-1)^2(2 - 2)^2 = 0$,
$c_3 = (-1)^3(3 - 2)^2 = -1$,
$c_4 = (-1)^4(4 - 2)^2 = 4$
The first four terms are -1, 0, -1, and 4.
25. $a_1 = \frac{(-1)^{2(1)}}{1^2} = 1$, $a_2 = \frac{(-1)^{2(2)}}{2^2} = \frac{1}{4}$,
$a_3 = \frac{(-1)^{2(3)}}{3^2} = \frac{1}{9}$, $a_4 = \frac{(-1)^{2(4)}}{4^2} = \frac{1}{16}$
The first four terms are 1, $\frac{1}{4}, \frac{1}{9}$, and $\frac{1}{16}$.
27. This sequence is a sequence of odd integers starting at 1. Even integers are represented by $2n$ and odd integers are all one less than an even integer. Try the formula $a_n = 2n - 1$. To see that it is correct, find a_1 through a_5.
29. A sequence of alternating ones and negative ones is obtained by using a power of -1. Since we want a_1 to be positive when $n = 1$, we use the $n + 1$ power on -1. A

formula for the general term is
$a_n = (-1)^{n+1}$.

31. This sequence is a sequence of even integers starting at 0. Even integers can be generated by $2n$, but to make the first one 0 (when $n = 1$), we use the formula
$a_n = 2n - 2$.

33. This sequence is a sequence consisting of the positive integral multiples of 3. Try the formula $a_n = 3n$ to see that it generates the appropriate numbers.

35. Each term in this sequence is one larger than the corresponding term in the sequence of Exercise 31. Try the formula $a_n = 3n + 1$ to see that it generates the given sequence.

37. To get the alternating signs on the terms of the sequence, we use a power of -1. Next we observe that the numbers 1, 2, 4, 8, 16 are powers of 2. The formula $a_n = (-1)^n 2^{n-1}$ will produce the given sequence. Remember that n always starts at 1, so in this case we use $n - 1$ as the power of 2 to get $2^0 = 1$.

39. Notice that the numbers 0, 1, 4, 9, 16, . . . are the squares of the nonnegative integers: 0^2, 1^2, 2^2, 3^2, 4^2,.... We would use n^2 to generate squares, but since n starts at 1, we use
$a_n = (n - 1)^2$ to get the given sequence.

41. After the first penalty the ball is on the 4-yard line. After the second penalty the ball is on the 2-yard line. After the third penalty the ball is on the 1-yard line, and so on. The sequence of five terms is 4, 2, 1, $\frac{1}{2}$, $\frac{1}{4}$.

43. To find the amount of increase, we take 5% of $43,568 to get $2178 increase. We could just multiply $43,568 by 1.05 to obtain the new price. In either case the price the next year is $45,746. To find the price for the next year we multiply the last year's price by 1.05: 1.05($45,746) = $48,034. Repeating this process gives us the prices $45,746, $48,034, $50,435, $52,957, and $55,605.

45. Of the $1 million, 80% is respent in the community. So $800,000 is respent. Of the $800,000 we have 80% respent in the community. So $640,000 is respent in the community, and so on. The first four terms of the sequence are $1,000,000, $800,000, $640,000, $512,000.

47. Possible vertical repeats are 27 in., 13.5 in., 9 in., 6.75 in., and 5.4 in.

49. If we put the word great in front of the word grandparents 35 times, then we have 2^{37} of this type of relative. Use a calculator to find that $2^{37} \approx 137{,}438{,}953{,}472$. This is certainly larger than the present population of the earth.

53. a) Use a calculator to find
$$a_{100} = (0.999)^{100} = 0.9048,$$
$$a_{1000} = (0.999)^{1000} = 0.3677,$$
and
$$a_{10000} = (0.999)^{10000} = 0.00004517.$$
b) a_n approaches zero as n gets larger and larger.

12.2 WARM-UPS

1. True, because of the definition of series.
2. False, the sum of a series can be any real number.
3. False, there are 9 terms in 2^3 through 10^3.
4. False, because the terms in the first series are opposites of the terms in the second series.
5. False, the ninth term is
$$\frac{(-1)^9}{(9+1)(9+2)} = -\frac{1}{110}.$$
6. True, because the terms are -2 and 4 and the sum is 2.
7. True, because of the distributive property.
8. True, because the notation indicates to add the number 4 five times.
9. True, because $2i + 7i = 9i$.
10. False, because in the series on the left side the 1 is added in three times and in the series on the right side the 1 is only added in once.

12.2 EXERCISES

1. Summation notation provides a way of writing a sum without writing out all of the terms.
3. A series is the indicated sum of the terms of a sequence.

5. $\sum_{i=1}^{5} i = 1 + 2 + 3 + 4 + 5 = 15$

7. $\sum_{i=1}^{4} i^2 = 1^2 + 2^2 + 3^2 + 4^2$
$= 1 + 4 + 9 + 16 = 30$

9. $\sum_{j=0}^{5}(2j - 1) = (2 \cdot 0 - 1) + (2 \cdot 1 - 1)$
$+ (2 \cdot 2 - 1) + (2 \cdot 3 - 1)$
$+ (2 \cdot 4 - 1) + (2 \cdot 5 - 1)$
$= -1 + 1 + 3 + 5 + 7 + 9 = 24$

11. $\sum_{i=1}^{5} 2^{-i} = 2^{-1} + 2^{-2} + 2^{-3} + 2^{-4} + 2^{-5}$
$= \frac{1}{2} + \frac{1}{4} + \frac{1}{8} + \frac{1}{16} + \frac{1}{32} = \frac{31}{32}$

13. $\sum_{i=1}^{10} 5i^0 = 5(1)^0 + 5(2)^0 + 5(3)^0 + 5(4)^0$
$+ 5(5)^0 + 5(6)^0 + 5(7)^0$
$+ 5(8)^0 + 5(9)^0 + 5(10)^0$
$= 5 + 5 + 5 + 5 + 5 + 5 + 5 + 5 + 5 + 5$
$= 50$

15. $\sum_{i=1}^{3}(i - 3)(i + 1) = (1 - 3)(1 + 1)$
$+ (2 - 3)(2 + 1) + (3 - 3)(3 + 1)$
$= -4 + (-3) + 0 = -7$

17. $\sum_{j=1}^{10}(-1)^j$
$= (-1)^1 + (-1)^2 + ... + (-1)^{10}$
$= -1 + 1 - 1 + 1 - 1 + 1 - 1$
$+ 1 - 1 + 1$
$= 0$

19. The sum of the first six positive integers is written as $\sum_{i=1}^{6} i$. There are other ways to indicate this series in summation notation but this is the simplest.

21. To get the signs of the terms to alternate from positive to negative we use a power of -1. To get odd integers we use the formula $2i - 1$. So this series is written $\sum_{i=1}^{6}(-1)^i(2i - 1)$.

23. This series consists of the squares of the first six positive integers. It is written in summation notation as $\sum_{i=1}^{6} i^2$.

25. This series consists of the reciprocals of the positive integers. If the index i goes from 1 to 4, we must use $2 + i$ to get the numbers 3 through 6. So the series is written $\sum_{i=1}^{4} \frac{1}{2 + i}$.

27. The terms of this series are logarithms of positive integers. If the index i goes from 1 to 3, we must use $i + 1$ to get the numbers 2 through 4. The series is written $\sum_{i=1}^{3} \ln(i + 1)$.

29. Since the subscripts range from 1 through 4, we let i range from 1 through 4: $\sum_{i=1}^{4} a_i$.

31. The subscripts on x range from 3 through 50, so i ranges from 1 through 48: $\sum_{i=1}^{48} x_{i+2}$.

33. The subscripts on w range from 1 through n, so i ranges from 1 through n: $\sum_{i=1}^{n} w_i$.

35. $\sum_{i=1}^{6} x^i = x + x^2 + x^3 + x^4 + x^5 + x^6$

37. $\sum_{j=0}^{3}(-1)^j x_j$
$= (-1)^0 x_0 + (-1)^1 x_1 + (-1)^2 x_2 + (-1)^3 x_3$
$= x_0 - x_1 + x_2 - x_3$

39. $\sum_{i=1}^{3} ix^i = 1x^1 + 2x^2 + 3x^3$
$= x + 2x^2 + 3x^3$

41. If $j = 0$ when $i = 1$, then $i = j + 1$ and j ranges from 0 through 4: $\sum_{j=0}^{4}(j + 1)^2$

43. If $j = 1$ when $i = 0$, then $i = j - 1$ and j ranges from 1 through 13. If we substitute $i = j - 1$ into $2i - 1$ we get $2i - 1 = 2(j - 1) - 1 = 2j - 3$. So the series is written $\sum_{j=1}^{13}(2j - 3)$.

45. If $j = 1$ when $i = 4$, then $i = j + 3$ and j ranges from 1 through 5. Substitute $j + 3$ for i to get $\sum_{j=1}^{5} \frac{1}{j + 3}$.

47. If $j = 0$ when $i = 1$, then $i = j + 1$ and j ranges from 0 through 3. The exponent $2i + 3$ becomes $2(j + 1) + 3 = 2j + 5$. The series is written as $\sum_{j=0}^{3} x^{2j+5}$.

49. If $j = 0$ when $i = 1$, then $i = j + 1$ and j ranges from 0 through $n - 1$. Replacing i by

$j+1$ gives us the series $\sum_{j=0}^{n-1} x^{j+1}$.

51. On the first jump he has moved $\frac{1}{2}$ yard. On the second jump he moves $\frac{1}{4}$ yard. On the third jump he moves $\frac{1}{8}$ yard, and so on. To express the reciprocals of the powers of 2, we use 2^{-i}. His total movement after nine jumps is expressed as the series $\sum_{i=1}^{9} 2^{-i}$.

53. $\sum_{i=1}^{4} 1{,}000{,}000(0.8)^{i-1}$.

55. A sequence is basically a list of numbers. A series is the indicated sum of the terms of a sequence.

12.3 WARM-UPS

1. False, because the common difference is -2 $(1-3=-2)$.
2. False, because the difference between consecutive terms is not constant.
3. False, because the difference between the consecutive terms is sometimes 2 and sometimes -2.
4. False, because the nth term is given by $a_n = a_1 + (n-1)d$.
5. False, because the second must be 7.5, making $d = 2.5$ and the fourth term 12.5.
6. False, because if the first is 6 and the third is 2, the second must be 4 in order to have a common difference.
7. True, because that is the definition of arithmetic series.
8. True, because the series is $5+7+9+11+13$ and the common difference is 2.
9. True, because of the formula for the sum of an arithmetic series.
10. False, because there are 11 even integers from 8 through 28 inclusive and the sum is $\frac{11}{2}(8+28)$.

12.3 EXERCISES

1. An arithmetic sequence is one in which each term after the first is obtained by adding a fixed amount to the previous term.
3. An arithmetic series is an indicated sum of an arithmetic sequence.
5. The common difference is $d=2$ and the first term is $a_1 = 2$. Use the formula $a_n = a_1 + (n-1)d$ to find that the nth term is $a_n = 2 + (n-1)2 = 2n$.
7. The common difference is $d=6$ and the first term is $a_1 = 0$. Use the formula $a_n = a_1 + (n-1)d$ to find that the nth term is $a_n = 0 + (n-1)6 = 6n - 6$.
9. The common difference is $d=5$ and the first term is 7. The nth term is $a_n = 7 + (n-1)5 = 5n + 2$.
11. The common difference is $d=2$ and the first term is $a_1 = -4$. Use the formula $a_n = a_1 + (n-1)d$ to find that the nth term is $a_n = -4 + (n-1)2 = 2n - 6$.
13. The common difference is $d = 1 - 5 = -4$ and the first term is $a_1 = 5$. The nth term is $a_n = 5 + (n-1)(-4) = -4n + 9$.
15. The common difference is $d = -9 - (-2) = -7$ and the first term is -2. The nth term is $a_n = -2 + (n-1)(-7) = -7n + 5$.
17. The common difference is $d = -2.5 - (-3) = 0.5$ and the first term is -3. The nth term is $a_n = -3 + (n-1)(0.5) = 0.5n - 3.5$.
19. The common difference is $d = -6.5 - (-6) = -0.5$ and the first term is -6. The nth term is $a_n = -6 + (n-1)(-0.5) = -0.5n - 5.5$.
21. $a_1 = 9 + (1-1)4 = 9$,
$a_2 = 9 + (2-1)4 = 13$,
$a_3 = 9 + (3-1)4 = 17$,
$a_4 = 9 + (4-1)4 = 21$,
$a_5 = 9 + (5-1)4 = 25$
The first five terms of the arithmetic sequence are 9, 13, 17, 21, and 25.
23. $a_1 = 7 + (1-1)(-2) = 7$,
$a_2 = 7 + (2-1)(-2) = 5$,
$a_3 = 7 + (3-1)(-2) = 3$,
$a_4 = 7 + (4-1)(-2) = 1$,
$a_5 = 7 + (5-1)(-2) = -1$

The first five terms of the arithmetic sequence are 7, 5, 3, 1, and -1.

25. $a_1 = -4 + (1-1)3 = -4$,
$a_2 = -4 + (2-1)3 = -1$,
$a_3 = -4 + (3-1)3 = 2$,
$a_4 = -4 + (4-1)3 = 5$,
$a_5 = -4 + (5-1)3 = 8$
The first five terms of the arithmetic sequence are -4, -1, 2, 5, and 8.

27. $a_1 = -2 + (1-1)(-3) = -2$,
$a_2 = -2 + (2-1)(-3) = -5$,
$a_3 = -2 + (3-1)(-3) = -8$,
$a_4 = -2 + (4-1)(-3) = -11$,
$a_5 = -2 + (5-1)(-3) = -14$
The first five terms of the arithmetic sequence are -2, -5, -8, -11, and -14.

29. $a_1 = -4(1) - 3 = -7$,
$a_2 = -4(2) - 3 = -11$,
$a_3 = -4(3) - 3 = -15$,
$a_4 = -4(4) - 3 = -19$,
$a_5 = -4(5) - 3 = -23$
The first five terms of the arithmetic sequence are -7, -11, -15, -19, and -23.

31. $a_1 = 0.5(1) + 4 = 4.5$,
$a_2 = 0.5(2) + 4 = 5$, $a_3 = 0.5(3) + 4 = 5.5$,
$a_4 = 0.5(4) + 4 = 6$, $a_5 = 0.5(5) + 4 = 6.5$
The first five terms of the arithmetic sequence are 4.5, 5, 5.5, 6, and 6.5.

33. $a_1 = 20(1) + 1000 = 1020$,
$a_2 = 20(2) + 1000 = 1040$,
$a_3 = 20(3) + 1000 = 1060$,
$a_4 = 20(4) + 1000 = 1080$,
$a_5 = 20(5) + 1000 = 1100$
The first five terms of the arithmetic sequence are 1020, 1040, 1060, 1080, and 1100.

35. Use $a_1 = 9$, $n = 8$, and $d = 6$ in the formula $a_n = a_1 + (n-1)d$.
$$a_8 = 9 + (8-1)6 = 51$$

37. Use $a_1 = 6$, $a_{20} = 82$, and $n = 20$ in the formula $a_n = a_1 + (n-1)d$.
$82 = 6 + (20-1)d$
$82 = 6 + 19d$
$76 = 19d$
$4 = d$

39. Use $a_7 = 14$, $d = -2$, and $n = 7$ in the formula $a_n = a_1 + (n-1)d$.
$14 = a_1 + (7-1)(-2)$
$14 = a_1 - 12$
$26 = a_1$

41. From the fact that the fifth term is 13 and the first term is -3, we can find the common difference.
$13 = -3 + (5-1)d$
$13 = -3 + 4d$
$16 = 4d$
$4 = d$
Use $a_1 = -3$, $n = 6$, and $d = 4$ in the formula $a_n = a_1 + (n-1)d$.
$$a_6 = -3 + (6-1)4 = 17$$

43. Use $a_1 = 1$, $a_{48} = 48$, and $n = 48$ in the formula $S_n = \frac{n}{2}(a_1 + a_n)$.
$$S_{48} = \frac{48}{2}(1 + 48) = 1176$$

45. To find n, use $a_1 = 8$, $d = 2$, and $a_n = 36$ in the formula $a_n = a_1 + (n-1)d$.
$36 = 8 + (n-1)2$
$36 = 8 + 2n - 2$
$30 = 2n$
$15 = n$
Use $a_1 = 8$, $a_{15} = 36$, and $n = 15$ in the formula $S_n = \frac{n}{2}(a_1 + a_n)$.
$$S_{15} = \frac{15}{2}(8 + 36) = 330$$

47. To find n, use $a_1 = -1$, $d = -6$, and $a_n = -73$ in the formula $a_n = a_1 + (n-1)d$.
$-73 = -1 + (n-1)(-6)$
$-78 = -1 - 6n + 6$
$-78 = -6n$
$13 = n$
Use $a_1 = -1$, $a_{13} = -73$, and $n = 13$ in the formula $S_n = \frac{n}{2}(a_1 + a_n)$.
$$S_{13} = \frac{13}{2}(-1 + (-73)) = -481$$

49. To find n, use $a_1 = -6$, $d = 5$, and $a_n = 64$ in the formula $a_n = a_1 + (n-1)d$.
$64 = -6 + (n-1)5$
$64 = -6 + 5n - 5$
$75 = 5n$
$15 = n$
Use $a_1 = -6$, $a_{15} = 64$, and $n = 15$ in the formula $S_n = \frac{n}{2}(a_1 + a_n)$.
$$S_{15} = \frac{15}{2}(-6 + 64) = 435$$

51. To find n, use $a_1 = 20$, $d = -8$, and $a_n = -92$ in the formula $a_n = a_1 + (n-1)d$.
$-92 = 20 + (n-1)(-8)$
$-92 = 20 - 8n + 8$
$-120 = -8n$

$15 = n$

Use $a_1 = 20$, $a_{15} = -92$, and $n = 15$ in the formula $S_n = \frac{n}{2}(a_1 + a_n)$.

$$S_{15} = \frac{15}{2}(20 + (-92)) = -540$$

53. $\sum_{i=1}^{12}(3i - 7) = -4 + (-1) + ... + 29$

Use $a_1 = -4$, $a_{12} = 29$, and $n = 12$ in the formula $S_n = \frac{n}{2}(a_1 + a_n)$.

$$S_{12} = \frac{12}{2}(-4 + 29) = 150$$

55. $\sum_{i=1}^{11}(-5i + 2) = -3 + (-8) + ... + (-53)$

Use $a_1 = -3$, $a_{11} = -53$, and $n = 11$ in the formula $S_n = \frac{n}{2}(a_1 + a_n)$.

$$S_{11} = \frac{11}{2}(-3 + (-53)) = -308$$

57. Use $a_1 = \$22{,}000$, $n = 7$, and $d = \$500$ in the formula $a_n = a_1 + (n - 1)d$.

$$a_7 = 22{,}000 + (7 - 1)500 = \$25{,}000$$

59. The students read 5 pages the first day, 7 pages the second day, 9 pages the third day, and so on. To find the number they read on the 31st day, let $n = 31$, $d = 2$, and $a = 5$ in the formula $a_n = a_1 + (n - 1)d$.

$$a_{31} = 5 + (31 - 1)2 = 65$$

To find the sum $5 + 7 + 9 + ... + 65$, use $n = 31$, $a_1 = 5$, and $a_{31} = 65$ in the formula $S_n = \frac{n}{2}(a_1 + a_n)$.

$$S_{31} = \frac{31}{2}(5 + 65) = 1085$$

61. The only sequence that does not have a common difference is (b) and so it is not arithmetic.

12.4 WARM-UPS

1. False, because the ratio of two consecutive terms is not constant.
2. False, there is a common ratio of 2 between adjacent terms.
3. True, because the general form for a geometric sequence is $a_n = a_1 r^{n-1}$.
4. True, because if $n = 1$, then $3(2)^{-1+3} = 12$.
5. True, because $a_1 = 12$ and $a_2 = 6$ gives $r = 1/2$.
6. True, because of the definition of geometric series.
7. False, because we have a formula for the sum of a finite geometric series.
8. False, because $a_1 = 6$.
9. True, because this is the correct formula for the sum of all of the terms of an infinite geometric series with first term 10 and ratio 1/2.
10. False, because there is no sum for an infinite geometric series with a ratio of 2.

12.4 EXERCISES

1. A geometric sequence is one in which each term after the first is obtained by multiplying the preceding term by a constant.
3. A geometric series is an indicated sum of a geometric sequence.
5. The approximate value of r^n when n is large and $|r| < 1$ is 0.
7. Since the first term is 1 and the common ratio is 2, the nth term is $a_n = 1(2)^{n-1} = 2^{n-1}$.
9. Since the first term is $1/3$ and the common ratio is 3, the nth term is $a_n = \frac{1}{3}(3)^{n-1}$.
11. Since the first term is 64 and the common ratio is $1/8$, the nth term is $a_n = 64\left(\frac{1}{8}\right)^{n-1}$.
13. Since the first term is 8 and the common ratio is $-1/2$, the nth term is $a_n = 8\left(-\frac{1}{2}\right)^{n-1}$.
15. Since the first term is 2 and the common ratio is $-4/2 = -2$, the nth term is $a_n = 2(-2)^{n-1}$.
17. Since the first term is $-1/3$ and the common ratio is $(-1/4)/(-1/3) = 3/4$, the nth term is $a_n = -\frac{1}{3}\left(\frac{3}{4}\right)^{n-1}$.
19. $a_1 = 2(1/3)^{1-1} = 2$,
$a_2 = 2(1/3)^{2-1} = 2/3$,
$a_3 = 2(1/3)^{3-1} = 2/9$,
$a_4 = 2(1/3)^{4-1} = 2/27$,
$a_5 = 2(1/3)^{5-1} = 2/81$
The first 5 terms are 2, $\frac{2}{3}$, $\frac{2}{9}$, $\frac{2}{27}$, and $\frac{2}{81}$.
21. $a_1 = (-2)^{1-1} = 1$,
$a_2 = (-2)^{2-1} = -2$, $a_3 = (-2)^{3-1} = 4$,
$a_4 = (-2)^{4-1} = -8$, $a_5 = (-2)^{5-1} = 16$
The first 5 terms are 1, −2, 4, −8, and 16.

23. $a_1 = 2^{-1} = 1/2,$
$a_2 = 2^{-2} = 1/4, \quad a_3 = 2^{-3} = 1/8,$
$a_4 = 2^{-4} = 1/16, \quad a_5 = 2^{-5} = 1/32$
The first 5 terms are $\frac{1}{2}, \frac{1}{4}, \frac{1}{8}, \frac{1}{16}$, and $\frac{1}{32}$.

25. $a_1 = (0.78)^1 = 0.78,$
$a_2 = (0.78)^2 = 0.6084,$
$a_3 = (0.78)^3 = 0.4746,$
$a_4 = (0.78)^4 = 0.3702, a_5 = (0.78)^5 = 0.2887$
The first 5 terms are 0.78, 0.6084, 0.4746, 0.3702, and 0.2887.

27. Use $a_4 = 40$, $n = 4$, and $r = 2$ in the formula $a_n = a_1 r^{n-1}$.

$$40 = a_1(2)^{4-1}$$
$$40 = 8a_1$$
$$5 = a_1$$

29. Use $a_4 = 2/9$, $n = 4$, and $a_1 = 6$ in the formula $a_n = a_1 r^{n-1}$.

$$\frac{2}{9} = 6r^{4-1}$$
$$\frac{1}{27} = r^3$$
$$\frac{1}{3} = r$$

31. Use $r = 1/3$, $n = 4$, and $a_1 = -3$ in the formula $a_n = a_1 r^{n-1}$.

$$a_4 = -3\left(\frac{1}{3}\right)^{4-1} = -3\left(\frac{1}{27}\right) = -\frac{1}{9}$$

33. Use $r = 1/2$, $a_1 = 1/2$, and $a_n = 1/512$ in the formula $a_n = a_1 r^{n-1}$ to find n.

$$\frac{1}{512} = \frac{1}{2}\left(\frac{1}{2}\right)^{n-1}$$
$$\frac{1}{2^9} = \left(\frac{1}{2}\right)^n$$
$$n = 9$$

Use $n = 9$, $a_1 = 1/2$, and $r = 1/2$ in the formula $S_n = \dfrac{a_1(1 - r^n)}{1 - r}$.

$$S_9 = \frac{\frac{1}{2}\left(1 - \left(\frac{1}{2}\right)^9\right)}{1 - \frac{1}{2}} = 1 - \frac{1}{512} = \frac{511}{512}$$

35. Use $n = 5$, $a_1 = 1/2$, and $r = -1/2$ in the formula $S_n = \dfrac{a_1(1 - r^n)}{1 - r}$.

$$S_5 = \frac{\frac{1}{2}\left(1 - \left(-\frac{1}{2}\right)^5\right)}{1 - \left(-\frac{1}{2}\right)} = \frac{\frac{1}{2}(\frac{33}{32})}{\frac{3}{2}} = \frac{11}{32}$$

37. First determine the number of terms. Since $r = 2/3$, the nth term is $30\left(\frac{2}{3}\right)^{n-1}$.
Solve

$$30\left(\frac{2}{3}\right)^{n-1} = \frac{1280}{729}$$
$$\left(\frac{2}{3}\right)^{n-1} = \frac{128}{2187} = \left(\frac{2}{3}\right)^7$$
$$n - 1 = 7$$
$$n = 8$$

Use $n = 8$, $a_1 = 30$, and $r = 2/3$ in the formula $S_n = \dfrac{a_1(1 - r^n)}{1 - r}$.

$$S_8 = \frac{30\left(1 - \left(\frac{2}{3}\right)^8\right)}{1 - \left(\frac{2}{3}\right)} = \frac{30(\frac{6305}{6561})}{\frac{1}{3}}$$
$$= \frac{63050}{729} \approx 86.4883$$

39. $\displaystyle\sum_{i=1}^{10} 5(2)^{i-1} = S_{10} = \frac{5(1 - (2)^{10})}{1 - (2)}$
$$= \frac{5(-1023)}{-1} = 5115$$

41. $\displaystyle\sum_{i=1}^{6} (0.1)^i = S_6 = \frac{0.1(1 - (0.1)^6)}{1 - (0.1)}$
$$= \frac{0.1(0.999999)}{0.9} = 0.111111$$

43. $\displaystyle\sum_{i=1}^{6} 100(0.3)^i = S_6$
$$= \frac{100(0.3)(1 - (0.3)^6)}{1 - (0.3)}$$
$$= \frac{100(0.3)(1 - (0.3)^6)}{0.7} = 42.8259$$

45. Use $a_1 = 1/8$ and $r = 1/2$ in the formula for the sum of an infinite geometric series $S = \dfrac{a_1}{1 - r}$.

$$S = \frac{\frac{1}{8}}{1 - \frac{1}{2}} = \frac{\frac{1}{8}}{\frac{1}{2}} = \frac{1}{4}$$

47. Use $a_1 = 3$ and $r = 2/3$ in $S = \dfrac{a_1}{1 - r}$.

$$S = \frac{3}{1 - \frac{2}{3}} = \frac{3}{\frac{1}{3}} = 9$$

49. Use $a_1 = 4$ and $r = -1/2$ in $S = \dfrac{a_1}{1 - r}$.

$$S = \frac{4}{1 - (-\frac{1}{2})} = \frac{4}{\frac{3}{2}} = \frac{8}{3}$$

51. Use $a_1 = 0.3$ and $r = 0.3$ in $S = \dfrac{a_1}{1 - r}$.

$$S = \frac{0.3}{1 - 0.3} = \frac{0.3}{0.7} = \frac{3}{7}$$

53. Use $a_1 = 3$ and $r = 0.5$ in $S = \dfrac{a_1}{1 - r}$.

$$S = \frac{3}{1 - 0.5} = \frac{3}{0.5} = 6$$

55. Use $a_1 = 0.3$ and $r = 0.1$ in $S = \frac{a_1}{1-r}$.
$$S = \frac{0.3}{1-0.1} = \frac{0.3}{0.9} = \frac{1}{3}$$
57. Use $a_1 = 0.12$ and $r = 0.01$ in $S = \frac{a_1}{1-r}$.
$$S = \frac{0.12}{1-0.01} = \frac{0.12}{0.99} = \frac{12}{99} = \frac{4}{33}$$
59. We want the sum of the geometric series $2000(1.12)^{45} + 2000(1.12)^{44} + ... + 2000(1.12)$.
Note that the last deposit is made at the beginning of the 45th year and earns interest for only one year. Rewrite the series as $2000(1.12) + 2000(1.12)^2 + ... + 2000(1.12)^{45}$ where the $a_1 = 2000(1.12)$, $n = 45$, and $r = 1.12$.
$$S_{45} = \frac{2000(1.12)(1-(1.12)^{45})}{1-1.12} = \$3{,}042{,}435.27$$

61. We want the sum of the finite geometric series $1 + 2 + 4 + 8 + 16 + ... + 2^{30}$, which has 31 terms and a ratio of 2.
$$S_{31} = \frac{1(1-2^{31})}{1-2} = 2^{31} - 1 = 2{,}147{,}483{,}647 \text{ cents} = \$21{,}474{,}836.47$$
63. Use $r = 0.80$, $a_1 = 1{,}000{,}000$ in $S = \frac{a_1}{1-r}$.
$$S = \frac{1{,}000{,}000}{1-0.80} = \$5{,}000{,}000$$
65. Only sequence (d) is not geometric because it is the only one that does not have a constant ratio.
67. Use $a_1 = 24/100 = 0.24$ and $r = 1/100 = 0.01$ in the formula for $S = \frac{a_1}{1-r}$.
$$S = \frac{0.24}{1-0.01} = \frac{0.24}{0.99} = \frac{8}{33}$$

12.5 WARM-UPS

1. False, because there are 13 terms in a binomial to the 12th power.
2. False, because the 7th term has variable part a^6b^6.
3. False, because if $x = 1$ the equation is incorrect.
4. True, because the signs alternate in any expansion of a difference.
5. True, because we can obtain it from the 7th line.
6. True, because $1 + 4 + 6 + 4 + 1 = 2^4$.
7. True, because of the binomial theorem.
8. True, because $2^n = (1+1)^n$
$$= \sum_{i=0}^{n} \frac{n!}{(n-i)!i!} 1^{n-i} 1^i = \sum_{i=0}^{n} \frac{n!}{(n-i)!i!},$$
and the last sum is the sum of the coefficients in the nth row.
9. True, by definition of $0!$ and $1!$.
10. True, because $\frac{7 \cdot 6 \cdot 5 \cdot 4 \cdot 3 \cdot 2 \cdot 1}{5 \cdot 4 \cdot 3 \cdot 2 \cdot 1 \cdot 2 \cdot 1} = 21$

12.5 EXERCISES

1. The sum obtained for a power of a binomial is called a binomial expansion.
3. The expression $n!$ is the product of the positive integers from 1 through n.
5. $\frac{4!}{4!0!} = \frac{4 \cdot 3 \cdot 2 \cdot 1}{4 \cdot 3 \cdot 2 \cdot 1 \cdot 1} = 1$
7. $\frac{5!}{2!3!} = \frac{5 \cdot 4}{2} = 10$
9. $\frac{8!}{5!3!} = \frac{8 \cdot 7 \cdot 6}{3 \cdot 2 \cdot 1} = 56$
11. The coefficients in the 3rd row of Pascal's triangle are 1, 3, 3, 1. Use these coefficients with the pattern for the exponents.
$$(x+1)^3 = 1x^3 1^0 + 3x^2 1^1 + 3x 1^2 + 1x^0 1^3 = x^3 + 3x^2 + 3x + 1$$
13. The coefficients in the 3rd row of Pascal's triangle are 1, 3, 3, 1. Use these coefficients with the pattern for the exponents.
$$(a+2)^3 = 1a^3 2^0 + 3a^2 2^1 + 3a 2^2 + 1a^0 2^3 = a^3 + 6a^2 + 12a + 8$$
15. The coefficients in the 5th row of Pascal's triangle are 1, 5, 10, 10, 5, 1. Use these coefficients with the pattern for the exponents.
$$(r+t)^5 = r^5 + 5r^4t + 10r^3t^2 + 10r^2t^3 + 5rt^4 + t^5$$
17. The coefficients in the 3rd row are 1, 3, 3, 1. Use these coefficients with the pattern for the exponents, and alternate the signs.
$$(m-n)^3 = m^3 - 3m^2n + 3mn^2 - n^3$$
19. Use the coefficients 1, 3, 3, 1 and let $y = 2a$ in the binomial theorem.
$$(x+2a)^3 = 1x^3(2a)^0 + 3x^2(2a)^1 + 3x(2a)^2 + (2a)^3 = x^3 + 6ax^2 + 12a^2x + 8a^3$$
21. Use the coefficients 1, 4, 6, 4, 1 in the binomial theorem.

$$(x^2 - 2)^4 = (x^2)^4 - 4(x^2)^3 2 + 6(x^2)^2 2^2 - 4x^2 2^3 + 1(x^2)^0 2^4 = x^8 - 8x^6 + 24x^4 - 32x^2 + 16$$

23. Use the coefficients from the 7th line 1, 7, 21, 35, 35, 21, 7, 1 and alternate the signs of the terms.

$$(x - 1)^7 = x^7 - 7x^6 + 21x^5 - 35x^4 + 35x^3 - 21x^2 + 7x - 1$$

25. Use the binomial theorem to write the first 4 terms of $(a - 3b)^{12}$.

$$\frac{12!}{12!0!}a^{12}b^0 - \frac{12!}{11!1!}a^{11}b^1 + \frac{12!}{10!2!}a^{10}b^2 - \frac{12!}{9!3!}a^9b^3$$
$$= a^{12} - 36a^{11}b + 594a^{10}b^2 - 5940a^9b^3$$

27. Use the binomial theorem to write the first 4 terms of $(x^2 + 5)^9$.

$$\frac{9!}{9!0!}(x^2)^9 5^0 + \frac{9!}{8!1!}(x^2)^8 5^1 + \frac{9!}{7!2!}(x^2)^7 5^2 + \frac{9!}{6!3!}(x^2)^6 5^3$$
$$= x^{18} + 45x^{16} + 900x^{14} + 10500x^{12}$$

29. Use the binomial theorem to write the first 4 terms of $(x - 1)^{22}$.

$$\frac{22!}{22!0!}x^{22}1^0 - \frac{22!}{21!1!}x^{21}1^1 + \frac{22!}{20!2!}x^{20}1^2 - \frac{22!}{19!3!}x^{19}1^3$$
$$= x^{22} - 22x^{21} + 231x^{20} - 1540x^{19}$$

31. Use the binomial theorem to write the first 4 terms of $\left(\frac{x}{2} + \frac{y}{3}\right)^{10}$.

$$\frac{10!}{10!0!}\left(\frac{x}{2}\right)^{10}\left(\frac{y}{3}\right)^0 + \frac{10!}{9!1!}\left(\frac{x}{2}\right)^9\left(\frac{y}{3}\right)^1 + \frac{10!}{8!2!}\left(\frac{x}{2}\right)^8\left(\frac{y}{3}\right)^2 + \frac{10!}{7!3!}\left(\frac{x}{2}\right)^7\left(\frac{y}{3}\right)^3$$
$$= \frac{x^{10}}{1024} + \frac{5x^9y}{768} + \frac{5x^8y^2}{256} + \frac{5x^7y^3}{144}$$

33. Use the formula for the kth term of $(x + y)^n$ with $k = 6$ and $n = 13$.

$$\frac{13!}{(13 - 6 + 1)!(6 - 1)!}a^{13-6+1}w^{6-1} = \frac{13!}{8!5!}a^8w^5 = 1287a^8w^5$$

35. Use the formula for the kth term with $k = 8$ and $n = 16$.

$$\frac{16!}{(16 - 8 + 1)!(8 - 1)!}m^{16-8+1}(-n)^{8-1} = \frac{16!}{9!7!}m^9(-n)^7 = -11440m^9n^7$$

37. Use the formula for the kth term with $k = 4$ and $n = 8$.

$$\frac{8!}{(8 - 4 + 1)!(4 - 1)!}x^{8-4+1}(2y)^{4-1} = \frac{8!}{5!3!}x^5(2y)^3 = 56x^5 8y^3 = 448x^5y^3$$

39. Use the formula for the kth term with $k = 7$ and $n = 20$.

$$\frac{20!}{(20 - 7 + 1)!(7 - 1)!}(2a^2)^{20-7+1}b^{7-1} = \frac{20!}{14!6!}(2a^2)^{14}b^6 = 635{,}043{,}840a^{28}b^6$$

41. Use $n = 8$, $x = a$, and $y = b$ in the binomial theorem with summation notation.

$$(a + m)^8 = \sum_{i=0}^{8} \frac{8!}{(8 - i)!\,i!}\,a^{8-i}\,m^i$$

43. Use $n = 5$, $x = a$, and $y = -2x$ in the binomial theorem with summation notation.

$$(a + (-2x))^5 = \sum_{i=0}^{5} \frac{5!}{(5 - i)!\,i!}\,a^{5-i}\,(-2x)^i = \sum_{i=0}^{5} \frac{5!\,(-2)^i}{(5 - i)!\,i!}\,a^{5-i}\,x^i$$

45. $(a + (b + c))^3$

$$= a^3 + 3a^2(b + c) + 3a(b + c)^2 + (b + c)^3$$
$$= a^3 + 3a^2b + 3a^2c + 3ab^2 + 6abc + 3ac^2 + b^3 + 3b^2c + 3bc^2 + c^3$$
$$= a^3 + b^3 + c^3 + 3a^2b + 3a^2c + 3ab^2 + 3ac^2 + 3b^2c + 3bc^2 + 6abc$$

Chapter 12 Wrap-Up

Enriching Your Mathematical Word Power

1. a **2.** d **3.** c **4.** b **5.** a
6. c **7.** d **8.** b **9.** d **10.** a

CHAPTER 12 REVIEW

1. $a_1 = 1^3, a_2 = 2^3, a_3 = 3^3, a_4 = 4^3, a_5 = 5^3$

The terms of the sequence are 1, 8, 27, 64, 125.

3. $c_1 = (-1)^1(2 \cdot 1 - 3) = 1,$
$c_2 = (-1)^2(2 \cdot 2 - 3) = 1,$
$c_3 = (-1)^3(2 \cdot 3 - 3) = -3,$
$c_4 = (-1)^4(2 \cdot 4 - 3) = 5,$
$c_5 = (-1)^5(2 \cdot 5 - 3) = -7$
$c_6 = (-1)^6(2 \cdot 6 - 3) = 9$

The terms are 1, 1, −3, 5, −7, 9.

5. $a_1 = -\frac{1}{1} = -1, a_2 = -\frac{1}{2}, a_3 = -\frac{1}{3}$

The first three terms are $-1, -\frac{1}{2}, -\frac{1}{3}$.

7. $b_1 = \frac{(-1)^{2\cdot1}}{2\cdot1+1} = \frac{1}{3},\quad b_2 = \frac{(-1)^{2\cdot2}}{2\cdot2+1} = \frac{1}{5}$

$b_3 = \frac{(-1)^{2\cdot 3}}{2\cdot 3+1} = \frac{1}{7}$

The first three terms are $\frac{1}{3}$, $\frac{1}{5}$, and $\frac{1}{7}$.

9. $c_1 = \log_2(2^{1+3}) = \log_2(2^4) = 4$
$c_2 = \log_2(2^{2+3}) = \log_2(2^5) = 5$
$c_3 = \log_2(2^{3+3}) = \log_2(2^6) = 6$
The first three terms are 4, 5, and 6.

11. $\sum_{i=1}^{3} i^3 = 1^3 + 2^3 + 3^3 = 36$

13. $\sum_{n=1}^{5} n(n-1)$
$= 1(1-1) + 2(2-1) + 3(3-1) + 4(4-1) + 5(5-1)$
$= 0 + 2 + 6 + 12 + 20 = 40$

15. The terms in the series are reciprocals of even integers. Even integers are usually represent as $2i$, but to get 4 in the denominator when $i = 1$, we use $2(i+1)$.

$$\sum_{i=1}^{\infty} \frac{1}{2(i+1)}$$

17. The terms in this series are the squares of integers. Squares are usually represented as i^2, but to get the first term 0 when $i = 1$, we use $(i-1)^2$.

$$\sum_{i=1}^{\infty} (i-1)^2$$

19. To get alternating signs for the terms, we use a power of -1. If we use $(-1)^i$, then $i = 1$ makes the first term negative. So we use $(-1)^{i+1}$.

$$\sum_{i=1}^{\infty} (-1)^{i+1} x_i$$

21. $a_1 = 6 + (1-1)5 = 6$
Since the common difference is 5, the first four terms are 6, 11, 16, and 21.

23. $a_1 = -20 + (1-1)(-2) = -20$
Since the common difference is -2, the first four terms are -20, -22, -24, and -26.

25. $a_1 = 1000(1) + 2000 = 3000$
Since the common difference is 1000, the first four terms are 3000, 4000, 5000, and 6000.

27. Use $a_1 = 1/3$, $d = 1/3$, and the formula $a_n = a_1 + (n-1)d$.

$$a_n = \frac{1}{3} + (n-1)\frac{1}{3} = \frac{n}{3}$$

29. Use $a_1 = 2$, $d = 2$, and the formula $a_n = a_1 + (n-1)d$.

$$a_n = 2 + (n-1)(2) = 2n$$

31. Use $a_1 = 1$, $a_{24} = 24$, $n = 24$, and the formula $S_n = \frac{n}{2}(a_1 + a_n)$.

$$S_{24} = \frac{24}{2}(1+24) = 300$$

33. Use $a_1 = 1/6$, $d = 1/3$, $a_n = 11/2$, and the formula $a_n = a_1 + (n-1)d$ to find n.
$\frac{11}{2} = \frac{1}{6} + (n-1)\frac{1}{3}$
$33 = 1 + (n-1)2$
$32 = 2n - 2$
$34 = 2n$
$17 = n$
Now use $n = 17$, $a_1 = 1/6$, $a_{17} = 11/2$, and the formula $S_n = \frac{n}{2}(a_1 + a_n)$ to find the sum.

$$S_{17} = \frac{17}{2}\left(\frac{1}{6} + \frac{11}{2}\right) = \frac{17}{2}\left(\frac{34}{6}\right) = \frac{289}{6}$$

35. Use $a_1 = -1$, $a_7 = 11$, $n = 7$, and the formula $S_n = \frac{n}{2}(a_1 + a_n)$ to find the sum.

$$S_7 = \frac{7}{2}(-1+11) = 35$$

37. $a_1 = 3\left(\frac{1}{2}\right)^{1-1} = 3$, $a_2 = 3\left(\frac{1}{2}\right)^{2-1} = \frac{3}{2}$,
$a_3 = 3\left(\frac{1}{2}\right)^{3-1} = \frac{3}{4}$, $a_4 = 3\left(\frac{1}{2}\right)^{4-1} = \frac{3}{8}$
The first four terms are 3, $\frac{3}{2}$, $\frac{3}{4}$, and $\frac{3}{8}$.

39. $a_1 = 2^{1-1} = 1$, $a_2 = 2^{1-2} = \frac{1}{2}$
$a_3 = 2^{1-3} = \frac{1}{4}$, $a_4 = 2^{1-4} = \frac{1}{8}$
The first four terms are 1, $\frac{1}{2}$, $\frac{1}{4}$, and $\frac{1}{8}$.

41. $a_1 = 23(10)^{-2(1)} = 0.23$,
$a_2 = 23(10)^{-2(2)} = 0.0023$,
$a_3 = 23(10)^{-2(3)} = 0.000023$,
$a_4 = 23(10)^{-2(4)} = 0.00000023$
The first four terms of the geometric sequence are 0.23, 0.0023, 0.000023, and 0.00000023.

43. Use $a_1 = 1/2$, $r = 6$, and the formula $a_n = a_1 r^{n-1}$.

$$a_n = \frac{1}{2}(6)^{n-1}$$

45. Use $a_1 = 7/10$, $r = 1/10$, and the formula $a_n = a_1 r^{n-1}$.

$$a_n = \frac{7}{10}\left(\frac{1}{10}\right)^{n-1}$$

47. Use $a_1 = 1/3$, $r = 1/3$, $n = 4$, and the formula $S_n = \frac{a_1(1-r^n)}{1-r}$.

$$S_4 = \frac{\frac{1}{3}\left(1 - \left(\frac{1}{3}\right)^4\right)}{1 - \frac{1}{3}} = \frac{\frac{1}{3}\left(\frac{80}{81}\right)}{\frac{2}{3}} = \frac{40}{81}$$

49. Use $a_1 = 0.3, r = 0.1, n = 10$, and the formula $S_n = \frac{a_1(1-r^n)}{1-r}$.

$$S_{10} = \frac{0.3(1-(0.1)^{10})}{1-0.1} = \frac{0.3(0.9999999999)}{0.9}$$
$$= 0.3333333333$$

Your calculator may not give ten 3's after the decimal point, but doing this computation without a calculator does give ten 3's and this is the exact answer.

51. Use $a_1 = 1/4, r = 1/3$, and the formula for the sum of an infinite geometric series $S = \frac{a_1}{1-r}$.

$$S = \frac{\frac{1}{4}}{1-\frac{1}{3}} = \frac{\frac{1}{4}}{\frac{2}{3}} = \frac{3}{8}$$

53. Use $a_1 = 18$, r = 2/3, and the formula for the sum of an infinite geometric series S $= \frac{a_1}{1-r}$.

$$\mathrm{S} = \frac{18}{1-\frac{2}{3}} = \frac{18}{\frac{1}{3}} = 54$$

55. The coefficients for the fifth power of a binomial are 1, 5, 10, 10, 5, and 1.

$$(m+n)^5 = m^5 + 5m^4n + 10m^3n^2$$
$$+ 10m^2n^3 + 5mn^4 + n^5$$

57. The coefficients for the third power of a binomial are 1, 3, 3, and 1. Alternate the signs because it is a difference to a power.

$$(a^2-3b)^3 = 1(a^2)^3(3b)^0 - 3(a^2)^2(3b)^1$$
$$+ 3(a^2)^1(3b)^2 - 1(a^2)^0(3b)^3$$
$$= a^6 - 9a^4b + 27a^2b^2 - 27b^3$$

59. Use $n = 12$ and $k = 5$ in the formula for the kth term.

$$\frac{12!}{(12-5+1)!(5-1)!}x^{12-5+1}y^{5-1} = \frac{12!}{8!4!}x^8y^4$$
$$= 495x^8y^4$$

61. Use $n = 14$ and $k = 3$ in the formula for the kth term.

$$\frac{14!}{(14-3+1)!(3-1)!}(2a)^{14-3+1}(-b)^{3-1}$$
$$= \frac{14!}{12!2!}(2a)^{12}(-b)^2 = 372{,}736a^{12}b^2$$

63. Use the binomial theorem expressed in summation notation, with $n = 7$.

$$(a+w)^7 = \sum_{i=0}^{7} \frac{7!}{(7-i)!\,i!}\,a^{7-i}\,w^i$$

65. The sequence has neither a constant difference nor a constant ratio. So it is neither arithmetic nor geometric.

67. There is a constant difference of 3. So the sequence is an arithmetic sequence.

69. There is a constant difference of 2. So the sequence is an arithmetic sequence.

71. Use $a_1 = 6, n = 4, a_4 = 1/30$, and the formula $a_n = a_1r^{n-1}$.

$$\frac{1}{30} = 6r^{4-1}$$
$$\frac{1}{180} = r^3$$
$$r = \sqrt[3]{\frac{1}{180}} = \frac{1}{\sqrt[3]{180}} = \frac{\sqrt[3]{150}}{30}$$

73.
$$\sum_{i=1}^{5} \frac{(-1)^i}{i!} = \frac{(-1)^1}{1!} + \frac{(-1)^2}{2!} + \frac{(-1)^3}{3!}$$
$$+ \frac{(-1)^4}{4!} + \frac{(-1)^5}{5!}$$
$$= -1 + \frac{1}{2} - \frac{1}{6} + \frac{1}{24} - \frac{1}{120}$$

75. This is the summation notation for the binomial expansion of $(a+b)^5$.

$$\frac{5!}{5!0!}a^5b^0 + \frac{5!}{4!1!}a^4b^1 + \frac{5!}{3!2!}a^3b^2 + \frac{5!}{2!3!}a^2b^3$$
$$+ \frac{5!}{1!4!}a^1b^4 + \frac{5!}{0!5!}a^0b^5$$
$$= a^5 + 5a^4b + 10a^3b^2 + 10a^2b^3 + 5ab^4 + b^5$$

77. There are 26 terms because in the expansion of $(x+y)^n$ there are $n+1$ terms.

79. The first \$3000 earns interest for 16 years. The second \$3000 earns interest for 15 years, and so on. The last \$3000 earns interest for 1 year. The total in the account at the end of 16 years is the sum of the following series.

$3000(1.1) + 3000(1.1)^2 + ... + 3000(1.1)^{16}$

This is a geometric series with $n = 16$, $a_1 = 3000(1.1)$ and $r = 1.1$.

$$S_{16} = \frac{3000(1.1)(1-(1.1)^{16})}{1-1.1} = \$118{,}634.11$$

81. We compute a new balance 16 times by multiplying by $1 + 0.10$ each time.

$3000(1.10)^{16} = \$13{,}784.92$

CHAPTER 12 TEST

1. $a_1 = -10 + (1 - 1)6 = -10$
$a_2 = -10 + (2 - 1)6 = -4$
$a_3 = -10 + (3 - 1)6 = 2$
$a_4 = -10 + (4 - 1)6 = 8$
The first four terms are -10, -4, 2, and 8.

2. $a_1 = 5(0.1)^{1-1} = 5$
$a_2 = 5(0.1)^{2-1} = 0.5$
$a_3 = 5(0.1)^{3-1} = 0.05$
$a_4 = 5(0.1)^{4-1} = 0.005$
The first four terms are 5, 0.5, 0.05, and 0.005.

3. $a_1 = \frac{(-1)^1}{1!} = -1$, $a_2 = \frac{(-1)^2}{2!} = \frac{1}{2}$,
$a_3 = \frac{(-1)^3}{3!} = -\frac{1}{6}$, $a_4 = \frac{(-1)^4}{4!} = \frac{1}{24}$
The first four terms are -1, $\frac{1}{2}$, $-\frac{1}{6}$, and $\frac{1}{24}$.

4. $a_1 = \frac{2(1) - 1}{(1)^2} = 1$, $a_2 = \frac{2(2) - 1}{(2)^2} = \frac{3}{4}$,
$a_3 = \frac{2(3) - 1}{(3)^2} = \frac{5}{9}$, $a_4 = \frac{2(4) - 1}{(4)^2} = \frac{7}{16}$
The first four terms are 1, $\frac{3}{4}$, $\frac{5}{9}$, and $\frac{7}{16}$.

5. The sequence is an arithmetic sequence with $a_1 = 7$ and $d = -3$. So the general term is $a_n = 7 + (n - 1)(-3) = 7 - 3n + 3$
$= 10 - 3n$.

6. This sequence is a geometric sequence with $a_1 = -25$, and $r = -1/5$. So the general term is $a_n = -25\left(-\frac{1}{5}\right)^{n-1}$.

7. This sequence is a sequence of even integers, which we can represent as $2n$. To get the signs to alternate, we use a power of -1. So the general term is $a_n = (-1)^{n-1}2n$.

8. This sequence is a sequence of squares of the positive integers. So the general term is $a_n = n^2$.

9. $\sum_{i=1}^{5}(2i + 3) = 2(1) + 3 + 2(2) + 3$
$+ 2(3) + 3 + 2(4) + 3 + 2(5) + 3$
$= 5 + 7 + 9 + 11 + 13$

10. $\sum_{i=1}^{6} 5(2)^{i-1} = 5(2)^{1-1} + 5(2)^{2-1}$
$+ 5(2)^{3-1} + 5(2)^{4-1} + 5(2)^{5-1} + 5(2)^{6-1}$
$= 5 + 10 + 20 + 40 + 80 + 160$

11. $\sum_{i=0}^{4} \frac{4!}{(4-i)!i!} m^{4-i}q^i = \frac{4!}{4!0!}m^4q^0$
$+ \frac{4!}{3!1!}m^3q^1 + \frac{4!}{2!2!}m^2q^2$
$+ \frac{4!}{1!3!}m^1q^3 + \frac{4!}{0!4!}m^0q^4$
$= m^4 + 4m^3q + 6m^2q^2 + 4mq^3 + q^4$

12. Use $a_1 = 9$, $a_{20} = 66$, and $n = 20$ in the formula for the sum of an arithmetic series.
$S_{20} = \frac{20}{2}(9 + 66) = 10(75) = 750$

13. Use $a_1 = 10$, $n = 5$, and $r = 1/2$ in the formula for the sum of a finite geometric series.

$$S_5 = \frac{10\left(1 - \left(\frac{1}{2}\right)^5\right)}{1 - \frac{1}{2}} = \frac{10\left(\frac{31}{32}\right)}{\frac{1}{2}} = \frac{155}{8}$$

14. Use $a_1 = 0.35$ and $r = 0.93$ in the formula for the sum of an infinite geometric series.
$S = \frac{0.35}{1 - 0.93} = \frac{0.35}{0.07} = 5$

15. Use $a_1 = 2$, $a_{100} = 200$, and $n = 100$ in the formula for the sum of a finite arithmetic series.
$S_{100} = \frac{100}{2}(2 + 200) = 50(202) = 10{,}100$

16. Use $a_1 = 1/4$ and $r = 1/2$ in the formula for the sum of an infinite geometric series.
$S = \frac{1/4}{1 - 1/2} = \frac{1/4}{1/2} = \frac{1}{2}$

17. Use $a_1 = 2$, $r = 1/2$ and $a_n = 1/128$ to find n:
$2\left(\frac{1}{2}\right)^{n-1} = \frac{1}{128}$
$\left(\frac{1}{2}\right)^{n-1} = \frac{1}{256} = \frac{1}{2^8}$
$n - 1 = 8$
$n = 9$
Use $a_1 = 2$, $n = 9$, and $r = 1/2$ to find the sum of the 9 terms.
$S_9 = \frac{2(1 - (1/2)^9)}{1 - 1/2} = \frac{511}{128} \approx 3.9922$

18. Use $a_1 = 3$, $a_5 = 48$, $n = 5$ in the formula for the general term of a geometric sequence.
$48 = 3r^{5-1}$
$16 = r^4$
$\pm 2 = r$

19. Use $a_1 = 1$, $a_{12} = 122$, $n = 12$, and the formula for the general term of an arithmetic sequence.
$122 = 1 + (12 - 1)d$
$121 = 11d$
$11 = d$

20. Use $n = 15$ and $k = 5$ in the formula for the kth term of a binomial expansion.

$$\frac{15!}{(15-5+1)!(5-1)!}r^{15-5+1}(-t)^{5-1}$$
$$= \frac{15!}{11!4!}r^{11}t^4 = 1365r^{11}t^4$$

21. Use $n = 8$ and $k = 4$ in the formula for the kth term of a binomial expansion.

$$\frac{8!}{(8-4+1)!(4-1)!}(a^2)^{8-4+1}(-2b)^{4-1}$$
$$= \frac{8!}{5!3!}a^{10}(-2)^3b^3 = -448a^{10}b^3$$

22. $800(1.10)^1 + 800(1.10)^2 + \ldots. + 800(1.10)^{25}$
$$= \frac{800(1.10)(1-1.10^{25})}{1-1.10} = \$86{,}545.41$$

Making Connections

Chapters 1 - 12

1. $f(3) = 3^2 - 3 = 9 - 3 = 6$

2. $f(n) = n^2 - 3$

3. $f(x+h) = (x+h)^2 - 3$
$= x^2 + 2xh + h^2 - 3$

4. $f(x) - g(x) = x^2 - 3 - (2x - 1)$
$= x^2 - 2x - 2$

5. $g(f(3)) = g(6) = 2(6) - 1 = 11$

6. $(f \circ g)(2) = f(g(2)) = f(3) = 3^2 - 3 = 6$

7. $m(16) = \log_2(16) = 4$

8. $(h \circ m)(32) =$
$h(m(32)) = h(5) = 2^5 = 32$

9. $h(-1) = 2^{-1} = 1/2$

10. $h^{-1}(8) = \log_2(8) = 3$

11. $m^{-1}(0) = 2^0 = 1$

12. $(m \circ h)(x) = m(h(x)) = m(2^x)$
$= \log_2(2^x) = x$
So $(m \circ h)(x) = x$.

13. If y varies directly as x, then $y = kx$. Since $y = -6$ when $x = 4$, we have $-6 = 4k$, or $k = -3/2$. When $x = 9$, we can use the original formula with $k = -3/2$.
$$y = -\frac{3}{2}(9) = -\frac{27}{2}$$

14. If a varies inversely as b, then $a = k/b$. If $a = 2$ when $b = -4$, we can find k.
$$2 = \frac{k}{-4}$$
$$-8 = k$$
To find a, use $k = -8$ and $b = 3$.
$$a = \frac{-8}{3} = -\frac{8}{3}$$

15. If y varies directly as w and inversely as t, then $y = (kw)/t$. Use $y = 16$, $w = 3$, and $t = -4$ to find k.
$$16 = \frac{k(3)}{-4}$$
$$-64 = 3k$$
$$\frac{-64}{3} = k$$
To find y, use $k = -64/3$, $w = 2$, and $t = 3$.
$$y = -\frac{64(2)}{3(3)} = -\frac{128}{9}$$

16. If y varies jointly as h and the square of r, then $y = khr^2$. Use $y = 12$, $h = 2$, and $r = 3$ to find k.
$$12 = k(2)(3)^2$$
$$12 = 18k$$
$$\frac{2}{3} = k$$
To find y, use $k = 2/3$, $h = 6$, and $r = 2$.
$$y = \frac{2}{3}(6)(2)^2 = 16$$

17. The graph of $x > 3$ is the region to the right of the vertical line $x = 3$. The graph of $x + y < 0$ is the region below the line $y = -x$. The region to the right of $x = 3$ and below $y = -x$ is shown in the following graph.

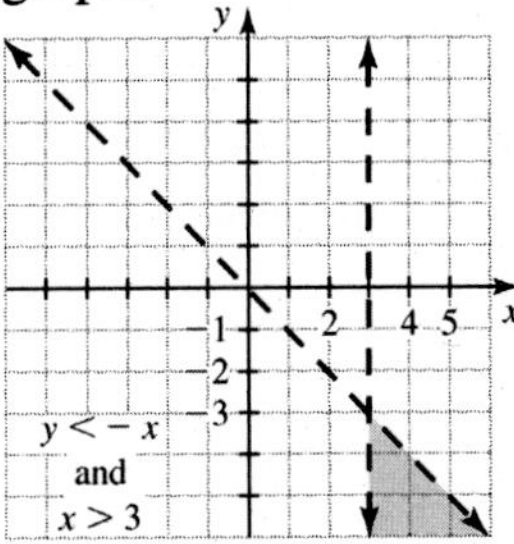

18. The inequality $|x - y| \geq 2$ is equivalent to the compound inequality $x - y \geq 2$ or $x - y \leq -2$. The graph of $x - y \geq 2$ is the region on or below the line $y = x - 2$. The graph of $x - y \leq -2$ is the region on or above the line $y = x + 2$. Since the word or is used, the graph of the compound inequality is the union of these two regions as shown in the following diagram.

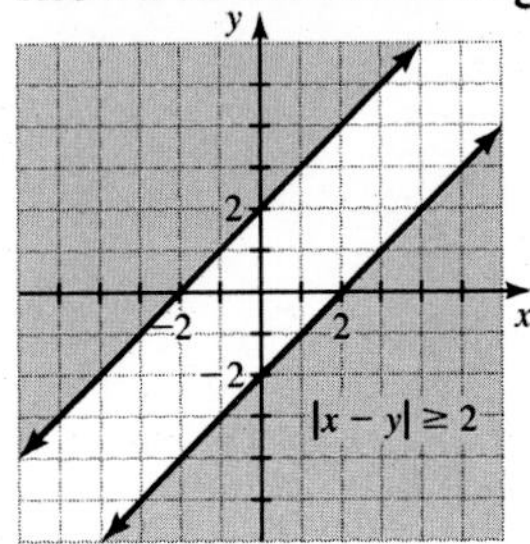

19. The graph of $y < -2x + 3$ is the region below the line $y = -2x + 3$. The graph of $y > 2^x$ is the region above the curve $y = 2^x$. Since the word and is used, the graph of the compound inequality is the intersection of these two regions, the points that lie above $y = 2^x$ and below $y = -2x + 3$. We could have used a test point in each of the four regions to see which region satisfies both inequalities.

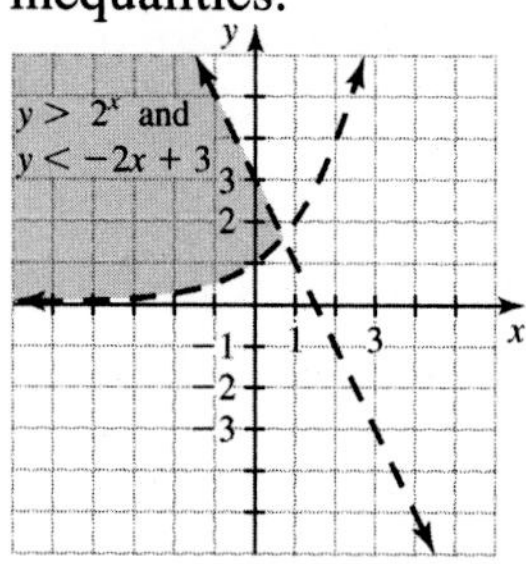

20. The inequality $|\, y + 2x \,| < 1$ is equivalent to $-1 < y + 2x < 1$. This inequality is also written as

$$y + 2x > -1 \quad \text{and} \quad y + 2x < 1$$
$$y > -2x - 1 \text{ and} \qquad y < -2x + 1$$

The graph of $y > -2x - 1$ is the region above the line $y = -2x - 1$. The graph of $y < -2x + 1$ is the region below the line $y = -2x + 1$. The points that satisfy both inequalities are the points that lie between these two parallel lines.

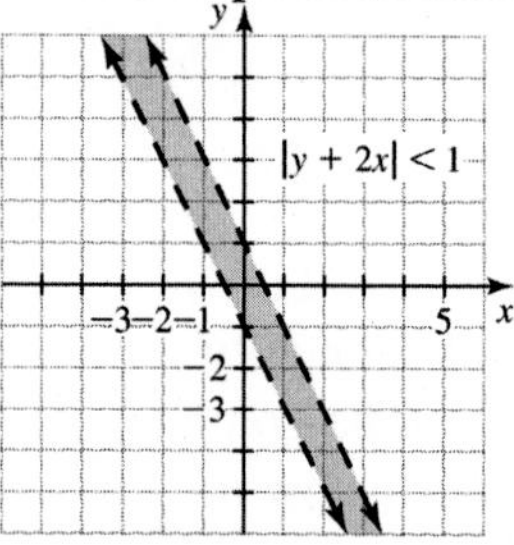

21. The graph of $x^2 + y^2 = 4$ is a circle of radius 2 centered at the origin. Since (0, 0) satisfies the inequality $x^2 + y^2 < 4$, we shade the region inside the circle.

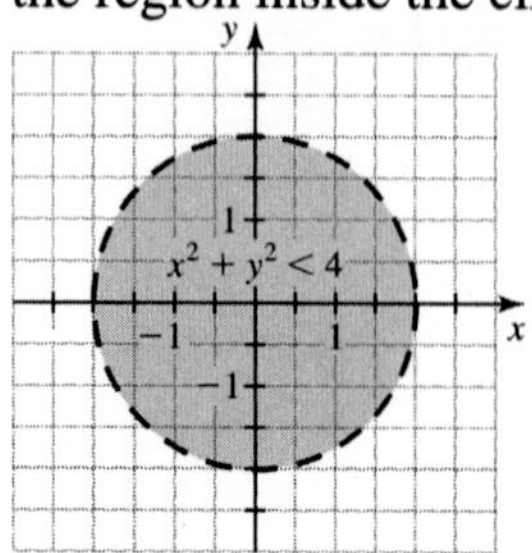

22. The graph of $x^2 - y^2 = 1$ is a hyperbola with x-intercepts $(-1, 0)$ and $(1, 0)$. The fundamental rectangle passes through the x-intercepts and (0, 1) and $(0, -1)$. Extend the diagonals for the asymptotes. The hyperbola opens to the left and right. Test a point in each region to see that only points in the region containing the origin satisfy the inequality.

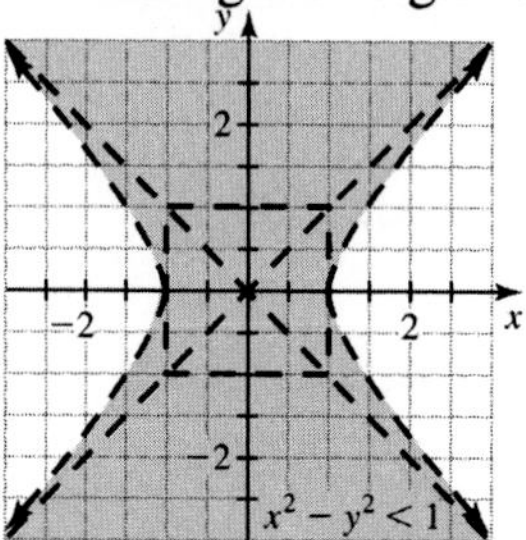

23. Graph the curve $y = \log_2(x)$ and shade the region below the curve to show the graph of $y < \log_2(x)$.

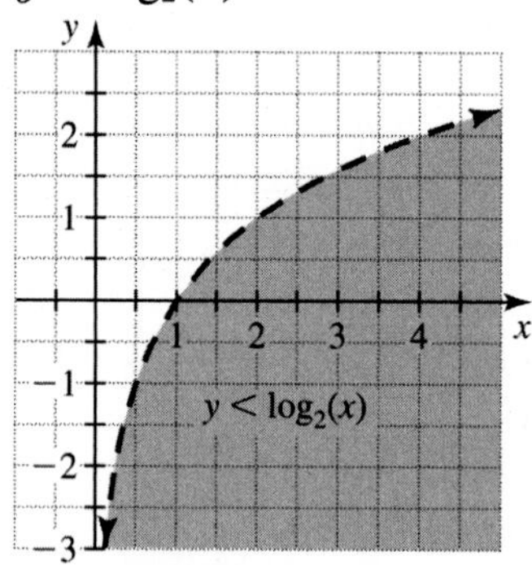

24. Write $x^2 + 2y < 4$ as $y < -\frac{1}{2}x^2 + 2$, to see that the boundary is a parabola opening downward with vertex at (0, 2).

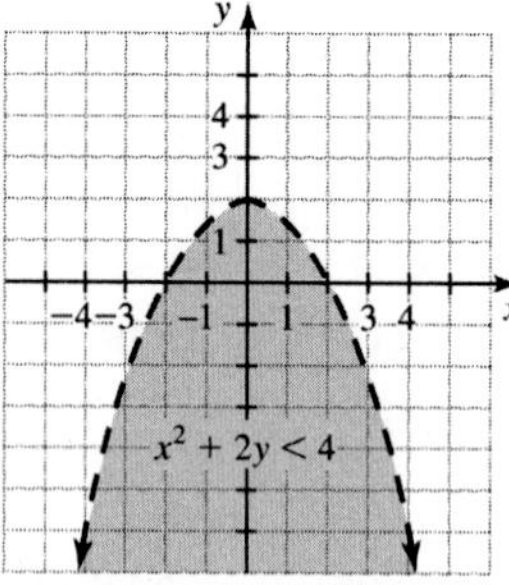

25. The graph of $\frac{x^2}{4} + \frac{y^2}{9} < 1$ is the region inside the ellipse $\frac{x^2}{4} + \frac{y^2}{9} = 1$. The graph of $y > x^2$ is the region above the parabola $y = x^2$. The points that satisfy the compound inequality are inside the ellipse and above the parabola as shown in the diagram.

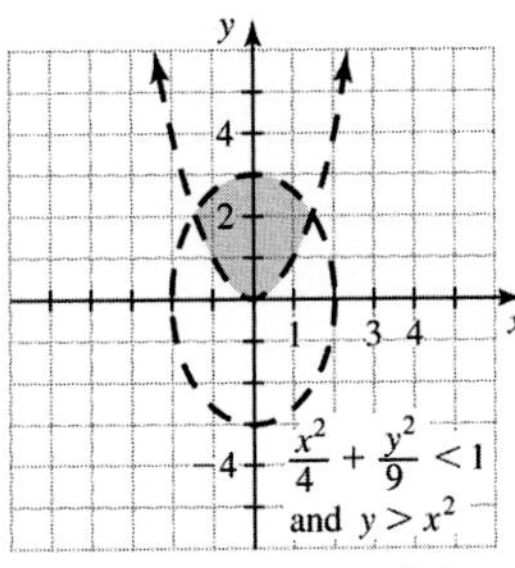

26. $\frac{a}{b}+\frac{b}{a}=\frac{a(a)}{b(a)}+\frac{b(b)}{a(b)}=\frac{a^2+b^2}{ab}$

27. $1-\frac{3}{y}=\frac{y}{y}-\frac{3}{y}=\frac{y-3}{y}$

28. $\frac{x-2}{x^2-9}-\frac{x-4}{x^2-2x-3}$

$$=\frac{(x-2)(x+1)}{(x-3)(x+3)(x+1)}-\frac{(x-4)(x+3)}{(x-3)(x+1)(x+3)}$$

$$=\frac{x^2-x-2}{(x-3)(x+3)(x+1)}-\frac{x^2-x-12}{(x-3)(x+3)(x+1)}$$

$$=\frac{10}{(x-3)(x+3)(x+1)}$$

29. $\frac{x^2-16}{2x+8}\cdot\frac{4x^2+16x+64}{x^3-16}$

$$\frac{(x-4)(x+4)}{2(x+4)}\cdot\frac{4(x^2+4x+16)}{x^3-16}$$

$$=\frac{2\,(x^3-64)}{x^3-16}$$

30. $\frac{(a^2b)^3}{(ab^2)^4}\cdot\frac{ab^3}{a^{-4}b^2}=\frac{a^6b^3}{a^4b^8}\cdot\frac{ab^3}{a^{-4}b^2}$

$$=\frac{a^7b^6}{b^{10}}=\frac{a^7}{b^4}$$

31. $\frac{x^2y}{(xy)^3}\div\frac{xy^2}{x^2y^4}=\frac{x^2y}{x^3y^3}\cdot\frac{x^2y^4}{xy^2}$

$$=\frac{x^4y^5}{x^4y^5}=1$$

32. $8^{2/3}=(\sqrt[3]{8})^2=2^2=4$

33. $16^{-5/4}=\frac{1}{(\sqrt[4]{16})^5}=\frac{1}{(2)^5}=\frac{1}{32}$

34. $-4^{1/2}=-\sqrt{4}=-2$

35. $27^{-2/3}=\frac{1}{(\sqrt[3]{27})^2}=\frac{1}{3^2}=\frac{1}{9}$

36. $-2^{-3}=-\frac{1}{2^3}=-\frac{1}{8}$

37. $2^{-3/5}\cdot 2^{-7/5}=2^{-10/5}=2^{-2}=\frac{1}{2^2}=\frac{1}{4}$

38. $5^{-2/3}\div 5^{1/3}=5^{-\frac{2}{3}-\frac{1}{3}}=5^{-1}=\frac{1}{5}$

39. $(9^{1/2}+4^{1/2})^2=(3+2)^2=5^2=25$

40. a) Age 4 years 3 months is 4.25 years.
$h(4.25)$

$$=79.041+6.39(4.25)-e^{3.261-0.993(4.25)}$$
$$=105.8\text{ cm}$$

$$105.8\text{ cm}\cdot\frac{1\text{ in}}{2.54\text{ cm}}=41.7\text{ in.}$$

c) Using a graphing calculator we get that a child who has a height of 80 cm has an age of 1.3 years

Appendix A

1. The perimeter is $3 + 4 + 5$ or 12 in.

2. The area is $\frac{1}{2} \cdot 12 \cdot 4$ or 24 ft^2.

3. Since the sum of the 3 angles of a triangle is 180°, then third angle is $180 - 30 - 90$ or 60°.

4. Since $A = \frac{1}{2}bh$, we have $36 = \frac{1}{2} \cdot 12h$ or $36 = 6h$, or $h = 6$ ft.

5. Since the side opposite 30° is one-half of the hypotenuse, the length of the hypotenuse is 20 cm.

6.
$$\begin{aligned} A &= \tfrac{1}{2}h(b_1 + b_2) \\ A &= \tfrac{1}{2}12(4 + 20) = 6(24) \\ &= 144 \text{ cm}^2 \end{aligned}$$

7. The area of a right triangle is one-half the product of the lengths of the legs.
$A = \frac{1}{2}(6)(8) = 24\text{ft}^2$

8. The hypotenuse of a right triangle is the longest side. So the hypotenuse is 13 ft.

9. Since $a^2 + b^2 = c^2$ for a right triangle, we have $a^2 + 40^2 = 50^2$ or $a^2 = 900$ and $a = 30$ cm.

10. Since $5^2 + 10^2 \neq 11^2$, the triangle is not a right triangle.

11. Since $7^2 + 24^2 = 15^2$, the triangle is a right triangle and its area is one-half the product of the lengths of the legs.
$A = \frac{1}{2}(7)(24) = 84 \text{ yd}^2$

12. Since the opposite side of a parallelogram are equal, the perimeter is $2(9) + 2(6)$ or 30 in.

13. The area of a parallelogram is bh. So the area is 32 ft^2.

14. Since a rhombus has four equal sides, it perimeter is 5(4) or 20 km.

15. The perimeter is $2L + 2W$ or $2(1.5) + 2(2)$ or 7 ft. The area is LW or $(1.5)(2)$ or 3 ft^2.

16.
$$\begin{aligned} P &= 2L + 2W \\ 60 &= 2L + 2(8) \\ 44 &= 2L \\ 22 &= L \end{aligned}$$
The length is 22 yd.

17. Since $A = \pi r^2$, $A = \pi(4)^2 \approx 50.3\,\text{ft}^2$.

18. Since $C = \pi d$, we have
$$C = \pi \cdot 12 \approx 37.7 \text{ ft}.$$

19. $V = \frac{1}{3}\pi r^2 h = \frac{1}{3}\pi(4)^2 9 \approx 150.80 \text{ cm}^3$

20. $S = \pi r\sqrt{r^2 + h^2} = \pi(12)\sqrt{12^2 + 20^2}$
$\approx 879.29\,\text{ft}^2$

21. $V = LWH = 12(6)(4) = 288\,\text{in.}^3$
$S = 2LW + 2LH + 2WH = 288\text{ in.}^2$

22.
$$\begin{aligned} V &= LWH \\ 120 &= 30h \\ 4 &= h \end{aligned}$$
The height is 4 cm.

23. $A = s^2 = 10^2 = 100 \text{ mi}^2$
$P = 4s = 4(10) = 40\,\text{mi}$

24.
$$\begin{aligned} A &= s^2 \\ 25 &= s^2 \\ 5 &= s \end{aligned}$$
$P = 4s = 4(5) = 20$ km

25.
$$\begin{aligned} P &= 4s \\ 26 &= 4s \\ s &= 6.5 \end{aligned}$$
$A = s^2 = 6.5^2 = 42.25 \text{ cm}^2$

26. $V = \frac{4}{3}\pi r^3 = \frac{4}{3}\pi(2)^3 \approx 33.510\,\text{ft}^3$
$S = 4\pi r^2 = 4\pi(2)^2 \approx 50.265 \text{ ft}^2$

27. $V = \pi r^2 h = \pi(2)^2 6 \approx 75.4 \text{ in.}^3$
$S = 2\pi rh + 2\pi r^2 = 2\pi(2)(6) + 2\pi(2)^2$
$\approx 100.5\,\text{in.}^2$

28. Since complementary angles have a sum of 90°, the other angle is $90 - 34$ or 56°.

29. Since an isosceles triangle has two equal sides, $x + 2(12) = 29$ or $x = 5$.
So the short side is 5 cm.

30. Since the corresponding sides of similar triangles are proportional,
$$\begin{aligned} \frac{10}{25} &= \frac{8}{x} \\ 10x &= 200 \\ x &= 20 \end{aligned}$$
Each side of the larger triangle is 2.5 times as large as the corresponding side in the smaller triangle. So the sides are 20 in. and $(2.5)(6)$ or 15 in.

31. Since supplementary angles have a sum of 180°, the other angle is $180 - 31$ or 149°.

32. Since all three sides of an equilateral triangle have the same length, the perimeter is 3(4) or 12 km.

33. The length of a side of an equilateral triangle is one-third the length of the perimeter. So the side has length $30/3$ or 10 yd.

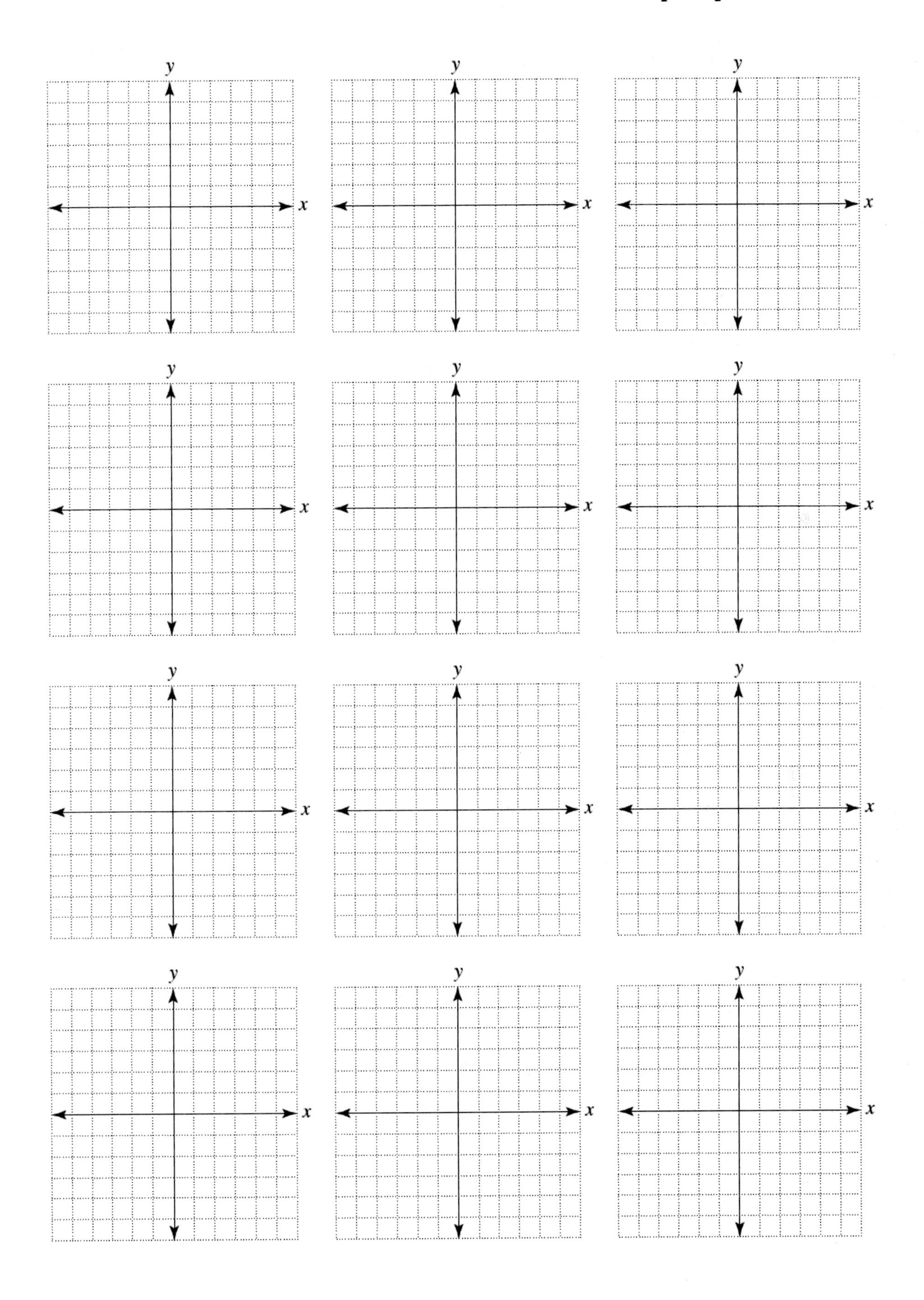
y
x

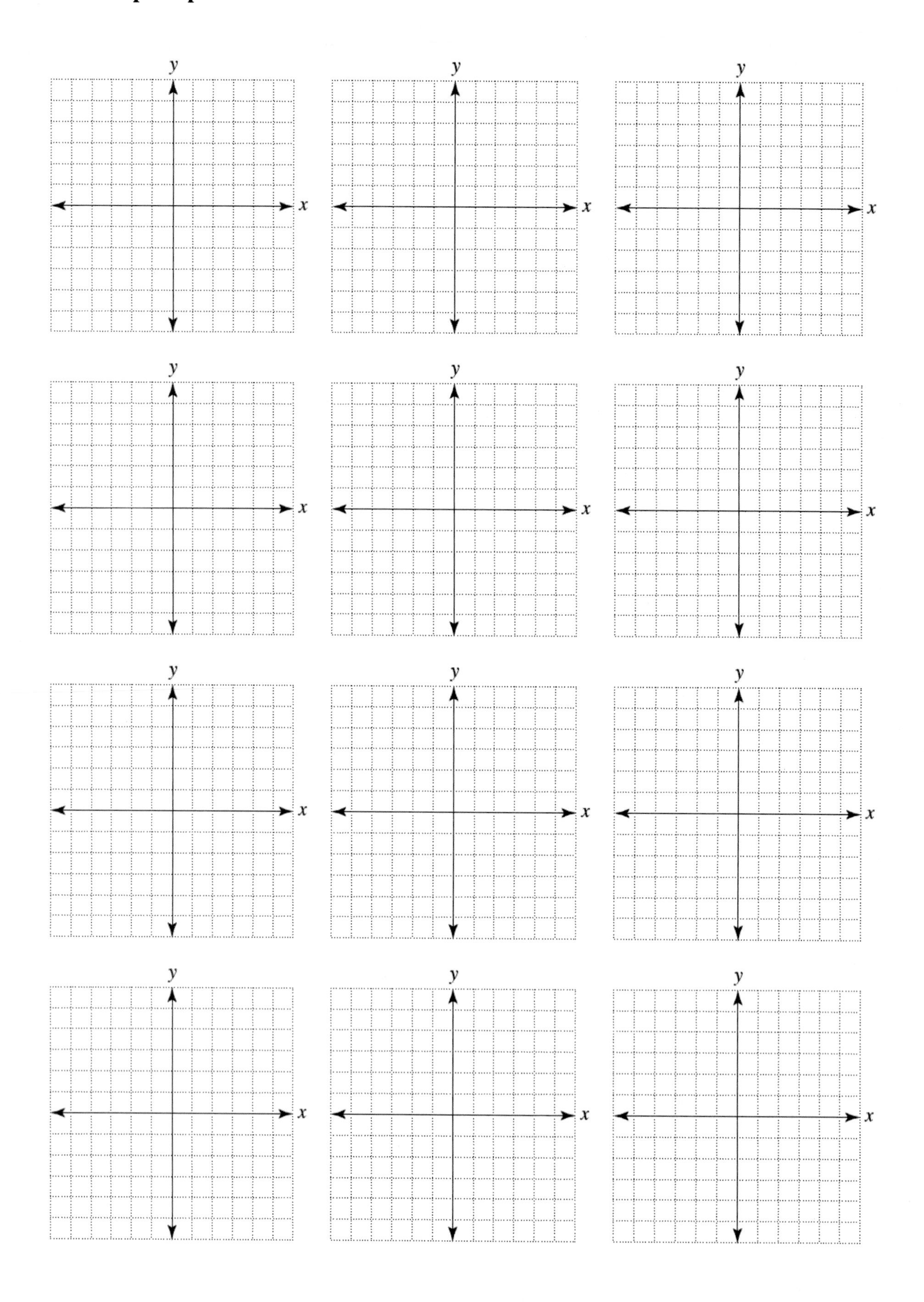
y
x
y
x
y
x
y
x
y
x
y
x
y
x
y
x
y
x
y
x
y
x
y
x

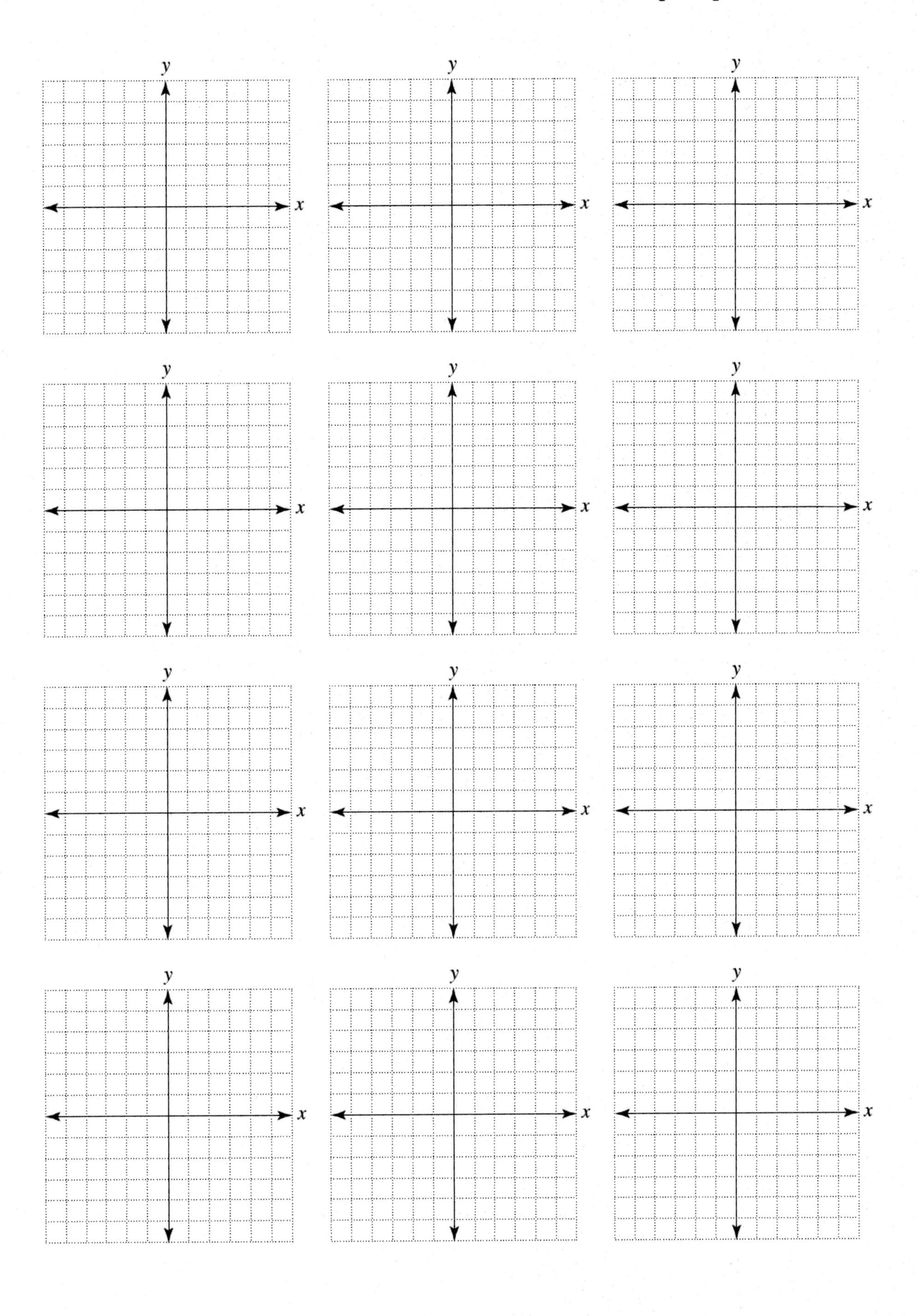
y
x
y
x
y
x
y
x
y
x
y
x
y
x
y
x
y
x
y
x
y
x
y
x

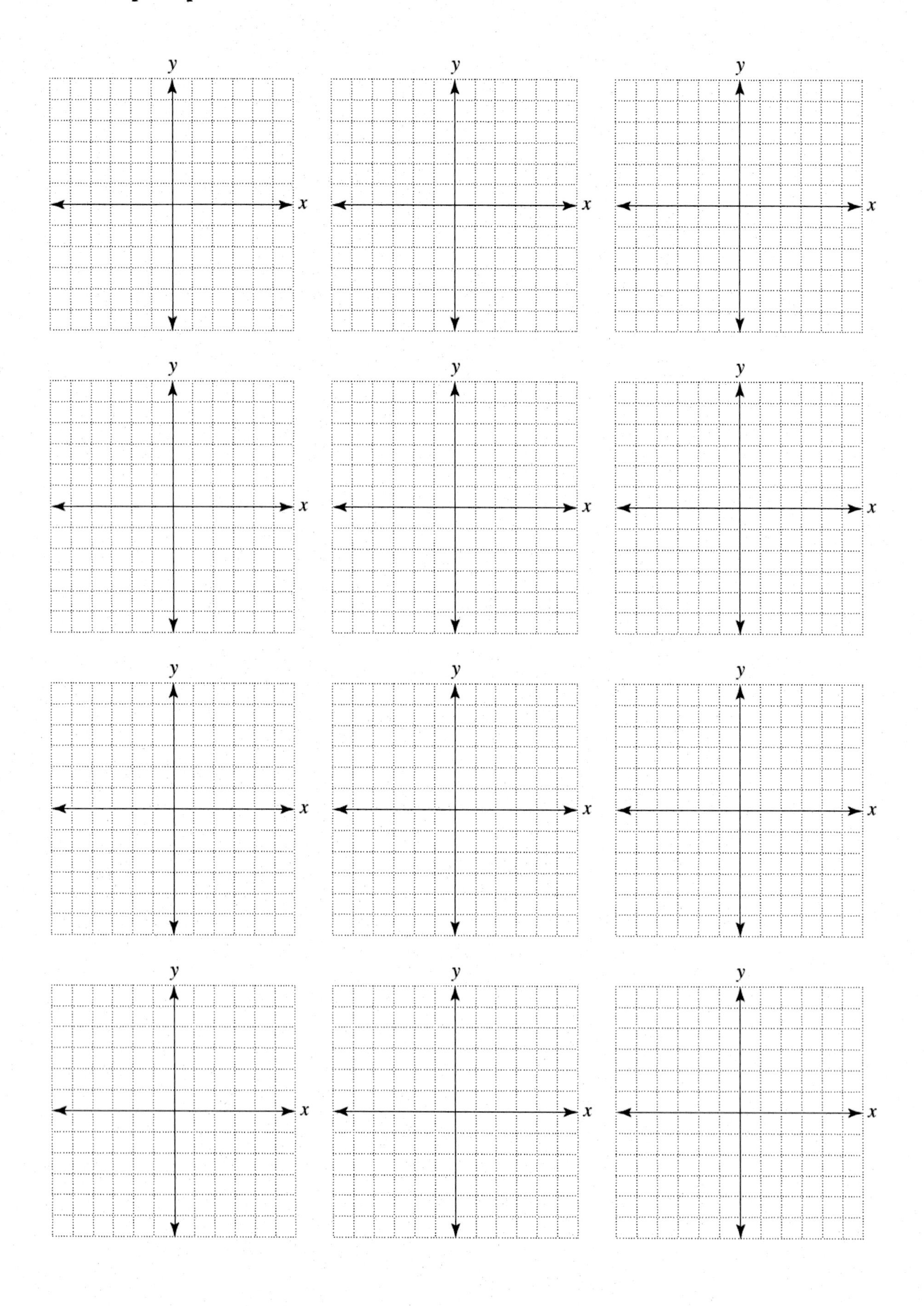
y
x
y
x
y
x
y
x
y
x
y
x
y
x
y
x
y
x
y
x
y
x
y
x

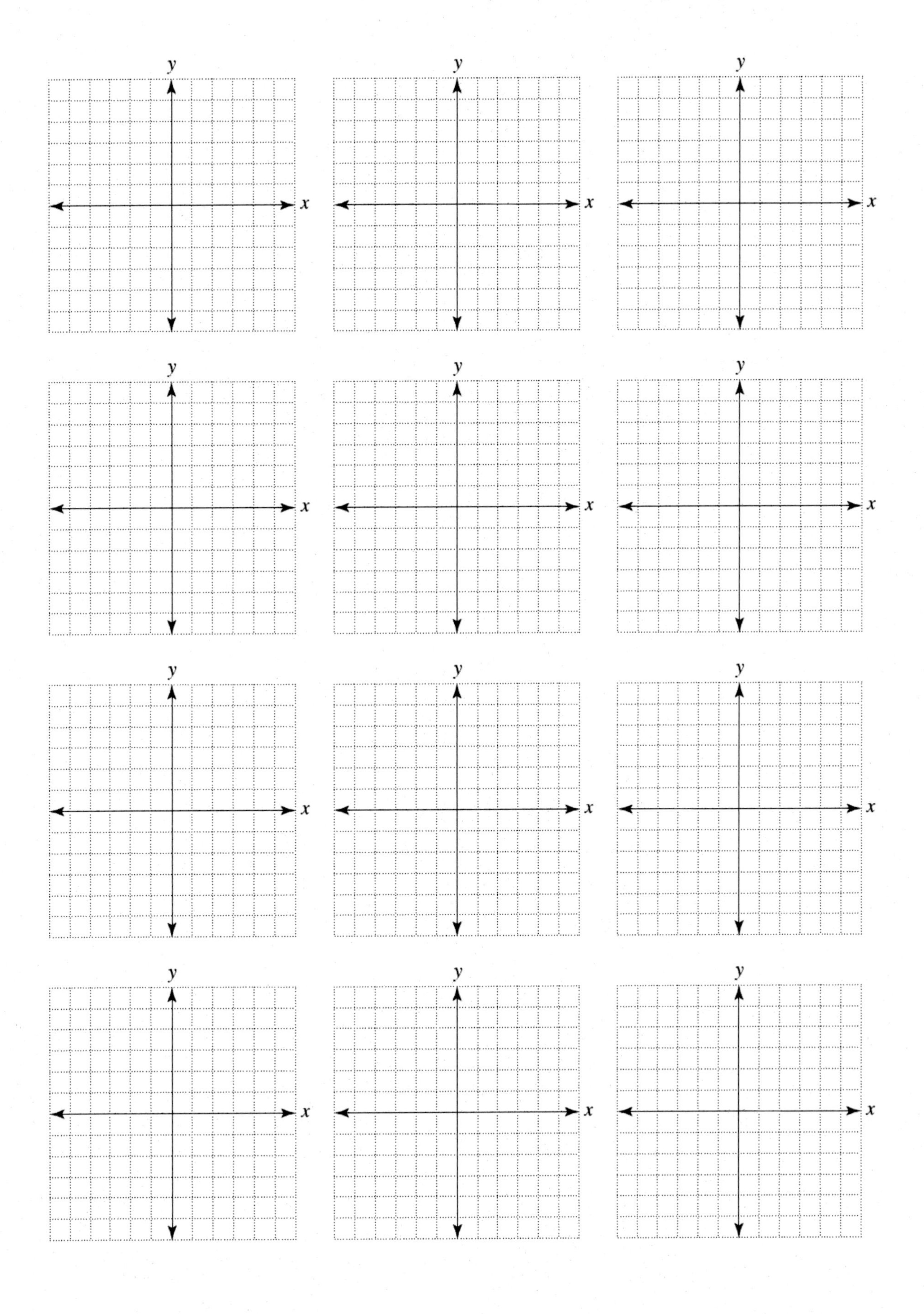
y
x
y
x
y
x
y
x
y
x
y
x
y
x
y
x
y
x
y
x
y
x
y
x

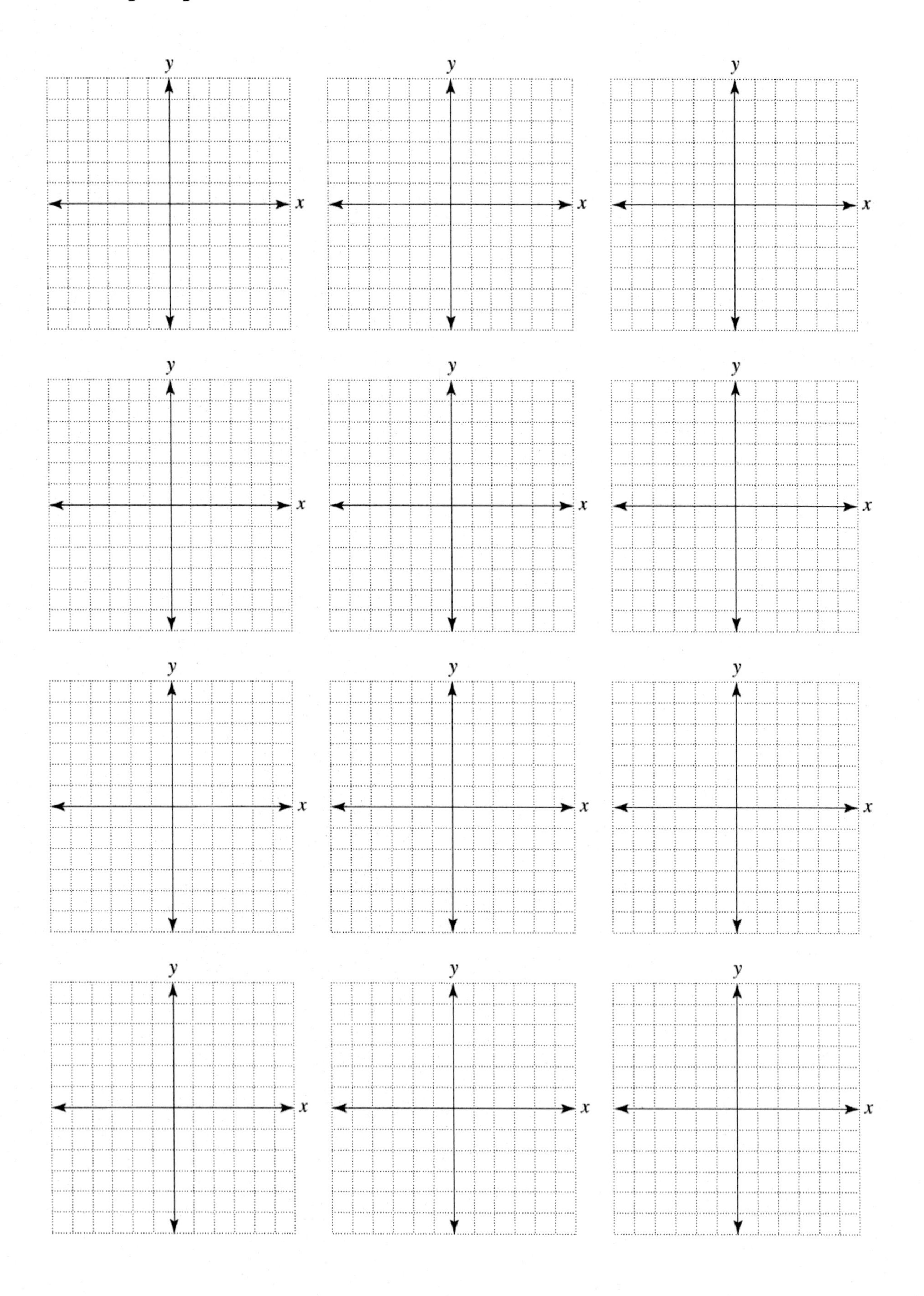

y
x
y
x
y
x
y
x
y
x
y
x
y
x
y
x
y
x
y
x
y
x
y
x

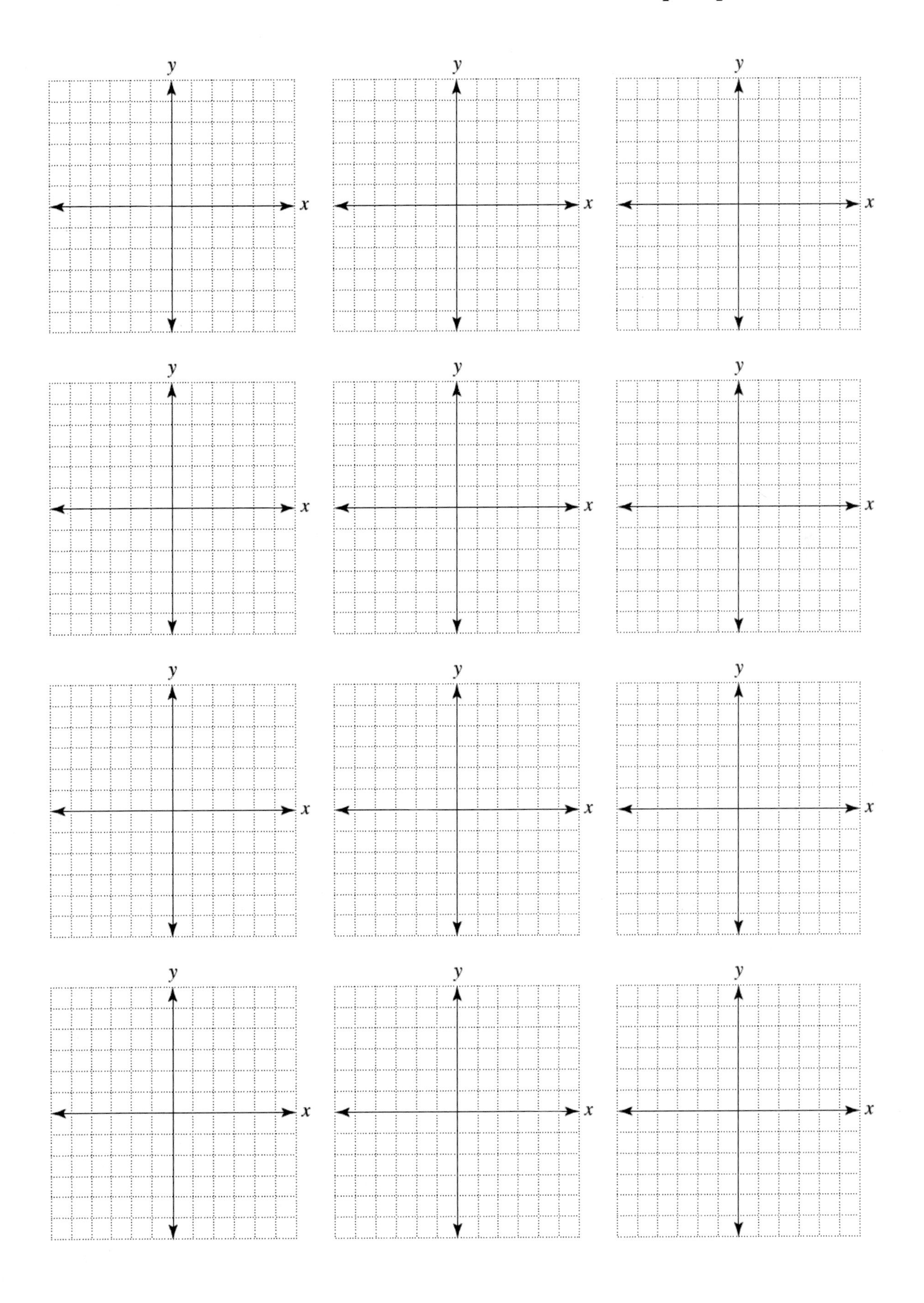
y
x

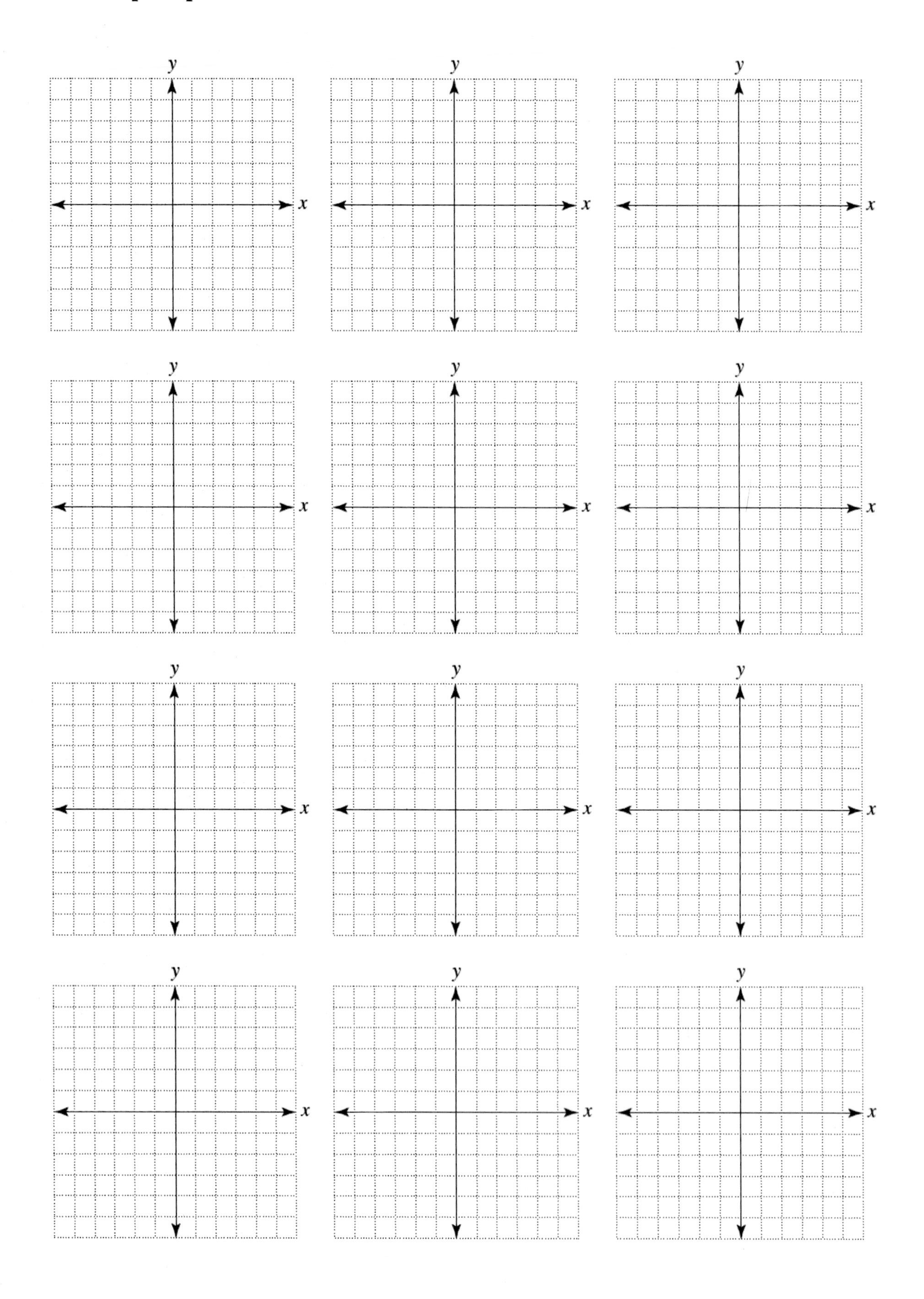
y
x
y
x
y
x
y
x
y
x
y
x
y
x
y
x
y
x
y
x
y
x
y
x

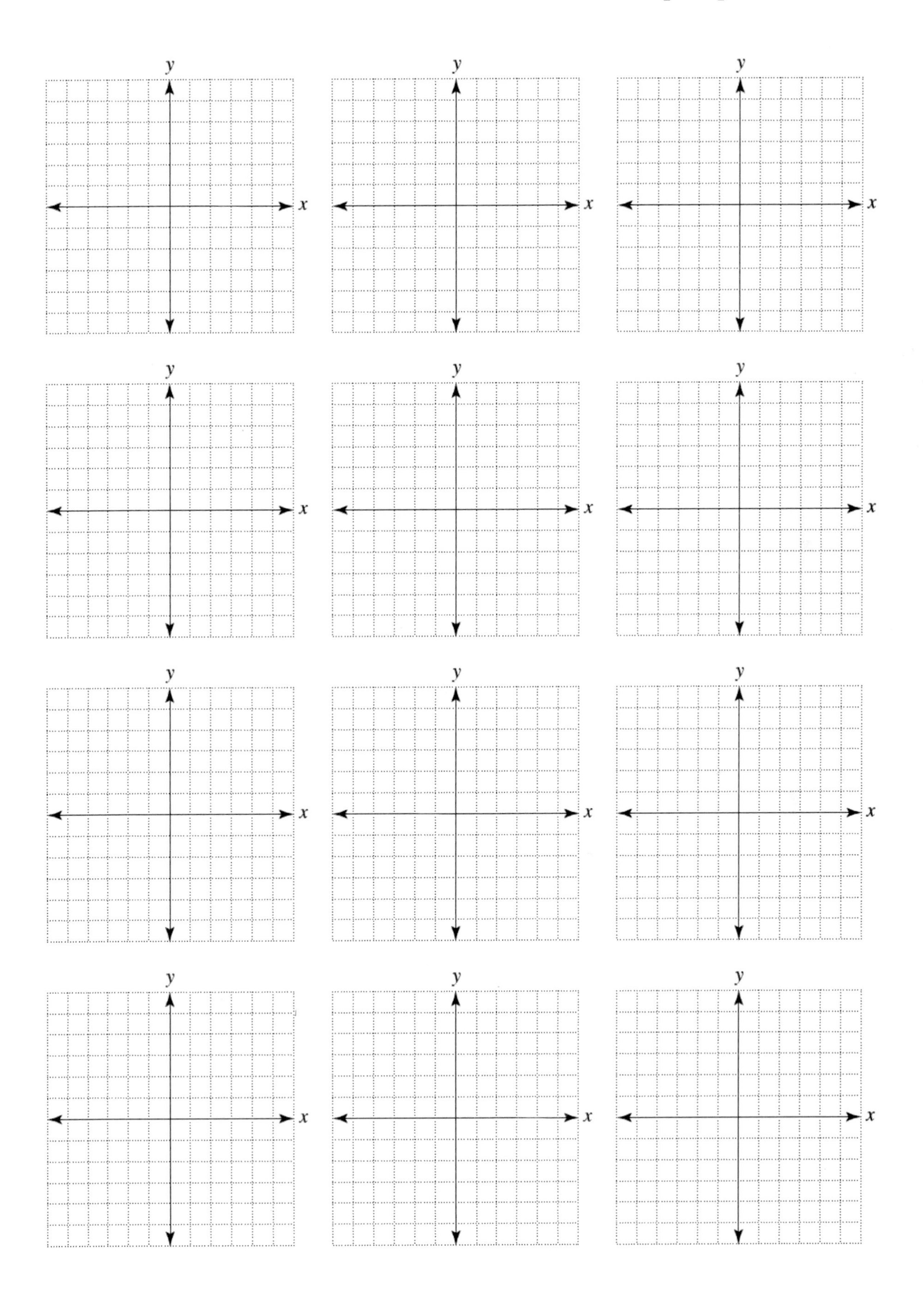

y
x

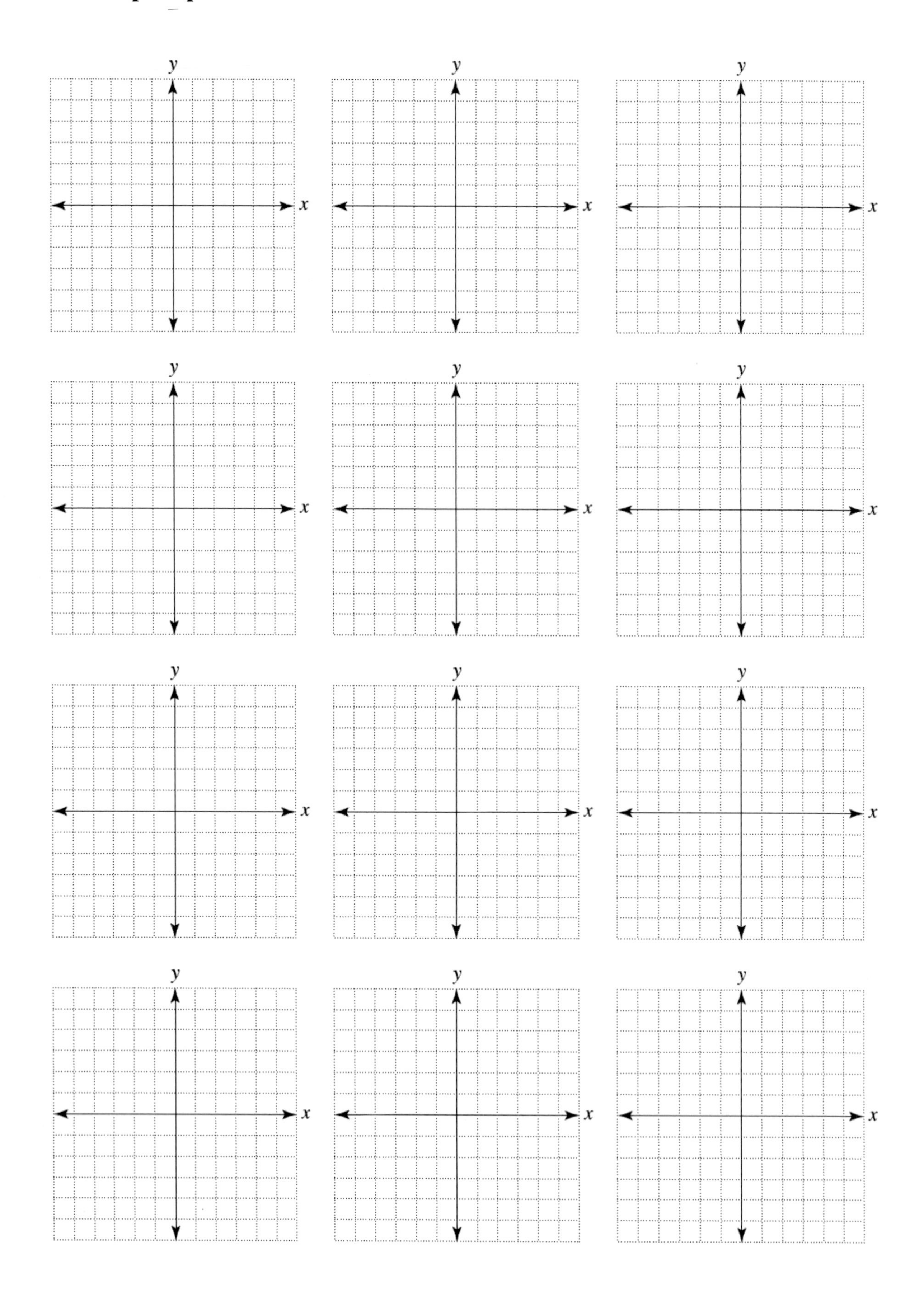
y
x
y
x
y
x
y
x
y
x
y
x
y
x
y
x
y
x
y
x
y
x
y
x

Notes

Notes